TP12010097
ENG

# Practical Applications in Digital Signal Processing

# Practical Applications in Digital Signal Processing

## Richard Newbold

PRENTICE
HALL

Upper Saddle River, NJ • Boston • Indianapolis • San Francisco
New York • Toronto • Montreal • London • Munich • Paris • Madrid
Capetown • Sydney • Tokyo • Singapore • Mexico City

Many of the designations used by manufacturers and sellers to distinguish their products are claimed as trademarks. Where those designations appear in this book, and the publisher was aware of a trademark claim, the designations have been printed with initial capital letters or in all capitals.

The author and publisher have taken care in the preparation of this book, but make no expressed or implied warranty of any kind and assume no responsibility for errors or omissions. No liability is assumed for incidental or consequential damages in connection with or arising out of the use of the information or programs contained herein.

The publisher offers excellent discounts on this book when ordered in quantity for bulk purchases or special sales, which may include electronic versions and/or custom covers and content particular to your business, training goals, marketing focus, and branding interests. For more information, please contact:

> U.S. Corporate and Government Sales
> (800) 382-3419
> corpsales@pearsontechgroup.com

For sales outside the United States please contact:

> International Sales
> international@pearson.com

Visit us on the Web: informit.com/ph

*Library of Congress Cataloging-in-Publication Data*
Newbold, Richard.
  Practical applications in digital signal processing / Richard Newbold.
    pages cm
  Includes bibliographical references and index.
  ISBN-13: 978-0-13-303838-5 (hardcover : alk. paper)
  ISBN-10: 0-13-303838-6 (hardcover : alk. paper)
  1. Signal processing—Digital techniques. 2. Electric filters, Digital. I. Title.
  TK5102.9.N49 2013
  621.382'2—dc23                                           2012024511

Copyright © 2013 Pearson Education, Inc.

ISBN-13: 978-0-13-303838-5
ISBN-10:    0-13-303838-6

Text printed in the United States on recycled paper at Edwards Brothers Malloy in Ann Arbor, Michigan.
First printing, October 2012

**Executive Editor**
Bernard Goodwin

**Managing Editor**
John Fuller

**Project Manager**
Caroline Senay

**Project Editor**
Scribe Inc.

**Copy Editor**
Scribe Inc.

**Indexer**
Scribe Inc.

**Proofreader**
Scribe Inc.

**Publishing Coordinator**
Michelle Housley

**Cover Designer**
Anne Jones

**Cover Art**
Laura Robbins

**Compositor**
Scribe Inc.

To my wife, Mary, who has always stood by me through thick and thin. To my son Shannon, a brilliant software design engineer, and to my son Daniel, an accomplished attorney, both of whom were self-sufficient right out of college.

# Contents

# Preface

I have spent more than 30 years toiling away as a digital hardware design engineer and as an unsophisticated self-taught software designer. Most of my software efforts were in support of my hardware designs and included endeavors such as bit-level simulations, microcode generation, assembly code, FORTRAN, C/C++, and writing Microsoft Windows application graphics-oriented test stations, which I utilized to verify the proper operation of my digital creations.

I began my digital design career when digital signal processing (DSP) was still in its infancy. In those days, all digital designs were implemented with small-scale integrated (SSI) circuits that weren't much more sophisticated than 4-bit adders and 8- to 1-bit multiplexers. The first company I worked for after graduation was heavily into the early phases of DSP.

DSP algorithms are for the most part dependent on repetitive multiplications and summation operations. The first digital multiplier I ever saw required an entire chassis of equipment to do a 16-by-16 multiplication. This multiplier consumed so much hardware that it was efficient to time-share it with other hardware that was engaged in processing independent tasks. Device propagation delays were so huge that building hardware systems that utilized a 5-MHz system clock was considered high tech.

To give some perspective about the state of the art at the time, the term *Silicon Valley* had not been coined yet. It was during this time that a little-known, small company that went by the name of Intel was operating out of a very tiny building located at 365 Middlefield Road in Mountain View, California. Intel had just introduced the world's first microprocessor. It was a 4-bit machine called the 4004 microcomputer. It was built under contract to the Nippon Calculating Machine Corporation in Tokyo, Japan. With the introduction of the 4004, the digital age changed gears. Digital technology soon began to evolve so quickly that hardware designed one year was almost obsolete by the next.

Program requirements always seemed to demand technology that wasn't developed yet. Design engineers were constantly tasked with implementing tomorrow's designs with today's technology. This struggle, in a large sense, fueled an atmosphere of intense research and development and drove the industry to continuously produce lower power, faster, and more complex devices and systems. Looking back, it seems like the world of DSP just exploded on all fronts. Start-up companies sprouted up in the Silicon Valley almost on a daily basis.

During this time, the science and technology of DSP grew and matured as integrated circuit manufacturers strived to produce higher speed signal processing components and lower power processors. Fusible link programmable logic devices were introduced, which quickly evolved into reprogrammable logic devices and, over time, evolved into field programmable gate arrays (FPGAs), complex programmable logic devices (CPLDs), and application-specific integrated circuits (ASICs), which are still in use today. Other companies began to prosper by serving as fabrication houses for extremely high-speed gallium arsenide and indium phosphide integrated circuits. They would teach engineers how to design using their processes and then fabricate their application-specific designs.

The design tools necessary to support the programming and testing of these complex devices have evolved into big-time software applications. FPGA companies are even taking most of the challenges out of DSP design by offering a library of DSP circuits called *cores* that can be incorporated into an FPGA design with a simple keystroke, without much knowledge on the designer's part of how these circuits operate.

During my 30-year career I have accumulated a fairly large library of DSP textbooks. With few exceptions, these books all cover the same basic topics. Different authors address the same subjects but each with their own unique approach. Reading several authors' treatment of the same subject helped me view DSP processing techniques from different perspectives and tended to fill a lot of the blanks in my understanding of the subject. These books were well written by astute people in the field, and they all provided an excellent technical baseline for DSP design.

However, there have been few textbooks written that deal specifically with the many DSP topics and algorithms that are commonly used in everyday applied DSP. As a rule, a good working knowledge of these applied DSP algorithms usually comes from word of mouth, design mentoring, and design experience. Over time, all design engineers accumulate (in their minds) a toolbox of circuits, procedures, algorithms, and techniques that are a product of years of long hours, a lot of sweat, tears, successes, failures, hand-wringing, and a fair amount of banging one's head against the wall. Unfortunately these toolboxes are not documented, and thus it is hard for other engineers to access the wealth of information contained within these toolboxes. Engineers for the most part are a secretive species and in their quest for job security are reluctant to publicize their hard-earned trade secrets.

There are many gray areas in DSP design that have not been addressed in detail by any of the engineering textbooks that I am familiar with. These gray areas usually don't address questions like *How do I design a circuit that will perform this or that critical DSP function?*

For example, no DSP textbook I am familiar with has discussed in detail applications that are heavy into the use of complex digital signals, the spectra of real and complex digital signals, the science of complex to real signal conversion, digital signal translation, or the concept of digital frequency synthesis.

I have not seen any text that provided a detailed analysis on how to design a numerically controlled oscillator (NCO) used in digital tuning applications, or how to design an elastic store memory used in pulse code modulation (PCM) multiplexing applications, or how to design a digital data locked loop (DLL) or a digital automatic gain control (dAGC).

Other design topics rarely discussed in application-oriented detail by the myriad of DSP books available today include applications of poly phase filters (PPF) and cascaded integrator comb (CIC) filters, and applications like digital channelizers, sometimes referred to as *transmultiplexers*. This versatile circuit is found in many applications, such as frequency division multiplex (FDM) to time division multiplex (TDM) conversion, mixing consoles, wideband scanners, and the processing of wideband intercepts in radio astronomy, to name just a few. All these subjects and more can be lumped into the general topic of *Practical Applications in Digital Signal Processing*.

## THE PURPOSE OF THIS BOOK

The purpose of this book is to unlock and dispense some of the contents of my own personal toolbox in the hope of filling in some of these DSP gray areas. It is my hope to provide a source of usable information and DSP design techniques suitable for use in real-world design applications.

There are a great many DSP textbooks that are considered bibles of the DSP design world. Many of these books, along with technical papers written by astute people in the field, are referenced within this book. It is not the intention of this book to repeat the work that has been done by so many previous authors. This book does not deal with the derivation and treatment of standard DSP concepts, which have been thoroughly addressed in great detail by many other authors. The sole purpose of this book is to serve as an application-oriented addendum to the many great DSP textbooks that have already been published.

## WHO SHOULD READ THIS BOOK

This book is not intended for a person with no previous DSP knowledge or experience. This book is intended for the undergraduate and graduate student who will soon enter the signal processing industry. It is also intended for the engineer already in the industry who has some experience in DSP design and who is now searching for additional information regarding the design and implementation of common but largely undocumented DSP hardware or software applications.

## HOW THIS BOOK IS ORGANIZED

This book is organized as a collection of tutorials on common DSP applications. The first four chapters are detailed reviews on the mathematical tools necessary to successfully analyze, design, and build complex digital processing systems. The remaining nine chapters provide detailed tutorials on independent signal processing applications commonly used in the industry. An appendix is included that provides an in-depth discussion on mixed language programming. The content of each chapter is summarized in the following sections.

### Chapter 1: Review of Digital Frequency

This chapter is a short tutorial on digital frequency and how it is related to the system sample rate. It shows how to mathematically represent the value of a particular digital frequency and how to determine the value of all the samples in a digital sinusoidal waveform.

### Chapter 2: Review of Complex Variables

This chapter presents a thorough review of the subject of complex variables. After reading this chapter, it is possible for a person with no prior experience to become proficient in the use of this valuable mathematical tool in the design and development of signal processing circuits and systems. The review starts by defining complex numbers and their properties and progresses all the way to a complete discussion of residue theory. The computation of residues provides the engineer an easy alternative to compute the impulse response of a digital system.

## Chapter 3: Review of the Fourier Transform

This chapter provides an in-depth review of the Fourier series and both the continuous and discrete Fourier transform (CFT and DFT, respectively). The discussion includes the derivation of transform properties, transform pairs, Parseval's theorem, and the derivation of energy and power spectral density (PSD) relationships. Attention is also given to the topic of spectral leakage, the band pass filter, and the low pass filter models of the DFT. Signal processing discussions include the use of windows, coherent and incoherent processing gain, and signal recognition. Even though this is an extensive review, it is written so that a reader without any background in the topics of Fourier series or Fourier transforms can proficiently use them when working with signal processing applications.

## Chapter 4: Review of the Z-Transform

This chapter provides a comprehensive review of the z-transform. Detailed discussions include the use of pole-zero diagrams, inverse z-transforms, convergence, and system stability. A person with no prior knowledge of z-transforms can, after reading this chapter, utilize the knowledge gained to analyze complex digital systems, thereby enabling them to derive a system frequency response, determine system stability, and compute a system impulse response. In addition, the reader will learn how to use the z-transform in real-world situations to modify existing designs to either enhance performance or alter the specifications for incorporation into other systems.

## Chapter 5: Finite Impulse Response Digital Filtering

The focus of this chapter is on the design of finite impulse response (FIR) digital filters. It is not my intent to repeat all of the excellent theoretical material that has already been published by so many astute authors. Almost all DSP texts devote substantial coverage to the history, theory, architecture, mathematics, and legacy design techniques of digital filters. Instead, the intent here is to concentrate solely on a single method for the design and implementation of some of the more common filter types. The purpose of this chapter is twofold. First, in order to establish a communication baseline, we will provide a very brief overview of digital filters. Second, we will demonstrate a computer-aided design methodology based on the Parks-McClellan optimal filter design program to implement several types of digital filters. A complete listing of this program is included in Appendix A.

## Chapter 6: Multirate Finite Impulse Response Filter Design

This chapter is a detailed discussion on the design of digital filters used to modify the sample rate of a signal. A designer is often faced with the task of either increasing or decreasing the sample rate of a signal by some integer or fractional amount. There are several methods that can be utilized to change the sample rate of a digital signal. All these methods involve the use of a digital filter, sometimes referred to as a *multirate filter*. Some multirate filters are better suited for specific rate change applications than others. In this chapter we will discuss three rate change methods that use the following three filter types:

1. *Poly phase filters*. The preferred method for moderate sized rate changes.
2. *Half band filters*. An efficient method for factor of two rate changes.
3. *CIC filters*. Computationally efficient filters for large rate changes.

## Chapter 7: Complex to Real Conversion

This chapter provides a detailed tutorial on the conversion of a complex signal to a real signal. This is a common signal processing function, yet material dealing with this very important topic is rarely found in engineering textbooks. A very good example of complex signal processing is seen in digital systems that employ a front-end tuner. These systems fall into a category that can be loosely categorized as "digital radio," in that an input wideband signal is tuned up or down in frequency and passed through a band pass or low pass filter to isolate some narrow band of interest. The mathematics of the tuning function converts the real input signal into a complex signal. The filtered narrow band signal is then processed in its complex form to implement whatever the particular application requires. After the intermediate processing is complete, the complex signal is generally converted back to real and provided as an output.

## Chapter 8: Digital Frequency Synthesis

There are numerous applications in the world of DSP that utilize a numerically controlled oscillator, or NCO. An NCO is a programmable oscillator that outputs a digital sinusoid at some user-specified frequency and phase. The sinusoid can be fixed at some programmed frequency, or it can be swept or hopped over a band of frequencies. The sinusoid can have a constant phase or it can be programmed to have multiple or switched phases. It can be a simple or a complex device, depending on the requirements of the application in which the NCO is used. A typical application utilizes the NCO to produce a programmable complex sinusoid to tune band pass signals down to base band for filtering and postprocessing, similar to the local oscillator in an AM radio. This chapter contains detailed figures that clearly illustrate both the design of the NCO and the workings of all the internal processing functions. Extensive simulations graphically illustrate the signals produced by the NCO.

## Chapter 9: Signal Tuning

This chapter provides a thorough discussion on the subject of signal tuning in both the continuous analog and discrete digital domains. It is often necessary when processing a signal to move it from one region of the frequency spectrum to another region. This is especially true when processing communications signals, where a band limited signal centered at frequency $f_1$ is tuned to another center frequency $f_2$ in order to simplify downstream processing. This chapter illustrates the methods used to translate the spectrum of real and complex signals both up and down in frequency.

## Chapter 10: Elastic Store Memory

During their careers, most engineers have designed interfaces between two or more data processing systems that utilized synchronous data streams. There are occasions, however, when a designer must interface two or more processing systems or data streams where the data rates are asynchronous to one another. For purposes of this chapter, the term *asynchronous* refers to the case where each data stream is time aligned to its own clock generated by an independent clock oscillator. The frequency and phase of each clocked data stream are similar but not necessarily identical. Each clock oscillator's output frequency uniquely varies over time and temperature. In many cases, these clocks may differ by as much as a few thousand hertz. In this chapter we illustrate how to synchronize these systems with an elastic store memory.

## Chapter 11: Digital Data Locked Loops

Suppose you are presented with a time division multiplex, or TDM, bit stream composed of a multiplex of two or more independent and originally asynchronous tributaries. How can we demultiplex these tributaries and synthesize an independent bit clock for each that is on average identical to its original premultiplex clock? This type of signal is similar to a high-level telephone PCM multiplex that carries several lower level tributaries. This is only one of many possible examples. The same question can be asked of any demultiplex processing where the multiplexed tributaries were originally asynchronous to one another. The answer requires utilizing a digital data locked loop, or DLL. The DLL is a fairly simple device that uses an elastic store memory to synthesize a bit stream clock and then synchronizes the demultiplexed bit stream or tributary with that clock, all with no prior knowledge of the original clock frequency. This chapter provides a thorough tutorial on how to design DLLs for just about any relevant application.

## Chapter 12: Channelized Filter Bank

This chapter presents a high-level functional discussion followed by an in-depth, detailed tutorial on the design of a digital channelizer, sometimes referred to as a *transmultiplexer*. As mentioned previously, this versatile circuit is found in many signal processing applications. The channelizer can easily replace hundreds of receivers with not much more than a single integrated circuit. In this chapter, we will design a working channelizer that simultaneously processes up to 2000 independent equal bandwidth signals.

## Chapter 13: Digital Automatic Gain Control

This chapter is a thorough discussion of a Type I and Type II digital automatic gain control, or dAGC. This subject matter is rarely covered in any engineering textbook available today, and if it is covered, it is usually given a cursory look amounting to not much more than a paragraph or two. In many electronic systems, one of the most important functions is automatic gain control (AGC). In general, an AGC is a nonlinear feedback circuit that if not designed properly can become unstable. The purpose of this chapter is to design a dAGC circuit; derive its operational parameters; simulate it; and then graphically illustrate the transient response, the steady state operation of the loop error, the loop gain, and the circuit output in response to various input signals and input signal perturbations.

## Appendix A: Mixed Language C/C++ FORTRAN Programming

Over the years, there is a good chance that engineers who have been in the business for a while have accumulated a few dusty, old FORTRAN programs, functions, or subroutines that represent some pretty valuable legacy code. If these coded routines weren't considered to be so valuable, the engineers more than likely would never have saved them. Typically, these routines represent a treasure chest of tested, debugged, and proven code that is still relevant in today's engineering environment. The one big problem is that most of the software today is developed in C or C++. If this is the predicament that you find yourself in, there is some good news and some bad news for you. The good news is there is a good chance that the program manager and design engineering staff has at their disposal a wealth of proven FORTRAN code. Incorporating this proven code into a project very well could result in a significant reduction in labor costs and a significant reduction in program schedule. The bad news, of course, is that C or C++ are today's preferred languages; therefore writing deliverable code in FORTRAN is really not a viable option. So if you are a program manager or a design engineer, what can you do in a situation such as this? One alternative is to build a mixed language program, where the bulk of the code including the main is written in C/C++ and linked with one or more valuable FORTRAN legacy functions and/or subroutines. This appendix is a tutorial on how to do just that.

# Acknowledgments

I extend my sincere appreciation to Bernard Goodwin, executive editor of Pearson North America, Prentice Hall Professional technical publications, for his generous help and support during the difficult task of introducing a new author to the world of technical publishing. I would also like to acknowledge Dr. John Treichler of Applied Signal Technology for his gracious permission to reference his original work on transmultiplexers.

I would like to thank Richard Lyons, author of *Understanding Digital Signal Processing* (Prentice Hall); David Myers; Jim Kemerling; Michael Myers; and C. Britton Rorabaugh, author of *Notes on Digital Signal Processing* (Prentice Hall), for their technical review of the manuscript.

I would like to give a long overdue thank you to Mike Tate of Electromagnetic Systems Laboratories, the absolute best technician I have ever worked with, who helped me successfully begin my career so many years ago. I would like to thank Tom Ranweiler of Northrop Grumman, the most technically astute software design engineer I have ever worked with, who was the person who helped me finish my career on a successful note with the design of a unique special-purpose signal processing system. I am also thankful for the opportunity to have worked alongside the brilliant systems engineer Dr. Pin-Wei Chen of Northrop Grumman who helped keep me involved in internal research and development design projects.

Finally I would like to extend my deep appreciation to my wife Mary for her patience, support, understanding, and encouragement throughout this very long project.

# About the Author

**Richard Newbold** received his B.S.E.E. and M.S.E.E. degrees in 1974 and 1978, respectively, and has spent more than 30 years as a digital hardware design engineer and self-taught software designer. His design experience includes special-purpose signal processing hardware and computers that processed real time wideband signals, direct sequence spread spectrum system processors, PCM multirate processing systems, high-speed signal processing systems implemented on special-purpose gallium arsenide ASICs, transmultiplexers, channelizers, multirate filters, tuners, frequency synthesizers, DLLs, synchronous digital hierarchy (SDH) demultiplexers, fractional resamplers, adaptive filters, elastic store memories, adaptive beam forming, asynchronous clock recovery, and fault tolerant signal processors. His software experience includes real time signal processing, bit-level hardware simulations, microcode and bit slice programming, assembly programming, FORTRAN, C/C++, and Microsoft Windows graphics-oriented test stations, which were used to bit-level simulate, graphically display, and verify the proper operation of his digital creations.

# CHAPTER ONE

# Review of Digital Frequency

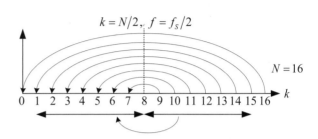

It is easy to mathematically represent an analog frequency on paper. The range of frequencies in the analog domain is both continuous and theoretically infinite. If we use the symbol $f_O$ to represent some arbitrary analog frequency all we need to do is to equate it with any one of an infinite number of available frequencies. We could, for example, choose $f_O$ to be equal to 23.456 Hz, or we could just as easily choose $f_O$ to be equal to 1.005 MHz. We could choose just about any other value to any precision that we can dream up. As long as we remain realistic, there is no limit on the values that $f_O$ can take on.

However, a digital system operates on digital data and generates digital results that are valid only at discrete increments of time equal to the period of the system sample clock. Therefore the value that a digitally generated discrete frequency can take on is a small subset of the range of values available to analog frequencies. The discrete frequency values within this subset are directly related to and dependent on the sample rate of the digital system clock.

This leads to some confusion when people deal with digital frequencies for the first time. Much of the confusion can be summed up with three frequently asked questions:

1. How do I define a digital frequency?
2. How do I mathematically represent a digital frequency?
3. How do I synthesize a digital frequency in hardware or software?

The scope of this chapter is to provide an answer for the first two of these questions. The answer to question number 3 requires its own chapter and is dealt with in detail in Chapter 8, "Digital Frequency Synthesis."

## 1.1  DEFINITIONS

In this chapter, we will make the following symbol definitions:

1. $f$ defines any arbitrary analog frequency in hertz.
2. $f_O$ defines a specific analog frequency in hertz.
3. $f_K$ defines a specific digital frequency in hertz.
4. $\omega_O$ defines a specific analog radian frequency in radians/second.
5. $\omega_K$ defines a specific digital radian frequency in radians/second.
6. $f_S$ defines the sample rate or the frequency of a digital system clock.
7. $T$ defines the period of the digital sample clock $T = 1/f_S$.

## 1.2  DEFINING DIGITAL FREQUENCIES

Unlike an analog frequency, a digitally generated frequency does not have infinite resolution. A digital frequency can only take on discrete values. A digital sine wave, for example, can only take on discrete values for frequency, phase, and amplitude. For the purposes of this chapter, the frequency resolution of a digitally generated sinusoid is limited by the period of the digital sample clock $T = 1/f_S$, and the precision of the sinusoidal waveform amplitude is limited by the bit width of each digital sample. Let us begin our discussion by considering a digital sinusoidal waveform.

We know that a sine wave has unity amplitude and is repetitive every $2\pi$ radians. As illustrated in Figure 1.1, we can draw a circle of radius 1, called the unit circle, and we can visualize a phasor of unity magnitude rotating around the unit circle at some fixed angular rate $\omega_K$. Every time the phasor makes a complete revolution around the unit circle, it has passed through $2\pi$ radians and has completed one cycle. We quantify the phasor's rotational speed as being $\omega_K$ radians per second. To get started, let us assign the label $C$ to this phasor. Since the phasor $C$ takes on values only at discrete instances of the sample period $T$, we can represent it as a function of discrete time by writing $C(nT)$ {for $n = 0, 1, 2, \cdots$}.

We can use some simple trigonometry to represent the phasor $C(nT)$ by its vertical and horizontal components, labeled $A(nT)$ and $B(nT)$ in Figure 1.1. The magnitude of $C$ is related to the value of its components by the Pythagorean theorem: $C(nT) = \sqrt{A^2(nT) + B^2(nT)}$.

We can see that as the phasor $C$ rotates around the unit circle, the magnitude of the $A$ component cyclically grows from 0 at 0 radians to +1 at $\pi/2$ radians. It then attenuates back to 0 at $\pi$ radians, grows to –1 at $3\pi/2$ radians, and finally attenuates back to a value of 0 as the phasor passes through

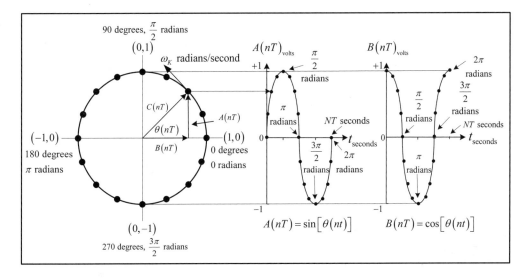

**Figure 1.1** Unit circle

$2\pi$ radians. Each time the phasor completes a rotation of $2\pi$ radians, the amplitude of the $A$ component traces out a sine wave and the amplitude of the $B$ component traces out a cosine wave, as illustrated in Figure 1.1.

Since we are dealing with a digital system, we know that the values of the phasor components $A$, $B$, and $C$, and the phasor phase angle $\theta$, shown in Figure 1.1, take on values only at discrete instants of time equal to the period of the sample frequency, or $T = 1/f_S$. Sequential sampling instants can be represented by the infinite series

$$1T,\ 2T,\ 3T, \cdots, nT, \cdots$$

We can represent any arbitrary sampling instant as $nT$ for $0 \le n < \infty$. Incorporating this notation into the unit circle phasor representation Figure 1.1, we see that the only values that can be represented by the phasor $C$, its vertical and horizontal components $A$ and $B$, and the phase angle $\theta$ are at the instants of time equal to $nT$. We know from trigonometry that

$$\sin(\theta) = \frac{A}{C} = \frac{A}{\sqrt{A^2 + B^2}}$$

and

$$\cos(\theta) = \frac{B}{C} = \frac{B}{\sqrt{A^2 + B^2}}$$

We also know that since we are working on the unit circle, the magnitude of $C = \sqrt{A^2 + B^2} = 1$. Therefore we can state that at any sampling instant $nT$,

$$A(nT) = \sin\left[\theta(nT)\right]$$

and

$$B(nT) = \cos\left[\theta(nT)\right]$$

### Equation 1.1

So far so good, but how do we quantify the discrete values taken on by the sinusoidal waveforms? Well, we can start by dividing the unit circle into $N$ equal arc segments, illustrated by the black dots on the unit circle in Figure 1.1. Each arch segment is a portion of the unit circle scribed by the tip of the phasor $C$ as the phase angle $\theta$ is incremented by $2\pi/N$ radians.

We will take this opportunity to coin a new and highly technical term. Let us define the angle $2\pi/N$ and the arch segment it describes as a *radian chunk*. The angle between the adjacent black dots on the unit circle in Figure 1.1 is equal to a radian chunk. The phasor $C(nT)$ and its components $A(nT)$ and $B(nT)$ can only be evaluated at each black dot corresponding to each radian chunk on the unit circle. Therefore the maximum number of samples that can represent a single cycle of a digital sine wave is equal to $N$.

If the digital oscillator that is generating the digital sinusoid is operating with a sample clock of $f_s$ Hz, then the phasor $C(nT)$ would rotate around the unit circle in discrete radian chunks of $2\pi/N$ at the clock rate of $f_s$ Hz. The lowest radian frequency of the digital oscillator can be mathematically defined as

$$\omega_K = \left(\frac{2\pi \text{ radians}}{N \text{ sample}}\right)\left(f_s \frac{\text{sample}}{\text{second}}\right) = \frac{2\pi}{N} f_s \text{ radians/second}$$

$$\omega_K = \frac{2\pi}{N} f_s \text{ radians/second}$$

### Equation 1.2

The lowest or fundamental frequency of the digital oscillator can be mathematically defined as

$$f_K = \left(\frac{1}{2\pi \text{ radians}}\right)\left(\frac{2\pi \text{ radians}}{N \text{ sample}}\right)\left(f_s \frac{\text{sample}}{\text{second}}\right) = \frac{1}{N} f_s \text{ second}^{-1} = \frac{1}{N} f_s \text{ Hz}$$

$$f_K = \frac{1}{N} f_s \text{ Hz}$$

**Equation 1.3**

If we think about it for a minute, Equation 1.2 and Equation 1.3 make sense. If the phasor points to each black dot on the unit circle for one sample period, it will take $N$ sample periods for it to move from dot zero to dot $N-1$. In doing so, it will make one revolution of the unit circle stopping once at each black dot in $N$ sample periods. It will take $N$ sample periods to trace out exactly one sinusoidal cycle. The total period for each cycle will be equal to $NT$ sec. The frequency of this sinusoidal cycle is then given by $f = 1/(NT \text{ second}) = (f_s/N)\text{Hz}$. Therefore the frequency of the sinusoids traced by the $A(nT)$ and $B(nT)$ components will be equal to $(f_s/N)\text{Hz}$.

Let us look at a very simple example. Suppose we have a digital oscillator clocked with a sample clock of $f_s = 32$ Hz, and suppose we decided that $N$ will be 16. The unit circle is subdivided into 16 equal arc lengths, giving us 16 equal radian chunks. The rotating phasor $C(nT)$ will be evaluated at 16 locations around the unit circle. This means there will be 16 samples per each period of the synthesized sinusoidal waveform. The digital radian frequency would be

$$\omega_K = \left(\frac{2\pi \text{ radians}}{16}\right)\left(\frac{32}{\text{sec}}\right) = 4\pi \text{ radians/second}$$

or, since $f_K = \omega_K/2\pi$, we can easily compute the digital oscillator frequency to be

$$f_K = \frac{\omega_K}{2\pi} = \left(\frac{4\pi \text{ radians/second}}{2\pi \text{ radians}}\right) = 2 \text{ second}^{-1} = 2 \text{ Hz}$$

In this simple example, 2 Hz is the lowest frequency other than zero that our simple digital oscillator can generate. This is based on the value of the sample frequency $f_s$ and our choice for the value of $N$. If, for the same sample rate, we had chosen $N$ to be a larger number, then the resolution of $f_K$ would have been greater. For example, if we had selected $N = 64$, the lowest frequency other than 0 Hz that our oscillator could produce would be

$$f_K = \frac{1}{N} f_s \text{ Hz} = \frac{1}{64} 32 \text{ Hz} = 0.5 \text{ Hz}$$

This is a good start, but a digital oscillator that can produce only a single frequency isn't as useful as an oscillator that can be programmed to produce any one of a whole range of discrete frequencies. Ideally, we would like to be able to program the digital oscillator to output any one of a wide range of discrete frequencies. We can achieve this enhancement with the addition of a multiplier "$k$" in Equation 1.2 and Equation 1.3. We can rewrite these equations to include the multiplier $k$ such that

$$\omega_K = \frac{2\pi k}{N} f_s \text{ radians/second}$$
$$\left\{ \text{where } k = 0, 1, 2, ... N/2 \right\}$$
$$f_K = \frac{k}{N} f_s \text{ Hz}$$

**Equation 1.4**

If $k = 1$, then the frequencies represented by Equation 1.4 are identical to those represented by Equation 1.2 and Equation 1.3. The value of $k$ can take on discrete integer values ranging from 0 to $N/2$. In our previous example, we set $f_s = 32$ Hz, and $N = 16$ so $k$ could take on values of $0, 1, 2, 3, 4, 5, 6, 7, 8$. All the possible frequency values that this example oscillator can take on are tabulated in Table 1.1.

As we can see from the table, this oscillator can be programmed to produce one of nine possible frequencies with a resolution of 2 Hz. The addition of the variable $k$ in Equation 1.4 causes the phasor $C(nT)$ to rotate around the unit circle in multiples of $2\pi/N$ radian chunks at a rate of $f_s$ sample per second. When $k$ is set to unity, the phasor will take on values at every black dot on the unit circle producing the oscillator's lowest or fundamental frequency. In this case, each cycle of the generated sinusoid will be composed of $N$ samples.

When $k$ is set to 2, the phase angle of the phasor $\theta(nT)$ will increase in increments of two radian chunks each tick of the sample clock. The phasor $C(nT)$ will take on the values of every second dot, and it will rotate around the unit circle twice as fast, producing a sine or cosine wave that is twice the fundamental frequency. In this case, each cycle of the sinusoid will be composed of half or $N/2$ samples. Similarly, when $k$ is set to 4 the phasor will travel around the unit circle at four times the fundamental rate, taking on values at every fourth dot to produce an output frequency that is four times the fundamental frequency. Each cycle, however, will be composed of $N/4$ number of samples per cycle.

**Table 1.1**   Example Digital Oscillator Frequencies

| $k$ | $\omega_k \dfrac{\text{radians}}{\text{second}}$ | $f_K$ Hz | Samples per cycle |
|---|---|---|---|
| 0 | 0.00 | 0 | — |
| 1 | $\pi/8$ | 2 | 16 |
| 2 | $2\pi/8$ | 4 | 8 |
| 3 | $3\pi/8$ | 6 | 16/3 |
| 4 | $4\pi/8$ | 8 | 4 |
| 5 | $5\pi/8$ | 10 | 16/5 |
| 6 | $6\pi/8$ | 12 | 16/6 |
| 7 | $7\pi/8$ | 14 | 16/7 |
| 8 | $8\pi/8$ | 16 | 2 |

When $k$ reaches its maximum value of $N/2$, which in this example is 8, there will be just two samples per sinusoidal period. The Nyquist rule states that two samples per cycle is the minimum number of samples allowed in order to be able to reconstruct an analog waveform from a digital waveform. This means that the highest frequency we can theoretically generate with a digital oscillator is half the sample rate or $f_s/2$. In our example, when $k = N/2 = 8$, we were able to generate a sinusoid of 16 Hz, which is exactly half the 32 Hz sample rate.

In Table 1.1, a few of the entries for the "Samples per cycle" column are in fractions. All this means is that each cycle of the sinusoid is generated using a different subset of the $N$ possible samples. That is, successive cycles of the sinusoidal waveform begin on a different sample value.

What happens if we continue to increase the value of $k$ beyond $N/2$ (which in our example is 8)? If we were to let $k = 9$, there would be less than two samples per cycle, the Nyquist rule would be violated, and the output waveform would take on the same frequency as if the value of $k$ had been set to 7. The resulting frequency is said to have been aliased or folded, about $f_s/2$. Undersampling an analog sinusoid with an analog to digital converter will cause the digital output sinusoid it to alias down in frequency. This is identical to setting the multiplication factor $k$ of a digital frequency to some number greater than $N/2$, which will result in the folding or aliasing about the Nyquist frequency of $f_s/2$.

In our example, if we were to say $k = 10$, the resulting digital frequency would be identical to that obtained by saying $k = 6$. If we choose $k = N - 1$

or 15 in our example, the resulting frequency would be identical to the case where $k = 1$. We can see that all frequencies above $k = 8$ are folded or aliased down in frequency about the so-called folding frequency of $f_s/2 = 16$ Hz. In general, the frequency $f_{N-K}$ aliases down to $f_K \{\text{for } (N/2) < k \leq (N-1)\}$.

A simple frequency folding diagram for the case where $N = 16$ is illustrated in Figure 1.2.

Care must be taken at the higher frequencies where the value of $k$ approaches the upper limit of $N/2$. On paper there is no problem as $k \rightarrow N/2$, but in a hardware or software implementation the samples must be carefully selected. An extreme example would be setting $k = N/2$ and using the samples at 0 and $\pi$ radians. This represents the optimum sample selection for a cosine wave, since the cosine sequence will take on the form

$$B(T) = \cos(n\pi) = +1, -1, +1, -1, \cdots \quad \{\text{for } n = 0, 1, 2, 3, \cdots\}$$

but these same samples will produce a DC output of 0 for a sine waveform

$$A(T) = \sin(n\pi) = 0, 0, 0, 0, \cdots \quad \{\text{for } n = 0, 1, 2, 3, \cdots\}$$

For this reason, in most designs dealing with narrow band signals, the minimum number of samples per cycle is usually held to some number around 2.5.

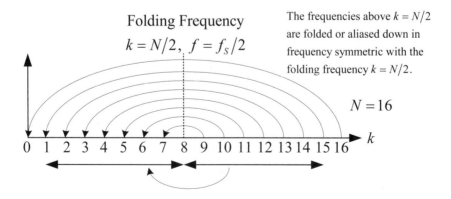

**Figure 1.2**   Frequency folding diagram

## 1.3   MATHEMATICAL REPRESENTATION OF DIGITAL FREQUENCIES

The notation in Equation 1.4 gives us a valid method to represent digital frequencies using a pencil and paper, but it doesn't help us when it comes to the generation of a digital sinusoidal waveform in hardware or software. In Equation 1.4, we explicitly included the sample rate represented by the term $f_s$. This is fine for mathematical computations on paper but the sample rate is a fixed entity that is already implicit in the operation of the digital hardware. It does not make sense to include the $f_s$ term in frequency synthesis because it is already present due to the fact that each digital computation takes place at the sample rate.

In addition, Equation 1.4 provides us with a single value for any particular frequency—something that we can write down on paper like the value 16 Hz or 48 Hz. This is not appropriate for the actual synthesis of a complete period of a sinusoidal frequency in hardware. In hardware or software, we need to generate all $N$ samples of a sinusoid at the sample rate $f_s$. To do that, we need to take into account the sample index $n$ of the generated sinusoid, and we need to normalize the sample frequency to 1. We normalize the frequency by dividing the frequency expression by $f_s$. In doing so, we can rewrite Equation 1.4 to include these changes and produce

$$\omega_K(n) = \frac{2\pi kn}{N} \text{ radians/second}$$
$$f_K(n) = \frac{kn}{N} \text{ Hz} \qquad \text{where} \begin{Bmatrix} k = 0,1,2,\ldots N/2 \\ n = 0,1,2,3,\ldots \infty \end{Bmatrix}$$

**Equation 1.5**

In Equation 1.5 we have normalized the sample frequency to 1. The normalized frequency of a digital waveform is usually expressed as a fraction given by $k/N$. To convert a normalized frequency back to an unnormalized frequency, we simply multiply by the sample rate, or

$$f_K = \frac{k}{N} f_s$$

Remember the sample rate is implicit in a hardware or software implementation. It is the rate at which the computations are performed (i.e., the rate at which the phasor advances between black dots on the unit circle).

For example, if our sample frequency $f_s = 256$ KHz, $k = 16$, and $N = 128$, then the normalized frequency would be expressed as   follows:

$f_K = k/N = 16/128 = 0.125$. The actual frequency in Hz can be determined by multiplying the fraction $k/N$ by the actual sample frequency $f_s$ or

$$f_K = \frac{k}{N} f_s = (0.125)(256 \text{ KHz}) = 32 \text{ KHz}$$

The function of the sample index $n$ in Equation 1.5 is to index successive samples and to advance the phasor around the unit circle by acting as an incrementing multiplier to the fixed value $k/N$. The term $nk/N$ can be thought of as an infinite series given by

$$\frac{k}{N}, \frac{2k}{N}, \frac{3k}{N}, \cdots, \frac{nk}{N}, \cdots$$

The correct way to generate a digital sinusoid waveform of unity magnitude and normalized frequency $k/N$ is given by

$$A(n) = \sin\left(\frac{2\pi k}{N} n\right)$$
$$\text{for} \left\{ \begin{array}{l} n = 0, 1, 2, \cdots \\ k = 0, 1, 2, \cdots, N/2 \end{array} \right\}$$
$$B(n) = \cos\left(\frac{2\pi k}{N} n\right)$$

**Equation 1.6**

It is important to remember that the term $\dfrac{2\pi}{N}$ is a radian chunk. The term $\dfrac{2\pi}{N} k$ is a $k$ multiple of a radian chunk. Since the sample rate is implicit to the hardware and since we are dealing with normalized values of frequency, we can now drop the index notation $nT$ as illustrated in Figure 1.1 and simply refer to the index as $n$ as we did in Equation 1.5 and Equation 1.6. The phasor $C(n)$ will rotate around the unit circle at the sample clock rate in increments of $2\pi k/N$ radian chunks. This means that the phasor sine and cosine components $A(n)$ and $B(n)$ will take on discrete values at each $k$th radian chunk and will do so at the clock rate.

Nothing is stopping us from writing down on a sheet of paper that the radian frequency associated with this radian chunk is $(2\pi k/N) f_s$, but keep in mind that this is just a number on a sheet of paper; it is not the sine or cosine argument that will trace out a sinusoidal waveform. This is where the sample index $n$ comes into play. The index increments by one every sample clock so the argument $\dfrac{2\pi}{N}nk$ takes on incremental $k$ radian chunk values with each clock period. This means that $\omega_k = (2\pi nk/N)$ will take on successive values of $0, 2\pi k/N, 4\pi k/N, 6\pi k/N, \cdots$ at the sample rate. This is the dynamic argument of the sine and cosine function. If we plot $A(n)$ and $B(n)$ for this dynamic argument as $n = 0,1,2,3,\cdots$, we will trace out a sine and a cosine waveform at the radian frequency $(2\pi k/N)f_s$.

For example, suppose we let $k = 1$, $N = 8$, and $f_s = 32\ \text{KHz}$. The sine argument would be implemented as

$$\frac{2\pi kn}{N} = \frac{2\pi n}{8} \quad \text{for } n = 0,1,2,3,\cdots$$

The sine waveform for the first eight values of the sample index $n$ is illustrated in Figure 1.3. The fundamental frequency of this sinusoid is computed by

$$\omega_K = \omega_1 \quad = \left(\frac{2\pi k}{N}\right)f_s \quad = \left(\frac{2\pi}{8}\right)32\ \text{KHz} \quad = 8\pi\ \text{K radians/second}$$

$$f_K = f_1 \quad = \left(\frac{k}{N}\right)f_s \quad = \left(\frac{1}{8}\right)32\ \text{KHz} \quad = 4\ \text{KHz}$$

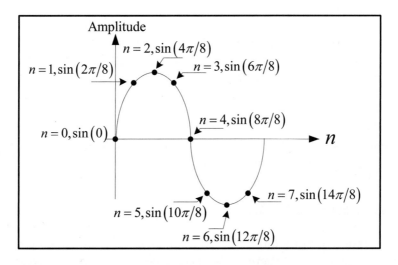

**Figure 1.3**  Mapped sine wave using a dynamic argument

The fundamental frequency is generated when $k = 1$. Multiples of the fundamental frequency are referred to as harmonic frequencies. Frequencies generated for $k = 2, 3, 4, \cdots, N/2$ are harmonics of the fundamental. When the multiplier $k = 0$, we end up with a frequency of 0 Hz, or DC.

We can represent the phase offset in a digital sinusoid by the expression

$$\sin\left(\frac{2\pi k n}{N} + \Phi\right) \text{ where } \Phi = \frac{2\pi p}{N} \quad 0 \le p \le N - 1$$

The phase offset can only take on values equal to $p$ radian chunks.

When implementing a digital sinusoidal generator in hardware, it is of paramount importance to remember the following seven points:

1. Since Equation 1.6 is computed every $T$ seconds, it implicitly includes the sample rate term $f_s$. This is because the index $n$ increments at the sample rate.

2. In normalized notation, the frequency of the sinusoid in Equation 1.6 is given by the ratio of $k/N$, where $0 \le k/N \le 0.5$.

3. The frequency resolution of the digital sinusoid waveforms $A(n)$ and $B(n)$ is determined by the size of $N$.

4. The phase resolution of the digital sinusoid waveforms $A(n)$ and $B(n)$ is determined by the size of $N$.

5. The precision of the amplitude of the sinusoidal waveform is determined by the bit width of the samples used to represent both $A(n)$ and $B(n)$.

6. The base or fundamental frequency of the sinusoids in Equation 1.6 is equal to $1/N$ normalized and $f_s/N$ unnormalized.

7. The set of discrete frequencies that can be output by a programmable oscillator is given by $k/N$ for $k = 0, 1, 2, \ldots, N/2$, which equates to a normalized frequency range of 0 to 0.5 in steps of $1/N$, or an actual frequency range of 0 Hz to $f_s/2$ Hz in steps of $f_s/N$.

We will show in Chapter 8, "Digital Frequency Synthesis," that the value of $k$ in item 7 can take on fractional values, which allows the engineer to design synthesizers that have much finer frequency resolution.

## 1.4 NORMALIZED FREQUENCY

The notation in Equation 1.4 is based on the concept of dividing the unit circle into $N$ equal arc segments. We can derive an equivalent expression simply by observing that the digital frequency

$$f_K = \frac{k}{N} f_S \quad \text{where } 0 \le k \le \frac{N}{2}$$

can be normalized by dividing both sides of the equation by the sample frequency $f_S$ to arrive at

$$\frac{f_K}{f_S} = \frac{k}{N} \quad \text{where } 0 \le f_K \le \frac{f_S}{2} \text{ and } 0 \le k \le \frac{N}{2}$$

**Equation 1.7**

Both sides of Equation 1.7 are now expressed as a fraction between 0 and 0.5. We can drop the subscript $K$ such that $f_K$ becomes $f$. When we do we can think of both $f$ and $f_S$ as being two analog frequencies whose ratio just happens to be $k/N$. This notation is useful if the unit circle is not specifically considered in the derivation of digital frequencies. The two methods of notation are equivalent, as illustrated in Equation 1.8.

$$\cos\left(\frac{2\pi f}{f_S} n\right) = \cos\left(\frac{2\pi k}{N} n\right) \quad \text{where } \left\{ \begin{array}{l} n = 0,1,2,3\cdots \\ \frac{f}{f_S} = \frac{k}{N} \end{array} \right\}$$

**Equation 1.8**

It's a matter of preference. Either notation is correct. This book uses both notations where appropriate.

## 1.5  REPRESENTATION OF DIGITAL FREQUENCIES

A digital frequency can be written on paper in units of Hz or in units of radians per second as

$$f_K = \frac{k}{N} f_S \quad \text{or} \quad \omega_K = \frac{2\pi k}{N} f_S$$

A more common method is to express a digital frequency as a fraction where the sampling frequency has been normalized to 1, such as

$$f_K = \frac{k}{N} \quad \text{or} \quad \omega_K = \frac{2\pi k}{N}$$

**Table 1.2**  Four Ways to Represent a Digital Frequency

|  |  | $k = 0$ | $k = 1$ | $k = 2$ | $\cdots$ | $k = N/2$ |
|---|---|---|---|---|---|---|
| Normalized radians | $\omega_k = \dfrac{2\pi k}{N}$ | 0 | $\dfrac{2\pi}{N}$ | $\dfrac{4\pi}{N}$ | $\cdots$ | $\pi$ |
| Radians | $\omega_k = \dfrac{2\pi k}{N} f_s$ | 0 | $\dfrac{2\pi}{N} f_s$ | $\dfrac{4\pi}{N} f_s$ | $\cdots$ | $\pi f_s$ |
| Normalized frequency | $f_k = \dfrac{k}{N}$ | 0 | $\dfrac{1}{N}$ | $\dfrac{2}{N}$ | $\cdots$ | 0.5 |
| Frequency | $f_k = \dfrac{k}{N} f_s$ | 0 | $\dfrac{f_s}{N}$ | $\dfrac{2 f_s}{N}$ | $\cdots$ | $\dfrac{f_s}{2}$ |

In all cases, the value of $k$ can range from 0 to $N/2$. Examples of the four representations of digital frequencies are illustrated in Table 1.2 for several values of $k$.

A frequency in the analog domain has infinite resolution and therefore can take on all possible values. A frequency in the digital domain can only take on specific discrete values, which are multiples of $f_s/N$. The value of an analog frequency can always be made to match exactly the value of a digital frequency, but since the value of the digital frequency does not have infinite resolution, the opposite is not true.

The amplitude of an analog frequency can take on an infinite number of different values, whereas the amplitude of a digital frequency can only take on discrete values. The number of amplitude values that can be represented by a digital sinusoid is dependent on the bit width of the individual digital samples. For example, suppose the bit width of a bipolar sinusoidal sequence is given by $B$. The sinusoidal amplitude can take on values equal to 0 and $\pm \left[ 1, 2, 3, \cdots, \left( 2^{B-1} - 1 \right) \right]$.

The sole purpose of this chapter is to introduce the mathematical representation of frequency in the digital domain. This book contains a great deal more information on the subject. For a detailed analysis and tutorial on the synthesis of digital frequencies the reader is encouraged to read Chapter 8, "Digital Frequency Synthesis."

# CHAPTER TWO

# Review of Complex Variables

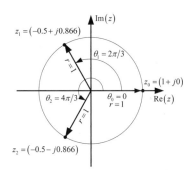

As the title of the chapter suggests, this is a review of complex variables. This review material is written with a significant amount of detail. Therefore it is possible for a reader with no previous exposure to the world of complex variables to gain a good understanding and to develop a working knowledge of the subject simply by reading this chapter. There have been many good textbooks written by astute authors that specifically deal with complex variables. A reader who wishes to obtain a more in depth treatment of the subject is directed to any or all of the excellent works listed in this chapter's references.

A function of a real variable such as $y = f(x) = 2x^2 + 4$ is an expression that computes a value for the real variable $y$ as a function of the value of the single real variable $x$. If we were to graph this expression, we would find that the independent variable $x$ is allowed to take on any value along the real or $x$-axis. When we insert this value for the variable $x$ into this expression we can compute the corresponding value of the dependent variable. This is why we say that $y$ is a function of $x$, expressed as $y = f(x)$. In a real function, the independent variable $x$ varies along a one-dimensional horizontal line called the $x$-axis, and the dependent variable $y$ varies along the one-dimensional vertical line called the $y$-axis.

The study of complex variables is an extension of the study of real variables. The standard convention is to let the symbol $z$ represent a complex number that is dependent upon two variables $x$ and $y$. An example of a complex variable is given as $z = x \pm jy$. Both $x$ and $y$ are real variables that can take on real values anywhere in the two-dimensional $x, y$ plane, referred to as the *complex plane*.

We introduce another complex variable with the standard notation $w$, which is analogous to the real variable $y$. This allows us to write an expression for complex variables that is similar to the expression we wrote for real variables—that is, $w = f(z) = f(x, y)$. To help clarify the notational concept, an example of an arbitrary real and an arbitrary complex function might be

$$y = f(x) = x^3 - 2x^2 + 5x + 3 \qquad \{\text{example function of a real variable}\}$$
$$w = f(z) = (x^2 - 2y) + j(3x + 4y^2) \quad \{\text{example function of a complex variable}\}$$

One major notational difference between a real and a complex number is that the complex number includes the pesky little term $j$. The $j$ term is a mathematical abstraction. It is assigned the value $j = \sqrt{-1}$, and its square is $j^2 = -1$. Typically, mathematicians use the term $i$, while engineers use the term $j$. In this book, we will for the most part stick with $j$.

This is a rather awkward number, and because it is difficult to conceptualize, mathematicians have given it the unfortunate name "imaginary" number. This is probably why they refer to it as $i$. Any expression associated with $j$ or $i$ is also labeled as imaginary. In our example of a complex function

$$w = f(z) = (x^2 - 2y) + j(3x + 4y^2)$$

the term $(x^2 - 2y)$ is called the real part of $f(z)$ and the term $(3x + 4y^2)$ is called the imaginary part of $f(z)$. However, both terms consist of real numbers and therefore the imaginary label is kind of misleading. From a graphical point of view, the real part of a complex number is generally associated with the real or $x$-axis, and the imaginary part of a complex number is generally associated with the imaginary or $y$-axis. For example, the complex number $z = 4 + j5$ is plotted as a point in the complex plane that is located 4 units in the $x$ direction and 5 units in the $y$ direction.

Complex numbers, complex functions, and complex computations are common in the world of electronic communications engineering. Engineers being extremely practical people have a solution to this imaginary notational quandary. Electronic engineers generally do not work with complex numbers; they usually work with complex electrical signals composed of a pair of electromagnetic or electrical waves that are $90°$ out of phase with one another. They refer to the real part of the complex signal as the "in-phase" component, and they refer to the imaginary part of the signal as the "quadrature" component.

For example, design engineers dealing with complex signals such as a signal output by a complex signal tuner, whether they be analog or digital, usually refer to the two signal components as the in-phase component and the quadrature component. In this regard, it is helpful to visualize the imaginary part of a complex number to be in "quadrature" with respect to the real or "in-phase" part.

The in-phase and quadrature notation works well and it does not infer any mystical, supernatural, or unrealizable properties to a number or signal (like the term *imaginary* does). Since engineers deal with both mathematicians and with other engineers within their industry on a daily basis, they tend to use these terms interchangeably. Therefore, in this book, we will use the terms *real* and *in phase* interchangeably, and we will also use the terms *imaginary* and *quadrature* interchangeably. The convention seems to be that when referring to mathematical expressions, we use the terms *real* and *imaginary*, but we use the terms *in phase* and *quadrature* when referring to signal components produced and processed by electronic circuits or digital hardware. In the study of complex variables, we make the following definitions with respect to $j$:

$$
\begin{aligned}
j &= \sqrt{-1} \\
j^2 &= -1 \\
\frac{1}{j} &= \frac{j}{j^2} = -j \\
j &= \frac{j^2}{j} = \frac{-1}{j}
\end{aligned}
$$

**Equation 2.1**

Complex numbers can be expressed in either Cartesian or in polar form. We begin our discussion of complex variables by defining the properties of each form.

## 2.1   CARTESIAN FORM OF COMPLEX NUMBERS

A complex number is an ordered pair of real numbers $x$ and $y$ such that $z = (x, y) = x \pm jy$. Given a complex variable $z = x + jy$ we can say that

$$
\begin{aligned}
\mathrm{Re}(z) &= x \quad \{\text{real part of } z\} \\
\mathrm{Im}(z) &= y \quad \{\text{imaginary part of } z\}
\end{aligned}
$$

An example of a complex point in Cartesian coordinates is illustrated in Figure 2.1.

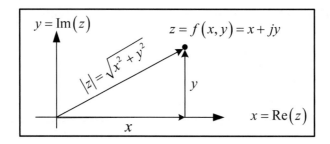

**Figure 2.1**   Complex point in Cartesian form

### 2.1.1   Properties of Complex Numbers Expressed in Cartesian Form

Given two complex numbers $z_1$ and $z_2$, such that

$$z_1 = x_1 + jy_1$$
$$z_2 = x_2 + jy_2$$

we can define the following properties:

**Equality**

$$z_1 = z_2 \text{ if } \left\{ x_1 = x_2 \text{ and } y_1 = y_2 \right\}$$

**Equation 2.2**

**Addition**

$$z_1 \pm z_2 = \left( x_1 \pm x_2 \right) + j \left( y_1 \pm y_2 \right)$$

**Equation 2.3**

**Multiplication**

$$z_1 z_2 = \left[ \left( x_1 x_2 - y_1 y_2 \right) + j \left( x_1 y_2 + x_2 y_1 \right) \right]$$

**Equation 2.4**

**Division**

$$\frac{z_1}{z_2} = \frac{z_1 \bar{z}_2}{z_2 \bar{z}_2} = \frac{\left( x_1 x_2 + y_1 y_2 \right) + j \left( x_2 y_1 - x_1 y_2 \right)}{x_2^2 + y_2^2} \quad \text{for } z_2 \neq 0$$

**Equation 2.5**

## Magnitude

$$|z| = \sqrt{x^2 + y^2}$$

### Equation 2.6

**Conjugates.** The conjugate of $z$ is represented by the notation $\bar{z}$. If a complex number is given by $z = x + jy$, then its conjugate is given by $\bar{z} = x - jy$. From this, we can develop several properties of complex numbers and their conjugates.

1. The conjugate of a sum is equal to the sum of conjugates $\overline{(z_1 + z_2)} = \bar{z}_1 + \bar{z}_2$.
   Proof: Given two complex numbers $z_1 = x_1 + jy_1$ and $z_2 = x_2 + jy_2$, then

$$\bar{z}_1 + \bar{z}_2 = (x_1 - jy_1) + (x_2 - jy_2)$$
$$= (x_1 + x_2) - j(y_1 + y_2)$$
$$= \overline{(x_1 + x_2) + j(y_1 + y_2)}$$
$$= \overline{z_1 + z_2}$$

We can therefore state that

$$\overline{(z_1 + z_2)} = \bar{z}_1 + \bar{z}_2$$

### Equation 2.7

and, by extension,

$$\overline{(z_1 + z_2 + z_3 + \cdots + z_n)} = \bar{z}_1 + \bar{z}_2 + \bar{z}_3 + \cdots + \bar{z}_n$$

### Equation 2.8

2. The conjugate of a product is equal to the product of the conjugates $\overline{z_1 z_2} = \bar{z}_1 \bar{z}_2$.
   Proof: Given two complex numbers $z_1 = x_1 + jy_1$ and $z_2 = x_2 + jy_2$, then

$$\bar{z}_1 \bar{z}_2 = (x_1 - jy_1)(x_2 - jy_2)$$
$$= (x_1 x_2 - y_1 y_2) - j(x_1 y_2 + x_2 y_1)$$
$$= \overline{(x_1 x_2 - y_1 y_2) + j(x_1 y_2 + x_2 y_1)}$$
$$= \overline{z_1 z_2}$$

We can therefore state that

$$\overline{z_1 z_2} = \overline{z}_1 \overline{z}_2$$

**Equation 2.9**

and, by extension,

$$\overline{\left(z_1 z_2 z_3 \cdots z_n\right)} = \overline{z}_1 \overline{z}_2 \overline{z}_3 \cdots \overline{z}_n$$

**Equation 2.10**

3. The conjugate of a quotient is equal to the quotient of conjugates $\overline{\left(\dfrac{z_1}{z_2}\right)} = \dfrac{\overline{z}_1}{\overline{z}_2}$.

   Proof: Given two complex numbers $z_1 = x_1 + jy_1$ and $z_2 = x_2 + jy_2$, then

$$\overline{\left(\frac{z_1}{z_2}\right)} = \overline{\left(\frac{x_1 + jy_1}{x_2 + jy_2}\right)\left(\frac{x_2 - jy_2}{x_2 - jy_2}\right)}$$

$$= \overline{\left(\frac{\left(x_1 x_2 + y_1 y_2\right) - j\left(x_1 y_2 - x_2 y_1\right)}{x_2^2 + y_2^2}\right)}$$

$$= \frac{\left(x_1 x_2 + y_1 y_2\right) + j\left(x_1 y_2 - x_2 y_1\right)}{x_2^2 + y_2^2}$$

and

$$\frac{\overline{z}_1}{\overline{z}_2} = \frac{\overline{\left(x_1 + jy_1\right)}}{\overline{\left(x_2 + jy_2\right)}}$$

$$= \left(\frac{x_1 - jy_1}{x_2 - jy_2}\right)\left(\frac{x_2 + jy_2}{x_2 + jy_2}\right)$$

$$= \frac{\left(x_1 x_2 + y_1 y_2\right) + j\left(x_1 y_2 - x_2 y_1\right)}{x_2^2 + y_2^2}$$

The two results are equal, so therefore we can state the following:

$$\overline{\left(\frac{z_1}{z_2}\right)} = \frac{\overline{z}_1}{\overline{z}_2}$$

**Equation 2.11**

4. A complex number summed with its conjugate $z + \overline{z} = 2\,\mathrm{Re}(z)$.
   Proof: Given a complex number $z = x + jy$ then

$$z + \overline{z} = (x + jy) + (x - jy) = 2x = 2\,\mathrm{Re}(z)$$

**Equation 2.12**

5. A complex number multiplied by its conjugate $z\,\overline{z} = |z|^2$.
   Proof: Given a complex number $z = x + jy$ then

$$z\,\overline{z} = (x + jy)(x - jy) = (x^2 + y^2) = |z|^2$$

**Equation 2.13**

## 2.2   POLAR FORM OF COMPLEX NUMBERS

Many times, computations involving complex variables are simplified by working with numbers formatted in polar form. This paragraph will discuss many of the properties of complex polar arithmetic.

We say that the modulus $r$ of the complex variable $z$ is the magnitude of $z$ or

$$r = |z| = \sqrt{x^2 + y^2}$$

**Equation 2.14**

This is the length or magnitude of a vector that extends from the origin of the complex plane to the point defined by $z$. We can also say that the argument of $z$ is the angle that the magnitude vector makes with the $x$-axis or

$$Arg(z) = \angle z = \theta = \tan^{-1}\left(\frac{y}{x}\right)$$

**Equation 2.15**

We can say that the complex variable $z$ is a function of the modulus and phase angle or $z = f(r, \theta)$. Both the magnitude and the argument of $z$ are illustrated in Figure 2.2.

Given the complex number defined by $z = x + jy$, we can derive the representation for $z$ in polar coordinates. Simple trigonometry tells us that

$$x = r\cos\theta$$
$$y = r\sin\theta$$

**Equation 2.16**

The magnitude of $z$ is $r = \sqrt{x^2 + y^2}$.

Proof: From Figure 2.2, we know that $x = r\cos\theta$ and $y = r\sin\theta$. It follows that

$$|z| = \sqrt{r^2\cos^2\theta + r^2\sin^2\theta} = \sqrt{r^2\left(\cos^2\theta + \sin^2\theta\right)} = \sqrt{r^2} = r$$

and

$$|z| = \sqrt{r^2\cos^2\theta + r^2\sin^2\theta} = \sqrt{x^2 + y^2}$$

and therefore

$$r = \sqrt{x^2 + y^2}$$

**Equation 2.17**

The argument of $z$ is

$$\theta = \tan^{-1}\left(\frac{y}{x}\right)$$

**Equation 2.18**

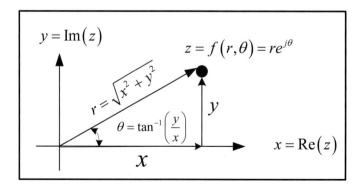

**Figure 2.2**   Complex point in polar form

Conversion between Cartesian to polar coordinates is given by

$$z = (x + jy) = r(\cos\theta + j\sin\theta)$$

**Equation 2.19**

## 2.2.1   Properties of Complex Numbers Expressed in Polar Form

Given two complex numbers expressed in polar coordinates

$$z_1 = r_1(\cos\theta_1 + j\sin\theta_1)$$
$$z_2 = r_2(\cos\theta_2 + j\sin\theta_2)$$

we can define the following properties.

**Equality**

$$z_1 = z_2 \text{ if } \{r_1 = r_2 \text{ and } \theta_1 = \theta_2\}$$

**Equation 2.20**

**Multiplication**

$$z_1 z_2 = r_1 r_2 \left[\cos(\theta_1 + \theta_2) + j\sin(\theta_1 + \theta_2)\right]$$

**Equation 2.21**

In this proof we will also show the following:

- The magnitude of a product is the product of the magnitudes.
- The argument of a product is the sum of arguments.

Proof:

$$
\begin{aligned}
z_1 z_2 &= r_1(\cos\theta_1 + j\sin\theta_1)r_2(\cos\theta_2 + j\sin\theta_2) \\
&= r_1 r_2 \left[(\cos\theta_1 \cos\theta_2 - \sin\theta_1 \sin\theta_2) + j(\sin\theta_1 \cos\theta_2 + \cos\theta_1 \sin\theta_2)\right] \\
&= r_1 r_2 \left[\cos(\theta_1 + \theta_2) + j\sin(\theta_1 + \theta_2)\right]
\end{aligned}
$$

Example: If

$$z_1 = 2\left(\cos 45° + j\sin 45°\right)$$

$$z_2 = 3\left(\cos 90° + j\sin 90°\right)$$

then

$$z_1 z_2 = 6\left(\cos 135° + j\sin 135°\right)$$

## Division

$$\frac{z_1}{z_2} = \frac{r_1}{r_2}\left[\cos\left(\theta_1 - \theta_2\right) + j\sin\left(\theta_1 - \theta_2\right)\right]$$

**Equation 2.22**

In this proof we will also show the following:

- The magnitude of a quotient is the quotient of the magnitudes.
- The argument of a quotient is the difference of arguments.

Proof:

$$\frac{z_1}{z_2} = \left[\frac{r_1\left(\cos\theta_1 + j\sin\theta_1\right)}{r_2\left(\cos\theta_2 + j\sin\theta_2\right)}\right]$$

$$= \frac{r_1\left(\cos\theta_1 + j\sin\theta_1\right)\left(\cos\theta_2 - j\sin\theta_2\right)}{r_2\left(\cos\theta_2 + j\sin\theta_2\right)\left(\cos\theta_2 - j\sin\theta_2\right)}$$

$$= \frac{r_1}{r_2}\left[\frac{\left(\cos\theta_1\cos\theta_2 + \sin\theta_1\sin\theta_2\right) + j\left(\sin\theta_1\cos\theta_2 - \sin\theta_2\cos\theta_1\right)}{\cos^2\theta_2 + \sin^2\theta_2}\right]$$

$$= \frac{r_1}{r_2}\left[\frac{\cos\left(\theta_1 - \theta_2\right) + j\sin\left(\theta_1 - \theta_2\right)}{1}\right]$$

$$= \frac{r_1}{r_2}\left[\cos\left(\theta_1 - \theta_2\right) + j\sin\left(\theta_1 - \theta_2\right)\right]$$

## Magnitude

$$r = \sqrt{x^2 + y^2}$$

**Equation 2.23**

Proof: The Pythagorean theorem states that for a right triangle the length or magnitude of the hypotenuse is equal to the square root of the sum the squares of the two sides. See Figure 2.2.

For some complex number $z = x + jy$ the magnitude or absolute value, or modulus is defined by

$$|z| = |x + jy| = r = \sqrt{x^2 + y^2}$$

**Powers**

$$z^n = r^n \left[ \cos(n\theta) + j \sin(n\theta) \right]$$

### Equation 2.24

Proof: If $z = r \left[ \cos(\theta) + j \sin(\theta) \right]$ and if $n$ is an integer such as 2, for example, then

$$
\begin{aligned}
z^2 \;&= r^2 \left[ \cos(\theta) + j \sin(\theta) \right]^2 \\
&= r^2 \left[ \cos^2(\theta) - \sin^2(\theta) + j2\cos(\theta)\sin(\theta) \right] \\
&= r^2 \left[ 2\cos^2(\theta) - 1 + j\sin(2\theta) \right] \\
&= r^2 \left[ \cos(2\theta) + j\sin(2\theta) \right]
\end{aligned}
$$

This result can be extended for all positive values of $n$. The multiplication property states that

$$z_1 z_2 = r_1 r_2 \left[ \cos(\theta_1 + \theta_2) + j \sin(\theta_1 + \theta_2) \right]$$

Therefore if $z = z_1 = z_2$ then

$$zz = z^2 = rr \left[ \cos(\theta + \theta) + j \sin(\theta + \theta) \right] = r^2 \left[ \cos(2\theta) + j \sin(2\theta) \right]$$

It follows then that

$$z_1 z_2 z_3 \cdots z_n = r_1 r_2 r_3 \cdots r_n \left[ \cos(\theta_1 + \theta_2 + \ldots \theta_n) + j \sin(\theta_1 + \theta_2 + \ldots \theta_n) \right]$$

If all the $z_i$ are equal then

$$z^n = r^n \left[ \cos(n\theta) + j \sin(n\theta) \right]$$

Since division and multiplication are inverse operations, it follows from Equation 2.21 and Equation 2.24 that reciprocals of complex numbers can be expressed as

$$z^{-1} = \frac{1}{z} = \frac{1}{r}\left[\cos(-\theta) + j\sin(-\theta)\right]$$
$$= \frac{1}{r}\left[\cos(\theta) - j\sin(\theta)\right]$$

This result can be expanded upon to define a complex number to a negative power $n$

$$z^{-n} = \frac{1}{z^n} = \frac{1}{r^n}\left[\cos(-n\theta) + j\sin(-n\theta)\right]$$
$$= \frac{1}{r^n}\left[\cos(n\theta) - j\sin(n\theta)\right]$$
$$= \left(\frac{1}{z}\right)^n$$

**Equation 2.25**

We will provide a much simpler one-line proof of the powers property when we discuss the exponential form of complex numbers.

**Dot Product.** Given two complex numbers $z_1 = x_1 + jy_1$ and $z_2 = x_2 + jy_2$, the dot product of the two complex numbers is given by

$$z_1 \cdot z_2 = |z_1||z_2|\cos\alpha = x_1x_2 + y_1y_2$$

where the angle $\alpha$ is the difference between the phase angles of $z_1$ and $z_2$. For example, suppose we have two vectors defined by the complex numbers $z_1$ and $z_2$, illustrated in Figure 2.3.

The phase angles for $z_1$ and $z_2$ are given by

$$\theta_1 = \tan^{-1}\left(\frac{4}{3}\right) = 53.1301°$$

$$\theta_2 = \tan^{-1}\left(\frac{1}{2}\right) = 26.5651°$$

The angle $\alpha = \theta_1 - \theta_2 = 26.5651°$. The dot product of $z_1$ and $z_2$ is then given by

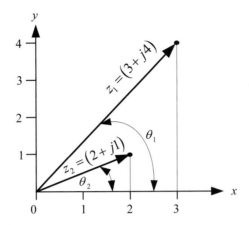

**Figure 2.3**   Dot product example

$$z_1 \cdot z_2 = |z_1||z_2|\cos\alpha = 5\sqrt{5}\cos\left(26.5651°\right) = 10$$

or alternatively the dot product can be computed by

$$z_1 \cdot z_2 = x_1 x_2 + y_1 y_2 = (3)(2)+(4)(1) = 10$$

## 2.3   ROOTS OF COMPLEX NUMBERS

The $n$ roots of a complex number $z$ are given by

$$z^{\frac{1}{n}} = \sqrt[n]{z} = \sqrt[n]{r}\left[\cos\left(\frac{\theta+2\pi k}{n}\right) + j\sin\left(\frac{\theta+2\pi k}{n}\right)\right] \quad \text{for} \quad k=0,1,2,3,\cdots n-1$$

**Equation 2.26**

Proof: Suppose we have a complex number $z_1$ that is raised to the $n$th power. Further suppose we equate this to some other complex number $z$ such that $z_1^n = z$. We can expand $z_1^n$ in polar form to get

$$z_1^n = r_1^n\left[\cos\left(n\theta_1\right) + j\sin\left(n\theta_1\right)\right]$$

**Equation 2.27**

Similarly we can expand $z$ in polar form to get $z = r\left[\cos(\theta) + j\sin(\theta)\right]$. Equating the two expressions gives us

$$z_1^n = z$$

**Equation 2.28**

$$r_1^n\left[\cos(n\theta_1) + j\sin(n\theta_1)\right] = r\left[\cos(\theta) + j\sin(\theta)\right]$$

**Equation 2.29**

In order for the expressions on both sides of Equation 2.29 to be equal, the modulus of each expression must be equal. We equate the modulus of each expression and take the $n$th root of both sides to get

$$r_1 = \sqrt[n]{r}$$

**Equation 2.30**

Also, for the equality in Equation 2.29 to hold, then the following relationships must hold:

$$\cos(n\theta_1) = \cos(\theta)$$
$$\sin(n\theta_1) = \sin(\theta)$$

Since the sine and cosine functions are periodic in $2\pi$ the previous two equations are true if and only if

$$n\theta_1 = \theta + 2\pi k \quad \text{for } k = 0,1,2,\ldots$$

This means that

$$\theta_1 = \frac{\theta + 2\pi k}{n} \quad \text{for } k = 0,1,2,\ldots$$

The values of $\theta_1$ are identical for any two integers $k$ that are separated by a multiple of $n$, so there are only $n$ distinct solutions for $\theta_1$ and therefore there are $n$ distinct roots for $z_1$. We can then write

$$\theta_1 = \frac{\theta + 2\pi k}{n} \quad \text{for } k = 0,1,2,\ldots,n-1$$

The $n$ roots are spaced at $n$ equal arc intervals around a circle of radius $\sqrt[n]{r}$ in the complex plane. The length of each arc interval is determined by the arc angle

$$\frac{\theta + 2\pi k}{n} \quad \{\text{for } k = 0, 1, 2, 3, \ldots n - 1\}$$

**Equation 2.31**

The root at $k = 0$ is called the principle root. For a general complex number of the form $z = x + jy$, we can state that the $n$ roots of $z$ can be computed by

$$z^{\frac{1}{n}} = \sqrt[n]{z} = \sqrt[n]{r}\left[\cos\left(\frac{\theta + 2\pi k}{n}\right) + j\sin\left(\frac{\theta + 2\pi k}{n}\right)\right] \quad \text{for } k = 0, 1, 2, 3, \cdots n - 1$$

These results can be extended to include a mix of roots and powers. For example,

$$z^{\frac{m}{n}} = \sqrt[n]{z^m} = \sqrt[n]{r^m}\left[\cos\left[\frac{m(\theta + 2\pi k)}{n}\right] + j\sin\left[\frac{m(\theta + 2\pi k)}{n}\right]\right] \quad \text{for } k = 0, 1, 2, 3, \cdots n - 1$$

**Equation 2.32**

The proof is left to the reader. In order to aid in the understanding of these concepts, this is probably a good time to stop and work a few examples.

**Example 1.** Find the two roots of $z$ when $z = j$. Let's begin by writing down what we know:

$$z = (x + jy) = (0 + j1) = j$$

The first thing we can do is compute the modulus or magnitude of $z$.

$$r = \sqrt{x^2 + y^2} = \sqrt{0^2 + 1^2} = 1$$

The next thing we do is to compute the argument of $z$.

$$\theta = \sin^{-1}\left(\frac{y}{r}\right) = \sin^{-1}\left(\frac{1}{1}\right) = \frac{\pi}{2}$$

In this example, does anyone know why we didn't compute the argument of $z$ using the typical expression

$$\theta = \tan^{-1}\left(\frac{y}{x}\right)?$$

The answer of course is that the value of $x$ is zero, and it is difficult to represent infinity on a calculator. Now we can use Equation 2.26, which is repeated here for clarity, to compute the two roots of $j$:

$$z^{\frac{1}{n}} = \sqrt[n]{z} = \sqrt[n]{r}\left[\cos\left(\frac{\theta + 2\pi k}{n}\right) + j\sin\left(\frac{\theta + 2\pi k}{n}\right)\right] \quad \text{for} \quad k = 0,1,2,3,\cdots n-1$$

If we substitute in the values we computed for $r$ and $\theta$ we arrive at

$$\sqrt{z} = \sqrt{j} \quad = \sqrt{r}\left[\cos\left(\frac{\theta + 2\pi k}{2}\right) + j\sin\left(\frac{\theta + 2\pi k}{2}\right)\right] \quad \{\text{for } k = 0,1\}$$

$$= \sqrt{1}\left[\cos\left(\frac{\frac{\pi}{2} + 2\pi k}{2}\right) + j\sin\left(\frac{\frac{\pi}{2} + 2\pi k}{2}\right)\right]$$

$$= \left[\cos\left(\frac{\pi}{4} + \pi k\right) + j\sin\left(\frac{\pi}{4} + \pi k\right)\right]$$

Now we can compute the two roots of $z = j$ to be

$$k = 0 \implies principle\ root \quad = \cos\left(\frac{\pi}{4}\right) + j\sin\left(\frac{\pi}{4}\right) \quad = (0.707 + j0.707)$$

$$k = 1 \implies root \quad = \cos\left(\frac{5\pi}{4}\right) + j\sin\left(\frac{5\pi}{4}\right) \quad = (-0.707 - j0.707)$$

The two roots of $j$ are illustrated in Figure 2.4. The first thing we notice from the figure is the two roots are equally spaced in the complex plane around the circle of radius $\sqrt{r}$. Both root values are at the tip of a vector of length $\sqrt{r}$ that begins at the origin and is offset from the real axis by $\theta_0$ and $\theta_1$ radians, respectively.

We should check our work to verify we got the right answer, so let's square each of the two roots we computed to see what we end up with.

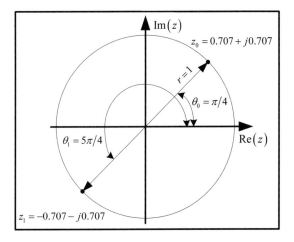

**Figure 2.4**  Two roots of the complex number $z = 0 + j$

$$k = 0 \quad \Rightarrow principle\ root \quad = \left(0.707 + j0.707\right)^2 \quad = \left(0.5 + j0.5 + j0.5 - 0.5\right) = j$$
$$k = 1 \quad \Rightarrow root \quad = \left(-0.707 - j0.707\right)^2 \quad = \left(0.5 + j0.5 + j0.5 - 0.5\right) = j$$

So both roots check out correctly and our computations were correct.

**Example 2.** Find the three roots of $z = 1$. In other words, find $\sqrt[3]{z} = \sqrt[3]{1}$. Once again, we write down the expanded version of $z$ to get

$$z = \left(x + jy\right) = \left(1 + j0\right) = 1$$

Then we compute the modulus or magnitude of $z$ to be

$$r = \sqrt{x^2 + y^2} = \sqrt{1^2 + 0^2} = 1$$

Next we compute the argument of $z$

$$\theta = \tan^{-1}\left(\frac{y}{x}\right) = \tan^{-1}\left(\frac{0}{1}\right) = 0$$

Now we can use Equation 2.26 to compute the three roots of 1:

$$\sqrt[3]{z} = \sqrt[3]{r}\left[\cos\left(\frac{\theta+2\pi k}{3}\right)+j\sin\left(\frac{\theta+2\pi k}{3}\right)\right] \quad \text{for } k=0,1,2$$

$$= \sqrt[3]{1}\left[\cos\left(\frac{0+2\pi k}{3}\right)+j\sin\left(\frac{0+2\pi k}{3}\right)\right]$$

$$= \left[\cos\left(\frac{2\pi k}{3}\right)+j\sin\left(\frac{2\pi k}{3}\right)\right]$$

The three roots of $z=1$ are computed to be

$$k=0 \quad \Rightarrow principle\ root \quad = \cos(0)+j\sin(0) \qquad = (1+j0)$$

$$k=1 \quad \Rightarrow root \qquad\qquad = \cos\left(\frac{2\pi}{3}\right)+j\sin\left(\frac{2\pi}{3}\right) \quad = (-0.5+j0.866)$$

$$k=2 \quad \Rightarrow root \qquad\qquad = \cos\left(\frac{4\pi}{3}\right)+j\sin\left(\frac{4\pi}{3}\right) \quad = (-0.5-j0.866)$$

The three roots of $z=1+j0$ are plotted in Figure 2.5. Note that all the roots are equally spaced around the circle of radius 1 in the complex plane.

Once again, we will check our work by raising each the three roots to the third power and verifying that the resulting answer is equal to 1. In this example we will express the argument of $z$ in degrees as opposed to radians.

$$k=0 \quad \Rightarrow principle\ root \quad = (1+j0)^3$$
$$= 1$$
$$k=1 \quad \Rightarrow root \qquad\qquad = (-0.5+j0.866)^3$$
$$= \left(\sqrt{.5^2+0.866^2}\right)^3\left[\cos(3\cdot120°)+j\sin(3\cdot120°)\right]$$
$$= (1)^3\left[\cos(360°)+j\sin(360°)\right]$$
$$= 1$$
$$k=2 \quad \Rightarrow root \qquad\qquad = (-0.5-j0.866)^3$$
$$= \left(\sqrt{.5^2+0.866^2}\right)^3\left[\cos(3\cdot240°)+j\sin(3\cdot240°)\right]$$
$$= (1)^3\left[\cos(720°)+j\sin(720°)\right]$$
$$= 1$$

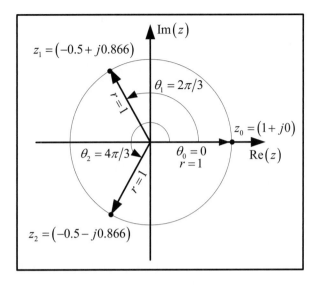

**Figure 2.5**   Three roots of the complex number $z = 1 + j0$

Everything checks out, so we are good to go.

**Example 3.** Find the 2 roots of $z$ when $z = 1 + j$. Without comment, we repeat all the steps outlined in the previous two examples.

$$z = (x + jy) = (1 + j1)$$

$$r = \sqrt{x^2 + y^2} = \sqrt{1^2 + 1^2} = \sqrt{2}$$

$$\theta = \tan^{-1}\left(\frac{y}{x}\right) = \tan^{-1}\left(\frac{1}{1}\right) = \frac{\pi}{4}$$

$$\sqrt{z} = \sqrt{r}\left[\cos\left(\frac{\theta + 2\pi k}{2}\right) + j\sin\left(\frac{\theta + 2\pi k}{2}\right)\right] \qquad \text{for } k = 0,1$$

$$= (2)^{\frac{1}{4}}\left[\cos\left(\frac{\frac{\pi}{4} + 2\pi k}{2}\right) + j\sin\left(\frac{\frac{\pi}{4} + 2\pi k}{2}\right)\right]$$

$$= (2)^{\frac{1}{4}}\left[\cos\left(\frac{\pi}{8} + \pi k\right) + j\sin\left(\frac{\pi}{8} + \pi k\right)\right]$$

The two roots of $z = 1 + j$ are computed to be

$$k = 0 \Rightarrow \text{principle root} = 2^{\frac{1}{4}}\left[\cos\left(\frac{\pi}{8}\right) + j\sin\left(\frac{\pi}{8}\right)\right] = (1.0987 + j0.4551)$$

$$k = 1 \Rightarrow \text{root} = 2^{\frac{1}{4}}\left[\cos\left(\frac{9\pi}{8}\right) + j\sin\left(\frac{9\pi}{8}\right)\right] = (-1.0987 - j0.4551)$$

The roots of $z = 1 + j$ are plotted in Figure 2.6.

Now we check our work to verify our computations are correct. This time we will work using Cartesian coordinates.

$$k = 0 \Rightarrow \text{principle root} = (1.0987 + j0.4551)^2$$
$$= 1.2071 + j0.5 + j0.5 - 0.2071$$
$$= 1 + j1$$

$$k = 1 \Rightarrow \text{root} = (-1.0987 - j0.4551)^2$$
$$= 1.2071 + j0.5 + j0.5 - 0.2071$$
$$= 1 + j1$$

The roots are valid, so we are done.

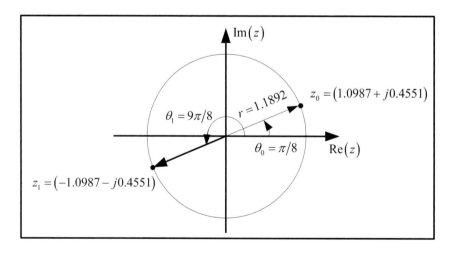

**Figure 2.6**  Two roots of the complex number $z = 1 + j$

## 2.4   ABSOLUTE VALUE OF COMPLEX NUMBERS

Given a complex number $z = x + jy$, we can state the following properties:

- The absolute value or magnitude of $z$ is given by $|z| = \sqrt{x^2 + y^2}$.
- The magnitude squared of $z$ is given by $|z|^2 = x^2 + y^2$.
- The magnitude of $z$ is always greater than or equal to zero. That is $|z| \geq 0$.
- The product of $z$ and its conjugate is the magnitude squared:

$$z\bar{z} = (x + jy)(x - jy) = x^2 + y^2 = |z|^2$$

- The absolute value of a product is the product of absolute values $|z_1 z_2|^2 = |z_1|^2 |z_2|^2$.

  Proof:

$$
\begin{aligned}
|z_1 z_2|^2 &= (z_1 z_2)\left(\overline{z_1 z_2}\right) \\
&= (z_1 z_2)\left(\bar{z}_1 \bar{z}_2\right) \\
&= (z_1 \bar{z}_1)(z_2 \bar{z}_2) \\
&= |z_1|^2 |z_2|^2
\end{aligned}
$$

- The absolute value of a quotient is equal to the quotient of absolute values

$$\left|\frac{z_1}{z_2}\right| = \frac{|z_1|}{|z_2|}$$

Proof: Let $z_3 = z_1 z_2$ and $z_2 = z_3/z_1$. It follows that

$$|z_2| = \left|\frac{z_3}{z_1}\right|$$

Also,

$$|z_1 z_2| = |z_1||z_2| = |z_3|$$

which means that

$$|z_2| = \frac{|z_3|}{|z_1|}$$

From these results, we can state that

$$\left|z_2\right| = \left|\frac{z_3}{z_1}\right| = \frac{\left|z_3\right|}{\left|z_1\right|} \quad \text{or} \quad \left|\frac{z_3}{z_1}\right| = \frac{\left|z_3\right|}{\left|z_1\right|}$$

- The distance from $z_1$ to $z_2$ is equal to $\left|z_1 - z_2\right|$.
- The absolute value of the difference is greater than or equal to the difference of the absolute values

$$\left|z_1 - z_2\right| \geq \left|z_1\right| - \left|z_2\right|$$

- The absolute value of the sum is less than or equal to the sum of the absolute values

$$\left|z_1 + z_2\right| \leq \left|z_1\right| + \left|z_2\right|$$

This result can be extended to

$$\left|z_1 + z_2 + z_2 + \cdots + z_n\right| \leq \left|z_1\right| + \left|z_2\right| + \left|z_3\right| + \cdots + \left|z_n\right|$$

- Real numbers can be ordered such that

$$x_1 < x_2 < x_3 < \cdots < x_n$$

Complex numbers can be ordered such that

$$\left|z_1\right| < \left|z_2\right| < \left|z_3\right| < \cdots < \left|z_n\right|$$

## 2.5  EXPONENTIAL FORM OF COMPLEX NUMBERS

The complex function $f(z) = e^z$ can be expressed as

$$f(z) = e^z = e^{x \pm jy} = e^x \left( \cos y \pm j \sin y \right)$$

The magnitude of $f(z)$ is given as

$$\left| f(z) \right| = \left| e^z \right| = \sqrt{\left( e^x \right)^2 \cos^2 y + \left( e^x \right)^2 \sin^2 y}$$

or

$$\left| e^z \right| = e^x$$

The complex variable $z$ can be expanded, first into its polar form and then into its exponential form so that

$$z = r \left( \cos \theta \pm j \sin \theta \right) = re^{\pm j\theta}$$

Suppose we have two complex numbers in exponential form, $z_1 = r_1 e^{j\theta_1}$ and $z_2 = r_2 e^{j\theta_2}$. We can state several basic properties with regard to $z_1$ and $z_2$:

- Multiplication of complex exponentials:

$$z_1 z_2 = r_1 e^{j\theta_1} r_2 e^{j\theta_2} = r_1 r_2 e^{j(\theta_1 + \theta_2)}$$

- Division of complex exponentials:

$$\frac{r_1 e^{j\theta_1}}{r_2 e^{j\theta_2}} = \frac{r_1}{r_2} e^{j(\theta_1 - \theta_2)}$$

- Powers of complex exponentials:

$$z^n = r^n e^{jn\theta}$$
$$z^{-n} = r^{-n} e^{-jn\theta}$$

- Roots of complex exponentials:

$$z^{\frac{1}{n}} = r^{\frac{1}{n}} e^{j\left( \frac{\theta + 2\pi k}{n} \right)} \quad \text{for } k=0,1,2,\cdots n-1$$

## 2.6  GRAPHS OF THE COMPLEX VARIABLE *z*

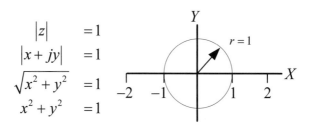

$$|z| = 1$$
$$|x + jy| = 1$$
$$\sqrt{x^2 + y^2} = 1$$
$$x^2 + y^2 = 1$$

**Figure 2.7**   Unit circle centered at the origin with a radius of 1

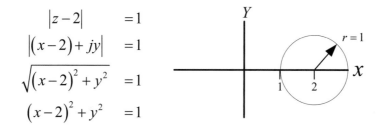

$$|z - 2| = 1$$
$$|(x - 2) + jy| = 1$$
$$\sqrt{(x - 2)^2 + y^2} = 1$$
$$(x - 2)^2 + y^2 = 1$$

**Figure 2.8**   Circle centered at $(x, y) = (2, 0)$ with a radius of 1

Note that in Figure 2.8, the expression $(x - 2)^2 + y^2 = 1$ has the form of the standard equation of a circle $(x - h)^2 + (y - k)^2 = r^2$, where the center is located at $(h, k)$ and where the radius is $r$.

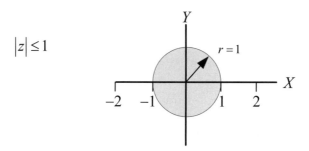

$$|z| \leq 1$$

**Figure 2.9**   Unit circle plus the area inside centered at the origin with a radius of 1, $|z| \leq 1$

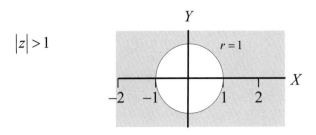

**Figure 2.10**   Everything outside of unit circle $|z| \geq 1$

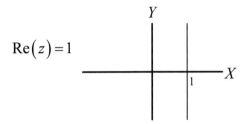

**Figure 2.11**   Straight vertical line passing through $x = 1$: Re(z) = 1

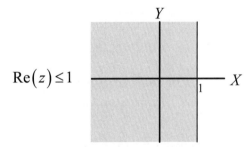

**Figure 2.12**   Region including and to the left of $x = 1$, Re(z) $\leq 1$

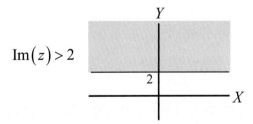

**Figure 2.13**   Region above the line $y = 2$, Im(z) > 2

## 2.7  LIMITS

A function $f(z)$ is continuous at a point $z = z_0$ if the following three conditions are met:

1. $\lim\limits_{z \to z_0} f(z)$ exists

2. $f(z_0)$ exists

3. $\lim\limits_{z \to z_0} f(z) = f(z_0)$

The limit of a function of a complex variable can be better understood if we first view it as the limit of real functions of two real variables. Let $f(z) = u(x, y) + jv(x, y)$, where $z = x + jy$ and $z_0 = x_0 + jy_0$. Then we can say that the limit of $f(z)$ exists if

$$\lim\limits_{z \to z_0} f(z) = f(z_0) = u_0 + jv_0$$

This can only be true if

$$\lim\limits_{\substack{x \to x_0 \\ y \to y_0}} u(x, y) = u_0 \text{ and } \lim\limits_{\substack{x \to x_0 \\ y \to y_0}} v(x, y) = v_0$$

The limit properties of complex functions are identical to those for real functions in that if the $\lim\limits_{z \to z_0} f_1(z) = w_1$ and $\lim\limits_{z \to z_0} f_2(z) = w_2$ then

- $\lim\limits_{z \to z_0}\left[ f_1(z) \pm f_2(z) \right] = \lim\limits_{z \to z_0} f_1(z) \pm \lim\limits_{z \to z_0} f_2(z) = w_1 \pm w_2$

- $\lim\limits_{z \to z_0}\left[ f_1(z) f_2(z) \right] = \left[ \lim\limits_{z \to z_0} f_1(z) \right]\left[ \lim\limits_{z \to z_0} f_2(z) \right] = w_1 w_2$

- $\lim\limits_{z \to z_0}\left[ \dfrac{f(z_1)}{f(z_2)} \right] = \dfrac{\lim\limits_{z \to z_0} f(z_1)}{\lim\limits_{z \to z_0} f(z_2)} = \dfrac{w_1}{w_2}$   for $w_2 \neq 0$

If $P(z)$ is a complex polynomial given by $P(z) = a_0 + a_1 z + a_2 z^2 + a_3 z^3 + \dots$, then

$$\lim_{z \to z_0} P(z) = P(z_0) = a_0 + a_1 z_0 + a_2 z_0^2 + a_3 z_0^3 + \ldots$$

## 2.8   ANALYTIC FUNCTIONS

In the world of complex variables, the term *analytic function* is used over and over again. The term *analytic* is important because it reveals a lot of important information about a function. There exists a plethora of text in the literature that defines the term analytic as it relates to complex variables, but it can all be boiled down to one simple but very important paragraph, as follows:

A function $f(z)$ is analytic in a region $R$ of the complex plane if its derivative exists and if it is continuous at every point in $R$. A function $f(z)$ is said to be analytic at some point $z_0$ if there exists a neighborhood $|z - z_0| < \delta$ within which the derivative of $f(z)$ exists.

- If $f(z)$ is analytic in some region $R$ of the complex plane, then all higher order derivatives of $f(z)$ are also analytic within $R$.
- If two functions $f(z)$ and $g(z)$ are analytic in a region $R$ of the complex plane, then
  - the sum and difference $f(z) \pm g(z)$ is also analytic in $R$,
  - the product $f(z)g(z)$ is also analytic in $R$, and
  - the quotient $f(z)/g(z)$ is also analytic in $R$ if $g(z)$ does not equal 0 anywhere in $R$.
- All trigonometric functions are analytic: $\sin z, \ \cos z, \cdots$etc.
- Polynomials are analytic everywhere:

$$w = f(z) = a_0 z^n + a_1 z^{n-1} + a_2 z^{n-2} + \cdots + a_{n-1} z + a_0 \ \text{ for } n \geq 0$$

- Rational algebraic functions such as

$$f(z) = \frac{P_n(z)}{Q_m(z)}$$

  are analytic everywhere except where $Q_m(z) = 0$.
- Exponential functions such as $w = e^z = e^{x+jy} = e^x \left( \cos y + j \sin y \right)$ are analytic everywhere.

The reader may ask, *How do I determine whether or not a function is analytic?* This is a very good question and we will answer this later when we see that a function that satisfies the Cauchy-Riemann conditions is said to be analytic.

## 2.9  SINGULARITY

If a function $f(z)$ is analytic for all points within a region $R$ with exception of the point $z_0$, then we define $z_0$ to be a singularity of $f(z)$. For a rational algebraic function such as

$$f(z) = P_n(z)/Q_m(z)$$

singularities are most commonly seen when $Q_m(z)$ takes on a value of zero. These types of singularities are typically referred to as *poles*. We will discuss the subject of poles in great depth later in this chapter.

## 2.10  ENTIRE FUNCTIONS

A function that is analytic over the entire finite range of the complex plane is called an *entire function*. Examples of entire functions are polynomials such as $\sin z$, $\cos z$, $e^z$, and $7z^3 + 3z^2 + 2z + 5 - j6$.

## 2.11  THE COMPLEX NUMBER $\infty$

The complex number $\infty$ is used a great deal in the mathematics of complex numbers, especially when dealing with limits. We treat this number in the complex number system via the use of a set of rules [1]:

- $z \pm \infty = \infty \quad \{z \neq \infty\}$

- $\dfrac{z}{0} = \infty \quad \{z \neq 0\}$

- $z \cdot \infty = \infty \quad \{z \neq 0\}$

- $\dfrac{z}{\infty} = 0 \quad \{z \neq \infty\}$

- $\dfrac{\infty}{z} = \infty \quad \{z \neq \infty\}$

Expressions in $z$ that are not defined are $\infty + \infty$, $\infty - \infty$, and $\infty/\infty$.

## 2.12 COMPLEX DIFFERENTIATION

Computing the derivative of a complex function is similar to computing the derivative of a real function. We define the derivative of $f(z)$ to be

$$\frac{d}{dz} f(z) = \lim_{\Delta z \to 0} \left[ \frac{f(z + \Delta z) - f(z)}{\Delta z} \right]$$

**Equation 2.33**

provided that the limit is the same for all paths by which $\Delta z \to 0$. For example, Figure 2.14 illustrates just a few of all the possible paths that can be taken from $z + \Delta z$ to $z$ as $\Delta z \to 0$. We can expand the term $\Delta z$ to be $\Delta z = \Delta x + j\Delta y$. In path P2 in Figure 2.14, for example, $\Delta z \to 0$ by first holding $\Delta x$ constant as $\Delta y \to 0$ and then holding $\Delta y$ constant as $\Delta x \to 0$. Path P1 achieves the same results in the exact opposite manner. Paths P3, P4, and P5 all take different routes. The point is that the derivative of a complex function exists only if the derivative in Equation 2.33 produces the same result for all possible paths $P_n$. As we will see later, functions of a complex variable that are differentiable satisfy the Cauchy-Riemann equations.

### 2.12.1 Rules of Differentiation

The rules for the differentiation of a complex function are similar to the rules for the differentiation of a real function. A brief review of these rules is as follows:

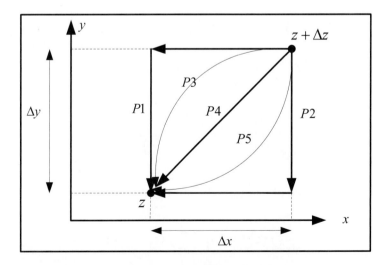

**Figure 2.14**  Example paths from $z + \Delta z$ to $z$

- Derivative of a complex constant $k = (a \pm jb)$ {where $a$ and $b$ are constants}

$$\frac{d}{dz}k = 0$$

- Derivative of a complex function

$$\frac{d}{dz}\left(kf(z)\right) = k\frac{d}{dz}f(z)$$

- Derivative of a complex polynomial

$$\frac{d}{dz}\left(z^n\right) = nz^{n-1}$$

An example of a complex function and its derivative are

$$f(z) \quad = z^3 + 3z^2 + (5 + j2)z + 9$$
$$\frac{d}{dz}f(z) \quad = 3z^2 + 6z + (5 + j2)$$

- Derivatives of a product of complex functions

$$\frac{d}{dz}\left\{f(z)g(z)\right\} = g(z)\frac{d}{dz}f(z) + f(z)\frac{d}{dz}g(z)$$

- Derivative of a quotient of complex functions

$$\frac{d}{dz}\left\{\frac{f(z)}{g(z)}\right\} = \frac{g(z)\dfrac{d}{dz}f(z) - f(z)\dfrac{d}{dz}g(z)}{g^2(z)}$$

- The derivative of a sum is the sum of derivatives

$$\frac{d}{dz}\left\{f(z) + g(z)\right\} = \frac{d}{dz}f(z) + \frac{d}{dz}g(z)$$

- The derivative of a difference is the difference of the derivatives

$$\frac{d}{dz}\{f(z)-g(z)\} = \frac{d}{dz}f(z) - \frac{d}{dz}g(z)$$

- L'Hôpital's rule

If $f(z)$ and $g(z)$ are analytic at point $z_0$ and if $f(z)/g(z)$ is in an indeterminate form such as

$$\lim_{z\to z_0} f(z) = \lim_{z\to z_0} g(z) = 0$$

Then

$$\lim_{z\to z_0}\left\{\frac{f(z)}{g(z)}\right\} = \lim_{z\to z_0}\left\{\frac{\dfrac{d}{dz}f(z)}{\dfrac{d}{dz}g(z)}\right\} = \lim_{z\to z_0}\left\{\frac{f'(z)}{g'(z)}\right\}$$

**Table 2.1**   Table of Common Complex Derivatives

| | |
|---|---|
| $\dfrac{d}{dz}(k) = 0$    for $k$ a complex constant $a \pm jb$ | $\dfrac{d}{dz}\csc^{-1}z = \dfrac{-1}{z\sqrt{z^2-1}}$ |
| $\dfrac{d}{dz}(z^n) = nz^{n-1}$ | $\dfrac{d}{dz}\sinh z = \cosh z$ |
| $\dfrac{d}{dz}(e^z) = e^z$ | $\dfrac{d}{dz}\cosh z = \sinh z$ |
| $\dfrac{d}{dz}\sin z = \cos z$ | $\dfrac{d}{dz}\ln z = \dfrac{1}{z}$ |
| $\dfrac{d}{dz}\cos z = -\sin z$ | $\dfrac{d}{dz}\log_p z = \dfrac{\log_p e}{z}$ |

*(continued on next page)*

**Table 2.1**  Table of Common Complex Derivatives (continued)

| | |
|---|---|
| $\dfrac{d}{dz}\tan z = \sec^2 z$ | $\dfrac{d}{dz}\tanh z = \operatorname{sech}^2 z$ |
| $\dfrac{d}{dz}\cot z = -\csc^2 z$ | $\dfrac{d}{dz}\coth z = -\operatorname{csch}^2 z$ |
| $\dfrac{d}{dz}\sec z = \sec z \tan z$ | $\dfrac{d}{dz}\operatorname{sech} z = -\operatorname{sech} z \tanh z$ |
| $\dfrac{d}{dz}\csc z = -\csc z \cot z$ | $\dfrac{d}{dz}\operatorname{csch} z = -\operatorname{csch} z \coth z$ |
| $\dfrac{d}{dz}\sin^{-1} z = \dfrac{1}{\sqrt{1-z^2}}$ | $\dfrac{d}{dz}\sinh^{1} z = \dfrac{1}{\sqrt{1+z^2}}$ |
| $\dfrac{d}{dz}\cos^{-1} z = \dfrac{-1}{\sqrt{1-z^2}}$ | $\dfrac{d}{dz}\cosh^{-1} z = \dfrac{1}{\sqrt{z^2-1}}$ |
| $\dfrac{d}{dz}\tan^{-1} z = \dfrac{1}{1+z^2}$ | $\dfrac{d}{dz}\tanh^{-1} z = \dfrac{1}{1-z^2}$ |
| $\dfrac{d}{dz}\cot^{-1} z = \dfrac{-1}{1+z^2}$ | $\dfrac{d}{dz}\coth^{-1} z = \dfrac{1}{1-z^2}$ |
| $\dfrac{d}{dz}\sec^{-1} z = \dfrac{1}{z\sqrt{z^2-1}}$ | $\dfrac{d}{dz}\operatorname{sech}^{-1} z = \dfrac{-1}{z\sqrt{1-z^2}}$ |
| $\dfrac{d}{dz}\left(a^z\right) = a^z \ln a$ | $\dfrac{d}{dz}\operatorname{csch}^{-1} z = \dfrac{-1}{z\sqrt{z^2+1}}$ |

If $\dfrac{d}{dz}f(z)$ and $\dfrac{d}{dz}g(z)$ are both zero, then the rule may be extended by taking higher order derivatives.

## 2.13  CAUCHY-RIEMANN EQUATIONS

The Cauchy-Riemann equations are extremely powerful tools that allow us to determine whether or not a function $f(z)$ is analytic. If we write $w = f(z)$, where $w = u + jv$, then

$$u + jv = f(z) = f(x + jy) = \text{Re}(f(z)) + \text{Im}(f(z))$$

That is, we can separate a complex function into its real and imaginary parts. If we assign the real parts to a function $u(x, y)$, and if we assign the imaginary parts to a function $v(x, y)$, we can rewrite the original complex function as

$$f(z) = u(x, y) + jv(x, y)$$

Both $u(x, y)$ and $v(x, y)$ are real functions of real variables $x$ and $y$. As an example, let

$$w = f(z) = z^2$$

We can expand this function to include its $x$ and $y$ components to get

$$f(z) = u(x, y) + jv(x, y) = z^2 = (x + jy)^2 = x^2 - y^2 + j2xy$$

Equating the real and imaginary parts we arrive at

$$u(x, y) \;\; = x^2 - y^2$$
$$v(x, y) \;\; = 2xy$$

What if we desired to compute the derivative of $w = f(z) = z^2$? How will we know in advance that the derivative exists? We can establish the conditions for the existence of a derivative at some point $z_0$ by the following. Given the functions

$$w \;\; = u + jv \;\; \Rightarrow \;\; \Delta w = \Delta u + j\Delta v$$
$$z \;\; = x + jy \;\; \Rightarrow \;\; \Delta z = \Delta x + j\Delta y$$

if $w = f(z) = u + jv$, then

$$\frac{d}{dz}f(z) = \frac{dw}{dz} = \lim_{\Delta z \to 0}\left\{\frac{\Delta w}{\Delta z}\right\} = \lim_{\substack{\Delta x \to 0 \\ \Delta y \to 0}}\left\{\frac{\Delta u + j\Delta v}{\Delta x + j\Delta y}\right\}$$

If we choose a path for $\Delta z = (\Delta x + j\Delta y) \to 0$ such that $\Delta x \to 0$ first followed by $\Delta y \to 0$, then

$$\frac{d}{dz}f(z) = \lim_{\Delta y \to 0}\left\{\frac{\Delta u + j\Delta v}{j\Delta y}\right\} = \lim_{\Delta y \to 0}\left\{\frac{\Delta v}{\Delta y} - j\frac{\Delta u}{\Delta y}\right\} = \frac{\partial v}{\partial y} - j\frac{\partial u}{\partial y}$$

**Equation 2.34**

If we choose a path for $\Delta z = (\Delta x + j\Delta y) \to 0$ such that $\Delta y \to 0$ first followed by $\Delta x \to 0$, then

$$\frac{d}{dz}f(z) = \lim_{\Delta x \to 0}\left\{\frac{\Delta u + j\Delta v}{\Delta x}\right\} = \lim_{\Delta x \to 0}\left\{\frac{\Delta u}{\Delta x} + j\frac{\Delta v}{\Delta x}\right\} = \frac{\partial u}{\partial x} + j\frac{\partial v}{\partial x}$$

**Equation 2.35**

We previously stated that for a derivative of a complex function to exist we must get the same results no matter what path we take as $\Delta z \to 0$. Therefore if our derivative exists, the results we obtained for the two paths we took in Equation 2.34 and Equation 2.35 must be equal. That is,

$$\left\{\frac{\partial v}{\partial y} - j\frac{\partial u}{\partial y}\right\} = \left\{\frac{\partial u}{\partial x} + j\frac{\partial v}{\partial x}\right\}$$

**Equation 2.36**

If we equate the real and imaginary parts of Equation 2.36, we end up with the Cauchy-Riemann equations, as shown in Equation 2.37. The Cauchy-Riemann equations tell us that if the following conditions are satisfied, then the function $f(z)$ is analytic and therefore has first order and higher derivatives at the point $z_0$.

$$\frac{\delta u}{\delta x} = \frac{\delta v}{\delta y}$$

$$\frac{\delta u}{\delta y} = -\frac{\delta v}{\delta x}$$

**Equation 2.37**

For example, let's return to the function $f(z) = z^2$ and suppose we wish to determine if it is analytic. The first thing we do is expand the function in terms of its $x$ and $y$ components. When we do we end up with

$$f(z) = z^2 = (x + jy)^2 = (x^2 - y^2) + j(2xy) = u + jv$$
$$\rightarrow \quad u = (x^2 - y^2)$$
$$\rightarrow \quad v = (2xy)$$

$$\frac{\delta u}{\delta x} = \frac{\delta v}{\delta y} = 2x$$

$$\frac{\delta u}{\delta y} = -\frac{\delta v}{\delta x} = -2y$$

The Cauchy-Riemann equations are satisfied, so we determine that $f(z)$ is indeed analytic.

We now present, without derivation, the equivalent Cauchy-Riemann equations in polar form. In this case $f(z) = u + jv$ is expressed in polar form as

$$f(z) = u(r, \theta) + jv(r, \theta)$$

The polar form of the Cauchy-Riemann equations are given by

$$\frac{\delta u}{\delta r} = \frac{1}{r}\frac{\delta v}{\delta \theta}$$

$$\frac{\delta v}{\delta r} = -\frac{1}{r}\frac{\delta u}{\delta \theta}$$

**Equation 2.38**

For example, suppose we are given the function $f(z) = \sqrt{z}$ and we wish to determine if this function is analytic. In polar form the variable $z$ is expressed as

$$z = re^{j\theta} \quad \text{for } \{r > 0, \text{ and } 0 \le \theta < 2\pi\}$$

We compute the functions $u(r, \theta)$ and $v(r, \theta)$ to be

$$f(z) = \sqrt{z}$$

$$= \sqrt{re^{j\theta}} = \sqrt{re^{j\frac{\theta}{2}}}$$

$$= \sqrt{r}\cos\left(\frac{\theta}{2}\right) + j\sqrt{r}\sin\left(\frac{\theta}{2}\right)$$

$$u(r,\theta) = \sqrt{r}\cos\left(\frac{\theta}{2}\right)$$

$$v(r,\theta) = \sqrt{r}\sin\left(\frac{\theta}{2}\right)$$

Now we apply the conditions imposed by the Cauchy-Riemann equations to arrive at

$$\frac{\delta u}{\delta r} = \frac{1}{r}\frac{\delta v}{\delta\theta} = \frac{1}{2\sqrt{r}}\cos\left(\frac{\theta}{2}\right)$$

$$\frac{\delta v}{\delta r} = -\frac{1}{r}\frac{\delta u}{\delta\theta} = \frac{1}{2\sqrt{r}}\sin\left(\frac{\theta}{2}\right)$$

The polar form of the Cauchy-Riemann equations is satisfied so the function $f(z) = \sqrt{z}$ is analytic.

Let's look at another example:

$$f(z) = e^z = e^{x+jy} = e^x\left(\cos y + j\sin y\right) = \left(u + jv\right)$$

Equating real and imaginary parts, we get

$$u = e^x\cos y$$

$$v = e^x\sin y$$

The Cauchy-Riemann equations produce

$$\frac{\partial u}{\partial x} = \frac{\partial v}{\partial y} \quad \frac{\partial\left(e^x\cos y\right)}{\partial x} = \frac{\partial\left(e^x\sin y\right)}{\partial y} \quad = e^x\cos y$$

$$\frac{\partial u}{\partial y} = -\frac{\partial v}{\partial x} \quad \frac{\partial\left(e^x\cos y\right)}{\partial y} = -\frac{\partial\left(e^x\sin y\right)}{\partial x} \quad = -e^x\sin y$$

The Cauchy-Riemann equations are satisfied, so the function $f(z) = e^z$ is analytic.

If you are planning to utilize the mathematics of complex variables for any future engineering endeavors, it is highly recommended that the Cauchy-Riemann equations be understood and kept near the top of your memory stack.

## 2.14  SIMPLY CONNECTED REGION

A region is simply connected if any closed path in the region can be contracted to a single point without leaving the region. This is an important concept to remember later on when we deal with line integrals.

## 2.15  CONTOURS

A contour is a continuous curve that is composed of a finite number of smooth arcs. A contour might be defined as a curve that consists of a starting point $z_1$, an end point $z_2$, and all points on the curve that lie in between. In general, this contour is given the label $C$. A contour of length $L \neq 0$ that does not intersect itself and has the property that the start and end point are the same is known as a *closed contour*.

The boundaries of a circle, a square, or a triangle are three examples of a closed contour. Some examples of various contours are illustrated in Figure 2.15. Contour $C_1$ is an example of a piecewise continuous curve, $C_2$ is a continuous curve, and $C_3$ and $C_4$ are examples of closed contours.

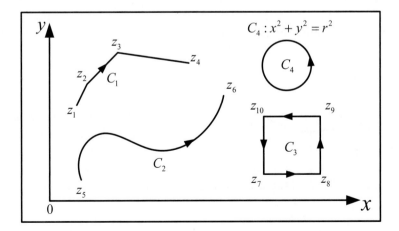

**Figure 2.15**  Examples of contours

## 2.16  LINE INTEGRALS

The integral of a function $f(z)$ from a point $z = z_1$ to a point $z = z_2$ in the complex plane is evaluated at all points along a path or contour $C$ that extends from $z_1$ to $z_2$. Therefore this integral is called a *line integral*. The computed value of a line integral is in general dependent on the choice of the contour and upon the function itself. There are, however, conditions where the value of a line integral is independent of the contour. We will discuss these special cases a bit later.

We can visualize the complex $z$ plane as having an infinite number of points. When we evaluate the line integral of $f(z)$ we evaluate it only along those points that fall on the defined contour $C$. If the path of integration is restricted to the defined contour, and if the integrand is restricted to be a continuous function on that contour, then the line integral may also be called a contour integral, and the integration is referred to as contour integration. In this book we will use the terms *line integral* and *contour integral* interchangeably.

Many readers have a difficult time visualizing exactly what a line integral is. We can aid in this visualization by first reviewing the definite integral that we all know from basic calculus. A definite integral computes the area in two-dimensional space that lies between a function $f(x)$ and the $x$-axis from the point A to the point B, as shown in Figure 2.16. The $x$-axis is sometimes referred to as the *base*, and the curve described by $f(x)$ is sometimes referred to as the *ceiling*. The area between the ceiling and the base is sometimes referred to as the *curtain*. The definite integral computes the area of the curtain.

In three-dimensional space, a function might be written as $z = f(x, y)$. We can think of this function as a three-dimensional surface that is hanging out somewhere in a coordinate system defined by an $x$-, $y$-, and $z$-axis. Instead of computing the area under the three-dimensional surface described by $z = f(x, y)$ and the $x$-axis, the contour integral computes the area of the

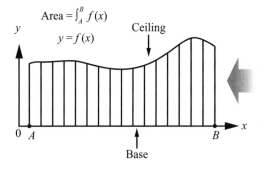

The definite integral computes the area between the function $y = f(x)$ and the $x$-axis from point $A$ to point $B$. This area is sometimes called the *curtain* simply because it looks like a shower curtain.

**Figure 2.16**   Definite integral summation curtain

curtain that extends from the contour $C$ and the intersection of the curtain with the surface. This intersection is called the *ceiling line* and the contour is sometimes called the *base*. This concept is illustrated in Figure 2.17. You can think of the ceiling line as the shadow cast on the three-dimensional surface produced when a light is projected from directly below the contour. So all we are doing in the line integral world is replacing $f(x)$ with $f(x,y)$, and we are replacing the x-axis with a contour $C$. The line integral computes the area of the curtain that extends from that contour to the ceiling line on the surface, defined by $z = f(x,y)$.

A function $f(z)$ that is integrated over a given contour is expressed as

$$\int_C f(z)\, dz$$

The $C$ at the bottom of the integral identifies a particular path or contour over which the function is integrated. If the integration takes place over a closed contour, then the integral is expressed as

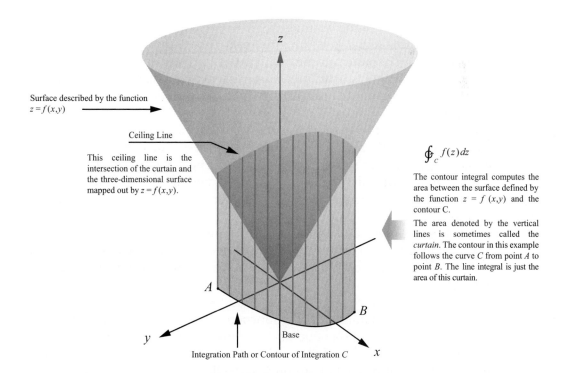

Surface described by the function
$z = f(x,y)$

Ceiling Line

This ceiling line is the intersection of the curtain and the three-dimensional surface mapped out by $z = f(x,y)$.

$$\oint_C f(z)\,dz$$

The contour integral computes the area between the surface defined by the function $z = f(x,y)$ and the contour C.

The area denoted by the vertical lines is sometimes called the *curtain*. The contour in this example follows the curve $C$ from point $A$ to point $B$. The line integral is just the area of this curtain.

$z$

$A$

$B$

$y$

Base

$x$

Integration Path or Contour of Integration $C$

**Figure 2.17**  Contour integral summation curtain

$$\oint_C f(z)\, dz$$

The circular, counterclockwise arrow indicates that the direction of integration along the closed contour is in the positive or counterclockwise direction. If the circular arrow pointed in the clockwise direction, then this would indicate that the direction of integration along the closed contour is in the negative or clockwise direction:

$$\oint_C f(z)\,dz$$

We will explain contours and the importance of integrating in the positive and negative direction a bit later.

## 2.17  REAL LINE INTEGRALS

When evaluating complex line integrals, we often break them down into an expression composed of real line integrals and then evaluate the real integrals. For this reason, it is necessary to spend some time developing a working knowledge of real line integrals.

A real line integral is nothing more than an integral very similar to a definite integral in basic calculus, with one exception. A definite integral in real variable calculus is usually evaluated along the $x$-axis from some point $x_1$ to some other point $x_2$, or along the $y$ axis from some point $y_1$ to some other point $y_2$.

A real line integral is an integral that is evaluated along some path, curve, or contour $C$ from some starting point $A$ to some end point $B$. The terms *path*, *curve*, and *contour* are used interchangeably to describe the path of integration. The only subject matter that is new here in regard to regular definite integrals is the method used to integrate along a contour instead of the $x$- or $y$-axis.

Let us begin by defining two real functions of $x$ and $y$ that we will call $P(x,y)$ and $Q(x,y)$. Furthermore we will define these functions to be continuous at all points along some contour $C$. This is pretty simple to visualize. If, for example, the functions $P$ and $Q$ are continuous for all points $x$ and $y$, then these functions are certainly continuous for the points $x$ and $y$ that happen to fall on the contour $C$. If the functions $P$ and $Q$ are continuous only within some smaller subset of points (say those contained within some region $R$) and if the contour $C$ is also completely contained within this same region, then the functions are continuous on $C$ as well.

Suppose we had some arbitrary integral such as

$$\int \left\{ \left( 6xy^3 \right) \, dx + \left( 9x^2y^2 \right) dy \right\}$$

Further suppose we wanted to integrate this integral along some contour $C$. In this case the integral would be written

$$\int_C \left\{ \left( 6xy^3 \right) \, dx + \left( 9x^2y^2 \right) dy \right\}$$

We would call this integral a *line integral*. In the process of performing the integration we would refer to the computations as *contour integration*. We can replace the line integral term $6xy^3$ with the more general term $P(x,y)$ and the line integral term $9x^2y^2$ with the general term $Q(x,y)$. In doing so, we have the general definition of a real line integral, written as

$$L = \int_C \left\{ P(x,y) \, dx + Q(x,y) \, dy \right\}$$

In this equation we have assigned to $L$ the value of the line integral. It is standard to write this integral in the shorthand notation

$$L = \int_C \left\{ P dx + Q dy \right\}$$

**Equation 2.39**

## 2.17.1 Properties of Real Line Integrals

Two versions of an identical contour are illustrated in Figure 2.18. Both curves have an arrow that indicates the direction we take along the contour when performing a contour integration. In part A of the figure, the arrow is oriented in the counterclockwise or positive direction. In part B of the figure, the arrow is oriented in the clockwise or negative direction. If we integrate some real function $f$ along each of these curves, we can represent the path taken in part A of the figure as $(ba)$ and that path in part B of the figure as $(ab)$.

When we write the line integrals for these two cases, we need to take into account the value of the line integral over path $(ba)$ is the negative of the line integral over path $(ab)$. This means the sign is reversed when the direction of integration along the contour is reversed. We can represent this as

$$\int_{ba} f \, ds = -\int_{ab} f \, ds$$

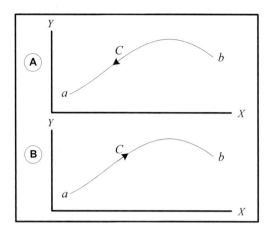

**Figure 2.18**  Example contour

where $ds$ is the infinitesimal arc length along the contour between points $a$ and $b$. Using the standard line integral notation, this becomes

$$\int_{ba}\{Pdx+Qdy\}=-\int_{ab}\{Pdx+Qdy\}$$

We can make several simple observations from Equation 2.39 (which is repeated here for clarity):

$$L=\int_{C}\{Pdx+Qdy\}$$

- If a contour $C$ is a straight line that is parallel to the y axis, then $dx=0$ and $L=\int_{C}Qdy$.
- If a contour $C$ is a straight line that is parallel to the x axis then $dy=0$ and $L=\int_{C}Pdx$.
- If a contour $C$ is a concatenation of two shorter contours $C_1$ and $C_2$, then

$$L=\int_{C}\{Pdx+Qdy\}=\int_{C1}\{Pdx+Qdy\}+\int_{C2}\{Pdx+Qdy\}$$

A contour must be continuous and single valued (i.e., it has only a single value $y$ for a single value $x$). If a contour is not single valued, such as a circle, then the contour must be segmented into two or more sections, $C=C_1C_2C_3\cdots C_K$, such that all the segments $C_K$ are single valued. For example, a circular contour may be segmented into two hemispheres and a contour integration performed on each hemisphere.

## 2.17.2   Real Line Integral Examples

Now is a pretty good time to reinforce the previous text with a few examples.

**Example 1.** Perform the integration of the line integral

$$L = \int_C f(x,y)\,ds = \int_C (x+y^2)\,dx + (x^2-y)\,dy$$

**Equation 2.40**

on the contour $C: y = x + 3$, as shown in Figure 2.19.

   The first thing we want to do is to equate this line integral with the standard form line integral of Equation 2.39 to get

$$L = \int_C \{P\,dx + Q\,dy\} = \int_C (x+y^2)\,dx + (x^2-y)\,dy$$

**Equation 2.41**

   If we equate the expressions associated with the $dx$ and $dy$ terms in Equation 2.41, we end up with

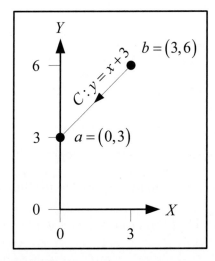

**Figure 2.19**   Path of integration for example 1

$$P = \left( x + y^2 \right)$$
$$Q = \left( x^2 - y \right)$$

**Equation 2.42**

Neither Equation 2.41 nor the expressions for $P$ and $Q$ in Equation 2.42 are important for this particular example. We need to draw attention to this procedure here, and will continue to do so in future examples, only because it should become second nature to you, especially when dealing with exact differentials and their properties, such as path-independent line integrals. These are all important topics that we will discuss later.

To evaluate this line integral, we need to incorporate the contour into the integral. We do so by first noting that the path is given by

$$C:\; y = x + 3, \text{ therefore } dy = dx$$

and the range of $x$ along the path is given by $0 \le x \le 3$. We incorporate the contour by substituting $x + 3$ for all occurrences of $y$ and $dx$ for all occurrences of $dy$ in Equation 2.40 and then integrating in the counterclockwise direction of the contour shown in Figure 2.19 (i.e., from $x = 3 \rightarrow 0$). When these substitutions are made and the limits of integration are established, we end up with

$$
\begin{aligned}
L &= \int_{x=3}^{x=0} \left( x + (x+3)^2 \right) dx + \left( x^2 - (x+3) \right) dx \\
&= 2 \int_{x=3}^{x=0} \left( x^2 + 3x + 3 \right) dx \\
&= 2 \left[ \frac{x^3}{3} + \frac{3x^2}{2} + 3x \right]_{x=3}^{x=0} = -63
\end{aligned}
$$

**Example 2.** Suppose we wished to integrate the function $f(x, y) = x + y^2$ along the contour shown in part A of Figure 2.20 (i.e., $C:\; x^2 + y^2 = 1$ for $0 \le x \le 1$). Thus we can write the expression for this contour integration as

$$L = \int_C f(x, y) \, ds = \int_C \left( x + y^2 \right) dx$$

The next thing we do is equate our line integral with the standard form line integral of Equation 2.39 to get

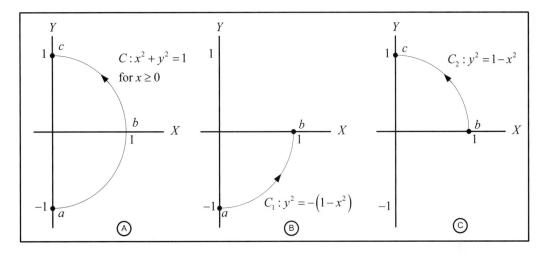

**Figure 2.20**   Path of integration for example 2

$$L = \int_C \{Pdx + Qdy\} = \int_C (x + y^2)\, dx$$

Equating the line integral components shows that $P = x + y^2$ and $Q = 0$. Therefore our line integral takes the form of

$$L = \int_C \{Pdx\} = \int_C (x + y^2)\, dx$$

Now we need to incorporate the path of integration into the line integral. We do that by noting the path $C: \; x^2 + y^2 = 1$ for $0 \le x \le 1$, a half circle of radius 1. We can express the path equation as a function of $x$ so that

$$y^2 = 1 - x^2$$

The first thing we see about the proposed path of integration is that $y$ is not single valued. Thus for every value of $x$, except $x = 1$, $y$ can take on two different values, $y = \pm\sqrt{1 - x^2}$. We therefore need to segment the path into two separate sections $C_1$ and $C_2$. These two segments are illustrated in parts B and C of Figure 2.20. We compute the segmented paths to be

$$C_1 : y^2 = -(1 - x^2)$$
$$C_2 : y^2 = +(1 - x^2)$$

So now our line integral can be rewritten as

$$L = \int_{C_1} \left( x + y^2 \right) dx + \int_{C2} \left( x + y^2 \right) dx$$

Now that our line integral expression takes into account the two paths, we can incorporate the path into the integral by substituting the path expression into the integral expression. We substitute $-\left(1 - x^2\right)$ for $y^2$ in path $C_1$ and integrate in the counterclockwise direction as $x$ moves from $0 \rightarrow 1$, and we substitute $1 - x^2$ for $y^2$ in path $C_2$ and integrate in the counterclockwise direction as $x$ moves from $1 \rightarrow 0$. Incorporating these changes causes the line integral to take the form

$$
\begin{aligned}
L &= \int_{x=0}^{1} \left( x - \left( 1 - x^2 \right) \right) dx + \int_{x=1}^{0} \left( x + \left( 1 - x^2 \right) \right) dx \\
&= \int_{x=0}^{1} \left( x^2 + x - 1 \right) dx + \int_{x=1}^{0} \left( -x^2 + x + 1 \right) dx \\
&= \int_{x=0}^{1} \left( x^2 + x - 1 \right) dx - \int_{x=0}^{1} \left( -x^2 + x + 1 \right) dx \\
&= 2 \int_{x=0}^{1} \left( x^2 - 1 \right) dx
\end{aligned}
$$

Note that we reversed the path of the second integral by changing the limits of integration from $x = 1 \rightarrow 0$ to $x = 0 \rightarrow 1$. When the path direction is reversed, the value of the integral is the negative of the original integral, so we keep our bookkeeping straight with the addition of a negative sign in front of the integral. After we perform the simple integration, we get

$$L = 2 \left( \frac{x^3}{3} - x \right) \Big|_{x=0}^{x=1} = 2 \left( \frac{1}{3} - 1 \right) = -\frac{4}{3}$$

This example was pretty straightforward. The important information to remember from this example is the method by which the path of integration was incorporated into the contour integration.

**Example 3.** In the previous example, we segmented the path into two pieces and performed the contour integration along path $C_1$, between the two end points $a = (0, -1)$ and $b = (1, 0)$, and then along the path $C_2$, between the two end points $b = (1, 0)$ and $c = (0, 1)$. Both path segments $C_1$ and $C_2$ were circular arcs. Now let's observe the result of performing the contour integration of the same function between the same end points but this time along a different path $C$. Once again we begin with the function $f(x,y) = x + y^2$. This time we wish to integrate along the contour illustrated in Figure 2.21 part A. We segment the path $C$ into two concatenated paths $C_1$ and $C_2$, as shown Figure 2.21 parts B and C. The two paths are expressed as

$$C_1 : y = \quad x - 1$$
$$C_2 : y = -x + 1$$

Once again our line integral takes the form of

$$L = \int_C \{Pdx\} = \int_C (x + y^2) dx$$

Now we need to incorporate the path of integration into the line integral. We accomplish this by realizing the path $C$ is the concatenation of the two path segments $C_1$ and $C_2$. So we will need to break our line integral into two separate integrals, one for each path segment, to get

$$L = \int_{C_1} (x + y^2) dx + \int_{C2} (x + y^2) dx$$

Now we can incorporate the path into the integral by substituting path expressions into the integral expression. We substitute $(x - 1)^2$ for $y^2$ in the first integral over $C_1$ and $(-x + 1)^2$ for $y^2$ in the second integral over $C_2$. Then we integrate over the range of $x$. Making these substitutions, our line integral becomes

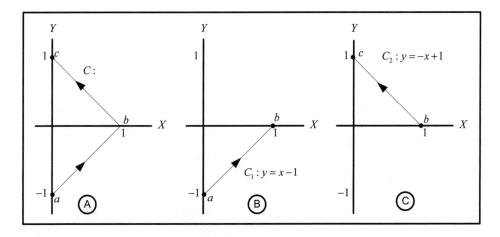

**Figure 2.21**   Path of integration for example 3

$$
\begin{aligned}
L &= \int_{x=0}^{1} \left( x + (x-1)^2 \right) dx + \int_{x=1}^{0} \left( x + (-x+1)^2 \right) dx \\
&= \int_{x=0}^{1} \left( x^2 - x + 1 \right) dx + \int_{x=1}^{0} \left( x^2 - x + 1 \right) dx \\
&= \int_{x=0}^{1} \left( x^2 - x + 1 \right) dx - \int_{x=0}^{1} \left( x^2 - x + 1 \right) dx \\
&= \int_{x=0}^{1} (0)\, dx = 0
\end{aligned}
$$

It is important to note that the two results we obtained in example 2 and example 3 (by performing a contour integration of the same function, between the same end points but over different paths) were not identical. In general, the value of a line integral is dependent on the path of integration. In other words, integrating along different paths between the same end points does not always give identical results. This makes sense because the ceiling line is different for a different contour and therefore the curtain has a different area.

We will discuss the very important special case where line integrals are path independent later. At that time, we will discuss the conditions whereby a contour integration between the same two end points produces the same result no matter what the path is.

**Example 4.** In the next two examples, we will demonstrate the evaluation of a line integral around a closed contour. Remember that the line integral symbol for a function $f(x, y)$ evaluated counterclockwise along a closed contour is given by $\oint_C f(x, y)$ and is considered to be the positive direction. The same function evaluated in the clockwise or negative direction on a closed contour is given by $\oint f(x, y)$. In this example, let's evaluate the line integral $L = \oint_C (x - y) dx + (2xy) dy$ over the triangular contour illustrated in Figure 2.22. It is clear from the figure that this contour $C$ is not single valued and that it lends itself well to decomposition into three concatenated subcontours $C_1, C_2$, and $C_3$. We will evaluate $L$ with three separate line integrals over three separate contours.

Before we begin, let's equate our line integral with the standard form line integral of Equation 2.39 to get

$$
L = \oint_C \{ P dx + Q dy \} = \oint_C (x - y) dx + (2xy) dy
$$

Equating the line integral components shows that

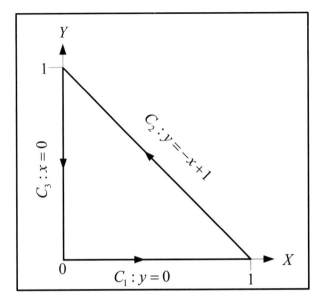

**Figure 2.22**   Path of integration for example 4

$$P = x - y$$
$$Q = 2xy$$

Once again, this identification of the functions $P$ and $Q$ is not important in this example. We won't use this information here. The expressions, however, are true and play a huge role when we begin to discuss exact differentials and associated properties, such as path-independent line integrals. Until then, we're still trying to instill through repetition the underlying method for the evaluation of all line integrals and the identification of path-independent line integrals.

So let's begin the evaluation of the line integral by evaluating each of the three contour integration components: $L = L_1 + L_2 + L_3$.

Contour $C_1$ : The line integral $L_1$ and the contour $C_1$ are given by

$$L_1 = \int_{C_1} (x - y)\, dx + (2xy)\, dy$$
$$C_1 : y = 0 \implies dy = 0$$

We incorporate the contour into the line integral by substituting $0$ for both $y$ and $dy$ into the line integral equation and integrating from $x = 0 \rightarrow 1$. This results in

$$L_1 = \int\limits_{x=0}^{x=1} x \; dx = \frac{x^2}{2}\bigg|_{x=0}^{x=1} = \frac{1}{2}$$

Contour $C_2$ : The line integral $L_2$ and the contour $C_2$ are given by

$$L_2 = \int_{C_2}(x-y)dx + (2xy)dy$$

$$C_2 : y = -x + 1 \;\; \Rightarrow \;\; dy = -dx$$

We incorporate the contour into the line integral by substituting $-x+1$ for $y$ and $-dx$ for $dy$ into the line integral equation and integrating in the path direction from $x = 1 \rightarrow 0$. When we do, we end up with

$$L_2 \;=\; \int\limits_{x=1}^{x=0}\big(x-(-x+1)\big)dx - \big(2x(-x+1)\big)dx$$

$$=\; \int\limits_{x=1}^{x=0}\big(2x^2 - 1\big)dx$$

$$=\; \left[\frac{2}{3}x^3 - x\right]_{x=1}^{x=0} = 0 - \left[\frac{2}{3} - 1\right] = \frac{1}{3}$$

Contour $C_3$ : The line integral $L_3$ and the contour $C_3$ are given by

$$L_3 = \int_{C_3}(x-y)dx + (2xy)dy$$

$$C_3 : x = 0 \;\; \Rightarrow \;\; dx = 0$$

This one is easy. Both $x$ and $dx$ are zero; therefore the entire integral is zero and $L_3 = 0$.

The line integral evaluates to

$$L = L_1 + L_2 + L_3 = \frac{1}{2} + \frac{1}{3} + 0 = \frac{5}{6}$$

**Example 5**. For our last example, we would like to evaluate the line integral $\int_C y \; dx$ on the circular contour $C : x^2 + y^2 = 4$, as illustrated in Figure 2.23.

Once again, we will need to break the contour into two separate paths. We will define these paths to be

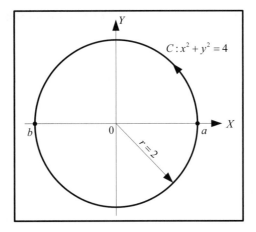

**Figure 2.23** Path of integration for example 5

$C_1:$ $y = +\sqrt{4 - x^2}$ from $\{x = +2 \text{ to } x = -2\}$ Top half circle

$C_2:$ $y = -\sqrt{4 - x^2}$ from $\{x = -2 \text{ to } x = +2\}$ Bottom half circle

Now we will compute the line integral via two separate integrals such that $L = L_1 + L_2$:

Contour $C_1$ : The line integral $L_1$ and the contour $C_1$ are given by

$$L_1 = \int_{C_1} (y) dx$$
$$C_1 : y = +\sqrt{4 - x^2}$$

Contour $C_2$ : The line integral $L_2$ and the contour $C_2$ are given by

$$L_2 = \int_{C_2} (y) dx$$
$$C_2 : y = -\sqrt{4 - x^2}$$

Now we can evaluate the line integral as

$$L = L_1 + L_2 = \int_{C_1} (y) dx + \int_{C_2} (y) dx$$

We incorporate the contour into the line integral by making the substitutions for the variable $y$ to get

$$L = \int_{x=2}^{x=-2} \left( \sqrt{4-x^2} \right) dx + \int_{x=-2}^{x=2} \left( -\sqrt{4-x^2} \right) dx$$

We can reduce this expression somewhat by reversing the path of integration in the second integral to get

$$\begin{aligned} L &= \int_{x=2}^{x=-2} \left( \sqrt{4-x^2} \right) dx - \int_{x=2}^{x=-2} \left( -\sqrt{4-x^2} \right) dx \\ &= 2 \int_{x=2}^{x=-2} \left( \sqrt{4-x^2} \right) dx \end{aligned}$$

We can solve this integral by trigonometric substitution. If we let $x = 2\sin\theta$, then $dx = 2\cos\theta\, d\theta$ and $x^2 = 4\sin^2\theta$. If we make these substitutions, we also need to modify the limits of integration from the variable $x$ to the variable $\theta$. When $x = 2$, $y = 0$, so the phase angle $\theta = 0$. When $x = -2$, $y = 0$, so the phase angle $\theta = \pi$.

Making these substitutions gives us the new expression for our line integral:

$$L = 2 \int_{\theta=0}^{\theta=\pi} \left( \sqrt{4-4\sin^2\theta} \right) 2\cos\theta\, d\theta$$

We can do some basic manipulations on this integral to arrive at

$$\begin{aligned} L &= 8 \int_{\theta=0}^{\theta=\pi} \left( \sqrt{1-\sin^2\theta} \right) \cos\theta\, d\theta \\ &= 8 \int_{\theta=0}^{\theta=\pi} \left( \sqrt{\cos^2\theta} \right) \cos\theta\, d\theta \\ &= 8 \int_{\theta=0}^{\theta=\pi} \cos^2\theta\, d\theta \end{aligned}$$

We can use the trigonometric identity $\cos^2\theta = \dfrac{1}{2}(1+\cos 2\theta)$ to simplify the integrand. When we do, we arrive at

$$L = 8 \int_{\theta=0}^{\theta=\pi} \frac{1}{2}(1 + \cos 2\theta) \ d\theta$$

$$= 4 \int_{\theta=0}^{\theta=\pi} (1 + \cos 2\theta) \ d\theta$$

$$= 4 \left[ \theta + \frac{1}{2} \sin 2\theta \right]_{\theta=0}^{\theta=\pi}$$

$$= 4 \left[ (\pi + 0) - (0) \right]$$

This gives us the final result of $L = 4\pi$.

### 2.17.3  Line Integrals Evaluated in Terms of Arc Length

Another method to evaluate line integrals is to move along the contour in increments of arc length. Suppose we wish to perform a contour integration of the function $f(x,y)$ along the contour $C$ between the points $P_1$ and $P_2$, as illustrated in Figure 2.24. Thus we wish to perform the contour integration

$$L = \int_C f(x,y) \ ds$$

In this context, $S$ is the arc length, $\delta s$ is a straight line that estimates a small change in arc length, and $ds$ is the differential arc length. We can approximate the contour $C$ by a series of straight lines $\delta s$ that are concatenated

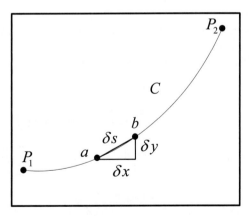

**Figure 2.24**  Length of arc along a contour

along the curve from point $P_1$ to $P_2$. Each line segment $\delta s$ can be broken down into its $x$ and $y$ components, $\delta x$ and $\delta y$. The magnitude of each line segment $\delta s$ is given by $\delta s = \sqrt{\delta x^2 + \delta y^2}$.

We can divide both sides by $\delta x$ to get

$$\frac{\delta s}{\delta x} = \sqrt{1 + \left(\frac{\delta y}{\delta x}\right)^2}$$

Now, in the limit as $\delta x \to 0$ and $\delta y \to 0$, we end up with

$$\frac{ds}{dx} = \sqrt{1 + \left(\frac{dy}{dx}\right)^2} \quad \text{or} \quad ds = \sqrt{1 + \left(\frac{dy}{dx}\right)^2}\, dx$$

If we make this substitution in the line integral for $ds$, we end up with

$$L = \int_C f(x,y)\, ds \;=\; \int_C f(x,y)\sqrt{1 + \left(\frac{dy}{dx}\right)^2}\, dx$$

or simply

$$L = \int_C f(x,y)\sqrt{1 + \left(\frac{dy}{dx}\right)^2}\, dx$$

**Equation 2.43**

**Example.** Evaluate $L = \int_C (4x + 3xy)\, ds$ on the contour $C : y = \frac{1}{2}x$, as illustrated in Figure 2.25.

The first thing we do is write down the base expression

$$L = \int_C f(x,y)\, ds \;=\; \int_C f(x,y)\sqrt{1 + \left(\frac{dy}{dx}\right)^2}\, dx$$

Then we make the appropriate substitutions to get

$$L = \int_C (4x + 3xy)\frac{1}{2}\sqrt{5}\, dx$$

We know that $y = \frac{1}{2}x$, so we can make this substitution and transform the entire expression in terms of $x$ only.

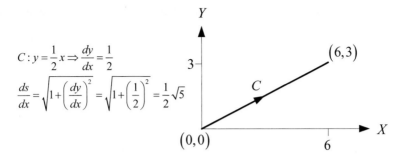

**Figure 2.25**  Line integral arc length

$$L = \int_C \left(4x + 3x\frac{x}{2}\right)\frac{1}{2}\sqrt{5}\ dx = \frac{\sqrt{5}}{2}\int_C \left(\frac{3}{2}x^2 + 4x\right)dx$$

Now we simply evaluate the integral to get

$$
\begin{aligned}
L &= \frac{\sqrt{5}}{2}\int_{x=0}^{x=6}\left[\frac{3}{2}x^2 + 4x\right]dx \\
&= \frac{\sqrt{5}}{2}\left[\frac{3}{6}x^3 + \frac{4}{2}x^2\right]_{x=0}^{x=6} \\
&= \frac{\sqrt{5}}{2}\left[\frac{3}{6}216 + \frac{4}{2}36\right] \\
&= \frac{\sqrt{5}}{2}[108 + 72] \\
&= \sqrt{5}\,(90)
\end{aligned}
$$

### 2.17.4  Line Integrals Evaluated in Terms of Parametric Equations

Say the variables $x$ and $y$ are expressed as a function of some other variable such as $t$. For example,

$$x = f_1(t)$$
$$y = f_2(t)$$

Then we say that $x$ and $y$ are in parametric form. Suppose we wish to perform a contour integration of the function $f(x(t),\ y(t))$ along the contour $C$

between the points $P_1$ and $P_2$, as illustrated in Figure 2.24. The line integral would take the form

$$L = \int_C f(x, y)\, ds$$

**Equation 2.44**

As mentioned previously, we can approximate the contour $C$ by a series of straight lines $\delta s$ that are concatenated along the curve from point $P_1$ to $P_2$. Each line segment $\delta s$ can be broken down into its $x$ and $y$ components $\delta x$ and $\delta y$. We can then state that

$$ds^2 = dx^2 + dy^2$$

The change in arc length $s$ with respect to $t$ is then given by

$$\frac{ds^2}{dt^2} = \frac{dx^2}{dt^2} + \frac{dy^2}{dt^2}$$

We take the square root of both sides of the above equation to get

$$\frac{ds}{dt} = \sqrt{\left(\frac{dx}{dt}\right)^2 + \left(\frac{dy}{dt}\right)^2}$$

Multiplying both sides of the equation by $dt$ gives us

$$ds = \sqrt{\left(\frac{dx}{dt}\right)^2 + \left(\frac{dy}{dt}\right)^2}\, dt$$

**Equation 2.45**

Substitution of Equation 2.45 into Equation 2.44 results in the parametric form of a line integral.

$$L = \int_C f(x, y) \sqrt{\left(\frac{dx}{dt}\right)^2 + \left(\frac{dy}{dt}\right)^2}\, dt$$

**Equation 2.46**

**Example**. Evaluate the line integral $L = \int_C (2x + 3y)\, ds$.
The contour $C$ is defined as

$$C: \quad x = \sin t \quad \Rightarrow \quad \frac{dx}{dt} = \cos t$$
$$\{t = 0 \text{ to } t = \pi/2\}$$
$$y = \cos t \quad \Rightarrow \quad \frac{dy}{dt} = -\sin t$$

$$f(x,y) \quad = 2x + 3y \qquad\qquad\qquad = 2\sin t + 3\cos t$$

$$ds \quad = \sqrt{\left(\frac{dx}{dt}\right)^2 + \left(\frac{dy}{dt}\right)^2}\, dt \quad = \sqrt{\sin^2 t + \cos^2 t}\; dt$$

So if we put all this together we get

$$L = \int_C f(x,y) \sqrt{\left(\frac{dx}{dt}\right)^2 + \left(\frac{dy}{dt}\right)^2}\, dt \quad = \int_{t=0}^{t=\frac{\pi}{2}} (2\sin t + 3\cos t)\sqrt{\sin^2 t + \cos^2 t}\; dt$$

Since $\sqrt{\sin^2 t + \cos^2 t} = 1$, we can rewrite this equation and solve to get

$$
\begin{aligned}
L \quad &= \int_{t=0}^{t=\frac{\pi}{2}} (2\sin t + 3\cos t)\, dt \\
&= \left[-2\cos t + 3\sin t\right]_{t=0}^{t=\pi/2} \\
&= \left(-2\cos\frac{\pi}{2} + 3\sin\frac{\pi}{2}\right) - (-2\cos 0 + 3\sin 0) \\
&= 5
\end{aligned}
$$

## 2.17.5   Exact Differentials

In calculus, it is a convention to express small differences in some variable through the use of what are called *differentials*. If we are given a function $z = f(x,y)$, we use the differential of $z$ to compute the small change in $z$ due to a small change in either or both $x$ and $y$. The differential of $z$ is given by

$$dz = \frac{\partial z}{\partial x} dx + \frac{\partial z}{\partial y} dy$$

**Equation 2.47**

As an example, the differential of the function $z = x^2 + y^2$ is given by

$$\begin{aligned} dz \ \ &= \frac{\partial z}{\partial x} dx + \frac{\partial z}{\partial y} dy \\ &= 2x \ dx + 2y \ dy \end{aligned}$$

One of the more important properties of line integrals is the property of exact differentials. Let's equate our newly defined differential with the component terms in the definition of the line integral that was presented in Equation 2.39. The line integral is repeated here for clarity:

$$L = \int_C \left\{ Pdx + Qdy \right\}$$

Doing so results in

$$dz = \frac{\partial z}{\partial x} dx + \frac{\partial z}{\partial y} dy = Pdx + Qdy$$

**Equation 2.48**

This gives us two new and important relationships:

$$Pdx = \frac{\partial z}{\partial x} dx \ \ \Rightarrow \ \ P = \frac{\partial z}{\partial x}$$

$$Qdy = \frac{\partial z}{\partial y} dy \ \ \Rightarrow \ \ Q = \frac{\partial z}{\partial y}$$

Now, what happens if we take the partial of $P$ with respect to $y$ and the partial of $Q$ with respect to $x$? In doing so, we end up with

$$\frac{\partial P}{\partial y} = \frac{\partial z^2}{\partial x \partial y}$$

$$\frac{\partial Q}{\partial x} = \frac{\partial z^2}{\partial x \partial y}$$

The terms on the right side of the two equations are equal. That is,

$$\frac{\partial z^2}{\partial x \partial y} = \frac{\partial z^2}{\partial x \partial y}$$

Therefore, this implies that

$$\frac{\partial P}{\partial y} = \frac{\partial Q}{\partial x}$$

**Equation 2.49**

A differential $dz = Pdx + Qdy$ is considered to be an exact differential if Equation 2.49 holds true. This is extremely important. Why? Because if the integrand $Pdx + Qdy$ of a line integral $L = \int_C \{Pdx + Qdy\}$ is an exact differential, then the line integral between two points evaluates to the same result, independent of the contour or path of integration. This is a very compelling statement. We will see why when we discuss path-independent line integrals in the next section.

## 2.17.6   Line Integrals That Are Path Independent

The importance of exact differentials is put into perspective by the following two statements. If $Pdx + Qdy$ is an exact differential, then

1. The line integral $\int_C \{Pdx + Qdy\}$ is independent of the path of integration

2. The contour integral $\oint_C \{Pdx + Qdy\}$ is zero when the path $C$ is a closed contour

**Equation 2.50**

The two items in Equation 2.50 are very powerful statements. Together, these two properties make our lives a whole lot simpler when we get into the important study of residue theory. These two statements are so important that you should put both near the top of your memory stack for quick retrieval.

Let's look at an example. Suppose we were tasked with computing the line integral of the function $f(x, y) = \frac{1}{6}x^3 y^2$ over some contour $C$. The first thing we could do is express this function in the standard line integral form given by Equation 2.39 and repeated here for clarity:

$$L = \int_C \{Pdx + Qdy\}$$

We know from previous work that if $z = f(x,y)$, then

$$P = \frac{\partial z}{\partial x} = \frac{\partial f(x,y)}{\partial x} = \frac{\partial\left(1/6\,x^3 y^2\right)}{\partial x} = \frac{1}{2}x^2 y^2$$

$$Q = \frac{\partial z}{\partial y} = \frac{\partial f(x,y)}{\partial y} = \frac{\partial\left(1/6\,x^3 y^2\right)}{\partial y} = \frac{1}{3}x^3 y$$

**Equation 2.51**

Therefore we can determine the integrand of the line integral expression to be

$$Pdx + Qdy = \left(\frac{1}{2}x^2 y^2\right)dx + \left(\frac{1}{3}x^3 y\right)dy$$

The expression $Pdx + Qdy$ is an exact differential if Equation 2.49, repeated here for clarity, holds true:

$$\frac{\partial P}{\partial y} = \frac{\partial Q}{\partial x}$$

We perform the indicated partial differentiation on Equation 2.51 to produce

$$\frac{\partial P}{\partial y} = \frac{\partial}{\partial y}\left(\frac{1}{2}x^2 y^2\right) = x^2 y$$

$$\frac{\partial Q}{\partial x} = \frac{\partial}{\partial x}\left(\frac{1}{3}x^3 y\right) = x^2 y$$

The two expressions are equal. Therefore the integrand

$$Pdx + Qdy = \left(\frac{1}{2}x^2 y^2\right)dx + \left(\frac{1}{3}x^3 y\right)dy$$

is an exact differential and therefore the line integral

$$\int_C Pdx + Qdy = \int_C \left(\frac{1}{2}x^2 y^2\right)dx + \left(\frac{1}{3}x^3 y\right)dy$$

is path independent, and contour integration of this integrand will evaluate to the same result between the same two end points no matter what the path. This is a very significant statement. Furthermore, if the path $C$ is a closed contour such as a circle, a square, an ellipse, or any other closed path, then this line integral will evaluate to 0. That is,

$$L = \oint_C P\,dx + Q\,dy = \oint_C \left(\frac{1}{2}x^2 y^2\right)dx + \left(\frac{1}{3}x^3 y\right)dy = \ 0 \ \left\{\text{ for } C : \text{a closed contour}\right\}$$

This statement is one of the most powerful tools available in the science of complex variables. Let's prove by example the two incredible statements listed in Equation 2.50.

**Statement 1.** If $P\,dx + Q\,dy$ is an exact differential, then $\int_C \{P\,dx + Q\,dy\}$ is independent of the path of integration.

Let's evaluate the line integral $L$ that we just discussed over two different contours that connect the same end points and see what we get. The two contours are shown as path 1 and path 2 in Figure 2.26. First, we will perform a contour integration over the curve defined as path 1:

$$C : y = x \ \Rightarrow \ dy = dx, \ \text{ for } 0 \le x \le 1$$

We incorporate the path into the integral by substituting $x$ for $y$ and $dx$ for $dy$. We then evaluate the line integral over path 1 to get

$$
\begin{aligned}
L \ &= \ \int_C \left(\frac{1}{2}x^2 y^2\right)dx + \left(\frac{1}{3}x^3 y\right)dy \\
&= \ \int_{x=0}^{x=1} \left(\frac{1}{2}x^2 x^2\right)dx + \left(\frac{1}{3}x^3 x\right)dx \\
&= \ \frac{5}{6} \int_{x=0}^{x=1} \left(x^4\right)dx \\
&= \ \left(\frac{5}{6}\right)\left(\frac{1}{5}\right)x^5 \Big|_{x=0}^{x=1} \ = \ \frac{1}{6}
\end{aligned}
$$

Now we evaluate the line integral over path 2. The contour laid out by path 2 is composed of two curves labeled $C_1$ and $C_2$, as shown in Figure 2.26. The parameters for each of the two paths are

$$C_1 : y = 0 \ \Rightarrow \ dy = 0 \ \text{ for } 0 \le x \le 1$$
$$C_2 : x = 1 \ \Rightarrow \ dx = 0 \ \text{ for } 0 \le y \le 1$$

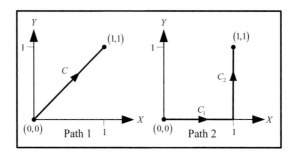

**Figure 2.26** Two different paths between the same end points

Our line integral will be the sum of two line integrals, one for each path or $L = L_1 + L_2$. We evaluate the line integral $L_1$ to be

$$
\begin{aligned}
L_1 &= \int_{C_1} \left( \frac{1}{2} x^2 y^2 \right) dx + \left( \frac{1}{3} x^3 y \right) dy \\
&= \int_{x=0}^{x=1} \left( \frac{1}{2} x^2 (0)^2 \right) dx + \left( \frac{1}{3} x^3 (0) \right) (0) \\
&= 0
\end{aligned}
$$

We evaluate the line integral $L_2$ to be

$$
\begin{aligned}
L_2 &= \int_{C_2} \left( \frac{1}{2} x^2 y^2 \right) dx + \left( \frac{1}{3} x^3 y \right) dy \\
&= \int_{y=0}^{y=1} \left( \frac{1}{2} (1)^2 y^2 \right) (0) + \left( \frac{1}{3} (1)^3 y \right) dy \\
&= \frac{1}{3} \int_{y=0}^{y=1} y \, dy \\
&= \frac{1}{3} \frac{y^2}{2} \Big|_{y=0}^{y=1} = \frac{1}{6}
\end{aligned}
$$

If we add the results for $L_1$ and $L_2$, we get

$$
L = L_1 + L_2 = \left( 0 + \frac{1}{6} \right) = \frac{1}{6}
$$

This is the same result for path 1. The reader is invited to confirm this powerful concept by performing several contour integrations of this function between the same two end points but using different paths.

**Statement 2:** If $Pdx + Qdy$ is an exact differential, then $\oint_C \{Pdx + Qdy\}$ is zero when the path $C$ is a closed contour.

This time we will evaluate the line integral on a closed contour. Note the change in the integral symbol to reflect this:

$$L = \oint_C \left(\frac{1}{2}x^2y^2\right)dx + \left(\frac{1}{3}x^3y\right)dy$$

For simplicity, let's use the unit circle $x^2 + y^2 = 1$ as the closed contour of integration. Since the unit circle is not single valued, we will need to partition the circle into two hemispherical paths, as illustrated in Figure 2.27. We choose to designate contour $C_1$ as the top half of the circular contour and we choose to designate contour $C_2$ as the bottom half of the circular contour. From the contour equation and from Figure 2.27, we can describe the original contour and the two paths as

$$C: \quad y^2 = 1 - x^2$$
$$C_1: \quad y = +\sqrt{1-x^2} \quad \text{for } x = +1 \rightarrow -1$$
$$C_2: \quad y = -\sqrt{1-x^2} \quad \text{for } x = -1 \rightarrow +1$$

When we set up the line integral, we will want to eliminate the $y$ terms from the integrand and express it in terms of $x$ only. To do that, we will need to determine the expression for $dy$. We begin by introducing a new variable $u$. We let

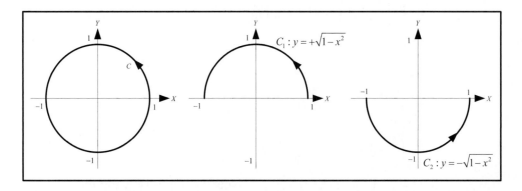

**Figure 2.27**   Closed contour integration

$$u = 1 - x^2 \implies du = -2x \, dx$$

Then for the path $C_1$

$$y = \sqrt{1 - x^2} = \sqrt{u} = u^{1/2}$$

$$dy = \frac{1}{2} u^{-1/2} du = \frac{1}{2} \frac{-2x}{\sqrt{1 - x^2}} dx = \frac{-x}{\sqrt{1 - x^2}} dx$$

And for the path $C_2$

$$y = -\sqrt{1 - x^2} = -\sqrt{u} = -u^{1/2}$$

$$dy = -\frac{1}{2} u^{-1/2} du = \frac{1}{2} \frac{2x}{\sqrt{1 - x^2}} dx = \frac{x}{\sqrt{1 - x^2}} dx$$

We will compute the line integral as the sum of two line integrals $L = L_1 + L_2$. We evaluate the line integral $L_1$ to be

$$L_1 = \int_{C_1} \left( \frac{1}{2} x^2 y^2 \right) dx + \left( \frac{1}{3} x^3 y \right) dy$$

$$= \int_{x=1}^{x=-1} \left( \frac{1}{2} x^2 \left( \sqrt{1 - x^2} \right)^2 \right) dx + \left( \frac{1}{3} x^3 \left( \sqrt{1 - x^2} \right) \right) \left( \frac{-x}{\sqrt{1 - x^2}} \right) dx$$

$$= \int_{x=1}^{x=-1} \left[ -\frac{1}{2} x^4 - \frac{1}{3} x^4 + \frac{1}{2} x^2 \right] dx$$

We evaluate the line integral $L_2$ to be

$$L_2 = \int_{C2} \left( \frac{1}{2} x^2 y^2 \right) dx + \left( \frac{1}{3} x^3 y \right) dy$$

$$= \int_{x=-1}^{x=1} \left( \frac{1}{2} x^2 \left( -\sqrt{1 - x^2} \right)^2 \right) dx + \left( \frac{1}{3} x^3 \left( -\sqrt{1 - x^2} \right) \right) \left( \frac{x}{\sqrt{1 - x^2}} \right)$$

$$= \int_{x=-1}^{x=1} \left[ \frac{1}{2} x^2 \left( 1 - x^2 \right) \right] dx - \frac{1}{3} x^4 dx$$

$$= \int_{x=-1}^{x=1} \left\{ -\frac{1}{2} x^4 - \frac{1}{3} x^4 + \frac{1}{2} x^2 \right\} dx$$

$$= -\int_{x=1}^{x=-1} \left\{ -\frac{1}{2} x^4 - \frac{1}{3} x^4 + \frac{1}{2} x^2 \right\} dx \quad \begin{Bmatrix} \text{Note the minus sign and the} \\ \text{path of integration is reversed} \end{Bmatrix}$$

We add the two integrals to get $L = L_1 + L_2 = L_1 - L_1 = 0$. Stated another way, the line integral

$$L = \oint_C \left(\frac{1}{2}x^2y^2\right)dx + \left(\frac{1}{3}x^3y\right)dy = 0 \quad \{\text{for } C: \ x^2 + y^2 = 1\}$$

This is a very neat result. The value of a line integral of a function that is an exact differential, evaluated on a closed path or closed contour, is zero.

Now that we know this, we can avoid a great deal of computation down the road when we discuss residue theory. We will be able to write down some pretty fancy equations and know in advance that they evaluate to zero. This is a hint of things to come: residue theory is a great tool for an engineer to have at his or her disposal when using z-transforms in a digital signal processing (DSP) design environment.

### 2.17.7   Green's Theorem

Let $C$ be a single closed contour such that a line parallel to the $x$- or $y$-axis will intersect the curve in at most two points. Furthermore, suppose we have two functions $P(x,y)$ and $Q(x,y)$ that are finite and continuous in some region $R$ bounded by the contour $C$. Suppose also that the first partial derivatives of the functions exist and are also finite and continuous in the same region $R$ and bounded by the contour $C$. When this is the case, Green's theorem says that

$$\oint_C (Pdx + Qdy) = \iint_R \left(\frac{\delta Q}{\delta x} - \frac{\delta P}{\delta y}\right)dxdy$$

**Equation 2.52**

Equation 2.52 says that the double integral over the region $R$ in the complex plane is equal to the line integral over the contour $C$ that encloses the region. An example region $R$ and its closed contour $C$ are illustrated in Figure 2.28.

Another way of saying this is that a double integral over the plane region $R$ can be transformed into a line integral over the contour $C$ that bounds the region. This gives us an option when solving line integrals. As we have seen, solving the line integral

$$\oint_C (Pdx + Qdy)$$

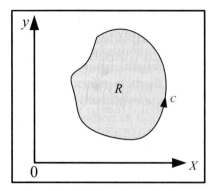

**Figure 2.28**   Region in the complex plane bounded by some contour c

can at times be a little tedious. Green's theorem gives us an alternate method to solve the same line integral that is oftentimes a bit simpler by evaluating

$$L = \oint_C \left( Pdx + Qdy \right) = \iint_R \left( \frac{\delta Q}{\delta x} - \frac{\delta P}{\delta y} \right) dxdy$$

**Example 1**. Suppose we wished to integrate the functions

$$P(x,y) = 4xy^3$$
$$Q(x,y) = 6x^2 y^2$$

over the closed contour $C : x^2 + y^2 = 1$, as illustrated in Figure 2.29.

We opt to use Green's theorem to perform the contour integration

$$L = \oint_C \left( Pdx + Qdy \right) = \iint_R \left( \frac{\delta Q}{\delta x} - \frac{\delta P}{\delta y} \right) dxdy$$

Evaluation of the line integral on the left is possible, but as we have seen, it can get a bit cumbersome, mainly because we would need to break the contour into two halves and solve two line integrals. So we decide to use the right double integral to evaluate the line integral. We can easily calculate that

$$\frac{\delta Q}{\delta x} = \frac{\delta}{\delta x} \left( 6x^2 y^2 \right) = 12xy^2$$
$$\frac{\delta P}{\delta y} = \frac{\delta}{\delta y} \left( 4xy^3 \right) = 12xy^2$$

Substituting these values into the right side of Green's equation, we get

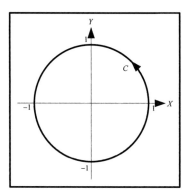

**Figure 2.29**   Circular contour

$$\iint_R \left( \frac{\delta Q}{\delta x} - \frac{\delta P}{\delta y} \right) dxdy = \iint_R \left( 12xy^2 - 12xy^2 \right) dxdy = 0$$

This particular line integral evaluation turned out to be a whole lot simpler than if we had chosen to evaluate the integral using $\oint_C \left( Pdx + Qdy \right)$. We should have known in advance that the integral would evaluate to zero since the integrand was an exact differential. How did we know this? We knew it because

$$\frac{\delta Q}{\delta x} = \frac{\delta P}{\delta y}$$

and the integral of an exact differential on a closed contour is equal to zero. The thought of how much work we just saved ourselves by using Green's theorem is exhilarating. Hopefully this exciting example jump-started your heart, increased your blood flow, and brought you out of a state of severe drowsiness. Stay tuned—there is more exciting stuff to come!

**Example 2.** In this example, we want to compute the line integral of

$$P(x, y) = 2x - y$$
$$Q(x, y) = x + 3y$$

over the closed contour $C : x^2 + 4y^2 = 4$, which is an ellipse. We start off with Green's equation:

$$L = \oint_C \left( Pdx + Qdy \right) = \iint_R \left( \frac{\delta Q}{\delta x} - \frac{\delta P}{\delta y} \right) dxdy$$

We decide to use the right side of the equation to compute the line integral. So we proceed with the calculations as

$$\frac{\delta Q}{\delta x} = \frac{\delta}{\delta x}(x+3y) = 1$$

$$\frac{\delta P}{\delta y} = \frac{\delta}{\delta y}(2x-y) = -1$$

Inserting these results into the right side of Green's equation gives us

$$L = \iint_R \left(\frac{\delta Q}{\delta x} - \frac{\delta P}{\delta y}\right) dxdy$$

$$= \iint_R \left(1-(-1)\right) dxdy$$

$$= 2\iint_R dx\,dy$$

You can recognize from basic calculus that $\iint_R dx\,dy$ is just the area of the region $R$. So $2\iint_R dx\,dy = 2A$, where $A$ is the area of the region $R$ bounded by the contour $C$. The contour that we are integrating over is the ellipse $x^2 + 4y^2 = 4$. The general form of an ellipse is given by

$$\frac{x^2}{a^2} + \frac{y^2}{b^2} = 1$$

The area of an ellipse is given by $A = \pi ab$. Hence, for our contour, $C: x^2 + 4y^2 = 4$, $a = 2$, and $b = 1$. Therefore

$$2\iint_R dx\,dy = 2A = 2\left[\pi(2)(1)\right] = 4\pi$$

You should readily see from this example that there are times when Green's theorem really simplifies the computation of line integrals.

**Example 3.** In this example, suppose

$$P(x,y) = x^2 + 2y^2$$
$$Q(x,y) = 0$$

The contour $C$ is a square described by the four vertexes illustrated in Figure 2.30. We begin by writing down Green's equation:

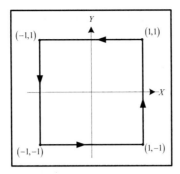

**Figure 2.30**   Contour of integration

$$L = \oint_C \left( P dx + Q dy \right) = \iint_R \left( \frac{\delta Q}{\delta x} - \frac{\delta P}{\delta y} \right) dx dy$$

Now we compute:

$$\frac{\partial P}{\partial y} = \frac{\partial}{\partial y} \left( x^2 + 2y^2 \right) = 4y$$

$$\frac{\partial Q}{\partial x} = 0$$

We evaluate the line integral using the right side of Green's equation:

$$
\begin{aligned}
L \ \ &= \iint_R \left( \frac{\delta Q}{\delta x} - \frac{\delta P}{\delta y} \right) dx dy \\
&= \iint_R \left( 0 - 4y \right) dx dy \\
&= -4 \int_{y=-1}^{y=1} \int_{x=-1}^{x=1} y \ dx \, dy \\
&= -4 \int_{y=-1}^{y=1} y \left[ x \right]_{x=-1}^{x=1} dy \\
&= -8 \int_{y=-1}^{y=1} y \, dy \\
&= -4 \, y^2 \Big|_{y=-1}^{y=1} \\
&= -4 \left[ (1)^2 - (-1)^2 \right] \\
&= 0
\end{aligned}
$$

In this particular problem, using the right side of Green's equation resulted in a significant reduction in computational work, as opposed to computing four separate line integrals (one for each path), as would be necessary if we had used the left side of the equation.

## 2.18  COMPLEX LINE INTEGRALS

The mathematics of complex line integrals is not much different from the mathematics of real line integrals. The main difference is that the computations are twice as lengthy. This is mainly due to the fact that a complex integral is usually broken down into two real integrals and then processed in the same manner as we discussed in the preceding text on real line integrals.

Suppose we have a function $f(z)$ of the complex variable $z = x + jy$ and we define the quantity $dz = dx + jdy$. Further, suppose that we have a complex function $w = u(x, y) + jv(x, y)$, where both $u(x, y)$ and $v(x, y)$ are real functions of real variables $x$ and $y$. We usually shorten the notation of $u(x, y)$ to $u$ and $v(x, y)$ to $v$, so now we can write

$$\int w \, dz = \int f(z) \, dz$$
$$= \int (u + jv)(dx + jdy)$$
$$= \int (udx - vdy) + j \int (vdx + udy)$$

**Equation 2.53**

Therefore the complex integral $\int w \, dz$ expands into two integrals, both of which are composed of real functions of real variables. Both integrals in Equation 2.53 are of the standard real line integral form

$$\int (Pdx + Qdy)$$

This will turn out to be a very convenient relationship. We will expand on this a bit later.

Suppose we have a complex variable $z$ that traverses a prescribed contour or path $C$ from one end point $z_1$ to another end point $z_2$, as illustrated in Figure 2.31. Further suppose that at every point along the contour the variable $z$ is associated with and evaluated by some function $f(z)$. We describe the process of integrating or summing all the evaluations of $f(z)$ along the contour between $z_1$ and $z_2$ by the notation

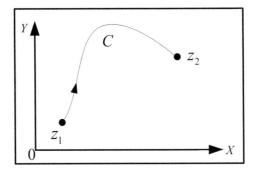

**Figure 2.31**  Example of a contour in the z-plane

$$\int_C f(z)\, dz$$

We call this integral a *line integral*, a *complex line integral*, or equivalently, a *contour integral* or *complex contour integral*.

It is critical to understand that the z-plane contains an infinite number of complex points $z_n$. The contour $C$ is made up of a very small subset of all the complex points in the plane. When we compute an integral of a function $f(z)$ along the path $C$, we are evaluating the function $f(z)$ only at the complex points that lie on the contour $C$. Many times, this simple fact is taken for granted and not specifically addressed. Therefore, the student new to the mathematics of complex variables sometimes gets confused as to what is actually going on in a line integral.

To further clarify this important point, suppose for simplicity that the contour $C$ is composed of $N$ continuous complex points out of the infinite number of points available in the z-plane. Then when we compute the line integral of $f(z)$ along $C$, we will end up evaluating the function at exactly the same $N$ complex points in the z-plane. Remembering this simple fact will avoid a great deal of confusion in later discussions.

Before we proceed any further, we need to define some basic properties of a complex line integral. If

$$w = f(z) = u(x, y) + jv(x, y) = u + jv$$

then the following is true:

1. The complex line integral can be expressed in terms of real integrals, such as

$$\int_C f(z)\ dz\ = \int_C (u + jv)(dx + jdy)$$

$$= \int_C (u\ dx - v\ dy) + j\int_C (v\ dx + u\ dy)$$

2. The line integral of the sum of two functions is equal to the sum of line integrals

$$\int_C (f(z) + g(z))\ dz = \int_C f(z)\ dz + \int_C g(z)\ dz$$

3. A complex constant $k$ can be brought outside of a line integral

$$\int_C kf(z)\ dz = k\int_C f(z)\ dz$$

4. A line integral evaluated in the counterclockwise direction on a contour from points $a$ to $b$ equals the negative of the same line integral evaluated in the clockwise direction from points $b$ to $a$:

$$\int_a^b f(z)\ dz = -\int_b^a f(z)\ dz$$

5. If the contour $C$ is composed of two separate but connected contours such as $C_1$, which connects some point $z_1$ to $z_2$, and $C_2$, which connects point $z_2$ to some other point $z_3$, then $C$ is a concatenation of $C_1$ and $C_2$ such that

$$\int_C f(z)\ dz = \int_{C1} f(z)\ dz + \int_{C2} f(z)\ dz$$

### 2.18.1   Path-Dependent Complex Line Integrals

As was the case for real line integrals, complex line integrals can be either path dependent or path independent. We will begin by looking at a few examples of path-dependent complex line integrals.

**Example 1.** Given the function $f(z) = \bar{z}$, compute the line integral on the path $C : |z - 2| = 3$ as shown in Figure 2.32. For academic purposes, let's make the assumption that the start and end point of integration on the circular contour is the point $(2, j3)$, as shown in the figure. As we will see, the selection of this point does not enter into the computations at all. We use this particular

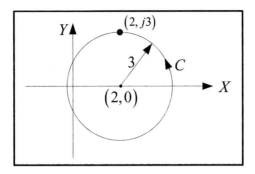

**Figure 2.32**   Contour of integration

point only to reference the contour of this example to the contour of the next example.

The first thing we do is write down the line integral. Doing so gives us

$$L = \oint_C \overline{z}\, dz \quad = \oint_C (x - jy)(dx + jdy)$$
$$= \oint_C x\, dx + y\, dy + j\oint_C x\, dy - y\, dx$$
$$= L_1 + jL_2$$

**Equation 2.54**

We have split the complex line integral $L$ into two separate real line integrals $L_1$ and $L_2$. Now let's evaluate each of these two integrals separately. We will begin with integral $L_1$. We can equate this integral to the standard form real line integral and in doing so take advantage of Green's theorem.

$$L_1 = \oint_C x\, dx + y\, dy = \oint_C (Pdx + Qdy) = \iint_R \left( \frac{\delta Q}{\delta x} - \frac{\delta P}{\delta y} \right) dxdy$$

In the interest of finding a simple solution, we will work with the right side of Green's equation, which gives us

$$L_1 = \oint_C x\, dx + y\, dy = \iint_R \left( \frac{\delta Q}{\delta x} - \frac{\delta P}{\delta y} \right) dxdy$$

We calculate the partials to be

$$\frac{\partial P}{\partial y} = \frac{\partial x}{\partial y} = 0$$

$$\frac{\partial Q}{\partial x} = \frac{\partial y}{\partial x} = 0$$

Upon substitution, we find that

$$L_1 = \iint_R \left( \frac{\delta Q}{\delta x} - \frac{\delta P}{\delta y} \right) dxdy = \iint_R (0-0)\, dxdy = 0$$

Now we will evaluate the integral $L_2$:

$$L_2 = \oint_C x\, dy - y\, dx = \oint_C (Pdx + Qdy) = \iint_R \left( \frac{\delta Q}{\delta x} - \frac{\delta P}{\delta y} \right) dxdy$$

We interchange terms in the left-hand integral to get it into the standard form of $\oint_C (Pdx + Qdy)$, which results in

$$L_2 = \oint_C -y\, dx + x\, dy = \oint_C (Pdx + Qdy) = \iint_R \left( \frac{\delta Q}{\delta x} - \frac{\delta P}{\delta y} \right) dxdy$$

Now we calculate the partials to be

$$\frac{\partial P}{\partial y} = \frac{\partial}{\partial y}(-y) = -1$$

$$\frac{\partial Q}{\partial x} = \frac{\partial}{\partial x}(x) = 1$$

Upon substitution of the partials, we find that

$$L_2 = \iint_R \left( \frac{\delta Q}{\delta x} - \frac{\delta P}{\delta y} \right) dxdy = \iint_R (1-(-1))\, dxdy = 2A$$

This is where $A$ is the area of the region $R$ enclosed by the contour $C$. The region $R$ is a circle. The area of a circle is $A = \pi r^2$, so the integral $L_2$ evaluates to $2A = 2\pi r^2$. The radius of the circle is $r = 3$, so $L_2 = 2\pi (3)^2 = 18\pi$. Now we finish the calculations by replacing both $L_1$ and $L_2$ with their computed values to get $L = (L_1 + jL_2) = (0 + j18\pi)$. Hence $\oint_C \bar{z} = (0 + j18\pi)$.

**Example 2**. The integral in this example is the same as the one used in the previous example. But the contour of integration is different. The new contour of integration is the 6 by 6 square illustrated in Figure 2.33. We derived the contour in this example by scribing a square superimposed on and tangent to the circle contour used in the previous example. This square is 6 units on each side. The contour start and end point is $(2, j3)$, which is the same point as on the circular contour in the previous example. Therefore the start and end points of the two contours can be considered to be the same.

The evaluation of the line integral in this example will be similar the previous example up to the point where the contour is considered. Therefore, rather than repeat the same computations, we can state that the line integral is the sum of two integrals:

$$L = L_1 + jL_2$$

Remember in the previous example that

$$L_1 = 0$$
$$L_2 = 2\iint_R dx\, dy = 2A$$

This is where $A$ is the area of the region $R$ bounded by the contour $C$. The area of the square contour is simply $A = 6 \times 6 = 36$ so that

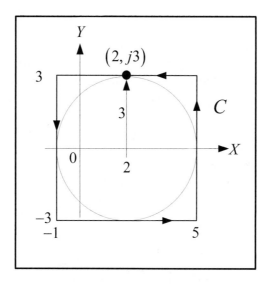

**Figure 2.33**   Contour of integration

$$L_2 = 2(36) = 72$$

Summing the two integrals gives

$$L = (L_1 + jL_2) = (0 + j72)$$

Hence, in this case,

$$\oint_C \bar{z} = (0 + j72)$$

We see in the previous two examples that integrating between the same end points but on different paths produced different results. This is because the line integral is path dependent. Also notice how in the two examples we utilized the standard form line integral expression and Green's theorem for real line integrals. We evaluate complex line integrals by breaking them down into real line integrals. The previous discussion of real line integrals therefore applies to the mathematics of complex line integrals as well.

We should notice how much simpler the computations were in the previous two examples due to the fact that we used Green's theorem. This doesn't happen 100% of the time, but it does happen often enough to warrant the energy expended on its use.

## 2.18.2  Path-Independent Complex Line Integrals

We have stated previously that the value of an integral is generally dependent on the contour over which it is evaluated. However, there exists a special set of line integrals that are indeed path independent. That is, they evaluate to the same value without regard for the integration path, or contour $C$. We can identify these line integrals by the following equations. If we can express a complex line integral $\int_C f(z)\ dz$ in terms of real functions of the form

$$\int_C f(z)\ dz = \int_C \{P_1 dx + Q_1 dy\} + j\int_C \{P_2 dx + Q_2 dy\}$$

**Equation 2.55**

we can say that the line integral is independent of path in a simply connected region if

$$\frac{\delta P_1}{\delta y} = \frac{\delta Q_1}{\delta x} \quad \text{and} \quad \frac{\delta P_2}{\delta y} = \frac{\delta Q_2}{\delta x}$$

**Equation 2.56**

If Equation 2.56 holds true, then we can say that the line integral $\int_C f(z)\,dz$ is path independent. It can be evaluated over any contour that connects the same end points and produce the exact same result. This is important. This means that if a line integral is path independent, we can evaluate using methods that are similar to those used for a standard, run-of-the-mill definite integral in calculus. If the end points of any arbitrary contour $C$ are $z_0$ and $z_1$ and if the line integral is path independent, then we can state

$$\int_C f(z)\,dz = \int_{z_0}^{z_1} f(z)\,dz$$

**Equation 2.57**

Stated another way, if $f(z)$ is analytic in region $R$ and if $f(z)$ is path independent, then

$$\int_{z_0}^{z_1} \frac{d}{dz} f(z) = \int_{z_0}^{z_1} f'(z)\,dz = f(z)\Big|_{z_0}^{z_1} = f(z_1) - f(z_0)$$

**Equation 2.58**

Equation 2.58 is true on every path in the region $R$ from $z_0$ to $z_1$. This is a powerful statement. Put this statement in your memory stack for immediate recall later. Right now, let's reinforce the importance of Equation 2.57 with a very simple example.

As an example, suppose we have a complex function $f(z) = z^2$ and we wish to evaluate this function along the straight line contour $C: y = x$ from the point $z = 0$ to the point $z = (1 + j)$, as shown in Figure 2.34.

The first thing that we need to do is to write down all the information necessary to solve this problem.

$$
\begin{aligned}
f(z) \;&= z^2 & dz &= dx + j\,dy \\
&= (x + jy)^2 & C&: \; y = x \Rightarrow dy = dx \\
&= x^2 - y^2 + j2xy &
\end{aligned}
$$

Now we break down the complex function into its component parts

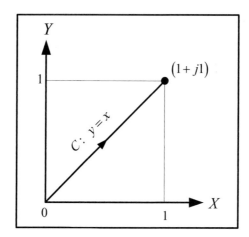

**Figure 2.34**  Example path of integration

$$L = \int_C f(z)\,dz \quad = \int_C z^2\,dz$$

$$= \int_C (x+jy)^2(dx+jdy)$$

$$= \int_C (x^2 - y^2 + j2xy)(dx+jdy)$$

$$L = \int_C f(z)\,dz = \int_C (x^2 - y^2)\,dx - (2xy)\,dy + j\int_C (2xy)\,dx + (x^2 - y^2)\,dy$$

**Equation 2.59**

Equation 2.59 is the breakdown of the function $f(z)\,dz$ into its general real and imaginary components. Now we need to modify these components so that we can make the integration of $f(z)$ specific to the contour of integration. This part is simple. If we look at Figure 2.34, we can see that the limit of integration for the variable $x$ ranges from $x = 0$ to $x = 1$. Since the contour is given by $y = x$, and since $dy = dx$, we can incorporate the contour into Equation 2.59 by substituting $x$ for $y$ and $dx$ for $dy$. This will give us an integral that is expressed in terms of $x$ only. We then can integrate from $x = 0$ to $x = 1$. Doing so gives us

$$L = \int_C f(z)\,dz \quad = \int_{x=0}^{1} \left\{ (x^2 - x^2)\,dx - (2x^2)\,dx \right\} + j\int_{x=0}^{1} \left\{ (2x^2)\,dx + (x^2 - x^2)\,dx \right\}$$

$$= \int_{x=0}^{1} \left\{ -(2x^2)\,dx \right\} + j\int_{x=0}^{1} \left\{ (2x^2)\,dx \right\}$$

Now we can evaluate the line integral by performing the routine integration to get

$$L = \left( -\frac{2}{3}x^3 + j\frac{2}{3}x^3 \right)\Bigg|_{x=0}^{x=1} = \frac{2}{3}(-1+j)$$

We have now computed the value of the line integral over a specific path $C$. Now let's go back and see if this particular line integral is path independent and if so demonstrate the validity of Equation 2.57. The first thing we need to do is to compare the components of Equation 2.59 to Equation 2.55. Equation 2.59 is repeated here for clarity:

$$L = \int_C f(z)\, dz = \int_C (x^2 - y^2)\, dx - (2xy)\, dy + j\int_C (2xy)\; dx + (x^2 - y^2)\, dy$$

Equation 2.55 is also repeated here for clarity:

$$L = \int_C f(z)\; dz = \int_C \{P_1 dx + Q_1 dy\} + j\int_C \{P_2 dx + Q_2 dy\}$$

Comparing the two equations, we can make the following associations:

$$P_1 = x^2 - y^2 \text{ and } Q_1 = -2xy$$

and

$$P_2 = 2xy \text{ and } Q_2 = x^2 - y^2$$

The line integral is path independent if it passes the test defined by Equation 2.56 (repeated here):

$$\frac{\delta P_1}{\delta y} = \frac{\delta Q_1}{\delta x} \text{ and } \frac{\delta P_2}{\delta y} = \frac{\delta Q_2}{\delta x}$$

Applying this test gives us the results

$$\frac{\delta P_1}{\delta y} = \frac{\delta Q_1}{\delta x} \Rightarrow \frac{\delta(x^2 - y^2)}{\delta y} = \frac{\delta(-2xy)}{\delta x} \Rightarrow -2y = -2y$$

and

$$\frac{\delta P_2}{\delta y} = \frac{\delta Q_2}{\delta x} \Rightarrow \frac{\delta(2xy)}{\delta y} = \frac{\delta(x^2 - y^2)}{\delta x} \Rightarrow 2x = 2x$$

The line integral passed the test. Therefore the line integral should evaluate to the same value no matter what path we take between the two points $z_0$ and $z_1$, and therefore Equation 2.57 and Equation 2.58 should hold true. Let's see if this is correct. We will integrate directly between the end points, as suggested by Equation 2.57, to get

$$
\begin{aligned}
L = \int_C f(z)\, dz &= \int_{z_0}^{z_1} f(z)\, dz \\
&= \int_{z=0}^{z=(1+j)} z^2\, dz \\
&= \frac{1}{3} z^3 \Big|_{z=0}^{z=(1+j)} \\
&= \frac{1}{3}(1+j)^3 \\
&= \frac{1}{3}(1+2j-1+j-2-j) = \frac{2}{3}(-1+j)
\end{aligned}
$$

The results are identical to those achieved by integrating $f(z)=z^2$ along the contour defined by $C$.

A detail-oriented engineer reading this text might respond to these results by saying, "Big deal! The path of integration was linear so the contour used by the line integral was identical to the path taken by evaluating the definite integral." Let's try and remove your doubts by evaluating the same function over a second contour and see what results we end up with. To wrap this up, let's repeat the integration of $f(z)=z^2$ along the parabolic contour $C: y=x^2$ illustrated in Figure 2.35. Once again we begin by writing down all the facts:

$$
\begin{aligned}
f(z) &= z^2 \\
&= (x+jy)^2 \qquad\qquad\quad C: \quad \begin{array}{l} dz = dx + jdy \\ y = x^2 \Rightarrow dy = 2x\, dx \end{array} \\
&= x^2 - y^2 + j2xy
\end{aligned}
$$

We break down the function in the same manner as before to get

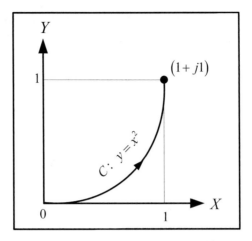

**Figure 2.35**   Path of integration

$$L = \int_C f(z)\, dz \quad = \int_C z^2\, dz$$

$$= \int_C (x + jy)^2 (dx + jdy)$$

$$= \int_C (x^2 - y^2 + j2xy)(dx + jdy)$$

$$L = \int_C f(z)\, dz = \int_C (x^2 - y^2)\, dx - (2xy)\, dy + j \int_C (2xy)\ dx + (x^2 - y^2)\, dy$$

and then we incorporate the contour into the breakdown by making the substitutions $y = x^2$ and $dy = 2x\, dx$ to get

$$L = \int_C f(z)\, dz \quad = \int_{x=0}^{1} \left\{ (x^2 - x^4)\, dx - (2x^3) 2x\, dx \right\} + j \int_{x=0}^{1} \left\{ (2x^3)\ dx + (x^2 - x^4) 2x\, dx \right\}$$

$$= \int_{x=0}^{1} \left\{ (x^2 - 5x^4)\, dx \right\} + j \int_{x=0}^{1} \left\{ (4x^3 - 2x^5)\ dx \right\}$$

Now we evaluate this line integral from $x = 0$ to $x = 1$ to get

$$L = \left( \frac{x^3}{3} - \frac{5x^5}{5} \right) + j \left( \frac{4x^4}{4} - \frac{2x^6}{6} \right) \Bigg|_{x=0}^{x=1} = \left( \frac{1}{3} - 1 \right) + j \left( 1 - \frac{1}{3} \right) = \frac{2}{3}(-1 + j)$$

We arrive at the same value using a different path. The information imparted to us by Equation 2.55, Equation 2.56, and Equation 2.57 is important and is a powerful tool in the mathematics of complex variables.

## 2.19  CAUCHY'S THEOREM

If a function $f(z)$ is analytic in a simply connected region $R$ then

$$\oint_C f(z)\, dz = 0$$

on every closed path $C$ within the region $R$. This is a fundamental theorem in the mathematics of complex variables. Because of the importance of this theorem, we will invest some time here to prove it.

Suppose we have a complex function $f(z)$ where $z = x + jy$ and $dz = dx + jdy$. Further suppose we have a second function $w$ that is a complex function of $x$ and $y$, such that $w = u(x,y) + jv(x,y) = u + jv$. If we equate the two we get

$$w = f(z) = (x + jy) = (u + jv)$$

Then, by substitution, we can write

$$\int_C f(z)\,dz \;=\; \int (u + jv)(dx + jdy)$$
$$= \int (u\, dx - v\, dy) + j \int (v\, dx + u\, dy)$$

Expressed in terms of a closed contour,

$$L \;=\; \oint_C f(z)\,dz \;=\; \oint_C (u\, dx - v\, dy) \;+\; j\oint_C (v\, dx + u\, dy)$$
$$= L_1 \qquad\qquad\qquad + jL_2$$

Now we apply Green's theorem on each of the two right-side integrals to get

$$L_1 = \oint_C (u\, dx - v\, dy) = \oint_C (Pdx + Qdy) = \iint_R \left( \frac{\delta Q}{\delta x} - \frac{\delta P}{\delta y} \right) dxdy$$

We choose to use the right-hand side of Green's theorem to evaluate $L_1$, which gives us

$$L_1 = \oint_C (u\ dx - v\ dy) = \iint_R \left( \frac{\delta Q}{\delta x} - \frac{\delta P}{\delta y} \right) dxdy = \iint_R \left( -\frac{\delta v}{\delta x} - \frac{\delta u}{\delta y} \right) dxdy$$

We know that the function $f(z)$ is analytic in the region $R$ in which it is evaluated. Therefore we can call upon the Cauchy-Riemann equations discussed earlier to help us evaluate $L_1$. The Cauchy-Riemann equations tell us that if the function $f(z)$ is analytic, then the following conditions are true:

$$\frac{\delta u}{\delta x} = \frac{\delta v}{\delta y}$$

$$\frac{\delta u}{\delta y} = -\frac{\delta v}{\delta x}$$

Therefore, by the Cauchy-Riemann equations, we know that $L_1$ evaluates to 0 because

$$\frac{\delta u}{\delta y} = -\frac{\delta v}{\delta x}$$

Let's go ahead and compute $L_1$ anyway. When we do we end up with

$$L_1 = \oint_C (u\ dx - v\ dy) = \iint_R \left( -\frac{\delta v}{\delta x} - \frac{\delta u}{\delta y} \right) dxdy = \iint_R (0)\ dxdy = 0$$

We evaluate the second integral $L_2$ in much the same manner:

$$L_2 = \oint_C (v\ dx + u\ dy) = \oint_C (Pdx + Qdy) = \iint_R \left( \frac{\delta Q}{\delta x} - \frac{\delta P}{\delta y} \right) dxdy$$

We choose to evaluate $L_2$ using the right side of Green's theorem, giving us

$$L_2 = \oint_C (v\ dx + u\ dy) = \iint_R \left( \frac{\delta u}{\delta x} - \frac{\delta v}{\delta y} \right) dxdy$$

By the Cauchy-Riemann equations, we see that $L_2 = 0$ because

$$\frac{\delta u}{\delta x} = \frac{\delta v}{\delta y}$$

We will compute $L_2$ anyway. When we do, we end up with

$$L_2 = \oint_C \left(v\,dx + u\,dy\right) = \iint_R \left(\frac{\delta u}{\delta x} - \frac{\delta v}{\delta y}\right) dxdy = \iint_R \left(0\right) dxdy = 0$$

Therefore

$$\oint_C f(z)\,dz = \left(L_1 + jL_2\right) = \left(0 + j0\right) = 0$$

The reader might ask the question, "How can Cauchy's theorem help us?" The answer is, it can reduce computations by a significant amount, thereby saving time and reducing the chance of computational error, as the following example will verify.

**Example 1.** Let us demonstrate the power of Cauchy's Theorem by turning a difficult problem into a trivial problem. Suppose we would like to compute the line integral of $f(z) = 1/(z - z_0)$ over the path $C$ illustrated in Figure 2.36. That is, we wish to compute

$$\oint_C f(z)\,dz = \oint_C \frac{dz}{\left(z - z_0\right)}$$

The function does have a singularity at $z = z_0$, but that singularity is outside the contour of integration. Therefore the function $f(z) = 1/(z - z_0)$ is analytic everywhere on the contour $C$ and within the region $R$ that is bounded by $C$. Therefore we can state with authority and without getting bogged down in any complex computations that

$$\oint_C \frac{dz}{\left(z - z_0\right)} = 0$$

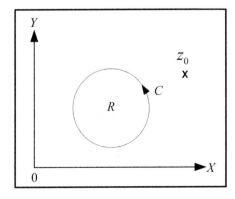

**Figure 2.36**   Cauchy's theorem contour, example 1

This is a simple example but it demonstrates how we can use Cauchy's theorem to break down complex problems into far simpler ones by eliminating many of the computations involved.

Before we invest time in other examples, it will benefit us to investigate the use of this theorem a bit further. In the previous example, we dealt with a function that was analytic in a specific region $R$ on and within the contour $C$. The function's one singularity, $z_0$, was not contained within the region, which made the evaluation of the line integral simple. This leads us to the obvious question, "What would have been the result if the singular point $z_0$ was located within the contour of integration?" This is a great question and I'm glad that I asked it. Now let's spend some time on the answer. We'll start by looking at the same problem as the previous example, but in this case we see that the contour $C$ contains a singular point $z_0$, as illustrated in Figure 2.37. The circular contour is described mathematically as $|z - z_0| = r$.

In the figure, we have drawn a vector from the origin to the point $z_0$ and labeled it $z_0$. We also have drawn a vector from the origin to some arbitrary point on the circular contour and labeled it $z$. No matter how the vector $z$ is drawn, as long as it originates at the origin and ends somewhere on the circle represented by path $C$, the magnitude of the difference between the two vectors $z$ and $z_0$ is always equal to $r$, the radius of the circle. The vector representing the difference is drawn connecting the end points of the vectors $z_0$ and $z$ and labeled $r$.

As the vector $z$ moves around the circle, it will continuously scribe a constant magnitude vector $r$ as it moves through the entire contour of $2\pi$ radians. We can express the vector $z$ in polar coordinates through the following equations. We begin with the original contour expressed as

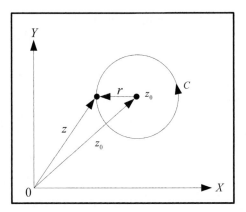

**Figure 2.37**   Contour that encloses a singularity

$$|z - z_0| = r$$

The quantity $r$ is a complex number so if we remove the magnitude notation, we can express $r$ in its polar notation to get

$$z - z_0 = re^{j\theta}$$

Now we add the constant $z_0$ to the both sides of the equation to get

$$z = z_0 + re^{j\theta}$$

The derivative of $z$ is given by

$$dz = jre^{j\theta}d\theta$$

The original line integral is expressed as

$$\oint_C \frac{dz}{z - z_0}$$

We can make these substitutions to convert the integral from Cartesian to polar coordinates. When we do, we arrive at

$$\oint_C \frac{dz}{z - z_0} = \oint_C \frac{jre^{j\theta}d\theta}{re^{j\theta}} = j\oint_C d\theta = j\int_{\theta=0}^{\theta=2\pi} d\theta$$

Notice that we have changed the limits of integration consistent with the polar notation so that $0 \le \theta \le 2\pi$. We can evaluate the final integral to get

$$j\int_{\theta=0}^{\theta=2\pi} d\theta = 2\pi j$$

Therefore the line integral around a closed contour that contains a singularity is given by

$$\oint_C \frac{dz}{z - z_0} = 2\pi j$$

If we choose to reverse the path of integration, then the integral will evaluate to its negative, or

$$\oint \frac{dz}{z-z_0} = -2\pi j$$

It should be noted that these results are expressed in engineering notation. Mathematicians usually replace the term $j$ with the term $i$ so that the results expressed in mathematician notation are

$$\oint_C \frac{dz}{z-z_0} = 2\pi i \qquad \text{and} \qquad \oint \frac{dz}{z-z_0} = -2\pi i$$

I have run into many mathematicians who are adamant about this notation. I remember a few of my old undergrad math professors getting visibly angry when they saw me using the $j$ notation on their classroom chalkboards. Their comments usually fell into the category of "you damn engineers!" I guess it's kind of like the age-old argument of "should the toilet seat be left up or down?" Personally I can handle it both ways. Either notation is fine with me. Just to prove it, we will use the $i$ notation for the remainder of this section.

Getting back to Cauchy's theorem, it looks like we can embellish it somewhat (at the risk of being the victim of serious mathematician wrath) by stating that if a function $f(z)$ is analytic in some region $R$, then

$$\oint_C f(z)\, dz \quad \begin{cases} = 0 \text{ if the contour } C \text{ contains no singularities} \\ = 2\pi i \text{ if the contour } C \text{ contains exactly one singularity} \end{cases}$$

This is true on every closed path $C$ within the region $R$.

Now is a good time to illustrate Cauchy's theorem with a few examples. Note how Cauchy's theorem simplifies calculation-intense problems.

**Example 2**. Given the line integral

$$\oint_C \frac{dz}{z-4}$$

the first thing we note is that $f(z)$ has a singularity at $z = 4$. Let us evaluate it over three different contours, as illustrated in Figure 2.38. We use the power of Cauchy's theorem to solve these three problems simply by inspection:

- Part A  $C:|z|=1 \quad \Rightarrow \quad \oint_C \frac{dz}{z-4} = 0 \quad \{z_0 \text{ not within contour } C\}$

- Part B  $C:|z-1|=2 \quad \Rightarrow \quad \oint_C \frac{dz}{z-4} = 0 \quad \{z_0 \text{ not within contour } C\}$

- Part C  $C:|z|=5 \quad \Rightarrow \quad \oint_C \frac{dz}{z-4} = 2\pi i \quad \{z_0 \text{ is within contour } C\}$

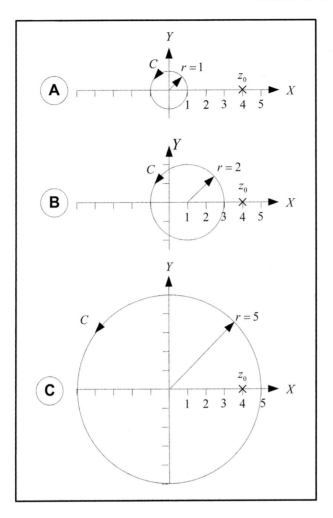

**Figure 2.38**   Cauchy's theorem contour, example 2

**Example 3**. Evaluate the line integral

$$\oint_C f(z)\,dz \;=\; \oint_C \frac{dz}{z^2 - 1} \quad C:|z| = 2$$

The first thing we do is factor $f(z)$ to get

$$f(z) = \frac{1}{z^2 - 1} = \frac{1}{(z+1)(z-1)}$$

The next thing we want to do is to split $f(z)$ into two parts, using the method of partial fractions:

$$f(z) = \frac{1}{(z+1)(z-1)} = \frac{A}{z+1} + \frac{B}{z-1}$$

We can solve for the coefficients $A$ and $B$ by

$$A = (z+1)f(z)\Big|_{z=-1} = -\frac{1}{2}$$

$$B = (z-1)f(z)\Big|_{z=+1} = +\frac{1}{2}$$

Substituting these coefficients into the partial fraction equation and computing the line integral of each fraction give us

$$\oint_C f(z)\,dz = -\frac{1}{2}\oint_C \frac{dz}{z+1} + \frac{1}{2}\oint_C \frac{dz}{z-1}$$

$$= -\frac{1}{2}2\pi i + \frac{1}{2}2\pi i$$

$$= 0$$

Since the contour of integration is a circle centered at the origin and of radius 2, the two singularities at $\pm 1$ are contained within the bounds of $C$. Therefore, by Cauchy's theorem, both integrals evaluate to $2\pi i$ times the constant outside the integrals.

**Example 4.** Evaluate the line integral

$$\oint_C f(z)\,dz = \oint_C \frac{2z+1}{z^2+z}\,dz$$

for each of the four cases where the contour of integration is given as

$$C_1 : \ |z| = \frac{1}{4}$$

$$C_2 : \ \left|z - \frac{1}{2}\right| = \frac{1}{4}$$

$$C_3 : \ |z - i| = \frac{1}{4}$$

$$C_4 : \ |z| = 2$$

The first thing we do is to factor $f(z)$ to get

$$f(z) = \frac{2z+1}{z^2+z} = \frac{2z+1}{z(z+1)}$$

The next thing we do is split $f(z)$ into its partial fractions

$$f(z) = \frac{2z+1}{z^2+z} = \frac{2z+1}{z(z+1)} = \frac{A}{z} + \frac{B}{z+1}$$

Now we solve for the partial fraction coefficients $A$ and $B$

$$A = (z)f(z)\big|_{z=0} = z\frac{2z+1}{z(z+1)}\bigg|_{z=0} = 1$$

$$B = (z+1)f(z)\big|_{z=-1} = (z+1)\frac{2z+1}{z(z+1)}\bigg|_{z=-1} = 1$$

Substituting the coefficients into the partial fraction equation and computing the line integral for each fraction give us

$$\oint_C f(z)\,dz = \oint_C \frac{dz}{z} + \oint_C \frac{dz}{z+1}$$

The line integral has singularities at $z=0$ and $z=-1$. Knowing the contours of integration, we can solve for the four cases by inspection using Cauchy's theorem:

$$
\begin{aligned}
\text{For } C_1 \quad \oint_C f(z)\,dz &= 2\pi i + 0 &&= 2\pi i \\
\text{For } C_2 \quad \oint_C f(z)\,dz &= 0 + 0 &&= 0 \\
\text{For } C_3 \quad \oint_C f(z)\,dz &= 0 + 0 &&= 0 \\
\text{For } C_4 \quad \oint_C f(z)\,dz &= 2\pi i + 2\pi i &&= 4\pi i
\end{aligned}
$$

In this example, $f(z)$ was a pretty hairy expression. We used the power of Cauchy's theorem to solve it for four different contours simply by inspection. This is amazing stuff, but as we will see, it gets even better.

**Example 5**. If a function $f(z)$ is analytic at all points interior to and on a closed contour $C$, then

$$\oint_C f(z)\, dz = 0$$

Given the region $R$ illustrated in Figure 2.39, that consists of the area between two concentric circular contours given by $C_1 : |z| = 1$ and $C_2 : |z| = 2$, where $C_1$ is scribed in the negative or clockwise direction and $C_2$ is scribed in the positive or counterclockwise direction. The function

$$f(z) = \frac{1}{z^2 (z^2 + 9)}$$

with singularities at $z = 0$ and $z = \pm 3j$ is analytic everywhere except at the points $z = 0$ and $z = \pm 3j$. Since these points lie outside the region,

$$\oint_R f(z)\, dz = \oint_R \frac{dz}{z^2 (z^2 + 9)} = 0$$

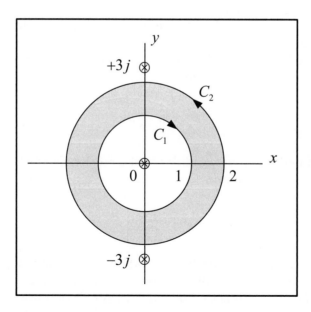

**Figure 2.39** Cauchy's theorem contour, example 5

Wow! What a time-saver Cauchy's theorem has proved to be. A problem like this one could have been computationally intensive. And just when you think you have this stuff understood, this book comes along and says, "Hey, wait a minute, there is more stuff to add to your memory stack!" The additional stuff we are going to add next may take you by surprise, but when we're done, we will have an even more powerful set of tools for solving complex line integrals.

Let's take the next step on this journey. Take a look at the line integral

$$L = \oint_C \frac{1}{z} \, dz$$

The function $f(z) = 1/z$ has a singularity at $z = 0$. It is analytic at all other points in the $z$-plane. Let's evaluate this line integral on the curve $C_1 : |z| = 5$, which is a circle centered at 0 with radius 5. The singularity is within the region bounded by $C$, and therefore by Cauchy's theorem the line integral evaluates to $2\pi i$. Now let's evaluate the same integral on the curve $C_2 : |z - 5| = 2$, which is a circle centered at $z = 5$ with a radius of $2$. In this case, the singularity at $z = 0$ is not contained within the region bounded by $C$ and right away the reader should be able to tell by Cauchy's theorem that the integral evaluates to 0.

What do you think would happen to the evaluation of the line integral if we modify the integrand such that $f(z) = (1/z^n)$ to produce a multiple order singularity at $z = 0$? We already know the answer for the case where $n = 1$ because we just discussed it. Now we shall see what interesting things happen when $n$ takes on the values of $2, 3, 4, 5, \cdots$. Consider the integral

$$L = \oint_C \frac{1}{z^n} \, dz$$

There is a multiple order singularity at $z = 0$. By Cauchy's theorem we know that if the closed contour $C$ does not enclose $z = 0$, then

$$L = \oint_C \frac{1}{z^n} \, dz = 0$$

If the contour does enclose the origin, then what will the integral evaluate to? Let's evaluate the integral and see. We will make this problem simpler by using polar coordinates. We make the conversion from Cartesian to polar by recognizing

$$z = re^{j\theta} \quad dz = jre^{j\theta} d\theta \quad z^n = r^n e^{jn\theta}$$

If we make these substitutions to the integral, we end up with

$$L = \oint_C \frac{1}{z^n} \, dz = \int_{\theta=0}^{\theta=2\pi} \frac{jre^{j\theta} \, d\theta}{r^n e^{jn\theta}} = \frac{j}{r^{n-1}} \int_{\theta=0}^{\theta=2\pi} e^{-j(n-1)\theta} \, d\theta$$

We evaluate the integral to get

$$L = \frac{-1}{(n-1)r^{n-1}} \left[ e^{-j(n-1)\theta} \right]_{\theta=0}^{\theta=2\pi} = \frac{-1}{(n-1)r^{n-1}} \left[ 1 - 1 \right] = 0$$

What is this? It is a line integral whose singularity lies within the contour of integration that evaluates to 0, as opposed to some multiple of $2\pi j$ (or $2\pi i$ if you prefer). What goes on here? Well, we first note that the singularity is a multiple-order singularity. We also note that the integral evaluates to 0 for $n \neq 1$. This unique property can be generalized to give us another valuable and powerful set of rules to use in the evaluation of line integrals. Before we can state these rules, it is first necessary to expand this generalization to functions other than $1/z$.

Suppose we have two multiple order singularities in the z-plane, at $z = a$ and $z = b$, represented by the functions

$$f(z) = \frac{1}{(z-a)^n} \quad \text{and} \quad f(z) = \frac{1}{(z-b)^n}$$

The contour is illustrated in Figure 2.40. The singularity at $a$ lies within the contour $C$ and the singularity at $b$ does not. Therefore the line integral

$$\oint_C f(z) = \oint_C \frac{1}{(z-b)^n} \, dz = 0$$

Since the singularity at $z = a$ lies within the contour, the line integral

$$\oint_C f(z) = \oint_C \frac{1}{(z-a)^n} \, dz$$

needs to be evaluated. We can make this evaluation simple or we can make it difficult. Obviously, it makes more sense to take the simple approach. We will do that by letting

$$w = (z-a) \quad \Rightarrow \quad dw = dz$$

If we make this substitution, we get

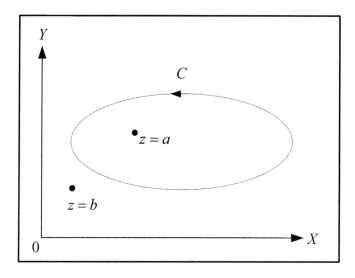

**Figure 2.40**  Multiple order singularities

$$\oint_C f(z) = \oint_C \frac{1}{(z-a)^n} \, dz = \oint_C \frac{1}{w^n} \, dw$$

We have already shown that for $n = 1$ the integral evaluates to

$$\oint_C f(z) = \oint_C \frac{1}{w} \, dw = 2\pi i$$

We have also seen that for $n = 2, 3, 4, \cdots$ the integral evaluates to

$$\oint_C f(z) = \oint_C \frac{1}{w^n} \, dw = 0$$

Therefore, we can state some new and powerful rules for evaluating line integrals. These rules are

$$\oint_C f(z) \, dz = \oint_C \frac{1}{(z-a)^n} \, dz = \begin{cases} 0 & \text{for } n \neq 1 \quad \text{and C does enclose } z = a \\ 0 & \text{and C does not enclose } z = a \\ 2\pi i & \text{for } n = 1 \quad \text{and C does enclose } z = a \end{cases}$$

**Equation 2.60**

The rules laid out in Equation 2.60 are simple and they are important. They are powerful tools in line integral evaluation. Remember them. These rules will come into play when we discuss residue theory.

## 2.20 TABLE OF COMMON INTEGRALS

If $f(z)$ is analytic, then complex integrals can be evaluated in much the same way as real integrals. Table 2.2 lists a few of the more common integrals without inclusion of the integration constant.

## 2.21 CAUCHY'S INTEGRAL

Suppose we have region $R$ within which the function $f(z)$ is analytic everywhere within and on a closed contour $C_1$ with exception of the singular point $z_0$, as shown in Figure 2.41. We have shown previously that the line integral

$$L = \oint_{C_1} \frac{dz}{z - z_0}$$

evaluated on the contour $C_1$ will result in the a value of $2\pi i$. That is, Cauchy's theorem tells us that

$$L = \oint_{C_1} \frac{dz}{z - z_0} = 2\pi i$$

We know by Cauchy's theorem that we will get the exact same result, should we evaluate the line integral on any closed contour that includes the point $z_0$. For example, we would get the same result if we evaluated the line integral on contours $C_2, C_3,$ or $C_4$, as shown in Figure 2.41. That is,

$$L = \oint_{C_1} \frac{dz}{z - z_0} = \oint_{C_2} \frac{dz}{z - z_0} = \oint_{C_3} \frac{dz}{z - z_0} = \oint_{C_4} \frac{dz}{z - z_0} = 2\pi i$$

Now suppose we wished to evaluate the line integral

$$L = \oint_C \frac{f(z)}{z - z_0} dz$$

Visualize the radius of the contour $C$ getting smaller and smaller, resulting in the region within $C$ also getting smaller and smaller so that the contour collapses to include only the single point $z_0$. The function $f(z)$ inside the contour would then evaluate to $f(z_0)$. We could then write the line integral as

$$L = \oint_C \frac{f(z_0)}{z - z_0} dz$$

**Table 2.2**   Table of Common Complex Integrals

$$\int z^n \, dz = \frac{z^{n+1}}{n+1} \quad n \ne -1$$

$$\int \sec z \tan z \, dz = \sec z$$

$$\int \frac{dz}{z} = \ln z$$

$$\int \csc z \cot z \, dz = -\csc z$$

$$\int e^z \, dz = e^z$$

$$\int \frac{dz}{\sqrt{z^2 \pm a^2}} = \ln\left(z + \sqrt{z^2 \pm a^2}\right)$$

$$\int a^z \, dz = \frac{a^z}{\ln a}$$

$$\int \frac{dz}{z^2 + a^2} = \frac{1}{a} \tan^{-1}\left(\frac{z}{a}\right)$$

$$\int \sin z \, dz = -\cos z$$

$$\int \frac{dz}{z^2 - a^2} = \frac{1}{2a} \ln\left\{\frac{z-a}{z+a}\right\}$$

$$\int \cos z \, dz = \sin z$$

$$\int \frac{dz}{\sqrt{a^2 - z^2}} = \sin^{-1}\frac{z}{a}$$

$$\int \tan z \, dz = -\ln \cos z$$

$$\int \frac{dz}{z\sqrt{a^2 \pm z^2}} = \frac{1}{a} \ln\left(\frac{z}{a + \sqrt{a^2 \pm z^2}}\right)$$

$$\int \cot z \, dz = \ln \sin z$$

$$\int \frac{dz}{z\sqrt{z^2 - a^2}} = \frac{1}{a} \cos^{-1}\frac{a}{z}$$

$$\int \sec z \, dz = \ln\left(\sec z + \tan z\right)$$

$$\int \sqrt{z^2 \pm a^2} \, dz = \frac{z}{2}\sqrt{z^2 \pm a^2}$$

$$\int \csc z \, dz = \ln\left(\csc z - \cot z\right)$$

$$\int \sqrt{a^2 - z^2} \, dz = \frac{z}{2}\sqrt{a^2 - z^2} + \frac{a^2}{2} \sin^{-1}\frac{z}{a}$$

$$\int \sec^2 z \, dz = \tan z$$

$$\int e^{az} \sin bz \, dz = \frac{e^{az}\left(a \sin bz - b \cos bz\right)}{a^2 + b^2}$$

$$\int \csc^2 z \, dz = -\cot z$$

$$\int e^{az} \cos bz \, dz = \frac{e^{az}\left(a \cos bz - b \sin bz\right)}{a^2 + b^2}$$

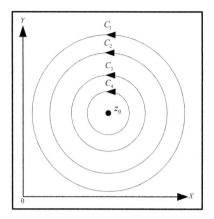

**Figure 2.41**  Different contours of integration

Since $f(z_0)$ is a constant, it can be brought outside the integral, giving us

$$L = f(z_0) \oint_C \frac{dz}{z - z_0} = (2\pi i) f(z_0)$$

Now if we rearrange terms, we get Cauchy's integral formula:

$$f(z_0) = \frac{1}{2\pi i} \oint_C \frac{f(z)}{z - z_0}\, dz$$

**Equation 2.61**

Boy this Cauchy fellow was a busy guy. So far in this chapter we have discussed three of his works. His full name was Augustin-Louis Cauchy. Cauchy was born in Paris, France, on August 21, 1789. He wasn't held in high regard by his fellow scientists due to his rigid religious posture. He did, however, churn out paper after paper that forever impacted the world of mathematics. I, for one, am very appreciative of his work. His work has made the lives of mathematicians and engineers a great deal easier.

Let us pause here for a few quick examples using Cauchy's integral formula.

**Example 1**. Suppose that

$$f(z) = \frac{z}{4 - z^2}$$

Further suppose that the contour of integration is the unit circle $C : |z| = 1$ and that there is a singularity within that contour at $z_0 = \dfrac{i}{2}$. The function

$f(z)$ has singularities at $z = \pm 2$, but both lie outside the contour of integration and therefore are not important here since $f(z)$ is analytic within the contour. We evaluate the line integral of $f(z)$ on the contour $C$ by rewriting Cauchy's integral formula as

$$\oint_C \frac{f(z)}{z - z_0} \, dz = 2\pi i \left[ f(z) \right]_{z = z_0}$$

We can start by filling in the blanks of Cauchy's integral formula to get

$$L = \oint_C \frac{f(z)}{z - z_0} \, dz = \oint_C \frac{f(z)}{(z - i/2)} \, dz = 2\pi i \left[ f(z) \right]_{z = \frac{i}{2}}$$

Now we evaluate $f(z)$ at $z = i/2$ to get

$$L = 2\pi i \left[ f(z) \right]_{z = \frac{i}{2}} = 2\pi i \left[ \frac{z}{4 - z^2} \right]_{z = \frac{i}{2}} = 2\pi i \left[ \frac{\frac{i}{2}}{4 - \left( \frac{i}{2} \right)^2} \right] = -\frac{4\pi}{17}$$

**Example 2.** Evaluate $f(z) = z^2 + z + 1$ on the contour $C : |z| = 2$ that contains a singularity at $z = 1$. We write the line integral as

$$L = \oint_C \frac{f(z)}{z - z_0} \, dz = 2\pi i \left[ f(z) \right]_{z=1} = 2\pi i \left[ z^2 + z + 1 \right]_{z=1} = 6\pi i$$

**Example 3.** Evaluate the function $f(z) = \frac{z+1}{z-7}$ on the contour $C : |z| = 5$ for $z_0 = 6$:

$$L = \oint_C \frac{f(z)}{z - z_0} \, dz = f(z_0) \oint_C \frac{dz}{z - 6}$$

We pop our memory stack, remembering the rules laid out in Equation 2.60, and realize that since $z_0$ is not within the contour of integration, the line integral evaluates to

$$L = f(z_0)(0) = 0$$

**Example 4.** Evaluate $L = \oint_C \frac{e^z}{z - 3} \, dz$ where $f(z) = e^z$ and $z_0 = 3$ for the three contours illustrated in Figure 2.42.

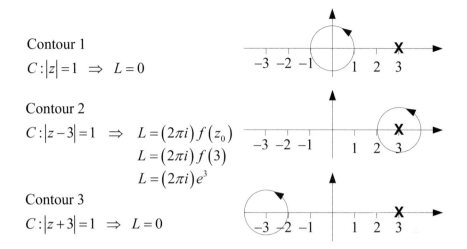

Contour 1

$C:|z|=1 \implies L=0$

Contour 2

$C:|z-3|=1 \implies L=(2\pi i)f(z_0)$
$L=(2\pi i)f(3)$
$L=(2\pi i)e^3$

Contour 3

$C:|z+3|=1 \implies L=0$

**Figure 2.42** Contours of integration, example 4

**Example 5.** Evaluate $L=\oint_C \dfrac{\cos(3z)}{[z+\pi/2]}\,dz$ where $f(z)=\cos(3z)$ and $z_0=-\pi/2$ for the two contours illustrated in Figure 2.43.

**Example 6.** Evaluate $L=\oint_C \dfrac{(z+1)^2}{z-i}\,dz$ where $f(z)=(z+1)^2$ and $z_0=i$ for the four contours illustrated in Figure 2.44.

There is one more item we need to take care of before we leave the subject of Cauchy's integral: we need to present without proof the expressions for the derivatives of Cauchy's integral. We will need these in our discussions on residue theory. Given the function

$$f(z_0)=\frac{1}{2\pi i}\oint_C \frac{f(z)}{z-z_0}\,dz$$

we can write down the first through $n$th derivatives as

$$\frac{d}{dz}f(z_0) \quad =f'(z_0) \quad =\frac{1}{2\pi i}\oint_C \frac{f(z)}{(z-z_0)^2}\,dz$$

$$\frac{d^2}{dz}f(z_0) \quad =f''(z_0) \quad =\frac{2!}{2\pi i}\oint_C \frac{f(z)}{(z-z_0)^3}\,dz$$

$$\vdots \qquad\qquad \vdots \qquad\qquad \vdots$$

$$\frac{d^n}{dz}f(z_0) \quad =f^n(z_0) \quad =\frac{n!}{2\pi i}\oint_C \frac{f(z)}{(z-z_0)^{n+1}}\,dz$$

**Equation 2.62**

Contour 1

$$C : |z| = 1 \quad \Rightarrow \quad L = 0$$

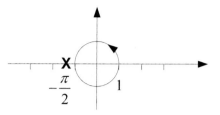

Contour 2

$$C : |z| = 2 \quad \Rightarrow \quad L = (2\pi i) f(z_0)$$
$$L = (2\pi i) f(-\pi/2)$$
$$L = (2\pi i) \cos(-3\pi/2)$$
$$L = (2\pi i) 0 = 0$$

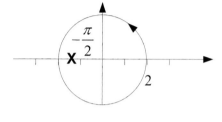

**Figure 2.43**   Contours of integration, example 5

Contour 1

$$C : |z| = 2 \quad \Rightarrow \quad L = (2\pi i) f(z_0)$$
$$L = (2\pi i) f(i)$$
$$L = (2\pi i)(1+i)^2$$
$$L = (2\pi i)(2i)$$
$$L = -4\pi$$

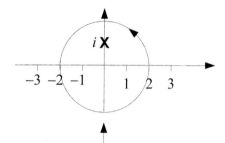

Contour 2

$$C : |z| = \frac{1}{2} \quad \Rightarrow \quad L = 0$$

Contour 3

$$C : |z - 1| = 1 \quad \Rightarrow \quad L = 0$$

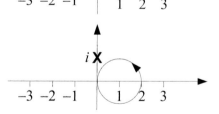

Contour 4

$$C : |z - i| = 1 \quad \Rightarrow \quad L = -4\pi$$

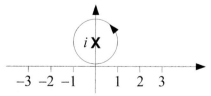

**Figure 2.44**   Contours of integration, example 6

The reader should know the general expression for the $n$th derivative. We will end up using this expression liberally.

**Example 7.** Evaluate $L = \oint_C \dfrac{e^z}{z^2} dz$ where $f(z) = e^z$ and $z_0 = 0$ for each of the four contours given in this example. We note that the line integral has a denominator that is squared. This sure looks like it fits the expression for the first derivative of $f(z_0)$ that was in Equation 2.62. So let's do some computing.

- *Contour 1*: $|z| = 2$. We know that $z_0$ lies within the contour, so we will start by writing down the expression for the first derivative of $f(z_0)$ and then fill in the blanks.

$$\frac{d}{dz} f(z_0) = f'(z_0) = \frac{1}{2\pi i} \oint_C \frac{f(z)}{(z - z_0)^2} dz$$

We rearrange terms to get the expression into a bit more palatable form and end up with

$$L = \oint_C \frac{f(z)}{(z - z_0)^2} dz = (2\pi i) f'(z_0)$$

Now we substitute line integral values to get

$$L = \oint_C \frac{f(z)}{(z - z_0)^2} dz = \oint_C \frac{e^z}{z^2} dz = (2\pi i) \frac{d}{dz} e^z \bigg|_{z=0}$$

After evaluation at $z_0 = 0$ we get the final result $L = 2\pi i$.

- *Contour 2*: $|z - 1| = 1/2$. The singularity at $z_0 = 0$ is not included within the contour, so the line integral evaluates to $L = 0$.
- *Contour 3*: $|z + i| = 1/4$. The singularity at $z_0 = 0$ is not included within the contour, so the line integral evaluates to $L = 0$.
- *Contour 4*: $|z + 1| = 1/2$. The singularity at $z_0 = 0$ is not included within the contour, so the line integral evaluates to $L = 0$.

**Example 8.** Evaluate $L = \oint_C \dfrac{e^z}{z^3} dz$ where $f(z) = e^z$ and $z_0 = 0$ for the contour $|z| = 2$. This time, we note that the line integral has a denominator that is cubed. This fits the expression for the second derivative of $f(z_0)$ that was shown in Equation 2.62. We know that $z_0$ lies within the contour so we will start by writing down the expression for the second derivative of $f(z_0)$ and then work from there.

$$\frac{d^2}{dz} f(z_0) = f''(z_0) = \frac{2!}{2\pi i} \oint_C \frac{f(z)}{(z-z_0)^3} \, dz$$

As before, we rearrange the expression so it suits our problem. In doing so we end up with

$$L = \oint_C \frac{f(z)}{(z-z_0)^3} \, dz = \frac{2\pi i}{2!} f''(z_0)$$

Upon substitution of the line integral values, we evaluate and get

$$L = \oint_C \frac{e^z}{z^3} \, dz = \frac{2\pi i}{2!} f''(0) = \frac{2\pi i}{2!} \frac{d^2}{dz^2} e^z \bigg|_{z=0} = \pi i$$

**Example 9.** Evaluate $L = \oint_C \frac{z^3}{(z+i)^4} dz$ where $f(z) = z^3$ and $z_0 = -i$ on the contours $|z| = 2$ and $|z| = 1/2$:

$$L = \oint_C \frac{f(z)}{(z-z_0)^4} \, dz = \frac{2\pi i}{3!} f'''(z_0) = \frac{2\pi i}{3!} \frac{d^3}{dz^3} z^3 \bigg|_{z=-i}$$

On the contour $|z| = 2$ the line integral evaluates to

$$L = \frac{2\pi i}{6}(6) = 2\pi i$$

If the integral is evaluated on the contour $|z| = 1/2$ the singularity at $z_0 = -i$ is not enclosed by $C$, and therefore the line integral evaluates to $L = 0$.

**Example 10.** Let's work an example that looks to be a bit more difficult. Evaluate the line integral

$$L = \oint_C \frac{z^3}{(z+1)(z-i)^2} \, dz$$

on the contour $C: |z| = 2$, as illustrated in Figure 2.45, part A. It is easy to see from the integral and from the figure that there is a single order singularity at $z = -1$ and a multiple singularity of order 2 at $z = i$.

Before we proceed, why don't we stop for a moment and think of what we know about Cauchy's integral. We know that if $f(z)$ is analytic on and

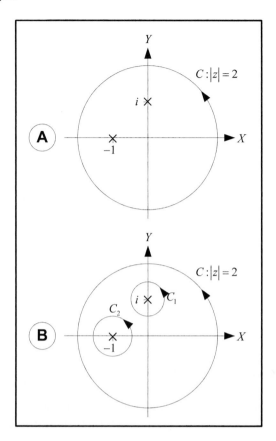

**Figure 2.45** Equivalent contours of integration

in some closed contour $C$, and if there are no singularities within the contour, then the line integral will evaluate to zero. Think outside the box for a moment and view the region within the contour that we just described as having no contribution to the evaluation. Its contribution is zero. In our example, however, we have two singularities that will contribute some value to the evaluation. So what do we do?

We know that the region within the contour $C:|z|=2$ will have a non-zero contribution to the line integral evaluation only at the two singularities—all others will contribute zero. Knowing this, we can split the original integral into two separate integrals. One line integral will be evaluated on contour $C_1$, and the second line integral will be evaluated on contour $C_2$, as shown in Figure 2.45, part B.

That is we can split the integral to obtain

$$L = \oint_C \frac{z^3}{(z+1)(z-i)^2} dz = \oint_{C_1} f_1(z) + \oint_{C_2} f_2(z)$$

Our objective is to derive the function $f_1(z)$ so that it is integrated on contour $C_1$ that encloses the singularity $z = i$ and to derive the function $f_2(z)$ so that it is integrated on contour $C_2$ that encloses the singularity $z = -1$. We accomplish this by using a bit of slick mathematical manipulation, as follows:

$$L = \oint_C \frac{z^3}{(z+1)(z-i)^2} dz = \oint_{C_1} \frac{z^3/(z+1)}{(z-i)^2} dz + \oint_{C_2} \frac{z^3/(z-i)^2}{(z+1)} dz$$

where

$$f_1(z) = z^3/(z+1)$$
$$f_2(z) = z^3/(z-i)^2$$

The alert reader might say that $f_1(z)$ has a singularity at $z = -1$. This is true, but it does not reside in the arbitrarily small contour $C_1$. The same can be said about $f_2(z)$. It has a singularity at $z = i$, but that singularity is not enclosed within the contour $C_2$. Now that we have rewritten the expression for our line integral, we can evaluate it using all we have learned about Cauchy's integral. Let us begin by representing the summation of integrals as $L = L_1 + L_2$ and then evaluating $L_1$ and $L_2$ separately. Evaluation of the integral $L_1$ produces

$$L_1 = \oint_{C_1} \frac{z^3/(z+1)}{(z-i)^2} dz = (2\pi i) \frac{d}{dz} f_1(z) \bigg|_{z=z_0} = (2\pi i) \frac{d}{dz} \left( \frac{z^3}{(z+1)} \right) \bigg|_{z=i} = (-3\pi - 2\pi i)$$

Evaluation of integral $L_2$ produces

$$L_2 = \oint_{C_2} \frac{z^3/(z-i)^2}{(z+1)} dz = (2\pi i) f_2(z) \big|_{z=z_0} = (2\pi i) \frac{z^3}{(z-i)^2} \bigg|_{z=-1} = -\pi$$

Evaluation of the integral $L$ is just the sum of $L_1$ and $L_2$, which is

$$L = L_1 + L_2 = (-3\pi - 2\pi i) + (-\pi) = (-4\pi - 2\pi i)$$

**Example 11.** Evaluate the line integral $L = \oint_C \frac{z^4 + z}{(z+i)^2} dz$ on the contour $C: |z| = 2$. The integrand tells us that we have a singularity of order two at

$z = -i$ and that both are enclosed within the contour of integration. From Equation 2.62 we find the expression that matches our line integral and we write it down as

$$\oint_C \frac{f(z)}{(z-z_0)^2}\, dz = (2\pi i)\frac{d}{dz} f(z_0)$$

Now we can easily setup the evaluation as

$$\oint_C \frac{f(z)}{(z-z_0)^2}\, dz = \oint_C \frac{z^4+z}{(z-i)^2}\, dz = (2\pi i)\frac{d}{dz}(z^4+z)\Big|_{z=i}$$

We can evaluate the integral as follows:

$$(2\pi i)\frac{d}{dz}(z^4+z)\Big|_{z=i} = (2\pi i)(4z^3+1)\Big|_{z=i} = (2\pi i)(1-4i) = (8\pi+2\pi i)$$

**Example 12.** Why don't we do one more example for the road? This time I'll leave some of the details to you, the reader. Evaluate the line integral

$$L = \oint_C \frac{z^2+1}{(z-1)(z+1)}\, dz$$

on the contour $C:|z|=2$.

Without a whole lot of jibber jabber, we can split the integral into two parts over two different contours and write

$$\begin{aligned}
L &= \oint_C \frac{z^2+1}{(z-1)(z+1)}\, dz \\
&= \oint_{C_1} \frac{(z^2+1)/(z+1)}{(z-1)}\, dz + \oint_{C_2} \frac{(z^2+1)/(z-1)}{(z+1)}\, dz \\
&= (2\pi i)\frac{(z^2+1)}{(z+1)}\Big|_{z=1} + (2\pi i)\frac{(z^2+1)}{(z-1)}\Big|_{z=-1} \\
&= 2\pi i - 2\pi i \\
&= 0
\end{aligned}$$

In this case, contours $C_1$ and $C_2$ are made to be arbitrarily small so that they enclose the singularities at $z=1$ and $z=-1$, respectively.

Sadly we have reached the end of the examples. The one thing the reader should take away from all this is that use of the Cauchy's integral sure makes short work of these relatively complicated problems.

## 2.22   RESIDUE THEORY

A function can be expanded into a series by several methods. In the study of calculus we learned several methods of series expansion. We learned about the Maclaurin series expansion, the Taylor series expansion, and the Laurent series expansion. If your memory is a bit foggy on these subjects, please consult references [1], [2], and [3]. The full treatment of these series expansion methods is beyond the scope of this book.

However, in the study of residues, it will be necessary to briefly mention the basic form of the Laurent series. A function $f(z)$ that is singular at the point $z = z_0$ can be expanded into a Laurent series of the form

$$f(z) = \sum_{n=-\infty}^{+\infty} a_n (z - z_0)^n = \cdots + \frac{a_{-2}}{(z - z_0)^2} + \frac{a_{-1}}{(z - z_0)} + a_0 + a_1 (z - z_0) + a_2 (z - z_0)^2 + \cdots$$

If we were to compute the line integral of the function $f(z)$ that is singular at a point $z = z_0$ within the contour $C$, we could write

$$L = \oint_C f(z)\, dz = \oint_C \left\{ \cdots + \frac{a_{-2}}{(z - z_0)^2} + \frac{a_{-1}}{(z - z_0)} + a_0 + a_1 (z - z_0) + a_2 (z - z_0)^2 + \cdots \right\} dz$$

Evaluating each term of the line integral would be a daunting task, and if there should be an infinite number of them, then the task would be impossible. Just how in the heck are we supposed to evaluate this humongous line integral? Well, a fellow named Cauchy has already provided a means to rescue us from attempting to solve this seemingly impossible problem. Remember the embellished Cauchy's integral of Equation 2.60 (repeated here for clarity):

$$\oint_C f(z)\, dz = \oint_C \frac{1}{(z - z_0)^n}\, dz = \begin{cases} 0 & \text{for } n \neq 1 \\ 0 & C \text{ does not enclose } z = z_0 \\ 2\pi i & \text{for } n = 1 \text{ and } C \text{ does enclose } z = z_0 \end{cases}$$

If we apply Cauchy's integral to the solution of each term in our line integral $L$ we discover that every term in the line integral equals zero with the exception of the $a_{-1}$ term. Therefore, if we evaluate the line integral, we end up with

$$L = \oint_C f(z)\, dz = \cdots 0 + 0 + 0 + \cdots + \oint_C \frac{a_{-1}}{(z - z_0)}\, dz + \cdots + 0 + 0 + 0 + \cdots = (2\pi i)\, a_{-1}$$

In other words

$$L = \oint_C f(z)\, dz = (2\pi i)\, a_{-1}$$

**Equation 2.63**

Since the term $a_{-1}$ is all that is left when the line integral of the Laurent series is evaluated, it is aptly called the *residue*. This is a significant result. To help bring home the importance of this result, let us repeat Equation 2.63 but this time with a little embellishment:

$$L = \oint_C f(z)\, dz = (2\pi i)(\text{Line Integral Residue}) = (2\pi i)\, a_{-1}$$

In order to demonstrate this method of solving line integrals, we will present an example, first using Cauchy's integral method and then second using the residue method of evaluation.

**Example 1.** Evaluate the line integral $\oint_C \dfrac{z}{(z+2)(z+4)}\, dz$ on the contour $C : |z| = 3$ using Cauchy's integral method.

The first thing we note is that there are two singularities located at $z = -2$ and $z = -4$. The singularity at $z = -4$ is not enclosed by the contour $C$. Therefore we can rewrite the integral as

$$L = \oint_C \frac{f(z)}{(z - z_0)} = \oint_C \frac{z/(z+4)}{(z+2)}\, dz = (2\pi i)\left[\frac{z}{(z+4)}\right]_{z=-2} = -2\pi i$$

Now let's evaluate the same line integral using the residue method. The first thing we do is expand $f(z)$ into a Laurent series. When we do, we end up with

$$f(z) = \frac{z}{(z+2)(z+4)}$$

$$= \frac{2}{z+4} - \frac{1}{z+2}$$

$$= \left\{ \cdots + \frac{8}{z^4} - \frac{4}{z^3} + \frac{2}{z^2} - \frac{1}{z} + \frac{1}{2} - \frac{z}{8} + \frac{z^2}{32} - \frac{z^3}{128} + \cdots \right\}$$

Looking at the series we see that the $a_{-1}$ term is $-1$. Therefore the evaluation of the line integral is

$$L = \oint_C \frac{z}{(z+2)(z+4)} dz = (2\pi i) a_{-1} = -2\pi i$$

As should be expected, we end up with the same result. I already know that you are disappointed at the complexity of the residue method. The Laurent expansion is time consuming, tedious, and subject to computational error. So what's the point?

The point is twofold. Point number one, I need to show the classical tedious method in order for us to fully appreciate a much simpler method to compute residues. Point number two, we need to see that there is some substance behind residue theory in order to validate it in our own minds. Once that is accomplished, we can show a simpler method to obtain the same result.

In the simpler method, we do not need to compute the Laurent series in order to compute the residue of a line integral. This is one of the reasons why we didn't spend a great deal of time discussing the Laurent series. Fortunately, a much simpler procedure to compute the $a_{-1}$ term exists. The residue method has some attributes that make it an extremely powerful DSP tool, as we will see later.

For a function $f(z)$ that is analytic both inside and on a simple closed contour $C$ except for some interior point $z_0$, at which there exists a singularity of order $n$, we can say that the residue at that singular point is computed by

$$a_{-1} = R = \lim_{z \to z_0} \left[ \frac{1}{(n-1)!} \frac{d^{n-1}}{dz^{n-1}} \left( (z-z_0)^n f(z) \right) \right]$$

**Equation 2.64**

where $n$ is the order of the singularity. Note that the term $R$ is used to represent the value of the residue since it is so much easier to write than $a_{-1}$. Equation 2.64 is important. Push it into your memory stack for quick recall. To demonstrate this simpler method of computing residues, let's evaluate the same line integral as before, only this time using Equation 2.64.

**Example 2.** Evaluate the line integral $\oint_C \frac{z}{(z+2)(z+4)} dz$ on the contour $C : |z| = 3$ using Equation 2.64.

Once again we know that the singularity is at $z = -2$. Also since the singularity is of the first order, then $n = 1$. Therefore we can write

$$a_{-1} = R = \lim_{z \to -2} \left[ \frac{1}{(1-1)!}(z+2)\frac{z}{(z+2)(z+4)} \right] = \lim_{z \to -2} \left[ \frac{z}{(z+4)} \right] = \frac{-2}{2} = -1$$

We know from Equation 2.63 that a line integral evaluates to

$$L = \oint_C f(z)\, dz = (2\pi i)\, a_{-1} = (2\pi i)\left( \text{Residue at } z_0 \right)$$

Therefore we can say that

$$L = \oint_C f(z)\, dz = (2\pi i)\, a_{-1} = (2\pi i)(-1) = -2\pi i$$

The power of the residue method lies in the fact that it allows us to solve a line integral with several singularities within the contour of integration quite simply. Suppose we have a line integral $\oint_C f(z)\, dz$ that is evaluated on some contour C. Further suppose that this contour encloses more than one singularity, as illustrated in Figure 2.46. Let us represent the $a_{-1}$ residue term for the singularity $z_0$ by $R_0$ and the $a_{-1}$ residue term for the singularity $z_1$ by $R_1$ and so on until we get to the $n$th singularity where the $a_{-1}$ residue term for the singularity $z_n$ is represented by $R_n$. The integral is taken counterclockwise around the contour $C$, and the individual residues $R_k \{ \text{for } 0 \le k \le n \}$ are computed using Equation 2.64.

Armed with this information, we can write the pretty amazing expression given by Equation 2.65:

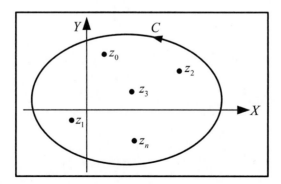

**Figure 2.46** Multiple singularities

$$\oint_C f(z)dz \quad = 2\pi i \left( R_0 + R_1 + R_2 + \cdots + R_n \right)$$
$$= (2\pi i) \sum \left( \text{Residues Enclosed by } C \right)$$

**Equation 2.65**

This is powerful stuff. Any reader who is involved in DSP and is using the z-transform to either design or analyze a complex digital circuit will fall in love with the residue method. In circuit theory, singularities are called *poles*. Locations where the circuit response is zero are called *zeros*. When designing or analyzing the frequency response or stability of a digital circuit using the z-transform, one can easily derive the circuit time domain impulse response using the method of residues.

This procedure is described in detail later in this book in Chapter 4, "Review of the Z-Transform." For now, let us look at a few examples.

**Example 3**. Evaluate $\oint_C \dfrac{e^z}{z(z-2)^2}$ on the contour $C:|z|=5$. There is a first order singularity or pole at $z=0$ and a second order singularity or pole at $z=2$. We can evaluate this integral by solving for the two residues:

$$\oint_C \frac{e^z}{z(z-2)^2} = 2\pi i \left( R_0 + R_1 \right)$$

Residue $R_0$ at $z=0$ is computed as

$$R_0 = \lim_{z \to z_0} \left[ \frac{1}{(n-1)!} \frac{d^{n-1}}{dz^{n-1}} \left( (z-z_0)^n f(z) \right) \right]$$

$$= \lim_{z \to 0} \left[ (z) \frac{e^z}{z(z-2)^2} \right] \quad = \frac{1}{4}$$

Residue $R_1$ at $z=2$ is computed as

$$R_1 = \lim_{z \to z} \left[ \frac{1}{(n-1)!} \frac{d^{n-1}}{dz^{n-1}} \left( (z-z_0)^n f(z) \right) \right]$$

$$= \lim_{z \to 2} \left[ \frac{1}{(2-1)!} \frac{d}{dz} \left( (z-2)^2 \frac{e^z}{z(z-2)^2} \right) \right]$$

$$= \lim_{z \to 2} \left[ \frac{e^z (z-1)}{z^2} \right] \quad = \frac{e^2}{4}$$

The final evaluation is computed as

$$\oint_C \frac{e^z}{z(z-2)^2} = 2\pi i\left(\frac{1}{4}+\frac{e^2}{4}\right) = \frac{\pi i}{2}\left(1+e^2\right)$$

**Example 4.** Evaluate $L=\oint_C \dfrac{1}{z^3\left(z-1\right)^2}\,dz$ on the contour $C:\left|z\right|=2.$

We can see that we have a third order singularity or pole at $z=0$ and a second order pole at $z=1$. Both of these poles are enclosed by the contour $C$. Let's begin by computing the residue $R_0$ for the third order pole $z=0$:

$$
\begin{aligned}
R_0 &= \lim_{z\to z_0}\left[\frac{1}{(n-1)!}\frac{d^{n-1}}{dz^{-1}}\left(\left(z-z_0\right)^n f\left(z\right)\right)\right]\\[2mm]
&= \lim_{z\to 0}\left[\frac{1}{(3-1)!}\frac{d^2}{dz^2}\left((z)^3\,\frac{1}{z^3\left(z-1\right)^2}\right)\right]\\[2mm]
&= \lim_{z\to 0}\left[\frac{1}{(2)!}\frac{d^2}{dz^2}\left(\frac{1}{\left(z-1\right)^2}\right)\right]\\[2mm]
&= \lim_{z\to 0}\left[\frac{1}{(2)!}\left(\frac{6}{\left(z-1\right)^4}\right)\right]\\[2mm]
&= 3
\end{aligned}
$$

Now let's compute the residue $R_1$ for the second order pole $z=1$:

$$
\begin{aligned}
R_1 &= \lim_{z\to z_0}\left[\frac{1}{(n-1)!}\frac{d^{n-1}}{dz^{-1}}\left(\left(z-z_0\right)^n f\left(z\right)\right)\right]\\[2mm]
&= \lim_{z\to 1}\left[\frac{1}{(2-1)!}\frac{d}{dz}\left(\left(z-1\right)^2\,\frac{1}{z^3\left(z-1\right)^2}\right)\right]\\[2mm]
&= \lim_{z\to 1}\left[\frac{1}{(1)!}\frac{d}{dz}\left(\frac{1}{(z)^3}\right)\right]\\[2mm]
&= \lim_{z\to 1}\left[\left(-\frac{3}{z^4}\right)\right]\\[2mm]
&= -3
\end{aligned}
$$

All that is left to do is to add the residues and multiply by $2\pi i$ to get

$$L = \oint_C \frac{1}{z^3(z-1)^2}\,dz = 2\pi i(R_0 + R_1) = 2\pi i(3-3) = 0$$

**Example 5.** Evaluate $\oint_C \frac{\cos z}{z^2 - 4\pi z + 3\pi^2}\,dz$ on the contour $C : |z| = 4\pi$. The first thing we need to do is factor the denominator to get

$$\oint_C \frac{\cos z}{z^2 - 4\pi z + 3\pi^2}\,dz = \oint_C \frac{\cos z}{(z-\pi)(z-3\pi)}\,dz$$

After factorization, we can see that we have two singularities and both are within the contour of integration. So we set up the evaluation by writing

$$L = \oint_C \frac{\cos z}{(z-\pi)(z-3\pi)}\,dz = 2\pi i(R_0 + R_1)$$

We compute the residue $R_0$ and get

$$R_0 = \lim_{z \to \pi}\left[\frac{\cos z}{(z-3\pi)}\right] = \frac{-1}{-2\pi} = \frac{1}{2\pi}$$

Next we compute the residue $R_1$ to get

$$R_1 = \lim_{z \to 3\pi}\left[\frac{\cos z}{(z-\pi)}\right] = \frac{-1}{2\pi} = -\frac{1}{2\pi}$$

Now we finish the evaluation by summing the residues and multiplying by $2\pi i$ to get

$$L = \oint_C \frac{\cos z}{(z-\pi)(z-3\pi)}\,dz = 2\pi i\left(\frac{1}{2\pi} - \frac{1}{2\pi}\right) = 0$$

**Example 6.** Find the residue at the singular point for $f(z) = \frac{\cos z}{z^5}$. In this example, let's think outside the box a bit. Let's not get into a rigid mode of thinking. In the interest of remaining flexible, we will take another route to this solution. We will expand $\cos z$ into its Laurent series about the singular point $z = 0$.

$$f(z) \quad = \frac{\cos z}{z^5}$$

$$= \frac{1}{z^5}\left[ 1 - \frac{z^2}{2!} + \frac{z^4}{4!} - \frac{z^6}{6!} + \cdots \right]$$

$$= \left[ \frac{1}{z^5} - \frac{1}{(2!)z^3} + \frac{1}{(4!)z} - \frac{z^1}{(6!)} + \cdots \right]$$

We notice from the series expansion that the $a_{-1}$ term (which is the residue term) is $1/4!$. Therefore the residue is $a_1 = R = 1/4!$ or $R = 1/24$. Therefore $L = 2\pi i/24 = \pi i/12$.

That's all there is in this review of complex variables, folks. We will be applying some of these concepts when we deal with the subject of the z-transform in Chapter 4.

## 2.23  REFERENCES

[1] Wilfred Kaplan. *Introduction to Analytic Functions*. Palo Alto, CA: Addison-Wesley, 1996.

[2] K. A. Stroud and Dexter Booth. *Complex Variables*. New York: Industrial Press, 2008.

[3] Ruel V. Churchill. *Complex Variables and Applications*. New York: McGraw-Hill, 1960.

[4] Murray R. Spiegel. *Schaum's Outlines: Complex Variables (with an Introduction to Conformal Mapping)*. 32nd printing. New York: McGraw-Hill, 1964 (1999).

[5] Mark Ablowitz and Athanassios Fokas. *Complex Variables and Applications*. 2nd ed. New York: Cambridge University Press, 2003.

# CHAPTER THREE

# Review of the Fourier Transform

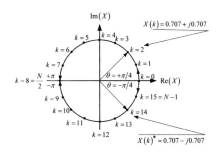

This chapter is not intended to be a detailed treatise on the Fourier transform. There have been a number of excellent books written that deal with this subject in great detail. This chapter is intended to be an in-depth review of the Fourier transform and its properties so that we may have a foundation on which to base our later discussions dealing with applications such as digital signal spectra and digital signal tuning. It is expected that the reader already has prior knowledge of and experience with the Fourier transform. If not, this review is detailed enough to give you a good understanding of the subject. If you are interested in a more thorough treatment on this subject, references [16], [17], and [18], listed at the end of this chapter, are recommended.

## 3.1  A BRIEF REVIEW OF THE FOURIER SERIES

Let's assume that $f(t)$ is a periodic function with period $T$ that is compliant with the set of three restrictions:

1. $f(t)$ has a finite number of maxima and minima within the period $T$ and all maxima and minima are of finite magnitude.
2. $f(t)$ has a finite number of discontinuities within the period $T$.
3. The integral of the absolute value of $f(t)$ over its period is finite:

$$\int_0^T |f(t)|\,dt < \infty$$

These three conditions are referred to as the *Dirichlet conditions* and are sufficient to guarantee that the Fourier series of $f(t)$ exists.

Any *periodic* waveform that meets the Dirichlet conditions can be represented by a constant term and the sum of an infinite number of sine and/or cosine terms that make up a mathematical series called the *Fourier series*. From this point on, unless specifically stated otherwise, we will assume that any function we refer to will be a function that meets the Dirichlet conditions. The Fourier series is a tool that allows an engineer to break down a periodic waveform into the sum of its individual frequency components. This provides the engineer with an extremely powerful tool to aid in the analysis, design, and test of most systems.

The general form for what is termed the *trigonometric Fourier series* is given by

$$f(t) = a_0 + \sum_{n=1}^{\infty} a_n \cos(n\omega_0 t) + \sum_{n=1}^{\infty} b_n \sin(n\omega_0 t)$$

**Equation 3.1**

Equation 3.1 tells us that any function $f(t)$ that meets the Dirichlet conditions can be broken down into the sum of its DC component, its fundamental frequency, and an infinite number of harmonic frequencies.

The series in Equation 3.1 represents the sum of a constant $a_0$ and an infinite collection of sinusoidal and/or cosinusoidal coefficient terms with amplitudes $a_n$ and $b_n$, corresponding to frequencies $n\omega_0$ where $n = 1, 2, 3, \cdots, \infty$. The $a_0$ term is the average value of the waveform. From an electronic engineering perspective, if $f(t)$ were a voltage waveform, then $a_0$ would be described as the value of the DC component of that waveform. The frequency $\omega_0$ corresponding to $n = 1$ is the fundamental frequency, and the frequencies at $n\omega_0$ for $n = 2, 3, 4, \cdots$ are referred to as the harmonic frequencies of the fundamental.

The derivation of the Fourier series of a function $f(t)$ requires that we compute the coefficients $a_0$, $a_n$, and $b_n$. It turns out that there are a few trigonometric identities that will significantly simplify these computations. Let's begin with a review of these identities.

We note that the integral over one period of a sine or cosine function is zero. Therefore we can write

$$\int_{t=0}^{t=T} \cos n\omega_0 t \ dt = \int_{t=t_0}^{t=t_0+T} \cos n\omega_0 t \ dt = 0$$

$$\int_{t=0}^{t=T} \sin n\omega_0 t \ dt = \int_{t=t_0}^{t=t_0+T} \sin n\omega_0 t \ dt = 0$$

**Equation 3.2**

This fact is easily seen since both the sine and cosine waveforms are symmetric and have equal positive and negative values over the duration of one period. We also note that

$$\int_{t=0}^{t=T} \cos^2 n\omega_0 t \ dt = \frac{1}{2} \int_{t=0}^{t=T} \left(1 + \cos 2n\omega_0 t\right) dt = \frac{T}{2}$$

$$\int_{t=0}^{t=T} \sin^2 n\omega_0 t \ dt = \frac{1}{2} \int_{t=0}^{t=T} \left(1 - \cos 2n\omega_0 t\right) dt = \frac{T}{2}$$

**Equation 3.3**

Finally, we note one other important property of sinusoids that are integrated over one period:

$$\int_0^T \cos(n\omega_0 t)\sin(m\omega_0 t) \ dt = 0 \quad \text{for all } m \text{ and } n$$

$$\int_0^T \cos(n\omega_0 t)\cos(m\omega_0 t) \ dt = 0 \quad \text{for } m \neq n$$

$$\int_0^T \sin(n\omega_0 t)\sin(m\omega_0 t) \ dt = 0 \quad \text{for } m \neq n$$

**Equation 3.4**

### 3.1.1  Trigonometric Fourier Series

The identities listed in Equation 3.2, Equation 3.3, and Equation 3.4 play a significant role in simplifying the derivation of a Fourier series for any periodic function. Now that we have a few key identities at our disposal, let's compute the value of $a_0$ in Equation 3.1, which is repeated here for clarity:

$$f(t) = a_O + \sum_{n=1}^{\infty} a_n \cos(n\omega_0 t) + \sum_{n=1}^{\infty} b_n \sin(n\omega_0 t)$$

If we integrate both sides of Equation 3.1 over a single period, we can write

$$\int_{t=0}^{t=T} f(t) dt = \int_{t=0}^{t=T} a_0 \ dt + \int_{t=0}^{t=T} \left[ \sum_{n=1}^{\infty} a_n \cos(n\omega_0 t) + \sum_{n=1}^{\infty} b_n \sin(n\omega_0 t) \right] dt$$

Now we interchange the order of summation and integration on the right-hand side to get

$$\int_{t=0}^{t=T} f(t) dt = \int_{t=0}^{t=T} a_0 \ dt + \sum_{n=1}^{\infty} \left\{ \int_{t=0}^{t=T} \left[ a_n \cos(n\omega_0 t) \right] dt + \int_{t=0}^{t=T} \left[ b_n \sin(n\omega_0 t) \right] dt \right\}$$

We can then use the identities listed in Equation 3.2 to determine that

$$\int_{t=0}^{t=T} f(t)\,dt = \int_{t=0}^{t=T} a_0\,dt + \sum_{n=1}^{\infty}\{0+0\} = \int_{t=0}^{t=T} a_0\,dt = a_0 T$$

We finish by rearranging terms to get

$$a_0 = \frac{1}{T}\int_{t=0}^{t=T} f(t)\,dt$$

**Equation 3.5**

It is easy to see that the term $a_0$ is just the average value over one period of the function $f(t)$. Since all periods are equal, $a_0$ is also the average of any multiple period of $f(t)$.

Now let's turn our attention to the computation of the Fourier series $a_n$ terms. We do this by multiplying both sides of Equation 3.1 by $\cos m\omega_0 t$ and integrating over one period such that

$$\int_{t=0}^{t=T} f(t)\cos(m\omega_0 t)\,dt \;=\; \int_{t=0}^{t=T} a_0\cos(m\omega_0 t)\; dt$$
$$+\int_{t=0}^{t=T}\left\{\cos(m\omega_0 t)\left[\sum_{n=1}^{\infty} a_n\cos(n\omega_0 t)+\sum_{n=1}^{\infty} b_n\sin(n\omega_0 t)\right]\right\}dt$$

Once again, we can interchange the operations of summation and integration to get

$$\int_{t=0}^{t=T} f(t)\cos(m\omega_0 t)\,dt \;=\; \int_{t=0}^{t=T} a_0\cos(m\omega_0 t)\; dt$$
$$+\sum_{n=1}^{\infty}\left[\int_{t=0}^{t=T} a_n\cos(n\omega_0 t)\cos(m\omega_0 t)\; dt + \int_{t=0}^{t=T} b_n\sin(n\omega_0 t)\cos(m\omega_0 t)\; dt\right]$$

We can see from the identities in Equation 3.2 that the integral of the $a_0$ term over one period is zero, which leaves us with

$$\int_{t=0}^{t=T} f(t)\cos(m\omega_0 t)\,dt = 0 + \sum_{n=1}^{\infty}\left[\int_{t=0}^{t=T} a_n\cos(n\omega_0 t)\cos(m\omega_0 t)\; dt + \int_{t=0}^{t=T} b_n\sin(n\omega_0 t)\cos(m\omega_0 t)\; dt\right]$$

We can also see from the identities in Equation 3.4 that

$$\int_0^T \sin(n\omega_0 t)\cos(m\omega_0 t)\ dt = 0 \ \text{ for all } m \text{ and } n$$

Therefore the integral of all the $b_n$ terms is zero, which reduces the integration to

$$\int_{t=0}^{t=T} f(t)\cos(m\omega_0 t)dt = \sum_{n=1}^{\infty}\left[\int_{t=0}^{t=T} a_n\cos(n\omega_0 t)\cos(m\omega_0 t)\ dt\right]$$

We also note from the identities in Equation 3.4 that

$$\int_0^T \cos(n\omega_0 t)\cos(m\omega_0 t)\ dt = 0 \ \ \text{ for } m \neq n$$

This means that every one of the infinite number of terms in the right-side integral is zero with the exception of the case where $m = n$. This leaves us with

$$\int_{t=0}^{t=T} f(t)\cos(n\omega_0 t)dt = \int_{t=0}^{t=T} a_n\cos(n\omega_0 t)\cos(n\omega_0 t)\ dt = \int_{t=0}^{t=T} a_n\cos^2(n\omega_0 t)\ dt$$

We know from the identities in Equation 3.3 that

$$\int_{t=0}^{t=T} \cos^2(n\omega_0 t)\ dt = \frac{T}{2}$$

Upon substitution we get

$$\int_{t=0}^{t=T} f(t)\cos(n\omega_0 t)dt = a_n\frac{T}{2}$$

By rearranging the terms, we arrive at the expression we use to compute the $a_n$ terms of the Fourier series:

$$a_n = \frac{2}{T}\int_{t=0}^{t=T} f(t)\cos(n\omega_0 t)dt \ \text{ for } n = 1,2,3,\cdots$$

**Equation 3.6**

We determine the expression for the computation of the $b_n$ terms just as we did for the $a_n$ terms, but this time we multiply the function $f(t)$ by $\sin(m\omega_0 t)$. For completeness, and since this is important, we will perform the full derivation of the expression for evaluating the $b_n$ terms. We begin with Equation 3.1, which is repeated here for clarity:

$$f(t) = a_O + \sum_{n=1}^{\infty} a_n \cos(n\omega_O t) + \sum_{n=1}^{\infty} b_n \sin(n\omega_O t)$$

We multiply both sides of Equation 3.1 by $\sin m\omega_0 t$ and integrate over one period such that

$$\int_{t=0}^{t=T} f(t)\sin(m\omega_O t)dt \;=\; \int_{t=0}^{t=T} a_O \sin(m\omega_O t)\; dt$$

$$+ \int_{t=0}^{t=T} \left\{ \sin(m\omega_O t)\left[ \sum_{n=1}^{\infty} a_n \cos(n\omega_O t) + \sum_{n=1}^{\infty} b_n \sin(n\omega_O t) \right] \right\} dt$$

Interchanging the operations of summation and integration, we get

$$\int_{t=0}^{t=T} f(t)\sin(m\omega_O t)dt \;=\; \int_{t=0}^{t=T} a_O \sin(m\omega_O t)\; dt$$

$$+ \sum_{n=1}^{\infty} \left[ \int_{t=0}^{t=T} a_n \cos(n\omega_O t)\sin(m\omega_O t)\; dt + \int_{t=0}^{t=T} b_n \sin(n\omega_O t)\sin(m\omega_O t)\; dt \right]$$

We can see from the identity in Equation 3.2 that the integral of the $a_0$ term is zero, which gives us

$$\int_{t=0}^{t=T} f(t)\sin(m\omega_O t)dt = 0 + \sum_{n=1}^{\infty} \left[ \int_{t=0}^{t=T} a_n \cos(n\omega_O t)\sin(m\omega_O t)\; dt + \int_{t=0}^{t=T} b_n \sin(n\omega_O t)\sin(m\omega_O t)\; dt \right]$$

We can also see from the identities in Equation 3.4 that

$$\int_{0}^{T} \cos(n\omega_O t)\sin(m\omega_O t)\; dt = 0 \text{ for all } m \text{ and } n$$

Therefore the integral of all the $a_n$ terms is zero, which reduces the integration to

$$\int_{t=0}^{t=T} f(t)\sin(m\omega_O t)dt = \sum_{n=1}^{\infty} \left[ \int_{t=0}^{t=T} b_n \sin(n\omega_O t)\sin(m\omega_O t)\; dt \right]$$

We also note from the identities in Equation 3.4 that

$$\int_0^T \sin(n\omega_0 t)\sin(m\omega_0 t)\ dt = 0 \ \text{ for } \ m \neq n$$

This means that every one of the infinite number of terms in the right-side integral is zero with the exception of the case where $m = n$. This leaves us with

$$\int_{t=0}^{t=T} f(t)\sin(m\omega_0 t)dt = \int_{t=0}^{t=T} b_n \sin(n\omega_0 t)\sin(n\omega_0 t)\ dt = \int_{t=0}^{t=T} b_n \sin^2(n\omega_0 t)\ dt$$

We know from Equation 3.3 that

$$\int_{t=0}^{t=T} \sin^2 n\omega_0 t\ dt = \frac{T}{2}$$

Upon substitution, we get

$$\int_{t=0}^{t=T} f(t)\sin(n\omega_0 t)dt = b_n \frac{T}{2}$$

By rearranging the terms, we arrive at the expression we use to compute the $b_n$ terms of the Fourier series:

$$b_n = \frac{2}{T}\int_{t=0}^{t=T} f(t)\sin(n\omega_0 t)dt \ \text{ for } n = 1, 2, 3, \cdots$$

**Equation 3.7**

Since the sine and cosine terms are periodic, we can integrate over any arbitrary number of periods and get the same results. It is common practice to integrate over the period that is centered at zero and extends from $-T/2$ to $+T/2$. When we do this we can say that the coefficients $a_0, a_n,$ and $b_n$ are determined by

$$a_O = \frac{1}{T}\int_{-T/2}^{T/2} f(t)\ dt$$

**Equation 3.8**

$$a_n = \frac{2}{T}\int_{-T/2}^{T/2} f(t)\cos(n\omega_0 t)dt \ \text{ for } n = 1, 2, 3, \cdots$$

$$b_n = \frac{2}{T}\int_{-T/2}^{T/2} f(t)\sin(n\omega_0 t)dt \ \text{ for } n = 1, 2, 3, \cdots$$

**Equation 3.9**

where

- $f(t)$ is the periodic waveform under study;
- $T$ is the period of the waveform;
- $n$ is the sequential index ranging from 1 to $\infty$;
- $\omega_O$ is the fundamental radian frequency of $f(t)$, $\omega_O = 2\pi f_O = 2\pi/T$; and
- $n\omega_O$ is the nth harmonic of the fundamental frequency $\omega_O$.

The coefficient $a_O$ in Equation 3.1 represents the average value of the waveform $f(t)$. It may be helpful to visualize the coefficients $a_n$ and $b_n$ as the magnitude of a correlation score between the waveform $f(t)$ and the nth harmonic frequency $\cos(n\omega_O)$ and $\sin(n\omega_O)$, respectively.

The computation of the coefficients $a_n$ and $b_n$ is carried out via Equation 3.9. The computations can be simplified significantly if the function $f(t)$ possesses either even or odd symmetry:

- If $f(t)$ is an even function such that $f(t) = f(-t)$, then $b_n = 0$ and

$$a_n = \frac{4}{T} \int_{t=0}^{t=\frac{T}{2}} f(t)\cos(n\omega_O t)\,dt$$

- If $f(t)$ is an odd function such that $f(t) = -f(-t)$, then $a_n = 0$ and

$$b_n = \frac{4}{T} \int_{t=0}^{t=\frac{T}{2}} f(t)\sin(n\omega_O t)\,dt$$

The expression given by Equation 3.1 for the infinite series produces a one-sided signal spectrum in that all the component frequencies are positive. Although not physically realizable, it is sometimes mathematically efficient to view the Fourier series as a two-sided spectrum and include the negative frequencies as well. This is done by halving the magnitude of the coefficients $a_n$ and $b_n$ and folding a mirror image of the positive spectra about the $y$-axis to graphically represent the negative frequencies.

Now that we have established a theoretical platform on which to stand, it is a good time to stop for a moment, catch our breath, and work through some examples.

**Example 1.** Compute the Fourier series for the pulsed waveform illustrated in Figure 3.1. We see from the figure that the waveform is periodic and that its amplitude switches between zero and $A$. Because we will get the same results if we integrate over any integer number of periods, we only need to concern ourselves with one period.

The period of the waveform is $T$, and it maintains an amplitude of $A$ for a length of time equal to $\tau$ and is zero for the remainder of the period. We choose to view the waveform as even symmetric about the $y$-axis, which means the period we're interested in extends from $-T/2$ to $T/2$.

The Fourier series for this waveform will take the general form of Equation 3.1, which is repeated here for clarity:

$$f(t) = a_O + \sum_{n=1}^{\infty} a_n \cos(n\omega_O t) + \sum_{n=1}^{\infty} b_n \sin(n\omega_O t)$$

In order for us to compute the Fourier series, we only need to fill in the blanks of this equation by computing the coefficients $a_0$, $a_n$, and $b_n$. Let's begin by computing the coefficient $a_0$. We know from Equation 3.8 that

$$a_O = \frac{1}{T} \int_{-T/2}^{T/2} f(t)\,dt$$

Since $f(t) = A$ between $-\tau/2$ and $+\tau/2$ and $f(t) = 0$ during the remainder of the period, we can write

$$a_O = \frac{1}{T} \int_{-T/2}^{T/2} f(t)\,dt = \frac{1}{T} \int_{-\tau/2}^{\tau/2} A\,dt = \frac{A}{T} t \Big|_{t=-\tau/2}^{t=+\tau/2} = \frac{A\tau}{T}$$

Thus we see that the $a_0$ term is just the average value of the waveform over one period. If all periods are identical (as is the case here) and if the

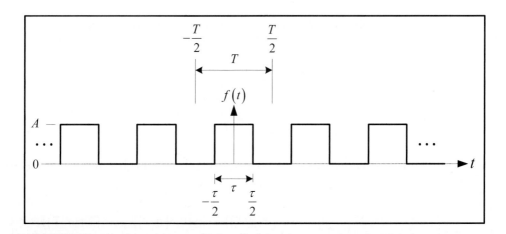

**Figure 3.1**   Square wave

waveform is composed of an integer number of periods, then $a_0$ is also the average value over any number of integer periods of the waveform.

Now lets us compute the $a_n$ coefficients. We know from Equation 3.9 that

$$a_n = \frac{2}{T} \int_{-T/2}^{T/2} f(t) \cos(n\omega_0 t) dt \quad \text{for } n = 1, 2, 3, \cdots$$

Therefore

$$a_n = \frac{2}{T} \int_{-\tau/2}^{\tau/2} A \cos(n\omega_0 t) dt$$

$$= \frac{2A}{T} \left( \frac{\sin(n\omega_0 t)}{n\omega_0} \right) \Bigg|_{t=-\frac{\tau}{2}}^{t=+\frac{\tau}{2}}$$

$$= \frac{2A}{T} \left( \frac{\sin(n\omega_0 \tau/2)}{n\omega_0} - \frac{\sin(-n\omega_0 \tau/2)}{n\omega_0} \right)$$

$$= \frac{4A}{T} \left( \frac{\sin(n\omega_0 \tau/2)}{n\omega_0} \right)$$

We can simplify things a bit if we multiply both numerator and denominator by $\tau$ to get

$$a_n = 4A \frac{\tau}{T} \left( \frac{\sin(n\omega_0 \tau/2)}{n\omega_0 \tau} \right)$$

We know that $\omega_0 = 2\pi f_0 = \frac{2\pi}{T}$, so upon substitution we get

$$a_n = 4A \frac{\tau}{T} \left( \frac{\sin\left( n \frac{2\pi}{T} \tau/2 \right)}{n \frac{2\pi}{T} \tau} \right)$$

Then after cancellation of terms and dividing both numerator and denominator by 2 we end up with the final result of

$$a_n = 2A \frac{\tau}{T} \left( \frac{\sin(n\pi(\tau/T))}{n\pi(\tau/T)} \right) = 2A \frac{\tau}{T} \left[ \frac{\sin(\pi(\tau/T))}{\pi(\tau/T)} + \frac{\sin(2\pi(\tau/T))}{2\pi(\tau/T)} + \frac{\sin(3\pi(\tau/T))}{3\pi(\tau/T)} + \cdots \right]$$

for $n = 1, 2, 3, \cdots$. Now let's compute the $b_n$ coefficients. We know from Equation 3.9 that

$$b_n = \frac{2}{T} \int_{-T/2}^{T/2} f(t) \sin(n\omega_O t) dt \quad \text{for } n = 1, 2, 3, \cdots$$

We also know that if $f(t)$ is an even function such that $f(t) = f(-t)$, then all of the $b_n$ coefficients equal zero. So there is no need to compute them. By knowing the symmetry property we could have saved ourselves a bit of computational work. For demonstration purposes, however, we choose to verify this symmetry property by computing the $b_n$ coefficients. In doing so we get

$$
\begin{aligned}
b_n &= \frac{2}{T} \int_{-T/2}^{T/2} f(t) \sin(n\omega_O t) dt \\
&= \frac{2}{T} \int_{-\tau/2}^{\tau/2} A \sin(n\omega_O t) dt \\
&= -\frac{2A}{T} \frac{\cos(n\omega_O t)}{n\omega_O} \bigg|_{t=-\frac{\tau}{2}}^{t=+\frac{\tau}{2}} \\
&= -\frac{2A}{T} \left[ \frac{\cos(n\omega_O(\tau/2))}{n\omega_O} - \frac{\cos(n\omega_O(-\tau/2))}{n\omega_O} \right] \\
&= -\frac{2A}{T}[0]
\end{aligned}
$$

So, as we predicted, the $b_n$ coefficients are zero for all $n$. So now we go back to the original Fourier series equation and plug in our results to get

$$f(t) = a_O + \sum_{n=1}^{\infty} a_n \cos(n\omega_O t) + \sum_{n=1}^{\infty} b_n \sin(n\omega_O t)$$

$$f(t) = \left( \frac{A\tau}{T} \right) + 2A \frac{\tau}{T} \sum_{n=1}^{\infty} \left( \frac{\sin(n\pi(\tau/T))}{n\pi(\tau/T)} \right) \cos(n\omega_O t)$$

**Equation 3.10**

We interpret Equation 3.10 to be a waveform composed of the following:

1. A DC or average term of amplitude $\dfrac{A\tau}{T}$

2. Amplitude terms that take on a discrete $\sin x/x$ envelope as $n$ is incremented from $1 \to \infty$

$$2A\frac{\tau}{T}\left[\frac{\sin\left(n\pi\left(\tau/T\right)\right)}{n\pi\left(\tau/T\right)}\right]$$

3. A frequency term $\cos\left(n\omega_o t\right)$ that, as $n$ is incremented, occurs at multiples of the fundamental frequency $\omega_O$

The on and off times, or duty cycle, of the pulse train is controlled by the ratio of $\tau/T$. For example, the Fourier series of a square wave would be the result of setting $\tau = T/2$.

Equation 3.10 looks a little formidable in its compact form, so let's expand the series to the point where we get the DC term, the fundamental, and the first two harmonics:

$$f(t) = \left(\frac{A\tau}{T}\right)$$

$$+2A\frac{\tau}{T}\left[\frac{\sin\left(\pi\left(\tau/T\right)\right)}{\pi\left(\tau/T\right)}\cos\left(\omega_o t\right) + \frac{\sin\left(2\pi\left(\tau/T\right)\right)}{2\pi\left(\tau/T\right)}\cos\left(2\omega_o t\right) + \frac{\sin\left(3\pi\left(\tau/T\right)\right)}{3\pi\left(\tau/T\right)}\cos\left(3\omega_o t\right) + \cdots\right]$$

**Equation 3.11**

The theoretically infinite Fourier series in Equation 3.11 is the expanded series or general expression for a pulsed waveform with duty cycle $\tau/T$ and even symmetry.

The resultant Fourier series for the square wave in Figure 3.1 is clear and compact. Each term in the series represents the amplitude of a particular discrete frequency that makes up the composite signal. We note that the square wave is composed of a DC term, a fundamental frequency, and an infinite number of harmonics. If we add all the infinite number of terms we will get a square wave identical to the one illustrated in Figure 3.1.

We can use Equation 3.10 or Equation 3.11 to produce the Fourier series for specific waveforms, as we will see in the following examples.

**Example 2**. Use Equation 3.11 to compute the Fourier series of a pulsed waveform of amplitude $A = 1$, with a duty cycle of 10%, where $\tau = T/10$. We replace $A$ and $\tau/T$ in the equation with their specified values and we end up with

$$f(t) = \left(\frac{1}{10}\right) + \frac{1}{5}\left[\frac{\sin(\pi/10)}{(\pi/10)}\cos(\omega_o t) + \frac{\sin(2\pi/10)}{2\pi/10}\cos(2\omega_o t) + \frac{\sin(3\pi/10)}{3\pi/10}\cos(3\omega_o t) + \cdots\right]$$

$$= \left(\frac{1}{10}\right) + \frac{1}{5}\left[\frac{10}{\pi}\sin(\pi/10)\cos(\omega_o t) + \frac{10}{2\pi}\sin(2\pi/10)\cos(2\omega_o t) + \frac{10}{3\pi}\sin(3\pi/10)\cos(3\omega_o t) + \cdots\right]$$

$$= \left(\frac{1}{10}\right) + \frac{10}{5\pi}\left[\sin(\pi/10)\frac{\cos(\omega_o t)}{1} + \sin(2\pi/10)\frac{\cos(2\omega_o t)}{2} + \sin(3\pi/10)\frac{\cos(3\omega_o t)}{3} + \cdots\right]$$

$$= 0.1 + 0.1967\cos(\omega_o t) + 0.1871\cos(2\omega_o t) + 0.1717\cos(3\omega_o t) + 0.1514\cos(4\omega_o t) + \cdots$$

The amplitude of the DC term and the first 47 harmonics of the Fourier series are computed and listed in Table 3.1. The envelope of the spectral lines produced by the Fourier series takes on a $\sin x/x$ shape with zero crossings at multiples of $n = 10$. This is because the $\sin x/x$ term is zero whenever its

**Table 3.1**   DC Term and First 47 Harmonics of a 10% Duty Cycle Pulsed Waveform

| $n$ | Amplitude | $n$ | Amplitude | $n$ | Amplitude |
|---|---|---|---|---|---|
| 0 | 0.1000 | 16 | −0.0378 | 32 | −0.0117 |
| 1 | 0.1967 | 17 | −0.0303 | 33 | −0.0156 |
| 2 | 0.1871 | 18 | −0.0208 | 34 | −0.0178 |
| 3 | 0.1717 | 19 | −0.0104 | 35 | −0.0182 |
| 4 | 0.1514 | 20 | 0.0000 | 36 | −0.0168 |
| 5 | 0.1273 | 21 | 0.0094 | 37 | −0.0139 |
| 6 | 0.1009 | 22 | 0.0170 | 38 | −0.0098 |
| 7 | 0.0736 | 23 | 0.0224 | 39 | −0.0050 |
| 8 | 0.0468 | 24 | 0.0252 | 40 | 0.0000 |
| 9 | 0.0219 | 25 | 0.0255 | 41 | 0.0048 |
| 10 | 0.0000 | 26 | 0.0233 | 42 | 0.0089 |
| 11 | −0.0179 | 27 | 0.0191 | 43 | 0.0120 |
| 12 | −0.0312 | 28 | 0.0134 | 44 | 0.0138 |
| 13 | −0.0396 | 29 | 0.0068 | 45 | 0.0141 |
| 14 | −0.0432 | 30 | 0.0000 | 46 | 0.0132 |
| 15 | −0.0424 | 31 | −0.0063 | 47 | 0.0110 |

argument equals multiples of $\pi$, or whenever $n\pi(\tau/T) = k\pi$ for $k = 1, 2, 3, \cdots$. When we rearrange the terms, we get $n = (T/\tau)k = k/\tau = 10k$ for $k = 1, 2, 3, \cdots$.

The plot of Fourier series is illustrated in Figure 3.2. The DC term, the fundamental frequency, and the first 47 harmonics are shown. Note the $\sin x/x$ envelope. Also note that the nulls in the envelope occur at multiples of $1/\tau$. In this example, $\tau = 0.1$, so $1/\tau = 10$ and therefore the envelope nulls occur at multiples of 10. The spacing between the spectral lines is $1/T$ Hz, which makes sense since each spectral line is at a multiple of the fundamental frequency $f_0 = 1/T$.

**Example 3.** Use Equation 3.10 or Equation 3.11 to compute the Fourier series of an even symmetric pulsed waveform of amplitude $A = 1$, with a 50% duty cycle, or $\tau = T/2$.

$$
\begin{aligned}
f(t) &= \left(\frac{A}{2}\right) + 2A\frac{\tau}{T}\left[\frac{\sin(\pi/2)}{(\pi/2)}\cos(\omega_0 t) + \frac{\sin(2\pi)}{2\pi/2}\cos(2\omega_0 t) + \frac{\sin(3\pi/2)}{3\pi/2}\cos(3\omega_0 t) + \cdots\right] \\
&= \left(\frac{A}{2}\right) + 2\frac{A}{2}\left[\frac{2}{\pi}\cos(\omega_0 t) + 0 - \frac{2}{3\pi}\cos(3\omega_0 t) + 0 + \frac{2}{5\pi}\cos(5\omega_0 t) + 0 + \cdots\right] \\
&= \left(\frac{A}{2}\right) + \frac{2A}{\pi}\left[\frac{\cos(\omega_0 t)}{1} - \frac{\cos(3\omega_0 t)}{3} + \frac{\cos(5\omega_0 t)}{5} + \cdots\right] \\
&= 0.5 + 0.6366\cos(\omega_0 t) - 0.2122\cos(3\omega_0 t) + 0.1273\cos(5\omega_0 t) + \cdots
\end{aligned}
$$

The amplitudes for the DC term and the first 47 harmonics are computed and listed in Table 3.2.

The spectrum for the 50% duty cycle even symmetric pulsed waveform is illustrated in Figure 3.3. Note the $\sin x/x$ envelope still exists, but the nulls are much closer together. The nulls in the envelope occur at intervals of $1/\tau$. In this example, the period $T = 1$, so $\tau = T/2 = 1/2$ and $n/\tau = 2, 4, 6, \cdots$.

It is interesting to note that as $\tau \to 0$, the waveform duty cycle approaches 0%, and every term including the DC term in the series approaches 0. This makes sense because in this case the waveform would approach zero as well. As $\tau \to T$, the waveform duty cycle approaches 100%, the fundamental and harmonic terms approach 0, and the DC term approaches the amplitude $A$. This also makes sense since the pulsed waveform would be morphed into a straight line as $\tau \to T$ and would be equal to the DC level of amplitude $A$.

**Example 4.** Compute the Fourier series for the periodic triangular waveform illustrated in Figure 3.4. We choose to compute the Fourier series over the period that is centered about the $y$-axis. The waveform over this period

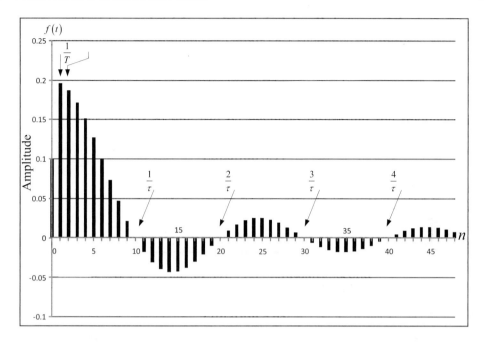

**Figure 3.2**   Fourier series of a 10% duty cycle pulsed waveform

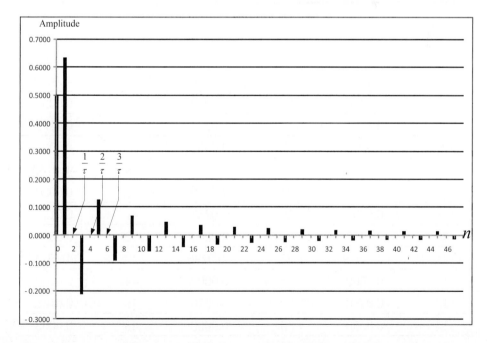

**Figure 3.3**   Fourier series for 50% duty cycle pulsed waveform

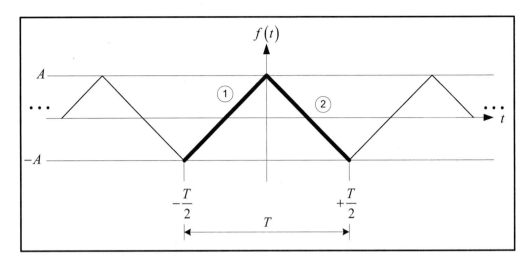

**Figure 3.4** Even symmetric triangular waveform

**Table 3.2** DC Term and First 47 Harmonics of a 50% Duty Cycle Pulsed Waveform

| $n$ | Amplitude | $n$ | Amplitude | $n$ | Amplitude |
|---|---|---|---|---|---|
| 0  | 0.5000  | 16 | 0.0000  | 32 | 0.0000  |
| 1  | 0.6366  | 17 | 0.0374  | 33 | 0.0193  |
| 2  | 0.0000  | 18 | 0.0000  | 34 | 0.0000  |
| 3  | −0.2122 | 19 | −0.0335 | 35 | −0.0182 |
| 4  | 0.0000  | 20 | 0.0000  | 36 | 0.0000  |
| 5  | 0.1273  | 21 | 0.0303  | 37 | 0.0172  |
| 6  | 0.0000  | 22 | 0.0000  | 38 | 0.0000  |
| 7  | −0.0909 | 23 | −0.0277 | 39 | −0.0163 |
| 8  | 0.0000  | 24 | 0.0000  | 40 | 0.0000  |
| 9  | 0.0707  | 25 | 0.0255  | 41 | 0.0155  |
| 10 | 0.0000  | 26 | 0.0000  | 42 | 0.0000  |
| 11 | −0.0579 | 27 | −0.0236 | 43 | −0.0148 |
| 12 | 0.0000  | 28 | 0.0000  | 44 | 0.0000  |
| 13 | 0.0490  | 29 | 0.0220  | 45 | 0.0141  |
| 14 | 0.0000  | 30 | 0.0000  | 46 | 0.0000  |
| 15 | −0.0424 | 31 | −0.0205 | 47 | −0.0135 |

is highlighted for illustrative purposes. We can immediately see that the waveform is even symmetric since $f(t) = f(-t)$. This tells us that all the $b_n$ coefficients will be equal to zero and therefore need not be computed. So by inspection we have eliminated approximately a third of the computations. We also note that the waveform is symmetric about the x-axis, so the DC or $a_0$ term will also be equal to zero. This will eliminate another third of the computations. For demonstration purposes, we will go ahead and compute the $a_0$ term and verify that it is indeed zero. Verification that the $b_n$ coefficients are all zero is left to the reader.

You may note from the figure that we have divided the waveform into two line segments labeled 1 and 2. We will break the problem up into two integrals—one for each segment. We derive expressions for each of the two line segments with the help of the straight line equation $y = mt + b$:

Segment 1:   $m = \dfrac{2A}{T/2} = \dfrac{4A}{T}, \quad b = A, \quad \text{and } y = \dfrac{4A}{T}t + A \quad \left\{ \text{for } -\dfrac{T}{2} \leq t \leq 0 \right\}$

Segment 2:   $m = \dfrac{-2A}{T/2} = \dfrac{-4A}{T}, \quad b = A, \quad \text{and } y = -\dfrac{4A}{T}t + A \quad \left\{ \text{for } 0 \leq t \leq +\dfrac{T}{2} \right\}$

Now let's compute the Fourier series $a_0$ term:

$$a_O = \frac{1}{T} \int_{-T/2}^{T/2} f(t)\,dt = \frac{1}{T} \int_{-T/2}^{0} \left( \frac{4A}{T}t + A \right) dt + \frac{1}{T} \int_{0}^{+T/2} \left( -\frac{4A}{T}t + A \right) dt$$

The computations for $a_0$ yield

$$a_0 = \frac{1}{T} \int_{-T/2}^{0} \left( \frac{4A}{T}t + A \right) dt + \frac{1}{T} \int_{0}^{+T/2} \left( \frac{-4A}{T}t + A \right) dt$$

$$= \frac{1}{T} \left[ \frac{2A}{T}t^2 + At \right]_{-T/2}^{0} + \frac{1}{T} \left[ \frac{-2A}{T}t^2 + At \right]_{0}^{+T/2}$$

$$= \left[ \frac{A}{2} - \frac{A}{2} \right] + \left[ -\frac{A}{2} + \frac{A}{2} \right]$$

$$= 0$$

So now we know that the DC or average term is zero and that all the $b_n$ coefficients are zero. The only thing left to do is to compute the $a_n$ coefficients. We begin by writing the expression used to compute the $a_n$ coefficients:

$$a_n = \frac{2}{T} \int_{-T/2}^{T/2} f(t) \cos(n\omega_0 t)\,dt \quad \text{for } n = 1, 2, 3, \cdots$$

We split the computation into two integrals, one for each curve segment, to get

$$a_n = \frac{2}{T}\int_{-T/2}^{0}\left(\frac{4A}{T}t + A\right)\cos\left(n\omega_o t\right)dt \; + \; \frac{2}{T}\int_{0}^{+T/2}\left(-\frac{4A}{T}t + A\right)\cos\left(n\omega_o t\right)dt$$

We can break each of the two integrals into two smaller integrals to get

$$\begin{aligned}
a_n &= \frac{8A}{T^2}\int_{-T/2}^{0} t\cos\left(n\omega_o t\right)dt + \frac{2A}{T}\int_{-T/2}^{0}\cos\left(n\omega_o t\right)dt \\
&\quad - \frac{8A}{T^2}\int_{0}^{+T/2} t\cos\left(n\omega_o t\right)dt + \frac{2A}{T}\int_{0}^{+T/2}\cos\left(n\omega_o t\right)dt
\end{aligned}$$

Now we can evaluate each of the four integrals separately. We can easily see that the two integrals on the right evaluate to zero, as follows. The complete evaluation of these integrals is left as an exercise for the reader.

$$\frac{2A}{T}\int_{-T/2}^{0}\cos\left(n\omega_o t\right)dt = +A\frac{\sin\left(n\pi\right)}{n\pi} = 0 \quad \text{for n=1,2,3,}\cdots$$

$$\frac{2A}{T}\int_{0}^{+T/2}\cos\left(n\omega_o t\right)dt = -A\frac{\sin\left(n\pi\right)}{n\pi} = 0 \quad \text{for n=1,2,3,}\cdots$$

This leaves us with two integrals to evaluate for $a_n$:

$$a_n = \frac{8A}{T^2}\int_{-T/2}^{0} t\cos\left(n\omega_o t\right)dt - \frac{8A}{T^2}\int_{0}^{+T/2} t\cos\left(n\omega_o t\right)dt$$

Both of these integrals contain the semi-nasty terms $t\cos\left(n\omega_o t\right)$ and must be integrated by parts. Doing so gives us

$$a_n = \frac{8A}{T^2}\left[\frac{\cos n\omega_o t}{\left(n\omega_o\right)^2} + \frac{t}{n\omega_o}\sin n\omega_o t\right]_{t=-\frac{T}{2}}^{t=0} - \frac{8A}{T^2}\left[\frac{\cos n\omega_o t}{\left(n\omega_o\right)^2} + \frac{t}{n\omega_o}\sin n\omega_o t\right]_{t=0}^{t=\frac{T}{2}}$$

We replace $\omega_0$ by its value of $\omega_0 = 2\pi/T$ and evaluate this expression at the limits to get

$$a_n = \frac{2A}{\pi^2 n^2}\left[1 - \cos\left(n\pi\right)\right] - \frac{2A}{\pi^2 n^2}\left[\cos\left(n\pi\right) - 1\right]$$

Rearranging and combining terms gives us

$$a_n = \frac{4A}{\pi^2 n^2}\left[1 - \cos(n\pi)\right]$$

**Equation 3.12**

If we take a look at Equation 3.12 as the index $n \to 1, 2, 3, \cdots$, we see that

$$a_n = \frac{4A}{\pi^2 n^2}\left[1 - \cos(n\pi)\right] \quad \left\{ \begin{array}{l} = 0 \ \text{for n even and n} \neq 0 \\[2ex] = \dfrac{8A}{\pi^2 n^2} \ \text{for n odd} \end{array} \right\}$$

Now that we know the values for $a_0, a_n,$ and $b_n$, we can substitute back into the Fourier expansion in Equation 3.1, which is repeated here for clarity:

$$f(t) = a_O + \sum_{n=1}^{\infty} a_n \cos(n\omega_0 t) + \sum_{n=1}^{\infty} b_n \sin(n\omega_0 t)$$

We then get the final result:

$$f(t) = \sum_{n=1}^{\infty} \frac{8A}{\pi^2 n^2} \cos(n\omega_0 t) \ \text{ for } n \text{ odd}$$

We expand this series out to the first five terms to get

$$f(t) = \frac{8A}{\pi^2}\left[\cos\omega_0 t + \frac{\cos 3\omega_0 t}{3^2} + \frac{\cos 5\omega_0 t}{5^2} + \frac{\cos 7\omega_0 t}{7^2} + \frac{\cos 9\omega_0 t}{9^2} + \cdots\right]$$

The computed values of the first 51 terms are listed in Table 3.3. We see that the triangular waveform is composed of an infinite sum of all the odd harmonic frequencies. We also see that the amplitude of each term diminishes as the square of the index term $n$. In this example, the sum of the first five harmonic terms is probably of sufficient accuracy to represent the triangle waveform. The plot of the Fourier series for this example is illustrated in Figure 3.5, which shows the relative significance of successive terms in the series. The total contribution of the sixth term and those that follow are minimal.

## 3.1.2   Compact Fourier Series Representation

The trigonometric form of the Fourier series in Equation 3.1 can be expressed in a compact form that is more suitable for the cases where neither the $a_n$ nor the $b_n$ coefficients are equal to zero. The sum of two sinusoids of the

**Table 3.3**  DC Term and First 47 Harmonics of a Triangle Waveform

| $n$ | Amplitude | $n$ | Amplitude | $n$ | Amplitude |
|---|---|---|---|---|---|
| 0 | 0 | 31 | 0.0008435 | 63 | 0.000204 |
| 1 | 0.8105695 | 33 | 0.0007443 | 65 | 0.000192 |
| 3 | 0.0900633 | 35 | 0.0006617 | 67 | 0.000181 |
| 5 | 0.0324228 | 37 | 0.0005921 | 69 | 0.000170 |
| 7 | 0.0165422 | 39 | 0.0005329 | 71 | 0.000161 |
| 9 | 0.010007 | 41 | 0.0004822 | 73 | 0.000152 |
| 11 | 0.0066989 | 43 | 0.0004384 | 75 | 0.000144 |
| 13 | 0.0047963 | 45 | 0.0004003 | 77 | 0.000137 |
| 15 | 0.0036025 | 47 | 0.0003669 | 79 | 0.000130 |
| 17 | 0.0028047 | 49 | 0.0003376 | 81 | 0.000124 |
| 19 | 0.0022453 | 51 | 0.0003116 | 83 | 0.000118 |
| 21 | 0.001838 | 53 | 0.0002886 | 85 | 0.000112 |
| 23 | 0.0015323 | 55 | 0.000268 | 87 | 0.000107 |
| 25 | 0.0012969 | 57 | 0.0002495 | 89 | 0.000102 |
| 27 | 0.0011119 | 59 | 0.0002329 | 91 | 0.000000 |
| 29 | 0.0009638 | 61 | 0.0002178 | 93 | 0.000000 |
| 30 | 0.0009006 | 62 | 0.0002109 | 94 | 0.000000 |

same frequency is a single waveform of the same frequency. The summed waveform, however, will have a new magnitude and phase due to the contributions of the two original sinusoids. We can express the sum of any two Fourier series terms of the same index $n$ as

$$c_n \cos(n\omega_0 t + \phi_n) = a_n \cos n\omega_0 t + b_n \sin n\omega_0 t$$

Similarly, we can separately combine all the sinusoids in the trigonometric form of the Fourier series with coefficients $a_n$ and $b_n$ and express them as a cosine wave at the same frequency with a new magnitude and with some phase offset. This compact form of the Fourier series is given by

$$f(t) = c_0 + \sum_{n=1}^{n=\infty} c_n \cos(n\omega_0 t + \phi_n)$$

**Equation 3.13**

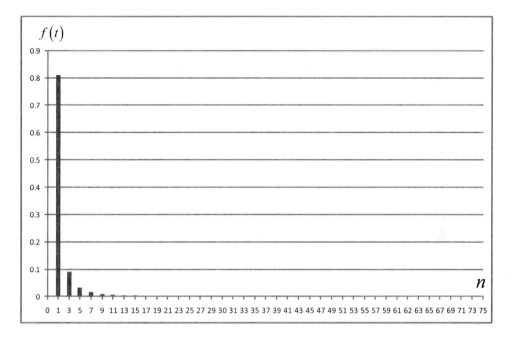

**Figure 3.5** Fourier series even symmetric triangular wave

where

$$c_0 = a_0$$

$$c_n = \sqrt{a_n^2 + b_n^2}$$

$$\phi_n = \tan^{-1}\left(-b_n/a_n\right)$$

### 3.1.3   Parseval's Theorem for the Fourier Series

Suppose we wish to measure the average power in some arbitrary periodic waveform. We could do so by integrating the square of the magnitude of the function over a single period and then normalizing the result to the length of the period. This is expressed as

$$P_{avg} = \frac{1}{T}\int_{t=0}^{t=T}\left|f\left(t\right)\right|^2\,dt \quad \left\{\text{for complex } f\left(t\right)\right\}$$

**Equation 3.14**

This expression assumes that $f\left(t\right)$ is a complex function and therefore $\left|f\left(t\right)\right|^2 = f\left(t\right)f^*\left(t\right)$. If the function $f\left(t\right)$ is a real function, then the absolute value operator can be dropped and we end up with

$$P_{avg} = \frac{1}{T} \int_{t=0}^{t=T} f(t)^2 \, dt \quad \{\text{for real } f(t)\}$$

The average power of a periodic function can also be obtained directly from the Fourier series coefficients $c_n$. Thus

$$P_{avg} = \sum_{n=-\infty}^{\infty} |c_n|^2 \quad \{\text{for a two sided series}\}$$

**Equation 3.15**

where

$$c_0^2 = a_0^2$$

$$|c_n|^2 = \left(a_n^2 + b_n^2\right)$$

Parseval's theorem states that the average power measured for a periodic time domain waveform is equal to the sum of the powers contributed by each term in the Fourier series for that waveform. Therefore we can equate Equation 3.14 and Equation 3.15 to get the equation that is known as Parseval's theorem.

$$P_{avg} = \frac{1}{T} \int_{t=0}^{t=T} |f(t)|^2 \, dt = \sum_{n=-\infty}^{\infty} |c_n|^2 \quad \{\text{for a two sided series}\}$$

**Equation 3.16**

This makes sense. The power of a periodic signal should be the same no matter if it is calculated in terms of the waveform as a whole or in terms of the sum of its components.

The Fourier series in Equation 3.16 has symmetric positive and negative harmonics that range from $-\infty < n < +\infty$ and is known as a two-sided series. A Fourier series that ranges from $1 \le n < +\infty$ has only positive harmonics and is known as a single-sided series. The amplitude of the single-sided coefficients is twice that of the double-sided series. Therefore, when computing an average power using a single-sided Fourier series, we use the expression

$$P_{avg} = \frac{1}{T} \int_{t=0}^{t=T} |f(t)|^2 \, dt = c_0^2 + \frac{1}{2} \sum_{n=1}^{\infty} |c_n|^2 \quad \{\text{for a single sided series}\}$$

**Equation 3.17**

**Example 1.** Compute the average power of the waveform $f(t) = D + A\cos(2\pi f_0 t)$. We note that the signal contains a DC component $D$ and a sinusoidal component of amplitude $A$. First, we will compute the power using the signal as a whole in the time domain, and then we will compute the power using the Fourier series coefficients. Computing the average power in the time domain gives us

$$
\begin{aligned}
P_{avg} &= \frac{1}{T}\int_0^T |f(t)|^2\, dt \\
&= \frac{1}{T}\int_0^T \left[D + A\cos(2\pi f_0 t)\right]^2 dt \\
&= \frac{1}{T}\int_0^T \left[D^2 + 2DA\cos(2\pi f_0 t) + A^2\cos^2(2\pi f_0 t)\right] dt \\
&= \frac{1}{T}\left[D^2 T + \frac{2DA}{2\pi f_0}\sin(2\pi f_0 T) + \frac{A^2 T}{2} + \frac{1}{4\pi f_0}\sin(4\pi f_0 T)\right] \\
&= D^2 + 0 + \frac{A^2}{2} + 0 \quad \{\text{The sin terms equal 0 because } T = 1/f_0\} \\
&= D^2 + \frac{A^2}{2}
\end{aligned}
$$

Now let's compute the Fourier series representation of the periodic waveform $f(t)$. We begin by writing down the trigonometric form of the Fourier series

$$
f(t) = a_O + \sum_{n=1}^{\infty} a_n \cos(n\omega_O t) + \sum_{n=1}^{\infty} b_n \sin(n\omega_O t)
$$

We can reduce our computations substantially by noting that the function $f(t)$ is even symmetric. The $b_n$ coefficients of an even symmetric function are all zero, so our Fourier series equation becomes

$$
f(t) = a_O + \sum_{n=1}^{\infty} a_n \cos(n\omega_O t)
$$

Remember that the coefficients are computed by using the following relations:

$$
a_0 = \frac{1}{T}\int_{-T/2}^{T/2} f(t)\, dt
$$

$$
a_n = \frac{2}{T}\int_{-T/2}^{T/2} f(t)\cos(n\omega_O t)\, dt \quad \text{for } n = 1, 2, 3, \cdots
$$

$$
b_n = \frac{2}{T}\int_{-T/2}^{T/2} f(t)\sin(n\omega_O t)\, dt \quad \text{for } n = 1, 2, 3, \cdots
$$

So let's compute the coefficient $a_0$:

$$a_O = \frac{1}{T}\int_0^T f(t)\,dt$$

$$= \frac{1}{T}\int_0^T \left[D + A\cos(2\pi f_0 t)\right]dt$$

$$= \frac{1}{T}\left[Dt + \frac{A}{2\pi f_0}\sin(2\pi f_0 t)\right]_0^T$$

$$= \frac{1}{T}\left[DT + \frac{A}{2\pi f_0}\sin(2\pi f_0 T)\right]$$

$$= D + 0 \quad \{\text{Because } T = 1/f_0\}$$

$$= D$$

This makes sense because both the $a_0$ and $D$ terms are DC terms. Now let's compute the coefficients $a_n$:

$$a_n = \frac{2}{T}\int_0^T f(t)\cos(2\pi n f_0 t)\,dt \quad \text{for } n = 1, 2, 3, \cdots$$

$$= \frac{2}{T}\int_0^T \left[D + A\cos(2\pi f_0 t)\right]\cos(2\pi n f_0 t)\,dt$$

$$= \frac{2}{T}\int_0^T D\cos(2\pi n f_0 t)\,dt + \frac{2}{T}\int_0^T A\cos(2\pi f_0 t)\cos(2\pi n f_0 t)\,dt$$

Now remember from Equation 3.4 that

$$\int_0^T \cos(n\omega_0 t)\cos(m\omega_0 t)\,dt = 0 \quad \text{for } m \neq n$$

This means that the second integral in the computation of $a_n$ is nonzero only for the case where $m = n$. In this example, $m = n$ only if $n = 1$, so we can reduce the second integral to get

$$a_1 = \frac{2}{T}\int_0^T D\cos(2\pi n f_0 t)\,dt + \frac{2}{T}\int_0^T A\cos^2(2\pi f_0 t)\,dt$$

$$= \frac{2}{T}\int_0^T D\cos(2\pi n f_0 t)\,dt + \frac{2}{T}\int_0^T \frac{A}{2}\left[1 + \cos(4\pi f_0 t)\right]dt$$

$$= \frac{2D}{T}\left[\frac{\sin(2\pi n f_0 t)}{(2\pi n f_0)}\right]_0^T + \frac{A}{T}\left[t + \frac{\sin(4\pi n f_0 t)}{(4\pi n f_0)}\right]_0^T$$

$$= A$$

Since the coefficients $b_n = 0$ the resultant Fourier coefficients are computed to be

$$a_0 = D$$
$$a_1 = A$$

We recall that the general form of the Fourier coefficients are given by

$$c_0^2 = a_0^2$$
$$c_n^2 = \left( a_n^2 + b_n^2 \right)$$

The computed one-sided Fourier series for this example is illustrated in part A of Figure 3.6. The corresponding two-sided Fourier series for this example is illustrated in part B of the same diagram.

We can now use the one-sided Parseval's equation to compute the average power in $f(t)$:

$$P_{avg} = c_0^2 + \frac{1}{2}\sum_{n=1}^{\infty}\left|c_n\right|^2 = c_0^2 + \frac{1}{2}\sum_{n=1}^{1}\left|c_n\right|^2 = \left( c_0^2 + \frac{1}{2}c_1^2 \right) = D^2 + \frac{A^2}{2}$$

If we had chosen to represent $f(t)$ as a two-sided Fourier series, we could then use the two-sided version of Parseval's equation to compute the average power in $f(t)$:

$$P_{avg} = \sum_{n=-\infty}^{+\infty}\left|c_n\right|^2 = \sum_{n=-1}^{+1}\left|c_n\right|^2 = \left( c_{-1}^2 + c_0^2 + c_1^2 \right) = \left( \frac{A^2}{4} + D^2 + \frac{A^2}{4} \right) = D^2 + \frac{A^2}{2}$$

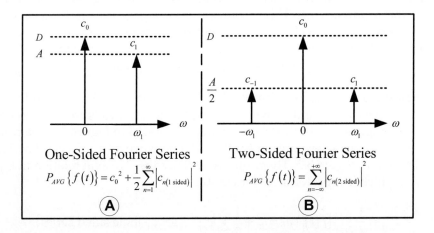

One-Sided Fourier Series
$$P_{AVG}\left\{ f(t) \right\} = c_0^2 + \frac{1}{2}\sum_{n=1}^{\infty}\left|c_{n(1\,sided)}\right|^2$$

**A**

Two-Sided Fourier Series
$$P_{AVG}\left\{ f(t) \right\} = \sum_{n=-\infty}^{+\infty}\left|c_{n(2\,sided)}\right|^2$$

**B**

**Figure 3.6**   One-sided and two-sided Fourier series diagram for example 1

**Example 2**. If we compute the average power in the periodic pulsed waveform illustrated in Figure 3.1, where the pulse amplitude $A = 1$ and the duty cycle $\tau/T = 0.1$, we get

$$P_{avg} = \frac{1}{T}\int_{t=0}^{t=T}\left|f(t)\right|^2\,dt \;=\; \frac{1}{T}\int_{t=-\frac{\tau}{2}}^{t=+\frac{\tau}{2}}A^2\,dt \;=\; A^2\frac{\tau}{T} \;=\; (1)^2\frac{1}{10} \;=\; 0.1$$

Now if we compute the average power by summing the single-sided Fourier series components of the square wave illustrated in Figure 3.2, we get

$$P_{avg} = c_0^2 + \frac{1}{2}\sum_{n=1}^{\infty}\left|c_n\right|^2$$

Obviously we cannot add an infinite number of terms, but we can add the most significant terms and get a very close approximation to the exact average power. For example, if we sum the DC term and the first 200 harmonics, we get

$$P_{avg} = 0.094493$$

The difference between the infinitely precise calculation of the time domain average power and the average power computed from the sum of the first 200 Fourier coefficients is

$$0.100000 - 0.094493 = 0.005507$$

Using only the first 200 Fourier series harmonics, we ended up with a result that was 94.5% of the calculated value. The remaining $200 \to \infty$ terms make up the residual 5.5%. This is probably not a good example to demonstrate power computations because the number of terms used is high, and we only reached 94.5% of the theoretical value. It does, however, point out that not all textbook examples or industry applications are ideal.

**Example 3**. Here is an example that is a bit more textbook friendly because it produces a nice and convenient result. Use Parseval's theorem to compute the average power of the triangle wave illustrated in Figure 3.4. We computed the Fourier series expansion for the triangle wave as a one-sided series, as illustrated in Figure 3.5. Therefore we will use the Parseval relation in Equation 3.17, which is repeated here for clarity:

$$P_{avg} = \frac{1}{T}\int_{t=0}^{t=T}\left|f(t)\right|^2\,dt = c_0^2 + \frac{1}{2}\sum_{n=1}^{\infty}\left|c_n\right|^2 \quad \left\{\text{for a single sided series}\right\}$$

Recall that the triangle waveform was broken into two line segments and that each line segment was integrated separately:

Segment 1:  $y = \dfrac{4A}{T}t + A \qquad \left\{ \text{for } -\dfrac{T}{2} \le t \le 0 \right\}$

Segment 2:  $y = -\dfrac{4A}{T}t + A \qquad \left\{ \text{for } 0 \le t \le +\dfrac{T}{2} \right\}$

We compute the average power of the time domain waveform first:

$$P_{avg} = \frac{1}{T}\int_{t=0}^{t=T}|f(t)|^2\,dt = \frac{1}{T}\int_{-\frac{T}{2}}^{0}\left(\frac{4A}{T}t + A\right)^2 dt \qquad\qquad +\frac{1}{T}\int_{0}^{+\frac{T}{2}}\left(-\frac{4A}{T}t + A\right)^2 dt$$

$$= \frac{1}{T}\left[\frac{16A^2}{T^2}\left(\frac{t^3}{3}\right) + \frac{8A^2}{T}\left(\frac{t^2}{2}\right) + A^2 t\right]_{-\frac{T}{2}}^{0} \quad +\frac{1}{T}\left[\frac{16A^2}{T^2}\left(\frac{t^3}{3}\right) - \frac{8A^2}{T}\left(\frac{t^2}{2}\right) + A^2 t\right]_{0}^{+\frac{T}{2}}$$

$$= \frac{4A^2}{3} - A^2 \quad = \frac{A^2}{3}$$

The amplitude of the triangle wave in the example was $A = 1$. Upon substitution we get

$$P_{avg} = \frac{A^2}{3} = \frac{1}{3} = 0.3333$$

Now we compute the average power using the Fourier series. In this case, we compute the average power using the one-sided version of Parseval's equation:

$$P_{avg} = c_0^2 + \frac{1}{2}\sum_{n=1}^{\infty}|c_n|^2$$

In this example we see that the spectral components of the Fourier series fall off very quickly, so most of the power is contained in the first few terms. As we see here, summing just the first five terms of Table 3.3 gives us an answer that agrees with the time domain calculations to four decimal places:

$$P_{avg} = c_0^2 + \frac{1}{2}\sum_{n=1}^{4}|c_n|^2 = (0)^2 + \frac{1}{2}\left[(0.810569)^2 + (0.090063)^2 + (0.032423)^2 + (0.016542)^2 + (0.010007)^2\right]$$

Performing these calculations gives us

$$P_{avg} = (0.328511 + 0.004056 + 0.000526 + 0.000137 + 0.0000501) = 0.3333$$

Adding additional terms in the calculation will produce results that are accurate to more decimal places. This is a good example of engineering approximation. For this example, we can write

$$P_{avg} = \frac{1}{T} \int_{t=0}^{t=T} |f(t)|^2 \, dt = c_0^2 + \frac{1}{2} \sum_{n=1}^{9} |c_n|^2 = 0.3333 \quad \{\text{for } n \text{ odd}, 1,3,5,7,\text{and } 9\}$$

Why is the upper limit on the summation equal to $9$ instead of $5$? Remember that the even terms in the Fourier series expansion were all zero, and therefore the first five nonzero terms are $n = 1,3,5,7,9$.

You may also ask the question, "What is the value of the Fourier series to me? How can I apply this theory in industry?" In many cases, a theory is taught in the classroom and we learn to apply this theory in industry. The application of the Fourier series might be one of those cases, but it does help to understand how the theory of the Fourier series is applicable to problems faced in everyday engineering.

The Fourier series allows us to determine the magnitude of the DC component, the fundamental component, and the harmonic components present in a given waveform. This capability is important in all fields of engineering. For example, if you happen to be designing a suspension bridge, wouldn't you like to be able to study, predict, and simulate all the harmonics of the motion waveform the bridge will experience due to some external forcing functions like wind, earthquakes, automobile traffic, or even marching bands? The mathematics of the Fourier series gives us a tool to do that.

How about the aerospace engineer who is tasked with analyzing the stress on the wings of a prototype aircraft due to vibration in response to a steady-state forcing function like continuous airflow or from impulse forces that occur in bumpy weather? This is an application that is well suited to Fourier series analysis.

As another example, the bandwidth of a system may alter the shape of some arbitrary waveform as it passes from input to output. Given the input waveform, how does the engineer predict what shape the output waveform will be? Well, if the engineer knows the system bandwidth characteristics and can expand the input waveform into its Fourier series, then he or she can accurately predict the shape of the output waveform simply by summing all the Fourier series components that fall within the bandwidth of the system and by ignoring all the series components that fall outside the pass band of the system.

Many times, engineers make calculations that are not exact but approximate. When dealing with real-world signals, systems, or data, many times it is impossible to compute an exact value for anything. In electrical engineering, for example, the additive noise picked up by an electrical signal as it passes through a transmission channel prevents an engineer from ever computing the exact response of some receiving circuit. He can calculate a response to be within some small statistical error bound though, and that is what is commonly done. Fortunately, in the great majority of cases, the error is usually negligible.

Engineers strive to make calculations that are accurate so that they do not exceed some predefined error bound. Many times an engineer will calculate system or statistical variables by considering only the first four or five significant terms of a Fourier series expansion. The terms in the series expansion of the triangle wave in the last example fall off in significance pretty rapidly. Using only the first few terms of that expansion will produce a very good estimate of the ideal value.

As another example, suppose an engineer wished to compute the power of a signal that had some arbitrary and complex waveform. He or she could expand the waveform into its Fourier series, as we did with the square wave and triangle wave. Now the engineer can use as many expansion terms in the series as necessary to compute an estimate of the signal power to within some error bound. Obviously, the more terms used, the greater the accuracy. It's a good bet that the engineer will not use the infinite number of terms that are available. He or she will use enough terms to provide a very close approximation to the ideal solution. In industry, where very few things are ideal, this is a valid approach.

## 3.2 A BRIEF REVIEW OF THE FOURIER TRANSFORM

We saw in the previous section that the Fourier series is a mathematical tool used to analyze periodic waveforms over a single period $T$. The Fourier series breaks the periodic time domain waveform into its component frequencies, consisting of the DC term, the fundamental, and all harmonics.

What happens when the period $T$ of a periodic waveform increases? We have seen that the frequency spacing between harmonics of the Fourier series line spectra was $1/T$. It can be easily seen that when the waveform period $T$ increases, the spacing between the line spectra decreases. In the limit as $T \to \infty$, the entire time domain waveform approaches that of a single period, and the spectral line spacing becomes infinitesimal resulting in a continuous frequency spectrum.

Suppose we had a periodic square wave of period $T$ and duty cycle $\tau/T$. Further suppose that we increase the period, so that $T \to \infty$. The end result is we would be left with an infinite length waveform consisting of single pulse of width $\tau$. The periodic waveform could then be viewed as a nonperiodic waveform. We compute the continuous spectrum of a nonperiodic waveform through the use of the Fourier transform.

A function $f(t)$ of the variable $t$ in the time domain and a function $F(f)$ of the variable $f$ in the frequency domain are said to be a Fourier transform pair if one function can be mathematically converted to the other function via the Fourier transform. For our purposes, we will utilize the symbol $\overset{f}{\leftrightarrow}$ to denote the relationship between a Fourier transform pair, such as

$$f(t) \overset{f}{\leftrightarrow} F(f)$$

The notation $F(f) = f\{f(t)\}$ is often used to indicate that $F(f)$ is the forward Fourier transform of $f(t)$, and the notation $f(t) = f^{-1}\{F(f)\}$ is used to indicate that $f(t)$ is the inverse Fourier transform of $F(f)$.

The definition of both the forward and inverse Fourier transform is fairly straightforward. The forward Fourier transform computes the frequency domain representation $F(f)$ of a time domain function $f(t)$. This computation is achieved via the expression

$$F(f) = \int_{-\infty}^{\infty} f(t)\, e^{-j2\pi ft}\, dt$$

**Equation 3.18**

The inverse Fourier transform computes the time domain representation $f(t)$ from its frequency domain representation $F(f)$ via the expression

$$f(t) = \int_{-\infty}^{\infty} F(f)\, e^{+j2\pi ft}\, df$$

**Equation 3.19**

If both $f(t)$ and $F(f)$ can be defined by Equation 3.18 and Equation 3.19, then we can state that $f(t)$ and $F(f)$ are a Fourier transform pair. We indicate this by the expression

$$f(t) \overset{f}{\leftrightarrow} F(f)$$

Many times it is convenient to modify the frequency variable $f$ and work with radian frequency $\omega = 2\pi f$. In this case, the forward Fourier transform is expressed as

$$F(\omega) = \int_{-\infty}^{+\infty} f(t)e^{-j\omega t}dt$$

**Equation 3.20**

We derive the inverse Fourier transform by making a simple substitution into Equation 3.19 by noting that $\omega = 2\pi f$ and therefore $d\omega = 2\pi\,df$ and $df = \dfrac{d\omega}{2\pi}$. Upon substitution, we end up with

$$f(t) = \frac{1}{2\pi}\int_{-\infty}^{+\infty} F(\omega)e^{+j\omega t}d\omega$$

**Equation 3.21**

Once again we can state that if Equation 3.20 and Equation 3.21 are valid then $f(t)$ and $F(\omega)$ are a Fourier transform pair and we can state that

$$f(t) \overset{f}{\leftrightarrow} F(\omega)$$

The reader can choose to work in either the $f$ or the $\omega$ domain—whichever is more suitable for the task at hand. In the text that follows, we will utilize both domain representations. The Fourier transform pairs discussed previously are tabulated and repeated here for clarity:

$$F(f) = \int_{-\infty}^{+\infty} f(t)e^{-j2\pi ft}\qquad \{\text{frequency domain}\}$$
$$f(t) = \int_{-\infty}^{+\infty} F(f)e^{+j2\pi ft}$$

$$F(\omega) = \int_{-\infty}^{+\infty} f(t)e^{-j\omega t}\qquad \{\text{radian frequency domain}\}$$
$$f(t) = \frac{1}{2\pi}\int_{-\infty}^{+\infty} F(\omega)e^{+j\omega t}$$

**Equation 3.22**

## 3.2.1 Fourier Transform of a Pulse Waveform

Let's begin our discussion on the Fourier transform by computing the frequency spectrum of the pulse waveform illustrated in Figure 3.7. The pulse has a width equal to $\tau$ and an amplitude equal to $A$. The waveform is mathematically defined as

$$f(t) = \begin{cases} A & \text{for } -\dfrac{\tau}{2} \leq t \leq +\dfrac{\tau}{2} \\[2mm] 0 & \text{for } |t| > \dfrac{\tau}{2} \end{cases}$$

This waveform, although trivial, plays a big role in signal processing. It is often referred to as a *rectangular window* or a *boxcar function*. We will see later that this waveform is important in many applications. In order to simplify our studies of the Fourier transform and how it is applied to the world of digital signal processing (DSP), we will need to thoroughly understand the frequency content of this simple pulse.

We can compute the Fourier transform of the pulse in Figure 3.7 by using the forward transform relation of Equation 3.18:

$$F(f) = \int_{-\infty}^{+\infty} f(t) e^{-j2\pi ft}\, dt$$

We make the substitutions for pulse width and amplitude and set up the Fourier integral to get

$$F(f) = A \int_{-\frac{\tau}{2}}^{+\frac{\tau}{2}} e^{-j2\pi ft}\, dt$$

Now we can break the exponential term into its sine and cosine components and perform the integration. In doing so, we perform the following steps

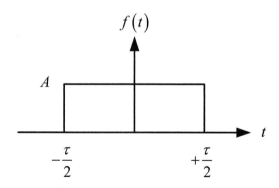

**Figure 3.7**  Pulse waveform

$$F(f) \quad = A \int_{-\frac{\tau}{2}}^{+\frac{\tau}{2}} e^{-j2\pi ft} \, dt$$

$$= A \int_{-\frac{\tau}{2}}^{+\frac{\tau}{2}} \left[ \cos(2\pi ft) - j\sin(2\pi ft) \right] dt$$

$$= A \left[ \frac{\sin(2\pi ft)}{2\pi f} + j \frac{\cos(2\pi ft)}{2\pi f} \right]_{t=-\frac{\tau}{2}}^{t=+\frac{\tau}{2}}$$

$$= A \left\{ \left( \frac{\sin(2\pi f \, \tau/2)}{2\pi f} + j \frac{\cos(2\pi f \, \tau/2)}{2\pi f} \right) - \left( \frac{\sin(2\pi f(-\tau/2))}{2\pi f} + j \frac{\cos(2\pi f(-\tau/2))}{2\pi f} \right) \right\}$$

$$= 2A \frac{\sin(\pi f \tau)}{2\pi f} + j0$$

$$= A\tau \frac{\sin(\pi f \tau)}{\pi f \tau}$$

We determine that the spectrum of the simple pulse is given by

$$F(f) = A\tau \frac{\sin(\pi f \tau)}{\pi f \tau}$$

**Equation 3.23**

As $f$ becomes small, $\sin(\pi f \tau) \to \pi f \tau$ and therefore $\sin(\pi f \tau)/\pi f \tau \to 1$. As a result, we see that the amplitude of the spectrum at DC is given by $F(0) = A\tau$. We also see that the amplitude of the spectrum at all other frequencies has a $\sin x/x$ response with nulls located at $n/\tau$ Hz $\{$for $n = 1, 2, 3, \cdots\}$. We know this because the term $\sin(\pi f \tau) = 0$ when its argument $\pi f \tau = n\pi \{$for $n = 1, 2, 3, \cdots\}$. Solving for $f$ gives us $f = n/\tau$. The magnitude spectrum given by Equation 3.23 is illustrated in Figure 3.8.

If we decide to work with radian frequency, then we could make the substitution of $\omega = 2\pi f \Rightarrow \omega/2 = \pi f$ into Equation 3.23 to get

$$F(\omega) = A\tau \frac{\sin(\omega \tau/2)}{\omega \tau/2}$$

In this particular case, the spectral nulls occur when $\omega \tau/2 = n\pi$ or $F(\omega) = 0$ at $\omega = 2n\pi/\tau$. The spectrum is illustrated in Figure 3.9.

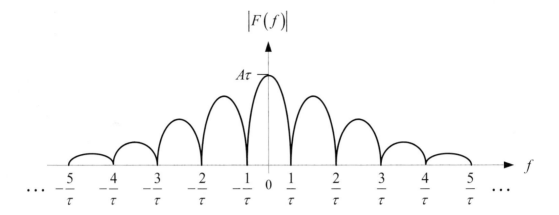

**Figure 3.8**  Frequency spectrum of a pulse of width

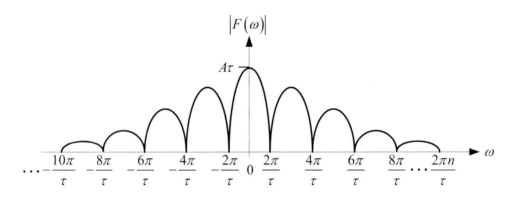

**Figure 3.9**  Radian frequency spectrum of a pulse of width

The spectrum for a rectangular pulse for various pulse widths is illustrated in Figure 3.10. In part A, we see a pulse whose amplitude equals $A$ and whose width is $\tau = 2$. The corresponding spectrum shows a DC amplitude of $A\tau = 2A$ and nulls at multiples of $1/\tau = 1/2$.

In part B, the pulse width is doubled to $\tau = 4$, which results in a compression of the spectrum such that the spectral nulls are now at multiples of $1/\tau = 1/4$ and the DC amplitude has doubled to $4A$.

In part C, the pulse width is widened to $\tau = 6$. The associated spectrum is narrowed such that the nulls now occur at multiples of $1/\tau = 1/6$ and the DC amplitude has increased to $6A$.

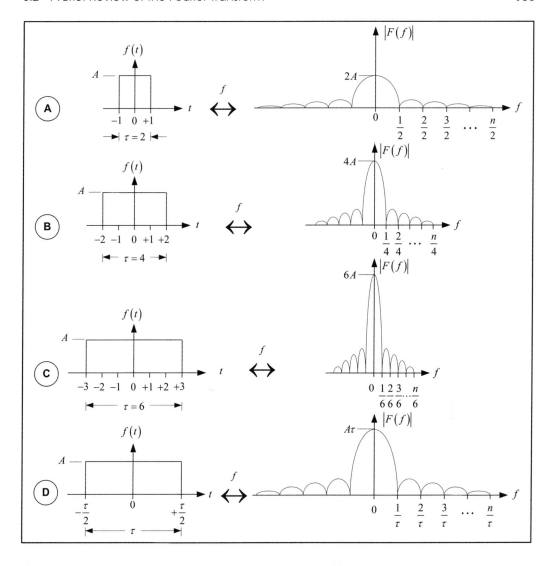

**Figure 3.10**   Time domain pulse width versus spectral nulls

In part D of the figure, we see the general expression for a pulse and its Fourier transform. In general, as the width of the pulse increases, the distance between zero crossings of the spectrum decreases and the width of the spectrum main lobe gets narrower. We can also see that the opposite is true. As the time domain pulse width gets narrower, the zero crossings of the spectrum spread out and the main lobe gets wider.

## 3.2.2  Fourier Transform Pairs

There are a handful of Fourier transform pairs that are considered the fundamental building blocks from which most other pairs are derived. These fundamental pairs consist of the time domain and frequency domain impulse functions and the sine and cosine functions.

### 3.2.2.1  Continuous Time Domain Unit Impulse

We define the continuous time domain unit impulse to be

$$\delta(t) \left\{ \begin{array}{ll} =1 & \text{for } t=0 \\ =0 & \text{for } t \neq 0 \end{array} \right\}$$

The Fourier transform of the unit impulse function is

$$f\{\delta(t)\} = \int_{t=-\infty}^{t=+\infty} \delta(t)e^{-j2\pi ft}\, dt = \delta(0)e^{-j0} = 1$$

This leads us to the important transform pair

$$\delta(t) \overset{f}{\leftrightarrow} 1$$

**Equation 3.24**

The unit impulse function shifted in time by $t_0$ is defined as

$$\delta(t-t_0) = \left\{ \begin{array}{ll} 1 & \text{for } t=t_0 \\ 0 & \text{for } t \neq t_0 \end{array} \right\}$$

$$\delta(t+t_0) = \left\{ \begin{array}{ll} 1 & \text{for } t=-t_0 \\ 0 & \text{for } t \neq -t_0 \end{array} \right\}$$

The Fourier transform of a shifted impulse function is given by

$$f\{\delta(t-t_0)\} = \int_{t=-\infty}^{t=+\infty} \delta(t-t_0)e^{-j2\pi ft}\, dt$$

If we let $p=t-t_0$, then $t=p+t_0$ and, since $t_0$ is a constant, $dp=dt$. Placing these substitutions into the previous equation gives us

$$f\{\delta(p)\} = \int_{p=-\infty}^{p=+\infty} \delta(p) e^{-j2\pi f(p+t_0)}\, dp$$

$$= e^{-j2\pi f t_0} \int_{p=-\infty}^{p=+\infty} \delta(p) e^{-j2\pi f(p)}\, dp$$

$$= e^{-j2\pi f t_0} \delta(0) e^{-j2\pi f(0)}$$

$$= e^{-j2\pi f t_0} \cdot 1 \cdot 1$$

$$= e^{-j2\pi f t_0}$$

Now we can substitute $t - t_0 = p$, to give us a fundamental Fourier transform pair:

$$\delta(t - t_0) \overset{f}{\leftrightarrow} e^{-j2\pi f t_0}$$

**Equation 3.25**

Similarly

$$f\{\delta(t + t_0)\} = \int_{t=-\infty}^{t=+\infty} \delta(t + t_0) e^{-j2\pi f t}\, dt = e^{+j2\pi f t_0}$$

Doing the same math as before, we derive the Fourier transform pair:

$$\delta(t + t_0) \overset{f}{\leftrightarrow} e^{+j2\pi f t_0}$$

**Equation 3.26**

### 3.2.2.2 Continuous Frequency Domain Unit Impulse

We define the continuous frequency domain unit impulse to be

$$\delta(f) \begin{cases} = 1 & \text{for } f = 0 \\ = 0 & \text{for } f \neq 0 \end{cases}$$

The inverse Fourier transform of the frequency impulse function is

$$f^{-1}\{\delta(f)\} = \int_{f=-\infty}^{f=+\infty} \delta(f) e^{+j2\pi f t}\, df = \delta(0) e^{+j0} = 1$$

This leads us to the transform pair

$$1 \overset{f}{\leftrightarrow} \delta(f)$$

**Equation 3.27**

The frequency domain unit impulse function shifted in frequency by $f_0$ is defined as

$$\delta(f - f_0) = \left\{ \begin{array}{ll} 1 & \text{for } f = f_0 \\ 0 & \text{for } f \neq f_0 \end{array} \right\}$$

$$\delta(f + f_0) = \left\{ \begin{array}{ll} 1 & \text{for } f = -f_0 \\ 0 & \text{for } f \neq -f_0 \end{array} \right\}$$

The inverse Fourier transform of a shifted impulse function is given by

$$f^{-1}\{\delta(f - f_0)\} = \int\limits_{f=-\infty}^{f=+\infty} \delta(f - f_0) e^{+j2\pi ft} \, df$$

If we let $p = f - f_0$, then $f = p + f_0$, and since $f_0$ is a constant, $df = dp$. Making these substitutions into the previous equation gives us

$$f^{-1}\{\delta(p)\} = \int\limits_{p=-\infty}^{p=+\infty} \delta(p) e^{+j2\pi(p+f_0)t} \, dp$$

$$= e^{+j2\pi f_0 t} \int\limits_{p=-\infty}^{p=+\infty} \delta(p) e^{+j2\pi pt} \, dp$$

$$= e^{+2\pi f_0 t} \delta(0) e^{+j2\pi 0t}$$

$$= e^{+j2\pi f_0 t} \cdot 1 \cdot 1$$

$$= e^{+j2\pi f_0 t}$$

Now if we substitute $f - f_0 = p$, we arrive at another fundamental Fourier transform pair:

$$e^{+j2\pi f_0 t} \overset{f}{\leftrightarrow} \delta(f - f_0)$$

**Equation 3.28**

Similarly,

$$f^{-1}\left\{\delta\left(f+f_0\right)\right\}=\int_{f=-\infty}^{f=+\infty}\delta\left(f+f_0\right)e^{+j2\pi f_0 t}\,dt\ =\ e^{-j2\pi f_0 t}$$

Doing the same math as before, we derive the Fourier transform pair:

$$e^{-j2\pi f_0 t}\ \overset{f}{\leftrightarrow}\ \delta\left(f+f_0\right)$$

**Equation 3.29**

### 3.2.2.3   Complex Exponential $e^{+j(2\pi f_0)t}$

Let's look at a frequency impulse response shifted in frequency by $f_0$, expressed as $\delta\left(f-f_0\right)$. We utilize the Fourier transform pair of Equation 3.28

$$e^{+j2\pi f_0 t}\ \overset{f}{\leftrightarrow}\ \delta\left(f-f_0\right)$$

to plot the line spectrum of $e^{j2\pi f_0 t}$ illustrated in Figure 3.11.

### 3.2.2.4   Complex Exponential $e^{-j(2\pi f_0)t}$

By utilizing the Fourier transform pair $e^{-j2\pi f_0 t}\ \overset{f}{\leftrightarrow}\ \delta\left(f+f_0\right)$ in Equation 3.29, we can similarly plot the line spectrum of $e^{-j(2\pi f_0)t}$ illustrated in Figure 3.12.

### 3.2.2.5   The Cosine Function $\cos\left(2\pi f_0 t\right)$

The spectrum of the function $f(t)=\cos\left(2\pi f_0 t\right)$ can be computed using a combination of Euler's identity and the Fourier transform identities that we derived in Equation 3.28 and Equation 3.29.

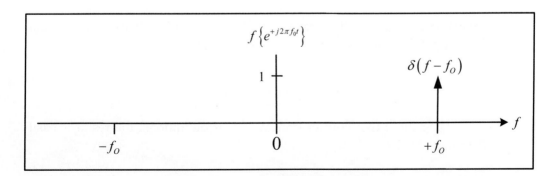

**Figure 3.11**   Positive exponential impulse spectrum

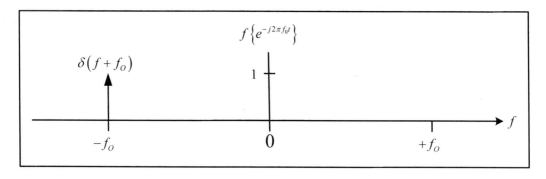

**Figure 3.12**   Negative exponential impulse spectrum

$$
\begin{aligned}
F(f) &= \int_{-\infty}^{\infty} f(t)e^{-j2\pi ft}\,dt \\
&= \int_{-\infty}^{\infty} \cos(2\pi f_0 t)e^{-j2\pi ft}\,dt \\
&= \int_{-\infty}^{\infty}\left[\frac{e^{j2\pi f_0 t}+e^{-j2\pi f_0 t}}{2}\right]e^{-j2\pi ft}\,dt \\
&= \frac{1}{2}\int_{-\infty}^{\infty}\left[e^{j2\pi f_0 t}\right]e^{-j2\pi ft}\,dt \;+\; \frac{1}{2}\int_{-\infty}^{\infty}\left[e^{-j2\pi f_0 t}\right]e^{-j2\pi ft}\,dt
\end{aligned}
$$

**Equation 3.30**

We recognize the first term in Equation 3.30 to be the Fourier transform of $e^{j2\pi f_0 t}$, which is defined by Equation 3.28:

$$
e^{j(2\pi f_0)t} \overset{f}{\leftrightarrow} \delta(f - f_0)
$$

We also recognize that the second term in Equation 3.30 is the Fourier transform of $e^{-j2\pi f_0 t}$, which is defined by Equation 3.29;

$$
e^{-j(2\pi f_0)t} \overset{f}{\leftrightarrow} \delta(f + f_0)
$$

Using these two Fourier transform pair definitions, the result is easily determined to be

$$
F(f) = \frac{1}{2}\{\delta(f - f_0) + \delta(f + f_0)\}
$$

Equation 3.31 defines the Fourier transform pair:

$$\cos(2\pi f_0 t) \overset{f}{\leftrightarrow} \frac{1}{2}\left\{\delta(f - f_0) + \delta(f + f_0)\right\}$$

**Equation 3.31**

The amplitude spectrum for $\cos(2\pi f_0 t)$ is illustrated in Figure 3.13.

Let's return to Equation 3.30 for a moment. We take the development of the equation through the first three steps as before and then rewrite the fourth step as the following:

$$
\begin{aligned}
F(f) &= \int_{-\infty}^{\infty} f(t)e^{-j2\pi ft}\,dt \\
&= \int_{-\infty}^{\infty} \cos(2\pi f_0 t)e^{-j2\pi ft}\,dt \\
&= \int_{-\infty}^{\infty}\left[\frac{e^{j2\pi f_0 t} + e^{-j2\pi f_0 t}}{2}\right]e^{-j2\pi ft}\,dt \\
&= \frac{1}{2}\int_{-\infty}^{\infty} e^{-j2\pi(f - f_0)t}\,dt + \frac{1}{2}\int_{-\infty}^{\infty} e^{-j2\pi(f + f_0)t}\,dt \\
&= \frac{1}{2}\left\{\delta(f - f_0) + \delta(f + f_0)\right\}
\end{aligned}
$$

The fourth step notation

$$\frac{1}{2}\int_{-\infty}^{\infty} e^{-j2\pi(f - f_0)t}\,dt + \frac{1}{2}\int_{-\infty}^{\infty} e^{-j2\pi(f + f_0)t}\,dt$$

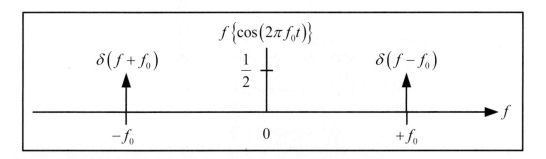

**Figure 3.13**   Line spectrum of cos(2πf₀t)

is the more common expression used when computing the Fourier transform of sinusoidal functions. We can split this expression into two integrals to get a pair of handy identities that are used all the time in signal processing applications:

$$\int_{-\infty}^{\infty} e^{-j2\pi(f-f_0)t}\, dt \;=\; \delta\big(f - f_0\big)$$

$$\int_{-\infty}^{\infty} e^{-j2\pi(f+f_0)t}\, dt = \delta\big(f + f_0\big)$$

**Equation 3.32**

### 3.2.2.6  The Sine Function $\sin\big(2\pi f_0 t\big)$

The spectrum of the function $f(t) = \sin\big(2\pi f_0 t\big)$ can be computed using a combination of Euler's identity and the Fourier transform identities that we derived in Equation 3.28 and Equation 3.29:

$$
\begin{aligned}
F(f) \quad &= \int_{-\infty}^{\infty} f(t) e^{-j2\pi ft}\, dt \\
&= \int_{-\infty}^{\infty} \sin\big(2\pi f_0 t\big) e^{-j2\pi ft}\, dt \\
&= \int_{-\infty}^{\infty} \left[ \frac{e^{j2\pi f_0 t} - e^{-j2\pi f_0 t}}{2j} \right] e^{-j2\pi ft}\, dt \\
&= \frac{1}{2j}\int_{-\infty}^{\infty} e^{-j2\pi(f-f_0)t}\, dt - \frac{1}{2j}\int_{-\infty}^{\infty} e^{-j2\pi(f+f_0)t}\, dt \\
&= \frac{1}{2j}\big\{\delta\big(f - f_0\big) - \delta\big(f + f_0\big)\big\}
\end{aligned}
$$

**Equation 3.33**

Equation 3.33 defines the Fourier transform pair:

$$\sin\big(2\pi f_0 t\big) \;\overset{f}{\leftrightarrow}\; \frac{1}{2j}\big\{\delta\big(f - f_0\big) - \delta\big(f + f_0\big)\big\}$$

**Equation 3.34**

We can modify Equation 3.34 by multiplying the right-hand side by $j/j$ to arrive at an alternate transform pair for $\sin\big(2\pi f_0 t\big)$:

$$F(f) = \frac{1}{2j}\left\{\delta(f - f_0) - \delta(f + f_0)\right\}$$

$$= \frac{j}{2j^2}\left\{\delta(f - f_0) - \delta(f + f_0)\right\}$$

$$= \frac{-j}{2}\left\{\delta(f - f_0) - \delta(f + f_0)\right\}$$

**Equation 3.35**

Equation 3.35 defines the alternate Fourier Transform pair:

$$\sin(2\pi f_0 t) \overset{f}{\leftrightarrow} \frac{-j}{2}\left\{\delta(f - f_0) - \delta(f + f_0)\right\}$$

**Equation 3.36**

The amplitude spectrum of $\sin(2\pi f_0 t)$, as defined by Equation 3.34, is illustrated in Figure 3.14. The amplitude spectrum representation of the alternate form of $\sin(2\pi f_0 t)$, as defined by Equation 3.36, is illustrated in Figure 3.15.

It is important that the reader understand that Figure 3.14 and Figure 3.15 are plots of spectral amplitude and not spectral magnitude. If the plots were spectral magnitude, then both impulses would be positive and the $j$ terms would vanish. It is also important to note that on the vertical axis $1/(2j) = -j/2$, which accounts for the $180°$ phase change between Figure 3.14 and Figure 3.15.

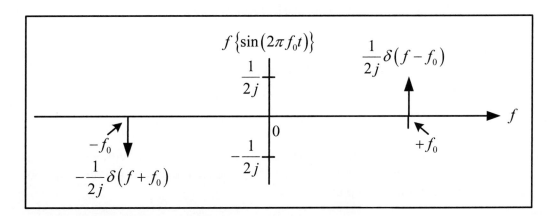

**Figure 3.14**   Line spectrum of $\sin(2\pi f_0 t)$

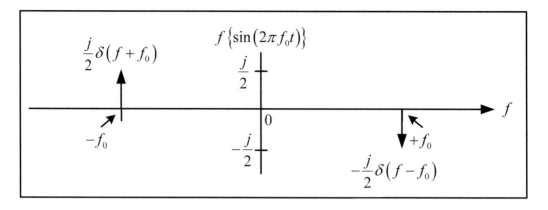

**Figure 3.15**   Alternate line spectrum of sin($2\pi f_0 t$)

### 3.2.2.7   Multiplication of a Real Signal $f(t)$ by $\cos(2\pi f_0 t)$

We define the function

$$y(t) = f(t)g(t)$$

where

$$g(t) = \cos(2\pi f_0 t)$$

Therefore we can write

$$y(t) = f(t)\cos(2\pi f_0 t)$$

Furthermore we define the spectrum of $y(t)$ as $Y(f)$. The two functions are then Fourier transform pairs denoted by

$$y(t) \overset{f}{\leftrightarrow} Y(f)$$

We define the function $f(t)$ as a band limited signal with a continuous frequency spectrum denoted by $F(f)$. These two functions are defined to be the Fourier transform pair

$$f(t) \overset{f}{\leftrightarrow} F(f)$$

The Fourier transform of $y(t)$ can be written as

$$f\{y(t)\} = \int_{-\infty}^{\infty} f(t)\cos(2\pi f_0 t)e^{-j2\pi ft}\,dt$$

Expansion of the cosine function by Euler's formula gives us

$$f\{y(t)\} = \int_{-\infty}^{\infty} f(t)\left[\frac{1}{2}e^{j2\pi f_0 t} + \frac{1}{2}e^{-j2\pi f_0 t}\right]e^{-j2\pi ft}\,dt$$

We can rearrange terms to get

$$f\{y(t)\} = \frac{1}{2}\int_{-\infty}^{\infty} f(t)e^{-j2\pi(f-f_0)t}\,dt \;+\; \frac{1}{2}\int_{-\infty}^{\infty} f(t)e^{-j2\pi(f+f_0)t}\,dt$$

**Equation 3.37**

We know that the Fourier transform of $f(t)$ is given by

$$F(f) = \int_{-\infty}^{\infty} f(t)e^{-j2\pi ft}\,dt$$

If we change the variable $f$ to $f - f_0$, we get

$$F(f - f_0) = \int_{-\infty}^{\infty} f(t)e^{-j2\pi(f-f_0)t}\,dt$$

**Equation 3.38**

Similarly, if we change the variable $f$ to $f + f_0$, we get

$$F(f + f_0) = \int_{-\infty}^{\infty} f(t)e^{-j2\pi(f+f_0)t}\,dt$$

**Equation 3.39**

Substituting Equation 3.38 and Equation 3.39 back into Equation 3.37, we arrive at

$$f\{y(t)\} = Y(f) = \frac{1}{2}\{F(f - f_0) \;+\; F(f + f_0)\}$$

This gives us the valuable Fourier Transform pair defined by Equation 3.40:

$$f(t)\cos(2\pi f_0 t) \overset{f}{\leftrightarrow} \frac{1}{2}\{F(f-f_0) + F(f+f_0)\}$$

**Equation 3.40**

This is important. Equation 3.40 tells us that if we multiply a band limited signal $f(t)$ by a cosine of frequency $f_0$, we can expect to see the spectrum of $f(t)$ translated to both $+f_0$ and $-f_0$, and both of these translations will be half the amplitude of the original.

The frequency domain representation of this translation is illustrated in Figure 3.16. Part A of the figure illustrates the band limited spectrum of the signal $f(t)$. Part B of the figure shows the spectrum of the cosine waveform $g(t) = \cos(2\pi f_0 t)$. Part C of the figure shows the amplitude spectrum of the product of $f(t)$ and $g(t)$. In the figure, we can clearly see that multiplication of the two signals in the time domain results in the individual spectra of the two signals being convolved in the frequency domain. That is, $Y(f) = F(f)*G(f)$.

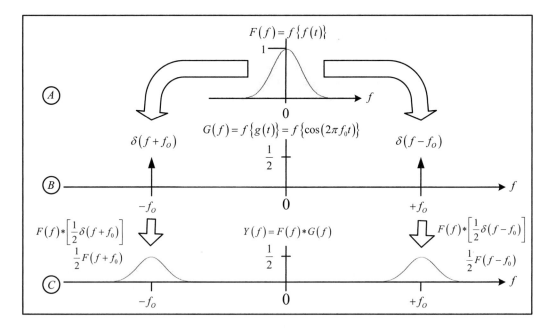

**Figure 3.16**   Multiplication of a band limited signal by $\cos(2\pi f_0 t)$

### 3.2.2.8 Multiplication of a Real Signal f(t) by sin(2πf₀t)

We can derive the Fourier transform pair for $f(t)\sin(2\pi f_0 t)$ using the same methods as multiplication by a cosine. In doing so, we arrive at

$$f(t)\sin(2\pi f_0 t) \overset{f}{\leftrightarrow} \frac{1}{2j}\{F(f-f_0) - F(f+f_0)\}$$

**Equation 3.41**

The frequency domain representation of this multiplication is illustrated in Figure 3.17.

Part A of the figure illustrates the band limited spectrum of the signal $f(t)$. Part B of the figure shows the spectrum of the sine waveform $g(t) = \sin(2\pi f_0 t)$. Part C of the figure clearly shows the amplitude spectrum of the product of $f(t)$ and $g(t)$ to be the convolution of the two individual signal spectra $Y(f) = F(f) * G(f)$.

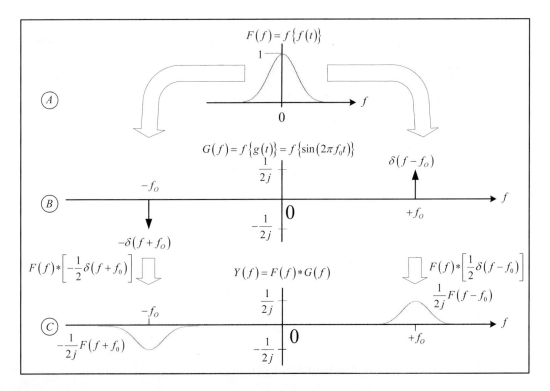

**Figure 3.17** Multiplication of a band limited signal by sin(2πf₀t)

### 3.2.2.9  Multiplication of a Real Signal $f(t)$ by $e^{+j2\pi f_0 t}$

We can easily derive another valuable transform pair by reworking Equation 3.38, which is repeated here for clarity:

$$F(f - f_0) = \int_{-\infty}^{\infty} f(t) e^{-j2\pi(f-f_0)t}\, dt$$

We can rearrange terms to produce

$$F(f - f_0) = \int_{-\infty}^{\infty} \left[ f(t) e^{+j2\pi f_0 t} \right] e^{-j2\pi ft}\, dt$$

The right-hand side of the equation is just the Fourier transform of $f(t)e^{+j2\pi f_0 t}$, giving us the Fourier transform pair in Equation 3.42:

$$f(t)e^{+j2\pi f_0 t} \overset{f}{\longleftrightarrow} F(f - f_0)$$

**Equation 3.42**

The transform pair of Equation 3.42 tells us that if we multiply a signal $f(t)$ whose spectrum is centered at 0 Hz and given by $F(f)$ by a positive complex exponential $e^{+j2\pi f_0 t}$, we will translate that spectrum up in frequency and it will be centered at the new frequency of $f_0$.

### 3.2.2.10  Multiplication of a Real Signal $f(t)$ by $e^{-j2\pi f_0 t}$

We can derive another Fourier transform pair by rewriting Equation 3.39, repeated here for clarity:

$$F(f + f_0) = \int_{-\infty}^{\infty} f(t) e^{-j2\pi(f+f_0)t}\, dt$$

We can rearrange terms to produce

$$F(f + f_0) = \int_{-\infty}^{\infty} \left[ f(t) e^{-j2\pi f_0 t} \right] e^{-j2\pi ft}\, dt$$

The right-hand side of the equation is just the Fourier transform of

$$f(t)e^{-j2\pi f_0 t}$$

This gives us the Fourier transform pair:

$$f(t)e^{-j2\pi f_0 t} \overset{f}{\leftrightarrow} F(f + f_0)$$

**Equation 3.43**

The transform pair of Equation 3.43 tells us that if we multiply a signal $f(t)$ whose spectrum is centered at 0 Hz and given by $F(f)$ by a negative complex exponential $e^{-j2\pi f_0 t}$, we will translate that spectrum down in frequency and it will be centered at the new frequency of $-f_0$.

### 3.2.2.11  Time Domain Multiplication $\overset{f}{\leftrightarrow}$ Frequency Domain Convolution

As we have seen in the previous two transform pairs, convolution plays a big part in the translation of signal spectra. Frequency translation involves the multiplication of two time domain signals such as a band limited signal $x(t)$ and a sinusoidal signal such as $h(t) = \sin(2\pi f_0 t)$. It turns out that the multiplication of two signals in the time domain results in the convolution of the two signal spectra in the frequency domain.

If, for example, we have two time domain signals $x(t)$ and $h(t)$ whose Fourier transforms are given by $f\{x(t)\} = X(f)$ and $f\{h(t)\} = H(f)$, then we can write the Fourier transform pair

$$x(t)h(t) \overset{f}{\leftrightarrow} X(f) * H(f)$$

**Equation 3.44**

The symbol $*$ is used to denote the convolution operation. We provide the proof of the Fourier transform pair given in Equation 3.44 by the following method. Let's define the convolution of $X(f)$ and $H(f)$ in the standard mathematical form given by

$$X(f) * H(f) = \int_{-\infty}^{\infty} X(u)H(f - u)\,du$$

**Equation 3.45**

The inverse Fourier transform of Equation 3.45 can be written as

$$f^{-1}\{X(f) * H(f)\} = \int_{-\infty}^{\infty} \left[ \int_{-\infty}^{\infty} X(u)H(f - u)\,du \right] e^{j2\pi ft}\,df$$

We can rearrange terms to get

$$f^{-1}\{X(f)*H(f)\} = \int_{-\infty}^{\infty} X(u)\left[\int_{-\infty}^{\infty} H(f-u)e^{j2\pi ft}\,df\right]du$$

**Equation 3.46**

We recognize that the term inside the brackets is just the inverse Fourier transform of the frequency shifted function $H(f-u)$. We already know from the Fourier transform pair given in Equation 3.42 that

$$h(t)e^{+j2\pi ut} \overset{f}{\longleftrightarrow} H(f-u)$$

so we can substitute this expression into Equation 3.46 to get

$$f^{-1}\{X(f)*H(f)\} = h(t)\left[\int_{-\infty}^{\infty} X(u)\,e^{+j2\pi ut}\,du\right]$$

If we replace $u$ with $f$ as the variable of integration, then $du = df$, and we can easily see that the quantity in the brackets is just the inverse Fourier transform of $x(t)$. Therefore

$$f^{-1}\{X(f)*H(f)\} = h(t)\left[\int_{-\infty}^{\infty} X(f)\,e^{j2\pi ft}\,df\right] = x(t)h(t)$$

We end up with the Fourier Transform pair:

$$h(t)x(t) \overset{f}{\longleftrightarrow} X(f)*H(f)$$

**Equation 3.47**

### 3.2.2.12   Time Domain Convolution $\overset{f}{\longleftrightarrow}$ Frequency Domain Multiplication

Convolution in the time domain results in multiplication in the frequency domain. This gives us another Fourier transform pair:

$$x(t)*h(t) \overset{f}{\longleftrightarrow} X(f)H(f)$$

**Equation 3.48**

where the $*$ symbol denotes the operation of convolution. The proof for the transform pair in Equation 3.48 is achieved by writing

$$y(t) = x(t) * h(t) = \int_{-\infty}^{\infty} x(\tau) h(t - \tau)\, d\tau$$

We compute the Fourier transform $Y(f) = f\{y(t)\}$ by

$$Y(f) = \int_{-\infty}^{\infty} y(t)\, e^{-j2\pi ft} dt = \int_{-\infty}^{\infty} \left[ \int_{-\infty}^{\infty} x(\tau) h(t - \tau) d\tau \right] e^{-j2\pi ft} dt$$

We can change the order of integration so that

$$Y(f) = \int_{-\infty}^{\infty} x(\tau) \left[ \int_{-\infty}^{\infty} h(t - \tau)\, e^{-j2\pi ft} dt \right] d\tau$$

Now let's define the variable $p = t - \tau$. This means that $t = p + \tau$, and since $\tau$ is a constant, $dt = dp$. If we make these substitutions in the previous equation, we get

$$\begin{aligned} Y(f) \quad &= \int_{-\infty}^{\infty} x(\tau) \left[ \int_{-\infty}^{\infty} h(p)\, e^{-j2\pi f(p+\tau)} dp \right] d\tau \\ &= \left[ \int_{-\infty}^{\infty} x(\tau) e^{-j2\pi f\tau}\, d\tau \right] \left[ \int_{-\infty}^{\infty} h(p)\, e^{-j2\pi fp} dp \right] \\ &= X(f)H(f) \end{aligned}$$

We end up with the Fourier transform pair:

$$x(t) * h(t) \overset{f}{\longleftrightarrow} X(f)H(f)$$

**Equation 3.49**

A few common Fourier transform pairs are listed in Table 3.4.

### 3.2.3 Fourier Transform Properties

A few of the important Fourier transform properties are listed in Table 3.5.

### 3.2.4 Parseval's Theorem for the Fourier Integral

If we are given the transform pairs

**Table 3.4**   Table of Continuous Fourier Transform Pairs

| Time domain signal | Fourier transform |
|:---:|:---:|
| $\delta(t)$ | $1$ |
| $1$ | $\delta(f)$ |
| $e^{j(2\pi f_0)t}$ | $\delta(f - f_0)$ |
| $e^{-j(2\pi f_0)t}$ | $\delta(f + f_0)$ |
| $\cos(2\pi f_0 t)$ | $\dfrac{1}{2}\{\delta(f - f_0) + \delta(f + f_0)\}$ |
| $\sin(2\pi f_0 t)$ | $\dfrac{1}{2j}\{\delta(f - f_0) - \delta(f + f_0)\}$ |
| $f(t)e^{+j(2\pi f_0)t}$ | $F(f - f_0)$ |
| $f(t)e^{-j(2\pi f_0)t}$ | $F(f + f_0)$ |
| $f(t)\cos(2\pi f_0 t)$ | $\dfrac{1}{2}\{F(f - f_0) + F(f + f_0)\}$ |
| $f(t)\sin(2\pi f_0 t)$ | $\dfrac{1}{2j}\{F(f - f_0) - F(f + f_0)\}$ |
| $x(t) * y(t)$ | $X(f)Y(f)$ |
| $x(t)y(t)$ | $X(f) * Y(f)$ |

$$f(t) \overset{f}{\leftrightarrow} F(\omega)$$

$$v(t) \overset{f}{\leftrightarrow} V(\omega)$$

then we can write

$$\int_{-\infty}^{+\infty} f(t)v(t)\,dt = \frac{1}{2\pi}\int_{-\infty}^{+\infty} F^*(\omega)\,V(\omega)\,d\omega$$

**Equation 3.50**

**Table 3.5**   Table of Fourier Transform Properties

| Property | Time domain signal | Fourier transform |
|---|---|---|
| Superposition | $ax(t) + by(t)$ | $aX(f) + bY(f)$ |
| Time delay | $x(t - t_0)$ | $X(f)e^{-j2\pi f t_0}$ |
| Scale change | $x(at)$ | $\|a\|^{-1} X(f/a) \quad a > 0$ |
| Time reversal | $x(-t)$ | $X(-f)$ |
| Frequency translation | $x(t)e^{\pm j2\pi f_0 t}$ | $X(f \mp f_0)$ |
| Modulation | $x(t)\cos(2\pi f_0 t)$ | $\dfrac{1}{2}\{X(f - f_0) + X(f + f_0)\}$ |
|  | $x(t)\sin(2\pi f_0 t)$ | $\dfrac{1}{2j}\{X(f - f_0) - X(f + f_0)\}$ |
| Convolution | $\int_{-\infty}^{+\infty} x(t - \tau)y(\tau)d\tau = x(t) * y(t)$ | $X(f)Y(f)$ |
| Multiplication | $x(t)y(t)$ | $\int_{-\infty}^{\infty} X(u)\, Y(f - u)du = X(f) * Y(f)$ |

where $F^*(\omega)$ is defined to be the conjugate of $F(\omega)$. Equation 3.50 is called *Parseval's theorem for the Fourier integral*. In the special case where $f(t) = v(t)$, Parseval's theorem can be written as

$$\int_{-\infty}^{+\infty} |f(t)|^2 dt = \frac{1}{2\pi}\int_{-\infty}^{+\infty} F^*(\omega)F(\omega)d\omega$$

We know from complex number theory that

$$F^*(\omega)F(\omega) = |F(\omega)|^2$$

Upon substitution, we get the more familiar expression of Parseval's theorem

$$\int_{-\infty}^{+\infty} |f(t)|^2 dt = \frac{1}{2\pi}\int_{-\infty}^{+\infty} |F(\omega)|^2 d\omega$$

**Equation 3.51**

If $\omega = 2\pi f$, then $d\omega = 2\pi df$, and we can make the substitution to convert Parseval's theorem from units of radian frequency to units of frequency. In doing so, we arrive at

$$\int_{-\infty}^{+\infty} \left| f(t) \right|^2 dt = \int_{-\infty}^{+\infty} \left| F(f) \right|^2 df$$

**Equation 3.52**

Equation 3.51 and Equation 3.52 are two extremely useful relations. Parseval is telling us that the total energy of a signal measured in the time domain is equal to the total energy of that same signal measured in the frequency domain. It makes sense to me! After all, the measurement of energy in a signal should be the same no matter which domain it is represented in.

### 3.2.5 Energy and Power Spectral Density

Suppose we are dealing with a voltage waveform whose amplitude is $A$ volts. Further suppose that this waveform is a pure cosine wave, such as the one illustrated in Figure 3.18. We know from past discussion that the frequency domain representation of this waveform is obtained via its Fourier transform and is given by

$$A\cos(2\pi f_0 t) \text{ volts} \overset{f}{\leftrightarrow} \frac{A}{2}\left[\delta(f - f_0) + \delta(f + f_0)\right] \text{ volts}$$

The frequency domain representation of this signal is also illustrated in Figure 3.18.

Ideally the bandwidth of the line spectra at $\delta(f - f_0)$ is exactly 1 Hz, and the bandwidth of the line spectra at $\delta(f + f_0)$ is also 1 Hz. The amplitude of both spectral lines is defined as $A/2$ volts.

We can say that the energy of this signal is given by $\left| F(f) \right|^2$. At the frequency $f = f_0$, $F(f_0) = A/2$ volts. The energy of this spectral component is

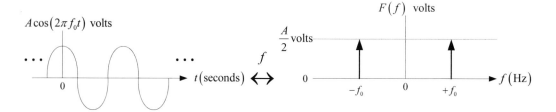

**Figure 3.18**    Cosine time/frequency representation

contained entirely within a 1 Hz bandwidth. Therefore the energy of this spectral component distributed across its 1 Hz bandwidth is given by

$$|F(+f_0)|^2 = \left(\frac{A/2 \text{ volts}}{1 \text{ Hz}}\right)^2 = \left(\frac{A}{2}\right)^2 \frac{\text{volt}^2}{\text{Hz}^2}$$

**Equation 3.53**

Similarly, the energy of the spectral component at $f = -f_0$ distributed across its 1 Hz bandwidth is given by

$$|F(-f_0)|^2 = \left(\frac{A/2 \text{ volts}}{1 \text{ Hz}}\right)^2 = \left(\frac{A}{2}\right)^2 \frac{\text{volt}^2}{\text{Hz}^2}$$

**Equation 3.54**

The total energy in this signal is just the sum of the two components or

$$E = |F(f)|^2 = \left[\left(\frac{A}{2}\right)^2 + \left(\frac{A}{2}\right)^2\right]\frac{\text{volt}^2}{\text{Hz}^2} = \frac{A^2}{2}\frac{\text{volt}^2}{\text{Hz}^2}$$

**Equation 3.55**

Now then, suppose we simplify the expressions in Equation 3.55 by replacing the term $A^2/2$ with the quantity $v^2$ volts. This gives us the more compact expression

$$|F(f)|^2 = v^2 \frac{\text{volt}^2}{\text{Hz}^2}$$

**Equation 3.56**

By convention we define the energy of a signal as the energy expended across a 1 Ω resistance $R$. So if our sinusoid is driving a 1 Ω load, then the energy into the load would be given by

$$|F(f)|^2 = \frac{v^2}{R}\frac{\text{volt}^2}{\Omega - \text{Hz}^2}$$

**Equation 3.57**

The "–" in the units of this expression is not a "minus" sign; it is a hyphen used to separate units. Even though the resistance $R$ is $1\,\Omega$, we include the $1\,\Omega$ resistance in Equation 3.57 to validate the addition of the unit $\Omega$ in the expression. This helps us to keep our unit bookkeeping straight.

Now then, suppose for a moment that we could somehow expand a continuous spectrum $F(f)$ so that we could easily observe the amplitude of individual frequencies within the spectrum. Further suppose that we could isolate a single frequency $f_n$ that had a magnitude of $v_n = A/\sqrt{2}$ volts, as illustrated in Figure 3.19.

We could represent the energy at this one particular frequency as

$$\left|F(f_n)\right|^2 = \frac{v_n^2}{R}\frac{\text{volt}^2}{\Omega-\text{Hz}^2}$$

We know that the unit of Hz is the reciprocal of the unit "seconds," so we can rewrite the units of this relation as

$$\left|F(f_n)\right|^2 = \frac{v_n^2}{R}\frac{\text{volt}^2-\text{seconds}}{\Omega-\text{Hz}}$$

**Equation 3.58**

At first glance, the units of $\left(\text{volt}^2-\text{seconds}\right)/\Omega$ seem to be a bit weird. Whatever these units may be, collectively they are relative to a single cycle or a single Hz. Let's do some exploring and see if we can figure this stuff out.

We know from basic electronics that a unit for measuring the quantity of electric charge is the coulomb $(Q)$, and it is defined by the amount of current $(I)$ that flows past some arbitrary plane each second $(t)$. In other words,

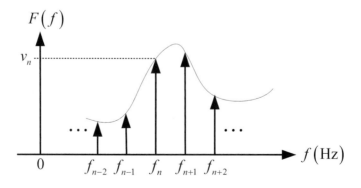

**Figure 3.19**   Theoretical expansion of a continuous spectrum

$Q = It$. We also know from basic electronics that a Joule $(J)$ is the amount of energy required to push one coulomb of charge across one volt $(V)$ of potential difference, or $J = QV$. Therefore we can combine the two relations and write

$$J = QV = (VI)t$$

In a resistive load, the current $I$ is defined by $I = V/R$, so by substitution

$$J = QV = (VI)t = \left(V\frac{V}{R}\right)t = \frac{V^2}{R}t \quad \frac{\text{volt}^2 - \text{seconds}}{\Omega}$$

So we see that the units $(\text{volt}^2 - \text{seconds})/\Omega$ are associated with Joules and hence correspond to units of energy. If we make the substitution of these units back into Equation 3.58, we end up with

$$\left|F(f_n)\right|^2 = \frac{v_n^2}{R}\left(\frac{\text{volt}^2 - \text{seconds}}{\Omega - \text{Hz}}\right) = \frac{v_n^2}{R}\left(\frac{\text{Joule}}{\text{Hz}}\right)$$

**Equation 3.59**

This relation is valid in a bandwidth over 1 Hz across a 1 $\Omega$ load. The convention is to always assume that the resistive load $R = 1\ \Omega$, so we can drop the $R$ term to get

$$\left|F(f_n)\right|^2 = v_n^2\left(\frac{\text{Joule}}{\text{Hz}}\right)$$

**Equation 3.60**

Furthermore, if we drop the subscript $n$ from $\left|F(f_n)\right|^2$, we end up with $\left|F(f)\right|^2$. Now instead of referring to a single isolated frequency $f_n$ we are referring to a continuous band of frequencies $f$. We can see that $\left|F(f)\right|^2$ has the units of Joule/Hz, or energy per Hz. We can interpret the term $\left|F(f)\right|^2$ to be the distribution of signal energy over some band of frequency. We call this the *energy spectral density* (ESD) of a signal. This is an important and powerful concept. It can be stated as

$$\text{ESD} = \left|F(f)\right|^2$$

**Equation 3.61**

Why is it powerful? One big reason is that we can compute the total energy $E$ of a signal over a continuous band of frequencies by integrating the signal energy density function to get

$$E = \int_{f1}^{f2} |F(f)|^2 \, df \text{ Joule}$$

We can therefore say that a signal $f(t)$ has an ESD given by $|F(f)|^2$.

Everyday communications signals are usually of unlimited duration and therefore the energy in the signal can be inconvenient to work with. Because of this we often work with average power instead. We derive the expression for average power as follows. The function $F(f)$ we have been discussing can be represented in the time domain by its inverse Fourier transform. That is,

$$f(t) \overset{f}{\leftrightarrow} F(f)$$

The average power of a signal $f(t)$ is given by

$$P_{AVG} = \lim_{T \to \infty} \frac{1}{T} \int_{-T/2}^{+T/2} |f(t)|^2 dt$$

We also know from Parseval's relation in Equation 3.52 that

$$\int_{-\infty}^{+\infty} |f(t)|^2 dt = \int_{-\infty}^{+\infty} |F(f)|^2 \, df$$

The average power of a signal then can be expressed as

$$P_{AVG} = \lim_{T \to \infty} \left\{ \frac{1}{T} \int_{-T/2}^{+T/2} |f(t)|^2 dt \right\} = \lim_{T \to \infty} \left\{ \frac{1}{T} \int_{-\infty}^{+\infty} |F(f)|^2 df \right\}$$

**Equation 3.62**

We can rewrite Equation 3.62 to get an expression that represents the average power of a signal computed in the frequency domain. In doing so, we arrive at

$$P_{AVG} = \left\{ \int_{-\infty}^{+\infty} \lim_{T \to \infty} \frac{|F(f)|^2}{T} \, df \right\}$$

The term $\lim\limits_{T\to\infty}\left\{|F(f)|^2/T\right\}$ is a measure of the average signal power distributed over frequency. This term is called the *power spectral density* (PSD) of the signal, sometimes represented by the notation $S(f)$. Therefore we can write

$$PSD = S(f) = \lim_{T\to\infty}\frac{|F(f)|^2}{T}$$

**Equation 3.63**

The units of power spectral density are derived from

$$\frac{|F(f)|^2}{T} = \frac{v^2}{T}\left(\frac{volt^2 - seconds}{\Omega - Hz}\right)\left(\frac{1}{seconds}\right) = \frac{v^2}{T}\left(\frac{Joule}{Hz - seconds}\right)$$

Once again, from basic electronics we know that the energy expenditure of 1 Joule/second is the equivalent to 1 watt of power. If we make this substitution, we can see that the units of power spectral density are given by watt/Hz and the average power for a signal is written as

$$P_{AVG} = \int_{-\infty}^{+\infty}\lim_{T\to\infty}\frac{1}{T}|F(f)|^2 df = \int_{-\infty}^{+\infty}S(f)df \text{ watts}$$

**Equation 3.64**

The power in a specific frequency band is computed by

$$P_{AVG} = \int_{f1}^{f2}S(f)df \quad \text{watts}$$

**Equation 3.65**

## 3.3  REVIEW OF THE DISCRETE FOURIER TRANSFORM (DFT)

The continuous Fourier transform (CFT) is an extremely powerful and useful tool in signal processing analysis. But the Fourier transform that we have discussed so far deals with continuous analog signals that have infinite precision in both the time and the frequency domains.

We cannot implement the CFT on a digital computer. This is due to the fact that computers deal with streams of digitized data samples captured at discrete intervals of time defined at the sample rate $f_S = 1/T_S$. The samples are quantized to discrete amplitudes with a resolution determined by the bit width of the computer memory. So right off the bat, we can see that the amplitude of the digitized signal cannot be described with infinite precision, and since the digitized signal is sampled at periodic intervals of time, its frequency cannot be described with infinite precision either. So we are in somewhat of a quandary. How can we compute the Fourier transform of a signal using a digital computer?

We solve this problem by using an algorithm called the discrete Fourier transform (DFT). The DFT is defined by the relation

$$X(k) = \sum_{n=0}^{N-1} x(n)\, e^{-j\frac{2\pi k}{N}n} \quad \text{for } k = 0,1,2,\cdots,N-1$$

**Equation 3.66**

where

$x(n)$    Sampled time domain signal

$X(k)$    Sampled frequency domain signal

$n$    Time domain sample index $n = 0,1,2,\cdots,N-1$

$k$    Frequency domain sample index $k = 0,1,2,\cdots,N-1$

$N$    Number of samples in both $x(n)$ and $X(k)$

For the discrete signal case, if Equation 3.66 exists for some $x(n)$, then we can say that $x(n)$ and $X(k)$ are a discrete Fourier transform pair and this relationship is written as

$$x(n) \overset{f}{\leftrightarrow} X(k)$$

Let's expand Equation 3.66 using Euler's formula to arrive at

$$X(k) = \sum_{n=0}^{N-1} x(n)\left[\cos\left(\frac{2\pi k}{N}\right)n - j\sin\left(\frac{2\pi k}{N}\right)n\right] \quad \text{for } k = 0,1,2,\cdots,N-1$$

**Equation 3.67**

In Equation 3.67 we can see that the $N$ sample input sequence $x(n)$ is multiplied by a complex sinusoid of radian frequency $2\pi k/N$. If we consider the input sequence $x(n)$ to be an $N$ sample vector, the computation of the DFT requires $x(n)$ to be multiplied by the $N$ sample complex sinusoid for a total of $N$ multiples. This vector multiplication is performed once for each value of the index $k$ for a total of $N$ times, bringing the total number of multiplies to $N^2$. Each vector multiply essentially correlates the input sequence against a complex sinusoid of normalized discrete frequency $2\pi k/N$. Successive vector multiplies are correlations against a complex sinusoid whose radian frequency increases in increments of $k/N$.

Each output $X(k)$ for $k = 0, 1, 2, \cdots, N-1$ is a single complex sample that represents the correlation score of the input sequence $x(n)$ correlated against the complex sinusoid at the discrete radian frequency of $2\pi k/N$. If, for example, the input signal was $N = 16$ samples in length and was a pure sinusoid at frequency $2\pi 4/16$, or $(k = 4)$, the DFT would produce a high correlation score for the output sample $X(4)$ and $X(12)$ and zero for all other values. Why do we get a high correlation when the value of $k = 12$? This is a good question, and the answer has to do with the interpretation of DFT frequency, which we will discuss next.

### 3.3.1  DFT Frequency

The concept of digital frequency is covered extensively in Chapter 1, "Review of Digital Frequency," and is utilized at length in digital signal tuning applications discussed in Chapter 9, "Digital Signal Tuning." If you have not already done so, please read Chapter 1 prior to proceeding. A short review of digital frequency is given here but only to simplify the discussion of the mechanics of the DFT.

Let's take a quick look at the sine and cosine terms and their argument in the DFT Equation 3.67:

$$\cos\left(\frac{2\pi k}{N}\right)n \qquad \sin\left(\frac{2\pi k}{N}\right)n$$

We see that the argument for both terms is $(2\pi k/N)n$. We can visualize a circle in the complex plane centered at the origin with radius 1, called the *unit circle*, as illustrated in Figure 3.20. The circumference of the circle is obtained by scribing an arc through an angle of $2\pi$ radians. If we segment the circle into smaller arcs of equal length by dividing $2\pi$ by $N$, then each arc length is scribed by an angle equal to $2\pi/N$ radians. This is illustrated in the figure by the equally spaced dots along the circumference of the unit circle. We can think of a phasor that originates at the origin of the unit circle and extends outward to one of the equally spaced dots. The phasor is rotating around the circle by jumping from dot to dot at the sample rate of the signal. The phasor is identified in the figure as $c(nT)$ and is composed of a real component $B(nT) = \cos(\theta(nT))$ and an imaginary component $A(nT) = \sin(\theta(nT))$.

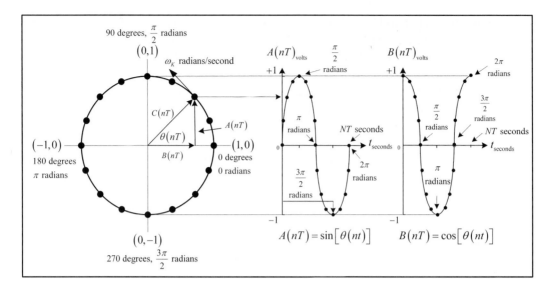

**Figure 3.20** Unit circle

As the phasor rotates around the unit circle, the sequence $B(nT)$ maps out the sample values of a cosine waveform and the sequence $A(nT)$ maps out the sample values of a sine waveform. These two mappings are illustrated in Figure 3.20.

In the case where the frequency index $k = 1$, both the sine and cosine argument take the form $(2\pi/N)n$. As the index $n$ advances through the range of $0, 1, 2, \cdots, N-1$, the argument increments in radian chunks equal to $(2\pi/N)$ radians, and the sine and cosine functions map out $N$ samples of a single sine and cosine waveform. It will take the phasor exactly $N$ sample clock periods to make one revolution of the unit circle. This will cause the sine and cosine functions to map out the lowest frequency possible other than DC. This is called the *fundamental frequency*. This sinusoidal frequency will contain exactly one cycle for every $N$ samples. The frequency of the fundamental will be equal to

$$f = \frac{f_S}{N} \text{ Hz} \text{ or } \omega = \frac{2\pi f_S}{N} \text{ radians/second}$$

where the term $f_S$ is the sample frequency.

When $k = 2$, the arguments will increment in radian chunks equal to $(2\pi \cdot 2/N)$, and the sine and cosine argument takes the form $(4\pi/N)n$. The phasor will rotate around the unit circle at twice the rate, skipping every second dot and mapping out a complex sinusoid at twice the fundamental frequency. This sinusoidal frequency will contain exactly two complete sine and cosine cycles for every $N$ samples and the output frequency will be equal to

$$f = \frac{2f_s}{N} \text{ Hz}$$

Some similar results can be obtained by setting the frequency index $k = 3, 4, 5, 6, \cdots, N-1$. Thus the argument will take on the general form

$$\frac{2\pi k}{N} n \text{ for } n = 0, 1, 2, \cdots, N-1$$

and the resulting discrete frequencies will take on the form

$$f = \frac{kf_s}{N} \text{ Hz or } \omega = \frac{2\pi k f_s}{N} \text{ radians/second}$$

**Equation 3.68**

When $k = N/2$, the sinusoidal frequency is

$$f = k\frac{f_s}{N} \text{ Hz} = \frac{N}{2}\frac{f_s}{N} \text{ Hz} = \frac{f_s}{2} \text{ Hz}$$

or

$$\omega = \frac{2\pi k f_s}{N} \text{ radians/second} = 2\pi \frac{N}{2}\frac{f_s}{N} \text{ radians/second} = \pi f_s \text{ radians/second}$$

This frequency consists of just two samples per cycle and is referred to as the *Nyquist frequency*.

It is the convention to normalize the expression for digital frequency by dividing by $f_s$. If we normalize Equation 3.68 to the sample rate, we end up with

$$f = \frac{k}{N} \text{ and } \omega = \frac{2\pi k}{N} \text{ for } k = 0, 1, 2, 3, \cdots, N-1$$

This fixes the range of normalized frequencies to be

$$0 \leq f \leq \frac{N/2}{N} = \frac{1}{2} \text{ and } 0 \leq \omega \leq \frac{(N/2)2\pi}{N} = \pi$$

with a frequency resolution of $1/N$ or $2\pi/N$, respectively. The maximum frequency possible is obtained when $k = N/2$. Nothing prevents us from making the frequency index $k$ larger, but we gain nothing from doing so. When $N/2 < k \leq N-1$, we end up with aliased frequencies. Aliased frequencies have less than two samples per cycle and appear to be identical to frequencies that have been "folded back" over the frequency $f_s/2$. In other words, the frequency corresponding to the index $k = N/2+1$ will look identical to the frequency corresponding to the index $k = N/2-1$. The frequencies corresponding to the index range $N/2+1 \leq k \leq N-1$ appear as identical or mirror frequencies corresponding to the index range $k = N/2-1$ down to 1. The normalized frequency $k = N/2$ and the frequency $f = f_s/2$ are referred to as *folding frequencies*.

A folded frequency axis is illustrated in Figure 3.21. In the figure, we can see that all frequencies above the folding frequency $f = f_s/2$ fold back and are indistinguishable from their mirror frequencies below the folding frequency. For example, as illustrated in the figure, a frequency at $f = f_s$ folds back to $f = 0$, $f = 3f_s/4$ folds back to $f = f_s/4$, and so forth.

The concept of folding or frequency aliasing is further illustrated in Figure 3.22. Here we have extended the frequency axis all the way to $3f_s$ and folded it every $f_s/2$ Hz. From the figure we can see that all the frequencies on the axis that passes through all the black dots fold back (or alias back) to the

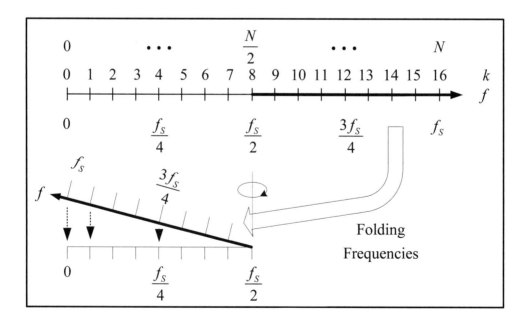

**Figure 3.21**  Frequency axis folding

base frequency of $f_s/8$. We can see from the figure that any number of under-sampled frequencies can fold back or alias into the spectrum between 0 and $f_s/2$. In the example shown in Figure 3.22, any frequency given by $nf_s \pm f_s/8$ will alias down to $f_s/8$. In general, frequencies given by $nf_s \pm f_s/M$ will alias down to $f_s/M$. Furthermore, the frequencies given by $nf_s - f_s/M$ are defined as being on an "odd fold" of the frequency axis, and the frequencies given by $nf_s + f_s/M$ are defined as being on an "even fold" of the frequency axis. It should be noted that a narrow band signal spectrum on odd folds will alias down frequency "inverted." In an inverted spectrum, the high and low frequencies of the band are reversed or swapped. That is, the original high frequencies now appear at the low end of the aliased spectrum, and the original low frequencies now appear on the high end of the aliased spectrum. A signal spectrum on even folds will alias down frequency "erect," where the frequencies are not reversed. The reader can verify this by unfolding the frequency axis in Figure 3.23 and observing the orientation of the narrow band spectrum represented by the triangle.

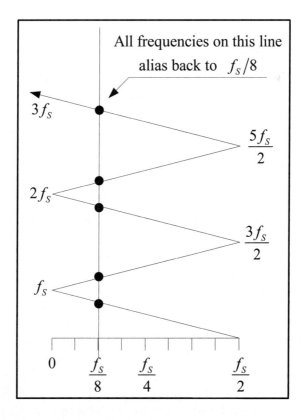

**Figure 3.22**   Extended frequency axis folding

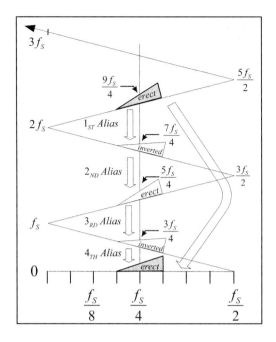

**Figure 3.23**   Aliasing tree diagram

This aliasing phenomenon is not always a bad thing. It can be put to practical use in many applications. For example, I have used aliasing to down convert certain pass band signals to base band. This is accomplished simply by choosing the correct rate to undersample an analog signal, causing it to alias down to base band, and then passing it through a low pass filter. As an example, suppose we had a narrow band analog signal centered at 2.25 MHz that we wish to digitize and then tune or down convert to a center frequency of $f_s/4 = 250$ KHz. We could choose to digitize the analog signal at a suitable sampling rate, follow that with a complex tuner to down convert the signal to 250 KHz, and then use a multirate filter to decimate the sample rate to 1 MHz. In doing so, we would end up with the narrow band signal of interest, centered at a quarter of the sample rate, as desired. However, we would have invested quite a bit of hardware to do it.

We could, however, create a "poor man's tuner" by using our A/D converter to do the tuning for us. For example, suppose we input this same analog narrow band signal to an A/D and set the sample rate to 1 MHz. The undersampled signal would cause the narrow band of interest to alias down in frequency four times and end up centered erect at 250 KHz, as illustrated in the aliasing tree in Figure 3.23. By undersampling the analog signal, we were able to save a lot of hardware. In this case, the low pass antialias filter that usually precedes an A/D would be replaced by a band pass filter that would isolate the input band of interest.

This is a signal processing trick that is useful in some but not all situations. For example, the signal could have aliased down inverted. In this case, we would need to flip the signal spectrum by multiplying by a sinusoid at frequency $f_S/2$. This is equivalent to multiplying the signal sequence by the repetitive sequence $\{+1,-1,+1,-1,\cdots\}$. This will spectrally invert a band limited signal centered at $f_S/4$. Another issue that could cause a problem is that the output sample rate might not match the sample rate of the follow on processing hardware. In this case, the poor man's tuner might not be a good fit for the application.

### 3.3.2   DFT Symmetry

Let's get back to the previous example where we processed an input signal vector consisting of a real sequence $x(n)$ that was $N = 16$ samples in length. The signal was a pure cosine at frequency $2\pi 4/16$, or $(k = 4)$. We said that the DFT would produce a high correlation score for the output sample $X(4)$ and $X(12)$ and zero for all other values.

Let's compute all 16 DFT outputs (correlation scores) and prove it. The input signal is illustrated in part A of Figure 3.24. We can write the values of each sample in the 16 sample vector as

$$x(n) = \quad 1 \quad 0 \quad -1 \quad 0 \quad 1 \quad 0 \quad -1 \quad 0 \quad 1 \quad 0 \quad -1 \quad 0 \quad 1 \quad 0 \quad -1 \quad 0$$

If we substitute this input vector into the DFT Equation 3.66, we get

$$X(k) \quad = \sum_{n=0}^{N-1} x(n)\, e^{-j\frac{2\pi k}{N}n} \quad \text{for } k = 0,1,2,\cdots,N-1$$

$$= \sum_{n=0}^{15} x(n)\, e^{-j\frac{2\pi k}{16}n} \quad \text{for } k = 0,1,2,\cdots,15$$

The first term $X(0)$ corresponds to the DC term. When $k = 0$, the exponential term reduces to unity, giving us

$$X(0) = \sum_{n=0}^{15} x(n) = \{1+0-1+0+1+0-1+0+1+0-1+0+1+0-1+0\} = 0$$

This makes sense. When we look at the time domain waveform in Figure 3.24 part A, we see that there is no DC component to the waveform, and therefore we expect the DFT result to be 0. All the odd terms of the input time series $x(n)$ equal zero, so there is no need to involve them in the rest of the DFT calculations. The DFT output for $k = 1$ is computed as

$$X(1) = \sum_{n=0}^{15} x(n)\, e^{-j\frac{\pi}{8}n} = \left\{ e^{-j\frac{\pi}{8}0} - e^{-j\frac{\pi}{8}2} + e^{-j\frac{\pi}{8}4} - e^{-j\frac{\pi}{8}6} + e^{-j\frac{\pi}{8}8} - e^{-j\frac{\pi}{8}10} + e^{-j\frac{\pi}{8}12} - e^{-j\frac{\pi}{8}14} \right\}$$

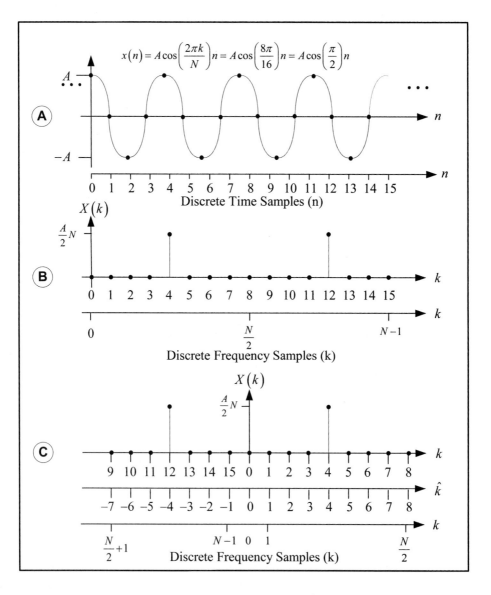

**Figure 3.24**  Discrete Fourier transform of a real input signal

By rearranging the terms to collect all the positive terms within brackets and all the negative terms within brackets, we end up with

$$X(1) = \sum_{n=0}^{15} x(n)\, e^{-j\frac{\pi}{8}n} = \left[ e^{-j\frac{\pi}{8}0} + e^{-j\frac{\pi}{8}4} + e^{-j\frac{\pi}{8}8} + e^{-j\frac{\pi}{8}12} \right] - \left[ e^{-j\frac{\pi}{8}2} + e^{-j\frac{\pi}{8}6} + e^{-j\frac{\pi}{8}10} + e^{-j\frac{\pi}{8}14} \right]$$

$$= \left[ e^{-j\frac{\pi}{8}0} + e^{-j\frac{\pi}{8}4} + e^{-j\frac{\pi}{8}8} + e^{-j\frac{\pi}{8}12} \right] - e^{-j\frac{\pi}{8}2} \left[ e^{-j\frac{\pi}{8}0} + e^{-j\frac{\pi}{8}4} + e^{-j\frac{\pi}{8}8} + e^{-j\frac{\pi}{8}12} \right]$$

$$= \left[ 1 - e^{-j\frac{\pi}{8}2} \right] \left[ e^{-j\frac{\pi}{8}0} + e^{-j\frac{\pi}{8}4} + e^{-j\frac{\pi}{8}8} + e^{-j\frac{\pi}{8}12} \right]$$

$$= \left[ 1 - e^{-j\frac{\pi}{8}2} \right][0] = 0$$

Another way to look at this is to remember that $e^{j\theta}$ is a complex number, and if we add all the terms in the positive bracket and subtract all the terms in the negative bracket, we get

$$\begin{bmatrix} +1 & +j0 \\ +0 & -j1 \\ -1 & +j0 \\ +0 & +j1 \end{bmatrix} - \begin{bmatrix} +0.707 & -j0.707 \\ -0.707 & -j0.707 \\ -0.707 & +j0.707 \\ +0.707 & +j0.707 \end{bmatrix} = 0 + j0 = 0$$

Thus we end up with $X(1) = 0$.

The DFT output for $k = 2$ is computed as

$$X(2) = \sum_{n=0}^{15} x(n)\, e^{-j\frac{2\pi}{8}n} = \left\{ e^{-j\frac{\pi}{4}0} - e^{-j\frac{\pi}{4}2} + e^{-j\frac{\pi}{4}4} - e^{-j\frac{\pi}{4}6} + e^{-j\frac{\pi}{4}8} - e^{-j\frac{\pi}{4}10} + e^{-j\frac{\pi}{4}12} - e^{-j\frac{\pi}{4}14} \right\}$$

By rearranging the terms to collect all the positive terms within brackets and all the negative terms within brackets, we end up with

$$X(2) = \sum_{n=0}^{15} x(n)\, e^{-j\frac{2\pi}{8}n} = \left[ e^{-j\frac{\pi}{4}0} + e^{-j\frac{\pi}{4}4} + e^{-j\frac{\pi}{4}8} + e^{-j\frac{\pi}{4}12} \right] - \left[ e^{-j\frac{\pi}{4}2} + e^{-j\frac{\pi}{4}6} + e^{-j\frac{\pi}{4}10} + e^{-j\frac{\pi}{4}14} \right]$$

$$= \left[ 1 - e^{-j\frac{\pi}{2}} \right] \left[ e^{-j\frac{\pi}{4}0} + e^{-j\frac{\pi}{4}4} + e^{-j\frac{\pi}{4}8} + e^{-j\frac{\pi}{4}12} \right] = \left[ 1 - e^{-j\frac{\pi}{2}} \right][0] = 0$$

We can also add all the terms in the positive bracket and subtract all the terms in the negative bracket to arrive at

$$\begin{bmatrix} +1 & -j0 \\ -1 & -j0 \\ +1 & -j0 \\ -1 & -j0 \end{bmatrix} - \begin{bmatrix} 0 & -j \\ 0 & +j \\ 0 & -j \\ 0 & +j \end{bmatrix} = 0 + j0 = 0$$

Thus we see that $X(2) = 0$.

The DFT output for $k = 3$ is computed as

$$X(3) = \sum_{n=0}^{15} x(n)\, e^{-j\frac{3\pi}{8}n} = \left\{ e^{-j\frac{3\pi}{8}0} - e^{-j\frac{3\pi}{8}2} + e^{-j\frac{3\pi}{8}4} - e^{-j\frac{3\pi}{8}6} + e^{-j\frac{3\pi}{8}8} - e^{-j\frac{3\pi}{8}10} + e^{-j\frac{3\pi}{8}12} - e^{-j\frac{3\pi}{8}14} \right\}$$

By rearranging terms to collect all the positive terms within brackets and all the negative terms within brackets, we end up with

$$X(3) = \sum_{n=0}^{15} x(n)\, e^{-j\frac{3\pi}{8}n} = \left[ e^{-j\frac{3\pi}{8}0} + e^{-j\frac{3\pi}{8}4} + e^{-j\frac{3\pi}{8}8} + e^{-j\frac{3\pi}{8}12} \right] - \left[ e^{-j\frac{3\pi}{8}2} + e^{-j\frac{3\pi}{8}6} + e^{-j\frac{3\pi}{8}10} + e^{-j\frac{3\pi}{8}14} \right]$$

If we add all the terms in the positive bracket and subtract all the terms in the negative bracket, we get

$$\begin{bmatrix} +1 & -j0 \\ +0 & +j1 \\ -1 & -j0 \\ +0 & -j1 \end{bmatrix} - \begin{bmatrix} -0.707 & -j0.707 \\ +0.707 & -j0.707 \\ +0.707 & +j0.707 \\ -0.707 & +j0.707 \end{bmatrix} = 0 + j0 = 0$$

Thus we see that $X(3) = 0$.

The DFT output for $k = 4$ is computed as

$$X(4) = \sum_{n=0}^{15} x(n)\, e^{-j\frac{4\pi}{8}n} = \left\{ e^{-j\frac{\pi}{2}0} - e^{-j\frac{\pi}{2}2} + e^{-j\frac{\pi}{2}4} - e^{-j\frac{\pi}{2}6} + e^{-j\frac{\pi}{2}8} - e^{-j\frac{\pi}{2}10} + e^{-j\frac{\pi}{2}12} - e^{-j\frac{\pi}{2}14} \right\}$$

By rearranging terms to collect all the positive terms within brackets and all the negative terms within brackets, we end up with

$$X(4) = \sum_{n=0}^{15} x(n)\, e^{-j\frac{4\pi}{8}n} = \left[ e^{-j\frac{\pi}{2}0} + e^{-j\frac{\pi}{2}4} + e^{-j\frac{\pi}{2}8} + e^{-j\frac{\pi}{2}12} \right] - \left[ e^{-j\frac{\pi}{2}2} + e^{-j\frac{\pi}{2}6} + e^{-j\frac{\pi}{2}10} + e^{-j\frac{\pi}{2}14} \right]$$

If we add all the terms in the positive bracket and subtract all the terms in the negative bracket, we get

$$\begin{bmatrix} +1 & -j0 \\ +1 & -j0 \\ +1 & -j0 \\ +1 & -j0 \end{bmatrix} - \begin{bmatrix} -1 & -j0 \\ -1 & -j0 \\ -1 & -j0 \\ -1 & -j0 \end{bmatrix} = [4-j0]-[-4-j0]=8$$

Thus we see that $X(4)=8$.

Similar calculations will show that

$$X(5)=X(6)=X(7)=X(8)=X(9)=X(10)=X(11)=X(13)=X(14)=X(15)=0$$

For demonstration purposes, we will calculate the $X(12)$ term. The DFT output for $k=12$ is computed as

$$X(12)=\sum_{n=0}^{15}x(n)\,e^{-j\frac{12\pi}{8}n}=\left\{e^{-j\frac{3\pi}{2}0}-e^{-j\frac{3\pi}{2}2}+e^{-j\frac{3\pi}{2}4}-e^{-j\frac{3\pi}{2}6}+e^{-j\frac{3\pi}{2}8}-e^{-j\frac{3\pi}{2}10}+e^{-j\frac{3\pi}{2}12}-e^{-j\frac{3\pi}{2}14}\right\}$$

By rearranging terms to collect all the positive terms within brackets and all the negative terms within brackets, we end up with

$$X(12)=\sum_{n=0}^{15}x(n)\,e^{-j\frac{12\pi}{8}n}=\left[e^{-j\frac{3\pi}{2}0}+e^{-j\frac{3\pi}{2}4}+e^{-j\frac{3\pi}{2}8}+e^{-j\frac{3\pi}{2}12}\right]-\left[e^{-j\frac{3\pi}{2}2}+e^{-j\frac{3\pi}{2}6}+e^{-j\frac{3\pi}{2}10}+e^{-j\frac{3\pi}{2}14}\right]$$

If we add all the terms in the positive bracket and subtract all the terms in the negative bracket we get

$$\begin{bmatrix} +1 & -j0 \\ +1 & -j0 \\ +1 & -j0 \\ +1 & -j0 \end{bmatrix} - \begin{bmatrix} -1 & -j0 \\ -1 & -j0 \\ -1 & -j0 \\ -1 & -j0 \end{bmatrix} = [4-j0]-[-4-j0]=8$$

Thus we get $X(12)=8$.

If we compare our computed results to Figure 3.24, part B, we see that the only elements in the DFT output that are nonzero are $X(4)$ and $X(12)$, and their amplitude is shown as being equal to $AN/2$. The amplitude of the time domain sinusoid in our example is set to $A=1$. The DFT length is $N=16$ samples, so we would expect to see a computed amplitude of $AN/2=(1)(16)/2=8$, which is exactly what we got.

The alert reader might ask the question, "When dealing with analog signals, we would expect to see that the amplitude of the CFT of a cosine wave is equal to $A/2$; why, then, is the amplitude for the DFT equal to $AN/2$?" If you happened to ask this question, then you asked a pretty darn good question. The answer will be provided shortly. Right now, however, we need to finish the discussion on DFT symmetry.

In general, if an $N$ sample input time series to a DFT is a real sequence, then the output of the DFT will be an $N$ sample complex sequence, as illustrated in Figure 3.25. The DFT is computed by correlating the input sequence with the complex exponential sequence $e^{-j\frac{2\pi kn}{N}}$. The exponential is divided into $N$ equal length arc segments on the unit circle. The correlations are computed between the input signal and the complex exponential at each of these arc segments or correlation points.

The correlation points at each arc segment on the unit circle from $0$ to $\pi$ are the complex conjugates of the correlation points at each arc segment on the unit circle from $0$ to $-\pi$. Therefore the magnitude of the DFT outputs obtained at the correlation points between $0$ to $-\pi$ will be identical and symmetric to those obtained at the correlation points between $0$ to $\pi$. For example, if we observe any two points of equal and opposite angles from $0$ degrees on the unit circle, as illustrated in Figure 3.26, we will see that the two points are the complex conjugates of one another. The magnitudes of these points will be equal and given by

$$|X(k)| = |X^*(k)|$$

In the figure, we specifically illustrate the points corresponding to $k = 2$ and $k = 14$ for an $N = 16$ point DFT. For an $N$ point DFT the computed values for the points from $1$ to $N/2-1$ are the complex conjugates of the computed values for the points from $N-1$ down to $N/2+1$. Thus, for an $N$ point DFT, $X^*(k) = X(N-k)$ for $k = 1, 2, \cdots (N/2-1)$. These two sets of points have identical magnitudes and therefore the second set is redundant and need not

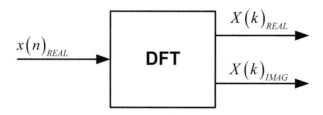

**Figure 3.25**  DFT real signal input complex signal output block diagram

be computed. We can get all the information we need from the DFT by computing only the first $N/2+1$ points, which includes the $X(0)$ or DC term.

There is one advantage to computing all $N$ points of a DFT and that is for purposes of display. For a real input signal, we can think of the values computed for the points $X(N-k)$ as being the negative frequencies associated with the signal spectrum. As illustrated in Figure 3.24, part C, we can slide the $X(N-k)$ values to the left of the origin and they take on the values associated with a signals' negative frequency content. This is commonly done to produce discrete spectral plots that mimic the spectral plots of continuous signals.

Another example is illustrated in Figure 3.27. Suppose the output of a 16-point DFT resembled part A of the figure. We can see that the horizontal axis for the DFT output in part A is calibrated using two different but equivalent units. The horizontal axis 1 in part A is calibrated by units of the discrete frequency index $k$ where $k=0,1,2,\cdots,15$. The calibration of axis 2 is in frequency expressed as a fraction of the sampling frequency $f_s$. In this calibration, we simply do the mapping

$$f = \frac{k}{N}f_s \ \{\text{for } k = 0,1,2,\cdots,N-1\}$$

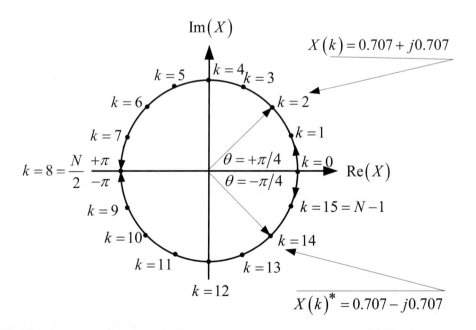

**Figure 3.26**   DFT unit circle for $N = 16$

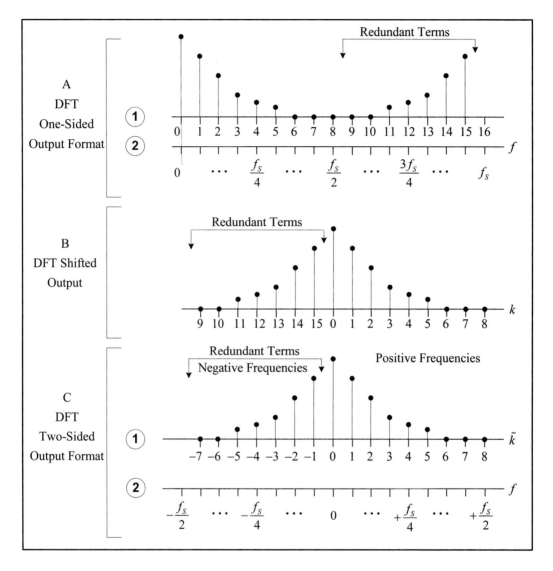

**Figure 3.27**   Example representations of a DFT output

Part B of the figure shows the DFT flipped output where the $X(N-k)$ terms are shifted to the left side of the origin. Note that the value of the frequency index $k$ has been retained for the shifted portion of the spectrum. Part C of the figure shows the two-sided spectral representation of the DFT. It also has two horizontal axes. The frequency index of the shifted $X(N-k)$ terms have been renumbered on axis 1 using negative values to represent their relative location on the horizontal axis. The bipolar frequency index for this representation is given by $\tilde{k}$. The second horizontal axis in part C is calibrated in frequency as before, this time using the mapping relation

$$f = \frac{\tilde{k}}{N} f_s \left\{ \text{for} \left[ -\frac{N}{2} + 1 \right] \le \tilde{k} \le \left[ -1 \right] \right\}$$

There are a couple of symmetry properties associated with the DFT that help simplify the computations:

- When the DFT input signal $x(n)$ is real and has even symmetry—that is, $x(n) = x(-n)$—then the complex DFT output $X(k)$ will be real and even.
- When the DFT input signal $x(n)$ is real and has odd symmetry—that is, $x(n) = -x(-n)$—then the complex DFT output $X(k)$ will be imaginary and odd.

These symmetry properties are illustrated in Figure 3.28.

We have taken a look at the case where the DFT input time series is real; now let's switch gears a bit and discuss the case where the DFT input is a complex time series. A good many signal processing architectures process complex signals. Why? There are two reasons that immediately come to mind:

1. Although it is not intuitive, processing complex signals in many instances is much simpler than processing real signals.

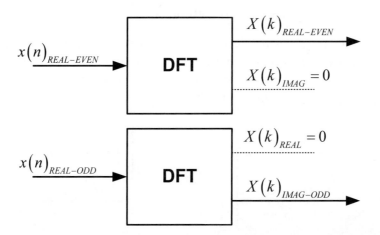

**Figure 3.28**   DFT symmetry properties

2. Many times, the first signal processing block in a system following the A/D converter is a digital tuner. The tuner inputs a real digital signal, down converts it to base band, and then sends it off for further processing. The tuner generally produces a complex signal at its output. Therefore the follow on processing must deal with a complex signal.

It makes sense to keep the signal in a complex format until it reaches the output stage of the processing system. At that point, the complex signal can be converted back to a real digital signal. The conversion of a signal from complex to real is fairly straightforward and is dealt with in Chapter 7, "Complex to Real Conversion." It is also nice to know that a complex signal has twice the Nyquist bandwidth as a real signal. The spectrum of a complex signal does not have the mirror image spectrum we associate with real signals. This interesting phenomenon will be demonstrated in the complex DFT discussion that follows.

Let us begin our discussion by looking at the input/output signal block diagram of the complex DFT shown in Figure 3.29. From the figure, it is clear that the input signal is a complex time series that has a real part and an imaginary part. The DFT outputs a complex signal that represents the spectrum of the input time series.

Let's begin by defining an example complex input signal. For simplicity of discussion, let's make the length of the signal $N = 16$. We can define the mathematical form of the input to be

$$x(n) = A\cos\left(\frac{2\pi m}{N}n\right) + jA\sin\left(\frac{2\pi m}{N}n\right) \quad \text{for } n = 0,1,2,\cdots,N-1$$

**Equation 3.69**

where $m$ is the frequency index and can take on values $0 \le m \le N-1$. Remember that for a real signal the range of $m$ is limited to $0 \le m \le N/2$ because larger values of $m$ cause frequency aliasing. So right away we can see one advantage of complex signals.

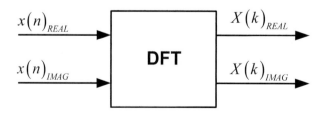

**Figure 3.29**   DFT complex signal input–complex signal output block diagram

Another advantage of complex signals is that we can simplify the signal processing mathematics by working with complex exponentials, as opposed to working with the sin and cosine functions. We can rewrite Equation 3.69 and express it in its exponential form as

$$x(n) = Ae^{+j\frac{2\pi m}{N}n}$$

Now we can write the equation for the DFT of $x(n)$ as

$$X(k) = \sum_{n=0}^{N-1} x(n)e^{-j\frac{2\pi k}{N}} = A\sum_{n=0}^{N-1} e^{+j\frac{2\pi m}{N}n}e^{-j\frac{2\pi k}{N}}$$

**Equation 3.70**

where the frequency index term for the input signal is $m$ and the frequency index term for the $N$ different correlation frequencies of the DFT is $k$. Let's stop for a moment and take a look at the complex input signal illustrated in Figure 3.30. Part A of the figure shows the real part of the input signal as being a cosine waveform of amplitude $A$. Part B of the figure shows the imaginary part of the input signal as being a sine wave, also of amplitude $A$. Since the example input signal is 16 samples in length, only the first 16 samples are shown. The reader should note that we have an integral number of cycles within the 16-point sample length. That is, there are 4 complete cycles of sine and cosine contained in the 16-point sequence length and therefore $m/N = 4/16 = 1/4$. We already know that the frequency of the sinusoidal waveforms in the figure is given by

$$f = \frac{m}{N}f_s$$

Therefore we can compute the frequency of the sinusoidal waveforms to be

$$f = \frac{4}{16}f_s = \frac{1}{4}f_s = \frac{f_s}{4}$$

We can simplify the DFT equation for this particular input signal by rewriting Equation 3.70 as

$$X(k) = A\sum_{n=0}^{N-1} e^{-j\frac{2\pi(k-m)}{N}n}$$

**Equation 3.71**

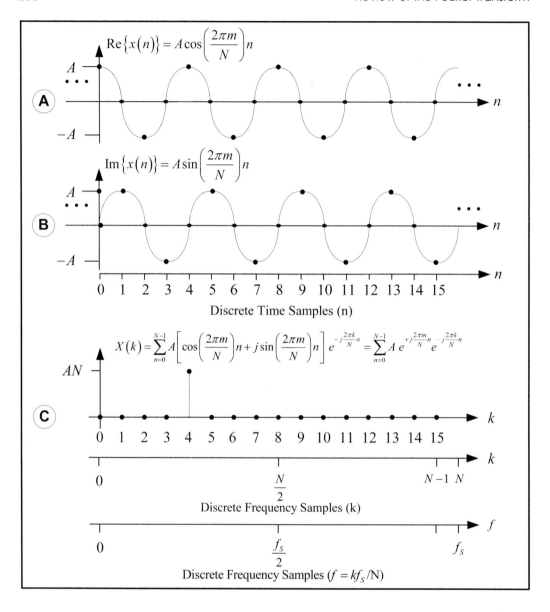

**Figure 3.30**   DFT of a complex input signal

This example is simple enough that we do not need to compute all 16 complex DFT output points. Instead, we will consider just two cases.

### 3.3.2.1   Case I: k = m

In this case, the input signal frequency is identical to one of the 16 correlation frequencies of the DFT. The complex exponential collapses to unity, and we can write

$$X(k) = X(m) = A \sum_{n=0}^{N-1} e^{-j\frac{2\pi(k-m)}{N}n} = A \sum_{n=0}^{N-1} e^{-j\frac{2\pi(0)}{N}n} = A \sum_{n=0}^{N-1}(1) = AN$$

Since the frequency of the input sinusoid is identical to one of the discrete frequencies of the DFT, we get a maximum correlation between the two. In our example, $N = 16$, and if the amplitude $A = 1$, then the output of the DFT at the point $k = m$ is $X(k) = AN = 16$. The reader should note from this example that if the complex input signal has an integer number of complete cycles within the DFT sample space, and if the amplitude of this signal is $A$, then the amplitude of the spectral line at the output of the DFT will be $AN$.

That was pretty simple. Now let's consider the remaining 15 correlation frequencies of this example DFT.

### 3.3.2.2   Case II: $k \neq m$

In this case, the input signal frequency is different from the correlation frequency of the DFT. We can utilize Equation 3.71 once again, which is repeated here for convenience:

$$X(k) = A \sum_{n=0}^{N-1} e^{-j\frac{2\pi(k-m)}{N}n}$$

If there are an integer number of cycles of the input waveform $x(n)$ contained in the $N$ points of the DFT, then we know from the trigonometric identities in Equation 3.4 that for $k \neq m$ the summation term will equal zero. Therefore we can write

$$X(k) = A \sum_{n=0}^{N-1} e^{-j\frac{2\pi(k-m)}{N}n} = 0 \qquad \text{for } k \neq m$$

**Equation 3.72**

This makes sense if we visualize the complex sinusoid as having an integral number of cycles within $N$ points and if we recall that the sinusoid is bipolar. The positive sinusoid values cancel the negative sinusoid values. The DFT of a complex sinusoid with $k$ complete cycles contained within the $N$ point sample space can be written by inspection as

$$X(m = k) = AN$$
$$X(m \neq k) = 0$$

The spectrum of this example complex input sinusoid is illustrated in part C of Figure 3.30. The reader should note that there is a single impulse located at $k = 4$ and all other discrete frequencies within the signal spectrum from $0$ to $f_s$ are zero. The reader may want to see some verification that Equation 3.72 is true. This can be proved simply enough by expanding the complex exponential into its sine and cosine components. In doing so, we end up with

$$X(k) \quad = A \sum_{n=0}^{N-1} e^{-j\frac{2\pi(k-m)}{N}n}$$

$$= A \sum_{n=0}^{N-1} \left[ \cos\left(\frac{2\pi m}{N}n\right) + j\sin\left(\frac{2\pi m}{N}n\right) \right] \left[ \cos\left(\frac{2\pi k}{N}n\right) - j\sin\left(\frac{2\pi k}{N}n\right) \right]$$

Now we perform the complex multiplication and get

$$X(k) = A \sum_{n=0}^{N-1} \left\{ \cos\left(\frac{2\pi m}{N}n\right)\cos\left(\frac{2\pi k}{N}n\right) + \sin\left(\frac{2\pi m}{N}n\right)\sin\left(\frac{2\pi k}{N}n\right) \right.$$

$$\left. -j\cos\left(\frac{2\pi m}{N}n\right)\sin\left(\frac{2\pi k}{N}n\right) + j\cos\left(\frac{2\pi k}{N}n\right)\sin\left(\frac{2\pi m}{N}n\right) \right\}$$

**Equation 3.73**

To finish this computation, we call on the help of the continuous form of the trigonometric identities of Equation 3.4. We convert these identities from the continuous form to the discrete form to obtain the identities

$$\sum_{n=0}^{N-1} \cos\left(\frac{2\pi m}{N}n\right)\sin\left(\frac{2\pi k}{N}n\right) = 0 \quad \text{for all } m \text{ and } k$$

$$\sum_{n=0}^{N-1} \cos\left(\frac{2\pi m}{N}n\right)\cos\left(\frac{2\pi k}{N}n\right) = 0 \quad \text{for } m \neq k$$

$$\sum_{n=0}^{N-1} \sin\left(\frac{2\pi m}{N}n\right)\sin\left(\frac{2\pi k}{N}n\right) = 0 \quad \text{for } m \neq k$$

**Equation 3.74**

Comparing the discrete form of the identities in Equation 3.74 to the expanded DFT expression in Equation 3.73, we can easily see that

$$X(k) = 0 + 0 - j0 + j0 \quad \{\text{for } m \neq k\}$$

The proof is in the pudding, so to speak. If the reader is still not convinced, it is easy to write an Excel spreadsheet to compute the sum given in

Equation 3.72. Just remember that a very small floating point number in Excel like 3.45E-14, for example, in a problem like this is due to the arithmetic noise caused by quantized data and is the equivalent of zero.

### 3.3.3   DFT Resolution

The resolution of a DFT simply means the smallest chunk of spectral bandwidth between adjacent bins in an $N$ point DFT. This is typically given by

$$f_{resolution} = \frac{f_S}{N}$$

So the resolution depends on the sample rate $f_S$ and the length of the DFT. If, for example, the input signal to a 512-point DFT was sampled at 512 KHz, the resolution would be

$$f_{resolution} = \frac{f_S}{N} = \frac{512 \text{ KHz}}{512} = 1 \text{ KHz}$$

In this example, each bin in the 512-point DFT has an ideal bandwidth of 1 KHz. The term *ideal* is used because, as we will soon see, the actual bandwidth of each DFT bin is determined by the type of preprocessing window used. A window is nothing more than a special sequence that is the same length as the data sequence processed by the DFT. The window sequence multiplies the data sequence to reduce or smooth time domain discontinuities at its end points. Some windows produce bin bandwidths that are more ideal than others. We will discuss window sequences later in this chapter.

### 3.3.4   Inverse DFT

The forward DFT is usually represented by

$$f\{x(n)\} = X(k) = \sum_{n=0}^{N-1} x(n) e^{-j\frac{2\pi k}{N}n} \quad \left\{ \begin{array}{l} \text{for } n = 0,1,2,3,\cdots,N-1 \\ \text{for } k = 0,1,2,3,\cdots,N-1 \end{array} \right\}$$

**Equation 3.75**

The inverse discrete Fourier transform (IDFT) is usually represented by

$$f^{-1}\{X(k)\} = x(n) = \sum_{k=0}^{N-1} X(k) e^{+j\frac{2\pi k}{N}n} \quad \left\{ \begin{array}{l} \text{for } n = 0,1,2,3,\cdots,N-1 \\ \text{for } k = 0,1,2,3,\cdots,N-1 \end{array} \right\}$$

**Equation 3.76**

The reader will note that the only difference between the forward and inverse transforms is the input sequence and the sign of the complex exponential. Both of these representations are correct and legitimate to use. Other forms of the inverse DFT are also used. Remember that the forward transform has a gain factor of $N$. Therefore some engineers and some software applications prefer to include a scale factor of $1/N$ for the inverse transform.

$$f^{-1}\{X(k)\} = x(n) = \frac{1}{N}\sum_{k=0}^{N-1}X(k)e^{+j\frac{2\pi k}{N}n} \quad \left\{ \begin{array}{l} \text{for } n = 0,1,2,3,\cdots,N-1 \\ \text{for } k = 0,1,2,3,\cdots,N-1 \end{array} \right\}$$

**Equation 3.77**

Since there is never 100% agreement on anything, some engineers and some software applications prefer to apply a scale factor to both the forward and inverse DFT such that

$$f^{-1}\{X(k)\} = x(n) = \frac{1}{\sqrt{N}}\sum_{k=0}^{N-1}X(k)e^{+j\frac{2\pi k}{N}n}$$

$$f\{x(n)\} = X(k) = \frac{1}{\sqrt{N}}\sum_{n=0}^{N-1}x(n)e^{-j\frac{2\pi k}{N}n}$$

**Equation 3.78**

Because we are usually more concerned about the relative magnitudes of a signal within each bin of the DFT, the scaling is usually not an issue for most applications.

### 3.3.5 DFT Representation

There are several acceptable methods of expressing a DFT in terms of the discrete frequencies that it produces. Two of the most common methods are discussed in the following sections.

### 3.3.5.1 Method 1

The first and most common method is to simply represent the output of each of the $N$ bins of the DFT as an instantaneous estimate of the discrete frequency within the input signal that falls within the bandwidth of that bin. This estimate is made for each frequency index $k$ where $0 \le k \le N-1$. This representation is the most familiar and is listed here:

$$H(k) = \sum_{n=0}^{N-1}h(n)\,e^{-j\left(\frac{2\pi}{N}\right)nk} \quad \text{for} \left\{ \begin{array}{l} 0 \le n \le N-1 \\ 0 \le k \le N-1 \end{array} \right\}$$

$$h(n) = \frac{1}{N} \sum_{k=0}^{N-1} H(k) \, e^{+j\left(\frac{2\pi}{N}\right)kn} \quad \text{for} \left\{ \begin{array}{l} 0 \le n \le N-1 \\ 0 \le k \le N-1 \end{array} \right\}$$

**Equation 3.79**

### 3.3.5.2 Method 2

When we think about discrete frequencies that are associated with DFT bins, we normally think in terms of

$$f = \frac{k}{N} f_s \; \{\text{unnormalized}\} \quad \text{or} \quad f = \frac{k}{N} \; \{\text{normalized to } f_s\}$$

The second method commonly used to represent the $N$ discrete frequency estimates produced by the DFT is by the normalized frequency $k/N$ for each of the $k$ bins. This representation is as follows:

$$H\left(\frac{k}{N}\right) = \sum_{n=0}^{N-1} h(n) \, e^{-j2\pi\left(\frac{k}{N}\right)n} \quad \text{for} \left\{ \begin{array}{l} 0 \le n \le N-1 \\ 0 \le k \le N-1 \end{array} \right\}$$

$$h(n) = \sum_{k=0}^{N-1} H\left(\frac{k}{N}\right) \, e^{+j2\pi\left(\frac{k}{N}\right)n} \quad \text{for} \left\{ \begin{array}{l} 0 \le n \le N-1 \\ 0 \le k \le N-1 \end{array} \right\}$$

**Equation 3.80**

The expressions given in Equation 3.80 may look a bit awkward at first, but they are useful when dealing with both the computations and the notation involved in the tuning of signal spectra up and down in frequency. We will use the notation of Equation 3.80 when we discuss discrete signal tuning in Chapter 9, "Signal Tuning."

### 3.3.6 DFT Discrete Power Spectrum

The $N$ output points of a DFT are many times referred to as *bins*. We use the terms *DFT point* and *DFT bin* interchangeably. Most of the time we like to postprocess the DFT output by computing the magnitude, the phase, and the power of the signal spectrum in each DFT bin. The complex output of some arbitrary DFT bin $k$, which represents the normalized frequency $k/N$, is expressed as a complex number in Cartesian coordinates as

$$X(k) = X(k)_{REAL} + j \, X(k)_{IMAG} \quad \{\text{method 1}\}$$

or equivalently

$$X\left(\frac{k}{N}\right) = X\left(\frac{k}{N}\right)_{REAL} + j\, X\left(\frac{k}{N}\right)_{IMAG} \quad \{\text{method 2}\}$$

The same complex number can be expressed in polar form as

$$X(k) = |X(k)| \angle\theta(k)$$

or equivalently

$$X\left(\frac{k}{N}\right) = \left|X\left(\frac{k}{N}\right)\right| \angle\theta(k)$$

where $|X(k)|$ is the magnitude and $\theta(k)$ is the phase of $X(k)$. If we use the notation of method 1, the magnitude of the $k$th bin is given by

$$|X(k)| = \sqrt{X(k)^2_{real} + X(k)^2_{imag}}$$

**Equation 3.81**

The phase is computed by

$$\theta(k) = \tan^{-1}\left(\frac{X(k)_{imag}}{X(k)_{real}}\right)$$

**Equation 3.82**

The power of the signal in the $k$th DFT bin is simply the square of the magnitude or

$$|X(k)|^2 \;=\; X(k)X^*(k) \;=\; X(k)^2_{real} + X(k)^2_{imag}$$

**Equation 3.83**

Expressed in units of decibels, the signal power is given by

$$|X(k)|^2 \;=\; 20Log_{10}\left(\sqrt{X(k)^2_{real} + X(k)^2_{imag}}\right) \;=\; 10Log_{10}\left(X(k)^2_{real} + X(k)^2_{imag}\right)$$

**Equation 3.84**

Usually, when plotting the spectral magnitude or power of a signal processed by a DFT, we normalize all the plot values to a magnitude of 1 or 0 db by computing the log of each bin magnitude relative to the peak bin magnitude:

$$|X(k)|^2 = 20Log_{10}\left[\frac{|X(k)|}{|X(k)|_{Max}}\right] = 10Log_{10}\left[\frac{|X(k)|^2}{|X(k)|^2_{Max}}\right]$$

**Equation 3.85**

### 3.3.7 DFT of a Pulsed Waveform

Let us compute the DFT of the unit pulse shown in Figure 3.31. This derivation will be similar to the derivation we did for the continuous time pulsed waveform. The major difference is that in this case we are dealing with a sampled waveform. The waveform that we are interested in is a sequence of samples labeled $x(n)$. The sequence is $N$ samples in length. Of the $N$ samples, $L$ samples have unity amplitude, and the remaining samples are zero. We begin by writing the equation of the DFT for $x(n)$:

$$X(k) = \sum_{n=0}^{N-1} x(n)\, e^{-j\left(\frac{2\pi k}{N}\right)n}$$

There is no need to involve the zero samples in our computation, so we can rewrite the equation as

$$X(k) = \sum_{n=0}^{L-1} (1)\, e^{-j\left(\frac{2\pi k}{N}\right)n}$$

We expand this summation to get

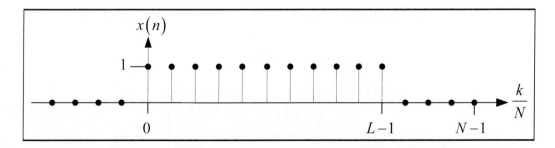

**Figure 3.31**   Discrete pulse waveform

$$X(k) = 1 + e^{-j\left(\frac{2\pi k}{N}\right)} + e^{-j\left(\frac{2\pi k}{N}\right)2} + e^{-j\left(\frac{2\pi k}{N}\right)3} \cdots + e^{-j\left(\frac{2\pi k}{N}\right)(L-1)}$$

**Equation 3.86**

Now if we multiply $X(k)$ by $e^{-j\left(\frac{2\pi k}{N}\right)}$ we get

$$X(k)e^{-j\left(\frac{2\pi k}{N}\right)} = e^{-j\left(\frac{2\pi k}{N}\right)} + e^{-j\left(\frac{2\pi k}{N}\right)2} + e^{-j\left(\frac{2\pi k}{N}\right)3} \cdots + e^{-j\left(\frac{2\pi k}{N}\right)(L-1)} + e^{-j\left(\frac{2\pi k}{N}\right)(L)}$$

**Equation 3.87**

If we subtract Equation 3.87 from Equation 3.86, we get

$$X(k) - X(k)e^{-j\left(\frac{2\pi k}{N}\right)} = 1 - e^{-j\left(\frac{2\pi k}{N}\right)(L)}$$

Rearranging the terms, we end up with

$$X(k)\left(1 - e^{-j\left(\frac{2\pi k}{N}\right)}\right) = 1 - e^{-j\left(\frac{2\pi k}{N}\right)(L)}$$

Dividing both sides by the quantity $1 - e^{-j\left(\frac{2\pi k}{N}\right)}$, we end up with

$$X(k) = \left[\frac{1 - e^{-j\left(\frac{2\pi k}{N}\right)(L)}}{1 - e^{-j\left(\frac{2\pi k}{N}\right)}}\right]$$

**Equation 3.88**

We can simplify Equation 3.88 by factoring the exponential terms to get

$$X(k) = \left[\frac{e^{-j\left(\frac{\pi k}{N}\right)(L)}\left(e^{+j\left(\frac{\pi k}{N}\right)(L)} - e^{-j\left(\frac{\pi k}{N}\right)(L)}\right)}{e^{-j\left(\frac{\pi k}{N}\right)}\left(e^{+j\left(\frac{\pi k}{N}\right)} - e^{-j\left(\frac{\pi k}{N}\right)}\right)}\right]$$

Now we can combine the factored terms to get

$$X(k) = e^{-j\left(\frac{\pi k}{N}\right)(L-1)} \left[ \frac{\left( e^{+j\left(\frac{\pi k}{N}\right)(L)} - e^{-j\left(\frac{\pi k}{N}\right)(L)} \right)}{\left( e^{+j\left(\frac{\pi k}{N}\right)} - e^{-j\left(\frac{\pi k}{N}\right)} \right)} \right]$$

The $e^{-j\left(\frac{\pi k}{N}\right)(L-1)}$ term is a phase term and the expression within brackets is the amplitude term for $X(k)$. We can rearrange Euler's well-known formula to help us clean up the amplitude term. In doing so, we get

$$\sin(\omega) = \frac{e^{+j\omega} - e^{-j\omega}}{2j} \quad \Rightarrow \quad 2j\sin(\omega) = e^{+j\omega} - e^{-j\omega}$$

Upon substitution, we arrive at

$$X(k) = e^{-j\left(\frac{\pi k}{N}\right)(L-1)} \left[ \frac{\sin\left(\frac{\pi k L}{N}\right)}{\sin\left(\frac{\pi k}{N}\right)} \right]$$

**Equation 3.89**

Now $X(k)$ is expressed in terms of its amplitude and phase. If we compute the magnitude of $X(k)$, we arrive at

$$|X(k)| = \left| \frac{\sin\left(\frac{\pi k L}{N}\right)}{\sin\left(\frac{\pi k}{N}\right)} \right|$$

and if we compute the phase we get

$$\angle\theta = e^{-j\left(\frac{\pi k}{N}\right)(L-1)}$$

Then we can write

$$X(k) = |X(k)|\angle\theta$$

From the expression in Equation 3.89, it appears that the magnitude response is similar to something like $\sin x / x$. We have a small problem left to clear up though: the magnitude response at $k=0$ is $0/0$, which is an indeterminate form. We could determine the response value at $k=0$ using l'Hôpital's rule, but we would like to use a simpler method and avoid all that pesky numerator and denominator differentiation stuff. Let's instead do something simple, like multiplying the numerator by

$$\left(\frac{\pi kL}{N}\bigg/\frac{\pi kL}{N}\right)$$

and multiplying the denominator by

$$\left(\frac{\pi k}{N}\bigg/\frac{\pi k}{N}\right)$$

In doing so, we get

$$|X(k)| = \left|\frac{\left[\left(\frac{\sin(\pi kL/N)}{\pi kL/N}\right)\pi kL/N\right]}{\left[\left(\frac{\sin(\pi k/N)}{\pi k/N}\right)(\pi k/N)\right]}\right| = L\left|\frac{\left[\left(\frac{\sin(\pi kL/N)}{\pi kL/N}\right)\right]}{\left[\left(\frac{\sin(\pi k/N)}{\pi k/N}\right)\right]}\right|$$

**Equation 3.90**

We can see from this expression that we now have a term in both the numerator and the denominator that takes on the familiar form of $\sin x/x$. And we know from basic calculus that in the limit as $x\to0$, $\sin x\to x$, and therefore $\sin x/x \to 1$. So let's see what happens when the take the limit of $|X(k)|$ as $k\to0$:

$$\lim_{k\to0}|X(k)| = \lim_{k\to0}L\left[\frac{\left(\frac{\sin(\pi kL/N)}{\pi kL/N}\right)}{\left(\frac{\sin(\pi k/N)}{\pi k/N}\right)}\right] = L\left(\frac{1}{1}\right) = L$$

Now we know that the amplitude at $k=0$ is the length of the unit pulse or $L$. Most of the time we like to normalize the frequency response by dividing by $L$ so that the maximum amplitude is one. After normalization, we end up with

$$|X(k)| = \frac{1}{L}\left[\frac{\sin\left(\frac{\pi kL}{N}\right)}{\sin\left(\frac{\pi k}{N}\right)}\right] \text{ and } \angle\theta = e^{-j\left(\frac{\pi k}{N}\right)(L-1)}$$

We can see that the normalized frequency magnitude response is unity at $k = 0$ and has zeros whenever

$$n\left(\frac{\pi k L}{N}\right) = 0, \text{ or when } \frac{\pi k L}{N} = m\pi \quad \{\text{for } m = 1, 2, 3, \cdots L - 1\}$$

If we rearrange terms, we can see that the frequency response zeros of the numerator term occur when the normalized frequency is

$$\frac{k}{N} = \frac{m}{L} \quad \{\text{for } m = 1, 2, 3, \cdots L - 1\}$$

If the reader so desires, the frequency response zeros can also be expressed in terms of normalized radians per second or

$$\frac{2\pi k}{N} = \frac{2\pi m}{L} \quad \{\text{for } m = 1, 2, 3, \cdots L - 1\}$$

The frequency response of the pulsed waveform is illustrated in Figure 3.32.

## 3.3.8 Properties of the DFT

There are several properties of the DFT that makes the visualization and the computation of a DFT a little easier. Most of these properties have to do with the even/odd symmetry of the DFT input and output signals. These properties are listed in Table 3.6.

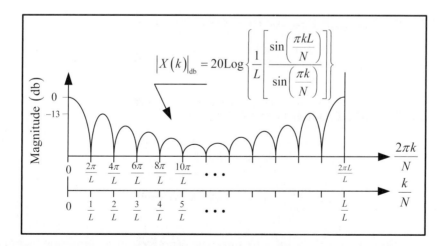

**Figure 3.32** Spectrum of a discrete pulse

**Table 3.6**   Properties of the DFT

| Time domain DFT input $x(n) = x_{REAL}(n) + jx_{IMAG}(n)$ | Frequency domain DFT output $X(k) = X_{REAL}(k) + jX_{IMAG}(k)$ |
|---|---|
| $x(n) = x_{REAL}(n)$ | $X(k) = X_{REAL+EVEN}(k) + jX_{IMAG+ODD}(k)$ |
| $x(n) = x_{IMAG}(n)$ | $X(k) = X_{REAL+ODD}(k) + jX_{IMAG+EVEN}(k)$ |
| $x(n) = x_{REAL+EVEN}(n) + jx_{IMAG+ODD}(n)$ | $X(k) = X_{REAL}(k)$ |
| $x(n) = x_{REAL+ODD}(n) + jx_{IMAG+EVEN}(n)$ | $X(k) = jX_{IMAG}(k)$ |
| $x(n) = x_{REAL+EVEN}(n)$ | $X(k) = X_{REAL+EVEN}(k)$ |
| $x(n) = x_{REAL+ODD}(n)$ | $X(k) = jX_{IMAG+ODD}(k)$ |
| $x(n) = x_{IMAG+EVEN}(n)$ | $X(k) = jX_{IMAG+EVEN}(k)$ |
| $x(n) = x_{IMAG+ODD}(n)$ | $X(k) = X_{REAL+ODD}(k)$ |
| $x(n) = x_{REAL+EVEN}(n) + jx_{IMAG+EVEN}(n)$ | $X(k) = X_{REAL+EVEN}(k) + jX_{IMAG+EVEN}(k)$ |
| $x(n) = x_{REAL+ODD}(n) + jx_{IMAG+ODD}(n)$ | $X(k) = X_{REAL+ODD}(k) + jX_{IMAG+ODD}(k)$ |

### 3.3.9   Discrete Fourier Transform Pairs

There exist several important time domain / frequency domain pairs that are useful when working with the DFT. These pairs are referred to as *discrete Fourier transform pairs*. Several of these pairs are listed in Table 3.7.

### 3.3.10   DFT Bin Band Pass Filter

What do we mean by a DFT "bin band pass filter"? Let's begin by revisiting Equation 3.66, which describes the DFT:

$$X(k) = \sum_{n=0}^{N-1} x(n) e^{-j\frac{2\pi k}{N}n} \quad \text{for } k = 0, 1, 2, \cdots, N-1$$

Sometimes it's helpful to step back and look at an equation as being the description of a signal processing system. We can look at Equation 3.66 from two different perspectives.

**Table 3.7**   Discrete Fourier Transform Pairs

| Property | DFT |
|---|---|
| DFT | $x(n) \overset{f}{\leftrightarrow} X(k)$ |
| Linearity | $ax(n) + by(n) \overset{f}{\leftrightarrow} aX(k) + bY(k)$ |
| Shift in time | $x(n-m) \overset{f}{\leftrightarrow} X(k)e^{-j\frac{2\pi m}{N}n}$ |
| Shift in frequency | $x(n)e^{\pm j\frac{2\pi m}{N}n} \overset{f}{\leftrightarrow} X(k \mp m)$ |
| Convolution | $y(n) = \sum_{m=0}^{N-1} x(m)h(n-m) = x(n) * h(n)$ |
| Time domain convolution | $x(n) * h(n) \overset{f}{\leftrightarrow} X(k)H(k)$ |
| Time domain multiply | $x(n)h(n) \overset{f}{\leftrightarrow} X(k) * H(k)$ |
| Unit impulse | $\Delta(n) \overset{f}{\leftrightarrow} 1$ |
| Complex conjugation | $x^*(n) \overset{f}{\leftrightarrow} X^*(-k)$ |
| Modulation | $x(n)\cos\left(\frac{2\pi k_0}{N}n\right) \overset{f}{\leftrightarrow} \frac{1}{2}\left[X(k+k_0) + X(k-k_0)\right]$ |
| Modulation | $x(n)\sin\left(\frac{2\pi k_0}{N}n\right) \overset{f}{\leftrightarrow} \frac{1}{2j}\left[X(k+k_0) - X(k-k_0)\right]$ |

The first perspective is that the $N$ point input sequence $x(n)$ is cross-correlated with $N$ complex sinusoid vectors. Each bin of the DFT output represents the correlation score of the input sequence with one of the $N$ complex sinusoids. This is the perspective that we have used previously.

A second perspective would be the input sequence $x(n)$ is down converted or tuned to base band $N$ times by $N$ equally spaced complex tuning frequencies and then filtered by a low pass filter to extract the spectral output associated with each DFT bin.

The reader might ask, "Where does the low pass filter come from?" The answer is, the $N$ point summation operator $\sum_{n=0}^{N-1}$ has a $\sin x / x$ "like" frequency response. The main lobe of this frequency response acts like a low pass filter. We can think of this filter as a DFT bin filter. As we will see later in this chapter, this is not an optimum filter by any means. As a matter of fact, it's a poor filter for most applications, but since it comes with the DFT, it is free, and it is sufficient for some uses. For precision spectral processing, however, this filter, inherent to the DFT, needs to be enhanced. The reader is encouraged to read Chapter 9, "Signal Tuning," and Chapter 12, "Channelized Filter Bank," for a more detailed discussion of this enhancement process.

### 3.3.11  DFT Leakage

The output of an $N$ point DFT consists of $N$ complex points or processed samples. Individual points or samples are often referred to as the output of a DFT "bin." Each bin is associated with a discrete DFT frequency index $k$, where $k = 0, 1, 2, \cdots, N-1$. The processed sample associated with any particular DFT bin is complex and represents the estimate of the magnitude and phase of all the signal spectral components that fall within the pass band of that particular bin's band pass filter.

#### 3.3.11.1   Ideal Case: No Leakage

In the ideal case, the spectral components of a signal will all have an integral number of complete periods contained within the $N$ samples of the DFT. In this case, each spectral component will have a frequency index $k$ that is an integer with the range $0 \le k \le N-1$.

For example, a signal may take the form of

$$x(n) = A_1 \cos\left(\frac{2\pi k_1}{N} n\right) + A_2 \cos\left(\frac{2\pi k_2}{N} n\right) + A_3 \cos\left(\frac{2\pi k_3}{N} n\right) \quad \text{for } 0 \le n \le N-1$$

**Equation 3.91**

In this ideal case, all the spectral energy associated with the real and even signal component

$$A_1 \cos\left(\frac{2\pi k_1}{N} n\right)$$

will reside entirely in the $k_1$th and $(N - k_1)$th frequency bin, and we would expect the magnitude of the DFT output in each of these two bins to be $A_1 N / 2$.

We would expect to see similar results for the other two cosine terms in this ideal signal. If, for example, we set the variables in Equation 3.91 as follows:

$$
\begin{aligned}
A_1 &= 1 & k_1 &= 8 \\
A_2 &= 2 & k_2 &= 12 \\
A_3 &= 3 & k_3 &= 16 \\
N &= 64 &
\end{aligned}
$$

then we will expect to see symmetrical line spectra located at

$$
\begin{aligned}
k = k_1 &= 8 \text{ and } & (N-k) &= 56 & Mag &= A_1N/2 = 32 \\
k = k_2 &= 12 \text{ and } & (N-k) &= 52 & Mag &= A_2N/2 = 64 \\
k = k_3 &= 16 \text{ and } & (N-k) &= 48 & Mag &= A_3N/2 = 96
\end{aligned}
$$

A simple Excel plot of this ideal DFT output signal is illustrated in Figure 3.33.

The situation described by Equation 3.91 and illustrated in Figure 3.33 represents the ideal case where the signal input to an $N$ point DFT contains sinusoidal components that are periodic in $N$ samples, and therefore all the energy associated with each sinusoid is contained in a symmetric pair of frequency impulses. This case does not arise too often in signal processing. The more common signal that is processed by a DFT is a wideband signal that

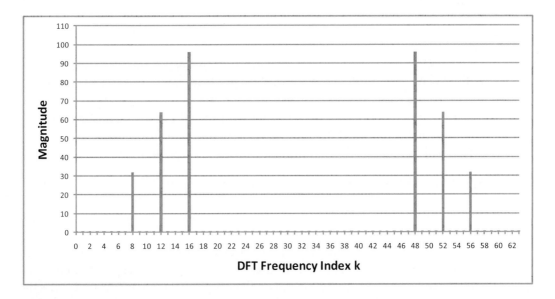

**Figure 3.33**   Ideal spectral case: No leakage

contains all frequencies within the band. The majority of these frequencies will not be ideal in that they will not be periodic within the $N$ point DFT sample space. This gives rise to a nasty DFT attribute called *leakage*, which we discuss next.

### 3.3.11.2  *Common Case: Spectral Leakage*

An $N$ point DFT can be thought of as a bank of $N$ adjacent band pass filters. In most applications, a DFT is used to process wideband signals. The band pass filters can be thought of as separating adjacent narrow bands of frequencies from the input wideband signal so that the energy within the narrow bandwidths can be computed and assigned to a DFT output bin. Expressed a different way, the signal energy that passes through each of these $N$ band pass filters can be computed as a spectral magnitude and is associated with the appropriate bin of the DFT. The magnitude response of each DFT bin band pass filter is given by

$$|H(\omega)| = \left\{ \frac{\sin\left(\dfrac{\omega N}{2}\right)}{\sin\left(\dfrac{\omega}{2}\right)} \right\}$$

The derivation of the DFT band pass filter response is discussed in detail in Chapter 12, "Channelized Filter Bank." The band pass filter frequency response is directly correlated with the method used to window or truncate the input signal such that a sequence of $N$ samples is extracted from a theoretically infinite signal stream. We will discuss the concept of windowing in the next section.

In an effort to aid the readers' understanding of this critical issue, we will present the concept of spectral leakage from two different perspectives. In both cases we will simply select, without modification, $N$ sequential samples from a sinusoidal signal as our DFT input. This is the equivalent of multiplying the input signal by a rectangular window of unity magnitude for $N$ sequential samples and zero magnitude for all other samples.

**Spectral leakage perspective 1.** We can think of the DFT as being a bank of adjacent band pass filters with each filter centered on a DFT output bin. Each of these band pass filters has an identical $\sin(\omega N/2)/\sin(\omega/2)$ frequency magnitude response.

For example, the $\sin(\omega N/2)/\sin(\omega/2)$ band pass filter response for bin 0 is illustrated in part A of Figure 3.34. The DFT bin 0 corresponds to the discrete frequency index $k = 0$ or $k/N = 0$ and is normally associated with the extraction of the DC component of any input signal. We can see from the figure that the band pass filter main lobe has a normalized bandwidth of $2/N$

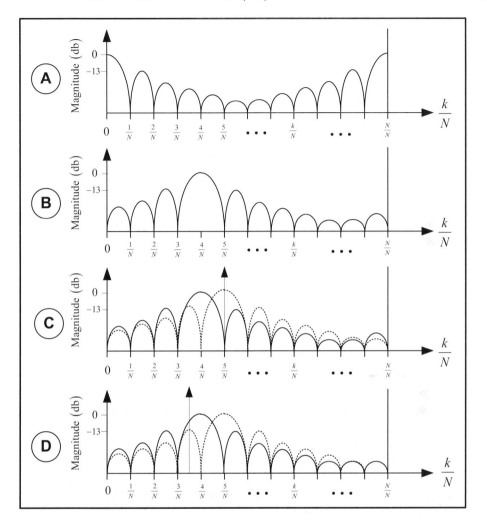

**Figure 3.34**    DFT inherent sin$x$/$x$ filter response

and all the filter response side lobes have a normalized bandwidth of $1/N$. The unnormalized bandwidths can be expressed in terms of the sample frequency as $2f_s/N$ and $f_s/N$, respectively.

We can also see from part A of the figure that the magnitude of the band pass filter first side lobe is only 13 db down from the main lobe magnitude. The attenuated magnitude of subsequent side lobes relative to the main lobe is not very impressive. For example, the second side lobe is down only 18 db from the main lobe and the third side lobe is only down 20 db.

In part B of the figure, we see the band pass filter response for the $k = 4th$ DFT bin. Note that the main lobe of the filter is centered at the normalized frequency of $4/N$, or equivalently at the unnormalized frequency of $4f_s/N$. Also note that the main lobe bandwidth overlaps into the two adjacent frequency bins. The filter side lobes, although attenuated in magnitude, overlap into all the $N$ DFT bins. This means that each bin filter will pass via its side lobes' attenuated energy associated with all other bins.

This band pass filter overlap presents a bit of a problem. In part C of the figure, we see the superposition of two adjacent bin band pass filters and their overlap. The two band pass filters correspond to the fourth and fifth DFT bins. The filter response centered on the $k = 4th$ bin is represented by a solid line. The filter response centered on the $k = 5th$ bin is represented by a dashed line. With this drawing, it is easy to visualize the overlapping frequency response of these two adjacent filters. In part C, we also show the line spectrum of a pure complex sinusoidal tone of discrete frequency $k = 5/N$. This is a sweet frequency in that it just so happens to have five complete periods within the $N$ sample DFT. (When working with DFTs, a frequency that falls directly into a single bin is often times called a *sweet frequency*.) Note that all the energy for this sinusoid falls directly in the center of the $k = 5th$ DFT bin. Also note that this tone also falls directly in the null of the filter response of the fourth DFT band pass filter response. In this example, we can see that the energy of this sinusoid will be confined to the fifth bin of the DFT on output, and no leakage of energy will be spread to the fourth DFT bin via the side lobes of the fourth bin filter. This is exactly the case for the spectral plot of the three sinusoids illustrated previously in Figure 3.33.

In part D of Figure 3.34, we see an example of a sinusoidal tone that does not fall directly on a DFT bin. In this case, the frequency of the tone is $k/N = 3.5/N$, which means that there is not an integer number of periods contained within the DFT $N$ point sample space. In this case, it is clear from the figure that the energy of the tone falls into the main lobe of the $k = 4th$ band pass filter bin and into the side lobe of the $k = 5th$ band pass filter bin. Therefore, in this simple example, both filters will pass tone energy, and we will see energy contributions due to this single tone in both DFT output bins. In the context of the DFT, this is not a sweet frequency.

Now let's look at a more general example—only this time we will consider the frequency response for all $N$ DFT band pass filters illustrated in Figure 3.35. In this figure, we see the frequency response of all $N$ DFT band pass filters superimposed onto the same plot. In this example, we see the line spectrum of a complex sinusoid at $k/N = 3/N$ and another at $k/N = 4.5/N$. It is easy to see that the tone at $3/N$ falls directly in the center of the main lobe of the third DFT band pass filter, and it falls directly in the nulls of the frequency response for all the other $N - 1$ band pass filters. The energy of this particular tone will be confined to the third DFT bin, and on output we will see only a single line spectrum located in the $k = 3$ DFT bin. In this example, the frequency $k/N = 3/N$ is a sweet frequency.

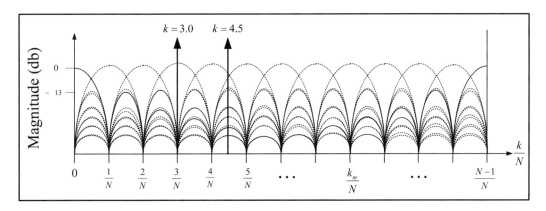

**Figure 3.35**   Adjacent band pass filter responses for an $N$ point DFT

The complex sinusoid at $4.5/N$, however, is a different story. We can clearly see that this tone falls in the main lobe of two adjacent band pass filters and in the side lobes of all the remaining $N-2$ band pass filters. Therefore, on output, the DFT will show energy from this tone distributed over all $N$ DFT bins. The magnitude of the signal energy that is leaked into adjacent bins is dependent upon the actual frequency of the tone, the bandwidth of the filter main lobe, and the amount of side lobe attenuation of that bins band pass filter. This spillage of energy from a single tone into all of the DFT bins is referred to as *spectral leakage*. When processing wideband signals with a DFT, the engineer needs to recognize that not all frequencies in the input signal are periodic in the N sample space of the DFT, and therefore they will make a contribution to the total leakage seen in the DFT output. Let's try and make this concept a bit clearer by presenting a couple of examples.

**Example 1**. This example shows an ideal case where no leakage occurs. Let's envision a real sinusoidal input signal with the DFT computed spectrum illustrated in Figure 3.36. Here we see the amplitude spectrum of a real cosine wave with a normalized frequency of $k_0/N$. The line spectrum is presented as a symmetrical two-sided signal spectrum where the DFT bins from $N/2+1$ to $N-1$ represent negative frequencies and are plotted on the left side of the origin. The normalized frequency $k_0/N$ is a sweet frequency because there is no leakage. Since there is no leakage, this means that this signal must have an integer number of periods contained within the $N$ sample time domain input to the DFT. All the energy from this input signal falls into exactly two DFT output bins: the $k_0$th bin and the $N-k_0$th bin. We can visualize the time domain signal for this example as the infinitely sampled sinusoid illustrated

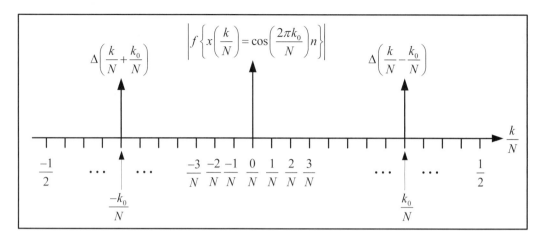

**Figure 3.36**   No spectral leakage

in Figure 3.37, part A. This sinusoid is at the normalized frequency of $k_0/N$ and is represented by

$$y(n) = \cos\left(\frac{2\pi k_0}{N} n\right)$$

The dots on the signal represent the digital sample instants. The DFT cannot process a signal that is infinite in length. Therefore we have to truncate the signal to some reasonable length equal to $N$ samples. We can visualize this process of truncation as multiplication of the input signal by a window function $w(n)$ that has unity amplitude for $N$ samples and zero amplitude for all other samples. This window function is illustrated in part B of Figure 3.37. The result of the multiplication is given by

$$\hat{y}(n) = w(n)\cos\left(\frac{2\pi k_0}{N} n\right)$$

The product of the two signals is illustrated in part C of the figure. It is this product that we input to the DFT for processing. It is important to note that in this case we were fortunate to pick a window length that was long enough to contain an integer number of complete signal periods. That is, the windowed signal contains no partial cycles. The DFT believes that the signal it is processing is periodic. It believes that the windowed signal it is process- ing repeats itself forever. If we concatenate the windowed signal to itself to form a larger signal, we can see that because the signal contains an integer

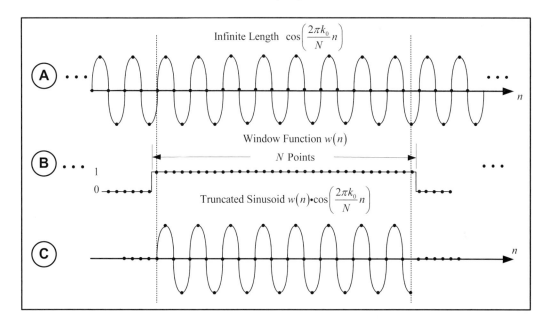

**Figure 3.37**   Truncated signal with an integer number of periods

number of signal periods, the concatenated sections butt together seamlessly with no discontinuities. The DFT sees this as an infinite sinusoidal waveform and produces the output illustrated in the simple Excel plot of Figure 3.38. The length of the fast Fourier transform (FFT) in this example is $N = 64$.

The amplitude of the sinusoid is $A = 1$ and the frequency is $k_0 = 16$. Note that the magnitude of the two spectral components is $AN/2 = (1)(64)/2 = 32$. Also note that since the input signal is real, the line spectra components occur at $k_0 = 16$ and at $k = N - k_0 = (64 - 16) = 48$ for a one-sided plot or at $k_0 = -16$ for a two-sided plot.

**Example 2**. This example shows a common case where spectral leakage occurs. Let's envision the more common case where the frequency of a sinusoidal signal falls between DFT bins, as illustrated in Figure 3.39. Here the normalized frequency of the cosine wave is $k_1/N$. Relative to the ideal case in the previous example, we can say that the frequency $k_1$ is equal to an ideal frequency $k_0$ plus some offset. We can think of our frequency as $k_1 = k_0 \pm \Delta_k$, where $\Delta_k$ is some small frequency increment less than one.

The spectrum of the sinusoidal frequency is shown in Figure 3.39 as a dotted line to indicate its location relative to the DFT output bins. Since the sinusoid falls between DFT bins, we know that we will see spectral leakage

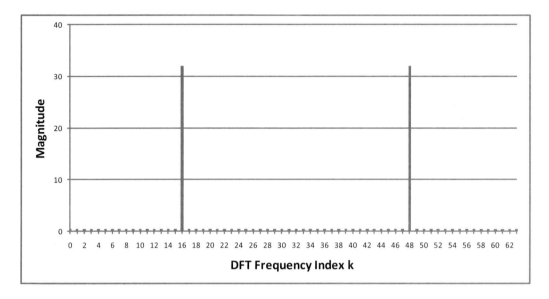

**Figure 3.38**   DFT of $w(n)\cdot\cos(2\pi k_0 n/N)$ for $k_0 = 16$

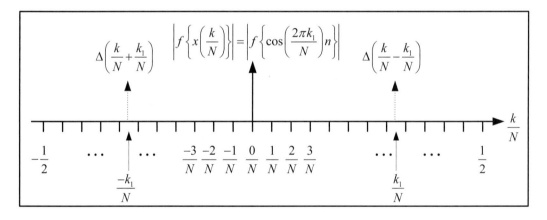

**Figure 3.39**   Ideal input signal spectrum falls between DFT bins

in the DFT output because the signal does not have an integer number of complete periods within the $N$ point DFT sample space.

The infinitely long sampled sinusoidal signal is illustrated in Figure 3.40, part A. In order to make this example an extension of example 1, the frequency of the sinusoid is selected to be $k_1 = k_0 + 0.25$, and therefore $k_1 = 16 + 0.25 = 16.25$. We can see from the figure that the sample instants are different from that of example 1. This is due to the slight increase in signal frequency.

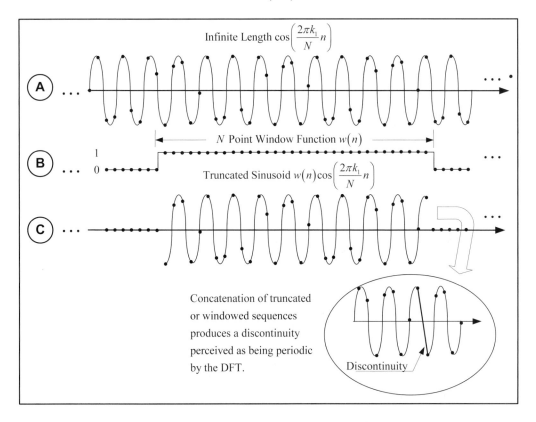

Infinite Length $\cos\left(\dfrac{2\pi k_1}{N}n\right)$

**A**

$N$ Point Window Function $w(n)$

**B**

Truncated Sinusoid $w(n)\cos\left(\dfrac{2\pi k_1}{N}n\right)$

**C**

Concatenation of truncated or windowed sequences produces a discontinuity perceived as being periodic by the DFT.

Discontinuity

**Figure 3.40** Truncated signal with a noninteger number of periods

In part B of the figure, we can see that we have the same 64 sample truncation window $w(n)$ as in example 1. This time, however, because the frequency of the signal has increased, we see that we do not end up with an integer number of cycles in the truncated signal. As a matter of fact, in this example, we end up with an additional 3/4 cycle. The truncated signal is illustrated in part C of the figure.

Remember that the DFT believes it is processing a signal that is periodic in $N$ samples. This time, if we concatenate the truncated signal processed by the DFT we immediately see that the extra portion of a cycle has created a discontinuity in the waveform that the DFT believes to be periodic. This sharp discontinuity in the time domain transforms to a whole bunch of high frequencies in the frequency domain. These extra frequencies appear in the DFT output as spectral leakage.

The spectral leakage caused by a discontinuity such as the one in Figure 3.40 is illustrated in Figure 3.41. In this figure, we see the location where the frequency $k_1/N$ would appear if the DFT had an infinite resolution. We also

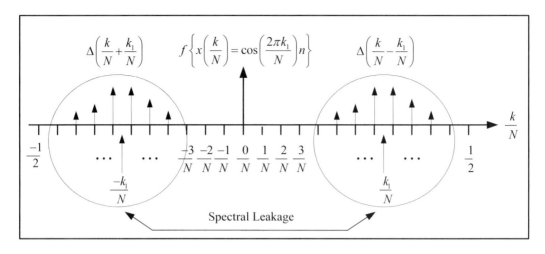

**Figure 3.41** Example of spectral leakage

see the spectral leakage in adjacent DFT bins caused by the sinusoid being detected by the side lobes of adjacent DFT band pass filters.

Three real examples follow that will help illustrate the concept of leakage. In all cases, the input signal to the DFT is a 64-sample real sinusoid.

**Example 3**. Remember when we set $k_0 = 16$ in the first example we got the clean single line spectra DFT output shown in Figure 3.38. If we increase the sinusoid frequency to $k_0 = 16.25$, we end up with a frequency that does not concentrate all of its energy in a single DFT bin. Instead, its energy is detected by the side lobes of adjacent DFT band pass filters, and we end up with the spectrum illustrated in Figure 3.42. We can see in the figure that more energy falls into the sixteenth bin than the seventeenth bin. The spectrum of this pure tone appears to be smeared across all the DFT output bins with the bulk of the energy captured in the sixteenth DFT bin. This makes sense because the sinusoidal energy is concentrated closer to the sixteenth bin than any other bin. This smearing or spectral leakage is an artifact of the DFT, and the reader should be well aware that some degree of leakage will always be present when processing wideband signals.

**Example 4**. If the sinusoidal frequency is increased to $k_0 = 16.50$, the line spectrum is positioned equally between bins $k = 16$ and $k = 17$. Equal energy is detected by the main lobes of both the $k = 16$ and $k = 17$ band pass filters. Each side lobe of the remaining filters detects attenuated but equal energy as well. The DFT in this case outputs a smeared but symmetrical line spectrum, as illustrated in Figure 3.43.

**Example 5**. Finally, if the frequency of the input signal is increased to $k_0 = 16.75$, the signal energy is closer to the seventeenth bin and further away from the sixteenth bin. The spectral plot illustrated in Figure 3.44 reflects this shift in signal energy, with the bulk of the energy being captured in the seventeenth bin of the DFT.

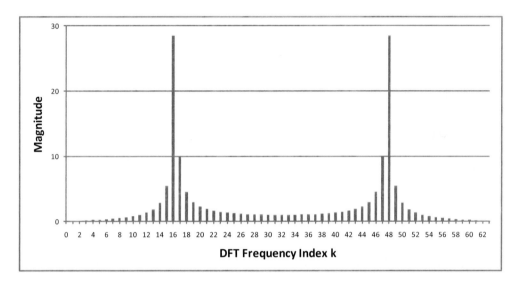

**Figure 3.42**   DFT of $w(n) \cdot \cos(2\pi k_0 n/N)$ for $k_0 = 16.25$

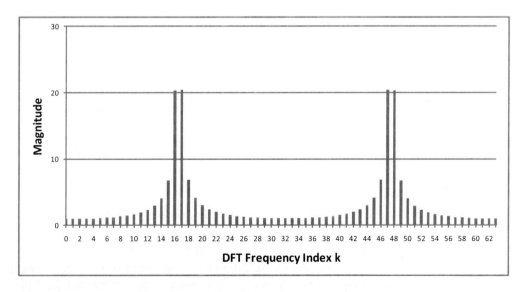

**Figure 3.43**   DFT of $w(n) \cdot \sin(2\pi k_0 n/N)$ for $k_0 = 16.50$

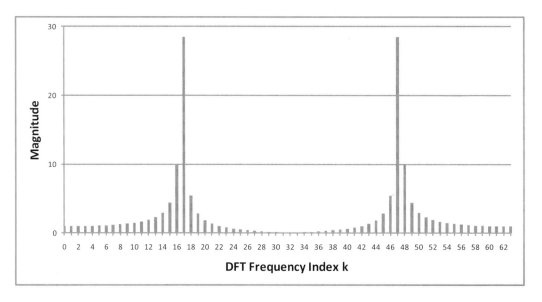

**Figure 3.44**   DFT of $w(n) \cdot \cos(2\pi k_0 n/N)$ for $k_0 = 16.75$

The reader should note that in almost all wideband applications, most if not all these frequencies will have a noninteger number of cycles contained within the DFT sample space. Therefore, most if not all these frequencies will contribute to the total spectral leakage.

There are methods that have been developed to reduce and/or minimize the effects of leakage. For instance, the sloppy $\sin(\omega N/2)/\sin(\omega/2)$ band pass filter magnitude response inherent to the DFT is a direct result of the rectangular window used to truncate the input signal to $N$ points.

One of the simplest methods to reduce leakage is through the use of a truncation window other than the rectangular window $w(n)$. Over the years, there has been a great deal of research by a host of brilliant people resulting in the development different truncation window functions other than the rectangular window. These window functions all share the same property of producing lower side lobes. The smaller side lobes reduce the effects of spectral leakage.

The world we live in is governed by the laws of physics, and because of that, one thing will always be true: you never get something for nothing. If you squeeze a balloon at one end, it pops out on the other end. It turns out that all truncation windows have a major trade-off. Window functions with lower side lobes and therefore greater leakage suppression have wider bandwidth main lobes and therefore have reduced spectral resolution. There is never a free lunch, especially in engineering and physics. Therefore, in selecting a specific window, the engineer must make an application-specific trade-off between leakage reduction and frequency resolution. The concept of windows is briefly discussed in the next section.

When making this trade-off, engineers can either preprocess the data with a standard window other than a rectangular window or they can opt to design their own window in the form of a finite impulse response (FIR) filter. When using the FIR method, it is possible to design a DFT with a bank of band pass filters that possess a flat in-band magnitude response, a steep transition band roll-off, and a 60 db or greater stop band attenuation so that interference from neighboring bins is negligible. This technique comes into play when we use a DFT to implement a precision digital filter bank or a digital channelizer. The procedure for designing a channelizer using an optimized DFT is discussed in detail in Chapter 12, "Channelized Filter Bank." In that chapter, the reader will learn how to preprocess DFT input data to produce nearly ideal DFT band pass filters and see how these filters work in a real-world signal processing application.

**Spectral leakage perspective 2.** In perspective 1's discussion of spectral leakage, we looked at the band pass filter response of each of the $N$ bins of a DFT. We also mentioned that the filter response was directly associated with the type of truncation window we happened to choose. In this second perspective, we will look at this filter response from a slightly different angle.

We know that the DFT equation is given by

$$X(f) = \sum_{n=0}^{N-1} x(n) \, e^{-j\left(\frac{2\pi k}{N}n\right)}$$

**Equation 3.92**

From this equation it is apparent that the DFT processes a sequence of length $N$. As far as the DFT is concerned, the input data is truncated so that all data samples before and after the $N$ point sequence are zero. In other words, it is exactly like we had multiplied the input sequence by some window $w(n)$, where

$$w(n) = \left\{ \begin{array}{l} 1 \text{ for } 0 \leq n \leq N-1 \\ 0 \text{ for all other } n \end{array} \right\}$$

The input sequence that is processed by the $N$ point DFT can be rewritten without modification of the output:

$$X(f) = \sum_{n=-\infty}^{+\infty} w(n)x(n) \, e^{-j\left(\frac{2\pi k}{N}n\right)} = \sum_{n=0}^{N-1} w(n)x(n) \, e^{-j\left(\frac{2\pi k}{N}n\right)} \text{ for } w(n) = \left\{ \begin{array}{l} 1 \text{ for } 0 \leq n \leq N-1 \\ 0 \text{ for all other } n \end{array} \right\}$$

**Equation 3.93**

All we are saying here is that we multiply the portion of the input sequence that is processed by the DFT by unity. The DFT expression in Equation 3.93 is identical to the DFT expression in Equation 3.92. The only difference is we have exposed a discrete signal processing function $w(n)$ that has up to this point been invisible to the reader. Identifying this invisible function is important because it is the determining factor in the magnitude of spectral leakage and in the determination of the spectral resolution produced at the output of a DFT.

A computerized graphic is worth several thousand words, so let's take a gander at Figure 3.45. In part A of the figure, we see a typical discrete rectangular window $w(n)$ that has unity magnitude for $N$ samples and is zero everywhere else. The continuous spectrum of this window is shown to the right. We can see that the spectrum of the window takes on the general shape of a $\sin x/x$ function. More precisely, it takes on a spectral envelope defined by $\sin(\omega N/2)/\sin(\omega/2)$. The output of the DFT is a sampled version of this continuous envelope.

It is important to note that the window spectrum consists of a single sample at zero frequency with magnitude equal to the peak of the main lobe. All other spectral samples for this window occur at response nulls and are therefore equivalent to zero.

In part B of the figure, we see the input signal $x(n)$, which is an infinite length sampled sinusoid of normalized frequency $f = 6/N$ or, if you wish, at a frequency of $f = 6f_s/N$. It turns out that this sinusoid is at a frequency so that exactly six complete periods are contained within the $N$ samples of the

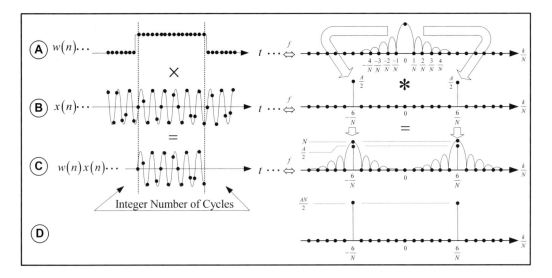

**Figure 3.45**  Processing an integral number of cycles with a DFT

window. The spectrum of this infinite length real sinusoid is shown to the right in part B. We note that all the sinusoid's energy falls directly in the DFT bins corresponding to the normalized frequencies $\pm 6/N$. The magnitude of each spectral line is $A/2$, where $A$ is the amplitude of the real sinusoid.

If we multiply the window sequence $w(n)$ in part A of the figure by the data sequence $x(n)$ in part B, we end up with the truncated time domain sequence shown in part C of Figure 3.45. We can see from this figure that there are an integer number of complete cycles contained in the truncated signal. The spectrum of the truncated sinusoid is shown on the right side of part C. We note that multiplication in the time domain is the equivalent of convolution in the frequency domain. The symbol used to denote convolution is $*$. In this example, we can say that

$$x(n)w(n) \overset{f}{\leftrightarrow} X(f) * W(f)$$

The spectrum of the rectangular window convolved with the spectrum of the infinite length sinusoid is illustrated in part C of the figure. One important fact to note here is that after the convolution of $x(n)$ and $w(n)$, the spectrum of the window is centered exactly on the frequency bins corresponding to $\pm 6/N$. Therefore all the sample values of the window with the exception of the sample at the window main lobe peak fall directly on the DFT bin nulls. The null samples make no contribution to the convolution, and therefore the output magnitude spectrum consists of just two line spectra, as shown in part D of the figure.

The convolution of the window spectrum that has a main lobe magnitude of $N$, with the sinusoid signal line spectrum of magnitude $A/2$, produces an output of magnitude $AN/2$. For this reason, many engineers scale the DFT output by $1/N$, so the DFT output mirrors the textbook analog Fourier transform output of $A/2$.

Now let's take a look at the case where a sinusoidal waveform is at a frequency such that an integer number of periods cannot be contained within the $N$ samples of the DFT window. This situation is illustrated in Figure 3.46. In this case, we have increased the normalized frequency of the sinusoid to be $f = 6.5/N$ or $f = 6.5 f_s/N$. Part A of the figure is identical to part A in the previous figure. The window is the same length and amplitude. The sinusoidal waveform has, however, increased in frequency from $6/N$ to $6.5/N$.

Part B of the figure shows the infinite length sampled time domain sinusoid. Part B also shows the ideal location of the line spectrum of this sinusoid superimposed upon the $N$ DFT bins. We can see that the line spectra fall between the sixth and seventh DFT bins. Therefore we will not get a perfect correlation between the input signal and any of the DFT's $N$ analysis sinusoids.

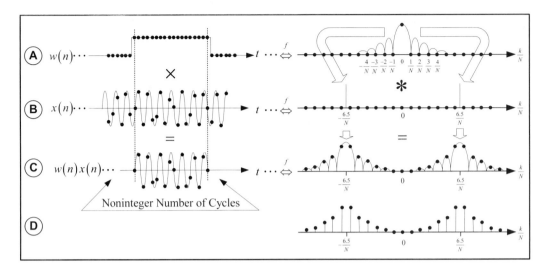

**Figure 3.46**   Processing a noninteger number of cycles with a DFT

As before, the multiplication of the input sequence $x(n)$ by the window sequence $w(n)$ produces a truncated sinusoid that is illustrated in part C of the figure. This time, however, the envelope of the rectangular window $\sin(\omega N/2)/\sin(\omega/2)$ spectrum is not centered on a DFT bin and, unlike the previous example, the nulls do not fall on bin centers and we get some partial correlation with all the DFT bins. The resulting output spectrum is illustrated in part D of the figure. The spectral leakage is very much like spectral smearing, and it is evident in even this simple example.

### 3.3.12   DFT Windowing

In the last section we discussed the nasty DFT attribute referred to as spectral leakage. We saw that the leakage is a natural artifact of a discrete process like a DFT that is constrained to operate on a finite number of quantized samples. We showed that by default a standalone DFT convolves the signal spectra with the spectrum of a rectangular window. The width of the main lobe and the magnitude of the side lobes of this window contribute significantly to the phenomena of DFT spectral smearing or leakage.

The reader can take some comfort in the fact that a great deal of work over the years has been aimed at the development of many different types of windows that can be used in place of the rectangular window. The number of windows in all the DSP literature are too numerous to mention in this short chapter. We will discuss the performance properties of five of the most common windows here. The reader is referred to references [1], [2], [14], and [15] for discussions of other window functions.

All windows perform the same signal processing function. That is, they are utilized to attenuate the discontinuity that occurs when an infinite length signal is truncated to a manageable number of samples. All windows come with design trade-off issues. The selection of an optimum window is dependent upon the application at hand. The windows we discuss here are chosen to illustrate the concept of windowing and to demonstrate the performance trade-offs that need to be made by the design engineer when selecting a window function. In this section, we will introduce the five common window types:

- Rectangular window
- Bartlett window
- Hanning window
- Hamming window
- Blackman window

In addition, this section will also discuss a special purpose high performance window: a FIR digital filter that, when used as a DFT preprocessor, produces a DFT output that is very close to an ideal bank of band pass filters. We will pay brief attention to this type of window in this chapter. The development and use of this window is discussed in detail in Chapter 12, "Channelized Filter Bank."

Way back in the early days of DSP there was a great deal of effort put forth in the development of the tools and the procedures utilized in the design of digital filters. Don't forget that in the early days, engineers were struggling to implement signal processing algorithms either on large mainframe computers like the IBM-360 or in special purpose digital hardware.

In those days, digital integrated circuits were extremely primitive compared to what is available today. Integrated circuits at the time were small scale integrated (SSI) circuits. They had huge propagation delays, limited fan out, and they consumed a tremendous amount of power. The good news was that in those days the rise and fall time of the signals output by one of one of these prehistoric integrated circuits was so lengthy that most engineers had never heard of terms like *ground bounce*. They never had to treat integrated circuit interconnects as though they were transmission lines, and they didn't worry about the physics of circuit board geometry and layout because the speed of the digital signals and their edge rates in those days was extremely slow. In 1975, for example, a state-of-the-art digital design ran at a clock rate of 5 MHz, and those circuit designs were tough because of the propagation delays. Complementary metal-oxide semiconductor (CMOS) devices at that time had a propagation delay on the order of 200 nanoseconds. It seems hard to believe, but it is true.

Thirty years later, designing galium arsenide gate arrays with at clock rates of 600 MHz was fairly common. That sounds a bit slow until you realize that the period of a 600 MHz clock is 1.67 nanoseconds. To put it in perspective, this means that at this clock rate, all arithmetic computational hardware and data processing hardware would input a new data sample and spit out a processed data sample every 1.67 billionths of a second. At these rates, the clock no longer resembles a textbook digital waveform. It looks very much like a sinusoid. For these circuits to work, the on chip drivers, the physical positioning of circuit macros, and the individual circuit loads must be tweaked on each and every circuit in order to add or delete picosecond propagation delays and minimize clock tree skew.

Anyway, the reason behind this long-winded history lesson is to put the importance of these windows in perspective. In the early days of DSP, there were many well-documented methods of designing a digital filter. None of them were optimum by today's standards. Most of those methods are obsolete now due to the rise of the high-speed digital computer, availability of lightning-fast hardware, and computer aided-design tools such as the Parks-McClellan FIR filter design algorithm, discussed in Chapter 5 "Finite Impulse Response Digital Filtering."

In those days, it was common for engineers to design an analog filter and then convert the analog impulse response of that filter to a digital filter impulse response using one of several techniques. One of those techniques was to generate an infinite filter impulse response and then truncate it to $N$ samples by multiplying the response with a window function. Another method was to use the window function itself as the impulse response in the digital filter design.

Efforts to optimize these techniques resulted in the development of many different kinds of window functions. These filter design techniques are pretty much a thing of the past and are rarely if ever used anymore. The window functions, however, are still around, they are still important, and are commonly used to minimize the errors associated with the truncation of digital signals prior to their being processed.

The DFT is a great example of the use of windows. A DFT processing system will usually preprocess the data sequence by multiplying it by some form of window function to reduce spectral leakage, modify spectral resolution, or maintain some trade-off position in between. So with this little bit of history behind us, let's begin our discussion of window functions.

### 3.3.12.1  Rectangular Window

Let's begin with the window function that we have already become familiar with: the rectangular window. The expression for a two-sided rectangular window is given by Equation 3.94. This is the cheapest window to implement since it involves multiplying the input DFT sequence by unity:

$$w(n) = 1 \quad \left\{ \text{for } -\frac{N-1}{2} \leq n \leq \frac{N-1}{2} \right\}$$

$$w(n) = 0 \quad \text{for all other values of } n$$

**Equation 3.94**

The spectrum of an $N = 16$ point rectangular window is illustrated in Figure 3.47. For convenience, the frequency axis of the spectral plot is calibrated in three different but equivalent scales. The first frequency axis is calibrated in normalized frequency and is expressed in units of $k/N$. The second axis is calibrated in units of normalized radian frequency $2\pi k/N$. The third axis is calibrated in units of fractional sample rate $f_s$.

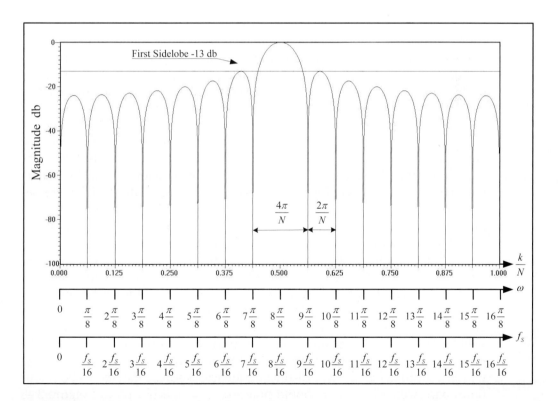

**Figure 3.47** Spectrum of a 16-point rectangular window

A brief inspection of Figure 3.47 shows us that the window main lobe has a bandwidth of $4\pi/N$ or $2f_s/N$, and the bandwidth of each side lobe is $2\pi/N$ or $f_s/N$. We also see that the magnitude of the first side lobe is only 13 db down from the peak magnitude of the main lobe. This spectral plot is a precise version of the rectangular window spectrum we referred to in the discussion of spectral leakage. The attenuation of side lobe magnitude for this window is small, and because of this the rectangular window is not a good window for applications that require minimal spectral leakage.

From this plot, it is easy for the reader to see that a convolution of this window spectrum with the line spectra of a sinusoidal waveform whose frequency does not fall precisely on a DFT bin will produce a great deal of spectral smearing or leakage.

The window spectrum is plotted in decibels with the magnitude of all points computed relative to the maximum value of the main lobe. Thus

$$\left| W\left(k\right) \right|_{db} = 20\mathrm{Log}_{10}\left( \frac{W\left(k\right)}{W\left(k\right)_{MAX}} \right) \ \text{ for } k = 0, 1, 2, 3, \cdots, N-1$$

**Equation 3.95**

Therefore the main lobe maxima relative to itself is 0 db. The actual amplitude of the rectangular window main lobe is equal to the number of points in the window or, in this case, $N = 16$. This is important to remember when analyzing the output of a DFT that uses a rectangular window.

What happens if we increase the value of $N$ by a factor of four? We can see what happens by observing the spectrum for an $N = 64$ point rectangular window illustrated in Figure 3.48. The plot indicates that the bandwidth of the main lobe and all the side lobes has decreased by a factor of four. Increasing the length of the DFT window decreases the main lobe width, which increases the frequency resolution of the window. Imagine the difference in (1) the output of a DFT of an off bin sinusoid using a 16-point rectangular window and (2) the output of the same DFT using a 64-point rectangular window.

With this image in your head, imagine the same comparison when performing a DFT on a signal that contains two sinusoids closely spaced in frequency. Because of the difference in main lobe width, the DFT using a 16-point window may not have the resolution necessary to detect two distinct and closely spaced frequencies. The energy of the two sinusoids could very well be smeared into one another when convolved with the wide main lobe of the window. Conversely, a larger DFT that uses the longer 64-point window may be able to detect the two distinct frequencies because the main lobe is four times narrower. The reader should note from the spectral plot of Figure 3.48

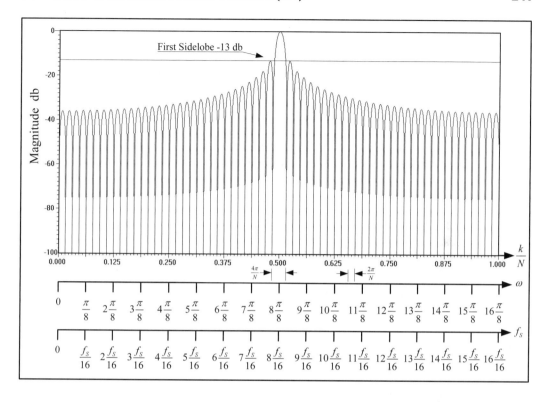

**Figure 3.48**   Spectrum of a 64-point rectangular window

that the length of the window has no effect on the side lobe magnitude. The magnitude of the first side lobe remains 13 db down from the main lobe, independent of the window length. Even though it is not shown in the magnitude plot, the reader should also note that the amplitude of the main lobe is now $N = 64$. The bandwidth of this main lobe has been decreased by a factor of four, but the amplitude has been increased by a factor of four.

### 3.3.12.2   Bartlett Window

We can reduce the magnitude of the window side lobes by designing windows that are tapered from the midpoint out to the two end points. This taper is designed to reduce the magnitude of signal discontinuity due to truncation by the window. From a frequency point of view, a less abrupt truncation introduces fewer high-frequency components.

The Bartlett window does just that. It is a window shaped like a triangle and is described by the expression given in Equation 3.96:

$$w(n) = 1 - \frac{2|n|}{N-1} \quad \left\{ \text{for} \quad -\frac{N-1}{2} \leq n \leq +\frac{N-1}{2} \right\}$$

$$w(n) = 0 \qquad \left\{ \text{for all other values of } n \quad \right\}$$

**Equation 3.96**

The length of the window in Equation 3.96 is $N$ samples. The spectrum of Bartlett window for $N = 64$ is illustrated in Figure 3.49. Here we can see that the magnitude of the first side lobe is 26 db down from the main lobe peak. This simple tapered window suppresses side lobe magnitude roughly by a factor of two relative to the rectangular window. Use of this window will considerably reduce the magnitude of DFT spectral leakage. The outlying side lobes decrease in magnitude to below –60 db at $\pm f_s/2$. The price we pay for this increase in leakage performance is a doubling of the main lobe bandwidth. We can see from the figure that the main lobe bandwidth for the

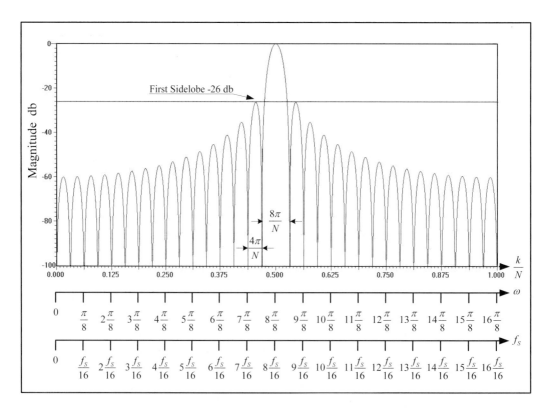

**Figure 3.49**    Spectrum of a 64-point Bartlett window

Bartlett window is $8\pi/N$. This means that the use of this window will reduce the frequency resolution of the DFT by a factor of two. That is, the ability of the DFT to distinguish between two closely spaced off bin sinusoids for the same length data sequence is only half that of the rectangular window. However, since the main lobe width is controlled by the length of the window the frequency, resolution can be increased by increasing the value of $N$.

The spectral amplitude of the Bartlett main lobe is $N/2$ or half that of the rectangular window. In this case, the main lobe amplitude right out of the 64-point DFT is 32. This makes sense because for the same length window, the area under a Bartlett window is half the area under a rectangular window.

### 3.3.12.3  Hanning Window

The Hanning window is named after Julius Von Hann (1839–1921), a professor of cosmic physics at the University of Vienna (1890–1910). The Hanning or Hann window is an $N$ point window that is shaped like a bell curve. It has a maximum amplitude of 1.0 with a raised cosine shaped taper to 0.0 at both ends of the window. The time domain expression for a two-sided Hanning window is given by

$$w(n) = 0.5\left(1+\cos\frac{2\pi n}{N-1}\right) \quad \left\{\text{for } -\frac{N-1}{2}\leq n\leq +\frac{N-1}{2}\right\}$$

$$w(n) = 0 \qquad\qquad\qquad \left\{\text{for all other values of } n \quad\right\}$$

**Equation 3.97**

The spectrum of a 64-point Hanning window is illustrated in Figure 3.50. Like the Bartlett window, the main lobe bandwidth is $8\pi/N$, but the magnitude of its first side lobe is down 31 db. The drop-off of outlying side lobes is below –100 db at $\pm f_s/2$, which is significantly better than either the rectangular or Bartlett window. The main lobe amplitude right out of the DFT is equal to $N/2$. For this example, the main lobe amplitude is 32 or one half that of the rectangular window.

### 3.3.12.4  Hamming Window

The Hamming window is named after its inventor, Richard Wesley Hamming, an American mathematician famous for his work in computer science. Besides the Hamming window, he is best known for his developmental work on error-detecting and error-correcting codes. The Hamming window is similar to the Hanning window in that it too has a raised cosine or bell shape in the time domain. Only, in this case, the raised cosine sits on a pedestal of height 0.54. The time domain expression for a two-sided Hamming window is given by

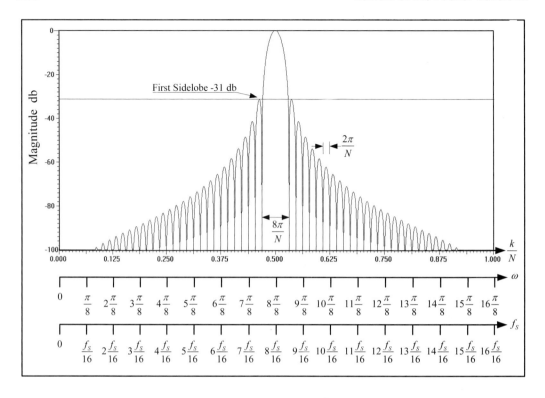

**Figure 3.50**   Spectrum of a 64-point Hanning window

$$w(n) = 0.54 + 0.46 \cos\left(\frac{2\pi n}{N-1}\right) \quad \left\{ \text{for} \quad -\frac{N-1}{2} \le n \le +\frac{N-1}{2} \right\}$$

$$w(n) = 0 \qquad\qquad\qquad\qquad \left\{ \text{for all other values of } n \quad \right\}$$

**Equation 3.98**

The spectrum of a $N = 64$ point Hamming window is illustrated in Figure 3.51. This window has a main lobe bandwidth of $8\pi/N$, as did the Bartlett and Hanning windows. In this case, however, the magnitude of the highest side lobe is 41 db down. The Hamming window has a 10 db advantage in the highest side lobe attenuation over the Hanning window, but the attenuation of outlying side lobes is approximately –52 db at $\pm f_S/2$, which far less than that of the Hanning window. The amplitude of the Hamming window spectral main lobe right out of the DFT is about 54% of the rectangular window main lobe, or approximately $N/1.8518$.

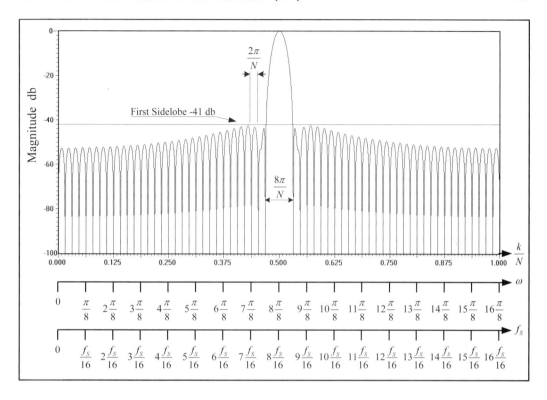

**Figure 3.51**   Spectrum of a 64-point Hamming window

### 3.3.12.5   Blackman Window

The Blackman window is named after Ralph Beebe Blackman, an American mathematician and engineer who was among the pioneers in the early days of signal processing. He worked at Bell Laboratories in the 1960s and authored several books on spectrum estimation. The time domain expression for a two-sided Blackman window is given by

$$w(n) = 0.42 + 0.5\cos\left(\frac{2\pi n}{N-1}\right) + 0.08\cos\left(\frac{4\pi n}{N-1}\right) \quad \left\{\text{for } -\frac{N-1}{2} \le n \le +\frac{N-1}{2}\right\}$$

$$w(n) = 0 \qquad\qquad\qquad\qquad\qquad\qquad \left\{\text{for all other values of } n \quad\right\}$$

**Equation 3.99**

The Blackman window gives the largest side lobe attenuation of all the previously mentioned windows. The first side lobe is down by a respectable 57 db. This is an excellent truncation window for leakage suppression. On the down side, however, the main lobe bandwidth is $12\pi/N$. When the Blackman window is convolved with the spectrum of a signal, its main lobe tends to smooth any spectral discontinuities, resulting in a reduction in spectral resolution. The spectrum of an $N = 64$ point Blackman window is illustrated in Figure 3.52. The main lobe width can be decreased by increasing the number of points $N$. The amplitude of the Blackman window spectral main lobe right out of the DFT is about 43% of the rectangular window main lobe, or approximately $N/2.3810$.

### 3.3.12.6   FIR Filter Window

The last window we will discuss is a window that the reader can design to optimize the band pass filter characteristics of an $N$ point DFT. This type of preprocessing window is utilized in applications such as digital channelizers because of its flat pass band, ultra sharp transition band, and 60 db or greater stop band attenuation. This type of window is the optimum DFT window because its main lobe bandwidth can be made very close to that of a single DFT bin, and the transition bands are narrow enough to minimize spectral leakage from adjacent DFT bins.

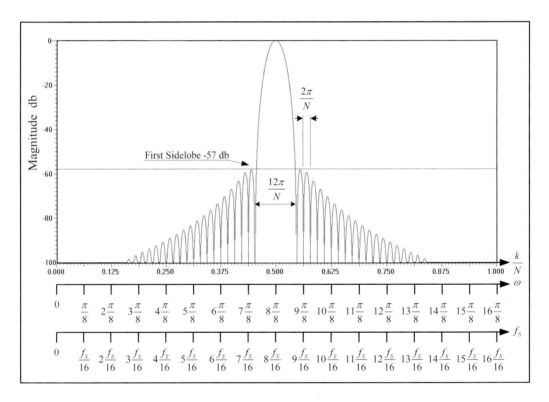

**Figure 3.52**   Spectrum of a 64-point Blackman window

We will briefly mention this window here so that we might give the reader some insight into more optimal window types. The design of this window is identical to the design of any FIR digital filter and can be done using the Parks-McClellan computer algorithm discussed in Chapter 5, "Finite Impulse Response Digital Filtering." The implementation of this window in concert with a DFT is discussed in detail in Chapter 12, "Channelized Filter Bank," where this type of preprocessing window enhances the DFT to produce an almost ideal bank of band pass filters.

Implementation of this window only requires that we design a FIR filter to meet the particular application specifications. The window preprocessing of data is accomplished through the use of a *poly phase FIR filter*. This is just a lot of fancy words for something quite simple, as the reader will see in Chapter 6, "Multirate Finite Impulse Response Filter Design," and Chapter 12, "Channelized Filter Bank."

In this example, we implement a filter that operates at a sample rate of 128 KHz, with a 3 db bandwidth of 3750 Hz, a 60 db bandwidth of 4500 Hz, and a stop band attenuation greater than 60 db. The filter bandwidth is approximately 4 KHz. The length of the DFT is 32 points. The DFT bin width is given by 128 KHz/32 = 4 KHz, which is equal to the filter bandwidth. The 4 KHz bandwidth filter or window is illustrated in Figure 3.53.

Assuming the signal we process is complex, meaning that it has a real part and an imaginary part, then we can implement a bank of 32 band pass filters using only the filter as a window coupled with an $N = 32$ point DFT. If the impulse response of this window is labeled $h(n)$ and if the signal being processed by the DFT is labeled $x(n)$, then we can process this signal and separate the 32 independent band pass signals simply by implementing the equation

$$y(k) = \sum_{n=0}^{31} x(n)h(n)e^{-j\left(\frac{2\pi k}{N}n\right)}$$

**Equation 3.100**

When computing a single DFT, implementing all 32 band pass filters is almost as simple as Equation 3.100 suggests. The issue with Equation 3.100 is that a filter with the tight specifications we described has an impulse response much larger than 32 coefficients. In addition, when computing successive DFTs to process continuous real time data, there are a few "gotchas" hidden in the implementation of this equation that the designer needs to be aware of. In addition, the hardware or software implementation of Equation 3.100 can be streamlined quite a bit using tricks of the trade to increase processing efficiency. These issues are all discussed in Chapter 12, "Channelized Filter Bank."

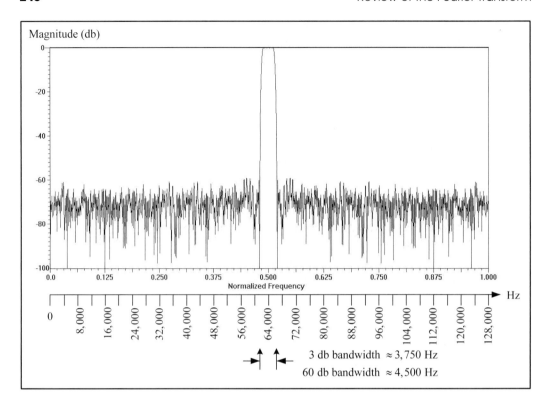

**Figure 3.53**   Window function constructed from a poly phase FIR filter

The overlay of the first 16 band pass filters in the bank of 32 is illustrated in Figure 3.54. We can clearly see from this figure how the frequency selectiveness of the DFT can be optimized for a specific application through the use of an appropriate window. The bandwidth of each band pass filter in Figure 3.54 closely approximates the DFT bin width. There is little overlap of the window main lobes, and the side lobes are all attenuated below 60 db.

The FIR-based window seen here is not a standard window like the Bartlett or Hamming window. We simply designed this window based on the specifications for a particular channelizer application. The reader can design his or her own FIR filter based window to suit the specifics of their application, combine it with a DFT, and produce an enhanced DFT similar to this one.

The techniques and procedures necessary to design a FIR filter are discussed in Chapter 5, "Finite Impulse Response Digital Filtering." You can purchase semiexpensive application software to design the filters; use your employers' design workstations, if any are available; or, if inspired, write your own software application using the FIR filter design software routine in Appendix A, "Mixed Language C/C++ FORTRAN Programming."

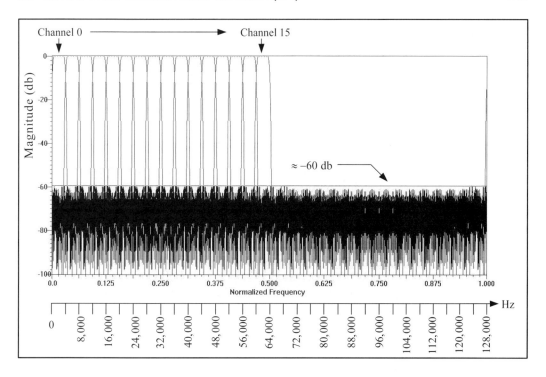

**Figure 3.54**   Plot of the first 16 frequency bins of a 32-point DFT

If the application that you are currently involved in could use an optimum DFT based real time channelizer such as the one illustrated in Figure 3.54 and if you view the figure as a possible design solution for that application, then it is strongly suggested that you read Chapter 12, "Channelized Filter Bank," to discover just how remarkably simple a channelizer such as this is to design and implement.

### 3.3.12.7   Window Processing Examples

Now that we have spent a great deal of time dealing with theoretical discussions on windows and their effect on a DFT output, let's take a look at some examples that one is likely to see in everyday engineering activities.

In the first example, we sample a 16 KHz sinusoid at 128 KHz and apply this to the input of a 1024-point DFT. We apply the signal directly to the DFT, which is the equivalent of preprocessing the signal by a rectangular window. The ratio of the signal frequency to the sample rate gives us the normalized frequency of the sinusoid:

$$16 \text{ KHz}/128 \text{ KHz} = 0.125$$

Expressed in units of the DFT frequency, we can see that

$$\frac{kf_s}{N} = 16,000 \text{ Hz or } k = \frac{N}{f_s} 16,000 \text{ Hz} = \frac{1024}{128,000} 16,000 = 128$$

The frequency index $k = 128$ is an integer; therefore the 16 KHz tone is an "on bin" frequency and there will be no spectral leakage. Since this sinusoid is real, we should expect to see an ideal line spectrum located at $k/N = 128/1024 = 0.125$ and at $1 - (k/N) = 1 - (128/1024) = 0.875$. These two line spectra can be interpreted as being equal to $\pm f_s/8$ on a two-sided representation of the DFT output. This is an example of a sweet frequency that is periodic in $N$ samples, resulting in no spectral leakage. The 1024-point DFT of this signal is illustrated in Figure 3.55.

The second example is identical to the first with the exception that we preprocessed the input signal with a 1024-point Blackman window. The results of the DFT in this case are illustrated in Figure 3.56. The reader should note that there is not a great deal of difference between the two examples, with the exception that the width of the line spectra has increased. This increased spectral width is due to the larger width of the Blackman window main lobe. The difference between the use of a rectangular and a Blackman window is not as evident here because we are processing an on bin or sweet frequency where there are no frequency discontinuities present. The most significant performance results are witnessed when processing off bin frequencies, as we shall see next.

In the third example, our processing is identical, but we change the frequency index $k$ slightly so that it becomes a noninteger. In this example, we set $k = 128.5$. We can calculate the input frequency as

$$f = \frac{kf_s}{N} = \frac{(128.5)(128,000)}{1024} \text{ Hz} = 16,062.5 \text{ Hz}$$

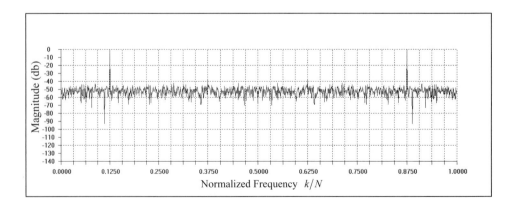

**Figure 3.55**   On bin sinusoid at 16 KHz preprocessed with a rectangular window

This is not a sweet frequency for this length DFT. The frequency index $k$ is not an integer. Therefore this is an off bin frequency, and we should expect to see spectral leakage. The DFT computed spectrum for this off bin sinusoid is illustrated in Figure 3.57. The input signal was processed with a rectangular window. Remember the spectrum of a rectangular window has the highest side lobes of all the windows we have discussed, and therefore it will be the source of the highest degree of leakage.

In the figure, we can see the peak spectra of the input sinusoid exactly where we would expect them to be. The slow roll-off skirts on either side of the peak spectra are a direct result of spectral leakage. If this is an example of a single sinusoid, what should we expect to see from a wideband signal that is composed of thousands of sinusoidal frequencies? Well, since most of the frequencies in a wideband signal are off bin frequencies relative to the length of the DFT, we would expect to see a smeared spectrum much worse that we see for a single sinusoid. The reader should be able to identify this signal processing phenomena and understand that it is not an accurate representation of the true signal spectrum. It is an artifact of the DFT.

Remember the purpose of windowing is to reduce the magnitude of frequency discontinuity experienced by off bin frequencies at the boundaries of the DFT. This has the effect of significantly reducing spectral leakage. In the fourth and final example, let's process the same off bin sinusoidal signal again, only this time we will preprocess the input signal with a Hanning window, as opposed to a rectangular window. The resulting signal spectrum is illustrated in Figure 3.58. A comparison of the spectrum in Figure 3.58 to the spectrum of Figure 3.57 clearly shows the processing enhancement that can come with the proper selection of a DFT window.

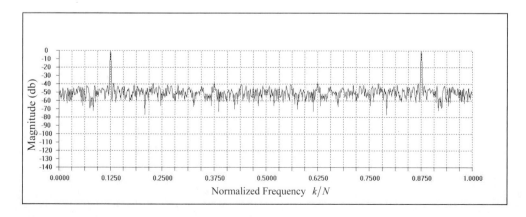

**Figure 3.56**   On bin sinusoid at 16 KHz preprocessed with a Blackman window

Preprocessing the input signal with a Hanning window resulted in the spectral leakage skirts being reduced essentially to the level of the noise floor, and a more accurate representation of the actual signal spectra has been achieved.

This is an example of DSP at its best. All you need to accomplish amazing feats like these is a knowledge of the processing tools and algorithms that are discussed in any one of the many DSP textbooks on the market today.

One of the best career moves a signal processing engineer can make is to either purchase or write a software application that allows him or her to perform system-level simulations. The results obtained from simulations with graphical interpretation of the data (such as the simple simulation used in the previous examples) provides an invaluable platform from which to observe the performance of any system that an engineer may design. The nice graphics make for an interesting and informative slide show at design presentations as well.

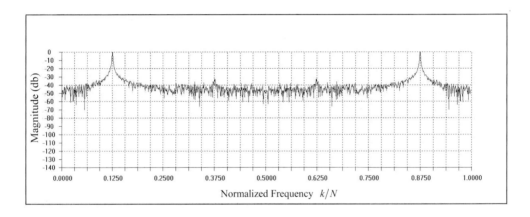

**Figure 3.57**  Off bin sinusoid at 16.0625 KHz with a rectangular window

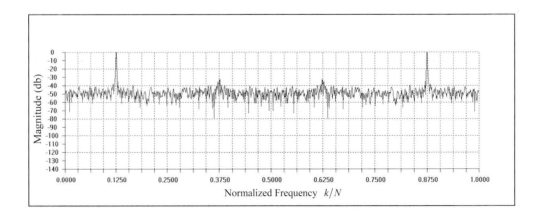

**Figure 3.58**  Off bin sinusoid at 16.0625 KHz preprocessed with a Hanning window

### 3.3.12.8   Window Summary

The parameters associated with different window types are listed in Table 3.8.

The most common methods of annotating the frequency axis for display of a DFT output are listed in Table 3.9.

**Table 3.8**   Common Window Parameters

| Window function | Approximate peak side lobe magnitude (db) | Approximate width of main lobe | Max DFT amplitude relative to rectangular window |
|---|---|---|---|
| Rectangular | −13 | $4\pi/N$ | 1 |
| Bartlett | −26 | $8\pi/N$ | 0.50 |
| Hanning | −31 | $8\pi/N$ | 0.50 |
| Hamming | −42 | $8\pi/N$ | 0.54 |
| Blackman | −57 | $12\pi/N$ | 0.42 |

**Table 3.9**   Frequency Axis Representations

| DFT frequency variable | Frequency | Two-sided frequency axis range | Frequency resolution | Spectrum repeats every |
|---|---|---|---|---|
| Frequency normalized to the sample rate $f_s$ | $\dfrac{k}{N}$  $-\dfrac{N}{2} \le k \le +\dfrac{N}{2}$ | −0.5 to +0.5 | $1/N$ | $N$ |
| Radian frequency $\omega$ normalized to the sample rate $f_s$ | $\dfrac{2\pi k}{N}$  $-\dfrac{N}{2} \le k \le +\dfrac{N}{2}$ | $-\pi$ to $+\pi$ | $2\pi/N$ | $2\pi$ |
| Fractional sample rate | $\dfrac{kf_s}{N}$  $-\dfrac{N}{2} \le k \le +\dfrac{N}{2}$ | $-f_s/2$ to $+f_s/2$ | $f_s/N$ | $f_s$ |

## 3.4  DFT PROCESSING GAIN

The DFT can be thought of as a bank of identical band pass filters. The bandwidth of each filter gets narrower as the length $N$ of the DFT increases. Suppose we have a signal composed of a pure sinusoid embedded in wideband noise. The signal to noise ratio (SNR) of this composite signal can be computed by

$$SNR = 10Log_{10}\left(\frac{Signal\ power}{Wideband\ noise\ power}\right)$$

Now suppose we pass this same signal through a narrow band filter of bandwidth $B$, much smaller than the noise bandwidth, centered on the sinusoidal frequency. The filter passes the sinusoid intact, and because of its narrow pass band, the filter only passes $B$ Hz of the wideband noise. In this case, only a fraction of the noise passes through the filter. This procedure increases the SNR considerably. This is exactly what happens in a DFT, only in this case there are $N$ adjacent band pass filters, each one centered on a DFT bin.

The amount of noise that passes through is determined by the bandwidth of the DFT band pass filter. The bandwidth of the filter pass band and the magnitude of the filter response side lobes are determined by the type of window chosen to preprocess the composite signal prior to the DFT. As the length of the DFT increases, the bandwidth of each filter decreases. The noise passing through each filter can now be referred to as *narrow band noise*. The narrow band noise power will be substantially less than that of the wideband noise. Therefore we can say that the SNR of a signal at the output of any DFT bin is given by

$$SNR = 10Log_{10}\left(\frac{Signal\ power}{Narrow\ band\ noise\ power}\right)$$

Since the filtered noise power is significantly reduced from that of the input wideband noise power, there is a gain in SNR at the output of the DFT. This gain is one component of the DFT processing gain. Processing gain is a powerful attribute of the DFT that allows us to "pull the signal out of the noise," as engineers like to say. This property makes the DFT a great tool for signal detection and identification.

This makes for a great midterm exam exercise. The course instructor might hand each student a CD with three signals completely buried in noise and ask them to detect, identify, and compute some statistics on each signal. The DFT would be just one of the tools at the student's disposal.

We can view the DFT as correlating both the sinusoidal input signal and the input noise against all $N$ of its analysis frequencies. If a real sinusoid is bin centered, then its maximum correlation score is equal to $(A/2)N$, while a bin-centered complex signal would have a maximum correlation score of $AN$. The noise is uncorrelated with the analysis frequencies and therefore has a low correlation score. This is a second component of the DFT processing gain. Since this processing gain is proportional to the length of the DFT, the gain can be quite large. If we process a signal with a DFT of length $N_1$ and process that same signal with a second DFT of length $N_2$, where $N_2 > N_1$, then the corresponding increase in processing gain due to the longer DFT is approximately

$$\text{DFT Processing Gain} \simeq 10\text{Log}_{10}\left(\frac{N_2}{N_1}\right)$$

**Equation 3.101**

This tells us that doubling the length of a DFT will produce a 3 db processing gain. This expression is only an approximation because of the following factors:

- We cannot accurately predict the exact power of $N$ samples of random noise.
- If the sinusoid is not periodic in the $N$ sample space of the DFT, then spectral leakage will reduce the overall signal power per bin and will raise the overall noise floor.
- Any windows that may be used to preprocess the composite signal prior to the DFT will affect the spectral leakage and thus affect the noise power.

We can, however, use Equation 3.101 as a rule of thumb to aid us in the determination of how to process a signal.

We can interpret the SNR of our composite signal to be the ratio of the signal power to the average noise power. If we compute the SNR in the frequency domain, then we would compute the average noise power by summing the noise magnitude for each bin of the DFT of length $N$ and then dividing by $N$:

$$
\begin{aligned}
N_P &= 20\text{Log}_{10}\left[\frac{1}{N}\sum_{n=0}^{N-1}\left(\sqrt{\text{noise}(n)^2_{REAL} + \text{noise}(n)^2_{IMAG}}\right)\right] \\
&= 10\text{Log}_{10}\left[\frac{1}{N}\sum_{n=0}^{N-1}\left(\text{noise}(n)^2_{REAL} + \text{noise}(n)^2_{IMAG}\right)\right]
\end{aligned}
$$

**Equation 3.102**

The term $\left( noise(n)_{REAL} + j\, noise(n)_{IMAG} \right)$ represents the noise contained within the $n$th bin of the DFT. The expression in Equation 3.102 represents the average noise power across all $N$ bins of a DFT.

For the case where the embedded sinusoid is bin centered, the signal power for $0 \leq n < N$ is just the peak value of the signal magnitude squared, or

$$S_P = 20\text{Log}_{10}\left[ \text{Maximum}\left\{ \sqrt{\text{signal}(n)_{REAL}^2 + \text{signal}(n)_{IMAG}^2} \right\}^2 \right]$$

$$= 10\text{Log}_{10}\left[ \text{Maximum}\left\{ \text{signal}(n)_{REAL}^2 + \text{signal}(n)_{IMAG}^2 \right\} \right]$$

**Equation 3.103**

This is true for this simple example because the signal will have a line spectrum that occupies just one DFT bin for a complex input signal or two DFT bins for a real input signal. Computing the signal power for an off bin sinusoid could be accomplished in the same manner, but remember that a portion of the signal power will be spread across adjacent DFT bins due to spectral leakage.

For the simple case where the signal is a pure on bin sinusoid, the SNR of the composite signal can be computed for $0 \leq n < N$ by

$$SNR = 10\text{Log}_{10}\left( \frac{\text{Peak signal power}}{\text{Average noise power}} \right)$$

$$= 10\text{Log}_{10}\left( \frac{\text{Max}\left\{ \text{signal}(n)_{REAL}^2 + \text{signal}(n)_{IMAG}^2 \right\}}{\dfrac{1}{N}\displaystyle\sum_{n=0}^{N-1}\left\{ \text{noise}(n)_{REAL}^2 + \text{noise}(n)_{IMAG}^2 \right\}} \right)$$

**Equation 3.104**

A variant on the SNR useful for many applications is to compute the SNR based on the ratio of peak signal power to peak noise power. This is expressed as

$$SNR = 10\text{Log}_{10}\left( \frac{\text{Peak signal power}}{\text{Peak noise power}} \right)$$

$$= 10\text{Log}_{10}\left( \frac{\text{Max}\left\{ \text{signal}(n)_{REAL}^2 + \text{signal}(n)_{IMAG}^2 \right\}}{\text{Max}\left\{ \text{noise}(n)_{REAL}^2 + \text{noise}(n)_{IMAG}^2 \right\}} \right)$$

**Equation 3.105**

In a more realistic case, where the signal is either an off bin sinusoid or is composed of many frequencies (such as would be the case with a narrow bandwidth signal), there will be spectral leakage. In this case, the common solution to computing the SNR is to establish a threshold $T$ such that all bins whose signal magnitude falls above the threshold are counted as signal and all bins whose signal magnitude falls below the threshold are counted as noise. In this case, the SNR is computed as

$$
\text{SNR} = 10\text{Log}_{10}\left(\frac{\dfrac{1}{K}\displaystyle\sum_{n=0}^{K-1}\left\{\text{signal}(n)_{REAL}^{2}+\text{signal}(n)_{IMAG}^{2}\right\}\geq T}{\dfrac{1}{M}\displaystyle\sum_{n=0}^{M-1}\left\{\text{signal}(n)_{REAL}^{2}+\text{signal}(n)_{IMAG}^{2}\right\}<T}\right)
$$

In this situation, using an $N$ point DFT, there would be $K$ samples above or equal to the threshold and $M$ samples below threshold, where $N = K + M$.

We can demonstrate the relative processing gain given in Equation 3.101 by starting with 1024 samples of a bin centered sinusoid with additive noise whose waveform is illustrated in Figure 3.59. We will compute the SNR using both the average noise and peak noise methods given in Equation 3.104 and Equation 3.105.

In this example, the sample rate is $f_S = 128$ KHz. The sinusoid frequency is $f_S/8 = 16$ KHz and has an amplitude equal to $10$. The random noise is bipolar wideband noise with a zero DC component and with a maximum possible amplitude of $10$.

Relevant signal parameters are computed and printed at the bottom of the figure. These are defined as follows:

- $s/a$ = the ratio of maximum signal and maximum possible noise amplitude
- $N$ = the number of samples being processed equals the length of the DFT

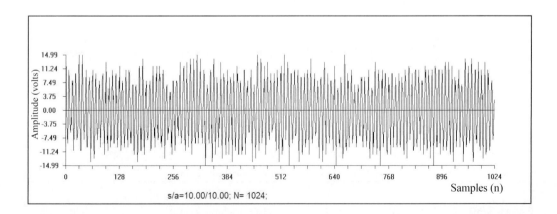

**Figure 3.59** Sinusoidal waveform with added noise

- aSNR = the computed average SNR in decibels
- pSNR = the computed peak SNR in decibels

We can see from Figure 3.59 that the additive noise has degraded the sinusoid considerably. If we were to pass this composite signal through a narrow band filter centered on $f_S/8$, we would eliminate much of the wideband noise and the amplitude distortion of the output sinusoidal waveform would improve considerably. This is exactly what happens inside the DFT. If we pre-process this signal with a rectangular window and apply it to a 1024-point DFT, we get the spectrum illustrated in Figure 3.60. We can see from the figure that the composite input signal is real because we have two symmetric line spectra located at the normalized frequencies $\pm f_S/8 = \pm 0.125$. The average SNR is computed to be 35.8 db, which correlates with the average noise level in the plot. The peak SNR is computed to be 26.1 db, which is substantially different than the average SNR. A horizontal cursor is placed 26.1 db down from the peak signal. We can see several points along the plot where the peak noise magnitude reaches this cursor. Both of these SNR values are valid and can be used in specific DSP applications.

Using the same time domain signal but doubling the number of input signal samples, we see the resulting spectrum for a 2048-point DFT in Figure 3.61. Here we see that for the same input signal our average SNR is computed to be 38.9 db, which is a gain of about 3.1 db over the 1024-point DFT case. The peak SNR is computed to be 29 db. Once again a horizontal cursor is slewed down to the peak SNR value, and we see at least one and perhaps two noise samples that equal the value of the cursor. If we use the same input signal but double the length of the DFT again to 4096 samples and compute the DFT, we get the spectrum illustrated in Figure 3.62.

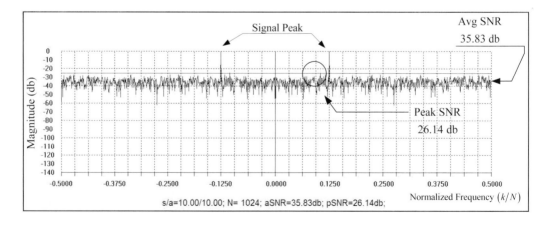

**Figure 3.60**   A 1024-point DFT of an $f_s/8$ sinusoid in additive noise

Here we measure the average SNR to be 41.9 db—a processing gain of 3 db above the case where the DFT was 2048 points long. The peak SNR is computed to be 31.5 db.

Finally, we double the length of the DFT again, this time to 8192 points. The spectrum for this case is illustrated in Figure 3.63. The average SNR has increased from the previous case by 3.1 db to 45.0 db, and the peak SNR has increased to 35.4 db.

It should be clear to the reader that the method used to compute the SNR produces significantly different results. Generally the computed value of SNR is used only as a reference that is specific to the computations associated with some particular application. As long as the reference and its methods

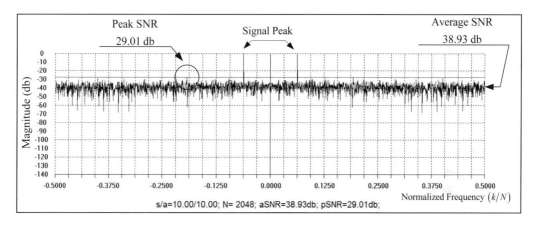

**Figure 3.61** A 2048-point DFT of an $f_s/8$ sinusoid in additive noise

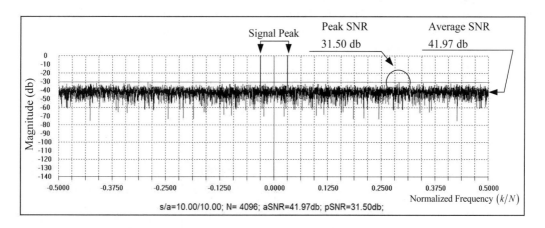

**Figure 3.62** A 4096-point DFT of an $f_s/8$ sinusoid in additive noise

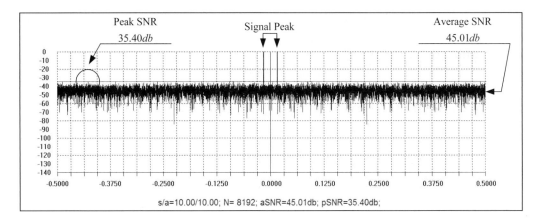

**Figure 3.63**   A 8192-point DFT of an $f_s/8$ sinusoid in additive noise

of computation are stable and consistent, then the interpretation of other computational values that are based on this reference will also be stable and consistent. The method used to compute SNR is open to interpretation and is ultimately defined by the application in which it is used.

### 3.4.1   Increasing the DFT Processing Gain by Averaging

We can increase the processing gain even further by averaging successive DFTs. This is referred to as *integration gain* and is the third component in the overall processing gain of a DFT. Assuming that the signal is present in all the waveform intervals where the DFT is computed, the average magnitude of the signal bins will remain constant. However, when averaged, the magnitude of the random noise will tend to cancel. This reduction of noise power increases the overall processed SNR. There are two methods of DFT averaging that are commonly used [15]: incoherent integration and coherent integration.

The incoherent integration method simply averages the magnitudes of individual bins for successive DFTs. Let's define $|F_n(k)|$ as the magnitude of the $k$th bin of the $n$th DFT in the average. Let's also define the DFT length to be $N$ points and say that we average $P$ DFTs. The averaging process can be described by

$$
\begin{aligned}
\left|\overline{F}(k)\right| &= \frac{1}{P}\sum_{n=0}^{P-1}\left|F_n(k)\right| \\
&= \frac{1}{P}\sum_{n=0}^{P-1}\sqrt{F_n(k)_{REAL}^2 + F_n(k)_{IMAG}^2}
\end{aligned}
\qquad
\left\{
\begin{array}{l}
\text{for } P \text{ DFTs} \\
\text{for } k = 0,1,2,3,\cdots,N-1 \text{ Bins per DFT}
\end{array}
\right\}
$$

**Equation 3.106**

Equation 3.106 tells us to add the contents of bin $k$ over $P$ DFTs and then divide the sum by $P$ and repeat this for all $N$ bins. Incoherent integration works well and is commonly used, but the reader should be cautioned that since the averaging is done on magnitudes, it is computed using all positive numbers.

The coherent integration method does not average magnitudes. In this method, we average the complex numbers out of each DFT bin. That is, we average all the real and imaginary parts of successive DFT bins. The real and imaginary parts are both positive and negative numbers and can tend to cancel when averaged.

The process is to first compute and separately sum the real parts and the imaginary parts for individual bins in successive DFTs and then divide by the number of DFTs. We can treat this as the sum of $P$ complex numbers divided by $P$:

$$\overline{F}(k) = \overline{F}(k)_{REAL} + j\,\overline{F}(k)_{IMAG} = \frac{1}{P}\sum_{n=0}^{P-1}\left(F_n(k)_{REAL} + j\,F_n(k)_{IMAG}\right) \left.\begin{array}{l} \text{for } P \text{ DFTs} \\ \text{for } k = 0, 1, \cdots, N-1 \text{ Bins per DFT} \end{array}\right\}$$

Then we compute the magnitude of the averaged complex number:

$$\left|\overline{F}(k)\right| = \sqrt{\overline{F}(k)_{REAL}^2 + \overline{F}(k)_{IMAG}^2}$$

Because coherent integration deals with bipolar numbers in most cases, the noise reduction produces a greater processing gain.

## 3.5   EXAMPLE DFT SIGNAL PROCESSING APPLICATION

How might we use a DFT in a signal processing system? That is, aside from using it to compute and view signal spectra, how else may a DFT be utilized?

Here is just one example. Suppose we were assigned the task of processing data streams that arrived from many different sources. Our objective is to process any or all the data streams and determine if any signals might be present. If a signal is identified, we would catalog it and compute some basic signal parameters such as signal center frequency, magnitude, and bandwidth so that other processing systems could extract the signal and perhaps demodulate it.

One solution to this task is to use the DFT as a front-end signal processor. We could use the processing gain of a single DFT and if necessary the additional integration gain by averaging several successive DFTs to pop the signal out of the noisy background. The processing scenario is illustrated in Figure 3.64 and might go something like this:

1. Capture a single DFT or average several successive DFTs to capitalize on the processing gain and the integration gain inherent in DFT averaging, thus popping the signal out of the noise.

2. Estimate the noise floor by computing the average noise power $N_{AVG}$, as given in Equation 3.102. In many applications, it is acceptable to compute the average noise power over a small segment of the signal spectra and then make the assumption that this average is representative of the actual average noise power over the entire spectrum from $f = 0$ to $f = f_s/2$. This technique is sometimes called *straight line averaging* in that the value calculated for $N_{AVG}$ over a portion of the spectrum is translated into a straight line across the spectrum. This process is illustrated in Figure 3.64. Here we compute the estimate of the average noise in the small sample space within the shaded box and make the assumption that this value is very close to the average noise in other parts of the spectrum.

3. Once the average noise power $N_{AVG}$ of Equation 3.102 is computed, we can add a programmable offset to this value and call the new value a *threshold*. In the figure, the value of $n$ db is added to the average noise power to produce a threshold $T = (N_{AVG} + n)$ *db*. As illustrated in the figure, this threshold is nothing more than a straight line at some computed power level that extends from $f = 0$ to $f = f_s/2$.

4. We now perform a bin by bin power comparison with the threshold value. All bin power measurements that are below the threshold are classified as noise or nonsignal bins. All bin magnitudes that are above the threshold are classified as signal.

5. Once we have an inventory of signal bins, we can easily compute approximate bandwidth, magnitude, and center frequency for each signal.

6. Using these parameters, we could pass this composite time domain signal through a tunable band pass filter to extract the signal of interest. Alternatively, we could down convert the center frequency of the detected signal to DC and then low pass filter the spectrum to extract the signal of interest. This later method is more common in DSP simply because it allows us to process the signal at a much lower sample rate.

This method of signal extraction works for both wideband and narrow band signals, as indicated in the figure. This example shows that DSP isn't as mysterious and difficult as it is sometimes perceived to be.

Any DSP application problem can be broken down into two phases: (1) the system engineering phase where the processing algorithm is thought out and developed and (2) the hardware/software engineering phase where the system design is actually implemented and tested.

A good systems engineer will have done a great deal of software simulations using real data to verify his algorithm prior to handing it off to the design engineers. Good hardware and software engineers will simulate their design to verify that it works and meets system performance specifications before they build it.

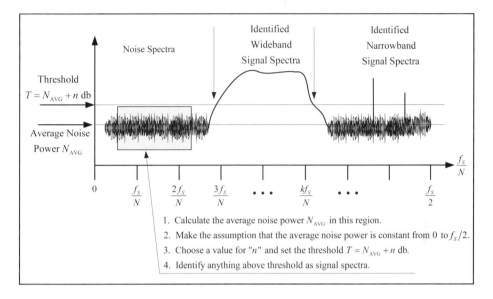

1. Calculate the average noise power $N_{AVG}$ in this region.
2. Make the assumption that the average noise power is constant from 0 to $f_S/2$.
3. Choose a value for "$n$" and set the threshold $T = N_{AVG} + n$ db.
4. Identify anything above threshold as signal spectra.

**Figure 3.64** Using the processing gain of the DFT to identify signal spectra

Hold on to your hats, folks, and get ready for the real world. Unfortunately, this simulation effort doesn't happen 100% of the time, and because of this, both program budgets and schedules suffer when the processing performance doesn't meet the program specifications. Some of this can be attributed to the lack of hours bid for engineering. When an engineer doesn't have enough labor hours in the program bid to do a thorough design, the first thing that gets eliminated is design simulation. When this happens, the design errors and subpar system performance issues aren't caught until the program test phase when all software is written and all the hardware is built and set in stone. Design changes to enhance performance, which could have been easily and cheaply implemented during the simulation phase of the design, now require the rewrite of a gazillion lines of code, the refabrication of circuit cards, and schedule delays that could take months. Cost and schedule overruns like these are the end result of the competitive bidding process where the low bid gets awarded the program contract. Most contractors are forced to "buy in" to win the contract and then hope to "gracefully overrun" during the life of the program. That said, it is usually true that the engineer who invests the time and effort to simulate his or her design will be the program star when his or her design works the first time it's tested or works after minimal tweaks.

## 3.6 DISCRETE TIME FOURIER TRANSFORM (DTFT)

Thus far we have used the DFT as a tool to estimate the discrete frequency response of a discrete time sequence. For a discrete time sequence of length $N$, the DFT will give us $N$ equally spaced estimates of the magnitude and phase of the frequency response for that sequence.

There is another tool available to the engineer that allows the computation of the continuous frequency response for a discrete time sequence. This tool is called the *discrete time Fourier transform* (DTFT). The DTFT is defined by the equation

$$X(\omega) = \sum_{n=-\infty}^{+\infty} x(n) e^{-j\omega n}$$

where $x(n)$ is a discrete sequence and $X(\omega)$ is a continuous frequency spectrum.

In order to achieve infinitely precise frequency resolution, the continuous frequency response requires that we process an infinite number of terms. This cannot be done on a computer or in special purpose hardware. It can only be computed by means of pencil and paper, using the mathematical concepts and procedures common to the study of infinite series. Even though the DTFT of a discrete sequence cannot be realized in any processing system, it is still a useful tool available to the engineer during the analysis and design phase of a program.

The DTFT of a discrete sequence is periodic in $2\pi$. This is obvious if we compute the DTFT at an offset of $2\pi$ and compare the results obtained from $|X(\omega)|$ and $|X(\omega+2\pi m)|$. For example,

$$
\begin{aligned}
X(\omega+2\pi m) &= \sum_{n=-\infty}^{+\infty} x(n) e^{-j(\omega+2\pi m)n} \\
&= \sum_{n=-\infty}^{+\infty} x(n) e^{-j\omega n} e^{-j(2\pi m)n} \\
&= X(\omega) \text{ since } e^{-j(2\pi m)n} = 1 \text{ for all integer values of } m
\end{aligned}
$$

Suppose we define the discrete sequence $x(n)$ as

$$x(n) = \left\{ \begin{array}{l} a^n \text{ for } n \geq 0 \\ 0 \text{ otherwise} \end{array} \right\} \text{ where } |a| < 1$$

We can compute the DTFT of the discrete sequence $x(n)$ by

$$X(\omega) = \sum_{n=-\infty}^{\infty} x(n) e^{-j\omega n} = \sum_{n=0}^{\infty} x(n) e^{-j\omega n} = \sum_{n=0}^{\infty} a^n e^{-j\omega n}$$

**Equation 3.107**

We begin by computing a closed form expression for the first $K$ terms of $X(\omega)$. The term $X_K(\omega)$ is used to designate the $K$ term expression:

$$X_K(\omega) = \sum_{n=0}^{K-1} a^n e^{-j\omega n}$$

**Equation 3.108**

We can expand this expression by writing all the terms to arrive at

$$X_K(\omega) = \sum_{n=0}^{K-1} a^n e^{-j\omega n} = 1 + ae^{-j\omega} + a^2 e^{-j2\omega} + \cdots a^{K-1} e^{-j2(K-1)\omega}$$

**Equation 3.109**

Now if we multiply both sides of this expression by $ae^{-j\omega}$, we get

$$ae^{-j\omega} X_K(\omega) = ae^{-j\omega} \sum_{n=0}^{K-1} a^n e^{-j\omega n} = ae^{-j\omega} + a^2 e^{-j2\omega} + \cdots a^{K-1} e^{-j2(K-1)\omega} + a^K e^{-jK\omega}$$

**Equation 3.110**

We can subtract Equation 3.110 from Equation 3.109 to get

$$
\begin{aligned}
X_K(\omega)\left(1 - ae^{-j\omega}\right) &= \sum_{n=0}^{K-1} a^n e^{-j\omega n}\left(1 - ae^{-j\omega}\right) \\
&= \left[1 + ae^{-j\omega} + a^2 e^{-j2\omega} + \cdots a^{K-1} e^{-j2(K-1)\omega}\right] - \left[ae^{-j\omega} + a^2 e^{-j2\omega} + \cdots a^{K-1} e^{-j2(K-1)\omega} + a^K e^{-jK\omega}\right] \\
&= 1 - a^K e^{-jK\omega}
\end{aligned}
$$

Now we divide both sides by $\left(1 - ae^{-j\omega}\right)$ to get the closed form expression for the first $K$ terms of $X(\omega)$. In doing so, we arrive at

$$X_K(\omega) = \frac{1 - a^K e^{-jK\omega}}{\left(1 - ae^{-j\omega}\right)}$$

It is a simple matter to compute the remaining terms of $X(\omega)$ by remembering that $|a| < 1$ and by letting $K \to \infty$. When we take these two facts into consideration, we can write

$$X(\omega) = \lim_{K \to \infty} X_K(\omega) = \lim_{K \to \infty} \left[\frac{1 - a^K e^{-jK\omega}}{\left(1 - ae^{-j\omega}\right)}\right] = \frac{1}{\left(1 - ae^{-j\omega}\right)}$$

Therefore the continuous frequency response of the discrete sequence $x(n)$ is given by

$$X(\omega) = \frac{1}{\left(1 - ae^{-j\omega}\right)}$$

**Equation 3.111**

We can expand Equation 3.111 by using Euler's identity to get

$$X(\omega) = \frac{1}{\left(1 - a\left(\cos\omega - j\sin\omega\right)\right)}$$

From this, we can compute the magnitude and phase of $X(\omega)$:

$$\left|X(\omega)\right| = \frac{1}{\sqrt{\left(1 - a\cos\omega\right)^2 + a^2\sin\omega^2}} = \frac{1}{\sqrt{1 + a^2 - 2a\cos\omega}}$$

$$\theta(\omega) = \tan^{-1}\left(\frac{a\sin\omega}{1 - a\cos\omega}\right)$$

**Equation 3.112**

We plot the magnitude of the continuous frequency response for the discrete sequence $x(n)$ by computing the values for $\left|X(\omega)\right|$ defined in Equation 3.112. In this example, we arbitrarily set the value for the variable $a$ to be $a = 0.5$. Figure 3.65 illustrates the continuous frequency response for $\omega = 2\pi f$ across the range of $0 \leq f \leq 2f_s$, where $f_s$ is the sample frequency of the

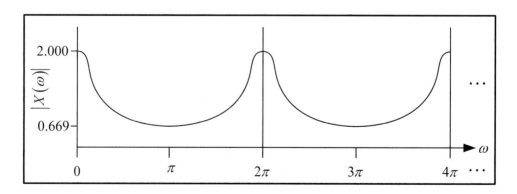

**Figure 3.65**   Continuous frequency response of a discrete sequence

discrete sequence. If the radian frequency $\omega$ is normalized to the sample rate $f_s$, this is the equivalent to the frequency range of $0 \le \omega \le 4\pi$.

We can clearly see from Equation 3.112 and Figure 3.65 that the frequency response is continuous and is periodic in $2\pi$. If we were to compute and plot the DFT of the same sequence and if we scaled the DTFT to the same frequency axis, overlaying the two, then we would see that the DFT is a sampled version of the DTFT.

The inverse of the DTFT is computed by

$$x(n) = \frac{1}{2\pi} \int_{-\pi}^{\pi} X(\omega) e^{j\omega n} \, d\omega$$

which is defined for the normalized frequency variable $\omega$.

## 3.7   FAST FOURIER TRANSFORM (FFT)

The DFT algorithm has existed for many years. As a matter of fact, the DFT goes way back to the days before the computer. There have always been applications that required the use of the DFT. There was a huge problem with the DFT in these days: the number of computations for an $N$ point transform was proportional to $N^2$. Since all the computations were done by hand, this became a serious problem. It could take months using an office calculating machine to compute a single DFT.

Even with the advent of the digital computer, the processing of data through a DFT was extremely time consuming. Fourier analysis was necessary for most design and research applications, but because it required so much time and effort, most people tried to avoid using the DFT. In the early 1960s, a fellow by the name of James W. Tukey was asked to come up with an algorithm to speed the processing of the DFT by making the computations more efficient. Tukey produced an efficient algorithm by drawing upon the work of people who had previously researched the problem. Once the algorithm was in place, another fellow by the name of J. W. Cooley was asked to program the algorithm. Together they published a paper in *Mathematics of Computation* in 1965. Their paper and their names soon became synonymous with the term *fast Fourier transform* (FFT).

The main attribute the FFT has over the DFT is that it computes the DFT using far fewer mathematical operations; therefore it is much faster. The number of computations required by the FFT to processes the same data as the DFT is proportional to $N\text{Log}_2 N$. This is significantly less than the $N^2$ computations required by the DFT.

Other than that, the FFT is essentially the same as the DFT, in that they both produce the same results. The FFT is just an algorithm that happens to be faster than the DFT. The principles of the Fourier theory are still the same. The FFT was a very big deal from the 1960s through the 1990s. However, with the tremendous increase in processing speed of today's computers, the processing time of a DFT has become much less of an issue.

The one drawback of the FFT is that it requires the length of the input data sequence to be a power of two. This can become a bit cumbersome because the choice of possible input sequence lengths differ from one another by a factor of two. This can cause the length of input sequences to grow rather quickly to an unmanageable number of samples. The DFT, on the other hand, can process any length sequence, thereby simplifying any Fourier analysis task. For example, the fact that a DFT can process any length sequence provides for a more accurate determination of the bin frequency resolution.

Over the years, many people have strived to improve upon the original Cooley-Tukey FFT algorithm. Richard C. Singleton has shown that data sequences of length other than a power of two can be processed by an FFT in his paper [19] that describes a mixed radix approach to computing the FFT.

Treatment of the FFT is not covered in this text because of the reasons mentioned previously and because there exists a plethora of detailed discussion on this algorithm in every DSP text written since 1965. Should the reader seek additional treatment of this subject, the industry bible on the FFT is the text authored by Oran Brigham [18].

## 3.8  REFERENCES

[1] Alan V. Oppenheim and Ronald W. Schafer. *Digital Signal Processing*. Englewood Cliffs, NJ: Prentice Hall, 1975.

[2] Alan V. Oppenheim and Ronald W. Schafer. *Discrete Time Signal Processing*. Englewood Cliffs, NJ: Prentice Hall, 1989.

[3] Douglas F. Elliott. *Handbook of Digital Signal Processing Engineering Applications*. San Diego, CA: Academic Press, 1987.

[4] Adrian Banner. *The Calculus Lifesaver*. Princeton, NJ: Princeton University Press, 2007.

[5] Charles Williams. *Designing Digital Filters*. Englewood Cliffs, NJ: Prentice Hall, 1986.

[6] Emmanuel C. Ifeachor and Barrie W. Jervis. *Digital Signal Processing: A Practical Approach*. Workingham, UK: Addison-Wesley, 1993.

[7] R. E. Ziemer and W. H. Tranter. *Principles of Communications*. Boston: Houghton Mifflin, 1976.

[8] Mischa Schwartz. *Information Transmission, Modulation and Noise*. 3rd ed. New York: McGraw-Hill, 1980.

[9]   Ferrel G. Stremler. *Introduction to Communication Systems*. 2nd ed. Boston: Addison-Wesley, 1982.

[10]  Philip F. Panter. *Modulation, Noise, and Spectral Analysis*. New York: McGraw-Hill, 1965.

[11]  Richard Shiavi. *An Introduction to the Analysis and Processing of Signals*. 2nd ed. New York: Academic Press, 1999.

[12]  Robert W. Ramirez. *The FFT Fundamentals and Concepts*. Englewood Cliffs, NJ: Prentice Hall, 1985.

[13]  Edward W. Kamen. *Introduction to Signals and Systems*. 2nd ed. New York: Macmillan, 1990.

[14]  Alan V. Oppenheim, Alan S. Willsky, and Ian T. Young. *Signals and Systems*. Englewood Cliffs, NJ: Prentice Hall, 1983.

[15]  Richard G. Lyons. *Understanding Digital Signal Processing*. 2nd ed. Englewood Cliffs, NJ: Prentice Hall, 2004.

[16]  Ron Bracewell. *The Fourier Transform and Its Applications*. San Francisco: McGraw-Hill, 1965.

[17]  Athanasios Papoulis. *The Fourier Integral and Its Applications*. San Francisco: McGraw-Hill, 1963.

[18]  Oran Brigham. *The Fast Fourier Transform*. Englewood Cliffs, NJ: Prentice Hall, 1974.

[19]  Richard C. Singleton. "An Algorithm for Computing the Mixed Radix Fast Fourier Transform." *IEEE Transactions on Audio and Electroacoustics*, vol. AU-17 (June 1969): 93–103.

[20]  C. S. Burrus, T. W. Parks, and J. F. Potts. *DFT/FFT and Convolution Algorithms*. New York: John Wiley and Sons, 1985.

# CHAPTER FOUR

# Review of the Z-Transform

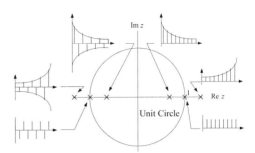

The z-transform plays a significant role in the design of digital hardware. In a great many instances, the mathematics of a discrete time system can be simplified through the use of the z-transform. The z-transform for discrete time signals is analogous to the Laplace transform for continuous time signals.

As we shall see, the z-transform is closely related to the Fourier transform and as a result allows us to easily compute the frequency response of a discrete time system. We can also use the transform to determine the stability of a digital system by investigating the poles of the system transfer function. This chapter provides a detailed review of the z-transform, its properties, and how it can be applied to the analysis of digital signal processing (DSP) functions. Although this review is extensive and is sufficient for most engineering applications, the reader can consult any of the references at the end of this chapter for a much more detailed treatment of the z-transform.

## 4.1 COMPLEX NUMBER REPRESENTATION

Let's begin this section with a few definitions. We will define the quantity $z$ to be a complex variable. As illustrated in Figure 4.1, the value of $z$ can be represented in either Cartesian coordinates or in polar coordinates.

In Cartesian coordinates the variable $z$ is a complex point in the z plane such that $z = a \pm jb$, where $a$ represents the real part of the variable and $b$ represents the imaginary part of the variable. The complex point $z$ is located at the tip of the hypotenuse of a right triangle whose two legs are measured to be $a$ units along the real axis and $b$ units along the imaginary axis.

**271**

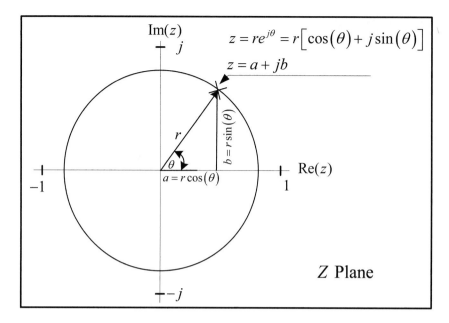

**Figure 4.1**   General complex point

In polar coordinates, the same complex variable $z$ can be represented as a point on the tip of a vector of length $r$ that is rotated $\theta$ radians from the real axis. The complex variable $z$ in polar representation takes the following form:

$$z = re^{\pm j\theta} = r\left[\cos(\theta) \pm j\sin(\theta)\right] \text{ where } \theta = \tan^{-1}(b/a)$$

Conversion between the two coordinate systems is achieved through the relationship $a = r\cos(\theta)$ and $b = r\sin(\theta)$. Because the variable $z$ is complex, it can be expressed in terms of its real and imaginary components. That is,

$$z = \text{Real}(z) + j\,\text{Imag}(z)$$

The magnitude of the complex variable $z$ is given by

$$|z| = \sqrt{\text{Real}(z)^2 + \text{Imag}(z)^2}$$

For the case illustrated in Figure 4.1, the magnitude expressed in Cartesian form is

$$|z| = \sqrt{a^2 + b^2}$$

The magnitude of the same point expressed in polar form is

$$|z| = \sqrt{\left(r\,\cos(\theta)\right)^2 + \left(r\,\sin(\theta)\right)^2} = \sqrt{r^2\left[\cos^2(\theta) + \sin^2(\theta)\right]} = \sqrt{r^2} = r$$

The phase of the complex variable $z$, sometimes referred to as the phase shift of $z$, is given by

$$\theta = \tan^{-1}\left(\frac{\text{Imag}(z)}{\text{Real}(z)}\right)$$

or, as illustrated in Figure 4.1,

$$\theta = \tan^{-1}\left(\frac{r\,\sin(\theta)}{r\,\cos(\theta)}\right) = \tan^{-1}\left(\frac{b}{a}\right)$$

Many times it is beneficial to express the complex variable $z$ in terms of power. The expression for power is just the square of the magnitude of $z$, or

$$\text{Power} = |z|^2 = \text{Real}(z)^2 + \text{Imag}(z)^2$$

The power of the complex point illustrated in Figure 4.1 is given by

$$\text{Power} = |z|^2 = \left(r\,\cos\theta\right)^2 + \left(r\,\sin\theta\right)^2 = a^2 + b^2 = r^2$$

The power expressed in units of decibels (db) is given by

$$\text{Power in db} = 10\text{Log}_{10}|z|^2 = 10\text{Log}_{10}\left[\text{Real}(z)^2 + \text{Imag}(z)^2\right] = 10\text{Log}_{10}\left[r^2\right]$$

Once again using the example in Figure 4.1, the power of the complex point represented by $z$ is given by

$$\text{Power in db} = 10\text{Log}_{10}|z|^2 = 10\text{Log}_{10}\left[a^2 + b^2\right] = 10\text{Log}_{10}\left[r^2\right]$$

We can see from Figure 4.1 that if the lengths of $a$ and $b$ change or if the angle $\theta$ changes such that $|z|$ remains constant, then the set of all possible values for the complex number $z$ lie on a circle centered at the origin with radius $r$. We also note that these values are repetitive in that they will repeat every time the angle $\theta$ rotates through $2\pi$ radians. If we imagine that $\theta$ changes in some uniform manner with respect to time, then $|z|$ will repetitively rotate around the circle of radius $r$ with a time varying phase angle described by $\theta(t)$. The angular rate at which the tip of the vector rotates is defined as $\omega = d\theta(t)/dt$.

A fixed point in the complex plane located on a circle of radius $r$ is defined in polar coordinates as $z = re^{j\theta}$. An angular rotation of that point along the same circle of radius $r$ is defined in polar coordinates as $z = re^{j\omega}$. We will discuss this further in the next section.

## 4.2  MECHANICS OF THE Z-TRANSFORM

The z-transform $H(z)$ of a discrete time sequence $h(n)$ is defined as

$$H(z) = \sum_{n=-\infty}^{\infty} h(n) z^{-n}$$

**Equation 4.1**

where $z$ is a complex variable defined in the z-plane as any one of the following three equivalents:

$$z = a \pm jb = r(\cos\theta \pm j\sin\theta) = re^{\pm j\theta}$$

The definition in Equation 4.1 is very similar to the definition of the Fourier transform of the same sequence, which is defined as

$$H(e^{j\omega}) = \sum_{n=-\infty}^{\infty} h(n) e^{-j\omega n}$$

**Equation 4.2**

It can be shown that the Fourier transform is a subset of the larger z-transform. For example, if we replace $z$ in Equation 4.1 with its polar coordinate representation $re^{j\omega}$, we get the expression

$$H(re^{j\omega}) = \sum_{n=-\infty}^{\infty} h(n)(re^{j\omega})^{-n} = \sum_{n=-\infty}^{\infty} r^{-n} h(n) e^{-j\omega n}$$

**Equation 4.3**

Equation 4.3 is identical to the Fourier transform representation, with the exception that the sequence $h(n)$ is multiplied by the sequence $r^{-n}$. If we set the value of $r$ to unity, then the z-transform of Equation 4.3 is identical to the Fourier transform given by Equation 4.2. That is, the z-transform and the discrete Fourier transform (DFT) are equivalent when the z-transform is evaluated on the unit circle, as illustrated in Figure 4.2. This is more easily recognized if we replace $\omega$ in Equation 4.2 with its discrete equivalent of $2\pi(k/N)$.

The frequency response of the discrete time system described by the sequence $h(n)$ can be obtained by evaluating the z-transform of the sequence on the unit circle. As we will see later, this is true only if the region of convergence (ROC) of the z-transform in question includes the unit circle. This is a very powerful statement because when the z-transform of a sequence is evaluated on the unit circle, we will obtain the same representation of that sequence in the frequency domain as we would with the DFT. This property illustrated by Equation 4.4 not only allows us to simplify the mathematics of a signal processing system by working in the z-domain but gives us a direct path to the frequency domain to analyze the frequency response of a discrete time system:

$$H(z)\Big|_{z=e^{j\omega}} = H\left(e^{j\omega}\right)$$

**Equation 4.4**

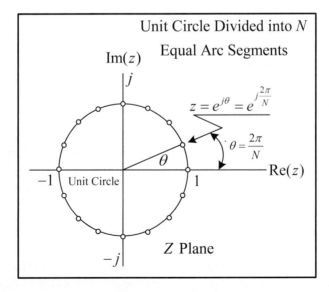

**Figure 4.2** Evaluation of the z-transform on the unit circle

In order to determine the frequency response of a digital system, we only need to evaluate Equation 4.4 on the range $0 \le \omega < 2\pi$. This is because the frequency response is cyclical and repeats itself every $2\pi$ radians. Equation 4.4 is a nifty relationship. It's a great way to illustrate the z-domain to frequency domain conversion concept. It also simplifies notation and is commonly used by everyone in the trade.

The expression in Equation 4.4 is compatible with pencil and paper computations because $\omega$ is a continuous term with infinite precision and can take on an infinite number of values between $0 \le \omega < 2\pi$. Therefore, with pencil and paper, we can determine the continuous frequency response representation of a digital system. Because the use of Equation 4.2 provides the engineer with the continuous frequency response of a discrete time system, it is sometimes referred to as a *discrete time Fourier transform* (DTFT). The DTFT is discussed in Chapter 3, "Review of the Fourier Transform." However, Equation 4.2 and Equation 4.4 are not compatible with computations on a digital computer. The reason is because $\omega$ is an analog term. It has infinite precision, and it takes on an infinite number of values within the range of $0 \le \omega < 2\pi$. In order to compute the frequency response of a digital system using a digital computer or special purpose hardware, we need to compute $H\left(e^{j\omega}\right)$ at discrete values of $\omega$.

Let's refer back to Figure 4.2, which illustrates the unit circle in the z-plane. We can partition the unit circle into $N$ equal arc segments so that each segment is defined by the angle $\theta$. In the figure, each arc segment is illustrated by the arc drawn between two adjacent dots on the unit circle. Now we can compute the value of $H\left(e^{j\omega}\right)$ at each segment dot to produce $N$ samples of the overall frequency response. The computations are carried out at each successive increment of the angle $\theta$. The total angle $\theta$ increases as it moves around the unit circle and is expressed as

$$\theta = \frac{2\pi}{N} n \quad \text{for } \left\{ 0 \le n < N-1 \right\}$$

The discrete values of the frequency response is therefore computed via

$$H(z)_{z=e^{j\omega}} = H\left(e^{j\omega}\right) \approx H\left(e^{j\theta}\right) = H\left(e^{j\frac{2\pi}{N}n}\right) \quad \text{for } \left\{ 0 \le n < N-1 \right\}$$

**Equation 4.5**

The approximation notation $(\approx)$ is used only to indicate that a finite number of all possible values of $H\left(e^{j\omega}\right)$ are computed. The value of that finite number is $N$.

Throughout this book we use the notation $h(n)$ to represent a discrete time series or sequence. This sequence is sometimes referred to as the impulse response of a system. We will use the notation $H(z)$ to represent the z-transform of the sequence $h(n)$. In addition, we will use the notation $\overset{z}{\leftrightarrow}$ to denote a z-transform pair. For example, the notation indicating that the sequences $h(n)$ and $H(z)$ are a z-transform pair is

$$h(n)\overset{z}{\leftrightarrow}H(z)$$

We will also use the z-transform operator $z\{h(n)\}$ as shorthand notation to describe the transformation from the discrete time domain to the z-domain. This notation is defined as

$$z\{h(n)\} = H(z) = \sum_{n=-\infty}^{\infty} h(n)z^{-n}$$

Notation that is commonly used in the industry for the inverse z-transform, or the transformation from the z-domain back to the discrete time domain, is

$$h(n) = z^{-1}\{H(z)\}$$

In both cases, the notation $z\{\ \}$ and $z^{-1}\{\ \}$ represent operators only. We will also use the accepted notation $H(e^{j\omega})$ to represent the frequency response of a digital system, keeping in mind that in doing so we are implicitly referring to the discrete frequency response defined in Equation 4.5.

## 4.3  LEFT-SIDED Z-TRANSFORM

Sequences that are 0 for all $n > N$ are termed *left-sided* or *noncausal* sequences. The left-sided z-transform of a sequence $h(n)$ is defined by Equation 4.6, where the summation range is $-\infty \leq n \leq N$:

$$H(z) = \sum_{n=-\infty}^{N} h(n)z^{-n}$$

### Equation 4.6

Depending upon the sequence, $N$ usually takes on a value of 0 or $-1$ such that

$$H(z) = \sum_{n=-\infty}^{-1} h(n) z^{-n} \quad \text{or} \quad H(z) = \sum_{n=-\infty}^{0} h(n) z^{-n}$$

## 4.4  RIGHT-SIDED Z-TRANSFORM

Sequences that are $0$ for all $n < N$ are termed *right-sided* or *causal* sequences. The right-sided z-transform of a sequence $h(n)$ is defined by Equation 4.7, where the summation range is $N \leq n \leq \infty$.

$$H(z) = \sum_{n=N}^{\infty} h(n) z^{-n}$$

### Equation 4.7

Depending upon the sequence, $N$ usually takes on the value of $0$ such that

$$H(z) = \sum_{n=0}^{\infty} h(n) z^{-n}$$

## 4.5  TWO-SIDED Z-TRANSFORM

Sequences that are nonzero on some arbitrary interval $-N < n < +M$ are referred to as *two-sided* sequences. The two-sided z-transform of a sequence $h(n)$ is defined by Equation 4.8:

$$H(z) = \sum_{n=-N}^{M} h(n) z^{-n}$$

### Equation 4.8

Most textbook two-sided transforms take on the limits of $-\infty$ to $+\infty$ such that

$$H(z) = \sum_{n=-\infty}^{\infty} h(n) z^{-n}$$

Many times, it is convenient to separate a two-sided sequence into its left-side and right-side components such as

$$h(n)_{TwoSided} = h(n)_{LeftSided} + h(n)_{RightSided}$$

$$\sum_{n=-\infty}^{\infty} h(n) = \sum_{n=-\infty}^{-1} h(n) + \sum_{n=0}^{\infty} h(n)$$

## 4.6 CONVERGENCE OF THE Z-TRANSFORM

If we have a sequence or series $H(z)$ that is either finite or infinite, we would like to know the limiting behavior of $H(z)$ as $z \to \infty$. Does the limit exist? Does the series grow without bound (in which case it is said to *diverge*)? Does the series tend to approach some finite value (in which case it is said to *converge*)? A system with a transfer function that diverges is unstable, and therefore the z-transform has meaning only if it converges.

The region of convergence, or ROC, of a z-transform is defined as that region in the z-plane where a series of the complex variable $z$ converges to some finite value. As an example, in the case of the right-sided z-transform, this means that

$$\left| H(z) \right| = \sum_{n=0}^{\infty} \left| h(n) z^{-n} \right| < \infty$$

The z-transform does not necessarily converge for all sequences or for all values of $z$. If $H(z)$ converges for some value of $z = re^{j\omega}$, then $H(z)$ will converge for all values of $z$ on the circle centered at the origin and defined by the radius $r$. Furthermore, this circle lies within the area of the z-plane defined as the ROC. Depending on the sequence, the convergence region may extend via ever-growing concentric circles to some other outer radius that defines a bounding circle of radius $r_O$. This bounding circle may include infinity. The convergence region may also extend via ever smaller concentric circles to some smaller inner bounding circle of radius $r_I$ that may include the origin. The annular area within the circles defined by the bounding radii $r_I$ and $r_O$ defines the convergence region. From Equation 4.3, we can see that if the ROC happens to include the unit circle scribed by $|z| = 1$, then the Fourier transform of the series exists and it also converges.

In determining the convergence region, it is useful to derive the factored closed form expression for a sequence. In general, this expression will take on the form

$$H(z) = \frac{N(z)}{D(z)} = \frac{(z - a_1)(z - a_2)(z - a_3) \cdots (z - a_N)}{(z - b_1)(z - b_2)(z - b_2) \cdots (z - b_M)}$$

where $a_n$ and $b_n$ are real or complex numbers and where $(z - a_n)$ and $(z - b_n)$ represent the zeros and the poles of $H(z)$, respectively. The convergence region cannot contain any poles, since by definition a function diverges toward infinity near a pole and therefore is not bounded. The poles of a sequence will oftentimes define a convergence region boundary. Boundaries can also be defined by either the origin or infinity.

You can determine if one or more poles or zeros exist at infinity simply by letting $z \to \infty$ and observing the behavior of $H(z)$. If the $\lim\limits_{z \to \infty} H(z) = 0$, then there exists at least one zero at infinity. On the other hand, if the $\lim\limits_{z \to \infty} H(z) = \infty$, then there exists at least one pole at infinity. If the resulting limit turns out to be an indeterminate form, then you must apply l'Hôpital's rule as many times as necessary to determine the result.

A simple guideline in this regard is as follows [6]:

- If the denominator of $H(z)$ is of higher order than the numerator, the z-transform has one or more zeros at infinity.
- If the denominator of $H(z)$ is of lower order than the numerator, then the z-transform has one or more poles at infinity.

### 4.6.1   Convergence of a Right-Sided Sequence

If $x(n)$ is a right-sided or causal sequence that is equal to $0$ for $n < N$, where the value of $N$ is usually chosen to be 0, the ROC will extend outward from the largest magnitude pole of $H(z)$ toward (and possibly including) $|z| = \infty$. The convergence region for a right-sided sequence is shown as the shaded area in Figure 4.3. This figure illustrates the case where the sequence has three real poles, labeled $P1$, $P2$, and $P3$, and one complex pole labeled $P4$ with its conjugate $P4^*$.

The boundary for the convergence region extends from the largest magnitude pole, in this case $P3$, out to (and possibly including) infinity. For clarity, we have also included in the figure the unit circle. In this example, since the unit circle is included in the ROC, the Fourier transform for this sequence also exists.

### 4.6.2   Convergence of a Left-Sided Sequence

If $x(n)$ is a left-sided or noncausal sequence that is equal to $0$ for $n > N$, the ROC will extend inward from the smallest magnitude pole of $H(z)$ toward (and possibly including) $|z| = 0$. For a left-sided sequence, the value of $N$ is usually chosen to be $-1$. The convergence region for a left-sided sequence is shown as the shaded area in Figure 4.4. This figure illustrates the case where the sequence again has three real poles labeled $P1$, $P2$, and $P3$, and one complex pole labeled $P4$ with its conjugate $P4^*$. The boundary for the convergence region extends from the smallest magnitude pole, in this case $P1$, inward to the origin. Again, for clarity, we have included in the figure the unit circle. The convergence region for this particular example does not contain the unit circle and therefore the Fourier transform for the system described by $H(z)$ does not exist.

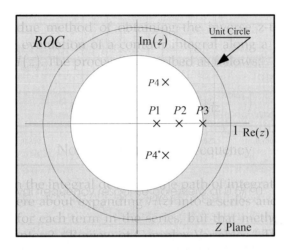

**Figure 4.3**  Convergence region for a right-sided sequence

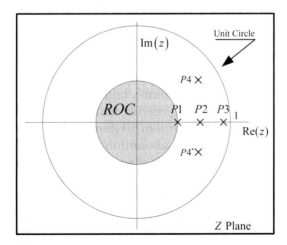

**Figure 4.4**  Convergence region for a left-sided sequence

### 4.6.3 Convergence of a Two-Sided Sequence

If $x(n)$ is a two-sided sequence, the ROC, should it exist, will consist of an annular ring whose width is bounded by at least two poles. We can think of a two-sided sequence as being the sum of a left-sided sequence and a right-sided sequence:

$$H(z) = H_L(z) + H_R(z)$$

The convergence region of the left-sided sequence $H_L(z)$ will extend from its smallest magnitude pole inward toward $|z| = 0$. The convergence region of the right-sided sequence $H_R(z)$ will extend from its largest magnitude pole outward toward $|z| = \infty$. The intersection of these two convergence regions will form an annular convergence ring and will represent the ROC for that sequence.

The convergence region for a two-sided sequence is illustrated as the shaded area in Figure 4.5. The sequence in this figure has four real poles, labeled $P_R1$, and $P_R2$, representing the two poles contributed by the right-sided sequence, and $P_L3$, and $P_L4$, representing the two poles contributed by the left-sided sequence. The figure illustrates that the convergence region is bounded by the largest magnitude pole $P_R2$ of the right-sided sequence and the smallest magnitude pole $P_L3$ of the left-sided sequence. Once again, for convenience, the unit circle is included in the figure. Since the unit circle lies within the ROC, this sequence has a Fourier transform.

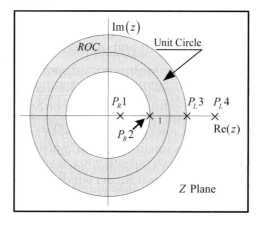

**Figure 4.5**   Convergence region for a two-sided sequence

### 4.6.4  **Test for Convergence**

The z-transform can be either a finite or an infinite series. As mentioned previously, the series has meaning only if it converges. There are several methods of testing whether or not a series converges. One such test is called the
*ratio test* and involves the absolute value of the ratio of successive terms of a
sequence. Let's assume we have an infinite sequence given by

$$\sum_{n=0}^{\infty} a_n = a_0 + a_1 + a_2 + \cdots a_k + a_{k+1} + \cdots$$

**Equation 4.9**

and we wish to determine if this sequence converges. We could form another
sequence $d_n$ that is composed of the ratio of successive terms of the original
sequence $a_n$, such that

$$d_n = \sum_{n=0}^{\infty} \frac{a_{n+1}}{a_n} = \frac{a_1}{a_0} + \frac{a_2}{a_1} + \frac{a_3}{a_2} + \cdots + \frac{a_{k+1}}{a_k} + \cdots$$

and then perform the ratio test for convergence on the new sequence. The
rule for the ratio test is if the sequence $d_n$ converges to a number less than 1,
then the original sequence $a_n$ also converges. We can define the ratio convergence test as

$$L = \lim_{n \to \infty} \left| \frac{a_{n+1}}{a_n} \right|$$

**Equation 4.10**

When we compute the limit in Equation 4.10, we can determine whether
or not the sequence converges by the following means:

$$\begin{cases} L < 1 & \text{The sequence converges} \\ L = 1 \text{ or does not exist} & \text{We don't know if the sequence converges or diverges} \\ L > 1 & \text{The sequence diverges} \end{cases}$$

If $L = 1$ or if $L$ does not exist, then the ratio test does not provide any information as to whether the sequence converges or diverges, and another method
of convergence test must be employed. There are a variety of convergence tests

that are suitable for different types of sequences. A few of these convergence tests are listed here and are defined in detail in [5]:

1. Comparison test
2. P-test
3. Root test
4. Integral test
5. Alternating series test

We will only discuss the ratio test here since it is best suited for the z-transform power series.

### 4.6.5  Example Convergence of a Right-Sided Sequence

For a right-sided sequence $h(n)$, the convergence region of the z-transform is bounded on the inside by the largest magnitude pole of $H(z)$ and can extend all the way to and include the circle defined by $|z| = \infty$. Let's illustrate this with an example. Suppose we have a z-transform of the form

$$H(z) = \sum_{n=0}^{\infty} a^n z^{-n} = 1 + az^{-1} + a^2 z^{-2} + a^3 z^{-3} + a^4 z^{-4} + \cdots$$

**Equation 4.11**

We need to determine if this sequence converges, and if so, identify its convergence boundaries. In other words, what is its ROC? To check for the convergence of $H(z)$, we use the ratio test:

$$L = \lim_{n \to \infty} \left| \frac{a^{n+1} z^{-(n+1)}}{a^n z^{-n}} \right| = \lim_{n \to \infty} \left| \frac{a^{n+1} z^{-n-1}}{a^n z^{-n}} \right| = |az^{-1}| = \left| \frac{a}{z} \right|$$

We know from the definition of the ratio test that the sequence converges only if $L < 1$, so we set the term

$$L = \left| \frac{a}{z} \right| < 1$$

The expression $L < 1$ is true only if $|z| > |a|$. Now that we know the sequence will converge if the ratio is less than 1, or equivalently $|z| > |a|$, let's compute the ROC and see if it agrees. Ideally, to do this computation, we would like to find a closed form expression for Equation 4.11 and then inspect its poles. Let's begin with Equation 4.11, which is repeated here for clarity:

$$\sum_{n=0}^{\infty} a^n z^{-n} = 1 + az^{-1} + a^2 z^{-2} + a^3 z^{-3} + a^4 z^{-4} + \cdots$$

If we multiply the z-transform sequence by the quantity $az^{-1}$, we will generate a second sequence given by

$$az^{-1} \sum_{n=0}^{\infty} a^n z^{-n} = az^{-1} + a^2 z^{-2} + a^3 z^{-3} + a^4 z^{-4} + \cdots$$

**Equation 4.12**

If we subtract Equation 4.12 from Equation 4.11, we obtain

$$\sum_{n=0}^{\infty} a^n z^{-n} - az^{-1} \sum_{n=0}^{\infty} a^n z^{-n} = \left\{ 1 + az^{-1} + a^2 z^{-2} + a^3 z^{-3} + a^4 z^{-4} + \cdots \right\} - \left\{ az^{-1} + a^2 z^{-2} + a^3 z^{-3} + a^4 z^{-4} + \cdots \right\}$$

This reduces to

$$\sum_{n=0}^{\infty} a^n z^{-n} \left( 1 - az^{-1} \right) = 1$$

If we divide both sides by the term $\left( 1 - az^{-1} \right)$ and then multiply both the numerator and denominator by $z$, we end up with

$$\sum_{n=0}^{\infty} a^n z^{-n} = \frac{1}{\left( 1 - az^{-1} \right)} = \frac{z}{z - a}$$

**Equation 4.13**

It is clear from Equation 4.13 that the closed form expression for $H(z)$ has a pole located at $z = a$. Since this is a right-sided sequence, we know that the ROC is bounded on the low side by the largest magnitude pole, which in this case is at $z = a$. We also know that the sequence converges for $|z| > |a|$. So the ratio test and the convergence calculation agree with one another, which is always a good thing. The graphical depiction of the ROC for $H(z)$ is illustrated as the shaded area in Figure 4.6. In the figure, we have shown that $|a| < 1$, which means the unit circle is included in the ROC and therefore the system Fourier transform exists.

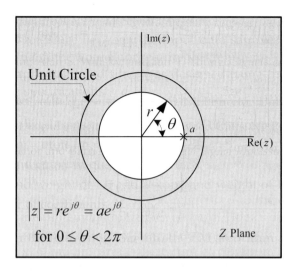

**Figure 4.6**  ROC for the right-sided $H(z)$

### 4.6.6  Example Convergence of a Left-Sided Sequence

Suppose we have a left-sided sequence given by $x(n) = -a^n u(-n-1)$. The z-transform of this sequence is computed using the same methods as the right-sided sequence, demonstrated as follows. The function $u(-n-1)$ is a unit step function that is defined to be

$$u(-n-1) = \left\{ \begin{array}{ll} 1 & \text{for } n \leq -1 \\ 0 & \text{for } n > -1 \end{array} \right\}$$

The z-transform can be computed as

$$\begin{aligned} X(z) &= -\sum_{n=-\infty}^{\infty} a^n u(-n-1) z^{-n} \\ &= -\sum_{n=-\infty}^{-1} a^n z^{-n} \\ &= -\sum_{n=1}^{\infty} a^{-n} z^{n} \\ &= 1 - \sum_{n=0}^{\infty} \left( a^{-1} z \right)^{n} \end{aligned}$$

**Equation 4.14**

The change in the lower limit of the summation index $n$, in Equation 4.14 from 1 to 0, results in the addition of a $-1$ term in the series. That term is accounted for and negated by the inclusion of a $+1$ term.

Temporarily disregarding the 1 term in Equation 4.14, we use the ratio test to see if the series term converges:

$$\lim_{n\to\infty}\left|\frac{a^{-n-1}z^{n+1}}{a^{-n}z^{n}}\right| < 1 \;\Rightarrow\; \lim_{n\to\infty}\left|\frac{z}{a}\right| < 1 \;\Rightarrow\; |z| < |a|$$

The series does indeed converge for $|z| < |a|$.

The closed form expression of Equation 4.14 is computed by first expanding the summation term to produce

$$\sum_{n=0}^{\infty}\left(a^{-1}z\right)^{n} \;=\; 1 + a^{-1}z + a^{-2}z^{2} + a^{-3}z^{3} + \cdots$$

**Equation 4.15**

If we multiply both sides of Equation 4.15 by the term $a^{-1}z$, we get

$$a^{-1}z\sum_{n=0}^{\infty}\left(a^{-1}z\right)^{n} \;=\; a^{-1}z + a^{-2}z^{2} + a^{-3}z^{3} + \cdots$$

**Equation 4.16**

If we subtract Equation 4.16 from Equation 4.15, we end up with

$$\sum_{n=0}^{\infty}\left(a^{-1}z\right)^{n} \;-\; a^{-1}z\sum_{n=0}^{\infty}\left(a^{-1}z\right)^{n} \;=\; 1$$

Now we can collect terms to get

$$\sum_{n=0}^{\infty}\left(a^{-1}z\right)^{n}\left(1 - a^{-1}z\right) = 1$$

$$\sum_{n=0}^{\infty}\left(a^{-1}z\right)^{n} \;=\; \frac{1}{\left(1 - a^{-1}z\right)}$$

If we include the 1 term from Equation 4.14, the closed form solution of the z-transform $X(z)$ is

$$X(z) = 1 - \sum_{n=0}^{\infty} \left(a^{-1}z\right)^n \quad = 1 - \frac{1}{\left(1 - a^{-1}z\right)}$$

$$= \frac{1}{\left(1 - az^{-1}\right)} \quad = \frac{z}{z - a}$$

We know that $X(z)$ has a pole at $z = a$, and we also know that $X(z)$ converges for $|z| < |a|$. Using these two pieces of information, we can draw the ROC for $X(z)$ as the shaded area illustrated in Figure 4.7. These results correlate with the fact that the ROC for a left-sided or noncausal sequence extends inward toward 0 from the smallest pole of the sequence.

### 4.6.7 Example Convergence of a Two-Sided Sequence

Now suppose we have a two-sided sequence defined by

$$X(z) = \sum_{n=-\infty}^{\infty} a^{|n|} z^{-n}$$

**Equation 4.17**

We can separate Equation 4.17 into its left-side and right-side components to produce

$$X(z) = \sum_{n=-\infty}^{-1} a^{-n} z^{-n} \;+\; \sum_{n=0}^{\infty} a^n z^{-n}$$

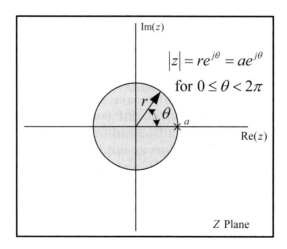

**Figure 4.7**   ROC for the left-sided $H(z)$

The details are left to the reader: Using the same procedure as we did for the left-sided sequence, we can determine the ROC for the term $\sum\limits_{n=-\infty}^{-1} a^{-n}z^{-n}$ to be $|z| < \left|\dfrac{1}{a}\right|$. Using the same procedure we did for the right-sided sequence, we can determine the ROC for the term $\sum\limits_{n=0}^{\infty} a^{n}z^{-n}$ to be $|z| > |a|$. We can also determine the closed form expression of $X(z)$ to be

$$X(z) = \frac{az}{1-az} + \frac{z}{z-a}$$

The pole of $X(z)$ that is contributed by the left-sided sequence is located at $\dfrac{1}{a}$ and the pole of $X(z)$ that is contributed by the right-sided sequence is located at $a$. The ROC for $X(z)$ is the annular ring defined by the intersection of the two convergence regions and is illustrated as the shaded area in Figure 4.8.

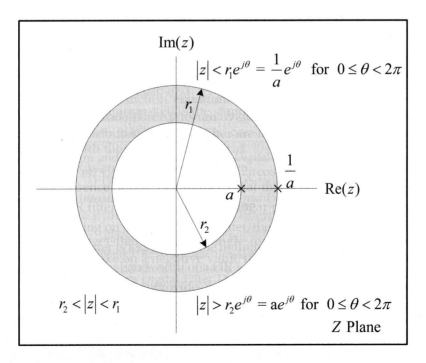

**Figure 4.8**   ROC for the two-sided *H(z)*

## 4.7   SYSTEM STABILITY

Let's assume that you have been tasked by your supervisor to design a recursive digital filter—that is, a filter with feedback, commonly referred to as an infinite impulse response (IIR) filter. These filters are very sensitive to coefficient values and, if not designed carefully, can be unstable.

You did your homework, you burned some midnight oil, you beat your head against the wall for a few hours, and finally you came up with a filter design whose impulse response is represented by the infinite sequence $h(n)$. Now you want to check the properties of your filter impulse response sequence and find the ROC. At this point, you may ask yourself, "Why in the blazes is the ROC of my filter impulse response important?" The answer is both very simple and very powerful.

You can use the z-transform of $h(n)$ to determine the stability of your filter or most other digital processing systems that you may design. It turns out that for linear time invariant (LTI) causal systems with right-sided sequences, if the unit circle is contained within the convergence region, then that system is defined to be stable.

This is a just a long-winded way of saying that if all the poles of a digital system lie within the unit circle, then that system is stable. Stated another way, the necessary and sufficient conditions for an LTI system to be stable require that its ROC contain the unit circle. Some authors state that a stable system is one that will produce a bounded output signal if the input signal is bounded. This is sometimes abbreviated BIBO for Bounded Input Bounded Output.

As an example, the poles plotted in Figure 4.3 for a right-sided sequence are all within the unit circle. Therefore the system that owns these poles is by definition stable.

Figure 4.9 and Figure 4.10 provide a conceptual illustration of the discrete time domain impulse response of a system versus the placement of that system's poles. The discrete time response versus pole placement for real poles is illustrated in Figure 4.9. We can see in this figure that the poles within the unit circle are correlated with a decaying response that approaches zero as the sample index $n \to \infty$. Poles on the right side of the $z$-plane produce unipolar sequences, while poles on the left side of the $z$-plane produce bipolar or alternating sequences.

Poles outside the unit circle are correlated with a time response that increases with sample index $n$. That is, the response grows without bound and is unstable. A simple example is a digital system whose output grows with time until the magnitude can no longer be represented in the fixed width bit field of the system. Once this occurs, the system produces nothing but random mumbo jumbo at its output. We all know that this can't be a good thing.

As illustrated in the figure, a positive real pole located on the unit circle is associated with a constant amplitude sequence. On the other hand, a negative real pole on the unit circle is associated with an alternating sequence of constant amplitude.

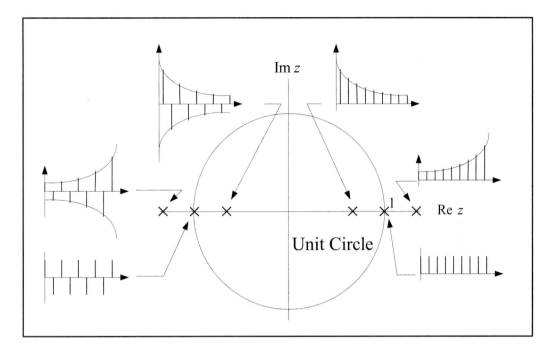

**Figure 4.9**   Pole locations versus time response for real poles

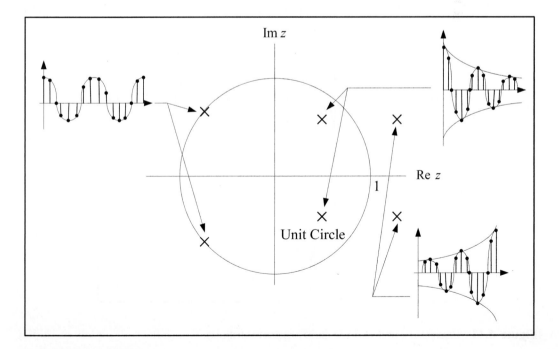

**Figure 4.10**   Pole locations versus time response for complex poles

The time response versus pole placement for complex poles is illustrated in Figure 4.10. We can see in the figure that the system time series behavior is sinusoidal with the same amplitude growth and decay properties as was observed for real poles.

Whenever you are working with a system pole-zero diagram, you should be able to visualize the general form of the time response for each real or complex pole.

## 4.8 PROPERTIES OF THE Z-TRANSFORM

### 4.8.1 The Unit Delay

We will begin our discussion on the properties of the z-transform, with the unit sample delay. The block diagram of a circuit that creates a unit delay is shown in Figure 4.11. Suppose we have an input sequence $x(n)$ that is equal to 0 for $n < 0$. Further suppose that this sequence is clocked through a shift register that delays the sequence by one clock period and produces an output sequence labeled $y(n)$. In part A of the figure, we see sequential samples of an input data sequence $x(n)$ being clocked into and out of a shift register. The register delays the input sequence $x(n)$ by one clock period so that on output the sequence notation is $x(n-1)$ relative to the input.

In part B of the figure, we see the same block diagram, but this time the term *shift register* has been replaced by the symbol $z^{-1}$. In the context of a data

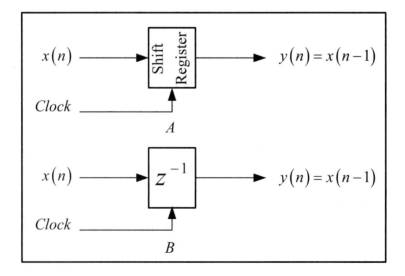

**Figure 4.11** Equivalent unit sample delay representations

delay, the term *shift register* and symbol $z^{-1}$ are equivalent. In signal process-ing notation, the block diagram representation in part B of the figure is identi-cal to that of part A.

In the complex z-domain, the notation $z^{-1}$ is mathematically identical to a single sample clock period delay. This is easily shown by the following: We see that the output sequence in part B of the figure is a sequence labeled $y(n)$ that is equal to the input sequence $x(n)$ delayed by one sample period such that

$$y(n) = x(n-1) \quad \text{where } x(n) = 0 \text{ for } n < 0$$

**Equation 4.18**

The z-transform of the sequence in Equation 4.18 is given by

$$Y(z) = \sum_{n=0}^{\infty} y(n) z^{-n} = \sum_{n=0}^{\infty} x(n-1) z^{-n}$$

**Equation 4.19**

We can simplify Equation 4.19 by substituting the index $m$ for the index $n-1$, so that $m = n-1$. With this substitution, we realize that $n = m+1$, and we recognize that when $n = 0$, $m = -1$. If we substitute these new indices back into the equation, we arrive at

$$Y(z) = \sum_{m=-1}^{\infty} x(m) z^{-(m+1)}$$

**Equation 4.20**

Note the range of summation has changed from $\{0 \text{ to } \infty\}$ to $\{-1 \text{ to } \infty\}$. Since we know that $x(m) = 0$ for $m < 0$, we can set the lower bound of the summation range to $m = 0$. Now we can rearrange the terms in Equation 4.20 to obtain

$$Y(z) = z^{-1} \left[ \sum_{m=0}^{\infty} x(m) z^{-m} \right]$$

The term in brackets is just the z-transform of $x(m)$ written as $x(m) \overset{z}{\longleftrightarrow} X(z)$. Therefore, the z-transform of a series delayed by one sample period is given by

$$Y(z) = z^{-1}X(z)$$

**Equation 4.21**

We can see from Equation 4.21 that the z-transform of the output sequence is equal to the z-transform of the input sequence multiplied by $z^{-1}$. In the z-domain, $z^{-1}$ represents the transform of a time domain unit delay equal to one clock period. This is why the term *shift register* in part A of Figure 4.11 is replaced by the symbol $z^{-1}$ in part B.

If we view the shift register as a system that operates on an input sequence $x(n)$ to produce an output sequence $y(n)$ and define the system function as $H(z)$, then we can describe the system transfer function for this shift register as

$$H(z) = \frac{Y(z)}{X(z)} = \frac{z^{-1}X(z)}{X(z)} = z^{-1}$$

This result can be extended to any number of delays. For example, suppose we had a series of shift registers as illustrated in Figure 4.12. In this figure, the input sequence $x(n)$ is clocked through a series of four shift registers. Each register delays the sequence by one clock period. The output of each register is tagged with its discrete time sequence notation and with its corresponding z-transform notation. It is clear from this figure that each successive delay multiplies the result by $z^{-1}$. The sequence $x(n)$ with z-transform $X(z)$ enters the register array and, four clock periods later, emerges at the output denoted by the sequence $y(n) = x(n-4)$ with a z-transform of $Y(z) = z^{-4}X(z)$. The four sample delay discussed here is a fairly simple example, but it clearly illustrates the z-transform delay property. As we shall see, the sample delay property of the z-transform is a handy tool that enables us to write by inspection the z-transform of a simple system either from the systems difference equation or from the system block diagram. For example, suppose we wish to determine the z-transform of the structure illustrated in Figure 4.13. The difference equation that defines this structure can be written by inspection of the figure as

$$y(n) = 1.0x(n) + 1.5x(n-1) + 0.5x(n-2) + 0.5y(n-1)$$

**Equation 4.22**

Using the unit delay property, the z-transform of this digital structure can be written from Equation 4.22 as

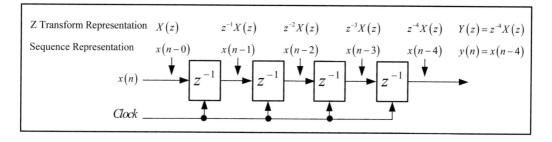

| Z Transform Representation | $X(z)$ | $z^{-1}X(z)$ | $z^{-2}X(z)$ | $z^{-3}X(z)$ | $z^{-4}X(z)$ | $Y(z)=z^{-4}X(z)$ |
|---|---|---|---|---|---|---|
| Sequence Representation | $x(n-0)$ | $x(n-1)$ | $x(n-2)$ | $x(n-3)$ | $x(n-4)$ | $y(n)=x(n-4)$ |

**Figure 4.12**   Four sample delay

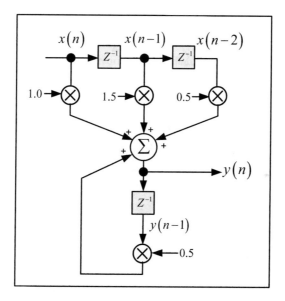

**Figure 4.13**   Typical recursive hardware structure

$$Y(z) = X(z) + 1.5z^{-1}X(z) + 0.5z^{-2}X(z) + 0.5z^{-1}Y(z)$$

We can rearrange and collect terms to get

$$Y(z)\left(1 - 0.5z^{-1}\right) = X(z)\left(1 + 1.5z^{-1} + 0.5z^{-2}\right)$$

The system transfer function is then obtained by dividing both sides of the equation by $X(z)$ and the term $\left(1 - 0.5z^{-1}\right)$ to get

$$\frac{Y(z)}{X(z)} = H(z) = \frac{\left(1 + 1.5z^{-1} + 0.5z^{-2}\right)}{\left(1 - 0.5z^{-1}\right)}$$

**Equation 4.23**

Many people do not like working in terms of inverse $z$'s, so we can multiply both the numerator and denominator of Equation 4.23 by $z^2$ to arrive at

$$H(z) = \frac{z^2 + 1.5z + 0.5}{z(z - 0.5)}$$

Now we can factor the numerator and rewrite the system function to expose the poles and zeros:

$$H(z) = \frac{(z + 1)(z + 0.5)}{z(z - 0.5)}$$

**Equation 4.24**

It is clear from Equation 4.24 that this example system has two real zeros, one at $z = -1$ and another at $z = -0.5$, and two real poles at $z = 0$ and $z = +0.5$. The poles and zeros are plotted relative to the unit circle in Figure 4.14. Once we have factored the transfer function $H(z)$, we can analyze the system and determine its response characteristics in the following ways:

- We can determine the system frequency response by evaluating $H(z)$ on the unit circle in the $z$-plane. This is accomplished by replacing the complex variable $z$ everywhere in $H(z)$ by the term $e^{j\omega}$ and computing the frequency response for $0 \le \omega < 2\pi$. We will demonstrate this in more detail later.
- We can compute the system ROC.
- We can plot the system function poles and zeros in the $z$-plane to verify that the system is stable. This is accomplished by verifying that all the poles lie within the unit circle.

We can see from the plot that the system is stable because the system poles are contained within the unit circle. Also, since the system is causal, we know that the ROC extends from the largest magnitude pole (in this case $z = 0.5$) in ever-expanding concentric circles to $|z| = \infty$. Since the ROC includes the unit circle, we have a correlating piece of information that confirms that

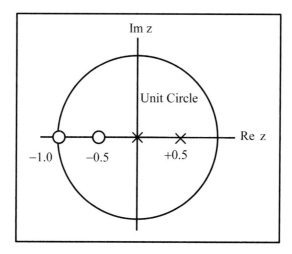

**Figure 4.14**   Pole-zero plot of a typical hardware structure

this system is stable, and we also know that the Fourier transform of this systems impulse response exists.

We can tell by inspection of the pole-zero diagram that the frequency response of this system is going to approximate a low pass characteristic. The location of the pole at 0.5 will cause the system response to have a fairly high magnitude around 0 radians or 0 Hz, and the occurrence of the 0 at $\pi$ radians will cause the system frequency response to be 0 at half the sample rate.

### 4.8.2   Linearity

Suppose we have two sequences $x(n)$ and $y(n)$ whose z-transforms are given by

- $z\{x(n)\} = X(z)$, which converges in the range $R_{X1} < |z| < R_{X2}$, and
- $z\{y(n)\} = Y(z)$, which converges in the range $R_{Y1} < |z| < R_{Y2}$.

If $a$ and $b$ are constants, then $z\{ax(n) + by(n)\} = aX(z) + bY(z)$, which converges for $R_1 < |z| < R_2$. The convergence region is as a minimum—the intersection of the two regions defined by $R_X$ and $R_Y$.

### 4.8.3   Shift

This property is a simple extension of the unit delay property. If $x(n) \overset{z}{\leftrightarrow} X(z)$, then we can define the z-transform pair:

$$x(n-k) \overset{z}{\leftrightarrow} z^{-k} X(z)$$

**Equation 4.25**

### 4.8.4  Left Shift of a Right-Sided Sequence

If $x(n) \overset{z}{\leftrightarrow} X(z)$, then

$$x(n+m) \overset{z}{\leftrightarrow} z^m \left[ X(z) - \sum_{n=0}^{m-1} x(n) z^{-n} \right]$$

**Equation 4.26**

Proof: If $y(n) = x(n+m)$, then $Y(z) = z\{x(n+m)\} = \sum_{n=0}^{\infty} x(n+m) z^{-n}$.

We change the index by letting $k = n + m$. Therefore $n = k - m$, and when $n = 0, k = m$, which gives

$$
\begin{aligned}
Y(z) &= \sum_{k=m}^{\infty} x(k) z^{-k+m} \\
&= z^m \sum_{k=m}^{\infty} x(k) z^{-k} \\
&= z^m \left[ \sum_{k=0}^{\infty} x(k) z^{-k} - \sum_{k=0}^{m-1} x(k) z^{-k} \right] \\
&= z^m \left[ X(z) - \sum_{k=0}^{m-1} x(k) z^{-k} \right]
\end{aligned}
$$

For example,

- $z\{x(n+1)\} = zX(z) - zx(0)$
- $z\{x(n+2)\} = z^2 X(z) - z^2 x(0) - zx(1)$
- $z\{x(n+3)\} = z^3 X(z) - z^3 x(0) - z^2 x(1) - zx(2)$

It is very important to avoid confusion by taking note of the $X(z)$ and $x(n)$ terms in these examples. $X(z)$ is the z-transform of $x(n)$, and $x(n)$ is the discrete time domain sequence.

## 4.8.5   Differentiation

If $x(n) \overset{z}{\leftrightarrow} X(z)$, then

$$nx(n) \overset{z}{\leftrightarrow} -z\frac{d}{dz}X(z)$$

### Equation 4.27

or the equivalent

$$-z\frac{d}{dz}X(z) = \sum_{n=-\infty}^{\infty} nx(n)z^{-n}$$

### Equation 4.28

The proof of this relationship is straightforward. For the z-transform sequence,

$$X(z) = \sum_{n=-\infty}^{\infty} x(n)z^{-n}$$

We differentiate $X(z)$ with respect to $z$ as follows:

$$\frac{d}{dz}X(z) = \frac{d}{dz}\left[\sum_{n=-\infty}^{\infty} x(n)z^{-n}\right]$$

$$= \sum_{n=-\infty}^{\infty} -nx(n)z^{-n-1}$$

We can put the equation into a more standard form by multiplying both sides by $-z$ to obtain

$$-z\frac{d}{dz}X(z) = -z\sum_{n=-\infty}^{\infty} -nx(n)z^{-n-1}$$

Now we can simplify by collecting terms to produce

$$-z\frac{d}{dz}X(z) = \sum_{n=-\infty}^{\infty} nx(n)z^{-n}$$

### Equation 4.29

## 4.8.6 Initial Value Theorem for a Right-Sided Sequence

If $x(n) \overset{z}{\leftrightarrow} X(z)$, then we can compute the initial value of the sequence $x(n)$ from its z-transform by

$$x(0) = \lim_{z \to \infty} X(z)$$

**Equation 4.30**

The proof of this is easily shown. If a sequence $x(n) = 0$ for $n < 0$, then the transform of $x(n)$ is the right-sided sequence

$$X(z) = \sum_{n=0}^{\infty} x(n) z^{-n}$$

When we take the limit as $z \to \infty$, we obtain

$$\lim_{z \to \infty} X(z) = \lim_{z \to \infty} \left[ \sum_{n=0}^{\infty} x(n) z^{-n} \right] = \lim_{z \to \infty} \left[ \frac{x(0)}{1} + \frac{x(1)}{z} + \frac{x(2)}{z^2} + \frac{x(3)}{z^3} + \cdots \right] = x(0)$$

## 4.8.7 Final Value Theorem for a Right-Sided Sequence

If $X(z)$ is the z-transform of a right-sided sequence $x(n)$ that is $0$ for $n < 0$, and if $x(n)$ approaches a finite constant as $n$ approaches $\infty$, then

$$\lim_{n \to \infty} x(n) = \lim_{z \to 1} (z - 1) X(z)$$

**Equation 4.31**

The proof is demonstrated by the following: Let's assume that we want to find the value for the sequence $y(n)$ as $n \to \infty$ where

$$y(n) = x(n+1) - x(n)$$

The z-transform is given by

$$Y(z) = \sum_{n=0}^{\infty} \left[ x(n+1) - x(n) \right] z^{-n}$$

We know from the z-transform left-shift and delay properties that

$$z\{x(n+1)\} = zX(z) - zx(0)$$

Substituting this property, we get

$$zX(z) - zx(0) - X(z) = \sum_{n=0}^{\infty} \left[ x(n+1) - x(n) \right] z^{-n}$$

Now we can collect terms and rearrange to get

$$(z-1)X(z) = zx(0) + \sum_{n=0}^{\infty} \left[ x(n+1) - x(n) \right] z^{-n}$$

We take the limit as $z \to 1$ on both sides of the equation:

$$\lim_{z \to 1} (z-1)X(z) = \lim_{z \to 1} \left\{ zx(0) + \sum_{n=0}^{\infty} \left[ x(n+1) - x(n) \right] z^{-n} \right\}$$

After taking the limit, the $z$ terms on the right-hand side of the equation vanish, leaving us with

$$\lim_{z \to 1} (z-1)X(z) = x(0) + \sum_{n=0}^{\infty} \left[ x(n+1) - x(n) \right]$$

If we expand the series on the right-hand side, we can see that

$$x(0) + \{x(1) + x(2) + x(3) + \cdots + x(\infty)\} - \{x(0) + x(1) + x(2) + x(3) + \cdots + x(\infty-1)\} = x(\infty)$$

which gives us our desired result:

$$\lim_{z \to 1} (z-1)X(z) = \lim_{n \to \infty} x(n)$$

### 4.8.8   Convolution of Sequences

If the series $y(n)$ is the convolution of two other series $x(n)$ and $h(n)$ given by

$$y(n) = \sum_{k=-\infty}^{\infty} x(k)h(n-k) = x(n) * h(n)$$

where the symbol $*$ is the convolution operator, we can state that

$$Y(z) = X(z)H(z)$$

This leads us to the z-transform pair

$$\sum_{k=-\infty}^{\infty} x(k)h(n-k) \overset{z}{\leftrightarrow} X(z)H(z)$$

or

$$x(n)*h(n) \overset{z}{\leftrightarrow} X(z)H(z)$$

## Equation 4.32

The proof is demonstrated by the following: if $x(n)$ is an input sequence to and $y(n)$ is the output sequence from some digital system, then the two sequences would be related by the impulse response function $h(n)$ of the system. The sequence $y(n)$ is determined to be the convolution of the input sequence $x(n)$ and the system impulse response sequence $h(n)$. In standard notation, the convolution between these two sequences is represented by

$$y(n) = h(n)*x(n)$$

To be more specific, the convolution of the two sequences is expressed as

$$y(n) = \sum_{k=-\infty}^{\infty} h(k)x(n-k)$$

The z-transform of $y(n)$ is computed by

$$Y(z) = \sum_{n=-\infty}^{\infty} \left[ \sum_{k=-\infty}^{\infty} h(k)x(n-k) \right] z^{-n}$$

We can interchange the summation order to arrive at

$$Y(z) = \sum_{k=-\infty}^{\infty} h(k) \left[ \sum_{n=-\infty}^{\infty} x(n-k) \right] z^{-n}$$

Let's replace the index $n - k$ with a new index $p = n - k$. Using this notation, we can also replace the index $n$ with $n = p + k$. After the substitution of indices, we obtain

$$Y(z) = \sum_{k=-\infty}^{\infty} h(k) \left[ \sum_{p=-\infty}^{\infty} x(p) \right] z^{-(p+k)}$$

Now we can rearrange terms to produce

$$Y(z) = \left[ \sum_{k=-\infty}^{\infty} h(k)\, z^{-k} \right] \left[ \sum_{n=-\infty}^{\infty} x(p) z^{-p} \right]$$

**Equation 4.33**

Comparing Equation 4.33 with the definition of the z-transform given in Equation 4.7, we see that the result is simply the product of the transforms of the two sequences.

$$Y(z) = H(z)X(z)$$

This relationship allows us to define the z-transform pair:

$$h(n) * x(n) \overset{z}{\longleftrightarrow} H(z)X(z)$$

### 4.8.9 Multiplication of Two Sequences

The z-transform of a product of two sequences $w(n) = x(n)y(n)$ is represented as

$$W(z) = \sum_{n=-\infty}^{+\infty} x(n)y(n)\, z^{-n}$$

We know from residue theory, discussed later in this chapter, that the inverse z-transform of $Y(z)$ is

$$y(n) = \frac{1}{2\pi j} \oint_C Y(v)\, v^{n-1} dv$$

Upon substitution, we arrive at

$$
\begin{aligned}
W(z) &= \frac{1}{2\pi j}\left[ \sum_{n=-\infty}^{+\infty} x(n) \oint_C Y(v)\, v^{n-1} dv \right] z^{-n} \\
&= \frac{1}{2\pi j}\left[ \sum_{n=-\infty}^{+\infty} x(n) \oint_C Y(v) \left(\frac{z}{v}\right)^{-n} v^{-1} dv \right] \\
&= \frac{1}{2\pi j} \oint_C \left[ \sum_{n=-\infty}^{+\infty} x(n) \left(\frac{z}{v}\right)^{-n} Y(v) v^{-1} dv \right] \\
&= \frac{1}{2\pi j} \oint_C \left[ \sum_{n=-\infty}^{+\infty} X\left(\frac{z}{v}\right) Y(v) v^{-1} dv \right]
\end{aligned}
$$

where the contour $C$ is the overlap region between $X(z/v)$ and $Y(v)$. Therefore, we write the z-transform pair as

$$
x(n)y(n) \overset{z}{\longleftrightarrow} \frac{1}{2\pi j} \oint_C X\left(\frac{z}{v}\right) Y(v) v^{-1} dv
$$

**Equation 4.34**

## 4.9 COMMON Z-TRANSFORM PAIRS

You have probably correctly assumed that there are a great many z-transform pairs that are commonly used in everyday engineering applications.
These transform pairs are particularly useful when computing the inverse z-
transform of a system transfer function—that is, computing sequence $h(n)$
from the transform $H(z)$. If the expression for $H(z)$ is fairly simple in that
it consists of a few zeros and or poles, then the computation of the inverse
transform $h(n)$ can be straightforward with the aid of a table of common
transform pairs. For more complex forms of $H(z)$, it may be necessary to utilize readily available computer programs to perform the computation. Table
4.1 lists some common z-transform pairs. As we will see later, when dealing
with inverse transform computations, the entries in this table will prove extremely handy. To illustrate the usefulness of Table 4.1, suppose we wish to
compute the z-transform of the discrete time series,

$$
h(n) = \sin(\omega n T)
$$

**Equation 4.35**

**Table 4.1          Some Common Z-Transform Pairs**

| Pair number | Discrete time sequence $h(n), \ n \geq 0$ | Z-transform $H(z)$ | ROC |
|:---:|:---:|:---:|:---:|
| 1 | $\Delta(n)$ | $1$ | All $z$ |
| 2 | $\Delta(n-m)$ | $z^{-m}$ | All $z$ except 0 if $m > 0$ $\infty$ if $m < 0$ |
| 3 | $k\Delta(n)$ | $k$ | All $z$ |
| 4 | $u(n)$ | $\dfrac{z}{z-1}$ | $|z| > 1$ |
| 5 | $-u(-n-1)$ | $\dfrac{z}{z-1}$ | $|z| < 1$ |
| 6 | $k$ | $\dfrac{kz}{z-1}$ | $|z| > 1$ |
| 7 | $kn$ | $\dfrac{kz}{(z-1)^2}$ | $|z| > 1$ |
| 8 | $kn^2$ | $\dfrac{kz(z+1)}{(z-1)^3}$ | $|z| > 1$ |
| 9 | $k\alpha^n$ | $\dfrac{kz}{z-\alpha}$ | $|z| > \alpha$ |
| 10 | $kn\alpha^n$ | $\dfrac{k\alpha z}{(z-\alpha)^2}$ | $|z| > \alpha$ |
| 11 | $\alpha^n \quad 0 \leq n \leq N-1$ $0 \quad$ Elsewhere | $\dfrac{1-\alpha^N z^{-N}}{1-\alpha z^{-1}}$ | $|z| > 0$ |
| 12 | $\alpha^n$ | $\dfrac{z}{z-\alpha}$ | $|z| > |\alpha|$ |

(*continued on next page*)

**Table 4.1      Some Common Z-Transform Pairs (continued)**

| Pair number | Discrete time sequence $h(n)$, $n \geq 0$ | Z-transform $H(z)$ | ROC |
|---|---|---|---|
| 13 | $\sin(\omega nT)$ | $\dfrac{z\sin(\omega T)}{z^2 - 2z\cos(\omega T) + 1}$ | $|z| > 1$ |
| 14 | $\cos(\omega nT)$ | $\dfrac{z(z - \cos(\omega T))}{z^2 - 2z\cos(\omega T) + 1}$ | $|z| > 1$ |
| 15 | $r^n \cos(\omega nT)$ | $\dfrac{1 - \left[r\cos(\omega T)\right]z^{-1}}{1 - \left[2r\cos(\omega T)\right]z^{-1} + r^2 z^{-2}}$ | $|z| > r$ |
| 16 | $r^n \sin(\omega nT)$ | $\dfrac{\left[r\sin(\omega T)\right]z^{-1}}{1 - \left[2r\cos(\omega T)\right]z^{-1} + r^2 z^{-2}}$ | $|z| > r$ |
| 17 | $e^{-\alpha n}\sin(\omega nT)$ | $\dfrac{ze^{-\alpha}\sin(\omega T)}{z^2 - 2ze^{-\alpha}\cos(\omega T) + e^{-2\alpha}}$ | |
| 18 | $e^{-\alpha n}\cos(\omega nT)$ | $\dfrac{2ze^{-\alpha}\left[ze^{\alpha} - \cos(\omega T)\right]}{z^2 - 2ze^{-\alpha}\cos(\omega T) + e^{-2\alpha}}$ | |
| 19 | $x(n)$ | $X(z)$ | |
| 20 | $x(n-k)$ | $z^{-k}X(z)$ | |

Whenever we get involved with derivations that include sinusoidal functions, it's a good bet that the computations can be simplified through the utilization of Euler's equation. For the case of the sinusoid, Euler's equation states

$$\sin\theta = \frac{e^{j\theta} - e^{-j\theta}}{2j}$$

We begin by rewriting Equation 4.35 using Euler's notation to get

$$h(n) = \sin(\omega nT) = \frac{1}{2j}\left[e^{j\omega nT} - e^{-j\omega nT}\right]$$

**Equation 4.36**

We will treat both terms in Equation 4.36 separately. We can go directly to Table 4.1 and find the z-transform for each term. From the table, we see the entry

$$k\alpha^n \overset{z}{\longleftrightarrow} \frac{kz}{(z-\alpha)}$$

In Equation 4.36, we can equate $k$ with the term $1/2j$, and we can equate $\alpha^n$ with the term $e^{j\omega nT}$. Making the substitutions, we get

$$h(n) = \sin(\omega nT) \overset{z}{\longleftrightarrow} H(z) = \frac{1}{2j}\left[\frac{z}{z - e^{+j\omega T}} - \frac{z}{z - e^{-j\omega T}}\right]$$

This is a correct representation of the z-transform, but perhaps we can manipulate the equation a bit to make it a bit more palatable. First, let's do the obvious and bring out the factor $z$ to get

$$H(z) = \frac{z}{2j}\left[\frac{1}{z - e^{+j\omega T}} - \frac{1}{z - e^{-j\omega T}}\right]$$

Now we cross-multiply to get a common denominator to produce

$$H(z) = \frac{z}{2j}\left[\frac{(z - e^{-j\omega T}) - (z - e^{+j\omega T})}{(z - e^{-j\omega T})(z - e^{+j\omega T})}\right]$$

Collecting terms in both the numerator and the denominator, we arrive at

$$H(z) = \frac{z}{2j}\left[\frac{z - e^{-j\omega T} - z + e^{+j\omega T}}{z^2 - ze^{-j\omega T} - ze^{+j\omega T} + e^{+j\omega T}e^{-j\omega T}}\right]$$

$$= \frac{z}{2j}\left[\frac{e^{+j\omega T} - e^{-j\omega T}}{z^2 - z(e^{+j\omega T} + e^{-j\omega T}) + 1}\right]$$

Now we can use Euler's equation in the numerator and in the denominator to get

$$H(z) = \frac{z}{2j}\left[\frac{2j(\sin \omega T)}{z^2 - 2z\cos \omega T + 1}\right]$$

Now we can cancel the $2j$ terms to get the final result:

$$H(z) = \left[ \frac{z \sin \omega T}{z^2 - 2z \cos \omega T + 1} \right]$$

**Equation 4.37**

Equation 4.37 is the z-transform of Equation 4.35, more formally stated as

$$\sin(\omega nT) \overset{z}{\longleftrightarrow} \left[ \frac{z \sin \omega T}{z^2 - 2z \cos \omega T + 1} \right]$$

The reader should verify this by checking the entry for $\sin(\omega nT)$ in Table 4.1. Computing z-transforms is fairly straightforward, and it is kind of a fun exercise. This method illustrates the computation of a z-transform with the aid of Table 4.1. The table helps simplify a seemingly complex task.

## 4.10  INVERSE Z-TRANSFORM

If $h(n)$ and $H(z)$ are a z-transform pair such that $h(n) \overset{z}{\leftrightarrow} H(z)$, then we can compute the discrete time sequence $h(n)$ from its z-transform by one of several mutually exclusive methods that are lumped together under an umbrella called the *inverse z-transform*.

We use the notation $h(n) = Z^{-1}\{H(z)\}$ to indicate that the discrete time sequence $h(n)$ is the inverse z-transform of sequence $H(z)$. The $Z^{-1}$ is mathematical notation only and is not to be confused with the $z^{-1}$ terms seen in a z-domain sequence. The following are the three most common methods of computing the inverse z-transform:

1. Power series expansion method
2. Partial fraction expansion method
3. Residue method

Each of the three methods is discussed separately in the sections that follow.

### 4.10.1  Power Series Expansion Method

If we are given the z-transform of a causal sequence expressed as a ratio of two polynomials in $z^{-1}$, such as

$$H(z) = \frac{Y(z)}{X(z)} = \frac{a_0 + a_1 z^{-1} + a_2 z^{-2} + a_3 z^{-3} + \cdots + a_k z^{-k}}{b_0 + b_1 z^{-1} + b_2 z^{-2} + b_3 z^{-3} + \cdots + b_m z^{-m}}$$

**Equation 4.38**

we can expand the sequence by long division into another series given by

$$H(z) = h(0) + h(1)z^{-1} + h(2)z^{-2} + h(3)z^{-3} + h(4)z^{-4} + \cdots = \sum_{n=0}^{\infty} h(n)z^{-n}$$

where the discrete time sequence $h(n)$ for $n = 0,1,2,\cdots$ is the sequence of coefficients of the $z^{-n}$ terms and is the inverse transform of $H(z)$. We will illustrate this with a very simple example. Suppose we had a z-transform representing the transfer function of a small system given by

$$H(z) = \frac{Y(z)}{X(z)} = \frac{0.5}{1 - 0.8z^{-1}}$$

**Equation 4.39**

We can determine the sequence $h(n)$ by dividing the numerator by the denominator

$$
\begin{array}{r}
0.5 + 0.4z^{1} + 0.32z^{2} + 0.256z^{3} + 0.2048z^{4} + \cdots \\
1 - 0.8z^{-1} \overline{)\, 0.5 \phantom{xxxxxxxxxxxxxxxxxxxxxxxxxxxxxxxxxx}}
\end{array}
$$

$$
\begin{array}{rl}
0.5 & -0.40z^{-1} \\
+0.40z^{-1} & -0.32z^{2} \\
& +0.32z^{2} \quad -0.256z^{3} \\
& \quad\quad\quad \vdots
\end{array}
$$

The resulting infinite sequence is

$$H(z) = 0.5 + 0.4z^{1} + 0.32z^{2} + 0.256z^{3} + 0.2048z^{4} + \cdots$$

**Equation 4.40**

We now determine by inspection of Equation 4.40 the discrete time sequence $h(n)$ for $n = 0,1,2,\cdots$, which is the inverse z-transform of $H(z)$:

$$
\begin{aligned}
h(0) &= \ 0.5 \\
h(1) &= \ 0.4 \\
h(2) &= \ 0.32 \\
h(3) &= \ 0.256 \\
h(4) &= \ 0.2048 \\
&\ \ \vdots \qquad \vdots
\end{aligned}
$$

An infinite series of this type is useful in many applications for approximating a solution to a problem by considering only the first $n$ significant terms. However, using the infinite series to design a piece of computational hardware really isn't of much use unless you can express the series in a closed form solution. So, just for the heck of it, let's convert the infinite sequence of Equation 4.40 back into its original closed form expression of Equation 4.39.

We begin with Equation 4.40. If we bring out a common factor of 0.5 we recognize that all the coefficients are powers of 0.8. This gives us

$$H(z) \quad = 0.5 + 0.4z^1 + 0.32z^2 + 0.256z^3 + 0.2048z^4 + \cdots$$

$$= 0.5\left(0.8^0 + 0.8^1 z^1 + 0.8^2 z^2 + 0.8^3 z^3 + 0.8^4 z^4 + \cdots\right)$$

Now we can rewrite the above equation to get

$$H(z) = 0.5\sum_{n=0}^{\infty} (0.8)^n z^{-n} = 0.5 + 0.40z^{-1} + 0.32z^{-2} + 0.256z^{-3} + \cdots$$

### Equation 4.41

If we multiply both sides of Equation 4.41 by the term $0.8z^{-1}$ we get the sequence

$$H(z)\left(0.8z^{-1}\right) = \left(0.8z^{-1}\right)0.5\sum_{n=0}^{\infty} (0.8)^n z^{-n} = 0.40z^{-1} + 0.32z^{-2} + 0.256z^{-3} + \cdots$$

### Equation 4.42

If we subtract Equation 4.42 from Equation 4.41, we arrive at

$$H(z)\left(1 - 0.8z^{-1}\right) \quad = \left\{0.5\sum_{n=0}^{\infty} (0.8)^n z^{-n} - \left(0.8z^{-1}\right)0.5\sum_{n=0}^{\infty} (0.8)^n z^{-n}\right\}$$

$$= \left\{0.5 + 0.4z^{-1} + 0.32z^{-2} + 0.256z^{-3} + \cdots\right\} - \left\{0.40z^{-1} + 0.32z^{-2} + 0.256z^{-3} + \cdots\right\}$$

$$= 0.5$$

Finally, we divide both sides by $\left(1 - 0.8z^{-1}\right)$ to arrive at the original closed form system transfer function equation, repeated here for clarity:

$$H(z) = \frac{Y(z)}{X(z)} = \frac{0.5}{\left(1 - 0.8z^{-1}\right)}$$

Now if we wish, we can compute the system difference equation by re-arranging terms to get

$$Y(z)(1-0.8z^{-1}) = 0.5X(z)$$

Solving for $Y(z)$, we get

$$Y(z) = 0.5X(z) + 0.8z^{-1}Y(z)$$

**Equation 4.43**

We can use the shift property of the z-transforms to convert Equation 4.43 from the z-domain into a difference equation in the discrete time domain. The resulting difference equation is given by

$$y(n) = 0.5x(n) + 0.8y(n-1)$$

**Equation 4.44**

Now we are at the point where we can draw the system block diagram. We do this by simple inspection of Equation 4.44. The result is the digital feedback system block diagram illustrated in Figure 4.15.

The underlying motivation for using the power series expansion method to compute inverse z-transforms is that it is a simple procedure. It is a useful tool for the simpler inverse z-transform computations, and it is well suited for computer computations. The major limitation of this method is that it does not necessarily lead to a closed form solution. As we shall see in the following sections, other methods such as the partial fraction expansion method and the residue method do lead to closed form solutions. The penalty we pay, however, is an increase in computational complexity.

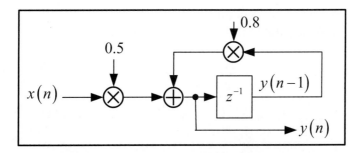

**Figure 4.15**  Block diagram representation of Equation 4.44

## 4.10.2 Partial Fraction Expansion Method

We can determine the inverse z-transform of a function by using the method of partial fraction expansion. In this method, we first expand the function into a sum of fractions called *partial fractions*. The inverse transform of each individual partial fraction is then determined from the table of transform pairs. The inverse transform for each partial fraction are then summed to produce the overall inverse z-transform. This is a preferred method as long as the denominator of the original function can be easily factored. The factorization of the denominator is the computational penalty we must pay in order to utilize this method. In most cases, the z-transform is seen as a ratio of polynomials either in $z$ or in $z^{-1}$:

$$H(z) = \frac{a_0 + a_1 z^{-1} + a_2 z^{-2} + \cdots + a_N z^{-N}}{b_0 + b_1 z^{-1} + b_2 z^{-2} + \cdots + b_M z^{-M}}$$

**Equation 4.45**

The poles of Equation 4.45 are obtained by factoring the denominator polynomial. In doing so, we may end up with any combination of real and complex poles. In addition, these poles might be either simple first order poles or they might be multiple order poles. We can expand a function of the type given in Equation 4.45 by following a simple set of five rules [7]:

**Rule 1.** If the poles of $H(z)$ are first order and if the order of the numerator and denominator are equal so that $N = M$, then $H(z)$ can be expanded by the following procedure: Factor the denominator to obtain a product of first order terms so that

$$H(z) = \frac{\text{Numerator}}{\left(1 - p_1 z^{-1}\right)\left(1 - p_2 z^{-1}\right)\left(1 - p_3 z^{-1}\right)\cdots\left(1 - p_M z^{-1}\right)}$$

**Equation 4.46**

Each term in the denominator represents a pole in the system transfer function $H(z)$. We separate the terms of Equation 4.46 into a collection of fractions:

$$
\begin{aligned}
H(z) &= A_0 + \frac{A_1}{1 - p_1 z^{-1}} + \frac{A_2}{1 - p_2 z^{-1}} + \cdots + \frac{A_M}{1 - p_M z^{-1}} \\
&= A_0 + \frac{A_1 z}{z - p_1} + \frac{A_2 z}{z - p_2} + \cdots + \frac{A_M z}{z - p_M} \\
&= A_0 + \sum_{K=1}^{M} \frac{A_K z}{z - p_K}
\end{aligned}
$$

**Equation 4.47**

where the numbers $A_1, A_2, A_3, \cdots A_K$ are unknown constants that must be determined. The numbers $p_1, p_2, p_3, \cdots p_K$ are the poles of $H(z)$. If the order of the numerator and denominator are equal and $N = M$, then the term $A_0$ is calculated to be $A_0 = a_N / b_M$.

**Rule 2.** If the poles of $H(z)$ are first order and if the order of the numerator is less than the order of the denominator and $N < M$, then the term $A_0 = 0$. The partial fraction expansion is still performed as defined by Equation 4.47.

**Rule 3.** If the poles of $H(z)$ are first order and if the order of the numerator is greater than the order of the denominator and $N > M$, then $H(z)$ must be reduced by long division until $N \leq M$. The remainder is then expanded according to Equation 4.47.

**Rule 4.** The coefficient $A_K$ associated with each pole in the partial fraction expansion is determined by multiplying both sides of the equation for $H(z)$ by the quantity

$$\frac{(z - p_K)}{z}$$

and evaluating the resulting expression at $z = p_K$. For example, if we have the function

$$H(z) = \frac{a_0 z^N + a_1 z^{N-1} + a_2 z^{N-2} + \cdots + a_N}{(z - p_1)(z - p_2) \cdots (z - p_K) \cdots (z - p_M)}$$

the first thing we do is rewrite it as a sum of fractions to get

$$H(z) = A_0 + \frac{A_1 z}{z - p_1} + \frac{A_2 z}{z - p_2} + \cdots + \frac{A_K z}{z - p_K} + \cdots + \frac{A_M z}{z - p_M}$$

Now if we wish to determine the partial fraction expansion coefficient $A_K$ corresponding to the pole at $z = p_K$, we would perform the following operation:

$$A_K = \frac{(z - p_K)}{z} H(z) \bigg|_{z = p_K} = \frac{(z - p_K)}{z} \left[ \frac{a_0 z^N + a_1 z^{N-1} + a_2 z^{N-2} + \cdots + a_N}{(z - p_1)(z - p_2) \cdots (z - p_K) \cdots (z - p_M)} \right] \bigg|_{z = p_K}$$

**Equation 4.48**

The proof of Equation 4.48 is easily shown as

$$A_K = \frac{(z - p_K)}{z} H(z)\Bigg|_{z=p_K} = \frac{(z - p_K)}{z}\left[\frac{a_0 z^N + a_1 z^{N-1} + a_2 z^{N-2} + \cdots + a_N}{(z - p_1)(z - p_2)\cdots(z - p_K)\cdots(z - p_M)}\right]\Bigg|_{z=p_K}$$

$$= \frac{(z - p_K)}{z}\left[A_0 + \frac{A_1 z}{z - p_1} + \frac{A_2 z}{z - p_2} + \cdots + \frac{A_K z}{z - p_K} + \cdots + \frac{A_M z}{z - p_M}\right]\Bigg|_{z=p_K}$$

$$= (z - p_K)\left[\frac{A_0}{z} + \frac{A_1}{z - p_1} + \frac{A_2}{z - p_2} + \cdots + \frac{A_K}{z - p_K} + \cdots + \frac{A_M}{z - p_M}\right]\Bigg|_{z=p_K}$$

$$= 0 + 0 + 0 + \cdots + A_K + \cdots + 0$$

**Rule 5.** If $H(z)$ contains multiple order poles, then the contribution to the overall partial fraction expansion due to those poles is determined separately by the following operation: If, for example, the factored version of $H(z)$ has an $n$th order pole at $z = p_K$ such that

$$H(z) = \frac{a_0 z^N + a_1 z^{N-1} + a_2 z^{N-2} + \cdots + a_N}{(z - p_1)(z - p_2)\cdots(z - p_K)^n \cdots(z - p_M)}$$

then the expansion associated for that pole will result in $n$ additional partial fraction terms of the form

$$\sum_{j=1}^{n} \frac{G_j}{(z - p_k)^j}$$

The $G_j$ terms are computed through the use of the following relation:

$$G_j = \frac{1}{(n-j)!}\frac{d^{n-j}}{dz^{n-j}}\left[\frac{(z - p_k)^n}{z} H(z)\right]\Bigg|_{z=p_K} \qquad \text{for } \{1 \le j \le n\}$$

**Equation 4.49**

Equation 4.48 and Equation 4.49 look a bit formidable, but as we shall see they are fairly simple to use. Our mission now is to demonstrate by example the method described previously for determining the partial fraction expansion of a function. Once the expansion has been achieved, we will compute the inverse z-transform for each fraction of that function.

### 4.10.2.1   First Order Real Poles

Suppose we have a function $H(z)$ that is the z-transform of any arbitrary system. In many cases, this will be the z-transform or transfer function of a digital recursive filter. If this were the case, we would probably like to compute the inverse transform of $H(z)$ in order to identify the filter coefficients $h(n)$. For other types of systems, we may be interested in determining the time domain behavior $h(n)$ in response to some input stimuli. To accomplish this, we would exploit the relation

$$h(n) = Z^{-1}\{H(z)\}$$

Suppose we would like to determine the inverse z-transform of a system described by the z-transform

$$H(z) = \frac{4z^2 - 1.25z}{z^2 - 0.75z + 0.125}$$

We can easily factor both the numerator and denominator to get

$$H(z) = \frac{4z^2 - 1.25z}{(z - 0.5)(z - 0.25)} \quad = \quad \left[\frac{z(4z - 1.25)}{(z - 0.5)(z - 0.25)}\right] = 4\left[\frac{z(z - 0.3125)}{(z - 0.5)(z - 0.25)}\right]$$

We can clearly see from the factored expression that the system has two poles located at $z = 0.5$ and $z = 0.25$, and it has two zeros located at $z = 0$ and $z = 0.3125$. We can also see that the system has a gain factor of four. The pole-zero diagram for this system is illustrated in Figure 4.16.

Let's continue by expressing $H(z)$ in terms of negative powers of $z$. We do this by dividing both the numerator and the denominator by $z^2$ to get

$$H(z) = \frac{4 - 1.25z^{-1}}{1 - 0.75z^{-1} + 0.125z^{-2}}$$

**Equation 4.50**

We see that $N < M$, so therefore by rule 2 the partial fraction expansion term $A_0 = 0$. We can expand Equation 4.50 according to rule 1 to get

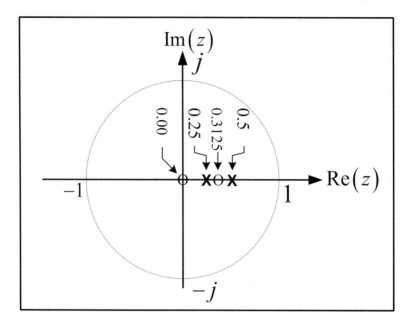

**Figure 4.16**  System pole-zero diagram

$$H(z) = \sum_{k=1}^{M} \frac{A_K z}{z - p_k} = \sum_{k=1}^{2} \frac{A_K z}{z - p_k} = \frac{A_1 z}{z - 0.5} + \frac{A_2 z}{z - 0.25}$$

**Equation 4.51**

We divide both sides of the equation by $z$ to produce

$$\frac{H(z)}{z} = \frac{A_1}{z - 0.5} + \frac{A_2}{z - 0.25}$$

   Now we can evaluate the partial fraction coefficients for each pole in the expansion by rule 4. In doing so, we get

$$A_1 = (z - 0.5) \left. \frac{H(z)}{z} \right|_{z=0.5} = \left. \frac{(z-0.5)}{z} \frac{(4z^2 - 1.25z)}{(z-0.5)(z-0.25)} \right|_{z=0.5} = \frac{0.75}{0.25} = 3$$

$$A_2 = (z - 0.25) \left. \frac{H(z)}{z} \right|_{z=0.25} = \left. \frac{(z-0.25)}{z} \frac{(4z^2 - 1.25z)}{(z-0.5)(z-0.25)} \right|_{z=0.25} = \frac{-0.25}{-0.25} = 1$$

We complete the partial fraction expansion by substituting the coefficients 3 and 1 for $A_1$ and $A_2$, respectively, in Equation 4.51 to get

$$H(z) = \frac{3z}{(z-0.5)} + \frac{z}{(z-0.25)}$$

**Equation 4.52**

We can now use Table 4.1 to find the inverse z-transform separately for each term in Equation 4.52. In the table, we find the following entry for the z-transform pair that fits our application:

$$k\alpha^n \overset{z}{\leftrightarrow} \frac{kz}{z-\alpha}$$

By substituting the values from our partial fraction terms into the z-transform pair in the table, we arrive at

$$H(z) = \left[ \frac{4 - 1.25z^{-1}}{1 - 0.75z^{-1} + 0.125z^{-2}} \right] = \left[ \frac{3z}{(z-0.5)} + \frac{z}{(z-0.25)} \right] \overset{z}{\leftrightarrow} h(n) = 3(0.5)^n + (0.25)^n$$

This is pretty simple stuff, isn't it? In this example, the sequence $h(n)$ could be interpreted as the impulse response of some hardware circuit or system. That is, if we input a unit impulse into a system consisting of a unit sample of magnitude 1 at time $t = 0$ and 0 for all time thereafter, then the system output would be identical to the time sequence described by $h(n)$.

There is still more that we can do, so let's not stop here. We can look at the hardware block diagram of this crazy system. How do we do that? Well, we simply use the unit delay property of z-transforms and convert the z-domain transfer function into the time domain difference equation. We then can draw the system block diagram directly from the difference equation. We begin with the system transfer function

$$H(z) = \frac{4 - 1.25z^{-1}}{1 - 0.75z^{-1} + 0.125z^{-2}}$$

**Equation 4.53**

We know that the transfer function $H(z)$ is the ratio of the z-transform of the system output $Y(z)$ to the z-transform of the system input $X(z)$. Thus we can rewrite Equation 4.53 as

$$H(z) = \frac{Y(z)}{X(z)} = \frac{4 - 1.25z^{-1}}{1 - 0.75z^{-1} + 0.125z^{-2}}$$

Now we can separate terms to get

$$Y(z)\left(1 - 0.75z^{-1} + 0.125z^{-2}\right) = X(z)\left(4 - 1.25z^{-1}\right)$$

$$Y(z) = X(z)\left(4 - 1.25z^{-1}\right) + Y(z)\left(+0.75z^{-1} - 0.125z^{-2}\right)$$

We employ the shift property of the z-transform $x(n - k) \overset{z}{\leftrightarrow} X(z)z^{-k}$ to get

$$y(n) = 4x(n) - 1.25x(n-1) + 0.75y(n-1) - 0.125y(n-2)$$

### Equation 4.54

The reader should note that if we insert a discrete impulse sequence $x(n)$, composed of a unit impulse at $n = 0$ and 0 at all other values, of $n$ into the difference equation given by Equation 4.54, then we will generate the exact same sequence as given by the impulse response we derived previously, repeated here for clarity:

$$h(n) = 3(0.5)^n + (0.25)^n$$

The block diagram of Equation 4.54 is easily drawn by inspection and illustrated in Figure 4.17. This is pretty powerful stuff, and as an added bonus, it's simple too. There is a lot more information that we can obtain about this system from its z-transform. For example, we can tell from the pole-zero diagram that the system is stable because all of its poles are contained within the unit circle. We also know that its ROC is within the range $0.5 < |z| \leq \infty$. Since the ROC contains the unit circle, we could use the power of the z-transform to compute the frequency response of this system. We can compute the frequency response by evaluating the system transfer function on the unit circle by replacing all the occurrences of $z$ with $e^{j\omega}$. Doing so gives us

$$H(z)\big|_{z=e^{j\omega}} = \frac{4 - 1.25e^{-j\omega}}{1 - 0.75e^{-j\omega} + 0.125e^{-j2\omega}}$$

We will pursue the frequency response topic a bit later, because right now we have other fish to fry. We still need to demonstrate the partial fraction expansion method for complex conjugate poles and higher order poles.

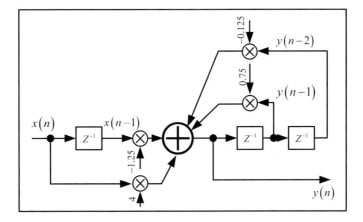

**Figure 4.17**  System block diagram

### 4.10.2.2  First Order Complex Conjugate Poles

Suppose we want to find the inverse z-transform of the following expression:

$$H(z) = \frac{z}{z^2 + 2z + 2}$$

**Equation 4.55**

If we divide both the numerator and the denominator by $z^2$, we will see that $N < M$. Therefore, by rule 2, the partial fraction expansion term $A_0 = 0$. We first factor the denominator of Equation 4.55 using the quadratic equation to get the roots of $z$:

$$z^2 + 2z + 2$$

$$z = \frac{-b \pm \sqrt{b^2 - 4ac}}{2a} = \frac{-2 \pm \sqrt{4 - 8}}{2} = (-1 \pm j)$$

The two roots are $z = -1 - j$ and $z = -1 + j$. These roots are the poles of $H(z)$. Now that we have the poles, we can expand Equation 4.55 according to rule 1 to get

$$H(z) = \sum_{k=1}^{M} \frac{A_K z}{z - p_k} = \sum_{k=1}^{2} \frac{A_K z}{z - p_k} = \frac{A_1 z}{z + 1 + j} + \frac{A_2 z}{z + 1 - j}$$

**Equation 4.56**

We divide both sides of the equation by $z$ to produce

$$\frac{H(z)}{z} = \frac{A_1}{z+1+j} + \frac{A_2}{z+1-j}$$

Now we can evaluate the partial fraction coefficients for each pole in the expansion by rule 4 to get

$$A_1 = (z+1+j)\frac{H(z)}{z}\bigg|_{z=(-1-j)} = \frac{(z+1+j)}{z}\left[\frac{z}{(z+1+j)(z+1-j)}\right]\bigg|_{z=(-1-j)} = \frac{1}{-2j} = \frac{j}{2}$$

$$A_2 = (z+1-j)\frac{H(z)}{z}\bigg|_{z=(-1+j)} = \frac{(z+1-j)}{z}\left[\frac{z}{(z+1+j)(z+1-j)}\right]\bigg|_{z=(-1+j)} = \frac{1}{2j} = \frac{-j}{2}$$

We can complete the partial fraction expansion by substituting the coefficients $j/2$ and $-j/2$ for $A_1$ and $A_2$, respectively, in Equation 4.56 to get

$$H(z) = \frac{(j/2)z}{z+1+j} + \frac{(-j/2)z}{z+1-j} = \frac{1}{2}\left[\frac{jz}{z+1+j} + \frac{-jz}{z+1-j}\right] = \frac{j}{2}\left[\frac{z}{z+1+j} - \frac{z}{z+1-j}\right]$$

**Equation 4.57**

Equation 4.57 is the partial fraction expansion of $H(z)$. We can look up the inverse transform for each of the two terms in the expansion in the transform Table 4.1. Once again, from the table, we see the format of the transform pair we need:

$$k(\alpha)^n \overset{z}{\longleftrightarrow} \frac{z}{z-\alpha}$$

**Equation 4.58**

Substituting the values in each of the two terms in Equation 4.57 into Equation 4.58, we get

$$h(n) = \frac{j}{2}(1+j)^n - \frac{j}{2}(1-j)^n$$

**Equation 4.59**

That's it, folks. Equation 4.59 is the inverse transform of Equation 4.57. This is indeed a valid inverse z-transform. However, it is not a satisfying result. Nobody likes imaginary operators like $j$ in their discrete time sequence. Results like these tend to cause ulcers and other undesirable by-products of stress. Perhaps if we played with the partial fraction expansion equation a little bit, we can get it into a more palatable form. Let's begin with the original partial fraction expansion of Equation 4.57:

$$H(z) = \frac{j}{2}\left[\frac{z}{z+1+j} - \frac{z}{z+1-j}\right]$$

We have a set of complex conjugate poles at $z = -1 \pm j$. Perhaps we will achieve more satisfying results if we work with this equation in polar form. We convert the two poles from Cartesian to polar form to get

$$z_1 = (-1-j) = \sqrt{2}e^{-j\left(\frac{3\pi}{4}\right)} = re^{-j\theta} \qquad \text{for } \left\{r = \sqrt{2}, \text{ and } \theta = 3\pi/4\right\}$$

$$z_2 = (-1+j) = \sqrt{2}e^{+j\left(\frac{3\pi}{4}\right)} = re^{+j\theta} = z_1^*$$

Note that in polar form the poles are still conjugates of one another. If we rewrite Equation 4.57 in polar form, we arrive at

$$H(z) = \frac{j}{2}\left(\frac{z}{z - re^{-j\theta}}\right) - \frac{j}{2}\left(\frac{z}{z - re^{+j\theta}}\right)$$

Now we can use the same transform pair from Table 4.1 to get the inverse transform—only this time in polar notation. The new discrete time sequence in polar notation is given by

$$h(n) = \frac{j}{2}\left(re^{-j\theta}\right)^n - \frac{j}{2}\left(re^{+j\theta}\right)^n$$

**Equation 4.60**

We still have the nasty $j$ in the equation, but this time we have an equation that looks a bit familiar and therefore it looks to have some promise in terms of simplification. Let's begin by rearranging and combining terms to get

$$h(n) = \frac{jr^n}{2}\left[\left(e^{-j\theta}\right)^n - \left(e^{+j\theta}\right)^n\right]$$

We can move the index $n$ inside the parenthesis to get

$$h(n) = \frac{jr^n}{2}\left[e^{-jn\theta} - e^{+jn\theta}\right]$$

If we multiply both the numerator and denominator by $2j$ and bring the minus sign out of the brackets, we arrive at

$$h(n) = \frac{-jr^n}{2}\left[\frac{e^{+jn\theta} - e^{-jn\theta}}{2j}\right]2j$$

We can combine terms and use Euler's equation to arrive at

$$h(n) = -j^2 r^n\left[\frac{e^{+jn\theta} - e^{-jn\theta}}{2j}\right] \;=\; r^n\left[\frac{e^{+jn\theta} - e^{-jn\theta}}{2j}\right] \;=\; r^n \sin(n\theta)$$

If we replace the magnitude term $r$ with $\sqrt{2}$ and the phase term $\theta$ with $3\pi/4$, we end up with an inverse transform that is a bit more pleasing to the eye. After all this fun work, we end up with the following:

$$h(n) = r^n \sin(n\theta) = \left(\sqrt{2}\right)^n \sin\left[(3\pi/4)n\right]$$

**Equation 4.61**

When we did the Cartesian to polar conversion, you probably noticed that the magnitude of the conjugate poles was greater than unity. Therefore the system described by the transfer function $H(z)$ should prove to be unstable because its poles are not contained within the unit circle. We can easily verify this by generating a bunch of values of $h(n)$ on an Excel spreadsheet. In this example, plots were made for 128 values of the index $n$. In order to increase the plot resolution and smooth the resulting curves, the index was incremented in steps of 0.1. Figure 4.18 shows the plot over 128 samples for the $\sin\left[(3\pi/4)n\right]$ term in $h(n)$. Figure 4.19 shows the plot over the same 128 samples of the magnitude or $r^n$ term in $h(n)$. The composite plot over the 128 samples of the complete discrete time domain sequence $h(n) = \left(\sqrt{2}\right)^n \sin\left[(3\pi/4)n\right]$ is illustrated in Figure 4.20.

Finally, just for the heck of it, Figure 4.21 shows the two terms plotted separately but superimposed on one another. The trend illustrated by the plots is clear. The discrete time sequence grows without bounds. It fails the convergence test, and the system is unstable.

The reader should compare these plots with the stability plots shown in Figure 4.9 and Figure 4.10. The complex pole in this example was on the left-hand side of the z-plane, and it was outside the unit circle. Therefore we should have envisioned a discrete sinusoidal time function with an increasing amplitude. When you are deriving the inverse z-transform, a comparison of the function poles with those shown in Figure 4.9 and Figure 4.10 should provide you will all the information you need to predict the behavior of the final discrete time domain sequence.

This is all very interesting stuff. So why stop here? Let's do one more example. In the third and last example, we will take a close look at how to handle second order poles in a partial fraction expansion.

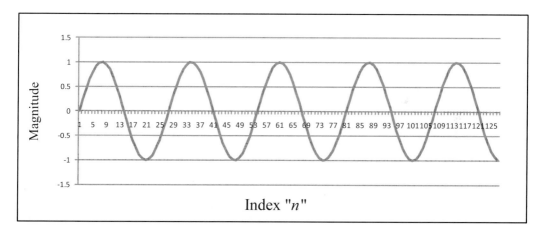

**Figure 4.18** Plot of the sine term for $h(n)$

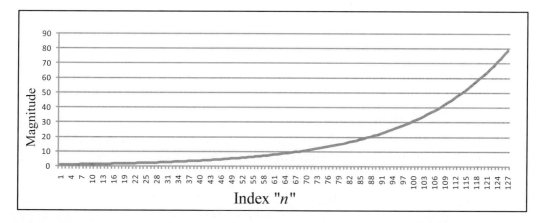

**Figure 4.19** Plot of the magnitude term for $h(n)$

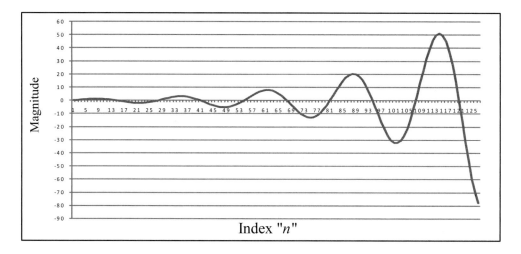

**Figure 4.20** Plot of the combined sine and magnitude terms of $h(n)$

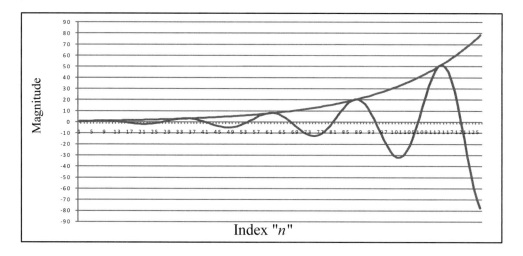

**Figure 4.21** Superimposed magnitude and sine terms of $h(n)$

### 4.10.2.3  Second Order Real Poles

Suppose that we wanted to use the partial fraction expansion method to determine the inverse z-transform of the system equation given by

$$H(z) = \frac{z^2}{(z-0.5)(z-1)^2}$$

**Equation 4.62**

The first thing to do is to run through the five partial fraction expansion rules discussed previously and follow the ones that are appropriate. Rule 1 tells us that if $M$ roots of the denominator are real, then we can expand the expression by

$$H(z) = A_0 + \sum_{K=1}^{M} \frac{A_K z}{z - p_K}$$

Rule 5 tells us that if $H(z)$ contains multiple order poles, then the contribution to the overall partial fraction expansion due to those poles is determined separately. If, for example, the factored version of $H(z)$ were to have an $n$th order pole at $z = p_K$ such that

$$H(z) = \frac{a_0 z^N + a_1 z^{N-1} + a_2 z^{N-2} + \cdots + a_N}{(z - p_1)(z - p_2)\cdots(z - p_K)^n \cdots (z - p_M)}$$

then the expansion associated with that pole will result in $n$ additional partial fraction terms of the form

$$\sum_{j=1}^{n} \frac{G_j}{(z - p_k)^j}$$

The $G_j$ terms are computed through the use of the following relation:

$$G_j = \frac{1}{(n-j)!} \frac{d^{n-j}}{dz^{n-j}} \left[ \frac{(z - p_k)^n}{z} H(z) \right]\Bigg|_{z = p_K} \quad \text{for } \{1 \leq j \leq n\}$$

As an example, take the function in $z$ that has one first order pole, one second order pole, one third order pole, such as

$$H(z) = \frac{z^2}{(z - p_K)(z - p_1)^2 (z - p_2)^3}$$

By rule 1 and rule 5, we know that the partial fraction expansion would take the form

$$H(z) = A_0 + \sum_{K=1}^{M=1} \frac{A_K z}{z - p_K} + \sum_{j=1}^{n=2} \frac{G_j z}{(z - p_1)^j} + \sum_{j=1}^{n=3} \frac{G_j z}{(z - p_2)^j}$$

**Equation 4.63**

We also know by rule 2 that the term $A_0 = 0$, since the order of the numerator is less than the order of the denominator, or $N < M$.

Now that we have mapped out our strategy, let's get going and compute the inverse transform of Equation 4.62 using the partial fraction expansion method. We begin with the z-transform:

$$H(z) = \frac{z^2}{(z-0.5)(z-1)^2}$$

We know from Equation 4.63 that the partial fraction expansion will be of the form

$$H(z) = \frac{A_1 z}{(z-0.5)} + \frac{G_1 z}{(z-1)} + \frac{G_2 z}{(z-1)^2}$$

**Equation 4.64**

so let's find the partial fraction expansion coefficient for the first order pole at $z = 0.5$:

$$A_1 = \frac{(z-0.5)}{z} H(z)\bigg|_{z=0.5} = \left[\frac{(z-0.5)}{z}\right]\left[\frac{z^2}{(z-0.5)(z-1)^2}\right]_{z=0.5} = \frac{z}{(z-1)^2}\bigg|_{z=0.5} = \frac{0.5}{0.25} = 2$$

Using rule 5, let's find the two partial fraction expansion coefficients for the second order pole at $z = 1$. The first coefficient $G_1$ is computed as

$$G_1 = \frac{1}{(2-1)!}\frac{d^1}{dz^1}\left\{\left[\frac{(z-1)^2}{z}\right]\left[\frac{z^2}{(z-0.5)(z-1)^2}\right]_{z=1}\right\}$$

$$= \frac{d}{dz}\left[\frac{z}{(z-0.5)}\right]_{z=1}$$

$$= \frac{(z-0.5)-z}{(z-0.5)^2}\bigg|_{z=1} = -2$$

The second coefficient $G_2$ is computed as

$$G_2 = \frac{1}{(2-2)!} \frac{d^0}{dz^0} \left\{ \left[ \frac{(z-1)^2}{z} \right] \left[ \frac{z^2}{(z-0.5)(z-1)^2} \right]_{z=1} \right\}$$

$$= \left[ \frac{z}{(z-0.5)} \right]_{z=1}$$

$$= \frac{z}{(z-0.5)} \bigg|_{=1} = 2$$

We obtain the partial fraction expansion by substitution of the newly computed coefficients back into Equation 4.63 to get

$$H(z) = \frac{A_1 z}{(z-0.5)} + \frac{G_1 z}{(z-1)} + \frac{G_2 z}{(z-1)^2}$$

$$H(z) = \frac{2z}{(z-0.5)} - \frac{2z}{(z-1)} + \frac{2z}{(z-1)^2}$$

**Equation 4.65**

We can complete the inverse z-transform by using the transform pairs in Table 4.1 to convert each of the three expansion terms in Equation 4.65. The inverse transform conversions for each of the three terms in the expansion are

1    $\dfrac{kz}{z-0.5} \overset{z}{\leftrightarrow} k\alpha^n \implies \dfrac{2z}{z-0.5} \overset{z}{\leftrightarrow} 2(0.5)^n$

2    $\dfrac{kz}{z-1} \overset{z}{\leftrightarrow} k \implies \dfrac{-2z}{z-1} \overset{z}{\leftrightarrow} -2$

3    $\dfrac{kz}{(z-1)^2} \overset{z}{\leftrightarrow} kn \implies \dfrac{2z}{(z-1)^2} \overset{z}{\leftrightarrow} 2n$

The transform of the three terms are combined to produce the composite inverse transform:

$$h(n) = 2(0.5)^n - 2 + 2n$$

Now we can factor out a 2 from the equation and collect terms to produce the final inverse z-transform equation given by

$$h(n) = 2\left[(0.5)^n + (n-1)\right]$$

The same procedure would be followed for the partial fraction expansion when dealing with an equation consisting of any number of higher order poles. The procedure is a bit time consuming, but it is also very straightforward. Just follow the rules and use the transform pair tables. For more complex transforms, it is probably wise to utilize purchased software applications that do the computations for you and spit out the results. This goes a long way toward preventing mathematical errors that always seem to occur when doing repetitive computations such as these.

The pros associated with the partial fraction expansion method of inverse z-transform calculation include the following:

- The partial fraction expansion method leads to a closed form solution.
- The partial fraction expansion method coefficients $A_K$ are easily mapped to the coefficients of parallel filter structures.
- The partial fraction expansion method is aided by the use of transform tables.

The cons associated with the partial fraction expansion method include the following:

- The denominator of the polynomial in $z$ needs to be factored in order to determine the poles of $H(z)$. This can on occasion be a difficult task.
- When the denominator of the polynomial contains multiple order poles, this method requires higher orders of differentiation, which creates additional computational complexity.

A rule of thumb for the use of the partial fraction expansion method is if the partial fractions are relatively straightforward and are listed in a table of transform pairs, then this may be the optimum method.

### 4.10.3  Residue Method

Suppose we have a function of $H(z)$ that is the z-transform of some circuit or system described by the discrete time function $h(n)$. In order to determine $h(n)$ we need to compute the inverse z-transform. So far we have explored two methods of computing the inverse z-transform: the power series method and the method of partial fraction expansion. Here we will explore a third method called the *residue method*.

There are several ways of computing residues. One method of computation utilizes the partial fraction expansion coefficients and produces the same results as the partial fraction expansion method. In this method, it is still necessary to first compute the expansion coefficients $A_K$ and $G_K$ and then perform a final computation, usually by using a table to convert the expansion terms from the z-domain to the discrete time domain. A second method actually computes the discrete time function $h(n)$ directly, thereby bypassing the use of a table. This second method of residue computation is significantly different from the partial fraction expansion method. This difference gives us another set of tools to derive inverse z-transforms, so this is the method we will discuss here.

The residue method of obtaining the inverse z-transform usually involves a long evaluation of a contour integral along a path that encloses all the poles of $H(z)$. The process is described as follows:

$$h(n) = \frac{1}{2\pi j} \oint_C H(z) \, z^{n-1} dz$$

### Equation 4.66

The $C$ in the integral describes the path of integration. We could go into great detail here about expanding $H(z)$ into a series and computing the contour integral for each term in the series, but that method is thoroughly discussed in Chapter 2, "Review of Complex Variables." The concept of contour integration is a very powerful tool used in the computation of residues. We will not repeat any of the contour integration material previously discussed in Chapter 2. Instead, we will introduce an equivalent method of implementing Equation 4.66.

We will, however, need to regurgitate a portion of Chapter 2 in order to define an *analytic function*. In the world of complex variables and z-transforms, the term *analytic function* is used over and over again. The term *analytic* is important because it reveals a lot of important information about a function. There exists a great amount of literature that defines the term *analytic* as it relates to complex variables and z-transforms, but it can all be boiled down to one simple but very important paragraph:

A function $H(z)$ is analytic in a region $R$ of the complex z-plane if its derivative exists and if it is continuous at every point in $R$. A function $H(z)$ is said to be analytic at some point $z_0$ if there exists a neighborhood $|z - z_0| < \delta$ within which the derivative of $H(z)$ exists. If $H(z)$ is analytic in some region $R$, then all higher order derivatives of $H(z)$ are also analytic within $R$.

If (1) the integration path $C$ of the contour integral in Equation 4.66 lies entirely within the ROC, (2) the contour contains a finite number of singularities (poles), and (3) $H(z)$ is analytic inside and on the closed contour except at the finite number of singularities, then we can state

$$h(n) = \frac{1}{2\pi j} \oint_C H(z) \, z^{n-1} dz = \overset{\text{Number of Residues}}{\underset{k=1}{\sum}} \left[ \text{Residue}_k \text{ of } z^{n-1} H(z) \right] \text{ at all singularities inside C}$$

**Equation 4.67**

In other words, we can compute $h(n)$ from $H(z)$ by adding the residues for each $m$th order pole of $z^{n-1}H(z)$. Trust me, this turns out to be a great deal less labor intensive than solving Equation 4.66. We can compute the residue of a simple pole by solving

$$\text{Res}\left[H(z)\right]_{z=p_k} = (z - p_k) z^{n-1} H(z)\Big|_{z=p_k}$$

**Equation 4.68**

We can compute the residue of an $m$th order pole by solving

$$\text{Res}\left[H(z)\right]_{z=p_k} = \frac{1}{(m-1)!} \frac{d^{m-1}}{dz^{m-1}} \left[ (z - p_k)^m z^{n-1} H(z) \right]_{z=p_k}$$

**Equation 4.69**

The reader can see that Equation 4.68 and Equation 4.69 are very similar to the equations used to compute the expansion coefficients for simple and multiple order poles in the partial fraction method. The difference is that with the residue equations shown here, we can realize the expression for $h(n)$ directly by computing the sum of the residues of $z^{n-1}H(z)$. Thus we do not need the services of a transform pair table. A few examples should make this statement a great deal clearer.

**Example 1.** This example deals with residues of first order poles. Suppose we wish use the residue method to find the inverse z-transform of

$$H(z) = \frac{z}{z^2 - 0.25z - 0.375}$$

The first order of business is to factor the denominator to determine the poles of $H(z)$. Factoring the denominator produces

$$H(z) = \frac{z}{(z - 0.75)(z + 0.5)}$$

Since we are dealing with two first order poles, we can compute the residues using Equation 4.68:

$$\text{Res}\Big[H(z)\Big]_{z=p_k} = (z-p_k)z^{n-1}H(z)\Big|_{z=p_k}$$

We can find the residue at the simple pole $z = 0.75$ from

$$
\begin{aligned}
\text{Res}\Big[H(z)\Big]_{z=0.75} &= (z-0.75)z^{n-1}\frac{z}{(z-0.75)(z+0.5)}\bigg|_{z=0.75} \\[2mm]
&= \frac{z^n}{(z+0.5)}\bigg|_{z=0.75} \\[2mm]
&= \frac{(0.75)^n}{1.25} \\[2mm]
&= \frac{4}{5}(0.75)^n
\end{aligned}
$$

We can find the residue at the simple pole $z = -0.5$ from

$$
\begin{aligned}
\text{Res}\Big[H(z)\Big]_{z=-0.5} &= (z+0.5)z^{n-1}\frac{z}{(z+0.5)(z-0.75)}\bigg|_{z=-0.5} \\[2mm]
&= \frac{z^n}{(z-0.75)}\bigg|_{z=-0.5} \\[2mm]
&= \frac{(-0.5)^n}{-1.25} \\[2mm]
&= -\frac{4}{5}(-0.5)^n
\end{aligned}
$$

Now we can compute the discrete time series $h(n)$ simply by summing the residues:

$$h(n) = \overset{\text{Number of Residues}}{\underset{k=1}{\sum}}\Big[\text{Residue}_k \text{ of } z^{n-1}H(z)\Big] = \frac{4}{5}(0.75)^n - \frac{4}{5}(-0.5)^n$$

Collecting terms and simplifying, we get the final result:

$$h(n) = \frac{4}{5}\Big[(0.75)^n - (-0.5)^n\Big]$$

The reader should note that we computed $h(n)$ directly and bypassed the previously necessary step of converting the expansion terms from the z-domain to the discrete time domain by using a table. The method of residues eliminates the need for tables (like Table 4.1). This could be a good or bad thing, depending on the situation.

**Example 2.** This example deals with residues of multiple order poles. Let's compute the inverse z-transform using the method of residues for the same $H(z)$ that we used in example 3 for the partial fraction expansion method. We should end up with the same result using the residue method. We begin with

$$H(z) = \frac{z^2}{(z-0.5)(z-1)^2}$$

Here we are dealing with a first order pole and second order pole. We will need to use Equation 4.68 to compute the residue of the first order pole, and we will employ the use of Equation 4.69 to compute the residue of the second order pole. We begin by finding the residue at the simple pole at $z = 0.5$ from

$$\text{Res}\big[H(z)\big]_{z=p_k} = (z-p_k)z^{n-1}H(z)\big|_{z=p_k}$$

$$\text{Res}\big[H(z)\big]_{z=0.5} = (z-0.5)z^{n-1}\frac{z^2}{(z-0.5)(z-1)^2}\bigg|_{z=0.5}$$

$$= \frac{z^{n+1}}{(z-1)^2}\bigg|_{z=0.5}$$

$$= \frac{(0.5)^{n+1}}{0.25} = \frac{(0.5)(0.5)^n}{0.25}$$

$$= 2(0.5)^n$$

We can find the residue at the second order pole at $z = 1$ from

$$\text{Res}\big[H(z)\big]_{z=p_k} = \frac{1}{(m-1)!}\frac{d^{m-1}}{dz^{m-1}}\Big[(z-p_k)^m z^{n-1}H(z)\Big]_{z=p_k}$$

$$\text{Res}\big[H(z)\big]_{z=1} = \frac{1}{(2-1)!}\frac{d^{2-1}}{dz^{2-1}}\left[(z-1)^2 z^{n-1}\frac{z^2}{(z-0.5)(z-1)^2}\right]_{z=1}$$

$$= \frac{d}{dz}\left[\frac{z^{n+1}}{(z-0.5)}\right]_{z=1}$$

$$= \left[\frac{(z-0.5)(n+1)z^n - z^{n+1}}{(z-0.5)^2}\right]_{z=1}$$

$$= \frac{0.5(n+1)-1}{0.25}$$

$$= 2(n-1)$$

Now that we have the residues, we can compute

$$h(n) = \sum_{k=1}^{\text{Number of Residues}} \big[\text{Residue}_k \text{ of } z^{n-1}H(z)\big] = 2(0.5)^n + 2(n-1)$$

Collecting terms and simplifying, we get the final result:

$$h(n) = 2\big[(0.5)^n + (n-1)\big]$$

We ended up with the same result without having to use z-transform tables. Note also that we did not have to compute two separate coefficients for the second order pole of $(z-1)^2$. In some instances, this reduces the amount of computation dramatically.

**Example 3**. This example deals with residues of second order exponential poles. Suppose we run into the system z-transform of the type

$$H(z) = \frac{5ze^{-\alpha}}{(z-e^\alpha)^2}$$

No problem. We can compute the inverse z-transform exactly as before. Since we have a second order pole, we begin with Equation 4.69:

$$\text{Res}\big[H(z)\big]_{z=p_k} = \frac{1}{(m-1)!} \frac{d^{m-1}}{dz^{m-1}}\Big[(z-p_k)^m z^{n-1} H(z)\Big]_{z=p_k}$$

$$\begin{aligned}
\text{Res}\big[H(z)\big]_{z=e^\alpha} &= \frac{d}{dz}\left[(z-e^\alpha)^2 z^{n-1} \frac{5ze^{-\alpha}}{(z-e^\alpha)^2}\right]_{z=e^\alpha} \\
&= \frac{d}{dz}\Big[5e^{-\alpha}z^n\Big]_{z=e^\alpha} \\
&= \Big[5e^{-\alpha}nz^{n-1}\Big]_{z=e^\alpha} \\
&= 5e^{-\alpha}n\big(e^{-\alpha}\big)^{n-1} \\
&= 5ne^{-\alpha n}
\end{aligned}$$

Therefore the system impulse response can be written as $h(n) = 5ne^{-\alpha n}$. Note that the residue method led us directly to the final solution and did not require the use of look up tables.

The pros associated with the residue method of inverse z-transform calculation include the following:

- The residue method leads to a closed form solution.
- The residue method does not require the use of transform tables.
- The residue method does not require computing $k$ separate coefficients for a pole of order $k$, which can result in significantly less computation.

The cons associated with the residue method include the following:

- As is the case with the partial fraction expansion method, the denominator of the polynomial in $z$ needs to be factored in order to determine the poles of $H(z)$. This can on occasion be a difficult task.
- As is the case with the partial fraction expansion method, when the denominator of the polynomial contains multiple order poles, this method requires higher orders of differentiation, which can be an additional computational burden.

## 4.11   POLE AND ZERO STANDARD FORM PLUG-IN EQUATIONS

Before we move on to some examples, it makes sense to define two standard form or "plug-in" equations that are useful time-savers when computing the frequency magnitude response of a system from its z-transform.

The first standard equation enables us to quickly plug in the values and write the frequency magnitude and phase response for a single real pole or single real zero. A single pole or zero is oftentimes referred to as a simple pole or zero. The second standard equation allows us to plug in the values and write the frequency magnitude and phase response for a complex pole or complex zero.

These two plug-in equations provide us with an efficient tool to quickly generate a frequency response for a system that consists of many simple and complex poles and zeros. We will derive the equations and illustrate them with a few simple examples. Later on, we will use these equations in a more complex example.

### 4.11.1   Standard Form for a Real Pole or Real Zero

Real poles and real zeros are always located on the real axis in the z-plane. An example of two real zeros and two real poles is shown in Figure 4.22, parts A and B, respectively. In this illustration, all the poles and zeros are located on the real axis, $a$ units from the origin of the z-plane. The poles are labeled $P_1$ and $P_2$, and the zeros are labeled $Z_1$ and $Z_2$. To avoid confusion, remember that the label $Z_K$ is purely a name assigned to the $k$th zero and should not be confused with the complex variable $z = x + jy$.

It is convention to represent the poles graphically by the character $X$ and to represent the zeros graphically by the character $O$. In polar form, a pole or zero can be represented as

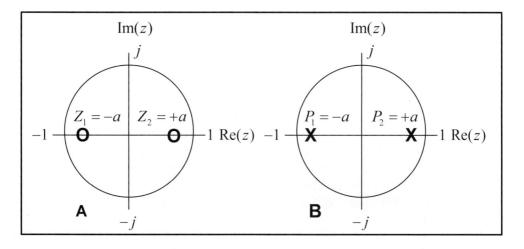

**Figure 4.22**   Example of real poles and real zeros

$$z = re^{\pm j\theta}$$

where $r$ is the magnitude equal to the length of the vector extending from the z-plane origin to the pole or zero and $\theta$ is the angle that vector makes with the real positive axis. In Cartesian form, the pole or zero can be represented as

$$z = r(\cos\theta \pm j\sin\theta) = a \pm jb$$

where

$$a = r\cos\theta$$
$$b = r\sin\theta$$

For real poles and zeros, the term $\theta = 0$ or $\pi$ radians, which fixes the term $b = 0$. Therefore, for real poles and zeros, $r = +a$ or $r = -a$ and the complex variable $z$ can only take on real values anywhere along the real axis. In most applications, we consider a real pole or real zero to be a factor of a larger polynomial in $z$ expressed as

$$H(z) = \frac{(z - z_1)(z - z_2)(z - z_3)\cdots(z - z_K)}{(z - p_1)(z - p_2)(z - p_3)\cdots(z - p_L)}$$

It is understood that the root of the factor $(z \pm z_n) = 0$ is $z = \mp z_n$ and corresponds to the location of a system zero. Likewise, the root of the factor $(z \pm p_n) = 0$ is $z = \mp p_n$ and corresponds to the location of a system pole.

Let us assume we have a system with a single pole or single zero located at $z = a$. This is the root of the polynomial $z - a = 0$, and the system function can be described by

$$H(z) = (z - a) \qquad \left\{ \begin{array}{l} \text{for the case where } a \text{ is the location of a zero} \\ \text{for the case where } a \text{ is the location of a pole} \end{array} \right\}$$
$$H(z) = 1/(z - a)$$

We wish to determine the frequency response of this function. To do so, we will need to evaluate it on the unit circle by substituting the complex variable $e^{j\omega}$ for the complex variable $z$ and then calculating $H(z)\big|_{z=e^{j\omega}} = H(e^{j\omega})$ for $\{0 \le \omega < 2\pi\}$. This is simply the discrete time Fourier transform (DTFT) of the discrete pole or zero. In the case of the real zero, we arrive at

$$H(z)\big|_{z=e^{j\omega}} = (z - a)\big|_{z=e^{j\omega}} = H(e^{j\omega}) = e^{j\omega} - a \qquad \{\text{for } 0 \le \omega < 2\pi\}$$

We can use Euler's formula to break this equation into its real and imaginary components to give us

$$H\left(e^{j\omega}\right) = \cos(\omega) - a + j\sin(\omega)$$

The magnitude of the frequency response is computed as

$$\left|H\left(e^{j\omega}\right)\right| = \sqrt{\left(\cos(\omega) - a\right)^2 + \sin(\omega)^2}$$

If we carry out the squaring operations under the radical and collect terms, we arrive at the standard plug-in equation for a real zero:

$$\left|H\left(e^{j\omega}\right)\right| = \sqrt{1 + a^2 - 2a\cos(\omega)}$$

**Equation 4.70**

The phase response of $H\left(e^{j\omega}\right)$ for a real zero or a real pole is computed as

$$\theta(\omega) = \tan^{-1}\left(\frac{\sin(\omega)}{\left(\cos(\omega) - a\right)}\right)$$

**Equation 4.71**

The inverse of Equation 4.70 is the standard plug-in form for a real pole:

$$\left|H\left(e^{j\omega}\right)\right| = \frac{1}{\sqrt{1 + a^2 - 2a\cos(\omega)}}$$

**Equation 4.72**

Don't forget that the phase angle for a pole is in the denominator of $H(z)$ and will take on a negative value if it is moved to the numerator:

$$H\left(e^{j\omega}\right) = \frac{r_Z e^{j\theta}}{r_P e^{j\psi}} = \left(\frac{r_Z}{r_P}\right)e^{j(\theta - \psi)}$$

Let's try a few of examples to see these standard equations for simple real poles or zeros in operation.

**Example 1.** This example shows a single real zero or pole located at $z = 1$. We know that $z = +1.0$. This places it directly on the unit circle at $\omega = 0$ or DC. We can calculate all the significant variables $r$, $\theta$, $a$, $2a$, and $a^2$:

$$
\left\{
\begin{array}{rl}
r = & 1.0 \\
\theta = & 0 \\
a = & 1.0 \\
2a = & 2.0 \\
a^2 = & 1.0
\end{array}
\right\}
$$

Plugging these values into Equation 4.70 gives us

$$\left|H\left(e^{j\omega}\right)\right| = \sqrt{1 + a^2 - 2a\cos(\omega)} = \sqrt{2 - 2\cos(\omega)} \qquad \left\{\text{for } 0 \leq \omega < 2\pi\right\}$$

The plot of $\left|H\left(e^{j\omega}\right)\right|$ $\left\{\text{for } 0 \leq \omega \leq \pi\right\}$ is illustrated in the Excel plot of Figure 4.23. The plot of $\omega$ for $\left\{\pi < \omega \leq 2\pi\right\}$ is the mirror image of Figure 4.23 and is not shown here.

We can clearly see from the figure that the magnitude of the frequency response for $H(z) = (z-1)$ is zero at $\omega = 0$ and tends to increase in magnitude to 2.0 as $\omega \rightarrow \pi$. It is clear that any zero located on the unit circle will cause the amplitude of the frequency response to be zero at the frequency corresponding to that zero location. In this example, the zero frequency response is at 0 Hz.

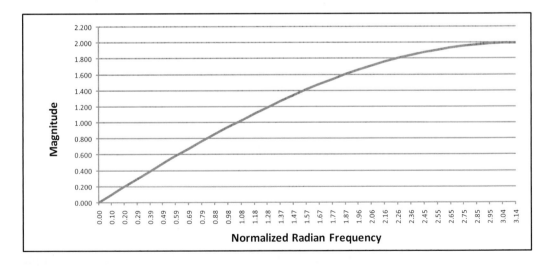

**Figure 4.23** Frequency response of a single zero function $H(z) = (z - 1)$

If we were dealing with a pole we would use the standard plug in equation

$$\left|H\left(e^{j\omega}\right)\right| = \frac{1}{\sqrt{1 + a^2 - 2a\cos(\omega)}}$$

When we plug in the values, we arrive at

$$\left|H\left(e^{j\omega}\right)\right| = \frac{1}{\sqrt{1 + a^2 - 2a\cos(\omega)}} = \frac{1}{\sqrt{2 - 2\cos(\omega)}} \qquad \left\{\text{for } 0 \le \omega < 2\pi\right\}$$

The plot of $\left|H\left(e^{j\omega}\right)\right| = \dfrac{1}{\sqrt{2 - 2\cos(\omega)}}$   $\left\{\text{for } 0 \le \omega < \pi\right\}$ is illustrated in Figure 4.24. Because the pole is the reciprocal of the zero, the system transfer function for the pole $H(z) = 1/(z-1)$ approaches $\infty$ as $z \to 1$ or equivalently as $\omega \to 0$, and it approaches $1/2$ as $\omega \to \pi$.

The reader should note that the $x$-axis in Figure 4.24 does not include $\omega = 0$ because

$$\left|H\left(e^{j\omega}\right)\right|\Big|_{\omega=0} = \infty$$

and we do not have enough paper to include that value in the plot. It is clear that any pole located on the unit circle will cause the amplitude of the frequency

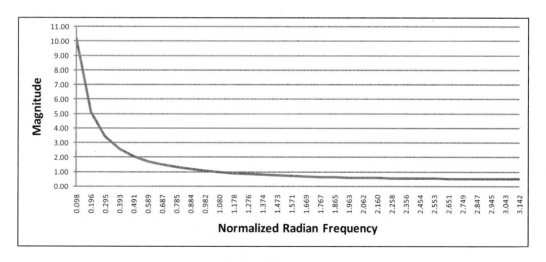

**Figure 4.24**   Frequency response of a single pole function $H(z) = 1/(z-1)$

response to approach ∞ at the frequency corresponding to that pole location. In this example that frequency is at $\omega = 0$. For LTI causal systems with right-sided sequences, if the unit circle is contained within the convergence region, then that system is defined as stable. Since this simple system response has a pole at $z = 1$, the unit circle is not contained within the ROC and the system that owns this pole and has no zeros to cancel it is not stable.

**Example 2.** This example shows a single real zero or pole located at z = 0.5. We know that $z = +0.5$, so we can calculate all the interesting variables $r$, $\theta$, $a$, $2a$, and $a^2$:

$$\left\{\begin{array}{rl} r = & 0.5 \\ \theta = & 0 \\ a = & 0.5 \\ 2a = & 1.0 \\ a^2 = & 0.25 \end{array}\right\}$$

Plugging these values into Equation 4.11 gives us

$$\left|H\left(e^{j\omega}\right)\right| = \sqrt{1+a^2 - 2a\cos(\omega)} = \sqrt{1+0.25 - \cos(\omega)} \qquad \{\text{for } 0 \le \omega < 2\pi\}$$

The frequency response of $\left|H\left(e^{j\omega}\right)\right| = \sqrt{1+0.25 - \cos(\omega)}$ $\{\text{for } 0 \le \omega < \pi\}$ is illustrated in the Excel plot of Figure 4.25. If this figure were the plot of a system transfer function zero $H(z) = (z-0.5)$, we can see that the zero attenuates the gain of the system by a factor of $1/2$ at $\omega = 0$. The gain of the circuit grows from that point to 1.5 at $\omega = \pi$. In between these two end points, the gain tends to increase monotonically in a sinusoidal-like curve. This plot takes on the characteristics of a high pass filter.

The frequency response for $H(z) = 1/(z-0.5)$ is plotted relative to $\omega$ in Figure 4.26 by using the plug-in equation

$$\left|H\left(e^{j\omega}\right)\right| = \frac{1}{\sqrt{1+0.25 - \cos(\omega)}} \qquad \{\text{for } 0 \le \omega < \pi\}$$

The gain of the system with this frequency response would be amplified by a factor of 2 at $\omega = 0$. We see this by setting $\omega = 0$ to get

$$\left|H\left(e^{j\omega}\right)\right|_{\omega=0} = \frac{1}{\sqrt{1+0.25 - \cos(0)}} = \frac{1}{\sqrt{0.25}} = 2.0$$

The gain contribution of this pole would continue to decrease as $\omega \to \pi$. When $\omega = \pi$, the magnitude would decrease to

$$\left|H\left(e^{j\omega}\right)\right|_{\omega=0} = \frac{1}{\sqrt{1+0.25-\cos(\pi)}} = \frac{1}{\sqrt{2.25}} = 0.666\cdots$$

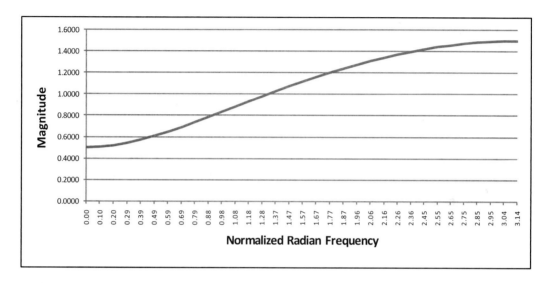

**Figure 4.25**   Frequency response of a real root function $H(z) = (z - 0.5)$

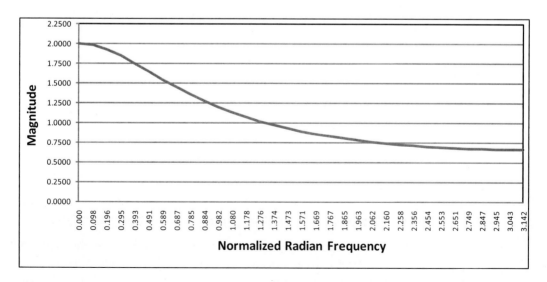

**Figure 4.26**   Frequency response of a real root function $H(z) = 1/(z - 0.5)$

We can see from the figure that this frequency response takes on the characteristics of a low pass filter.

**Example 3**. This example shows a single real zero or pole located at $z = -1.0$. We know that $z = -1.0$. This places it directly on the unit circle at $\omega = \pi$ or $f_s/2$. We can calculate all the variables $r$, $\theta$, $a$, $2a$, and $a^2$:

$$\left\{ \begin{array}{rc} r = & 1.0 \\ \theta = & \pi \\ a = & -1.0 \\ 2a = & -2.0 \\ a^2 = & 1.0 \end{array} \right\}$$

Plugging these values into Equation 4.11 gives us

$$\left|H\left(e^{j\omega}\right)\right| = \sqrt{1+a^2-2a\cos(\omega)} = \sqrt{2+2\cos(\omega)} \qquad \{\text{for } 0 \le \omega < 2\pi\}$$

The plot of $\left|H\left(e^{j\omega}\right)\right|$ $\{\text{for } 0 \le \omega < \pi\}$ is illustrated in the Excel plot of Figure 4.27.

The system function $H(z) = (z+1)$ has a zero at $z = -1$. The magnitude is 2 at $\omega = 0$ and tends to decrease in a sinusoidal decay to 0 as $\omega \to \pi$. Remember that a zero located on the unit circle will cause the frequency response to be zero at the frequency corresponding to that location. In this example, the frequency amplitude response of this system is exactly zero at $\omega = \pi$ or equivalently $f_s/2$.

The system function for a pole at $z = -1$ is $H(z) = 1/(z+1)$. For this case, we use the plug-in equation

$$\left|H\left(e^{j\omega}\right)\right| = \frac{1}{\sqrt{1+a^2-2a\cos(\omega)}} = \frac{1}{\sqrt{2+2\cos(\omega)}} \qquad \{\text{for } 0 \le \omega < 2\pi\}$$

The contribution to the overall system amplitude would be $1/2$ at $\omega = 0$ and would increase to $\infty$ as $\omega \to \pi$. A system with this pole and no zero to cancel it would be unstable.

### 4.11.2  Standard Form for a Complex Pole or Complex Zero

Complex poles and complex zeros can be located anywhere in the z-plane. Complex poles and zeros always occur in pairs called *conjugate pairs*. For example, a complex pole or zero may be located at $z = a + jb$. Its conjugate will be located at $z^* = a - jb$. The asterisk notation indicates that $z^*$ is the complex

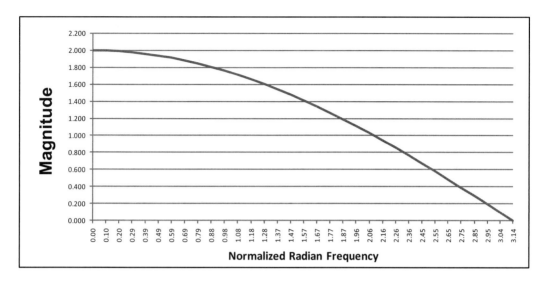

**Figure 4.27**   Frequency response of a single root function $H(z) = (z + 1)$

conjugate of $z$. In polar notation, the complex conjugate pairs can be represented as

$$z = re^{j\theta}, \ z^* = re^{-j\theta}$$

An example of two complex zeros and two complex poles is illustrated in Figure 4.28, parts A and B, respectively. In practice, the system transfer function $H(z)$ that contains only complex poles and zeros can be represented by

$$H(z) = \frac{Z_1 Z_1^* \cdots Z_K Z_K^*}{P_1 P_1^* \cdots P_M P_M^*} = \frac{\left(z - a_1 + jb_1\right)\left(z - a_1 - jb_1\right)\cdots\left(z - a_K + jb_K\right)\left(z - a_K - jb_K\right)}{\left(z - c_1 + jd_1\right)\left(z - c_1 - jd_1\right)\cdots\left(z - c_M + jd_M\right)\left(z - c_M - jd_M\right)}$$

The notation $Z_K Z_K^*$ is used only to identify each of the $k$ conjugate zeros of $H(z)$ and should not be confused with the complex variable $z = a \pm jb$. The notation $P_M P_M^*$ is used to identify each of the $M$ conjugate poles of $H(z)$.

Now it's time to derive the standard "plug-in" equation for complex conjugate poles and zeros. Once we are finished, we will be able inspect a pole-zero diagram and plug in the values to derive the frequency response of a system described by $H(z)$. Let's begin with a system equation $H(z)$ defined by either a single complex zero or a single complex pole such that

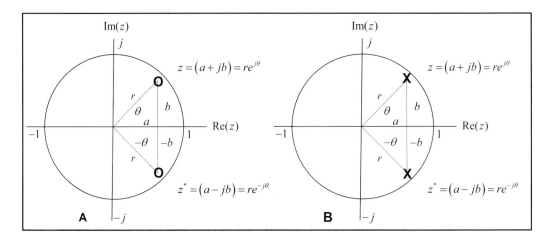

**Figure 4.28**  Complex pole and zero graphical representation

$$H(z) = (z - a + jb)(z - a - jb) \quad \{\text{for a system with one complex zero}\}$$

$$H(z) = \frac{1}{(z - a + jb)(z - a - jb)} \quad \{\text{for a system with one complex pole}\}$$

In either case, we are dealing with the exact same derivation. If we perform the multiplication of the two conjugates, we get

$$H(z) = (z - a + jb)(z - a - jb) = z^2 - 2az + a^2 + b^2$$

**Equation 4.73**

Now if we want to determine the frequency response of the complex conjugate pair, we need to substitute the complex quantity $e^{j\omega}$ for each $z$ in Equation 4.73, which gives us the relation

$$H(z)\big|_{z = e^{j\omega}} = H(e^{j\omega}) = e^{j2\omega} - 2ae^{j\omega} + a^2 + b^2$$

We can use Euler's formula to break this equation into its real and imaginary components to get

$$H(e^{j\omega}) = \left[\cos(2\omega) - 2a\cos(\omega) + a^2 + b^2\right] + j\left[\sin(2\omega) - 2a\sin(\omega)\right]$$

The magnitude of $H\left(e^{j\omega}\right)$ is computed as

$$\left|H\left(e^{j\omega}\right)\right| = \sqrt{\left[\cos(2\omega) - 2a\cos(\omega) + a^2 + b^2\right]^2 + \left[\sin(2\omega) - 2a\sin(\omega)\right]^2}$$

**Equation 4.74**

The phase of $H\left(e^{j\omega}\right)$ is computed as

$$\theta(\omega) = \tan^{-1}\left(\frac{\sin(2\omega) - 2a\sin(\omega)}{\cos(2\omega) - 2a\cos(\omega) + a^2 + b^2}\right)$$

**Equation 4.75**

Equation 4.74 and Equation 4.75 are the two plug-in frequency response equations for a complex conjugate pole or zero. Now let's demonstrate the use of these equations with a few examples.

**Example 1**. This example deals with a single complex zero or pole located at z = ± j. The pole-zero diagram for this example is illustrated in Figure 4.29. In this figure, only the complex zero is shown. The complex pole diagram would be identical. The only change would be that the Os would be replaced with Xs. This diagram corresponds to a system transfer function:

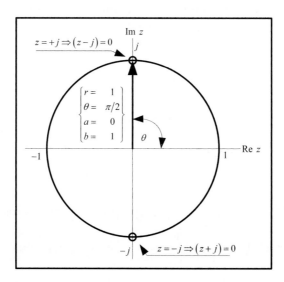

**Figure 4.29**    Zero diagram for (z − j)(z + j)

$$H(z) = (z - j)(z + j) \quad \{\text{for complex conjugate zeros}\}$$

$$H(z) = \frac{1}{(z - j)(z + j)} \quad \{\text{for complex conjugate poles}\}$$

This example is simple enough that we can fill out the variables $r$, $\theta$, $a$, $b$, $2a$, and $a^2 + b^2$ by inspection:

$$\left\{ \begin{array}{rc} r = & 1 \\ \theta = & \pi/2 \\ a = & 0 \\ b = & 1 \\ 2a = & 0 \\ a^2 + b^2 = & 1 \end{array} \right\}$$

Plugging these values into Equation 4.74, we can compute the frequency response as

$$\left| H\left(e^{j\omega}\right) \right| = \sqrt{\left[\cos(2\omega) - 2a\cos(\omega) + a^2 + b^2\right]^2 + \left[\sin(2\omega) - 2a\sin(\omega)\right]^2}$$

$$\left| H\left(e^{j\omega}\right) \right| = \sqrt{\left[\cos(2\omega) + 1\right]^2 + \left[\sin(2\omega)\right]^2} \quad \{\text{ for } 0 \le \omega < 2\pi\}$$

We can carry out the operations indicated under the radical to further reduce the magnitude equation to get

$$\left| H\left(e^{j\omega}\right) \right| = \sqrt{2 + 2\cos(2\omega)}$$

**Equation 4.76**

Depending upon whether this magnitude response is due to a pole or a zero, we can say that

$$\left| H\left(e^{j\omega}\right) \right| = \sqrt{2 + 2\cos(2\omega)} \quad \{\text{for conjugate zeros}\}$$

$$\left| H\left(e^{j\omega}\right) \right| = \frac{1}{\sqrt{2 + 2\cos(2\omega)}} \quad \{\text{for conjugate poles}\}$$

**Equation 4.77**

The Excel-generated plot of the magnitude frequency response of $H(e^{j\omega})$ in Equation 4.76 is shown in Figure 4.30. If we interpret this to be a plot of a complex zero, we see a magnitude of zero exactly at $\pi/2$ or equivalently at $f_s/4$, corresponding to the zero at $(z-j)$. If we would have continued the plot through $2\pi$ radians, we would have seen a second zero located at $3\pi/2$ or equivalently at $3f_s/4$, corresponding to the location of $(z+j)$, the conjugate zero. If we interpret this to be a plot of a complex pole, we would expect to see an infinite magnitude at $\pi/2$ and $3\pi/2$ or equivalently at $f_s/4$ and $3f_s/4$, corresponding to the poles at $(z-j)$ and $(z+j)$, respectively. This is true because these conjugate poles evaluate to 0 in the denominator of the transfer function $H(z)$ at $f_s/4$ and $3f_s/4$.

**Example 2.** This equation deals with single complex zero or pole located at $z = 0.8 \pm j0.5$.

The pole-zero diagram for this example is illustrated in Figure 4.31. In this figure, only the complex pole is shown. The complex zero diagram would be identical. The only change would be that the complex pole notation would change from Xs to Os. This diagram corresponds to a system transfer function of

$$H(z) = (z - 0.8 - j0.5)(z - 0.8 + j0.5) \quad \{\text{for complex conjugate zeros}\}$$

$$H(z) = \frac{1}{(z - 0.8 - j0.5)(z - 0.8 + j0.5)} \quad \{\text{for complex conjugate poles}\}$$

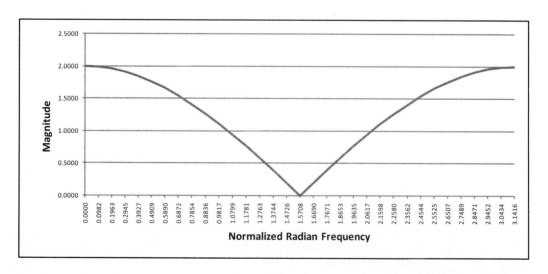

**Figure 4.30** Frequency response of the complex root function $H(z) = (Z = \pm j)$

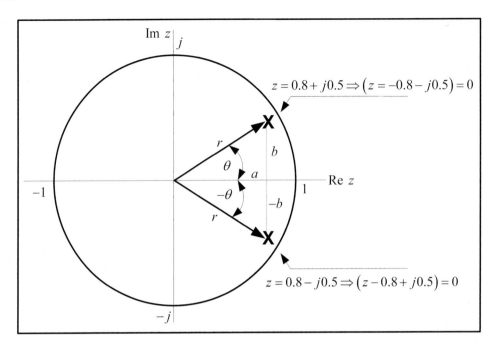

**Figure 4.31**    Pole-zero diagram for $(z - 0.8 + j0.5)(z - 0.8 - j0.5)$

This example takes a small amount of computation in order to define all the variables $r$, $\theta$, $a$, $b$, $2a$, and $a^2 + b^2$:

$$
\left\{
\begin{aligned}
r &= & \sqrt{0.8^2 + 0.5^2} & & = 0.9434 & \\
\theta &= & \tan^{-1}\left(\frac{0.5}{0.8}\right) = 32.0054^\circ & & = 0.5586 \text{ Radian} \\
a &= & 0.8 & \\
b &= & 0.5 & \\
2a &= & 1.6 & \\
a^2 + b^2 &= & 0.89 &
\end{aligned}
\right\}
$$

Plugging these values into Equation 4.74, we can compute the frequency response as

$$\left|H\left(e^{j\omega}\right)\right| = \sqrt{\left[\cos(2\omega) - 2a\cos(\omega) + a^2 + b^2\right]^2 + \left[\sin(2\omega) - 2a\sin(\omega)\right]^2}$$

$$\left|H\left(e^{j\omega}\right)\right| = \sqrt{\left[\cos(2\omega) - 1.6\cos(\omega) + 0.89\right]^2 + \left[\sin(2\omega) - 1.6\sin(\omega)\right]^2} \quad \{ \text{ for } 0 \le \omega < 2\pi\}$$

This equation looks formidable, but it can be solved in Excel very quickly. Depending upon whether this magnitude response is due to a pole or a zero, we can say that

$$\left|H\left(e^{j\omega}\right)\right| = \sqrt{\left[\cos\left(2\omega\right)-1.6\cos\left(\omega\right)+0.89\right]^2 + \left[\sin\left(2\omega\right)-1.6\sin\left(\omega\right)\right]^2} \quad \{\text{for conjugate zeros}\}$$

$$\left|H\left(e^{j\omega}\right)\right| = \frac{1}{\sqrt{\left[\cos\left(2\omega\right)-1.6\cos\left(\omega\right)+0.89\right]^2 + \left[\sin\left(2\omega\right)-1.6\sin\left(\omega\right)\right]^2}} \quad \{\text{for conjugate poles}\}$$

**Equation 4.78**

The Excel-generated plot of the magnitude frequency response of $H\left(e^{j\omega}\right)$ for the conjugate zero case in Equation 4.78 is shown in Figure 4.32. If we interpret this to be a plot of a complex zero, we see a minimal frequency response magnitude at $\theta = \tan^{-1}\left(\dfrac{b}{a}\right) = \tan^{-1}\left(\dfrac{0.5}{0.8}\right) = 0.5586$ radians or equivalently $f_s/11.25$, corresponding to the zero at $(z-0.8-j0.5)$. As a matter of fact, if we carry out the computations, we find the response magnitude at that frequency to be 0.0584. If we would have continued the plot through $2\pi$ radians, we would have also seen a second minimal frequency response magnitude corresponding to the location of the conjugate zero $(z-0.8+j0.5)$.

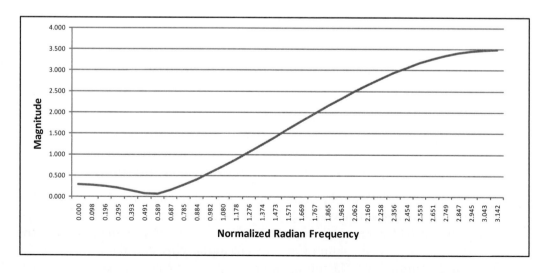

**Figure 4.32**  Frequency response of complex root function $H(z) = (z - 0.8 \pm j0.5)$

If we plot the response of the complex pole, we see a maximal frequency response magnitude at $\theta = \tan^{-1}\left(\dfrac{b}{a}\right) = \tan^{-1}\left(\dfrac{0.5}{0.8}\right) = 0.5586$ radian or equivalently $f_s/11.25$, corresponding to the pole at $(z - 0.8 - j0.5)$. This is true because this pole evaluates to a small number in the denominator of the transfer function $H(z)$. In this example, this number is on the order of 0.0584 and its reciprocal is 17.123.

### 4.11.3  Plug-In Equation Summary

Finding the frequency response of a system from its z-domain transfer function can look a bit difficult. Typical functions contain several real and complex poles and zeros. The trick is to divide the unit circle in to $N$ equally spaced points. Use the plug-in equations and an Excel spreadsheet to compute a frequency response column for each real or complex pole and zero separately. Each column will have $N$ entries. Then create another column that is the log magnitude of the product of all the zero columns divided by the product of all the pole columns. Once complete, you can use Excel to generate a plot of the magnitude column. It's that simple.

We would not have devoted so much time on this subject if we were not going to use it later. In the next section, we will use the techniques we just described to simplify the solution of a problem that is typical of those seen in the industry.

## 4.12  APPLICATIONS OF THE Z-TRANSFORM

After all this discussion, we would like to see the z-transform in action. In particular, we would like to see it used to solve a problem similar to what may be encountered in the industry. We will do that in the next two sections by discussing two scaled-down but realistic examples. The examples are scaled down only so the methods and procedures are clearly emphasized and are not lost in a pile of useless complexity.

### 4.12.1  Example Application 1: Recursive Accumulator

Let's consider the simple digital integrator illustrated in Figure 4.33. The block in the diagram labeled $Z^{-1}$ is nothing more than a shift register that delays the input by one clock period. As shown in the figure, the output of the integrator $y(n)$ is the sum of the current input $x(n)$ and the product of the previous output $y(n-1)$ multiplied by some coefficient $\beta$. This integrator is recognized as being nothing more than a simple recursive digital filter where

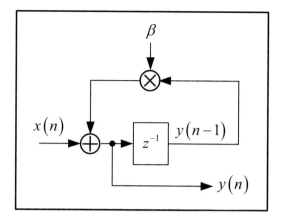

**Figure 4.33**   Simple integrator block diagram

$\beta$ is the only filter coefficient. The difference equation that describes this integrator is

$$y(n) = x(n) + \beta y(n-1)$$

**Equation 4.79**

Initially, we wish to compute the z-transform of this integrator and then determine the location of its pole(s) and/or zero(es) and from that determine its ROC. The z-transform is easily derived by inspection of Equation 4.79:

$$Y(z) = x(z) + \beta z^{-1} Y(z)$$

where we used the shift property of the z-transform to convert the term $\beta y(n-1)$ in the discrete time domain to $\beta z^{-1} Y(z)$ in the z-domain. Now we can rearrange and collect terms to produce

$$Y(z)(1 - \beta z^{-1}) = X(z)$$

$$Y(z) = \frac{1}{1 - \beta z^{-1}} X(z) = \left(\frac{z}{z - \beta}\right) X(z)$$

**Equation 4.80**

The transfer function $H(z)$ of the integrator is obtained by dividing both sides of Equation 4.80 by $X(z)$ to obtain

$$H(z) = \frac{Y(z)}{X(z)} = \frac{z}{z - \beta}$$

We know from previous examples that the term $z/(z-\beta)$ is the closed form expression for the sequence $\sum_{n=0}^{\infty} \beta^n z^{-n}$. If we apply the ratio test to this sequence, we can see that this sequence converges for

$$\lim_{n \to \infty} \left| \frac{\beta^{n+1} z^{-n-1}}{\beta^n z^{-n}} \right| < 1 \quad \text{for} \quad |z| > |\beta|$$

The integrator ROC lies within the range $\beta < |z| \leq \infty$ and is illustrated by the shaded area in Figure 4.34. From our previous discussion, we know that the sequence is a right-sided sequence, and we also know that if $\beta$ is less than unity, the ROC will contain the unit circle and therefore the circuit will be stable. This is a lot of powerful information to know, and it was very simple to obtain.

Now let's do some fun stuff! In order to demonstrate the power an engineer has at his disposal when using the z-transform, let's continue on with the design of this integrator. If we turn on the accumulator and let it run from now until infinity, the obvious question that might be asked is, "How many bits wide should the integrator shift register be in order to hold the infinite accumulation of some input $x(n)$ without ever overflowing?"

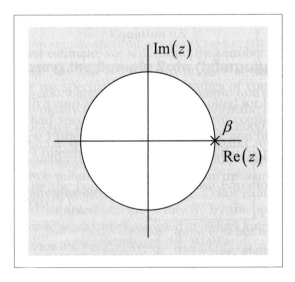

**Figure 4.34**  Convergence region for $y(n)$

This is a great question. On the surface, it seems like the bit width of the integrator shift register will require an infinite number of bits in order to avoid overflow. Is this really the case? The z-transform makes the solution to this difficult problem a great deal simpler than one might think. The final value property of the z-transform will make the tough solution to this problem easy. We know from Equation 4.80 that the z-transform of the output is related to the z-transform of the input by

$$Y(z) = \left(\frac{z}{z-\beta}\right)x(z)$$

**Equation 4.81**

Let's assume that the input data sequence $x(n)$ is an infinitely long sequence consisting of sequential samples equal to some constant value we will call $\alpha$. Our first objective is to derive the closed form expression for $X(z)$. We accomplish this by first expanding the z-transform of $x(n)$ to get

$$\begin{aligned} X(z) &= \sum_{n=0}^{\infty} x(n)z^{-1} \\ &= \sum_{n=0}^{\infty} \alpha z^{-1} \\ &= \alpha\left\{1 + z^{-1} + z^{-2} + z^{-3} + \cdots\right\} \end{aligned}$$

**Equation 4.82**

If we multiply both sides of Equation 4.82 by $z^{-1}$, we get

$$\begin{aligned} z^{-1}X(z) &= z^{-1}\sum_{n=0}^{\infty} \alpha z^{-1} \\ &= \alpha\left\{z^{-1} + z^{-2} + z^{-3} + \cdots\right\} \end{aligned}$$

**Equation 4.83**

If we subtract Equation 4.83 from Equation 4.82, we end up with

$$\left\{X(z) - z^{-1}X(z)\right\} = \left\{\sum_{n=0}^{\infty} \alpha z^{-1} - z^{-1}\sum_{n=0}^{\infty} \alpha z^{-1}\right\} = \alpha\left\{1 + z^{-1} + z^{-2} + z^{-3} + \cdots\right\} - \alpha\left\{z^{-1} + z^{-2} + z^{-3} + \cdots\right\}$$

The result of the subtraction is

$$X(z)\left(1-z^{-1}\right) = \sum_{n=0}^{\infty} \alpha z^{-1}\left(1-z^{-1}\right) = \alpha$$

If we rearrange and collect terms, we arrive at the closed form expression for $X(z)$:

$$X(z) = \sum_{n=0}^{\infty} \alpha z^{-1} = \frac{\alpha}{\left(1-z^{-1}\right)} = \frac{\alpha z}{z-1}$$

$$X(z) = \frac{\alpha z}{z-1}$$

**Equation 4.84**

Now that we know $X(z)$, we can substitute it back into Equation 4.81 to get

$$Y(z) = \left(\frac{z}{z-\beta}\right)X(z)$$

$$= \left(\frac{z}{z-\beta}\right)\left(\frac{\alpha z}{z-1}\right)$$

**Equation 4.85**

Remember that we are trying to determine the maximum bit width of the accumulator register so we can avoid overflow. To do this, we will use the final value theorem to determine the final value of the accumulated sum $y(n)$ as $n \to \infty$:

$$\lim_{n\to\infty} y(n) = \lim_{z\to 1}(z-1)Y(z)$$

$$= \lim_{z\to 1}(z-1)\left(\frac{z}{z-\beta}\right)\left(\frac{\alpha z}{z-1}\right)$$

$$= \lim_{z\to 1}\left(\frac{z}{z-\beta}\right)(\alpha z)$$

$$= \frac{\alpha}{1-\beta}$$

So after all the fun work is complete, we now have a firm result. Our calculations have shown that the maximum value that $y(n)$ can ever attain after an infinite number of accumulation cycles is

$$y(n)\Big|_{n\to\infty} = \frac{\alpha}{1-\beta}$$

**Equation 4.86**

Right away we know from Equation 4.86 that as long as $\beta \neq 1$, the infinite accumulation produces a $y(n)$ that has a finite value. This is starting to get interesting. But, wait, there's more!

Before we go any further, it is always wise to do a mathematical sanity check. We know that $\beta$ is a system pole. We also know that the z-transform is a right-sided sequence and therefore the ROC must contain the unit circle in order to guarantee stability.

Knowing this and looking at Equation 4.86, we know right from the start that $\beta$ must always be less than unity. As $\beta$ approaches unity, then the value of $y(n)$ approaches infinity, which would mean that our integrator register bit width would also approach infinity. However, as $\beta$ approaches zero, the output sequence $y(n)$ approaches the value of the input sequence $x(n) = \alpha$ and the integrator becomes nothing more than a simple wire that passes the input sequence directly to the output, unmodified. The output bit width would be the same as the input bit width. This is verified by the integrator block diagram in Figure 4.33. The figure clearly shows that if $\beta$ is equal to zero, the feedback path is broken and the input is summed with zero, resulting in $y(n) = x(n)$.

Now let's solve the original problem in the context of a real-world application. For this example, let's say that the input data $x(n)$ is an infinite data sequence generated by an 8-bit-wide analog to digital converter. Therefore the magnitude of the input is restricted to the range $0 \leq x(n) \leq 255$. In this example, we are making the assumption that we are dealing with unsigned numbers. If we happened to be dealing with a bipolar input data stream in 2s complement format, we would adjust the maximum range to be $-128 \leq x(n) \leq +127$.

We want to solve for the worst case condition, so let's use all positive numbers and set the value of the input data sequence to a constant stream of the largest magnitude samples possible, or $x(n) = \alpha = 255$. If we substitute this value into Equation 4.86, we get

$$y(n) = \frac{255}{1-\beta}$$

Now the worst case final value of $y(n)$ and therefore the width of the integrator register is determined only by the value of the coefficient $\beta$. If the input sequence $x(n)$ is constrained to be the maximum possible value over all time, then the maximum value that will be attained by the output sequence $y(n)$ will depend on the maximum value of $\beta$ or $\beta_{Max}$. We can now compute the value of $y(n)_{Max}$ and therefore the maximum width of the accumulator shift register:

$$y(n)_{Max} = \frac{255}{1 - \beta_{Max}} \quad \{0 \le \beta_{Max} < 1\}$$

The width of the accumulator register in terms of bits can be computed by

$$\text{Register Bit Width} = \log_2\left(y(n)_{Max}\right) = \log_2\left(\frac{255}{1 - \beta_{Max}}\right)$$

Table 4.2 summarizes the values for $y(n)_{Max}$ and the integrator register bit width for several values of $\beta$, ranging from 0 to 0.95. It's interesting to note that a $\beta$ with a value of 0.95 will require a register width of at least 13 bits to avoid overflow. It's also important to note that when $\beta$ is equal to zero, the required register width is only 8 bits, which is identical to the bit width of the input sequence $x(n)$.

We used the final value theorem of the z-transform to derive the worst case maximum accumulated value that the integrator register could ever attain. This value then enabled us to determine the maximum register bit width. The final value property of the z-transform really helped us turn a difficult problem into a simple one. Now that we have accomplished the seemingly impossible, the reader may still ask the question, "How can an integrator with a finite width register continue to accumulate data over an infinite amount of time and still not overflow?"

We will answer this follow-up question by looking at the integrator accumulation in the discrete time domain. The difference equation that describes the integrator was given by Equation 4.79 and is repeated here for clarity:

$$y(n) = x(n) + \beta y(n-1)$$

We can use Equation 4.79 to generate a history for the previous $N$ accumulations. As an example, the history for the previous five accumulations is shown in Equation 4.87. The history equations are listed 1 through 5: equation 1 represents the current output of the integrator, equation 2 represents the previous integrator output, equation 3 represents the output two clock ticks prior, and so on.

**Table 4.2**   Integrator Register Width versus the Integrator Coefficient

| β | $y(n)_{max}$ | Bit width | Next highest integer |
|---|---|---|---|
| 0.95 | 5100.00 | 12.316282 | 13 |
| 0.90 | 2550.00 | 11.316282 | 12 |
| 0.85 | 1700.00 | 10.731319 | 11 |
| 0.80 | 1275.00 | 10.316282 | 11 |
| 0.75 | 1020.00 | 9.994353 | 10 |
| 0.70 | 850.00 | 9.731319 | 10 |
| 0.65 | 728.57 | 9.508927 | 10 |
| 0.60 | 637.50 | 9.316282 | 10 |
| 0.55 | 566.67 | 9.146357 | 10 |
| 0.50 | 510.00 | 8.994353 | 9 |
| 0.45 | 463.64 | 8.856850 | 9 |
| 0.40 | 425.00 | 8.731319 | 9 |
| 0.35 | 392.31 | 8.615842 | 9 |
| 0.30 | 364.29 | 8.508927 | 9 |
| 0.25 | 340.00 | 8.409391 | 9 |
| 0.20 | 318.75 | 8.316282 | 9 |
| 0.15 | 300.00 | 8.228819 | 9 |
| 0.10 | 283.33 | 8.146357 | 9 |
| 0.05 | 268.42 | 8.068354 | 9 |
| 0.00 | 255.00 | 7.994353 | 8 |

$$
\begin{aligned}
1: \quad & y(n) & = x(n) \quad & + \beta y(n-1) \\
2: \quad & y(n-1) & = x(n-1) \quad & + \beta y(n-2) \\
3: \quad & y(n-2) & = x(n-2) \quad & + \beta y(n-3) \\
4: \quad & y(n-3) & = x(n-3) \quad & + \beta y(n-4) \\
5: \quad & y(n-4) & = x(n-4) \quad & + \beta y(n-5) \\
& \quad \vdots & \vdots \quad & \vdots
\end{aligned}
$$

**Equation 4.87**

If we substitute $y(n-1)$ given by the history equation 2 into the history equation 1, we get

$$y(n) = x(n) + \beta x(n-1) + \beta^2 y(n-2)$$

If we replace $y(n-2)$ by the expression in the history equation 3, we arrive at

$$y(n) = x(n) + \beta x(n-1) + \beta^2 x(n-2) + \beta^3 y(n-3)$$

If we replace $y(n-3)$ by the expression in the history equation 4, we get

$$y(n) = x(n) + \beta x(n-1) + \beta^2 x(n-2) + \beta^3 x(n-3) + \beta^4 y(n-4)$$

If we replace $y(n-4)$ by the expression in the history equation 5, we end up with

$$y(n) = x(n) + \beta x(n-1) + \beta^2 x(n-2) + \beta^3 x(n-3) + \beta^4 x(n-4) + \beta^5 y(n-5)$$

This replacement process can continue on without end. If we continue on as $k \rightarrow \infty$, the expansion will continue to grow and will take on the form of

$$y(n) = x(n) + \beta x(n-1) + \beta^2 x(n-2) + \beta^3 x(n-3) + \beta^4 x(n-5) + \cdots + \beta^k y(n-k)$$

**Equation 4.88**

Let's take a close look at Equation 4.88. We can see that each time a sample is recirculated around the integrator feedback loop, its contribution to the total sum is attenuated by a factor of $\beta$. If we had set the value of $\beta$ to be zero, then the only term output from the integrator is the current input $x(n)$ since all other terms would be zero. This is the equivalent of breaking the feedback loop. If we had set $\beta$ to a value of 0.1, then the contribution of each input to the total sum is reduced by a factor of 0.1 each time it recirculates around the feedback loop. That is, the contribution to the output of each previous input sample decays by a factor of $\beta$ each sample period. If $\beta$ is a large number but less than 1, the decay is slow. If $\beta$ is a small number much less than 1, the rate of decay is much faster. And if $\beta$ is greater than 1, the accumulation will grow without bound and the shift register will overflow. If $\beta > 1$, the integrator becomes an unstable system, which agrees with the pole-zero plot and the ROC calculations we performed using the integrator z-transform. So all the information we derived about the integrator from its z-transform agrees with the actual performance calculations. The track of an individual unit sample as it recirculates through the accumulator is illustrated in Table 4.3 for nine different values of $\beta$ versus the number of sample recirculations $k$.

As we move across the columns to the case where $\beta = 0.9$, we can see that the history samples tend to make a longer contribution to the total integrator sum. After 16 recirculations with a $\beta$ of 0.9, the sample has only attenuated by 80%. If we look at the case where $\beta = 0.1$, we can see that after only nine recirculations, the contribution of that unit sample to the running sum has decayed to zero above eight decimal places.

This was a fun and educational exercise. For those readers who see themselves designing infinite impulse response (IIR) filters in the future, you now know how to compute the minimum bit width for all registers within the filter. This is a very powerful piece of information.

All the bits and pieces we discussed in previous sections come together in this example. This simple exercise should turn on a light bulb somewhere. This z-transform theory we have been discussing is simple, but it is also a very powerful tool for signal processing engineers. The z-transform makes many of the difficult DSP computations easy.

We've had so much fun up to this point, why stop here? Let's continue on and use the power of the z-transform to compute the frequency response of our example integrator. We'll begin by revisiting Equation 4.81, repeated here for clarity:

$$Y(z) = \left( \frac{z}{z - \beta} \right) X(z)$$

We can compute the system transfer function $H(z)$ by dividing both sides of the equation by $X(z)$ to get

$$H(z) = \frac{Y(z)}{X(z)} = \left( \frac{z}{z - \beta} \right) = \left( \frac{1}{1 - \beta z^{-1}} \right)$$

**Equation 4.89**

If a right-sided causal sequence converges and the ROC includes the unit circle, then the Fourier transform for that sequence also converges and exists. We know that the system function for the integrator has a single pole at $|z| = |\beta|$, and we know that $\beta < 1$. Therefore we know that the ROC includes the unit circle and the function converges.

Since the ROC includes the unit circle, we can use the Fourier transform to compute the frequency response of our integrator. We accomplish this by

**Table 4.3**    Attenuation of History Samples versus the Feedback Coefficient Beta

| $k$ | $\beta^k = (0.1)^k$ | $\beta^k = (0.2)^k$ | $\beta^k = (0.3)^k$ | $\beta^k = (0.4)^k$ | $\beta^k = (0.5)^k$ | $\beta^k = (0.6)^k$ | $\beta^k = (0.7)^k$ | $\beta^k = (0.8)^k$ | $\beta^k = (0.9)^k$ |
|---|---|---|---|---|---|---|---|---|---|
| 0 | 1.0000000 | 1.0000000 | 1.0000000 | 1.0000000 | 1.0000000 | 1.0000000 | 1.0000000 | 1.0000000 | 1.00000000 |
| 1 | 0.1000000 | 0.2000000 | 0.3000000 | 0.4000000 | 0.5000000 | 0.6000000 | 0.7000000 | 0.8000000 | 0.90000000 |
| 2 | 0.0100000 | 0.0400000 | 0.0900000 | 0.1600000 | 0.2500000 | 0.3600000 | 0.4900000 | 0.6400000 | 0.81000000 |
| 3 | 0.0010000 | 0.0080000 | 0.0270000 | 0.0640000 | 0.1250000 | 0.2160000 | 0.3430000 | 0.5120000 | 0.72900000 |
| 4 | 0.0001000 | 0.0016000 | 0.0081000 | 0.0256000 | 0.0625000 | 0.1296000 | 0.2401000 | 0.4096000 | 0.65610000 |
| 5 | 0.0000100 | 0.0003200 | 0.0024300 | 0.0102400 | 0.0312500 | 0.0777600 | 0.1680700 | 0.3276800 | 0.59049000 |
| 6 | 0.0000010 | 0.0000640 | 0.0007290 | 0.0040960 | 0.0156250 | 0.0466560 | 0.1176490 | 0.2621440 | 0.53144100 |
| 7 | 0.0000001 | 0.0000128 | 0.0002187 | 0.0016384 | 0.0078125 | 0.0279936 | 0.0823543 | 0.2097152 | 0.47829690 |
| 8 | 0.0000000 | 0.0000025 | 0.0000656 | 0.0006553 | 0.0039062 | 0.0167961 | 0.0576480 | 0.1677721 | 0.43046721 |
| 9 | 0.0000000 | 0.0000005 | 0.0000196 | 0.0002621 | 0.0019531 | 0.0100777 | 0.0403536 | 0.1342177 | 0.38742049 |
| 10 | 0.0000000 | 0.0000001 | 0.0000059 | 0.0001048 | 0.0009765 | 0.0060466 | 0.0282475 | 0.1073741 | 0.34867844 |
| 11 | 0.0000000 | 0.0000000 | 0.0000017 | 0.0000419 | 0.0004882 | 0.0036279 | 0.0197732 | 0.0858993 | 0.31381060 |
| 12 | 0.0000000 | 0.0000000 | 0.0000005 | 0.0000167 | 0.0002441 | 0.0021767 | 0.0138412 | 0.0687194 | 0.28242954 |
| 13 | 0.0000000 | 0.0000000 | 0.0000001 | 0.0000067 | 0.0001220 | 0.0013060 | 0.0096889 | 0.0549755 | 0.25418658 |
| 14 | 0.0000000 | 0.0000000 | 0.0000000 | 0.0000026 | 0.0000610 | 0.0007836 | 0.0067822 | 0.0439804 | 0.22876792 |
| 15 | 0.0000000 | 0.0000000 | 0.0000000 | 0.0000010 | 0.0000305 | 0.0004701 | 0.0047475 | 0.0351843 | 0.20589113 |
| 16 | 0.0000000 | 0.0000000 | 0.0000000 | 0.0000004 | 0.0000152 | 0.0002821 | 0.0033232 | 0.0281475 | 0.18530202 |

replacing $z$ in Equation 4.89 with $e^{j\omega}$ and then evaluating the system function on the unit circle. So we begin with the system transfer function

$$H(z) = \left( \frac{1}{1 - \beta z^{-1}} \right)$$

Now we replace $z$ with $e^{j\omega}$ to obtain

$$H\left(e^{j\omega}\right) = \left( \frac{1}{1 - \beta e^{-j\omega}} \right)$$

We can expand the $e^{j\omega}$ term into its cosine and sine components and then collect real and imaginary terms to produce

$$H\left(e^{j\omega}\right) = \left( \frac{1}{1 - \beta(\cos\omega - j\sin\omega)} \right)$$

$$= \left( \frac{1}{(1 - \beta\cos\omega) - (j\beta\sin\omega)} \right)$$

Now we can compute the magnitude of $H\left(e^{j\omega}\right)$ by

$$\left| H\left(e^{j\omega}\right) \right| = \frac{1}{\sqrt{(1 - \beta\cos\omega)^2 + (-\beta\sin\omega)^2}}$$

If we carry out the operations indicated under the radical, we get

$$\left| H\left(e^{j\omega}\right) \right| = \frac{1}{\sqrt{1 - 2\beta\cos\omega + \beta^2\cos^2\omega + \beta^2\sin^2\omega}}$$

The terms under the radical collapse, and we end up with Equation 4.90, which expresses the magnitude of integrator frequency response as a function of $\omega$:

$$\left| H\left(e^{j\omega}\right) \right| = \frac{1}{\sqrt{1 + \beta^2 - 2\beta\cos\omega}}$$

**Equation 4.90**

The curves shown in Figure 4.35 are the result of plotting Equation 4.90 as a function of normalized frequency for eleven different values of $\beta$, ranging from 0.1 to 0.99. We can see from the figure that the integrator frequency response takes on a low pass filter characteristic. We can also see that the filter roll-off is dependent on the value of $\beta$.

It is interesting to note that the frequency response equation that we derived for this accumulator (Equation 4.90) is identical to the single pole plug-in equation we derived previously (Equation 4.70). We could have used the plug-in equation and avoided all the work, but for this example it was informative to perform the derivation.

The attenuation for each response is measured at a frequency equal to half the sample rate $(f_s/2)$, which is a normalized frequency of 0.5. Table 4.4 lists the measured attenuation at the normalized frequency of 0.5 versus the value of $\beta$. When $\beta$ equals zero, the integrator takes on the characteristics of an all pass filter, the output sequence $y(n)$ is equal to the input sequence $x(n)$, and the attenuation is zero. As $\beta$ becomes larger and approaches unity, the response roll-off increases, resulting in a larger attenuation, and the integrator takes on the characteristics of a narrow band low pass filter that attenuates

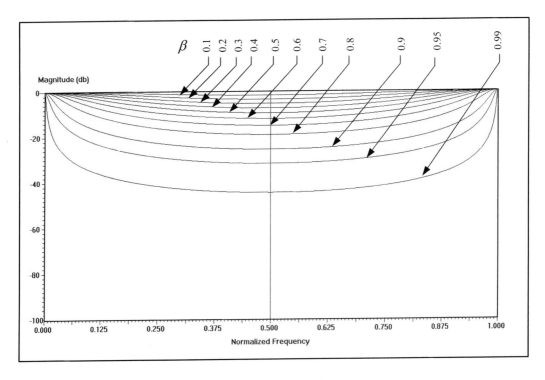

**Figure 4.35**   Integrator frequency response versus beta

**Table 4.4**  Integrator Attenuation versus Beta

| $\beta$ | Attenuation db |
|---|---|
| 0.00 | 0.00 |
| 0.10 | 1.60 |
| 0.20 | 3.40 |
| 0.30 | 5.20 |
| 0.40 | 7.20 |
| 0.50 | 9.40 |
| 0.60 | 12.00 |
| 0.70 | 15.00 |
| 0.80 | 19.00 |
| 0.90 | 25.40 |
| 0.95 | 31.60 |
| 0.99 | 44.40 |

high-frequency components from the input sequence. The integrator becomes more sluggish for larger values of $\beta$, in that it is less responsive to the higher input frequencies and tends to take on the role of a long term signal trend averager. An integrator similar to the one in this example with a large valued $\beta$ is sometimes used in frame pulse synchronizers to remove high-frequency pulse timing jitter. This jitter removal aids in locking to the framing pulse in pulse position modulated signals. It can also be used with a large $\beta$ to closely approximate the removal of the DC component from an input signal simply by subtracting the output from the input—that is, $w(n) = x(n) - y(n)$. This can be thought of as digital approximation of an analog capacitor. We will see a more efficient version of a digital capacitor later.

### 4.12.2  Example Application 2: Modifying an Existing Circuit

On many occasions an engineer may wish to modify the performance of an existing system. The goal might be to fine tune the operation of the system in order to bring it into compliance with some design specification or perhaps to change its operational specifications in a controlled manner so that it can be reused and interfaced with some other system. Oftentimes the motivation is to make corrections to mitigate the problems associated with an unsuspected or unplanned environmental interference, such as might be the case with interferring noise generated by power line radiation.

As happens on many occasions, an underpaid and overworked engineer might be tasked to modify an existing system design and is handed a pile of minimal documentation from which to proceed. In this case, the best course of action is to use the minimal documentation to reverse engineer the existing design, characterize it in the time and frequency domain, and then modify the existing design to meet the new specifications. The next example will pursue this line of thought. We will first introduce a simple system with minimal technical documentation that will represent the existing system design. We will then reverse engineer and characterize the system, and then we will make the modifications necessary to produce the desired system operation.

Let's suppose that there exists a large DSP system that is located in a nice air-conditioned computer room in a modern building located in downtown Mozambique. Some electricians were in the room recently, stringing wire to power another system soon to be installed in the same room. Shortly after the electricians finished their work and left, it was noticed that the performance of the signal processing system had degraded considerably. The problem has been traced to a low pass recursive digital filter located in the heart of the system.

Let's now suppose that your supervisor hands you an airplane ticket to Mozambique and a pole-zero diagram of the low pass recursive filter in question. He tells you to drop what you're doing, get on the next plane to Mozambique, troubleshoot a system you've never seen before, and fix the problem. Oh, and by the way, your travel allowance is limited to $30 a day and any expense above the allowance comes out of your pocket.

Clearly this is an undesirable assignment. Nevertheless, you are the junior engineer with minimal experience, and you need to keep your job at least one more year, so you agree. So, where do you start? How do you proceed? What is your plan? How are you going to live on $30 a day? You begin by looking at the only documentation your supervisor gave you: the pole-zero diagram of the low pass recursive digital filter.

Let's assume that the pole-zero diagram is the one illustrated in Figure 4.36. We can tell that the circuit does indeed resemble a low pass characteristic because of the pole-zero placement. There is a real pole located at $z = 0.9$, which means the circuit frequency response will have a fairly high gain at 0 Hz. In addition, there is a complex pole of magnitude 0.9 at $\theta = \pm 11.25$ degrees or $\pm \pi/16$ radians, and a complex pole at $\phi = \pm 70$ degrees or $\pm \pi/2.57$ radians that will tend to maintain the high gain. There is also a complex zero located at $z = \pm j$ and a real zero located at $z = -1$, which forces the filter frequency response to zero at a quarter the sample rate and half the sample rate (i.e., $f_s/4$ and $f_s/2$).

While on the lengthy airplane trip to Mozambique, you begin this design effort by reverse engineering the circuit using the pole-zero diagram as the starting point. The first thing you do is write down the z-transform of the

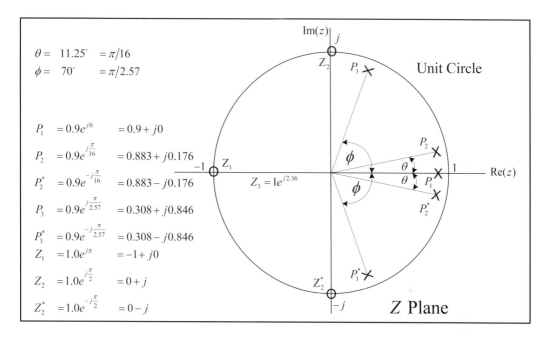

**Figure 4.36**  Low pass recursive filter pole-zero placement

filter transfer function, $H(z)$. We can do this by inspection of the pole-zero diagram. The resultant function is given by

$$H(z) = \frac{(z-j)(z+j)(z+1)}{(z-0.9)(z-0.883-j0.176)(z-0.883+j0.176)(z-0.308-j0.846)(z-0.308+j0.846)}$$

**Equation 4.91**

The transfer function looks a bit formidable, but it's really a piece of cake in disguise. We know from the pole placement that the circuit is stable, since all poles are contained within the unit circle and the ROC extends from $|z| = 0.9$ to $|z| = \infty$ and includes the unit circle. This makes sense since the filter had been working without incident for a long time.

Your next step in characterizing this filter is to determine its frequency response by replacing all occurrences of the complex variable $z$ in Equation 4.91 with the complex variable $e^{j\omega}$. You're not going to do this with a pencil and paper, so you pull out your laptop, open the Microsoft Excel application, and use the plug-in equations we developed previously to compute the frequency magnitude response of the individual poles and zeros of the transfer function. Each pole and zero will be evaluated individually at $N$ equally spaced points on the unit circle using the standard form plug-in equations. You divide the unit circle into $N$ equal arc segments. You evaluate each standard form equation for both poles and zeros at each of the $N$ equally spaced arc segments around the unit circle. The value of $\omega$ for each segment is computed by

$$\omega = \left(\frac{2\pi}{N}\right)k \quad \{\text{for } 0 \leq k \leq N-1\}$$

A total of $N$ values of the magnitude response for each pole and each zero are tabulated in a separate Excel spreadsheet column.

Larger values of $N$ will produce a finer resolution in the frequency response plots. For this example, we will set $N = 32$, which is sufficient to illustrate the procedure. It simplifies your task considerably if you work with each of the pole and zero factors in Equation 4.91 separately.

So you evaluate each real pole or real zero in the transfer function at $N$ values of $\omega$ using the standard form given in Equation 4.70, which is repeated here for clarity:

$$\left|H\left(e^{j\omega}\right)\right|_{ZERO} = \sqrt{1+a^2 - 2a\cos(\omega)} \qquad \left|H\left(e^{j\omega}\right)\right|_{POLE} = 1\Big/\sqrt{1+a^2 - 2a\cos(\omega)}$$

You evaluate each complex pole or complex zero in the transfer function at the same $N$ values of $\omega$ using the standard form given in Equation 4.74, which is also repeated here for clarity:

$$\left|H\left(e^{j\omega}\right)\right|_{ZERO} = \sqrt{\left[\cos(2\omega) - 2a\cos(\omega) + a^2 + b^2\right]^2 + \left[\sin(2\omega) - 2a\sin(\omega)\right]^2}$$

$$\left|H\left(e^{j\omega}\right)\right|_{POLE} = 1\Big/\sqrt{\left[\cos(2\omega) - 2a\cos(\omega) + a^2 + b^2\right]^2 + \left[\sin(2\omega) - 2a\sin(\omega)\right]^2}$$

The magnitude of the frequency response for each pole and each zero is tabulated for all $N$ values of $\omega$ and placed in a separate column of the spreadsheet. Once all of the pole-zero columns are computed, you end up with a bunch of columns, each with $N$ elements, in the spreadsheet that represent the magnitude response for each pole and zero in the system.

Now you simply multiply the zero columns together and divide by the product of the pole columns to produce another column in the spreadsheet that tabulates the $N$ different values of the magnitude versus the frequency of the composite $\left|H\left(e^{j\omega}\right)\right|$. Then you compute a final log magnitude column that tabulates $N$ entries of $20Log_{10}\left(\left|H\left(e^{j\omega}\right)\right|\Big/\left|H\left(e^{j\omega}\right)_{MAX}\right|\right)$. You use Excel to generate a plot of the log magnitude column. Now you can see for the first time the plot of the frequency response of the filter in question. That's the procedure, so let's walk through it step by step.

You begin by evaluating all the factors that make up the zeros of $H(z)$. That is,

$$(z-j)(z+j)(z+1)$$

You treat a complex conjugate pair as a single pole or zero and use the plug-in equations to generate the individual frequency responses.

**Factor 1.** The first factor is the magnitude response of the real zero $|(z+1)|_{z=e^{j\omega}}$. The zero occurs at $z=-1$. You identify the variables and use Equation 4.70 to generate the frequency response:

$$\left\{ \begin{array}{rl} r = & 1.0 \\ \theta = & 180 \\ a = & -1.0 \\ 2a = & -2.0 \\ a^2 = & 1.0 \end{array} \right\}$$

$$\left|H\left(e^{j\omega}\right)\right|_{(z+1)} = \sqrt{1+a^2-2a\cos(\omega)} = \sqrt{2+2\cos(\omega)}$$

**Factor 2.** The second factor is the magnitude response of the complex zero $|(z+j)(z-j)|_{z=e^{j\omega}}$. This corresponds to a complex zero at $z=\pm j$. You identify the variables and use Equation 4.74 to generate the frequency response:

$$\left\{ \begin{array}{rl} r = & 1 \\ \theta = & \pi/2 \\ a = & 0 \\ b = & 1 \\ 2a = & 0 \\ a^2+b^2 = & 1 \end{array} \right\}$$

$$\left|H\left(e^{j\omega}\right)\right| = \sqrt{\left[\cos(2\omega)-2a\cos(\omega)+a^2+b^2\right]^2+\left[\sin(2\omega)-2a\sin(\omega)\right]^2}$$

$$\left|H\left(e^{j\omega}\right)\right| = \sqrt{\left[\cos(2\omega)+1\right]^2+\left[\sin(2\omega)\right]^2}$$

$$\left|H\left(e^{j\omega}\right)\right| = \sqrt{2+2\cos(2\omega)}$$

Now you address the one real pole and the two complex poles in the system.

**Factor 3.** The third factor is the magnitude response of the real pole $|(z-0.9)|_{z=e^{j\omega}}$. The real pole occurs at $z=0.9$. You identify the variables and use Equation 4.70 to generate the frequency response:

$$\begin{Bmatrix} r = & 0.9 \\ \theta = & 0 \\ a = & 0.9 \\ 2a = & 1.8 \\ a^2 = & 0.81 \end{Bmatrix}$$

$$\left|H\left(e^{j\omega}\right)\right| = 1\big/\sqrt{1+a^2 - 2a\cos(\omega)} = 1\big/\sqrt{1.81 - 1.8\cos(\omega)}$$

**Factor 4.** The fourth factor is the magnitude response of the complex pole:

$$\left|(z-0.833 - j0.176)(z-0.833 + j0.176)\right|_{z=e^{j\omega}}$$

The complex pole occurs at $z=0.833 \pm j0.176$. You identify the variables and use Equation 4.74 to generate the frequency response:

$$\begin{Bmatrix} r = & \sqrt{0.883^2 + 0.176^2} & = 0.9 \\ \theta = & \tan^{-1}\left(\dfrac{0.176}{0.883}\right) = 11.25° & = 0.1963 \text{ Radian} \\ a = & 0.883 \\ b = & 0.176 \\ 2a = & 1.766 \\ a^2 + b^2 = & 0.81 \end{Bmatrix}$$

$$\left|H\left(e^{j\omega}\right)\right| = 1\big/\sqrt{\left[\cos(2\omega) - 2a\cos(\omega) + a^2 + b^2\right]^2 + \left[\sin(2\omega) - 2a\sin(\omega)\right]^2}$$

$$\left|H\left(e^{j\omega}\right)\right| = 1\big/\sqrt{\left[\cos(2\omega) - 1.76\cos(\omega) + 0.81\right]^2 + \left[\sin(2\omega) - 1.76\sin(\omega)\right]^2}$$

**Factor 5.** The fifth factor is the magnitude response of the complex pole

$$\left|(z-0.308 - j0.846)(z-0.308 + j0.846)\right|_{z=e^{j\omega}}$$

The complex pole occurs at $z = 0.308 \pm j0.846$. Once again, you identify the variables and use Equation 4.74 to generate the frequency response:

$$\left\{ \begin{array}{lll} r = & \sqrt{0.308^2 + 0.846^2} & = 0.9 \\[2mm] \theta = & \tan^{-1}\left(\dfrac{0.846}{0.308}\right) = 70° & = 1.22 \text{ Radian} \\[2mm] a = & 0.308 \\ b = & 0.846 \\ 2a = & 0.616 \\ a^2 + b^2 = & 0.81 \end{array} \right.$$

$$\left| H\left(e^{j\omega}\right) \right| = 1 \Big/ \sqrt{\left[\cos(2\omega) - 2a\cos(\omega) + a^2 + b^2\right]^2 + \left[\sin(2\omega) - 2a\sin(\omega)\right]^2}$$

$$\left| H\left(e^{j\omega}\right) \right| = 1 \Big/ \sqrt{\left[\cos(2\omega) - 0.616\cos(\omega) + 0.81\right]^2 + \left[\sin(2\omega) - 0.616\sin(\omega)\right]^2}$$

So now that all of the individual pole-zero magnitude computations are complete and tabulated in separate spreadsheet columns, you use the spreadsheet to compute the magnitude of the transfer function:

$$\left| H\left(e^{j\omega}\right) \right| = \left(\frac{(\text{Factor 1})(\text{Factor 2})}{(\text{Factor 3})(\text{Factor 4})(\text{Factor 5})}\right) \quad \text{for } 0 \le \omega < 2\pi$$

Then you determine the maximum value of all the magnitude computations, which gives you

$$\left| H\left(e^{j\omega}\right) \right|_{MAX} = \left(\frac{(\text{Factor 1})(\text{Factor 2})}{(\text{Factor 3})(\text{Factor 4})(\text{Factor 5})}\right)_{MAX} \quad \text{for } 0 \le \omega < 2\pi$$

You compute the normalized log magnitude of the frequency response by

$$\left| H\left(e^{j\omega}\right) \right|_{db} = 20 \text{Log}_{10}\left(\left| H\left(e^{j\omega}\right) \right| \Big/ \left| H\left(e^{j\omega}\right) \right|_{MAX}\right) \quad \text{for } 0 \le \omega < 2\pi$$

The example Excel spreadsheet for this problem is shown in Table 4.5. The reader does not necessarily have to spend any time studying this table. The important concept to grasp here is that by segmenting the task into smaller pieces, you can make finding the solution to tough problems simple. The beauty of this spreadsheet approach is that you can change the values for the poles and zeros and get an immediate update to the frequency response

**Table 4.5** Magnitude Tabulations for Poles and Zeros in the Recursive Filter Example

| | | Complex Poles | | Complex Zeros | | | REAL POLES | REAL ZEROS | Mag | Log 10 Mag |
|---|---|---|---|---|---|---|---|---|---|---|
| | | P1 | P2 | Z1 | | | P3 | Z2 | \|H(w)\| | |
| | R | 0.90 | 0.90 | 1.00 | | R | 0.90 | 1.00 | | |
| | Θ | 11.25 | 70.00 | 90.00 | | Θ | 0.00 | 180.00 | | |
| | a | 0.88 | 0.31 | 0.00 | | a | 0.90 | -1.00 | | |
| | b | 0.18 | 0.85 | 1.00 | | b | 0.00 | 0.00 | | |
| | 2a | 1.77 | 0.62 | 0.00 | | 2a | 1.80 | -2.00 | | |
| | $a^2+b^2$ | 0.81 | 0.81 | 1.00 | | $a^2$ | 0.81 | 1.00 | | |
| | | | | | | | | | | |
| n | ω | Mag | Mag | Mag | | ω | Mag | Mag | | |
| 0 | 0.00 | 0.04 | 1.19 | 2.0000 | | 0.00 | 0.10 | 2.00 | 751.14 | 0.00 |
| 1 | 0.10 | 0.04 | 1.19 | 1.9904 | | 0.10 | 0.14 | 2.00 | 607.19 | -1.85 |
| 2 | 0.20 | 0.04 | 1.16 | 1.9616 | | 0.20 | 0.21 | 1.99 | 415.65 | -5.14 |
| 3 | 0.29 | 0.06 | 1.12 | 1.9139 | | 0.29 | 0.30 | 1.98 | 177.66 | -12.52 |
| 4 | 0.39 | 0.12 | 1.06 | 1.8478 | | 0.39 | 0.38 | 1.96 | 75.51 | -19.95 |
| 5 | 0.49 | 0.19 | 0.98 | 1.7638 | | 0.49 | 0.47 | 1.94 | 38.49 | -25.81 |
| 6 | 0.59 | 0.28 | 0.90 | 1.6629 | | 0.59 | 0.56 | 1.91 | 22.59 | -30.44 |
| 7 | 0.69 | 0.39 | 0.79 | 1.5460 | | 0.69 | 0.65 | 1.88 | 14.72 | -34.16 |
| 8 | 0.79 | 0.50 | 0.68 | 1.4142 | | 0.79 | 0.73 | 1.85 | 10.44 | -37.14 |
| 9 | 0.88 | 0.63 | 0.55 | 1.2688 | | 0.88 | 0.82 | 1.81 | 8.01 | -39.44 |
| 10 | 0.98 | 0.78 | 0.42 | 1.1111 | | 0.98 | 0.90 | 1.76 | 6.67 | -41.03 |
| 11 | 1.08 | 0.93 | 0.29 | 0.9428 | | 1.08 | 0.98 | 1.72 | 6.12 | -41.78 |
| 12 | 1.18 | 1.09 | 0.19 | 0.7654 | | 1.18 | 1.06 | 1.66 | 5.77 | -42.29 |
| 13 | 1.28 | 1.25 | 0.20 | 0.5806 | | 1.28 | 1.13 | 1.61 | 3.23 | -47.33 |
| 14 | 1.37 | 1.42 | 0.32 | 0.3902 | | 1.37 | 1.21 | 1.55 | 1.09 | -56.77 |
| 15 | 1.47 | 1.60 | 0.48 | 0.1960 | | 1.47 | 1.28 | 1.48 | 0.30 | -68.04 |
| 16 | 1.57 | 1.78 | 0.64 | 0.0000 | | 1.57 | 1.35 | 1.41 | 0.00 | -150.00 |
| 17 | 1.67 | 1.95 | 0.82 | 0.1960 | | 1.67 | 1.41 | 1.34 | 0.12 | -76.12 |
| 18 | 1.77 | 2.13 | 0.99 | 0.3902 | | 1.77 | 1.47 | 1.27 | 0.16 | -73.40 |
| 19 | 1.87 | 2.30 | 1.16 | 0.5806 | | 1.87 | 1.53 | 1.19 | 0.17 | -72.88 |
| 20 | 1.96 | 2.46 | 1.32 | 0.7654 | | 1.96 | 1.58 | 1.11 | 0.17 | -73.14 |
| 21 | 2.06 | 2.62 | 1.48 | 0.9428 | | 2.06 | 1.63 | 1.03 | 0.15 | -73.81 |
| 22 | 2.16 | 2.78 | 1.63 | 1.1111 | | 2.16 | 1.68 | 0.94 | 0.14 | -74.70 |
| 23 | 2.26 | 2.92 | 1.77 | 1.2688 | | 2.26 | 1.72 | 0.86 | 0.12 | -75.77 |
| 24 | 2.36 | 3.05 | 1.90 | 1.4142 | | 2.36 | 1.76 | 0.77 | 0.11 | -76.97 |
| 25 | 2.45 | 3.17 | 2.02 | 1.5460 | | 2.45 | 1.79 | 0.67 | 0.09 | -78.33 |
| 26 | 2.55 | 3.27 | 2.12 | 1.6629 | | 2.55 | 1.82 | 0.58 | 0.08 | -79.85 |
| 27 | 2.65 | 3.36 | 2.21 | 1.7638 | | 2.65 | 1.84 | 0.49 | 0.06 | -81.60 |
| 28 | 2.75 | 3.44 | 2.29 | 1.8478 | | 2.75 | 1.86 | 0.39 | 0.05 | -83.68 |
| 29 | 2.85 | 3.50 | 2.35 | 1.9139 | | 2.85 | 1.88 | 0.29 | 0.04 | -86.30 |
| 30 | 2.95 | 3.54 | 2.39 | 1.9616 | | 2.95 | 1.89 | 0.20 | 0.02 | -89.90 |
| 31 | 3.04 | 3.57 | 2.42 | 1.9904 | | 3.04 | 1.90 | 0.10 | 0.01 | -95.97 |
| 32 | 3.14 | 3.58 | 2.43 | 2.0000 | | 3.14 | 1.90 | 0.00 | 0.00 | -150.00 |

plot. So it allows you to iteratively modify variables and instantly see the effects. The predicted frequency response plot for the troubled recursive filter is illustrated in Figure 4.37 for $0 \le \omega \le \pi$. This is what we would expect to see if we injected a swept digital sinusoid into the filter input, passed the filter output through a D/A converter, and then looked at the result on an analog spectrum analyzer that is displaying data in the *max hold* setting. We could also use a digital spectrum analyzer to inject pure sinusoids into the filter input, then measure and plot the magnitude of the filter output for each of the $N = 32$ frequencies that we originally used to mathematically evaluate the transfer function.

Does anyone know why we cannot inject a digital impulse into the filter, capture the impulse response $h(n)$ at the filter output, and then perform a DFT on $h(n)$ to determine the filter frequency response $H(K)$? In other words, what prevents us from performing the textbook Fourier transform pair $h(n) \overset{DFT}{\leftrightarrow} H(K)$?

The answer is we cannot do that because the filter we are dealing with is recursive and hence has an impulse response that is infinite. In order to perform a DFT, we would have to truncate the impulse response at some point, resulting in a seriously degraded spectral display. This method can work for short length finite impulse response (FIR) filters. However, this is still not a suitable technique for generating the frequency response of several FIR filters connected in cascade. Why is this? The reason is that the energy of the digital

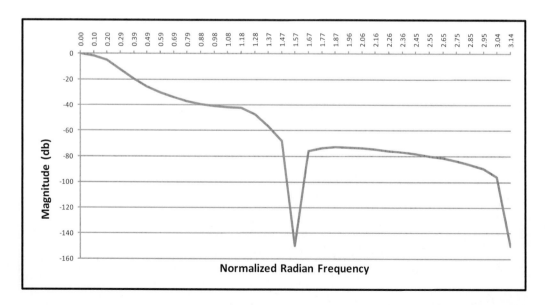

**Figure 4.37** Predicted frequency response of the recursive filter

impulse is spread across all the coefficients in the filter output. This energy is spread even further by each successive filter in the cascade. Finally, at the output of the last filter in the cascade, the power in original impulse has been spread and attenuated to the extent that it is either confined to the least significant bits (LSBs) of the cascade output or lost altogether, and hence it provides degraded results at best.

Getting back to the problem at hand, you arrive in Mozambique, armed with an idea of what the working filter's frequency response should look like. Now, when you start testing the hardware and view the spectrum analyzer measurements, you see the response you expected from your pole-zero computations, but you also see a large narrow band interfering signal at the radian frequency $\omega = 2.36$ radians/second, kind of like the one illustrated in Figure 4.38. This interference is somehow injected into and being passed by the filter side lobe, resulting in the degradation of the performance of the follow on system hardware. In the course of your testing, you determine that the sample rate of the filter is $f_s = 160$ Hz. The frequency of the narrow band interfering signal is then determined from simple mathematics:

$$\frac{f}{f_s} = \frac{\omega}{2\pi} \quad \text{or} \quad \frac{f}{160} = \frac{2.36}{6.28} \Rightarrow f = 60 \text{ Hz}$$

So it looks like the interfering signal is a 60-Hz tone that has somehow been injected into the system. The electricians that were stringing wire

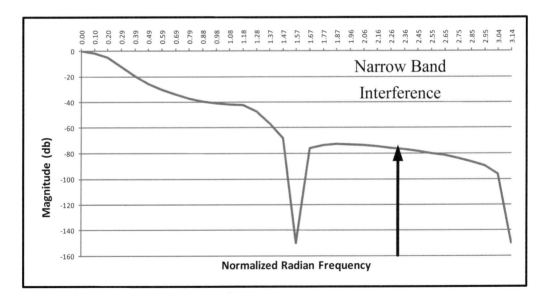

**Figure 4.38**   Spectrum of the recursive filter with narrow band interference

probably did not install power line filters, or they did not use shielded cable, or perhaps they didn't use proper grounding techniques. Who knows? It's really not your problem, but it is still imperative that the system get up and running as soon as possible. So you are still on the hook at the very least to design a temporary fix and save the day.

What do you do? Let's assume that the filter is not implemented in fixed hardware, but instead, it is implemented either in software or in a field programmable gate array (FPGA). Both of these mediums can be easily modified in the field, assuming you have the compilers readily available. If this is the case, then you are in luck. All you need to do is to add a complex zero $\left(Z_3 Z_3^*\right)$ to the filter transfer function located on the unit circle at 2.36 radians, which corresponds to 60 Hz. This will force the filter response side lobe to be zero at 60 Hz, therefore canceling out the interference, preventing it from being passed on to downstream hardware, and it will not adversely affect the oddly shaped, application-specific filter main lobe. If you can pull this off, people will celebrate and you will be a hero for a day.

The modified pole-zero diagram that includes the new complex zero is illustrated in Figure 4.39. Here we see that we added a complex zero at

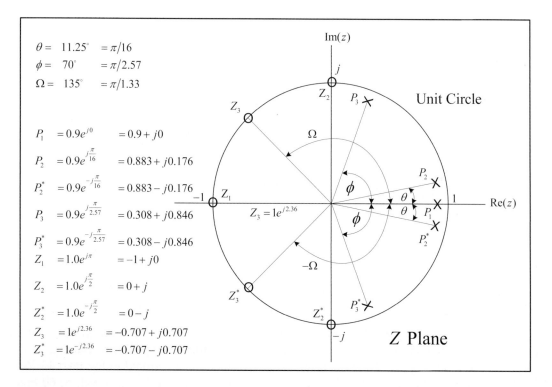

**Figure 4.39** Modified filter pole-zero diagram

$$z_3 = -0.707 + j0.707 \quad \text{and} \quad z_3^* = -0.707 - j0.707$$

which corresponds to $\Omega = 2.36$ radians, or $135°$. It also corresponds to a frequency of 60 Hz. This should place a null in the frequency response of the filter at 60 Hz, and the filter will not pass the unwanted interference. To keep your bookkeeping straight, you name the new complex zero "factor 6 " of the modified filter transfer function $H(e^{j\omega})$. You then compute its magnitude function using the same procedure as before and then enter it into a separate column in the filter Excel spreadsheet. The computations for factor 6 are as follows:

**Factor 6.** The sixth factor is a complex zero

$$\left| (z + 0.707 + j0.707)(z + 0.707 - j0.707) \right|_{z=e^{j\omega}}$$

The complex zero occurs at $z = -0.707 \pm j0.707$. As before, you identify the variables and use Equation 4.74 to generate the frequency response.

$$\left\{ \begin{array}{lll} r = & 1 & \\ \theta = & 135° & = 2.36 \text{ radian} \\ a = & -0.707 & \\ b = & 0.707 & \\ 2a = & 1.414 & \\ a^2 + b^2 = & 1 & \end{array} \right\}$$

$$\left| H(e^{j\omega}) \right| = \sqrt{\left[ \cos(2\omega) - 2a\cos(\omega) + a^2 + b^2 \right]^2 + \left[ \sin(2\omega) - 2a\sin(\omega) \right]^2}$$

$$\left| H(e^{j\omega}) \right| = \sqrt{\left[ \cos(2\omega) - 1.414\cos(\omega) + 1 \right]^2 + \left[ \sin(2\omega) - 1.414\sin(\omega) \right]^2}$$

When you add the new complex zero to the old filter transfer function, you get the new filter transfer function given by

$$\left| H(e^{j\omega}) \right|_{db} = 20\text{Log}_{10}\left( \frac{(\text{Factor 1})(\text{Factor 2})(\text{Factor 6})}{(\text{Factor 3})(\text{Factor 4})(\text{Factor 5})} \right) \quad \text{for } 0 \le \omega < 2\pi$$

Now you modify the code for the filter in the system software and/or filter FPGA to include this new complex zero and perform the spectrum analyzer measurements again. The new filter transfer function is illustrated in Figure 4.40. We see that the addition of a zero at 2.36 radians has produced a null at 60 Hz, which has effectively cancelled the interfering tone. If we had evaluated the

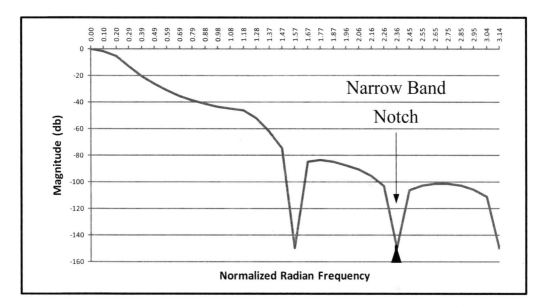

**Figure 4.40**  Interference removed by adding a transfer function complex zero

frequency response at more than 32 frequencies, we would have seen a much higher resolution plot. If we had used 128 frequencies, the notch would have been much narrower on the plot. Anyway, the temporary fix has worked and system is back up and running. Now you can fly home to a standing ovation from your peers.

We all know that this is just a band-aid fix because the source of the problem still exists. Once the source of the problem is found and is corrected by those responsible, it's a simple matter to reprogram the software/FPGA to reinstall the original filter. If you have documented your temporary design fix appropriately, any engineer or technician can be sent to the field to reverse the change. If you have not documented your fix, then you have what is termed in the trade as "job security"—at least until after you are sent back to the field to reverse the fix. Either way, your supervisor now sees you as a technical superman who can fix any tough problem that might occur in any other system the company is responsible for. The bad news is you will probably be tagged as the "go-to" guy whenever another nasty problem like this arises. Now you have to keep your bags packed.

### 4.12.3  Example Application: Digital DC Removal

At the end of the first example dealing with the digital integrator, we mentioned that the integrator could be used to approximate a digital capacitor. This was a true statement, but there are other more efficient implementations of a digital capacitor. One of these implementations is discussed next. The mathematical derivation is straightforward and identical to that of previous examples. Therefore it is left to the reader. The main interest of this example is to use a DC canceling circuit to demonstrate how the relative placement of poles and zeros affects the overall circuit frequency response. Let's begin with the z-domain transfer function of Equation 4.92 and its associated pole-zero diagram of Figure 4.41.

$$H(z) = \frac{z-1}{z-\alpha}$$

**Equation 4.92**

From the pole-zero diagram, we would expect to see a null in the frequency response at 0 radians or DC, due to the zero at $z = 1$. What makes this circuit interesting is the placement of the pole at $z = \alpha$. As $\alpha \to 1$, we would expect to see the frequency response gain at DC approach infinity. Since there is already a zero located at $z = 1$, theoretically the pole and zero will cancel each other when the two are superimposed, and the circuit gain will be unity for all frequencies. But what happens as the pole inches closer and closer to the zero? We know for sure that the gain is pinned to zero at DC, and we also know that as the pole moves closer to the zero, the gain of the circuit will tend to increase. The frequency response of the transfer function in Equation 4.92 for the case where the pole is fixed at $z = 0.5$ is illustrated in Figure 4.42. Ideally, we hope to see a frequency response that has null at DC and is flat at all other frequencies.

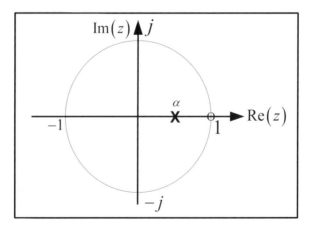

**Figure 4.41**  DC removal pole-zero diagram

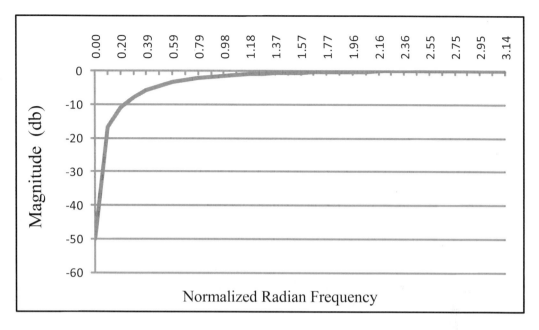

**Figure 4.42**   DC removal frequency response pole/zero at 0.5/1.0

In the figure, note the relatively large frequency response roll-off near DC. Although there is a huge attenuation at DC due to the zero placement, as we had predicted, we can see that there is also a significant attenuation on adjacent frequencies that extends well into the pass band. For example, there is a huge attenuation on the order of 10 db at 0.2 radians, and major attenuation continues to occur all the way to 1.18 radians and beyond. This means the frequency gain is attenuated all the way out to and beyond a quarter of the sample rate or $f_s/4$.

The pass band is far from being flat. Depending upon the application, this may or may not be a good thing. The odds are that this frequency response will degrade the performance of any follow on circuitry. What we would like to do is to increase the gain near DC so that we can minimize the band pass roll-off and maintain a flat gain that extends as close to DC as possible. We can increase the frequency response gain near DC by moving the pole at $z = \alpha$ closer to the unit circle such that $\alpha \rightarrow 1$.

For this example, we have exaggerated the pole placement by moving it from $z = 0.5$ to $z = 0.99$. The resulting frequency response is illustrated in Figure 4.43. The first thing we see in the figure is the absence of a long, slow roll-off of the near unity pass band. In its place, we see a sharp cutoff at approximately 0.1 radians and a flat pass band all the way out to half the sample

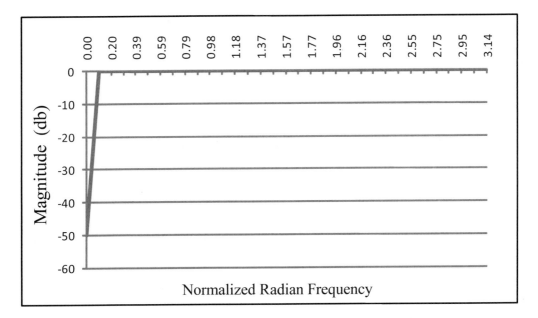

**Figure 4.43**  DC removal frequency response pole/zero at 0.99/1.0

rate, or $f_s/2$. This is quite an improvement. It should be noted that the width of the stop band at DC is shown in the plot as 0.1 radians only because this is the resolution of the plot. In actuality, the width is much narrower, as a higher resolution plot would show.

It is important to note that in the generation of this plot, the –50 db attenuation at DC (as shown in both figures) was artificially set to that value because the computation of $20\text{Log}_{10}(0)$ is impossible to compute. The arbitrary attenuation at DC shown in the plot was set to –50 db by the simple statement $\text{if}\big(\text{mag}(i) < -50\big)\ \text{mag}(i) = -50$, where $\text{mag}(i)$ is the magnitude function. This was done for a couple of reasons. First, setting the max attenuation at DC to a smaller number would give the false illusion of mediocre circuit performance. And second, setting the max attenuation at DC to a larger number would compress the plot, hiding any minute amplitude fluctuation detail in the pass band response.

As long as we have gone this far, we might as well show the circuit block diagram. Let's start with the circuit transfer function

$$H(z) = \frac{Y(z)}{X(z)} = \frac{z-1}{z-\alpha} = \frac{1-z^{-1}}{1-\alpha z^{-1}}$$

Collecting terms, we get

$$Y(z) = X(z)\left(1 - z^{-1}\right) + Y(z)\left(\alpha z^{-1}\right)$$

Using the unit delay property of z-transforms, we can, by inspection, convert between the z-domain and the discrete time domain to get

$$y(n) = x(n) - x(n-1) + \alpha y(n-1)$$

**Equation 4.93**

We can draw the circuit block diagram directly from Equation 4.93 to arrive at the diagram illustrated in Figure 4.44.

This is pretty cool stuff, but we need to stop here for a moment and inject some reality into the picture. When an engineer places poles and zeros around and within the unit circle, he needs to keep in mind that he is working within the confines of a digital system and therefore does not have infinite precision in their placement. The values of each pole and each zero will be quantized to the bit width $N$ of the digital number system used in the digital computations.

When we arbitrarily set the value of the real pole in the previous example to 0.99, the reader has to ask the following questions:

- Is the value 0.99 obtainable using $N$-bit-wide words?
- Is that value obtainable using the system floating point format?
- What is the pole/zero placement error introduced by quantization?

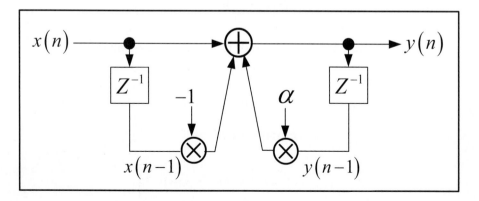

**Figure 4.44** DC removal circuit block diagram

These are great questions, and you should be proud if you asked any one of them. What if, for example, the quantization error caused the desired pole placement at 0.99 to actually be placed at 1.01? This would be disastrous because a system that is stable on paper where we have infinite precision becomes unstable when we build it in hardware or software. Why? Because we did not take into account the effects of quantization error on the pole-zero placement.

Now that we thoroughly understand the basics of z-transform theory and completely understand the powerful tool we have at our disposal, it is time to move on to bigger and better things in the following chapters. If you feel the need to fill in some blank spots in your grasp of the z-transform, there is a plethora of good information and deeper technical detail on this subject in the references listed at the end of this chapter.

## 4.13 SUMMARY OF USEFUL EQUATIONS

$$H(z) = \sum_{n=-\infty}^{\infty} h(n) z^{-n}$$

**Equation 4.1: Definition of the two-sided z-transform**

$$H(e^{j\omega}) = \sum_{n=-\infty}^{\infty} h(n) e^{-j\omega n}$$

**Equation 4.2: Definition of the Fourier transform**

$$H(z) = \sum_{n=0}^{\infty} h(n) z^{-n}$$

**Equation 4.7: Definition of the one-sided z-transform**

$$x(n-k) \overset{z}{\leftrightarrow} z^{-k} X(z)$$

**Equation 4.25: Shift**

$$x(n+m) \overset{z}{\leftrightarrow} z^{m} \left[ X(z) - \sum_{n=0}^{m-1} x(n) z^{-n} \right]$$

**Equation 4.26: Left shift of right-sided z-transform**

$$nx(n) \overset{z}{\leftrightarrow} -z\frac{d}{dz}X(z)$$

**Equation 4.27: Differentiation**

$$x(0) = \lim_{z \to \infty} X(z)$$

**Equation 4.30: Initial value theorem**

$$\lim_{n \to \infty} x(n) = \lim_{z \to 1} (z-1)X(z)$$

**Equation 4.31: Final value theorem**

$$x(n) * h(n) \overset{z}{\leftrightarrow} X(z)Y(z)$$

**Equation 4.32: Convolution of two sequences**

$$x(n)y(n) \overset{z}{\leftrightarrow} \frac{1}{2\pi j} \oint_C X\left(\frac{z}{v}\right) Y(v)v^{-1}dv$$

**Equation 4.34: Multiplication of two sequences**

$$\left|H(e^{j\omega})\right| = \sqrt{1 + a^2 - 2a\cos(\omega)}$$

**Equation 4.70: Real pole or zero plug ins**

$$\left|H(e^{j\omega})\right| = \sqrt{\left[\cos(2\omega) - 2a\cos(\omega) + a^2 + b^2\right]^2 + \left[\sin(2\omega) - 2a\sin(\omega)\right]^2}$$

**Equation 4.74: Complex pole or zero frequency response plug in**

## 4.14  REFERENCES

[1] Alan V. Oppenheim and Ronald W. Schafer. *Digital Signal Processing*. Englewood Cliffs, NJ: Prentice Hall, 1975.

[2] Alan V. Oppenheim and Ronald W. Schafer. *Discrete Time Signal Processing*. Englewood Cliffs, NJ: Prentice Hall, 1989.

[3] Douglas F. Elliott. *Handbook of Digital Signal Processing Engineering Applications*. San Diego, CA: Academic Press, 1987.

[4] Alan V. Oppenheim, Alan S. Willsky, and Ian T. Young. *Signals and Systems*. Englewood Cliffs, NJ: Prentice Hall, 1983.

[5] Adrian Banner. *The Calculus Lifesaver*. Princeton, NJ: Princeton University Press, 2007.

[6] Charles Williams. *Designing Digital Filters*. Englewood Cliffs, NJ: Prentice Hall, 1986.

[7] Emmanuel C. Ifeachor and Barrie W. Jervis. *Digital Signal Processing: A Practical Approach*. Workingham, UK: Addison-Wesley, 1993.

# CHAPTER FIVE

# Finite Impulse Response Digital Filtering

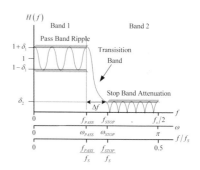

This chapter focuses on the design of finite impulse response (FIR) digital filters. It is not the author's intent to repeat all the excellent theoretical material
that has already been published by so many astute authors. Almost all of the
digital signal processing (DSP) texts devote substantial coverage to the history, theory, architecture, and mathematics of digital filters. A good selection
of these texts is listed at the end of this chapter in the reference list. The intent
here is to concentrate solely on a method for the design and implementation
of some of the more common filter types.

The purpose of this chapter is twofold: First, in order to establish a communication baseline, we will provide a very brief overview of digital filters.
Second, we will demonstrate a design methodology to implement several
types of digital filters.

There are two types of digital filters that are utilized most frequently in
industry: *finite impulse response* (FIR) filters and *infinite impulse response* (IIR)
filters. Over the years there have been many methods developed for the design of these types of filters. All methods have their advantages and disad-
vantages (and are documented extensively in this chapter's reference list).

Over the last several decades, engineers, scientists, and countless other
smart people have devoted an immeasurable amount of time and effort and
even whole careers toward the development of methods, procedures, and architectures for the design of IIR filters. Several methods have been developed,
such as the bilinear transform, impulse invariance, and the pole-zero placement method, to name just a few.

A great deal of research and work has also been devoted to the analysis,
design methods, and procedures for FIR filters. Two of the most familiar design
techniques for FIR filters are the tried-and-true window method, documented

in Ifeachor and Jervis [3], and the frequency sampling method, documented in Rabiner and Gold [9]. Both methods produce quality filters. Understanding the underlying concepts can provide a considerable amount of insight into the design of FIR filters. The reader who seeks an in-depth treatment of these two methods should consult the previously mentioned sources.

Although the window and frequency sampling methods were powerful and extremely useful in the early days of DSP, they are not the preferred methods in use today. The window method has the drawback that the cutoff frequencies cannot be precisely specified, and the frequency sampling method can become mathematically complex. For these reasons, many design engineers today consider these methods legacy techniques. More modern and efficient computer-aided FIR design methods such as the Parks-McClellan algorithm have since been developed and have become the industry standard. We will not deal with IIR filters in this book. Instead, we will devote our time to the modern day procedure for the design of FIR digital filters. We will concentrate our attention on the one technique that is currently in vogue today—the Parks-McClellan method of designing linear phase FIR filters—and we will present a step-by-step methodology for the design of these digital filters. In addition, for the reader's convenience, a full listing of the modified Parks-McClellan FORTRAN program along with a detailed description of how to compile and link it into a C/C++ main program is presented in Appendix A, "Mixed Language C/C++ FORTRAN Programming." The modifications to this program include converting the program into a FORTRAN subroutine and replacing the original program I/O with variables that are passed back to the C/C++ main through the call list.

## 5.1  REVIEW OF DIGITAL FIR FILTERS

The block diagram of a direct form FIR filter architecture is illustrated in Figure 5.1. The architecture is just a simple delay line that inputs a digitized data sequence $x(n)$. There is a multiplier that inputs the sequence of samples output by each stage in the delay line. Each multiplier forms the product of a delayed input signal sample $x(n-k)$ and a filter coefficient $h(k)$. The sample delay index is given by $k = 0,1,2,\cdots N-1$, where $N$ is the length of the filter. The individual products are summed to form the output sequence $y(n)$. The output $y(n)$ can be thought of as the dot product between the two vectors $x(n-k)$ and $h(k)$.

The discrete time equation that describes a FIR filter is

$$y(n) = \sum_{k=0}^{N-1} h(k)x(n-k) \quad n = 0,1,2,3,\cdots$$

**Equation 5.1**

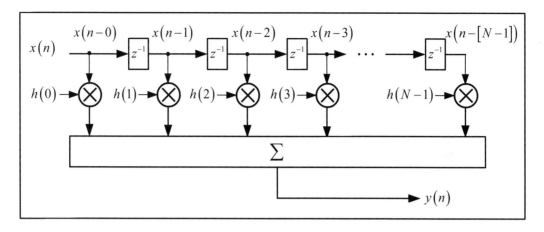

**Figure 5.1**   Direct form architecture of a FIR digital filter

Given the filter impulse response coefficients $h(k)$, we can determine the corresponding FIR filter transfer function equation:

$$H(z) = \sum_{k=0}^{N-1} h(k)\, z^{-k}$$

**Equation 5.2**

The transfer function equation is the z-transform of the FIR filter impulse response and is useful for analyzing the filter in the frequency domain. Some of the attributes of the FIR filter are as follows:

- The FIR is an all-zero no-pole filter.
- The absence of poles guarantees that this type of filter is always stable.
- The FIR filter can have perfectly linear phase and constant group delay.
- Depending on the length of the filter, the group delay for a FIR can be large.
- The FIR filter is less susceptible to finite word length effects than IIR filters.
- The transition bandwidth of a FIR varies inversely with the filter length $N$.
- The FIR architecture provides an efficient implementation of multirate filters.
- FIR filters are simple to implement.

To further comment on the last attribute, all DSP processors and most programmable logic devices such as field programmable gate arrays (FPGAs) have architectures that can efficiently realize FIR filters. In addition, the software development tools associated with FPGAs and application-specific integrated circuits (ASICs) generally contain scalable component libraries in support of various DSP architectures, which include FIR filters.

If the input sequence to a FIR filter is the unit impulse response defined as

$$\Delta(n) \quad = 1 \ \text{for } n = 0$$
$$= 0 \ \text{for } n \neq 0$$

then, as the impulse is clocked through the filter delay line, the filter output $y(n)$ will take on the sequence of values equal to the filter coefficients $h(k)$ for $k = 0, 1, 2, 3, \cdots, N-1$, followed by zeros thereafter. The filter coefficients therefore define the impulse response of the filter.

## 5.1.1 Linear Phase

One of the most powerful attributes of a FIR filter is that it can be designed with a perfectly linear phase response. When an input sequence $x(n)$ passes through a filter, its amplitude and phase are shaped by the amplitude and phase response characteristics of that filter. The frequency response of a FIR filter can be written as the product of its magnitude and phase components:

$$H\left(e^{j\omega}\right) = \left|H\left(e^{j\omega}\right)\right| e^{j\theta(\omega)}$$

The term $e^{j\theta(\omega)}$ is the phase response of the filter, where $\theta(\omega) = a - b\omega$, and both $a$ and $b$ are constants. These constants are defined by the filter being analyzed, as we will see later. The phase delay $P_D$ of a filter is the time delay that each frequency component within a composite signal experiences as it passes through the filter. The group delay $G_D$ of a filter is the average time delay of the composite signal as it passes through the filter. The definitions of phase delay and group delay are given by

$$P_D = -\theta(\omega)/\omega$$

**Equation 5.3**

$$G_D = -\frac{d\theta(\omega)}{d\omega}$$

**Equation 5.4**

It can be shown that a filter has a linear phase response if its phase response satisfies one of the two relationships, given in Equation 5.5 and Equation 5.6 [3],[9]:

$$\theta(\omega) = -b\omega$$

**Equation 5.5**

$$\theta(\omega) = a - b\omega$$

**Equation 5.6**

If the phase response of the filter meets the requirement of Equation 5.5, it will have a constant phase delay and a constant group delay [3],[9]. If this is the case, the filter impulse response will have even symmetry and will take the form of

$$h(n) = h(N-1-n) \quad \text{for} \quad \left\{ \begin{array}{ll} n = 0,1,2,\cdots(N-1)/2 & \text{N odd} \\ n = 0,1,2,\cdots(N/2-1) & \text{N even} \end{array} \right\}$$

where

$$b = \frac{(N-1)}{2}$$

**Equation 5.7**

If the phase response of the filter meets the requirement of Equation 5.6, it will only have a constant group delay [3],[9]. If this is the case, the filter impulse response will have odd symmetry and will take the form of

$$h(n) = -h(N-1-n) \quad \text{for} \quad \left\{ \begin{array}{ll} n = 0,1,2,\cdots(N-1)/2 & \text{N odd} \\ n = 0,1,2,\cdots N/2-1 & \text{N even} \end{array} \right\}$$

where $a = \dfrac{\pi}{2}$ and $b = \dfrac{(N-1)}{2}$.

## 5.1.2   Filter Types

The results of the previous section are important in that they define of the conditions that must be satisfied in order to design a linear phase FIR filter. The most important condition is the even or odd symmetry of the filter co-efficient vector $h(n)$. There are exactly four types of linear phase FIR filter architectures. These four are identified by the even/odd number of filter coefficients and by the even/odd symmetry of the filter impulse response.

The previous results also serve as the seed for the derivation of the mathematical form that describes these four filter types. The derivation is well documented in references [3] and [9] and is not repeated here. The result of that work, however, is presented in Table 5.1. The table shows the four possible types of linear phase FIR filters and their attributes.

The terms $\alpha(n)$ and $\beta(n)$ in Table 5.1 are the filter coefficients defined by

$$\alpha(0) = h(N-1)/2$$
$$\alpha(n) = 2h\left[(N-1)/2 - n\right]$$
$$\beta(n) = 2h(N/2 - n)$$

There is no need to memorize all the squiggly lines in Table 5.1. The key piece of information the reader should take away from the table is the relationship between the odd and even coefficients, the symmetry of the impulse response, and the critical zeros for each of the four types of linear phase filters. These are the important attributes that dictate the kind of linear phase FIR filters that can be implemented with each filter type.

The most general of the four filter types is filter type 1. It is implemented with an odd number of coefficients and has an even symmetric impulse

**Table 5.1**   The Four Types of Linear Phase FIR Filters

| Filter type | Number of coefficients | Frequency response $H(\omega)$ | Impulse response symmetry | Critical zeros |
|---|---|---|---|---|
| 1 | Odd | $e^{-j\omega(N-1)/2} \sum\limits_{n=0}^{(N-1)/2} \alpha(n)\cos(\omega n)$ | $h(n) = h(N-1-n)$ <br> *even* | None |
| 2 | Even | $e^{-j\omega(N-1)/2} \sum\limits_{n=1}^{N/2} \beta(n)\cos\left[\omega(n-1/2)\right]$ | $h(n) = h(N-1-n)$ <br> *even* | $f = 0.5$ |
| 3 | Odd | $e^{-j\left[\omega(N-1)/2 - \pi/2\right]} \sum\limits_{n=1}^{(N-1)/2} \alpha(n)\sin(\omega n)$ | $h(n) = -h(N-1-n)$ <br> *odd* | $f = 0.0$ <br> $f = 0.5$ |
| 4 | Even | $e^{-j\left[\omega(N-1)/2 - \pi/2\right]} \sum\limits_{n=1}^{N/2} \beta(n)\sin\left[\omega(n-1/2)\right]$ | $h(n) = -h(N-1-n)$ <br> *odd* | $f = 0.0$ |

response. A type 1 filter is suitable for the implementation of low pass, band pass, band stop, and high pass filters.

A type 2 filter is implemented with an even number of coefficients and has an even symmetric impulse response. A type 2 filter has a zero located at $\omega = \pi$, or $f = 0.5$, and therefore cannot be used for the implementation of high pass or band stop filters. This can be seen from the term $\cos\left[\omega(n-1/2)\right] = \cos\left[2\pi f(n-1/2)\right]$. When the normalized frequency is $f = 0.5$, the cosine argument reduces to $(\pi n - \pi/2)\mod 2\pi = \pm\pi/2$ for $n = 0,1,2,\cdots$, which of course reduces the cosine term to zero. Therefore $H(\omega)\big|_{\omega=\frac{\pi}{2},\ f=0.5} = 0$.

A type 3 filter is implemented with an odd number of filter coefficients and has an odd symmetric impulse response. A type 3 filter response has a 90° phase shift, and it has a zero located at both $f = 0$ and $f = 0.5$. This type of filter is not suited for either low pass, band stop, or high pass design. Because of the 90° phase shift, a type 3 filter is more suited for the implementation of differentiators or Hilbert transforms.

A type 4 filter is implemented with an even number of filter coefficients and has an odd symmetric impulse response. This type of filter also implements a 90° phase shift. It has a zero at $f = 0$, making it unsuited for the implementation of low pass and band stop filters. As is the case with a type 3 filter, a type 4 filter is better suited for the implementation of differentiators or Hilbert transforms.

### 5.1.3   Linear Phase FIR Filter Direct Form Architectures

We saw in the direct form FIR architecture of Figure 5.1 that $N$ coefficient multiplies were required for a filter of length $N$. If the filter has been designed for linear phase, then we can take advantage of the fact that the coefficient vector $h(n)$ is symmetric. That is, the filter impulse response $h(n) = h(N-1-n)$ has

$$N/2 \qquad\qquad \text{unique coefficients for N even}$$
$$(N-1)/2 \ + \ 1 \quad \text{unique coefficients for N odd}$$

Compared to the general direct form case, the linear phase FIR only requires half the coefficient memory and only needs to perform half the coefficient multiplies. The two architectures for the even/odd coefficient cases are shown in the following sections.

### 5.1.3.1   *Even Coefficient Architecture for Type 2 and Type 4 Filters*

The equation for the direct form architecture in Figure 5.1 is given as

$$y(n) = \sum_{k=0}^{N-1} h(k)x(n-k) \quad n = 0,1,2,3,\cdots$$

**Equation 5.8**

A type 2 filter has an even number of coefficients and an even symmetric impulse response and is given by

$$h(n) = h(N-1-n)) \quad \text{for } n = 0,1,2,\cdots N/2 - 1$$

If we insert the even symmetric impulse response into Equation 5.8, the equation for a type 2 FIR filter is

$$y(n) = \sum_{k=0}^{(N/2)-1} h(k)\{x[n-k] + x[n-(N-1-k)]\} \quad \text{for } n = 0,1,2,3,\cdots$$

**Equation 5.9**

A type 4 filter has an even number of coefficients and an odd symmetric impulse response and is given by

$$h(n) = -h(N-1-n)) \quad \text{for } n = 0,1,2,\cdots N/2 - 1$$

If we insert the odd symmetric impulse response into Equation 5.8, the equation for a type 4 FIR filter is

$$y(n) = \sum_{k=0}^{(N/2)-1} h(k)\{x[n-k] \ - \ x[n-(N-1-k)]\} \quad \text{for } n = 0,1,2,3,\cdots$$

**Equation 5.10**

The block diagram of the direct form implementation of a type 2 or type 4 linear phase FIR filter is illustrated in Figure 5.2. Note that coefficient symmetry has reduced the number of multiplies by a factor of two.

### 5.1.3.2   Odd Coefficient Architecture for Type 1 and Type 3 Filters

The equation for the direct form architecture in Figure 5.1 is repeated here for clarity:

$$y(n) = \sum_{k=0}^{N-1} h(k)x(n-k) \quad n = 0,1,2,3,\cdots$$

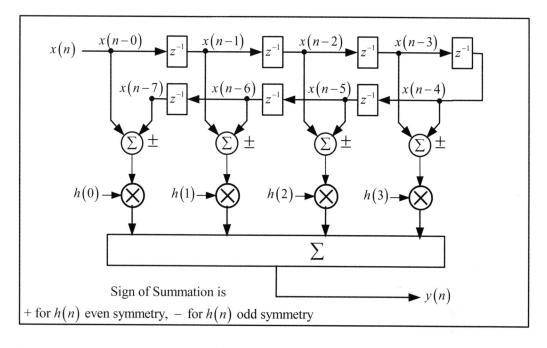

**Figure 5.2**   Direct form even coefficient FIR digital filter architecture

A type 1 filter has an odd number of coefficients and an even symmetric impulse response and is given by

$$h(n) = h(N-1-n) \text{ for } n = 0,1,2,\cdots(N-1)/2-1, \text{ and } h\left(\frac{N-1}{2}\right) \text{ for } n = (N-1)/2$$

If we insert the even symmetric impulse response into Equation 5.8, the equation for a type 1 FIR filter is

$$y(n) = h\left(\frac{N-1}{2}\right)x\left(n-\frac{N-1}{2}\right) + \sum_{k=0}^{\frac{N-1}{2}-1} h(k)\{x[n-k]+x[n-(N-1-k)]\} \text{ for } n = 0,1,2,3,\cdots$$

**Equation 5.11**

A type 3 filter has an odd number of coefficients and an odd symmetric impulse response and is given by

$$h(n) = -h(N-1-n) \text{ for } n = 0,1,2,\cdots(N-1)/2-1, \text{ and } h\left(\frac{N-1}{2}\right) \text{ for } n = (N-1)/2$$

If we insert the odd symmetric impulse response into Equation 5.8, the equation for a type 3 FIR filter is

$$y(n) = h\left(\frac{N-1}{2}\right)x\left(n - \frac{N-1}{2}\right) + \sum_{k=0}^{\frac{N-1}{2}-1} h(k)\left\{x[n-k] - x[n-(N-1-k)]\right\} \quad \text{for } n = 0,1,2,3,\cdots$$

**Equation 5.12**

The block diagram of the direct form implementation of a type 1 or type 3 linear phase FIR is illustrated in Figure 5.3.

## 5.2  PARKS-MCCLELLAN METHOD OF FIR FILTER DESIGN

In December 1973, Thomas W. Parks and James H. McClellan, while graduate students at Rice University, published a method for designing optimum linear phase FIR filters using the Remez exchange algorithm. This method, called the Parks-McClellan algorithm, rapidly became the preferred method of designing FIR filters and is still an industry standard today. Along with their algorithm, they published a FORTRAN program that input a set of user-defined

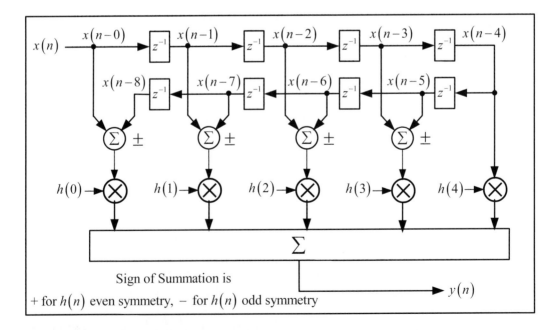

**Figure 5.3**   Direct form odd coefficient FIR digital filter architecture

filter specifications, computed the filter impulse response coefficients, and then returned these coefficients to the user. In this section, we will illustrate the design of FIR filters using this Parks-McClellan algorithm.

A copy of this Parks-McClellan FORTRAN program, modified to encapsulate it in the form of a FORTRAN subroutine, is included Appendix A, "Mixed Language C/C++ FORTRAN Programming." The routine has also been modified to eliminate the old FORTRAN WRITE and FORMAT statements. The routine output is now achieved by passing variables back to the C/C++ main through the call list for standard C/C++ disc file output.

Since C/C++ seems to be the preferred language in use today, it may seem odd that a FORTRAN program would be discussed in this book. The reasoning is that this program is so important it cannot be ignored. It makes sense to integrate this program into any DSP code written in C/C++ that computes the coefficients of FIR filters. Therefore Appendix A of this book includes a section that discusses the implementation of a mixed-language C/C++ / FORTRAN program. In particular, the discussion includes the procedure for calling the Parks-McClellan FORTRAN subroutine from a C/C++ main program. Details are provided for calling, argument passing, compiling, and linking a mixed language C/C++ / FORTRAN program. In addition, code is also provided in Appendix A for the C/C++ main that calls the Parks-McClellan subroutine, passes user-defined input variables, and processes the routine outputs to disk file.

From this point on, where it is convenient, the abbreviation PM will be used to refer to the longer term "Parks-McClellan." The intent of this book is to deal with DSP applications. It is not in the scope of this book to give a detailed analysis of the PM algorithm. Most of the references at the end of this chapter provide in-depth discussions on this subject. Rather than repeat the excellent work of these authors, it is the intention here to introduce the PM algorithm and provide detailed examples of how to use it.

### 5.2.1   Parks-McClellan Filter Design Procedure

Suppose we wish to design the coefficients $h(n)$ of a FIR low pass digital filter whose frequency response $H(f)$ conforms to some tolerance specification like the one illustrated in Figure 5.4. For convenience, the frequency axis in the figure is shown calibrated in terms of frequency $f$, radians $\omega$, and in normalized frequency $f/f_S$, where $f_S$ is the sample rate.

We specify the filter pass band to extend from 0 to $f_{PASS}$ Hz and the filter stop band to extend from $f_{STOP}$ to $f_S/2$ Hz. We should expect to see some ripple in both bands. Ideally we wish to approximate a value of unity for $H(f)$ in the filter pass band with a maximum error given by

$$\varepsilon(f) \le |\delta_1| \quad \text{for } 0 \le f \le f_{PASS}$$

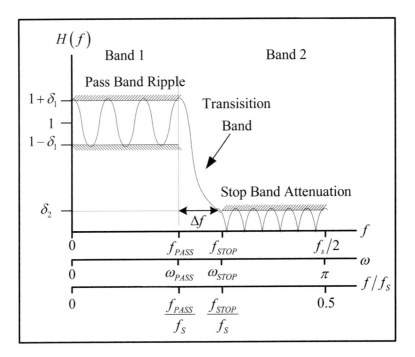

**Figure 5.4**   Tolerance specification for a low pass filter approximation

Therefore the filter response within the pass band is constrained to fall within the boundary

$$(1 - \delta_1) \le H(f) \le (1 + \delta_1) \quad \text{for } 0 \le f \le f_{PASS}$$

It is also desired to approximate a value of zero for the stop band of $H(f)$, with a maximum error given by $\varepsilon(f) \le \delta_2$ for $f_{STOP} \le f \le f_S/2$. The frequency response tolerance of this desired filter is shown in Figure 5.4. The ripple in the pass band oscillates between $1 - \delta_1$ and $1 + \delta_1$. In the stop band, the filter response attenuation lies between 0 and $\delta_2$. The difference between the ideal filter and the realizable filter can be represented by an error function

$$\varepsilon(f) = Wtx(f)\left[H(f)_{Desired} - H(f)_{Realizable}\right]$$

**Equation 5.13**

where $H(f)_{Desired}$ is the ideal or desired frequency response, $H(f)_{Realizable}$ is the actual or realizable frequency response, and $Wtx(f)$ is a weighting function that allows the designer to assign relative importance of the approximation error for different frequency response bands. The objective is to derive the filter coefficients $h(n)$ so that the value of the error $|\varepsilon(f)|$ in Equation 5.13 is minimized over all frequency pass bands and stop bands.

All the information we need to design a PM filter either is defined by or can be derived from the parameters illustrated in the tolerance specification in Figure 5.4. So that we may keep our variable bookkeeping straight, and so that we may avoid any additional modification to the PM algorithm, we will define the argument notation used by the PM subroutine. The PM FORTRAN subroutine call statement refers to the desired filter response $H(f)_{Desired}$ as FX, the desired weighting $Wtx(f)$ as WTX, the output array of realizable filter coefficients $H(f)_{Realizable}$ as H, and the number of coefficients $N$ as NFILT. The user does not have to input a great deal of information into the PM subroutine. The unmodified PM subroutine argument list is given by

```
SUBROUTINE PARK(NFILT,JTYPE,NBANDS,LGRID,EDGE,FX,WTX,H)
```

The modified PM subroutine argument list, as presented in Appendix A, "Mixed Language C/C++ FORTRAN Programming," is given by

```
SUBROUTINE PARK(NFILT,JTYPE,NBANDS,LGRID,EDGE,FX,WTX,H,
ENFCNS,NEG,NODD,NZ,KUP,EDEV,STATUS)
```

The difference between the unmodified and modified call list is that the unmodified PM subroutine dates back to 1973. At that time, the subroutine was a FORTRAN program, and its program outputs were to a line printer via standard FORTRAN WRITE and FORMAT statements. The modified PM subroutine has eliminated the old-style FORTRAN outputs in favor of passing output variables back to a C/C++ main program through the call list. The code for the C/C++ main that calls the modified PM subroutine and uses these extra variables in the call list is presented in Appendix A. For the purposes of this chapter, we will only deal with the variables in the unmodified call list because all the arguments, with the exception of the filter coefficient array H, are subroutine inputs used to specify the filter performance.

The filters are designed in this chapter using a C main program written in Visual C/C++ that runs in Microsoft Windows. The C main is written as a mixed language program that allows it to call the modified FORTRAN version of the PM subroutine. Windows is chosen because of the need to provide a graphical user interface (GUI) and because of the need to generate screen plots of the resultant filters. The definition of each argument in the unmodified call list is given in Table 5.2.

The filter design procedure is fairly simple and is itemized as follows:

1. Determine the filter sample rate $f_s$.
2. Specify all the frequency response band edges.

3. With the information from step 1 and step 2, compute the normalized frequency for each band edge. For example, for a simple low pass filter, this would consist of a pass band extending from 0.0 to a normalized pass band cutoff frequency $f_{PASS}/f_S$ and a normalized stop band frequency $f_{STOP}/f_S$ extending to $(f_S/2)/f_S$ or 0.5.

4. Specify the magnitude in decibels of the pass band ripple and the stop band attenuation.

**Table 5.2**  PM Algorithm Subroutine Arguments

| Argument | Type | Definition |
|---|---|---|
| NFILT | Integer | Specifies the length of the filter. It is the number of filter coefficients to compute. |
| JTYPE | Integer | Specifies the type of filter to design<br>1 = multiple pass band filter, (low pass, high pass, band pass, and band stop)<br>2 = differentiator<br>3 = Hilbert transformer |
| NBANDS | Integer | Specifies the number of filter bands. For example, a low pass filter would have two bands consisting of a pass band and a stop band. A band pass filter would have three bands consisting of a pass band and two stop bands and so on. |
| LGRID | Integer | Specifies the algorithm grid density. It defaults within the program to a value of 16. Typically a value of 16 or 32 is used. LGRID is the number of frequency points at which the algorithm checks for optimal error conditions that exist in the filter frequency response. |
| EDGE | Float array | An array that specifies the band edges of each pass band and stop band. Each edge is expressed in normalized frequency. The number of EDGE entries is equal to twice the number of bands or 2*NBANDS. |
| FX | Float array | An array that specifies the desired response for each band. Typical inputs are 1.0 for each pass band and 0.0 for each stop band. |
| WTX | Float array | An array that specifies the weight associated with each band. The weight is calculated based upon the $\delta_1, \delta_2, \delta_3 \cdots$ filter specifications. |
| H | Double precision array | An array of filter coefficients output from the subroutine. |

5. With the information from step 4, compute the values for $\delta_1, \delta_2 \cdots$.
6. With the information from step 5, estimate of the number of required filter coefficients.
7. With the information from step 5, compute the band weights.
8. Determine the desired filter response for each band. Usually this is as simple as assigning a 1 for a pass band and a 0 for a stop band.
9. Enter all these data into the argument list of the PM subroutine, call the routine, and compute the optimum set of filter impulse response coefficients.

Initially this design procedure may seem a bit intimidating, but it is very straightforward, and after a couple of filter designs, it will become second nature. All the steps in the procedure are discussed in detail the next section, but first we need to clarify steps 4, 5, 6, and 7.

In step 4 of the procedure, we want to specify the value in decibels of the pass band ripple error and the stop band attenuation. In step 5, we wish to use these values to compute the error bounds defined by $\delta_1, \delta_2 \cdots$. As an example, for the design of a two-band low pass filter, let's define the term *PBR* to represent the magnitude in decibels of the filter pass band ripple. Likewise, let's define the term *SBA* to represent the magnitude in decibels of the filter stop band attenuation. PBR and SBA are given by the following equations:

$$\text{PBR}\,(db) = 20\text{Log}_{10}\left(1+\delta_1\right)$$

**Equation 5.14**

$$\text{SBA}\,(db) = -20\text{Log}_{10}\left(\delta_2\right)$$

**Equation 5.15**

When we initially lay out the filter design specifications, we will assign desired values for PBR and SBA. Once we have decided on a value for PBR, we can rearrange the terms in Equation 5.14 to compute the value of $\delta_1$:

$$\delta_1 = 10^{\frac{\text{PBR}}{20}} - 1$$

**Equation 5.16**

Similarly, we can compute the value of $\delta_2$ by rearranging the terms in Equation 5.15 to get

$$\delta_2 = 10^{-\frac{SBA}{20}}$$

**Equation 5.17**

Once we know the values of $\delta_1$ and $\delta_2$, we can compute the estimated number of coefficients $N$ in step 6 and both the band weights in step 7. The weights for each of the two bands in the low pass frequency response are computed by

$$Wtx(f) = \begin{cases} \dfrac{\delta_1}{\delta_1} & \text{for } 0 \leq f \leq f_{PASS} \\ \dfrac{\delta_1}{\delta_2} & \text{for } f_{STOP} \leq f \leq f_s/2 \end{cases}$$

The deviation $\delta_1$ is usually the error in the filter pass band. In general, the weights for all bands in a multiband filter are given by the ratio of

$$Wtx(f) = \begin{cases} \dfrac{\delta_1}{\delta_1} & \text{Band 1} \\ \dfrac{\delta_1}{\delta_2} & \text{Band 2} \\ \vdots & \vdots \\ \dfrac{\delta_1}{\delta_L} & \text{Band } L \end{cases}$$

**Equation 5.18**

where $\delta_1$ is associated with the ripple specification of one of the filter pass bands. The reader should note that this procedure for deriving the weight values is not set in stone. The reader is by no means restricted to using this method. Other means of computing the weight values can be employed as the filter design application dictates.

The approximate number of coefficients required by the filter is obtained by using the following estimate presented by Kaiser [7]:

$$N = \frac{-20\text{Log}_{10}\left(\sqrt{\delta_1\delta_2}\right) - 13}{14.6\,\Delta f} + 1$$

**Equation 5.19**

where $\Delta f = f_{STOP} - f_{PASS}$ is the width of the transition band between the pass band and the stop band. Equation 5.19 is a fairly accurate first estimate for the number of filter impulse response coefficients. After the filter is designed, the number of coefficients may be adjusted upward or downward and the PM algorithm run again to iteratively achieve optimum performance. In most cases, only one additional iteration is ever necessary. Armed with this information, we will design some example filters.

## 5.2.2  PM Low Pass FIR Filter Design

Referring to the tolerance specification drawing in Figure 5.4, let's assume that our design specifications are the following:

- The filter sample rate is $f_S = 128$ KHz.
- The pass band extends from 0 Hz to 25.6 KHz.
- The stop band extends from 38.4 KHz to 64 KHz.
- The pass band ripple (PBR) error is chosen to be PBR = 0.01 db.
- The stop band attenuation (SBA) is chosen to be SBA = 60 db.
- The transition bandwidth $\Delta f = f_{STOP} - f_{PASS} = 12.8$ KHz.
- The hardware that implements the filter will utilize 16-bit integer coefficients.

**Step 1**. Determine the filter sample rate. A good percentage of the time, the sample rate is predetermined by system specifications. Other times, for more complex filters, the designer may have to specify a particular sample rate. In our example, the sample rate has been set at the system level to be $f_S = 128$ KHz.

**Step 2**. Specify the frequency response band edges. The band edges are pre-specified to be 0 Hz, 25.6 KHz, 38.4 KHz, and 64 KHz.

**Step 3**. Compute the normalized frequency response band edges. The normalized band edges are computed to be

$$\frac{0 \text{ Hz}}{128 \text{ KHz}}, \quad \frac{25.6 \text{ KHz}}{128 \text{ KHz}}, \quad \frac{38.4 \text{ KHz}}{128 \text{ KHz}}, \quad \frac{64 \text{ KHz}}{128 \text{ KHz}} = 0.0, 0.2, 0.3, 0.5$$

The normalized transition bandwidth $\Delta f$ is 12.8 KHz / 128 KHz = 0.1.

**Step 4**. Specify the pass band ripple and stop band attenuation. The PBR is specified to be 0.01 db. The SBA is specified to be 60 db.

**Step 5**. Compute the pass band and stop band error $\delta_1$ and $\delta_2$. We use Equation 5.16 to compute $\delta_1$:

$$\delta_1 = 10^{\frac{PBR}{20}} - 1$$
$$= 10^{\frac{0.01}{20}} - 1$$
$$= 0.0012$$

We use Equation 5.17 to compute $\delta_2$:

$$\delta_2 = 10^{-\frac{SBA}{20}}$$
$$= 10^{-\frac{60}{20}}$$
$$= 0.001$$

**Step 6**. Compute the estimated number of filter coefficients $N$. We use Equation 5.19 to compute the estimated value of $N$:

$$N = \frac{-20\mathrm{Log}_{10}\left(\sqrt{\delta_1 \delta_2}\right) - 13}{14.6\, \Delta f} + 1$$

$$= \frac{-20\mathrm{Log}_{10}\left(\sqrt{(0.0012)(0.001)}\right) - 13}{(14.6)\,(0.1)} + 1$$

$$= 32.65$$

For our first estimate, we will round the number of coefficients up to an even value of 34. The PM routine will produce an even symmetry impulse response. Since we specified an even number of coefficients, this will be a type 2 filter with a zero at $f = 0.5$, which is suited for low pass and band pass designs. If we had specified an odd number of coefficients, the filter would have been a type 1 filter, suitable for low, band, and high pass designs. Why? Because a type 1 filter has no inherent zeros in the frequency response.

**Step 7**. Compute the band weights. We use Equation 5.18 to compute the band weights $Wtx(f)$:

$$Wtx(f) = \begin{cases} \dfrac{\delta_1}{\delta_1} & \text{for } 0 \le f \le f_{PASS} \\[2ex] \dfrac{\delta_1}{\delta_2} & \text{for } f_{STOP} \le f \le f_s/2 \end{cases} = \begin{cases} \dfrac{0.0012}{0.0012} & \text{Band 1} \\[2ex] \dfrac{0.0012}{0.001} & \text{Band 2} \end{cases} = \begin{cases} 1 \\ 1.2 \end{cases}$$

**Step 8**. Determine the desired filter response. We want the ideal or desired response of the pass band to equal 1 and the ideal or desired response of the stop band to equal 0. So we set the desired response array $H(f)_{Desired} = FX[\ ] = 1, 0$.

**Step 9**. Enter the data into the PM program and compute the coefficients. We enter the following data into the PM algorithm argument list:

- **NFILT = 34** [Number of coefficients to compute = 34]
- **JTYPE = 1** [Filter Type = 1 for multi band filter]
- **NBANDS = 2** [Number of Bands = 2]
- **LGRID = 16** [Grid Size = 16]
- **EDGE = {0.0, 0.2, 0.3, 0.5}** [Band Edges = 0.0, 0.2, 0.3, 0.5]
- **FX = {1,0}** [Desired Response = 1, 0]
- **WTX = {1, 1.2 }** [Weights = 1, 1.2]
- Call the PM Routine
  CALL PARK(NFILT,JTYPE,NBANDS,LGRID,EDGE,FX,WTX,H)
- Read the **NFILT** computed coefficients from array **H**

The PM algorithm inputs these data and returns with a set of 34 double-precision floating point filter coefficients representing the impulse response of the filter. The PM-computed double-precision coefficients are then post-processed to scale them to 16-bit integers. The frequency response is computed by passing the array of 16-bit coefficients through a discrete Fourier transform (DFT). The frequency response of the specified filter is illustrated in Figure 5.5. It is interesting to see that there is a frequency zero located at $f_s/2$, which is expected for a type 2 filter.

The same routine that plots the frequency response shown in the figure also supports the placement of scrolling horizontal and vertical cursors for accurate parameter measurement purposes. The frequency response of Figure 5.5 is repeated in Figure 5.6, with the addition of the measurement cursors to better visualize the filter transition band and stop band attenuation.

The horizontal cursor in Figure 5.6 clearly shows that the stop band specification of 60 db was met with a 34-coefficient filter. The same filter with 33 coefficients produced an SBA of only 57.6 db. This clearly illustrates the need to perform several iterations of the filter design each time, slightly tweaking an input until the desired results are achieved. This usually requires the slight modification in the number of filter coefficients, which is expected since the original number of coefficients was only a close estimate. In this example, the addition of a single filter coefficient bought us 2.4 db of increased stop band performance. This example also clearly illustrates how close the filter performance is to the initial design specifications.

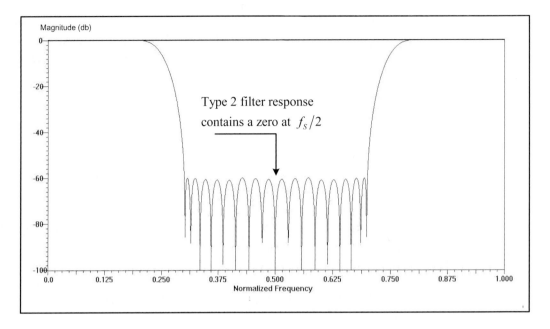

**Figure 5.5**   Low pass filter frequency response

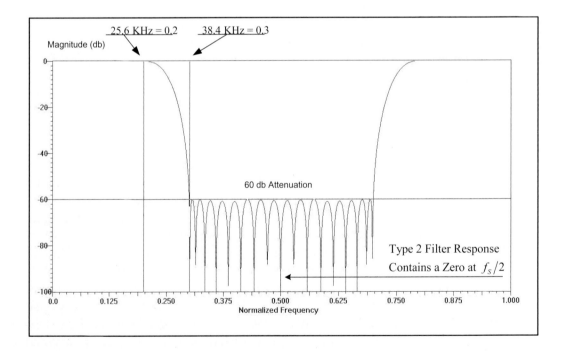

**Figure 5.6**   Low pass filter cursor measurements

The figure also shows two vertical cursors at the filter transition band edges. The example specified pass band edges of 0.0 and 0.2 and stop band edges of 0.3 and 0.5. The two vertical cursors correspond to the edges at 0.2 and 0.3. It is interesting to note that in this example the stop band attenuation of 60 db begins exactly at the cursor, marking the stop band edge of 0.3. This corresponds to the specified tolerance diagram illustrated in Figure 5.4.

A partial listing of the output data printed from the modified PM routine is illustrated in Table 5.3. In the table, the coefficients are represented as both double-precision floating point and 16-bit integer. Other data of interest are the tabulated input data consisting of band edges, desired response, band weighting, and the deviation in decibels of the band approximation error. Remember the example specified a PBR of 0.01 db and an SBA of 60 db. The tabulated results indicate that the filter design achieved 0.01 db and 60.32 db, respectively.

**Table 5.3**   Low Pass Filter PM Routine Output

Finite Impulse Response FIR
Linear Phase Digital Filter Design
Remez Exchange Algorithm
Band Pass Filter
Filter Length = 34
***** Impulse Response *****

| H(0)  | = | 0.00057940  | = | H(33) | = | 19 |
| H(1)  | = | -0.00143848 | = | H(32) | = | -46 |
| H(2)  | = | -0.00199142 | = | H(31) | = | -64 |
| H(3)  | = | 0.00300130  | = | H(30) | = | 98 |
| H(4)  | = | 0.00418997  | = | H(29) | = | 137 |
| H(5)  | = | -0.00610421 | = | H(28) | = | -199 |
| H(6)  | = | -0.00802244 | = | H(27) | = | -262 |
| H(7)  | = | 0.01107391  | = | H(26) | = | 363 |
| H(8)  | = | 0.01416825  | = | H(25) | = | 464 |
| H(9)  | = | -0.01899166 | = | H(24) | = | -621 |
| H(10) | = | -0.02417187 | = | H(23) | = | -791 |
| H(11) | = | 0.03223908  | = | H(22) | = | 1056 |
| H(12) | = | 0.04212628  | = | H(21) | = | 1380 |
| H(13) | = | -0.05858603 | = | H(20) | = | -1919 |
| H(14) | = | -0.08527554 | = | H(19) | = | -2793 |
| H(15) | = | 0.14773008  | = | H(18) | = | 4841 |
| H(16) | = | 0.44889525  | = | H(17) | = | 14709 |

|                 | Band 1      | Band 2       |
|-----------------|-------------|--------------|
| Lower Band Edge | 0.0000000   | 0.3000000    |
| Upper Band Edge | 0.2000000   | 0.5000000    |
| Desired Value   | 1.0000000   | 0.0000000    |
| Weighting       | 1.0000000   | 1.2000000    |
| Deviation       | 0.0011563   | 0.0009636    |
| Deviation(db)   | 0.0100375   | -60.3223428  |

It should be noted here that the coefficients in the previous example were transformed from floating point to 16-bit integer only to support a 16-bit integer hardware processor. Other hardware platforms such as floating point processors would more than likely utilize the unmodified double-precision PM coefficients. Also, the reader should take note that modifying the double-precision PM filter coefficients by converting them to K-bit integers will affect the filter performance. The modified performance should be compared to the original filter tolerance specifications, and corrections should be made if necessary.

In this particular example, the 16-bit integer coefficients produced a filter with an SBA of 60 db or 0.32 db worse than the attenuation produced by the double-precision coefficients. This is a fairly good comparison but is not the norm. If you are implementing a digital filter on a floating point processor, it stands to reason that the double-precision coefficients should be used.

### 5.2.3   PM Band Pass FIR Filter Design

Let's now turn our attention to the design of a band pass filter using the PM algorithm. The procedure is almost identical to that discussed for the design of a low pass filter. We will begin with a drawing that illustrates the tolerance specifications for a band pass filter design, illustrated in Figure 5.7. In this example, we have three bands to specify.

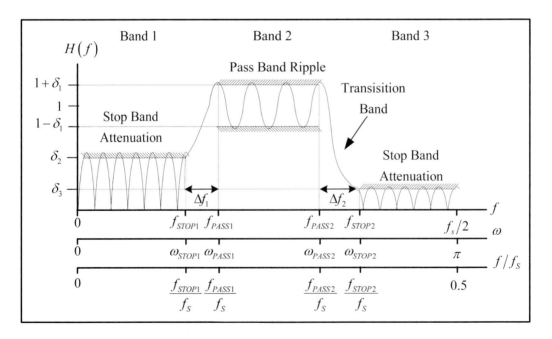

**Figure 5.7**   Tolerance specification for a band pass filter approximation

Band 2 is the desired pass band, and bands 1 and 3 are the desired stop bands. To make this a general example, the two stop bands have significantly different attenuation tolerance specifications. Band 2, the pass band, is approximated by a value of unity for $H(f)$ with a maximum error given by

$$\varepsilon(f) \le |\delta_1| \quad \text{for } f_{PASS1} \le f \le f_{PASS2}$$

Band 1, a stop band, is approximated by a value of zero for $H(f)$ with a maximum error given by

$$\varepsilon(f) \le \delta_2 \quad \text{for } 0 \le f \le f_{STOP1}$$

Band 3, the second stop band, is approximated by a value of zero for $H(f)$ with a maximum error given by

$$\varepsilon(f) \le \delta_3 \quad \text{for } f_{STOP2} \le f \le f_s/2$$

In this example, let's assume that the filter design specifications are as follows:

- The filter sample rate is $f_s = 128$ KHz.
- Stop band 1 extends from 0 Hz to 12.8 KHz.
- The pass band extends from 25.6 KHz to 38.4 KHz.
- Stop band 2 extends from 51.2 KHz to 64 KHz.
- The pass band ripple (PBR) is specified to be 0.01 db.
- The stop band 1 attenuation $SBA_1$ is specified to be 40 db.
- The stop band 2 attenuation $SBA_2$ is specified to be 50 db.
- The hardware that implements the filter will utilize 16-bit integer coefficients.

For the filter design, we can follow the same procedural steps that were outlined for the low pass filter design.

**Step 1.** Determine the filter sample rate. The sample rate is part of the specification and is $f_s = 128$ KHz.

**Step 2.** Specify the frequency response edges. The band edges are specified to be 0 Hz, 12.8 Hz, 25.6 KHz, 38.4KHz, 51.264 KHz, and 64 KHz.

**Step 3.** Compute the normalized frequency response edges. The normalized band edges are computed to be

$$\frac{0}{128 \text{ KHz}}, \quad \frac{12.8 \text{ KHz}}{128 \text{ KHz}}, \quad \frac{25.6 \text{ KHz}}{128 \text{ KHz}}, \quad \frac{38.4 \text{ KHz}}{128 \text{ KHz}}, \quad \frac{51.264 \text{ KHz}}{128 \text{ KHz}}, \quad \frac{64 \text{ KHz}}{128 \text{ KHz}}$$

or 0.0, 0.1, 0.2, 0.3, 0.4, and 0.5.

- Transition band 1 is normalized to $\dfrac{25.6 - 12.8}{128} = 0.1$.

- Transition band 2 is normalized to $\dfrac{51.2 - 38.4}{128} = 0.1$.

**Step 4**. Specify the pass band ripple and stop band attenuation. The PBR is specified to be 0.01 db, the $SBA_1$ is specified to be 40 db, and the $SBA_2$ is specified to be 50 db.

**Step 5**. Compute the values for $\delta_1$, $\delta_2$, and $\delta_3$. We use Equation 5.16 to compute $\delta_1$:

$$
\begin{aligned}
\delta_1 &= 10^{\frac{PBR}{20}} - 1 \\
&= 10^{\frac{0.01}{20}} - 1 \\
&= 0.0012
\end{aligned}
$$

We use Equation 5.17 to compute both $\delta_2$ and $\delta_3$:

$$
\begin{aligned}
\delta_2 &= 10^{-\frac{SBA_1}{20}} \qquad && \delta_3 = 10^{-\frac{SBA_2}{20}} \\
&= 10^{-\frac{40}{20}} \quad \text{and} \quad && \quad\ = 10^{-\frac{50}{20}} \\
&= 0.01 && \quad\ = 0.0032
\end{aligned}
$$

**Step 6**. Compute the estimated number of filter coefficients $N$. Here we have to diverge a bit from the low pass filter procedure. We would like to use Equation 5.19 to compute the estimated number of coefficients, but it is intended for the case where there are only two bands, such as would be the case for a low pass filter.

It turns out that we can still use that equation if we input variables that will give us the maximum number of coefficients. It seems intuitive that the worst-case tolerance specifications will produce the largest estimate of the number of filter coefficients $N$. For example, $N$ is proportional to the product of the $\delta$s for the pass band ripple and stop band attenuation. Therefore we should choose for the estimate of $N$ the two $\delta$s that represent the largest difference. In this example, the pass band $\delta_1$ and the 50-db stop band $\delta_3$

represent the largest difference. The value for $N$ is also inversely proportional to the width of the filter response transition band. Therefore we should use the narrowest transition bandwidth $\Delta f$ in the calculation. This will give us the maximum estimate of the value for $N$. We may have to tweak the value of $N$, but initially it will serve well as a pretty good first approximation.

In this example, $\delta_1$ and $\delta_3$ represent the largest separation. Both transition bands are equal to 0.01, so there is no choice to be made here. The calculation of $N$ then is given by

$$N = \frac{-20\text{Log}_{10}\left(\sqrt{\delta_1\delta_3}\right) - 13}{14.6\,\Delta f} + 1$$

$$= \frac{-20\text{Log}_{10}\left(\sqrt{(0.0012)(0.0032)}\right) - 13}{(14.6)\,(0.1)} + 1$$

$$= 29.19$$

We will round up this coefficient estimate to 30 and use it for our initial value of $N$. The PM routine will produce an even symmetry impulse response. Since we are using an even number of coefficients, this will be a type 2 filter. If we had used an odd number of coefficients, the filter type would have been a type 1 design. Both types are suited for a band pass filter.

If we would have used the smaller pass band and stop band error differential between $\delta_1$ and $\delta_2$ in the calculations, the estimated number of coefficients would have been 25.80 or 4 less than the worst case estimate.

**Step 7**. Compute the band weights. Since this example has three bands as opposed to two bands, this step is also a bit different than the design procedure for the low pass filter. Here we want to compute the ratio of the pass band $\delta$ to the two stop band $\delta$s. We will use Equation 5.18 to compute the band weights as follows:

$$Wtx(f) = \begin{cases} \dfrac{\delta_1}{\delta_2} & \text{for } 0 \le f \le f_{STOP1} \\[2mm] \dfrac{\delta_1}{\delta_1} & \text{for } f_{PASS1} \le f \le f_{PASS2} \\[2mm] \dfrac{\delta_1}{\delta_3} & \text{for } f_{STOP2} \le f \le \dfrac{f_S}{2} \end{cases} = \begin{cases} \dfrac{0.0012}{0.01} & \text{Band 1} \\[2mm] \dfrac{0.0012}{0.0012} & \text{Band 2} \\[2mm] \dfrac{0.0012}{0.0032} & \text{Band 3} \end{cases} = \begin{cases} 0.12 \\ 1 \\ 0.375 \end{cases}$$

**Step 8**. Determine the desired filter response. We set the desired response of the band pass filter to be 0, 1, 0 in bands 1, 2, and 3, respectively.

**Step 9**. Enter the data into the PM program and compute the coefficients. We enter the following data into the PM algorithm argument list:

- `NFILT = 30`   [Number of coefficients to compute = 30]
- `JTYPE = 1`   [Filter Type = 1 for multi band filter]
- `NBANDS = 3`   [Number of Bands = 3]
- `LGRID = 16`   [Grid Size = 16]
- `EDGE = {0.0, 0.1, 0.2, 0.3, 0.4, 0.5}`   [Band Edges = 0.0, 0.1, 0.2, 0.3, 0.4, 0.5]
- `FX = {0,1,0}`   [Desired Response = 0, 1, 0]
- `WTX = {0.12, 1, 0.375}`   [Weights = 0.12, 1, 0.375]
- Call the PM Routine
  `CALL PARK(NFILT,JTYPE,NBANDS,LGRID,EDGE,FX,WTX,H)`
- Read the `NFILT` computed coefficients from array `H`

The PM algorithm inputs the data and returns with a set of 30 double-precision floating point filter coefficients, representing the impulse response of the filter. The coefficients are then converted to 16-bit integer and then applied to a DFT to compute the filter frequency response shown in Figure 5.8. Notice that the SBA is different in stop band 1 and stop band 2, as specified. Figure 5.9 illustrates the same filter response but with the vertical cursors marking the boundaries of the first transition band and the horizontal cursor marking the stop band 1 attenuation. The display program is written in C++ and runs in Microsoft Windows, and it allows the user to slew the vertical cursors across the entire plot width. Here they have been slewed to the normalized frequencies of 0.1 and 0.2. The horizontal cursor has been slewed down to the peak of stop band 1 and measures an attenuation of 41.6 db. Don't forget that the plot used 16-bit integer coefficients. The PM program output of Table 5.4 lists the attenuation for the double-precision floating point coefficients as 41.726 db.

The PM program output in Figure 5.10 illustrates the vertical cursors marking the boundaries of the second transition band and the horizontal cursor marking the stop band 2 attenuation. The vertical cursors have been slewed to the normalized frequencies of 0.3 and 0.4. The horizontal cursor has been slewed down to the peak of stop band 2 and measures a value of 51.4 db. Table 5.4 lists the attenuation of stop band 2 as 51.738 db. The floating point and 16-bit integer filter coefficients are also listed in the PM output. The slight difference between the frequency response plot and the listing in Table 5.4 is attributed to the performance divergence between the 16-bit integer coefficients

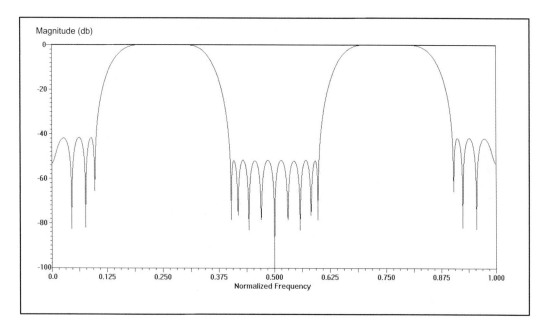

**Figure 5.8** Band pass filter frequency response

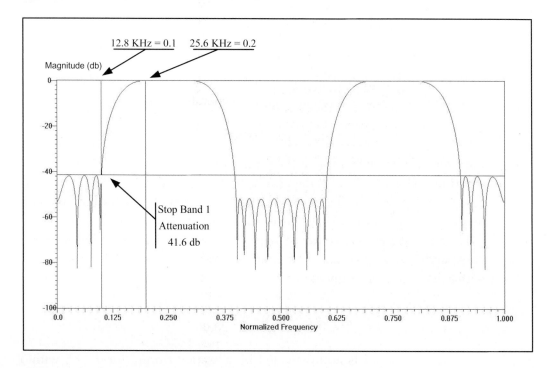

**Figure 5.9** Band pass filter lower band edge cursor measurements

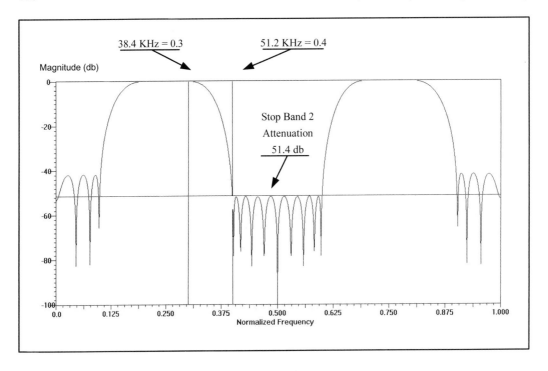

**Figure 5.10**   Band pass filter upper band edge cursor measurements

used to generate the plot and the double-precision floating point coefficients created by the PM algorithm and used to generate the table.

### 5.2.4   PM High Pass FIR Filter Design

Let's now direct our efforts to the design of a high pass filter using the PM algorithm. The procedure is identical to that discussed for the design of a low pass filter. Figure 5.11 illustrates the tolerance specifications for high pass filter design. In this example, we have only two bands to specify.

Band 1 is the filter response stop band, and band 2 is the filter response pass band. The stop band is approximated by a value of $H(f)$ equal to zero with a maximum error given by

$$\varepsilon(f) \le \delta_2 \quad \text{for } 0 \le f \le f_{STOP}$$

The pass band is approximated by a value of unity for $H(f)$, with a maximum error given by

**Table 5.4**   Band Pass Filter PM Routine Output

Finite Impulse Response FIR
Linear Phase Digital Filter Design
Remez Exchange Algorithm
Band Pass Filter
Filter Length = 30
\*\*\*\*\* Impulse Response \*\*\*\*\*

| | | | | | | |
|---|---|---|---|---|---|---|
| H(0) | = | 0.00240726 | = | H(29) | = | 79 |
| H(1) | = | -0.00133146 | = | H(28) | = | -43 |
| H(2) | = | 0.00693000 | = | H(27) | = | 227 |
| H(3) | = | 0.00723102 | = | H(26) | = | 237 |
| H(4) | = | -0.01171340 | = | H(25) | = | -383 |
| H(5) | = | 0.00012259 | = | H(24) | = | 4 |
| H(6) | = | -0.01780774 | = | H(23) | = | -583 |
| H(7) | = | -0.03092937 | = | H(22) | = | -1012 |
| H(8) | = | 0.04658752 | = | H(21) | = | 1527 |
| H(9) | = | 0.02583094 | = | H(20) | = | 846 |
| H(10) | = | 0.01430593 | = | H(19) | = | 469 |
| H(11) | = | 0.08138358 | = | H(18) | = | 2667 |
| H(12) | = | -0.17554559 | = | H(17) | = | -5751 |
| H(13) | = | -0.24090470 | = | H(16) | = | -7893 |
| H(14) | = | 0.29213624 | = | H(15) | = | 9573 |

| | Band 1 | Band 2 | Band 3 |
|---|---|---|---|
| Lower Band Edge | 0.0000000 | 0.2000000 | 0.4000000 |
| Upper Band Edge | 0.1000000 | 0.3000000 | 0.5000000 |
| Desired Value | 0.0000000 | 1.0000000 | 0.0000000 |
| Weighting | 0.1200000 | 1.0000000 | 0.3800000 |
| Deviation | 0.0081974 | 0.0009837 | 0.0025887 |
| Deviation(db) | -41.7264498 | 0.0085400 | -51.7384969 |

$$\varepsilon(f) \le |\delta_1| \quad \text{for } f_{PASS} \le f \le f_S/2$$

In this example, let's assume that the filter design specifications are as follows:

- The filter sample rate is $f_S = 64$ KHz.
- The stop band extends from 0 Hz to 20 KHz.
- The pass band extends from 25 KHz to 32 KHz.
- The PBR is specified to be 0.02 db.
- The SBA is specified to be 60 db.
- The transition bandwidth is 5 KHz.
- The hardware that implements the filter will utilize 16-bit integer coefficients.

For this design, we can follow the same procedural steps that were outlined for the low pass filter design.

**Step 1.** Determine the filter sample rate. The sample rate is part of the specification and is $f_S = 64$ KHz.

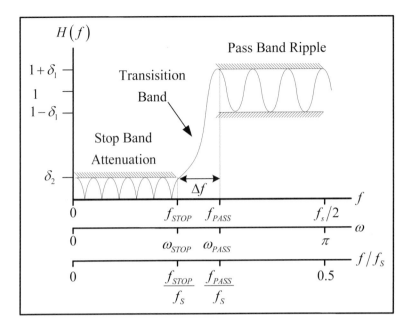

**Figure 5.11**   Tolerance specifications for a high pass filter approximation

**Step 2**. Specify the frequency response edges. The band edges are specified as 0 Hz, 20 KHz, 25 KHz, and 32 KHz.

**Step 3**. Compute the normalized frequency response edges. The normalized band edges are computed as

$$\frac{0 \text{ KHz}}{64 \text{ KHz}}, \quad \frac{20 \text{ KHz}}{64 \text{ KHz}}, \quad \frac{25 \text{ KHz}}{64 \text{ KHz}}, \quad \frac{32 \text{ KHz}}{64 \text{ KHz}} = 0.0, \ 0.3125, \ 0.3906, \ 0.5$$

The normalized transition band is $\Delta f = \dfrac{5 \text{ KHz}}{64 \text{ KHz}} = 0.0781$.

**Step 4**. Specify the pass band ripple and stop band attenuation. The PBR is specified as 0.02 db, and the SBA is specified as 60 db.

**Step 5**. Compute the values for $\delta_1$ and $\delta_2$. We use Equation 5.16 to compute $\delta_1$:

$$\delta_1 = 10^{\frac{PBR}{20}} - 1$$
$$= 10^{\frac{0.02}{20}} - 1$$
$$= 0.0023$$

We use Equation 5.17 to compute $\delta_2$:

$$\delta_2 = 10^{-\frac{SBA}{20}}$$
$$= 10^{-\frac{60}{20}}$$
$$= 0.001$$

**Step 6**. Compute the estimated number of filter coefficients $N$. We use Equation 5.19 to compute the estimated number of coefficients:

$$N = \frac{-20\text{Log}_{10}\left(\sqrt{\delta_1 \delta_2}\right) - 13}{14.6 \, \Delta f} + 1$$

$$= \frac{-20\text{Log}_{10}\left(\sqrt{(0.0023)(0.001)}\right) - 13}{(14.6) \, (0.0781)} + 1$$

$$= 39.0463$$

We will boost this estimate to the next odd number of 41 and use it for our initial value of $N$. The PM routine will produce an even symmetry impulse response. Since this is a high pass filter, we are constrained to use an odd number of filter coefficients and implement a type 1 filter, suited for low pass and band pass and high pass designs. Why? Because a type 2 filter frequency response has an inherent zero at half the sample rate or $f = 0.5$. If the algorithm worked at all, we would see a high pass frequency response that dropped to zero at $f = 0.5$. This would not be a good thing. As a matter of fact, if we try to use an even number of coefficients, the algorithm will produce a filter frequency response that looks more like random noise than a well-behaved high pass response. The reader should verify this by substituting an even value for the number of coefficients and see what the PM routine produces. It is always educational and satisfying to verify a textbook theory with real-world simulations.

**Step 7**. Compute the band weights. We use Equation 5.18 to compute the band weights as follows:

$$Wtx(f) = \begin{cases} \dfrac{\delta_1}{\delta_2} & \text{for } 0 \le f \le f_{STOP} \\[2ex] \dfrac{\delta_1}{\delta_1} & \text{for } f_{PASS} \le f \le f_s/2 \end{cases} = \begin{cases} \dfrac{0.0023}{0.001} & \text{for band 1} \\[2ex] \dfrac{0.0023}{0.0023} & \text{for band 2} \end{cases} = \begin{Bmatrix} 2.3 \\ 1 \end{Bmatrix}$$

**Step 8**. Determine the desired filter response. We set the desired response of the high pass filter to be 0 in band 1 and 1 in band 2, respectively.

**Step 9**. Enter the data into the PM program and compute the coefficients. We enter the following data into the PM algorithm argument list:

- **NFILT = 41** [Number of coefficients to compute = 41]
- **JTYPE = 1** [Filter Type = 1 for multi band filter]
- **NBANDS = 2** [Number of Bands = 2]
- **LGRID = 16** [Grid Size = 16]
- **EDGE = {0.0, 0.3125, 0.3906, 0.5}** [Band Edges = 0.0, 0.3125, 0.3906, 0.5]
- **FX = {0,1}** [Desired Response = 0, 1]
- **WTX = {2.3, 1 }** [Weights = 2.3, 1.0]
- Call the PM Routine
  CALL PARK(NFILT,JTYPE,NBANDS,LGRID,EDGE,FX,WTX,H)
- Read the **NFILT** computed coefficients from array **H**

The PM algorithm inputs these data and returns with a set of 41 double-precision floating point filter coefficients representing the impulse response of the filter. The coefficients are then converted to 16-bit integer and then applied to a DFT to compute the filter frequency response shown in Figure 5.12.

Figure 5.13 illustrates the same filter response but with the vertical cursors marking the boundaries of the transition band and the horizontal cursor marking the max value of the stop band attenuation. The vertical display cursors have been slewed to the normalized band edge frequencies of 0.3125 and 0.3906. From the plot, we can pick any two major ticks on the frequency axis and calculate the distance between minor ticks to be

$$\left[(0.375 - 0.250)/8\right]64\ \text{KHz} = 1000\ \text{Hz}$$

There are five minor ticks between the vertical cursors, which gives us a $5(1000\ \text{Hz})=5000\ \text{Hz}$ transition band, exactly as we had specified. The horizontal cursor has been slewed down to the specified level of the stop band of 60 db. We note that the first lobe of the stop band in the plot does not meet the original design stop band specifications. It is measured to be approximately 57 db. This clearly does not meet the original filter specifications. If we increase the number of filter coefficients from 41 to 43 and if we run the PM algorithm a second time, we produce the filter response illustrated in Figure 5.14. We can see in Figure 5.15 that the additional two filter coefficients bought us approximately 5 db of stop band attenuation. This beats

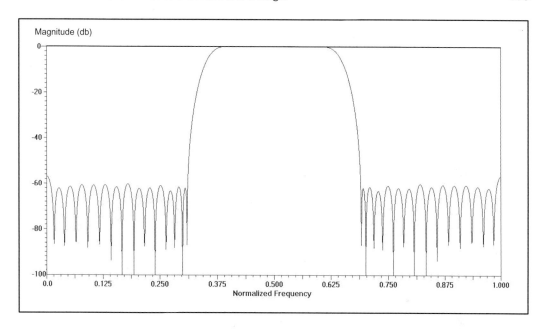

**Figure 5.12**   High pass filter frequency response: 41 coefficients

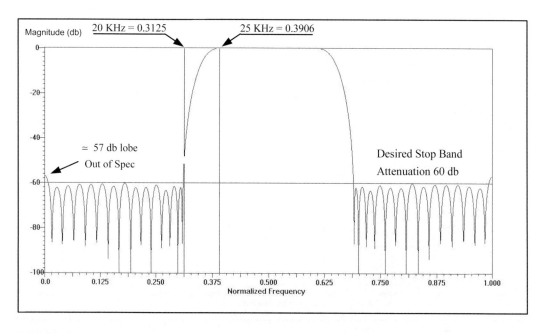

**Figure 5.13**   High pass filter cursor measurements: 41 coefficients

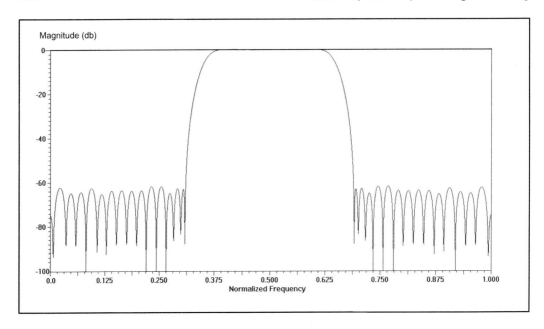

**Figure 5.14**   High pass filter frequency response: 43 coefficients

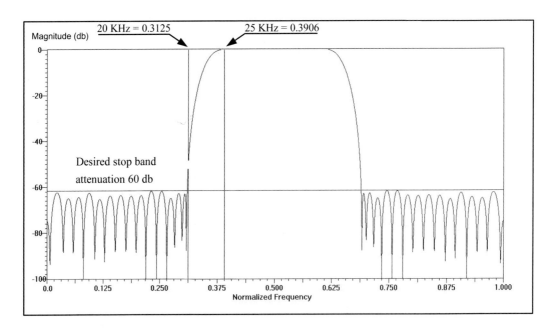

**Figure 5.15**   High pass filter cursor measurements: 43 coefficients

the original specification by almost 2 db. Again, this illustrates the point that it may be necessary to run the PM algorithm a couple of times to iteratively tweak the filter until the design specifications are met. More importantly, it points out the difference in performance between the double-precision floating point coefficients computed by the PM algorithm and the 16-bit integer coefficients used to generate the plots.

The filter coefficients produced by the PM subroutine are listed in Table 5.5. Once again, both the double-precision floating point and 16-bit integer coefficients are shown. We can see that the stop band attenuation for the floating point coefficients is –63 db, which is 3 db better than the design specification of –60 db. We also see that the pass band ripple is 0.013 db, which better than the design specification of 0.02 db.

**Table 5.5**   High Pass Filter PM Routine Output

```
Finite Impulse Response FIR
Linear Phase Digital Filter Design
Remez Exchange Algorithm
Band Pass Filter
Filter Length = 43
*****  Impulse Response  *****
         H(0)    =   -0.00035808   = H(42) =      -11
         H(1)    =   -0.00048401   = H(41) =      -15
         H(2)    =    0.00158592   = H(40) =       52
         H(3)    =   -0.00200285   = H(39) =      -65
         H(4)    =    0.00031580   = H(38) =       10
         H(5)    =    0.00320501   = H(37) =      105
         H(6)    =   -0.00565829   = H(36) =     -184
         H(7)    =    0.00333900   = H(35) =      109
         H(8)    =    0.00424400   = H(34) =      139
         H(9)    =   -0.01179996   = H(33) =     -386
        H(10)    =    0.01097573   = H(32) =      360
        H(11)    =    0.00199702   = H(31) =       65
        H(12)    =   -0.01979215   = H(30) =     -648
        H(13)    =    0.02616401   = H(29) =      857
        H(14)    =   -0.00841617   = H(28) =     -275
        H(15)    =   -0.02799640   = H(27) =     -916
        H(16)    =    0.05585843   = H(26) =     1830
        H(17)    =   -0.04063184   = H(25) =    -1330
        H(18)    =   -0.03420703   = H(24) =    -1120
        H(19)    =    0.14923144   = H(23) =     4890
        H(20)    =   -0.25432171   = H(22) =    -8333
        H(21)    =    0.29680749   = H(21) =     9726
```

|                  | Band 1       | Band 2      |
| ---------------- | ------------ | ----------- |
| Lower Band Edge  | 0.0000000    | 0.3900000   |
| Upper Band Edge  | 0.3100000    | 0.5000000   |
| Desired Value    | 0.0000000    | 1.0000000   |
| Weighting        | 2.3000000    | 1.0000000   |
| Deviation        | 0.0006968    | 0.0016026   |
| Deviation(db)    | -63.1381306  | 0.0139087   |

### 5.2.5   PM Stop Band FIR Filter Design

Let's go through the filter design procedure one more time. This time, we will dedicate our efforts to the design of a band stop filter using the PM algorithm. The procedure is identical to that discussed for the design of a band pass filter. Figure 5.16 illustrates the error tolerance for a band stop filter design. As was the case for the band pass filter, we have three bands to specify.

Band 2 is the filter response stop band, and bands 1 and 3 are the filter response pass bands. As in the previous examples, we wish to approximate the stop band setting $H(f)$ equal to zero with a maximum error given by

$$\varepsilon(f) \le \delta_2 \quad \text{for } f_{STOP1} \le f \le f_{STOP2}$$

The two pass bands are approximated by a value of unity for $H(f)$ with a maximum error given by

$$\varepsilon(f) \le |\delta_1| \quad \left\{ \begin{array}{l} \text{for } 0 \le f \le f_{PASS1} \\ \text{for } f_{PASS2} \le f \le f_s/2 \end{array} \right\}.$$

In this example, let's assume we are given a somewhat strange set of filter design specifications:

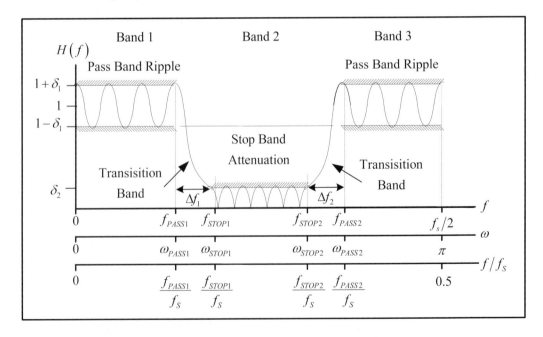

**Figure 5.16**   Tolerance specifications for a stop band filter approximation

- The filter sample rate is some strange number such as $f_s = 47.8$ KHz.
- The pass band 1 extends from 0 Hz to 9.560 KHz.
- The stop band extends from 11.472 KHz to 13.862 KHz.
- The pass band 2 extends from 15.744 KHz to 23.900 KHz.
- The ripple in both pass bands is specified to be PBR = 0.01 db.
- The side band attenuation is specified to be 60 db.
- Transition bandwidth $\Delta f_1 = 1.912$ KHz and $\Delta f_2 = 1.882$ KHz.
- The hardware that implements the filter will utilize 16-bit integer coefficients.

For this design, we can follow the same procedural steps that were outlined for the band pass filter design.

**Step 1.** Determine the filter sample rate. The sample rate was part of the specification and is $f_s = 47.8$ KHz.

**Step 2.** Specify the frequency response edges. The band edges are specified to be 0 Hz, 9.560 KHz, 11.472 KHz, 13.862 KHz, 15.744 KHz, and 23.900 KHz.

**Step 3.** Compute the normalized frequency response edges. The normalized band edges are computed as

$$\frac{0 \text{ KHz}}{47.8 \text{ KHz}}, \quad \frac{9.560 \text{ KHz}}{47.8 \text{ KHz}}, \quad \frac{11.472 \text{ KHz}}{47.8 \text{ KHz}}, \quad \frac{13.862 \text{ KHz}}{47.8 \text{ KHz}}, \quad \frac{15.744 \text{ KHz}}{47.8 \text{ KHz}}, \quad \frac{23.900 \text{ KHz}}{47.8 \text{ KHz}}$$

or 0.0, 0.2, 0.24, 0.29, 0.33, and 0.5.

- The normalized transition band $\Delta f_1 = \dfrac{1.9120 \text{ KHz}}{47.8 \text{ KHz}} = 0.04$.

- The normalized transition band $\Delta f_2 = \dfrac{1.882 \text{ KHz}}{47.8 \text{ KHz}} = 0.0394$.

**Step 4.** Specify the pass band ripple and stop band attenuation. The PBR for both pass bands was specified to be 0.01 db, and the SBA for the stop band was specified to be 60 db.

**Step 5.** Compute the values for $\delta_1$ and $\delta_2$. We use Equation 5.16 to compute $\delta_1$:

$$\delta_1 = 10^{\frac{\text{PBR}}{20}} - 1$$
$$= 10^{\frac{0.01}{20}} - 1$$
$$= 0.0012$$

We use Equation 5.17 to compute $\delta_2$:

$$\delta_2 = 10^{-\frac{SBA_1}{20}}$$

$$= 10^{-\frac{60}{20}}$$

$$= 0.001$$

**Step 6.** Compute the estimated number of filter coefficients $N$. Recognizing that the two pass bands are identically specified and that the two transition bands are of different widths, we can use Equation 5.19 to compute the estimated number of coefficients. To generate the worst-case maximum estimate, we will use the narrowest transition bandwidth of $\Delta f_2 = 1.882$ KHz $= 0.0394$ normalized:

$$N = \frac{-20 Log_{10}\left(\sqrt{\delta_1 \delta_2}\right) - 13}{14.6\ \Delta f_2} + 1$$

$$= \frac{-20 Log_{10}\left(\sqrt{(0.0012)(0.001)}\right) - 13}{(14.6)\ (0.0394)} + 1$$

$$= 81.3285$$

Just for the heck of it, let's round down this estimate to 81, use it for our initial value of $N$, and see what kind of performance we get. Keep in mind that we cannot use an even number of coefficients for this design because we need to implement a type 1 filter to avoid the zero at $f = 0.5$ inherent in type 2 filters.

**Step 7.** Compute the band weights. Here we want to compute the ratio of the pass band $\delta$s to the stop band $\delta$. We will use Equation 5.18 to compute the band weights as follows:

$$Wtx(f) = \begin{cases} \dfrac{\delta_1}{\delta_1} & \text{for } 0 \leq f \leq f_{PASS1} \\[2mm] \dfrac{\delta_1}{\delta_2} & \text{for } f_{STOP1} \leq f \leq f_{STOP2} \\[2mm] \dfrac{\delta_1}{\delta_1} & \text{for } f_{PASS2} \leq f \leq \dfrac{f_s}{2} \end{cases} = \begin{cases} \dfrac{0.0012}{0.0012} & \text{Band 1} \\[2mm] \dfrac{0.0012}{0.001} & \text{Band 2} \\[2mm] \dfrac{0.0012}{0.0012} & \text{Band 3} \end{cases} = \begin{cases} 1 \\ 1.2 \\ 1 \end{cases}$$

**Step 8.** Determine the desired filter response. We set the desired response of the band pass filter to be 1, 0, and 1 in bands 1, 2, and 3, respectively.

**Step 9**. Enter the data into the PM program and compute the coefficients. We enter the following data into the PM algorithm argument list:

- **NFILT = 81**  [Number of coefficients to compute = 81]
- **JTYPE = 1**  [Filter Type = 1 for multi band filter]
- **NBANDS = 3**  [Number of Bands = 3]
- **LGRID = 16**  [Grid Size = 16]
- **EDGE = {0.0, 0.2, 0.24, 0.29, 0.33, 0.5}**  [Edges = 0.0, 0.2, 0.24, 0.29, 0.33, 5]
- **FX = {1, 0, 1}**  [Desired Response = 1, 0, 1]
- **WTX = {1, 1.2, 1}**  [Weights = 1, 1.2, 1]
- Call the PM Routine
  CALL PARK(NFILT,JTYPE,NBANDS,LGRID,EDGE,FX,WTX,H)
- Read the **NFILT** computed coefficients from array **H**

The PM algorithm inputs the specified data and returns with a set of 81 double-precision floating point filter coefficients representing the impulse response of the filter. The coefficients are then converted to 16-bit integer and then applied to a DFT to compute the filter frequency response shown in Figure 5.17.

Figure 5.18 illustrates the same filter response but with the vertical cursors marking the boundaries of the first transition band and the horizontal

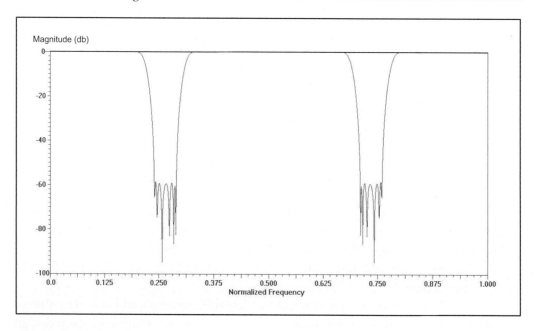

**Figure 5.17**  Stop band filter frequency response

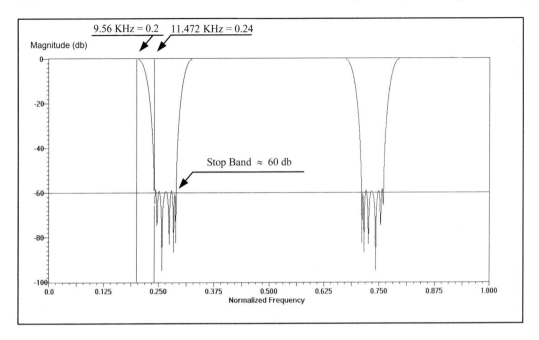

**Figure 5.18**  Stop band filter cursor measurements

cursor marking the stop band attenuation. The vertical cursors have been slewed to the normalized frequencies of 0.2 and 0.24. The horizontal cursor has been slewed down to 60 db. We can see that the stop band attenuation just barely misses the specification of 60 db. The PM program output illustrated in Table 5.6 lists the attenuation as –59.5 db. Strictly speaking, the filter illustrated in this figure is not within the original specification, and therefore our filter will not pass a customer design review. In addition, Table 5.6 shows a pass band ripple of 0.011 db that is very close to the original specification of 0.01 db, but it is still out of specification. Not to worry, though. We can easily gain another decibel or so of performance by increasing the length of the filter impulse response by a couple of coefficients. A second run of the PM algorithm showed that a filter with an impulse response of 83 coefficients easily beat the 60 db SBA specification with a figure of 61.5 db. The PBR figure of 0.008 beat the specification of 0.01 and the customer should be satisfied. The floating point and 16-bit integer filter coefficients are also listed in the PM output table for the 81 coefficient case.

From all of our previous examples, the reader should see the value of the PM filter design algorithm. It is far superior to any of the legacy design methods and tools that have preceded it.

**Table 5.6**   Stop Band Filter PM Routine Output

```
                        Finite Impulse Response FIR
                       Linear Phase Digital Filter Design
                          Remez Exchange Algorithm
                             Band Pass Filter
                            Filter Length = 81
                       *****  Impulse Response  *****
      H(0)  =   -0.00074406 =  H(80)   =        -23
      H(1)  =   -0.00084267 =  H(79)   =        -27
      H(2)  =    0.00153199 =  H(78)   =         50
      H(3)  =    0.00045085 =  H(77)   =         15
      H(4)  =   -0.00161495 =  H(76)   =        -52
      H(5)  =   -0.00022928 =  H(75)   =         -7
      H(6)  =    0.00090221 =  H(74)   =         30
      H(7)  =    0.00000484 =  H(73)   =          0
      H(8)  =    0.00122054 =  H(72)   =         40
      H(9)  =   -0.00064600 =  H(71)   =        -20
      H(10) =   -0.00410217 =  H(70)   =       -133
      H(11) =    0.00244161 =  H(69)   =         80
      H(12) =    0.00639588 =  H(68)   =        210
      H(13) =   -0.00474622 =  H(67)   =       -155
      H(14) =   -0.00656767 =  H(66)   =       -214
      H(15) =    0.00570910 =  H(65)   =        187
      H(16) =    0.00404272 =  H(64)   =        132
      H(17) =   -0.00304424 =  H(63)   =        -99
      H(18) =    0.00002863 =  H(62)   =          1
      H(19) =   -0.00444116 =  H(61)   =       -145
      H(20) =   -0.00303453 =  H(60)   =        -98
      H(21) =    0.01533118 =  H(59)   =        502
      H(22) =    0.00246415 =  H(58)   =         81
      H(23) =   -0.02521797 =  H(57)   =       -825
      H(24) =    0.00192627 =  H(56)   =         63
      H(25) =    0.02816321 =  H(55)   =        923
      H(26) =   -0.00667402 =  H(54)   =       -218
      H(27) =   -0.01966535 =  H(53)   =       -643
      H(28) =    0.00516911 =  H(52)   =        169
      H(29) =   -0.00029535 =  H(51)   =         -9
      H(30) =    0.00951524 =  H(50)   =        312
      H(31) =    0.02598102 =  H(49)   =        851
      H(32) =   -0.04064680 =  H(48)   =      -1331
      H(33) =   -0.04753868 =  H(47)   =      -1557
      H(34) =    0.08480638 =  H(46)   =       2779
      H(35) =    0.05518605 =  H(45)   =       1808
      H(36) =   -0.13176712 =  H(44)   =      -4317
      H(37) =   -0.04399511 =  H(43)   =      -1441
      H(38) =    0.16784017 =  H(42)   =       5500
      H(39) =    0.01675836 =  H(41)   =        549
      H(40) =    0.81861610 =  H(40)   =      26824
```

|                 | Band 1      | Band 2        | Band 3      |
|-----------------|-------------|---------------|-------------|
| Lower Band Edge | 0.0000000   | 0.2400000     | 0.3300000   |
| Upper Band Edge | 0.2000000   | 0.2900000     | 0.5000000   |
| Desired Value   | 1.0000000   | 0.0000000     | 1.0000000   |
| Weighting       | 1.0000000   | 1.2000000     | 1.0000000   |
| Deviation       | 0.0012716   | 0.0010597     | 0.0012716   |
| Deviation(db)   | 0.0110381   | -59.4965105   | 0.0110381   |

## 5.2.6   PM Band Weighting Example

The next two filters are presented only to show the relative importance of the weighting function and illustrate the effect that weight selection has on a filter design. Keep in mind that there are many different ways to compute the filter band weights. The method that this author has chosen represents only one. It is important to remember that the method used should be dependent on the application at hand and on the original filter specifications.

### 5.2.6.1  Case I

Figure 5.19 and the corresponding PM output Table 5.7 pertain to a simple low pass filter with a pass band weighting set to 10 and a stop band weighting set to 1. The elements in the PM weight array are given by $Wtx = \{10,1\}$. This says that the relative importance of the pass band approximation error is 10 times that of the stop band approximation error.

### 5.2.6.2  Case II

Figure 5.20 and the corresponding PM output Table 5.8 pertain to the same low pass filter, but this time the pass band weighting is set to 1 and the stop band weighing is set to 10. The PM weight array is then given by $Wtx = \{1,10\}$. Just the opposite is true for this filter, in that the relative importance of the stop band approximation error is now 10 times that of the pass band approximation error.

In comparing Figure 5.19 and Figure 5.20, it is clear that the filter in case I has less ripple in the pass band than the filter in case II. This is due to its pass band weighting being 10 times more important than the stop band weighting. If we compare Table 5.7 and Table 5.8, the pass band deviation for the filter in case I is 0.09 db compared to 0.9 db for the filter in case II. This factor of 10 difference is what we should expect.

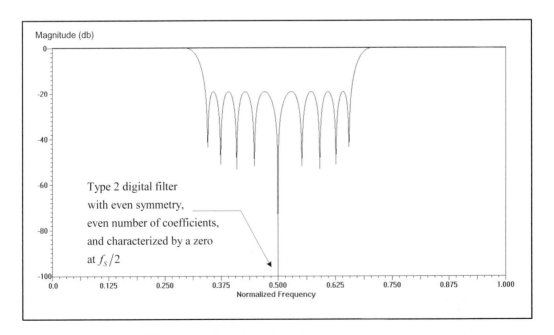

**Figure 5.19**   Case I pass band weight 10 / stop band weight 1

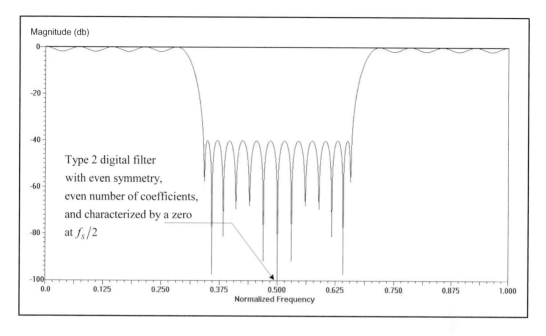

Magnitude (db)

Type 2 digital filter
with even symmetry,
even number of coefficients,
and characterized by a zero
at $f_S/2$

Normalized Frequency

**Figure 5.20** Case II pass band weight 1 / stop band weight 10

The frequency response for the case II filter shows a substantial pass band ripple. Similarly, the stop band for the filter presented in case II has a 39 db attenuation, as opposed to the 19 db side band attenuation for the case I filter. The 20-db difference is due to the 10-to-1 weight ratio for the stop band between the two filters.

## 5.3 PM IMPLEMENTATION OF HALF BAND FILTERS

There is a unique form of FIR filter called the *half band filter*. This filter is useful in multirate filtering applications where the sample rate of a signal is changed by a factor of two. The multirate application of this filter is discussed in greater detail in the next chapter. For now, let's limit our attention to understanding the special attributes of this filter and how it can be designed.

The frequency magnitude response of a half band filter is illustrated in Figure 5.21. Depending on the frequency scale we use, we note from the figure that the transition band is symmetrical about the frequency $f = 0.25$, $f = f_S/4$, or $\omega = \pi/2$ and that the amplitude of the frequency response at that frequency is equal to 0.5.

The bandwidth of the pass band and the stop band are equal. The filter stop band frequency is related to the filter pass band frequency by

**Table 5.7**   Case I Pass Band Weight 10 / Stop Band Weight 1

Finite Impulse Response FIR
Linear Phase Digital Filter Design
Remez Exchange Algorithm
Band Pass Filter
Filter Length = 32
\*\*\*\*\* Impulse Response \*\*\*\*\*

| | | | | | |
|---|---|---|---|---|---|
| H(0) | = | 0.01293945 | = H(31) = | | 424 |
| H(1) | = | -0.01314401 | = H(30) = | | -430 |
| H(2) | = | 0.00142900 | = H(29) = | | 47 |
| H(3) | = | 0.02068127 | = H(28) = | | 678 |
| H(4) | = | -0.03475920 | = H(27) = | | -1138 |
| H(5) | = | 0.02383450 | = H(26) = | | 781 |
| H(6) | = | 0.00660768 | = H(25) = | | 217 |
| H(7) | = | -0.02719479 | = H(24) = | | -890 |
| H(8) | = | 0.01155067 | = H(23) = | | 378 |
| H(9) | = | 0.03124887 | = H(22) = | | 1024 |
| H(10) | = | -0.05514724 | = H(21) = | | -1806 |
| H(11) | = | 0.01830116 | = H(20) = | | 600 |
| H(12) | = | 0.06564878 | = H(19) = | | 2151 |
| H(13) | = | -0.11572384 | = H(18) = | | -3791 |
| H(14) | = | 0.01715056 | = H(17) = | | 562 |
| H(15) | = | 0.54212691 | = H(16) = | | 17764 |

| | Band 1 | Band 2 |
|---|---|---|
| Lower Band Edge | 0.0000000 | 0.3400000 |
| Upper Band Edge | 0.3000000 | 0.5000000 |
| Desired Value | 1.0000000 | 0.0000000 |
| Weighting | 10.0000000 | 1.0000000 |
| Deviation | 0.0110995 | 0.1109954 |
| Deviation(db) | 0.0958782 | -19.0939015 |

$$f_{Stop} = \left(f_S/2 - f_{Pass}\right) = \left(0.5 - f_{Pass}/f_S\right)$$

A typical tolerance specification for a half band filter such as the one illustrated in Figure 5.22 shows that the pass band and the stop band are specified to have the same error tolerance—that is, $\delta_1 = \delta_2$. Therefore the weight values used by the PM program will both be equal to unity.

The half band filter has an even symmetrical impulse response in that $h(n) = h(N-1-n)$, and the number of filter coefficients is constrained to be odd. Therefore the half band filter is classified as a type 1 filter.

The characteristics of this filter are unique in that every even coefficient is equal to zero with the exception of the center coefficient $h\left(\dfrac{N-1}{2}\right)$, which is equal to 0.5. That is,

**Table 5.8** Case II Pass Band Weight 1 / Stop Band Weight 10

```
                    Finite Impulse Response FIR
                 Linear Phase Digital Filter Design
                    Remez Exchange Algorithm
                         Band Pass Filter
                        Filter Length = 32
                  ***** Impulse Response *****
       H(0)   =   -0.00661560   =  H(31) =      -216
       H(1)   =    0.02408035   =  H(30) =       789
       H(2)   =    0.04251486   =  H(29) =      1393
       H(3)   =   -0.00152996   =  H(28) =       -49
       H(4)   =   -0.01727738   =  H(27) =      -565
       H(5)   =    0.02293995   =  H(26) =       752
       H(6)   =    0.00005743   =  H(25) =         2
       H(7)   =   -0.02963165   =  H(24) =      -970
       H(8)   =    0.02811081   =  H(23) =       921
       H(9)   =    0.01410196   =  H(22) =       462
       H(10)  =   -0.05410218   =  H(21) =     -1772
       H(11)  =    0.03276661   =  H(20) =      1074
       H(12)  =    0.05398454   =  H(19) =      1769
       H(13)  =   -0.12195953   =  H(18) =     -3995
       H(14)  =    0.03530657   =  H(17) =      1157
       H(15)  =    0.53233766   =  H(16) =     17444
```

| | Band 1 | Band 2 |
|---|---|---|
| Lower Band Edge | 0.0000000 | 0.3400000 |
| Upper Band Edge | 0.3000000 | 0.5000000 |
| Desired Value | 1.0000000 | 0.0000000 |
| Weighting | 1.0000000 | 10.0000000 |
| Deviation | 0.1101689 | 0.0110169 |
| Deviation(db) | 0.9077809 | -39.1588223 |

$$
\begin{aligned}
h(n) &= 0.0 \\
&= 0.5 \quad \text{for} \\
h(n) &
\end{aligned}
\quad
\left\{
\begin{array}{l}
n = 0, 2, 4, 6, \cdots \\
n = (N-1)/2 \\
n = 1, 3, 5, 7, \cdots
\end{array}
\right\}
$$

Because of the symmetrical impulse response, the number of coefficient multiplies is reduced by a factor of two from that required for a direct form FIR filter. In addition, since every even coefficient is zero, we can further reduce the number of coefficient multiplies almost by a second factor of two. When the half band filter is implemented in the direct form odd coefficient architecture, then $(N-1)/4+1$ coefficient multiplies are necessary, which is significant reduction from the direct form transversal filter architecture. Back in the days when multipliers consumed a significant amount of hardware and

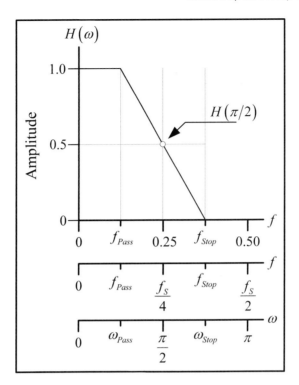

**Figure 5.21**   Half band filter symmetrical frequency response

power, any computational reduction was a major design goal. The half band filter was held in high regard by DSP design engineers because it significantly reduced the computational load.

This type of filter is useful in applications such as multirate processing, where minimum computation filters are used to halve the clock rate and reduce the signal bandwidth at the higher sample rates, which then enables the use of less complicated hardware to perform large-scale follow on sample rate decimation. Typically the half band filter would be the first filter in a two filter cascade and would be used to reduce the sample rate by a factor of 2. The follow on filter in the cascade could then operate at half the original sample rate to implement a more precise FIR filter with a much sharper roll-off. This type of architecture substantially reduces the filter design complexity.

The block diagram of a direct form linear phase half band filter is shown in Figure 5.23. This architecture was chosen specifically to illustrate the computational advantage this type of filter offers. Note that for this simple example of a nine-coefficient filter, only three multipliers are required. If the sample rate was slow enough, all the filter computations could be time-shared by a single multiplier.

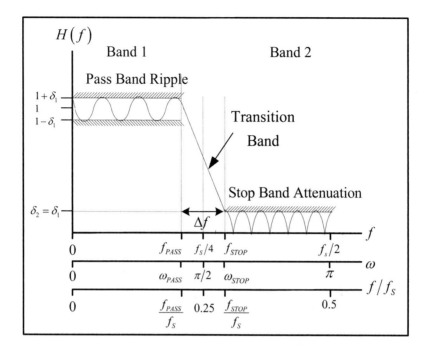

**Figure 5.22**   Tolerance specifications for a half band filter approximation

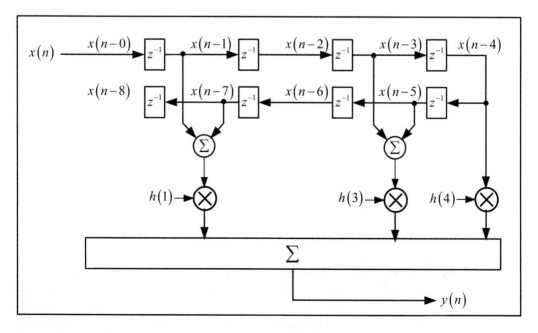

**Figure 5.23**   Direct form linear phase half band filter

For illustration purposes, the last delay register is included only to provide a visual reference between the input data samples and the filter coefficients. An actual filter would not require this register.

An example half band filter was designed using the PM algorithm with the following specifications:

- A normalized transition bandwidth of 0.06 symmetrical about $f_S/4$
- Pass band cutoff frequency $f_{Pass} = 0.22$
- Stop band cutoff frequency $f_{Stop} = 0.28$
- Pass band weight of 1.0 and a stop band weight of 1.0
- Number of coefficients $N = 65$

We enter the following data into the PM algorithm argument list:

- **NFILT = 65**  [Number of coefficients to compute = 65]
- **JTYPE = 1**  [Filter Type = 1 for multi band filter]
- **NBANDS = 2**  [Number of Bands = 2]
- **LGRID = 16**  [Grid Size = 16]
- **EDGE = {0.0, 0.22, 0.28, 0.5}**  [Band Edges = 0.0, 0.22, 0.28, 0.5]
- **FX = {1,0}**  [Desired Response = 1,0 ]
- **WTX = {1, 1}** [Weights = 1, 1]
- Call the PM Routine
  CALL PARK(NFILT,JTYPE,NBANDS,LGRID,EDGE,FX,WTX,H)
- Read the **NFILT** computed coefficients from array **H**

The PM algorithm inputs these data and returns with a set of 65 double-precision floating point filter coefficients representing the impulse response of the filter. The coefficients are then converted to 16-bit integer and then applied to a DFT to compute the half band filter frequency response shown in Figure 5.24.

Don't forget that the frequency response in the figure is plotted in decibels. Therefore the plot of the transition band midpoint is given by $H(\omega)\big|_{\omega=\pi/2} = 20\log_{10}(0.5) = -6$ db.

This particular half band filter required $(N-1)/4 + 1$ or $(65-1)/4+1 = 17$ multiplies per output point, as opposed to 65 multiplies for a direct form 1 implementation. So it's clear that this type of filter has its place in the world of DSP. Note in the coefficient listing presented in Table 5.9 that all of the even coefficients with the exception of coefficient $(N-1)/2 = 32$ are equal to zero for the integer coefficients and are not much more than computational noise for the floating point coefficients, which is really the equivalent of a zero.

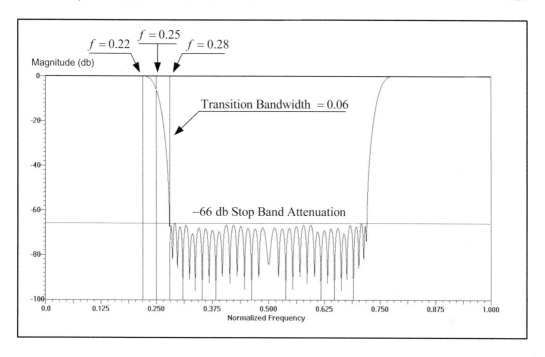

**Figure 5.24**  Half band filter frequency response

We will see in Chapter 6, "Multirate Finite Impulse Response Filter Design," that the half band filter can be used to reduce the sample rate of a signal by a factor of two. An example of a half band filter application might be a filter that precedes some DSP processor. The processor may be computationally overtaxed because of a high data sample rate. It might be possible to insert a half band filter between the input data and the processor. Because the half band filter has minimal computational requirements, it might be used to process the high sample rate input data sequence so that it can be decimated by a factor of two, providing for more computational headroom for the follow on signal processor.

Another example would be to cascade several of these half band filters in a string. Each filter would reduce the sequence bandwidth by a factor of two. This could reduce the amount of resources spent on computational hardware in the front end of a system where the input data rate is extremely high.

It is interesting at this point to put things in perspective. In the mid-1970s, a hardware multiplier used for processing data that was 8 to 16 bits wide filled an entire chassis that required a couple of feet of rack space. Because of its size and power consumption, it was most definitely time-shared between several mathematical processes. The first commercial LSI 16-by-16-bit multiplier chips were produced by TRW in the mid- to late-1970s.

**Table 5.9**  Half Band Filter PM Routine Output

```
                          Finite Impulse Response FIR
                         Linear Phase Digital Filter Design
                            Remez Exchange Algorithm
                               Band Pass Filter
                              Filter Length = 65
                        *****  Impulse Response  *****
            H(0)   =      0.00000009   =   H(64)  =           0
            H(1)   =     -0.00046530   =   H(63)  =         -14
            H(2)   =     -0.00000019   =   H(62)  =           0
            H(3)   =      0.00071205   =   H(61)  =          23
            H(4)   =      0.00000040   =   H(60)  =           0
            H(5)   =     -0.00123035   =   H(59)  =         -39
            H(6)   =     -0.00000088   =   H(58)  =           0
            H(7)   =      0.00197165   =   H(57)  =          65
            H(8)   =      0.00000133   =   H(56)  =           0
            H(9)   =     -0.00299475   =   H(55)  =         -97
            H(10)  =     -0.00000187   =   H(54)  =           0
            H(11)  =      0.00437039   =   H(53)  =         143
            H(12)  =      0.00000228   =   H(52)  =           0
            H(13)  =     -0.00618584   =   H(51)  =        -202
            H(14)  =     -0.00000289   =   H(50)  =           0
            H(15)  =      0.00855544   =   H(49)  =         280
            H(16)  =      0.00000353   =   H(48)  =           0
            H(17)  =     -0.01163979   =   H(47)  =        -380
            H(18)  =     -0.00000418   =   H(46)  =           0
            H(19)  =      0.01568522   =   H(45)  =         514
            H(20)  =      0.00000482   =   H(44)  =           0
            H(21)  =     -0.02110708   =   H(43)  =        -691
            H(22)  =     -0.00000535   =   H(42)  =           0
            H(23)  =      0.02868508   =   H(41)  =         940
            H(24)  =      0.00000590   =   H(40)  =           0
            H(25)  =     -0.04009562   =   H(39)  =       -1313
            H(26)  =     -0.00000600   =   H(38)  =           0
            H(27)  =      0.05972159   =   H(37)  =        1957
            H(28)  =      0.00000644   =   H(36)  =           0
            H(29)  =     -0.10369821   =   H(35)  =       -3397
            H(30)  =     -0.00000673   =   H(34)  =           0
            H(31)  =      0.31750144   =   H(33)  =       10404
            H(32)  =      0.50000658   =   H(32)  =       16384
```

|                 | Band 1      | Band 2       |
|-----------------|-------------|--------------|
| Lower Band Edge | 0.0000000   | 0.2800000    |
| Upper Band Edge | 0.2200000   | 0.5000000    |
| Desired Value   | 1.0000000   | 0.0000000    |
| Weighting       | 1.0000000   | 1.0000000    |
| Deviation       | 0.0004281   | 0.0004281    |
| Deviation(db)   | 0.0037179   | -67.3685306  |

These chips opened the door for multiply-intensive processing and signaled the beginning of a new era in DSP. These devices consumed about 6 square inches of circuit board real estate, and they burned about 5 watts each. The reader can see that in those days there was a great deal of incentive to develop DSP algorithms that minimized computations involving multiplication. I once saw a 17-by-17-inch circuit board with 64 of these 16-by-16 multipliers mounted on it. At 5 watts each, the multipliers consumed 320 watts total.

The original designer had used $1k\Omega$ pull down resistors on each of the multiplier's 48 input and output data pins, which burned another 77 watts. This board required its own dedicated power supply. The story goes that the engineers who came into work early in the morning would power up this board and use it to keep their coffee warm.

The half band filter, being a computationally efficient architecture, has been a staple in the design toolbox of many engineers. These days, this is not such an important issue. Today's FPGAs contain a wealth of resources, and the FPGA computer-aided design tools have libraries that contain not only individual low power multipliers but also building blocks for entire filter structures. They also provide scalable, completely predesigned DSP circuits sometimes called *cores* that the designer can add to a design with a simple click of the mouse. The old saying "here today, gone tomorrow" is especially true in the world of technology. Relative to technology, this saying is perhaps more aptly "here today, obsolete tomorrow." Because of the rate at which technology evolves, the electronic design engineer's career is a never-ending learning curve.

## 5.4 REFERENCES

[1] James H. McClellan, Thomas W. Parks, and Lawrence R. Rabiner. "A Computer Program for Designing Optimum FIR Linear Phase Digital Filters." *IEEE Transactions on Audio and Electroacoutsics*, vol. AU-21, no. 6 (1973).

[2] Paul M. Embree and Bruce Kimble. *C Language Algorithms for Digital Signal Processing*. Englewood Cliffs, NJ: Prentice Hall, 1991.

[3] Emmanuel C. Ifeachor and Barrie W. Jervis. *Digital Signal Processing: A Practical Approach*. Wokingham, UK: Addison-Wesley, 1993.

[4] Alan V. Oppenheim and Ronald W. Schafer. *Digital Signal Processing*. Englewood Cliffs, NJ: Prentice Hall, 1975.

[5] Steven A. Tretter. *Introduction to Discrete Time Signal Processing*. New York: John Wiley and Sons, 1976.

[6] P. P. Vaidyanathan. *Handbook of Digital Signal Processing Engineering Applications*. Edited by Douglas Elliott. San Diego, CA: Academic Press, 1987.

[7] J. F. Kaiser. "Non Recursive Digital Filter Design Using the Io-sinh Window Function." Proceedings of IEEE International Symposium of Circuits and Systems. April 1974.

[8] Charles S. Williams. *Designing Digital Filters*. Englewood Cliffs, NJ: Prentice Hall, 1986.

[9] Lawrence R. Rabiner and Bernard Gold. *Theory and Applications of Digital Signal Processing*. Englewood Cliffs, NJ: Prentice Hall, 1975.

[10] David R. Smith. *Digital Transmission Systems*. 2nd ed. New York: Chapman and Hall, 1993.

# CHAPTER SIX

# Multirate Finite Impulse Response Filter Design

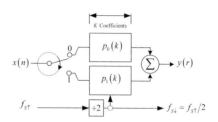

Many times it is necessary to change the sample rate of a digital sequence. A designer is often faced with the task of either increasing or decreasing the sample rate of a signal by some integer or fractional amount. A few examples include the following:

- Interfacing to a follow on system that operates with a slightly different sample clock
- Reducing the sample rate following the narrow band filtering of wide-band signals
- Implementing a zoom Fourier transform to increase frequency resolution
- Performing signal channelization to extract a bank of narrow band signals

There are several methods that can be utilized to change the sample rate of a digital signal. All these methods involve the use of a digital filter, sometimes referred to as a *multirate filter*. The term *multirate* refers to the fact that the filter is converting data from one sample rate to another sample rate. As you might expect, some multirate filters are better suited for specific rate change applications than others. In this chapter, we will discuss three rate change methods that use the following three filter types:

1. *Poly phase filters (PPFs)*. The preferred method for moderate-sized rate changes.
2. *Half band filters*. An efficient method for factor of two rate changes.
3. *Cascaded integrator comb (CIC) filters*. A computationally efficient filter for large rate changes.

In many applications, the multirate filter is used to perform the actual rate change, and a cleanup filter follows in cascade to enhance the roll-off of the transition band and remove residual spectral components in the stop band. This is particularly true when using CIC filters.

## 6.1  POLY PHASE FILTER (PPF)

There are many methods available to modify the sample rate of a signal. Perhaps one of the simpler, more efficient techniques is through the use of PPFs. The term *poly phase filter* (PPF) is a fancy name given to an ordinary everyday finite impulse response (FIR) filter that is implemented in a slightly different manner than a standard FIR filter.

A PPF can be thought of as a composite filter with impulse response $h(n)$, where $n = 0, 1, 2, \cdots, N-1$, that is segmented into $J$ separate shorter length filters operating in parallel. Each of the $J$ segments uses a subset of the composite set of filter coefficients. Each segment is called a *phase* and is represented by the coefficients $p_j(k)$ for $k = 0, 1, 2, \cdots, K-1$. The number of composite coefficients $N$ is chosen so that the number of coefficients per phase is an integer given by $K = N/J$.

The polyphase structures for implementing a filter that either increase the sample rate or decrease the sample rate are very similar and are discussed in the following sections. The terms usually given to describe a sample rate increase include *up sampling* and *interpolation*. The terms commonly associated with a sample rate decrease include *down sampling* and *decimation*. A filter structure that implements a sample rate change on an input signal usually operates at two different sample rates and is referred to as a *multirate filter*. A PPF is a class of multirate filters.

Before we begin our discussion on PPFs, let's first define a few variables that are commonly associated with multirate filters:

$f_{s\uparrow}$   The high sample rate input to or output from a multirate filter

$f_{s\downarrow}$   The low sample rate input to or output from a multirate filter

$x(n)$   The filter input data sequence

$y(r)$   The filter output data sequence

$h(n)$   The composite filter coefficient vector $n = 0, 1, 2, 3, \cdots, N-1$

$M$   The sample rate decimation factor $M = f_{s\uparrow(IN)}/f_{s\downarrow(OUT)}$ ; also equal to the $M$ phases of a polyphase decimation filter

$L$   The sample rate interpolation factor $L = f_{s\uparrow(OUT)}/f_{s\downarrow(IN)}$ ; also equal to the $L$ phases of a polyphase interpolation filter

$M\downarrow$   The symbol that represents a down sampling or decimation operation

$L\uparrow$   The symbol that represents an up sampling or interpolation operation

$N$   The number of coefficients in the composite multirate filter $h(n)$

$K$   The number of coefficients in each phase of a PPF

$n$   The index of the composite coefficient filter vector $n = 0,1,2,3,\cdots,N-1$; also equal to the sample index of the input data sequence $n = 0,1,2,3,\cdots$

$r$   The sample index of the output data sequence $r = 0,1,2,3,\cdots$

$k$   The index of PPF coefficients $k = 0,1,2,3,\cdots,K-1$

$m$   The $m$th phase of a polyphase decimation filter $m = 0,1,2,3,\cdots,M-1$

$l$   The $l$th phase of a polyphase interpolation filter $l = 0,1,2,3,\cdots,L-1$

$p_m(k)$   The coefficient vector of length $K$ for the $m$th phase of a decimation filter

$p_l(k)$   The coefficient vector of length $K$ for the $l$th phase of an interpolation filter

### 6.1.1   Decreasing the Sample Rate (Decimation)

Quite often it is necessary to decrease or decimate the sample rate at some point in a digital signal processing (DSP) system. Many times decimation is necessary in order to match the sample rate of a follow on processing system. Other times, decimation is performed to make the signal processing architecture computationally efficient, as might be the case when processing narrow band signals embedded in a wideband input. With today's hardware and software resources, modifying the sample rate within a processing system is commonplace and is fairly simple.

The operator notation that indicates a down sampling or decimation operation is usually given by the symbol $M\downarrow$. As an example, processing a discrete signal $x(n)$ sampled at a clock rate $f_S$ by the operator $M\downarrow$ would produce a new discrete signal $y(r)$ sampled at the clock rate of $f_S/M$, as illustrated in Figure 6.1.

When we decimate the sample rate of a signal, we can't just throw away the unwanted samples to reduce the sample rate. When the sample rate $f_S$ of a digital signal is decimated by a factor of $M$ such that $f_{s\downarrow} = f_{s\uparrow}/M$, we need to eliminate all frequencies that will alias below the new folding frequency of $f_{s\downarrow}/2$ and corrupt the new signal pass band. This is accomplished by a low pass filter. The typical symbol used in the industry to represent this decimation

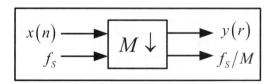

**Figure 6.1**   Down sampling operator

operation is a low pass filter followed by a down sample operator $M\downarrow$, as illustrated in Figure 6.2.

The antialiasing low pass filter that is inherent in the decimation operation is represented by the block with the filter impulse response $h(n)$. Although not completely accurate in terms of the actual implementation of down sampling hardware or software, the figure does convey all the information necessary to visualize and understand the process.

We could implement a brute force design to reduce the sample rate by a factor of $M$ by first low pass filtering the digital signal and then discarding $M-1$ of every $M$ samples of the filtered signal. This procedure will work and might actually be the procedure of choice when the input signal is already sampled at a low rate. If you think about it for a moment, it really does not make sense to waste the resources to compute the $M-1$ samples that we end up throwing away. Therefore if the input signal is sampled at a fairly high rate, it makes sense to merge the antialias filter function and the down sample function together so that the decimation filter uses fewer resources by

1. Operating at the reduced or decimated sample rate
2. Not computing the $M-1$ samples that end up getting thrown away

This makes for a computationally efficient and less complex design. We begin our discussion by looking at a digital signal $x(n)$ with a representative digital spectrum $X(\omega)$, illustrated in Figure 6.3. In part A of the figure, we see that the signal is sampled at a high sample rate given by $f_{s\uparrow}$. The two-sided spectrum extends all the way to the Nyquist folding frequency of $f_{s\uparrow}/2$. Now suppose our task is to reduce the sample rate of the signal $x(n)$ by a factor of two. In other words, we will decimate the signal by two. If we were to simply discard every other sample to reduce the sample rate by a factor of two, we would experience substantial aliasing, as indicated in part B of the figure. The hatched area indicates the frequencies that have been corrupted by aliasing. In this example, all the original frequencies have been corrupted and can never be reconstructed.

A more realistic method of decimating the sample rate of a signal is to filter the signal to remove all frequencies that will alias when the sample rate is reduced and then discard every other sample. The filtering process is

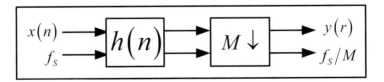

**Figure 6.2**  Functional representation of sample rate decimation

illustrated in part C of the figure and the spectrum of the decimated signal is illustrated in part D. In part D we see the spectrum relative to the low sample rate of $f_{s\downarrow} = f_{s\uparrow}/2$. Note that because the signal was prefiltered, the frequencies that would have aliased down and corrupted the output signal pass band have been removed and we experience no degradation in the filter output spectrum.

This method of decimation will work, and it has its merits for special applications, but it can be made a great deal more efficient, especially if the input sample rate $f_{s\uparrow}$ is a high frequency. The more practical method of decimating the sample rate of a signal is to perform the filtering and sample rate reduction simultaneously. Since the filter computations in this method are

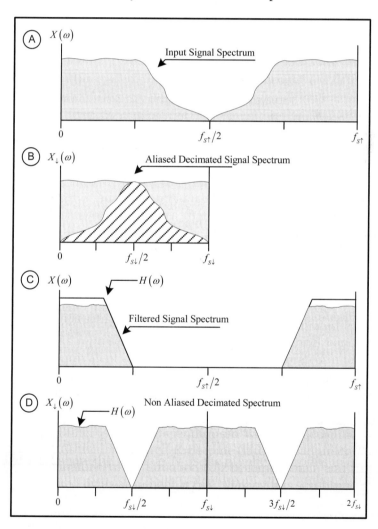

**Figure 6.3**   Decimate by two signal spectra

done at the low sample rate, this method has the benefit of reducing the computational rate and as a result reducing the amount of hardware necessary to perform the task.

Figure 6.4 shows a typical input data sequence $x(n)$ and its associated clock. Also shown below the input sequence is a second sequence $h(n)$ that represents the filter coefficient vector and one sample of the output sequence $y(r)$. For illustrative purposes, the filter vector in this example is only eight coefficients in length. The filter coefficient vector is shown aligned with an eight sample segment of the data sequence above it. The sum of products of the data vector segment and the coefficient vector produces the first output sample $y(r) = y(0)$. In this chapter, we will use the notation $y(0)$ for the first meaningful output sample produced by the filter, which will serve as our output signal starting point.

Thus the filter vector has the notation $h(n)|_{r=0}$. In a FIR filter, the individual data samples and coefficients that are aligned with one another are multiplied and then summed to produce the $r$th output sample of the sequence $y(r)$. This is sometimes referred to as a *dot product* between the data and coefficient vectors. In the figure, the dot product is performed between the input data and the filter coefficient vector to produce the output sample $y(0)$.

In a typical FIR filter, the coefficient vector is fixed and the input data sequence is clocked through a register delay line, one sample at a time. At each sample clock, the aligned data and filter coefficients are multiplied and summed to produce a new output sample. This continues forever, or until someone shuts off the power, whichever comes first. Using the data and coefficient notation of Figure 6.4, the output sequence $y(r)$ is computed:

$$y(r) = \sum_{n=0}^{N-1} h(n)x(r+n)$$

## Equation 6.1

For example, suppose the filter vector was $N = 7$ coefficients in length. The first three filter outputs would be computed by the following sequence:

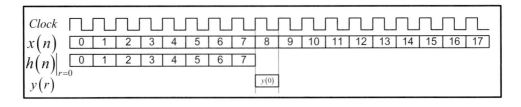

**Figure 6.4**  Input data stream, clock, filter coefficient vector, and output sample

$$y(0) \; = \sum_{n=0}^{6} h(n)x(0+n) \quad = h(0)x(0) \quad +h(1)x(1) \quad +h(2)x(2) \quad +h(3)x(3) \quad +h(4)x(4) \quad +h(5)x(5) \quad +h(6)x(6)$$

$$y(1) \; = \sum_{n=0}^{6} h(n)x(1+n) \quad = h(0)x(1) \quad +h(1)x(2) \quad +h(2)x(3) \quad +h(3)x(4) \quad +h(4)x(5) \quad +h(5)x(6) \quad +h(6)x(7)$$

$$y(2) \; = \sum_{n=0}^{6} h(n)x(2+n) \quad = h(0)x(2) \quad +h(1)x(3) \quad +h(2)x(4) \quad +h(3)x(5) \quad +h(4)x(6) \quad +h(5)x(7) \quad +h(6)x(8)$$

$$\vdots \qquad\qquad \vdots \qquad\qquad \vdots \qquad\quad \vdots \qquad\quad \vdots \qquad\quad \vdots \qquad\quad \vdots \qquad\quad \vdots$$

**Equation 6.2**

For our purposes, it is convenient to visualize the data sequence $x(n)$ to be infinitely long and fixed with respect to time. In this perspective, we visualize the coefficient vector $h(n)$ as sliding by the data sequence left to right, one sample per clock tick. This concept is analogous to the data sequence being an infinitely long straight road constructed from data samples placed end to end. The coefficient vector is analogous to a vehicle traveling along the road at a velocity equal to the sample rate. At each tick of the sample clock, a dot product is computed. Successive dot products produce an output sequence that we label $y(r)$.

Now let's suppose that we wished to use a filter of length $N = 8$ to process the data sequence such that the output sample rate is decimated by a factor of two. We could filter all the data and throw away every other output sample, but this requires the filter to perform multiply and accumulate operations to produce output samples that we will never use. That is, we would need to compute all the output samples and throw away every other one. This is a tremendous waste of computational power, and it only gets worse as the decimation rate increases.

A more efficient method is to compute only those output samples that are actually used. In other words, if the decimation factor is two, we only need to compute every other output sample, if the decimation factor is four, we only need to compute every fourth output sample, and so on. The benefit of this approach is that fewer computations are done, and they are done at the lower output clock rate, which is more efficient computationally and requires fewer hardware resources.

If we expand on Figure 6.4 to show the position of the sliding coefficient vector relative to the fixed data sequence at sample times $r = 0, 1, 2, 3, 4, \cdots$, we arrive at Figure 6.5. In this figure, we see the position of the coefficient vector relative to the data sequence for the first six output samples, and we see the first five samples of the filtered output sequence $\left[ y(r) \text{ for } r = 0 \rightarrow 4 \right]$. Note in this figure that since we are only computing every other output sample, successive $h(n)$ vectors are advanced in position relative to the data sequence by two samples. Also note that the period of each sample in the output sequence

$y(r)$ is now twice that of the input sample, indicating that the output sample rate is half the input sample rate.

When looking at a timing diagram such as the one in Figure 6.5, it is always a good idea to look for properties that can be exploited. In this case, it is clear from the timing diagram that the filter coefficient $h(0)$ is only used to multiply data samples

$$x(0), x(2), x(4), x(6), x(8), x(10), \cdots$$

Similarly the coefficient $h(1)$ is only used to multiply the data samples

$$x(1), x(3), x(5), x(7), x(9), x(11), \cdots$$

and so on. We can make comparable observations for all the filter coefficients. The trend we see here is that the even coefficients are restricted to the multiplication of all the even data samples, and the odd coefficients are restricted to the multiplication of all the odd data samples. This fact allows us to process the input signal with two phases of the original filter. Phase 0, represented by $p_0$, is a filter that is composed of all the even coefficients of the original composite filter, and phase 1, represented by $p_1$, is a filter that is composed of all the odd coefficients of the original composite filter.

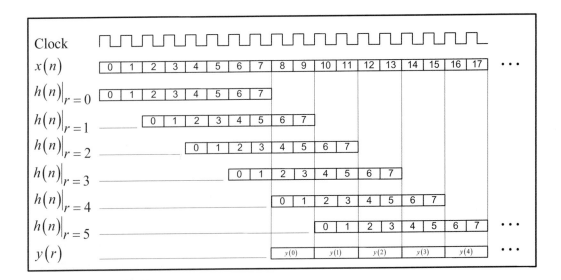

**Figure 6.5**   Decimate by two timing

Remember that in our simple example we are decimating a signal by a factor $M = 2$. Keeping with the spirit of learning from simple examples, let's suppose that we have already designed a filter of length $N = 8$ coefficients to eliminate the aliasing frequencies. Obviously, eight coefficients is a small number, but remember this example is supposed to be more instructive than realistic. We wish to implement this eight-coefficient decimate by two filter as a PPF. There are $M$ phases in a decimate by $M$ PPF, so the filter in our example will be implemented as an $M = 2$ phase filter. The number of coefficients for each phase in this example is given by

$$K = \frac{N}{M} = \frac{8}{2} = 4$$

The top-level block diagram of the resulting decimate by two PPFs is illustrated in Figure 6.6. Note in the figure that the input signal samples are applied to the input of the two filters $p_0$ and $p_1$ by a multiplexer modeled as a wiper arm. Initial conditions require that the wiper arm applies the first sample $x(0)$ to the phase $p_0$ filter first. The wiper arm then applies the second sample $x(1)$ to the phase $p_1$ filter, sample $x(2)$ to the phase $p_0$ filter, and so on. The wiper ports in the figure are labeled 0 and 1. The wipe on sequence is $0,1,0,1,\cdots$. This sequence may seem trivial now, but it is important for filters with more phases.

Each phase of the two-phase filter is $K = 4$ coefficients long. An output sample $y(r)$ is produced every time the wiper arm completes one revolution and inputs a sample to both filter phases. Note that the high rate input sequence clock $f_{s\uparrow}$ is divided by two to form the low rate output clock $f_{s\downarrow}$. The

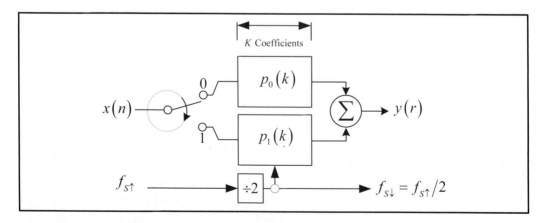

**Figure 6.6** Block diagram decimate by two PPF

low rate clock also is used to clock the internal operations within each of the two filter phases.

The operation of inputting two samples for every output sample accomplishes the decimation of the sample rate by a factor of two. The filter removes the aliasing frequencies, and the output sequence is a narrow band version of the input sequence at half the sample rate. The composite filter would probably be designed using the Parks-McClellan algorithm discussed in Chapter 5, "Finite Impulse Response Digital Filtering," and would be composed of a coefficient vector described by $h(n)$. The coefficients for each of the $M$ phases of the filter are derived from

$$p_m(k) = h(m+kM) \quad \left\{ \begin{array}{l} \text{for } m = 0,1,2,\cdots,M-1 \\ \text{for } k = 0,1,2,\cdots,K-1 \end{array} \right\}$$

**Equation 6.3**

In our simple example, the coefficients for each phase are determined to be

$$p_0(k) = \quad h(0) \quad h(2) \quad h(4) \quad h(6)$$
$$p_1(k) = \quad h(1) \quad h(3) \quad h(5) \quad h(7)$$

Using the notation in Figure 6.5, where the oldest data sequence sample is $x(0)$, we can derive the expression for the filtered and decimated output sequence $y(r)$:

$$y(r) = \sum_{n=0}^{N-1} h(n)x(rM+n) \quad \{\text{In this case } M = 2\}$$

**Equation 6.4**

Using Equation 6.4, we determine the multiply/accumulate (MAC) operations for the first three output samples in our example to be those illustrated in Figure 6.7.

We can see from Figure 6.7 that the polyphase architecture did not alter the mathematics of the filter process. It did, however, alter the input sample sequence alignment relative to the filter coefficient vector, which allowed us to process every other input data sequence vector. Note that in Figure 6.7 the filter coefficient $h(0)$ only processes the even data samples $x(0), x(2), x(4), \cdots$ and the coefficient $h(1)$ only processes the odd data samples $x(1), x(3), x(5), \cdots$, as stated earlier.

| $y(r)$ | 0 | 1 | 2 | 3 | 4 | 5 | 6 | 7 |
|---|---|---|---|---|---|---|---|---|
| | | | | $n$ | | | | |
| $y(0)=$ | $h(0)x(0)$ | $+h(1)x(1)$ | $+h(2)x(2)$ | $+h(3)x(3)$ | $+h(4)x(4)$ | $+h(5)x(5)$ | $+h(6)x(6)$ | $+h(7)x(7)$ |
| $y(1)=$ | $h(0)x(2)$ | $+h(1)x(3)$ | $+h(2)x(4)$ | $+h(3)x(5)$ | $+h(4)x(6)$ | $+h(5)x(7)$ | $+h(6)x(8)$ | $+h(7)x(9)$ |
| $y(2)=$ | $h(0)x(4)$ | $+h(1)x(5)$ | $+h(2)x(6)$ | $+h(3)x(7)$ | $+h(4)x(8)$ | $+h(5)x(9)$ | $+h(6)x(10)$ | $+h(7)x(11)$ |
| $\vdots$ | $\vdots$ | $\vdots$ | $\vdots$ | $\vdots$ | $\vdots$ | $\vdots$ | $\vdots$ | $\vdots$ |

**Figure 6.7**  First three output samples of a decimate by two filter

The data flow through each phase of the filter is functionally illustrated in Figure 6.8 for the first three output sample times, $r = 0,1,2$. We can see the filter coefficients in each of the two filter phases and the segment of the input data sequence being shifted through the filters at the three time increments. The detailed block diagram of our example decimate by two filters is shown in Figure 6.9.

This time the position of the wiper arm is shown relative to the input sample rate. On the sixth and seventh input clock ticks, the input samples $x(6)$ and $x(7)$ are wiped onto the two-filter delay lines and the first output point $y(0)$ is computed. From this point on, the wiper arm continues to apply samples of the input sequence and an output sample is computed at the end of each complete revolution of the wiper. Note again that the internal computations are carried out at the low output clock rate.

The reader should note that the two delay elements at the input to each of the two phases of the filter are included simply to clock and hold the input data samples from the wiper arm. If we were to eliminate these registers, then the input to the $h(6)$ and $h(7)$ multipliers would be transitory and the multiplier outputs would be corrupted.

The process for designing a filter that decimates the sample rate of a signal by any integer $M$ is identical to that discussed previously for the case where $M = 2$. For example, suppose you were tasked to design a low pass filter that will decimate the sample rate of an input signal by a factor of $M = 4$. The first thing to be done is to design an $N$ coefficient low pass filter, most likely using the Parks-McClellan algorithm, that processes the input signal to eliminate aliasing. The designer would choose the value of $N$ to be an integer multiple of $M = 4$ so that the number of coefficients $K$ in each phase of the resultant filter will be equal. If the design requires the number of filter coefficients to be $N = 128$, then we calculate the number of coefficients per phase as

$$K = N/M = 128/4 = 32$$

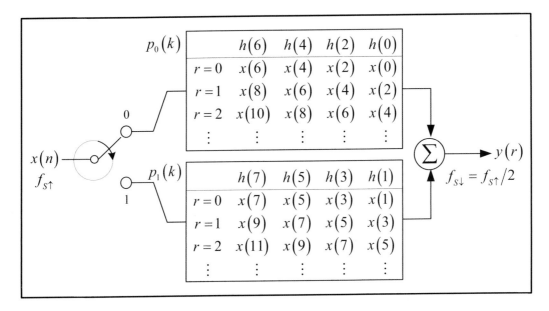

**Figure 6.8**  Two phase filter data flow

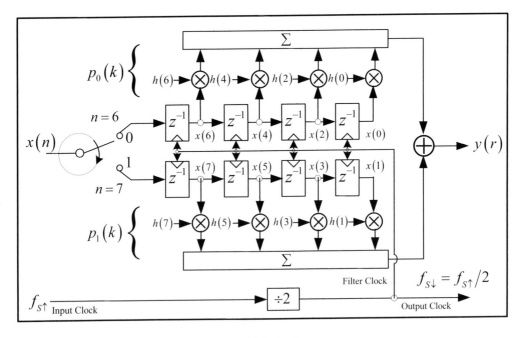

**Figure 6.9**  Decimate by two filter detailed block diagram

Since $M = 4$, we know that we will have a four-phase filter. We use Equation 6.3, repeated here for clarity, to determine the four sets of polyphase coefficients:

$$p_m(k) = h(m + kM) \quad \left\{ \begin{array}{l} \text{for } m = 0, 1, 2, \cdots, M-1 \\ \text{for } k = 0, 1, 2, \cdots, K-1 \end{array} \right\}$$

The coefficients for the four phases are computed as

$$
\begin{array}{rllllll}
p_0(k) = & h(0) & h(4) & h(8) & \cdots & h(124) \\
p_1(k) = & h(1) & h(5) & h(9) & \cdots & h(125) \\
p_2(k) = & h(2) & h(6) & h(10) & \cdots & h(126) \\
p_3(k) = & h(3) & h(7) & h(11) & \cdots & h(127)
\end{array}
$$

The resulting block diagram of the filter architecture is shown in Figure 6.10. The timing diagram for the decimate by four filter is illustrated in Figure 6.11.

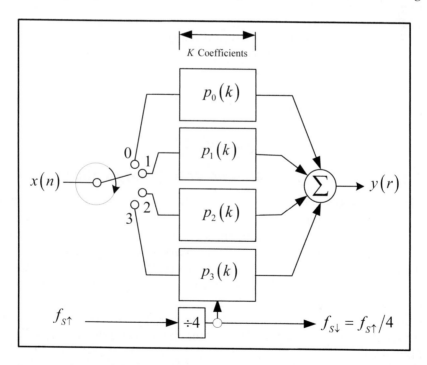

**Figure 6.10** Decimate by four filter block diagram

**Figure 6.11**   Timing diagram for the decimate by four filter

Note that successive coefficient vectors are offset in time from one another by four samples. Also note that coefficient $h(0)$ only operates on the data samples $x(0), x(4), x(8), x(12), \cdots$. The coefficient $h(1)$ only operates on the data samples $x(1), x(5), x(9), x(13), \cdots$, and so forth. These data sequences are generated and shifted into each of the four phase filters by the wiping action of the input multiplexer. The filtered output sequence is given by

$$y(r) = \sum_{n=0}^{N-1} h(n) x(rM + n) \qquad \{\text{In this case } M = 4\}$$

**Equation 6.5**

## 6.1.2  Increasing the Sample Rate (Interpolation)

Many times it is necessary to increase the sample rate of a system by some integer factor $L$. Typically this is necessary to match the data rate output by one system to the data rate input of another system. A good example is the creation of a wideband signal from a group of narrow band signals such as would be the case in a frequency division multiplex (FDM) processing system.

The operator notation that indicates an up sampling or interpolation function is usually given by the symbol $L\uparrow$. For example, processing a discrete signal $x(n)$ sampled at a clock rate $f_s$ by the operator $L\uparrow$ would produce a new discrete signal $y(r)$ sampled at a clock rate of $Lf_s$, as illustrated in Figure 6.12.

When the sample rate of a digital signal is interpolated, we need to account for the image frequencies that are created at multiples of the original sample rate. This is accomplished by low pass filtering the signal such that all

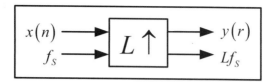

**Figure 6.12** Up sampling operator

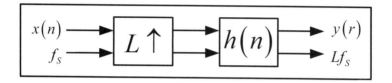

**Figure 6.13** Functional representation of sample rate interpolation

these image frequencies are eliminated. The typical symbol used in the industry to represent this interpolation operation includes the up sampling operator $L \uparrow$ and the low pass filter, as illustrated in Figure 6.13.

The anti-imaging low pass filter that is inherent in the interpolation operation is represented in the figure by the filter impulse response $h(n)$. This figure is not completely accurate in terms of representing the actual implementation of up sampling hardware or software. The figure does, however, convey all the information necessary to visualize and understand the process.

The process of up sampling or interpolating a signal is straightforward. We illustrate the procedure in Figure 6.14. In part A of the figure, we see a discrete signal with samples spaced every $T_{s\downarrow}$ second. The sample frequency for this signal is $1/T_{s\downarrow} = f_{s\downarrow}$. The spectrum of this discrete signal is shown as the Fourier transform of the time series.

In this example, we wish to increase the sample rate by a factor of $L = 4$. Functionally, this is accomplished by inserting three zero valued samples in between every adjacent input sample, as illustrated in part B of the figure. This increases the sample rate from $f_s$ to $4f_s$. The spectrum of this interpolated signal is shown as having replicated frequency bands that extend from 0 Hz to $f_{s\uparrow} = 4f_{s\downarrow}$ Hz. The replicated frequencies are called *image bands* and are represented with a dotted line.

The three image bands are centered at multiples of the original sample rate. We now apply a filter to the interpolated signal to eliminate these image bands. Part C of the figure illustrates the frequency response of the low pass anti-imaging filter $H(k)$. The frequency response is represented by a dashed line. It is designed to pass only the original signal spectrum.

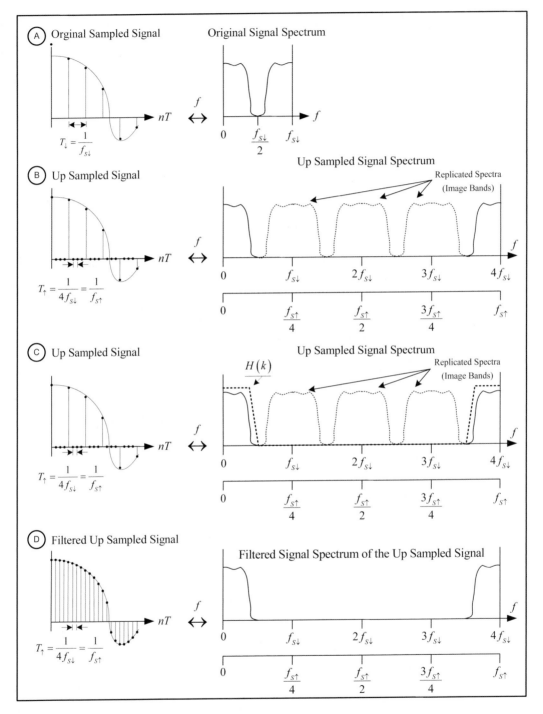

**Figure 6.14** Graphical illustration of signal interpolation by a factor of four

Part D of the figure shows the resultant filtered signal. We note that all the image bands have been filtered out. We also note that the zero samples of the filtered time series have been "interpolated" or averaged by the low pass filter so that their amplitude closely approximates the shape of the original time series. Although this method of interpolation will work, it is not the most efficient method because the filtering is done at the high sample rate after the zero samples have been added. Not only this, but 75% of all the computations performed in this particular interpolation example involve zero valued signal samples. This is a serious waste of processing resources, and it only gets worse as the interpolation factor increases. A better method is to manipulate the data and filter coefficients so that the insertion of zeros is not necessary. This will allow us to perform all the filter computations at the original low sample rate, and we need not waste computations on zero valued samples.

We look at the timing diagram of a simple interpolation filter to see if there are any attributes that we can exploit to simplify the design. Let's begin this discussion with the example timing diagram of Figure 6.15 for a simple filter that interpolates the signal by a factor of two. Once again it is beneficial to view the filter input data sequence $x(n)$ as an infinite sequence that

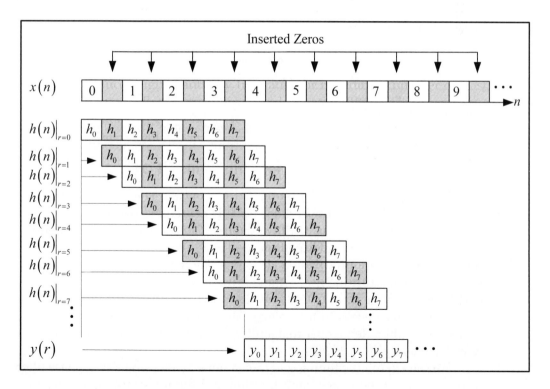

**Figure 6.15** Interpolate by a factor of two timing diagram

is fixed in time. In this visualization, the filter coefficient vector slides by the fixed data sequence one sample at a time at the low sample rate. The figure shows the position of the filter coefficient vector relative to the input data sequence for the first eight clock periods $(R = 0 \rightarrow 7)$. Note that for this example we have inserted a zero valued sample between adjacent input samples to increase the sample rate by a factor of two. Each zeroed data value is identified by a shaded sample.

Each clock period, a new output $y(r)$ is computed by multiplying the data vector by the time-aligned coefficient vector. Using the sequence notation of Figure 6.15, we can represent the filter calculations:

$$y(r) = \sum_{n=0}^{N-1} h(n) x(n + r)$$

**Equation 6.6**

We can see from the figure that we compute an output sample for each input sample. Because of zero insertion, the input sequence has been up sampled by a factor of two. Therefore the output sample rate is twice the rate of the original input signal sample rate.

We can see that at time $r = 0$, the coefficient vector and data vector are aligned so that the only coefficients involved in the computation of the output sample are the even coefficients: $h(0)$, $h(2)$, $h(4)$, and $h(6)$. The odd coefficients are multiplied by zero valued data samples and therefore make no contribution to the filter output.

We can also see from the timing diagram that at time $r = 1$ the only coefficients that are used to compute an output sample are the odd coefficients $h(1), h(3), h(5)$, and $h(7)$. The even coefficients are multiplied by zero valued data samples and therefore make no contribution to the filter output.

If we put on our engineering thinking caps for a moment, we realize that we can exploit this attribute because there is no need to perform the computations that involve zero value data samples. These calculations add nothing to the final output, so why expend resources to perform them? We can break the composite filter into two smaller filters: one that contains all the even filter coefficients, $h(0)$, $h(2)$, $h(4)$, and $h(6)$, and one that contains all of the odd filter coefficients, $h(1)$, $h(3)$, $h(5)$, and $h(7)$. Each of the smaller filters is referred to as a *phase*. In this interpolate by two example, we end up with a two-phase filter. Remember that a filter that is broken into several phases is referred to as a *poly phase filter*.

We can exploit the timing diagram symmetry by using a two-phase filter to perform both the image band filtering and the up sampling. Let us suppose that we designed the composite filter $h(n)$ using the Parks-McClellan algorithm to have $N$ coefficients. Being good engineers, we also made sure that $N$ is an integer multiple of the interpolation factor $L$.

We know that $L = 2$, so our filter will have two phases, $p_0$ and $p_1$. We also know that the number of coefficients in each phase is given by $K = N/L$. In our trivial example, we will let $N = 8$, so $K = 8/2 = 4$. The coefficients for each phase are determined by

$$p_l(k) = h(l + kL) \quad \left\{ \begin{array}{l} \text{for } l = 0,1,2,\cdots,L-1 \\ \text{for } k = 0,1,2,\cdots,K-1 \end{array} \right\}$$

**Equation 6.7**

Thus the coefficients for each of the two filter phases are

$$p_0(k) = \quad h(0) \quad h(2) \quad h(4) \quad h(6)$$
$$p_1(k) = \quad h(1) \quad h(3) \quad h(5) \quad h(7)$$

The block diagram of this simple filter is illustrated in Figure 6.16. The input signal sequence $x(n)$ is applied to both filters simultaneously. The output signal sequence $y(r)$ is the multiplex of the two filter outputs. This multiplex is illustrated in the figure as a wiper arm. If we look at the timing diagram of Figure 6.15, we see that the first output sample is computed with the coefficients of filter phase 0, or $p_0(k)$. Therefore the block diagram shows the output wiper rotating in a clockwise direction, with the initial position of the wiper blade at the output terminal 0 of filter $p_0(k)$. The output wiper arm completes one revolution, producing two output samples for each input

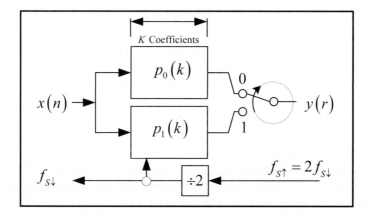

**Figure 6.16**   Interpolate by two PPF

sample. Therefore, this simple two-phase configuration interpolates the sample rate by a factor of two by outputting two samples for every input sample, and we did not have to add or waste any valuable resources to process any zero valued samples. The output data sequence for our simple example, where the number of composite filter coefficients is $N = 8$, the sample interpolation rate is $L = 2$, and the number of coefficients per phase is $K = 4$, is

$$
\begin{aligned}
y(0) &= x(0)h(0) \quad +x(1)h(2) \quad +x(2)h(4) \quad +x(3)h(6) \\
y(1) &= \quad\quad\quad\quad\; x(1)h(1) \quad +x(2)h(3) \quad +x(3)h(5) \quad +x(4)h(7) \\
y(2) &= \quad\quad\quad\quad\; x(1)h(0) \quad +x(2)h(2) \quad +x(3)h(4) \quad +x(4)h(6) \\
y(3) &= \quad\quad\quad\quad\quad\quad\quad\quad\; x(2)h(1) \quad +x(3)h(3) \quad +x(4)h(5) \quad +x(5)h(7) \\
y(4) &= \quad\quad\quad\quad\quad\quad\quad\quad\; x(2)h(0) \quad +x(3)h(2) \quad +x(4)h(4) \quad +x(5)h(6) \\
&\quad\; \vdots \quad\quad\quad\quad\quad\quad\quad\quad\quad\quad\quad\quad \vdots
\end{aligned}
$$

**Equation 6.8**

The block diagram of the generic architecture for our example two-phase filter is illustrated in Figure 6.17. Note that in this diagram the composite filter is divided into two phases, each with half the coefficients. The phase 0 filter $p_0(k)$ contains all the even coefficients, and the phase 1 filter $p_1(k)$ contains all the odd coefficients. We also see from the figure that the filter computations occur at the low clock rate. This is because we do not perform the computations for the "zero" data samples. Also note that the first output sample $y(0)$ is taken from the output of the phase 0 or $p_0(k)$ filter. The division of the system clock by two is illustrated in the block diagram as being implemented local to the filter. This is done only for clarity. A much better approach is to generate and distribute all the system clocks in a separate logic section within the system so that clock frequencies and relative phases can be precisely derived and distributed phase aligned through drivers that can adequately source both the clock load and clock transmission line.

Once again, the reader should note that the delay elements at the input to each of the two phases of the filter are included simply to reclock and hold the input data samples. If we were to eliminate these registers, then the input to the $h(6)$ and $h(7)$ multipliers might be transitory and the multiplier outputs would be corrupted.

If we take a second look at the filter outputs in Equation 6.8, we see that the data samples $x(n)$ processed by both filter phases are identical. We can see that the data in the filter registers remain constant for each complete cycle of the output wiper arm. This attribute allows us to remove half the data storage resisters in Figure 6.17. In doing so, we streamline the filter architecture somewhat and arrive at the new filter architecture illustrated in Figure 6.18. Because of the full

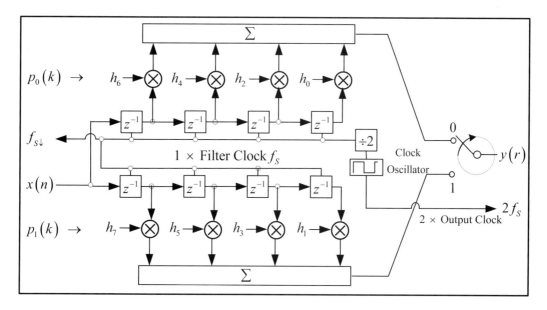

**Figure 6.17**  Block diagram of a interpolate by two PPF

use of hardware resources, this is the fastest polyphase interpolator architecture that can be implemented. An architecture of this type would be utilized for high sample rate cases where computational times are short and it is not possible to share hardware resources. After the filter delay line is initialized with input data samples, the sequence of operations is outlined as

| Sample clock | Operation | Output / Wiper arm |
|:---:|:---|:---:|
| | Compute | $y(0)/0$ |
| 0 | Clock in next input sample | |
| | Compute | $y(1)/1$ |
| | Compute | $y(2)/0$ |
| 1 | Clock in next input sample | |
| | Compute | $y(3)/1$ |
| | Compute | $y(4)/0$ |
| 2 | Clock in next input sample | |
| | Compute | $y(5)/1$ |
| | Compute | $y(6)/0$ |
| $\vdots$ | $\vdots$ | $\vdots$ |

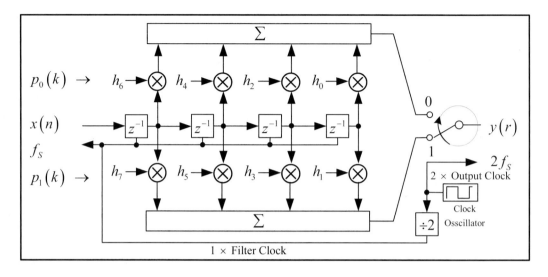

**Figure 6.18**  Interpolate by two example filter architecture

The output sample is taken from the indicated wiper arm post. If we have the luxury of working at lower data rates, then we may be able to time-share some of the workhorse hardware components such as the multipliers. An example of this type of architecture is shown in Figure 6.19. Here we have eliminated half of the multipliers and added a recirculating coefficient memory and address counter. This type of architecture is hardware efficient, but it does require that $L$ vector computations be performed sequentially for each input sample. This, of course, means that the filter will be $L$ times slower than its counterpart of Figure 6.18. A general form of this architecture for any interpolation value $L$ is illustrated in Figure 6.20. The good news is that this efficient model of a PPF pays big dividends in hardware resource savings as the interpolation ratio $L$ and the number of phase coefficients $K$ become large. For example, a 16-phase filter with 32 coefficients in each phase will require 512 multipliers in the full blown architecture but only 32 multipliers in the hardware efficient architecture. At low sample rates it may be possible to time-share these remaining multipliers as well, reducing the hardware resource count even further.

Let's look at another simple example. This example is intended to show that the design of any polyphase interpolation filter is a simple extension of the interpolate by two case we just discussed. We need to keep the examples simple only because the illustrations of the timing diagrams and filter architectures demand it. A larger filter capable of operating in a real application is designed and built exactly as these simple examples are described. For this next example, we will look at the implementation of an interpolate by four filter.

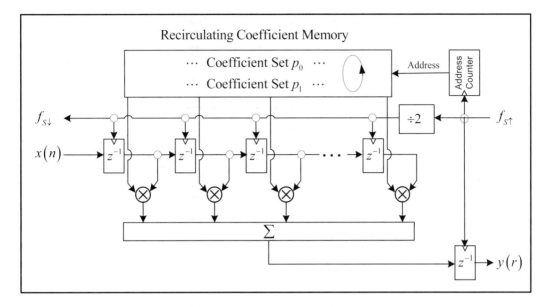

**Figure 6.19**   Hardware efficient implementation of a two-phase interpolation filter

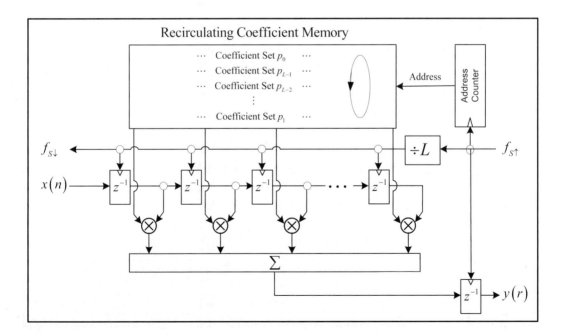

**Figure 6.20**   Hardware efficient multiphase interpolation filter architecture

Let's assume that we have an input data sequence $x(n)$ at some low sample rate $f_{s\downarrow}$ and we would like to up sample this signal by a factor of four and remove the three image bands that will result. The timing diagram of such a signal is illustrated in Figure 6.21.

Since our intention is to increase the sample rate by a factor of four, we can accomplish this by simply inserting three zero valued samples in between adjacent input signal samples. The input signal after the zero insertion is as shown in the figure. The zero samples in the input sequence are shown as shaded samples.

It is important to remember that the insertion of zeros is theoretically correct but does not actually occur. As with the case of the interpolate by two filter we discussed previously, we will implement this filter as a polyphase structure and we will simply not bother to perform the calculations that involve data samples that are zero inserts. The timing diagram shades all the zero data samples and all the unused filter coefficients for each successive dot product. As before, we will adopt the model of an infinite data sequence that is fixed in time. In this model, the filter coefficient vector slides by the data sequence at a rate of one sample per clock cycle. This is the visualization that is depicted in Figure 6.21.

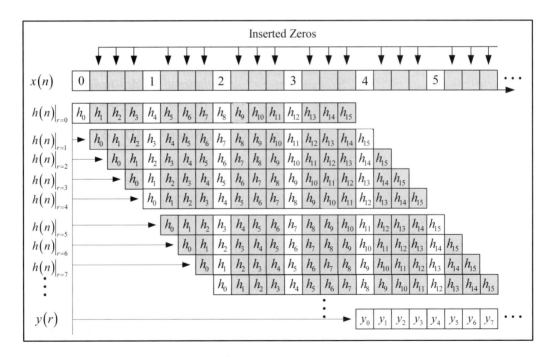

**Figure 6.21**   Interpolate by four input data and filter coefficients

Being astute engineers, we will implement this interpolate by four filter as a poly phase filter. Right away, we know that since we intend to up sample by a factor of $L = 4$, there will be four phases in the filter.

Our first task is to design a composite FIR filter that will pass the original signal bandwidth and attenuate the image bands that result when the sample rate is increased. There will be $L - 1 = 3$ image bands that will need to be filtered out. We will design the filter using the Parks-McClellan algorithm discussed previously. In that filter design, we will ensure that the number of filter coefficients $N$ is an integer multiple of the interpolation ratio $L$.

For our design example, we will assume that we were able to design a suitable filter with $N = 16$ coefficients. Now that we know $N$ and $L$, we can compute the number of coefficients $K$ required by each of the four filter phases. The value of $K$ is given by

$$K = \frac{N}{L} = \frac{16}{4} = 4$$

The rest of the design is straightforward. First we need to divide the composite filter coefficients into the groups used by each filter phase. We accomplish this feat through the following:

$$p_l(k) = h(l + kL) \quad \left\{ \begin{array}{l} \text{for } l = 0,1,2,\cdots,L-1 \\ \text{for } k = 0,1,2,\cdots,K-1 \end{array} \right\}$$

**Equation 6.9**

The coefficients for each of the four phases of the PPF are given by

$$
\begin{aligned}
p_0(k) &= & h(0) & \quad h(4) & \quad h(8) & \quad h(12) \\
p_1(k) &= & h(1) & \quad h(5) & \quad h(9) & \quad h(13) \\
p_2(k) &= & h(2) & \quad h(6) & \quad h(10) & \quad h(14) \\
p_3(k) &= & h(3) & \quad h(7) & \quad h(11) & \quad h(15)
\end{aligned}
$$

The block diagram of the four-phase filter is illustrated in Figure 6.22. The reader can see that the filter architecture is identical in structure to that described for the two-phase filter discussed earlier. The only difference is that this filter has four phases as opposed to two.

As before, the input signal is applied to all four filter phases so that each phase is operating on identical data samples. The filter output is a provided

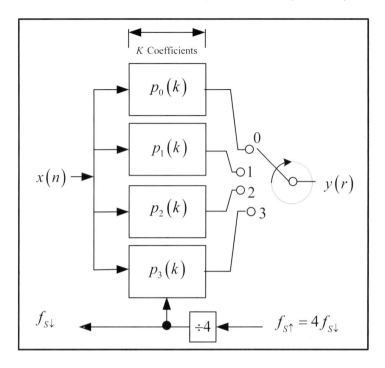

**Figure 6.22**   Block diagram of a four-phase interpolation filter

by a multiplexer, modeled as a rotating wiper arm. If you compare the wiper arm action with the timing diagram of Figure 6.21 and the coefficients generated by Equation 6.9, you can see that the wiping sequence is clockwise, beginning with the output of $p_0(k)$. The wiping sequence is

$$p_0(k), p_3(k), p_2(k), p_1(k), p_0(k), p_3(k), p_2(k), p_1(k), p_0(k), \cdots$$

This is pretty simple stuff. The architecture of a bigger interpolation filter is identical, differing only by the factors of $N, L$, and $K$. The first 13 output samples from the interpolate by four filter are computed and shown in Figure 6.23. Once again, the reader should note that the same input samples are used in the computations for all four filter phases.

If the period of the low sample rate is long enough, we can simplify the filter architecture by time-sharing the filter data delay line and the coefficient multipliers. The architecture of a hardware-efficient four-phase PPF is illustrated in Figure 6.24.

The sequence of operation of the four-phase efficient architecture is identical to that described previously for the two-phase architecture. We first initialize the filter by filling the delay line with input samples, and then we follow the periodic sequence given by

| Filter Phase | Output Sample | | | | |
|---|---|---|---|---|---|
| $p_0(k)$ | $y(0)$ | $h(0)x(0)$ | $+h(4)x(1)$ | $+h(8)x(2)$ | $+h(12)x(3)$ |
| $p_3(k)$ | $y(1)$ | $h(3)x(1)$ | $+h(7)x(2)$ | $+h(11)x(3)$ | $+h(15)x(4)$ |
| $p_2(k)$ | $y(2)$ | $h(2)x(1)$ | $+h(6)x(2)$ | $+h(10)x(3)$ | $+h(14)x(4)$ |
| $p_1(k)$ | $y(3)$ | $h(1)x(1)$ | $+h(5)x(2)$ | $+h(9)x(3)$ | $+h(13)x(4)$ |
| | | | | | |
| $p_0(k)$ | $y(4)$ | $h(0)x(1)$ | $+h(4)x(2)$ | $+h(8)x(3)$ | $+h(12)x(4)$ |
| $p_3(k)$ | $y(5)$ | $h(3)x(2)$ | $+h(7)x(3)$ | $+h(11)x(4)$ | $+h(15)x(5)$ |
| $p_2(k)$ | $y(6)$ | $h(2)x(2)$ | $+h(6)x(3)$ | $+h(10)x(4)$ | $+h(14)x(5)$ |
| $p_1(k)$ | $y(7)$ | $h(1)x(2)$ | $+h(5)x(3)$ | $+h(9)x(4)$ | $+h(13)x(5)$ |
| | | | | | |
| $p_0(k)$ | $y(8)$ | $h(0)x(2)$ | $+h(4)x(3)$ | $+h(8)x(4)$ | $+h(12)x(5)$ |
| $p_3(k)$ | $y(9)$ | $h(3)x(3)$ | $+h(7)x(4)$ | $+h(11)x(5)$ | $+h(15)x(6)$ |
| $p_2(k)$ | $y(10)$ | $h(2)x(3)$ | $+h(6)x(4)$ | $+h(10)x(5)$ | $+h(14)x(6)$ |
| $p_1(k)$ | $y(11)$ | $h(1)x(3)$ | $+h(5)x(4)$ | $+h(9)x(5)$ | $+h(13)x(6)$ |
| $\vdots$ | $\vdots$ | $\vdots$ | $\vdots$ | $\vdots$ | $\vdots$ |

**Figure 6.23**   First 13 output samples from the interpolate by four filter

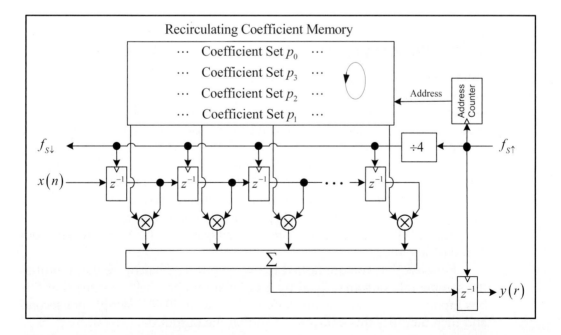

**Figure 6.24**   Efficient implementation of a four-phase interpolate by four filter

| Sample Clock | Operation | Output / Wiper Arm | Filter Phase |
|---|---|---|---|
| | Compute | $y(0)/0$ | $p_0(k)$ |
| 1 | Clock in next input sample | | |
| | Compute | $y(1)/3$ | $p_3(k)$ |
| | Compute | $y(2)/2$ | $p_2(k)$ |
| | Compute | $y(3)/1$ | $p_1(k)$ |
| | Compute | $y(4)/0$ | $p_0(k)$ |
| 2 | Clock in next input sample | | |
| | Compute | $y(5)/3$ | $p_3(k)$ |
| | Compute | $y(6)/2$ | $p_2(k)$ |
| | Compute | $y(7)/1$ | $p_1(k)$ |
| | Compute | $y(8)/0$ | $p_0(k)$ |
| 3 | Clock in next input sample | | |
| | Compute | $y(9)/3$ | $p_3(k)$ |
| | Compute | $y(10)/2$ | $p_2(k)$ |
| | Compute | $y(11)/1$ | $p_1(k)$ |
| | Compute | $y(12)/0$ | $p_0(k)$ |
| $\vdots$ | $\vdots$ | $\vdots$ | $\vdots$ |

## 6.1.3  Changing the Sample Rate by a Noninteger Value

We have discussed the methods used to both decrease the sample rate by an integer factor $M$ and to increase the sample rate by an integer factor $L$. Now we turn our attention to the methods used for changing the sample rate of a system by a noninteger value that can be expressed as a rational number (i.e., the ratio of two integers).

As we did before, we will define the rational sample rate change operator symbol as illustrated in Figure 6.25. This operator signifies the change of a data stream sample rate by a rational number $L/M$. As we can see from the figure, the input sequence $x(n)$ is sampled at a rate $f_s$, and the output sequence $y(r)$ is sampled at some noninteger rate $(L/M)f_s$, where both $L$ and $M$ are integers.

Suppose that you are tasked to design a box of hardware that modifies the sample rate of some signal by a rational number. Let's assume that this box inputs a serial data stream clocked at a rate of $20.48K$ sample per second and then modifies this data stream so that the output data stream is clocked

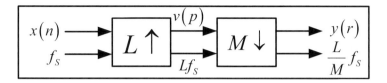

**Figure 6.25**   Rational sample rate change operator

at $19.20K$ sample per second. An application such as this is a fairly common operation in many disciplines, such as in the telecom world.

The ratio of the output sample rate to the input sample rate is $19,200/20,480 = 0.9375$. Initially this looks like a very difficult task. Perhaps in a career saving move, you could hand this task off to a junior engineer, thus keeping your path to management untarnished. But then you suddenly realize that the ratio of $19,200/20,480$ can be factored down to a simple ratio of $L/M = 15/16$. There is no reason the junior engineer should get the credit for such a complicated looking but simple design, so you decide that this task can be successfully implemented by designing a two-filter cascade. The first filter in the cascade would be an interpolation filter that would up sample the input signal by a factor of $L = 15$. The second filter in the cascade would be a decimation filter that would down sample the signal by a factor of $M = 16$.

This two filter cascade is very simple to implement utilizing a polyphase architecture for both filters. The functional configuration for this rational rate change filter is illustrated in Figure 6.26.

There are methods to significantly reduce the amount of hardware necessary to build this cascade, but these methods also increase the design complexity. Modern day field programmable gate arrays (FPGAs) can easily host a filter cascade such as the one in Figure 6.26, while utilizing only a very small portion of their onboard resources.

The reader should keep in mind that a good design is not necessarily the minimal design, nor is a good design always the most technically astute design. Many times, a good design is one that is simple, works reliably, and is completed on program schedule and under program budget. These important attributes of a good design are the direct result of employing minimal design complexity. One of the best tools in an engineer's toolbox is the philosophy of "keep it simple." However, should it be absolutely necessary, we can take steps to reduce the amount of hardware required to implement this cascade filter. We begin by expanding the rational rate change operator of Figure 6.25. If we expand the operator to include the associated interpolation and decimation low pass filters, we end up with the operator illustrated in Figure 6.27, where the impulse response of the interpolation low pass filter is given by

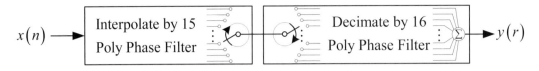

**Figure 6.26**  Cascade filter for a noninteger sample rate change

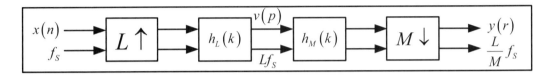

**Figure 6.27**  Expansion of the rational sample rate change operator

$h_L(k)$ and the impulse response of the decimation low pass filter is given by $h_M(k)$. We note from the figure that the signal out of the interpolation stage is termed $v(p)$, and it sampled at a rate of $Lf_s$. The sequence $v(p)$ is applied directly to the input of the decimation filter.

Because $h_L(k)$ and $h_M(k)$ are in cascade, they can be combined into a single composite filter $h_C(k)$. This results in the rational rate change operator illustrated in Figure 6.28. This configuration is different from the two independent multirate filters operating in cascade discussed previously. In that case, the computations in the interpolation filter occur at the low input sample rate of $f_s$, and the computations in the decimation filter occur at the low output sample rate of $(L/M)f_s$. The increase in computation rate for the filter $h_C(k)$ increases the complexity of the overall design. Care must also be taken in the design of $h_C(k)$ as well, because it now must include the attributes of both the interpolation and decimation filters. The selection of different coefficient "phases" from the composite coefficient vector adds additional design complexity. And, of course, the test and integration of the hardware will only go smoothly if the design is completely understood and thoroughly simulated beforehand.

Rabiner and Crochiere [2] have shown a method for implementing a composite rational rate change filter such as the one shown in Figure 6.28 using a direct form FIR filter structure with time varying filter coefficients. Their method gives us the advantage of performing the filter computations at the output sample rate of $(L/M)f_s$. If minimizing the hardware real estate consumption is a high design priority, the reader is encouraged to thoroughly investigate the architecture provided in reference [2].

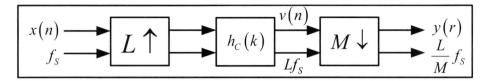

**Figure 6.28**   Functional representation of a rational sample rate change operator

## 6.2   HALF BAND FILTER

In Chapter 5, "Finite Impulse Response Digital Filtering," we discussed a unique form of FIR filter that is appropriate for multirate filtering applications where the sample rate of a signal is changed by a factor of two. This type of filter is called a *half band filter*. It has some appealing attributes that make it attractive for use in certain multirate applications.

The frequency domain amplitude response of a half band filter is illustrated in Figure 6.29. There are three frequency scales shown in the figure, to illustrate the equivalence of different frequency axis calibrations. Depending on the frequency scale the reader uses, we note from the figure that the filter transition band is symmetrical about the frequency $f = 0.25$, $f = f_S/4$, or $\omega = \pi/2$ and that the amplitude of the frequency response at that frequency is equal to 0.5. We also note that the pass band amplitude from 0 to $f_{PASS}$ Hz is 1.0. The bandwidth of the pass band and the stop band are equal. The filter stop band frequency $f_{STOP}$ is related to the filter pass band frequency by

$$
\begin{aligned}
f_{Stop} &= f_S/2 - f_{Pass} \\
&= 0.5 \; - f_{Pass}/f_S
\end{aligned}
$$

where $f_S$ is the sample rate. When we use the Parks-McClellan algorithm to design a filter with the frequency response shown here, we need to specify that the filter response is symmetric about $f_S/4$ and that the pass band ripple (PBR) $\delta_{PASS}$ and stop band attenuation (SBA) $\delta_{STOP}$ are equal. We also specify that the number of filter coefficients $N$ is an odd number. When we design a filter with the these specifications, the resultant filter impulse response $h(n)$ takes on a very interesting characteristic in that

$$
h(n) = \begin{cases}
0.5 & n = (N-1)/2 \\
0 & n = \text{even} \\
h(n) & n = \text{odd}
\end{cases}
$$

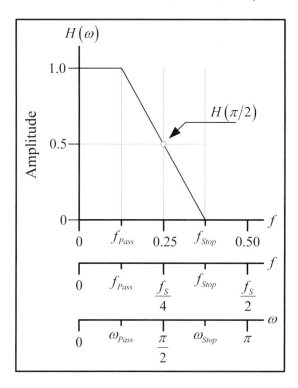

**Figure 6.29** Half band filter symmetrical frequency response

That is, the center coefficient $h\left(\dfrac{N-1}{2}\right)$ is equal to 0.5, and all even co-efficients are zero. The only nonzero coefficients other than the center coefficient are the odd coefficients. The half band filter design example discussed in Chapter 5, "Finite Impulse Response Digital Filtering," clearly illustrates these characteristics. This is very interesting: because all the even filter coefficients are zero, the number of filter computations can be reduced by half. In addition, if the filter is designed with symmetric coefficients, the computations can be halved once again. This makes for a very efficient filter from a purely computational point of view. A conceptual example of a 23-coefficient half band filter impulse response is illustrated in Figure 6.30.

If we choose our filter pass band carefully and choose the width of the filter transition bands carefully so that we minimize the effects of aliasing, then we can reduce the sample rate out of a half band filter by a factor of two. This can be as simple as computing every other sample of the output sequence. In this application, the filter is oftentimes referred to as a *decimate by two filter*.

If the design engineer is trying to minimize the number of filter coefficients utilized by a half band filter, the transition band can be quite wide. Therefore many times it is necessary to follow a half band filter or a cascade

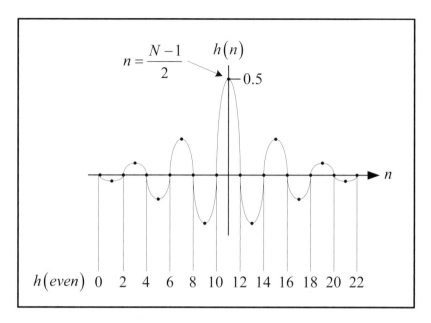

**Figure 6.30**   Half band filter impulse response

of half band filters with a cleanup FIR filter. The cleanup filter is a standard FIR filter. It typically has a narrow transition band, and it is used to eliminate residue frequencies left by the wide skirts of a half band filter.

Many times the half band filter function and the sample rate reduction function are merged into a single block diagram symbol such as illustrated in Figure 6.31. The figure illustrates an architecture composed of a decimate by two half band filter in cascade with a low pass FIR filter.

Changing the sample rate by multiple powers of two is just an extension of the decimate by two example. For example, the block diagram for a decimate by eight half band filter cascade is illustrated in Figure 6.32.

Let's take a look at the signal processing from the frequency domain perspective of a typical decimate by two half band filter modeled by the block diagram Figure 6.31. The frequency domain processing is shown in Figure 6.33.

In part A of the figure, we see a representative wideband signal spectrum that extends almost to the folding frequency of $f = 0.5$. The hatched area of the wideband spectrum represents the spectra of the desired narrow band signal that we wish to extract. The wideband signal spectrum is labeled $X(f)$. Its time domain counterpart $x(n)$ is shown on the right side of the diagram as the input to the half band filter / cleanup filter cascade.

Part B of the figure shows the same wideband spectra with the superimposed frequency response of the half band filter. We note from the figure that the filter transition band extends from $f_{PASS}$ to $f_{STOP}$ and is symmetric about $f = 0.25$. The amplitude of the filter response at this frequency is 0.5.

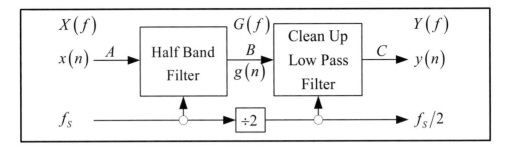

**Figure 6.31**    Half band / cleanup filter cascade

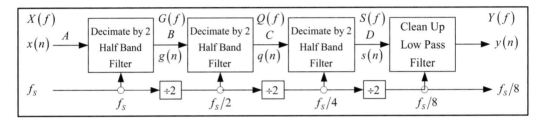

**Figure 6.32**    Decimate by eight half band filter cascade

In part C of the figure, we can see the wideband signal spectra after the sample rate has been reduced by a factor of two. We note that the cross-hatched area where spectrum aliasing occurs. Note that the aliasing does not extend into the pass band of interest. This is because we chose the filter pass band frequency $f_{PASS}$ to be reasonably greater than the desired pass band edge. This is an important point of designing a larger half band cascade. To simplify design constraints, the filter pass band edge for each successive half band filter in a cascade can be smaller than its predecessor. This is of no consequence as long as the pass band edge of the filter is greater than the pass band edge of the desired signal spectrum. The resulting spectrum of the filtered signal is labeled $G(f)$ and is shown on the right of the figure as being the input to the cleanup filter.

Part D of the figure illustrates the final spectrum at the output of the cleanup filter. This spectrum is labeled $Y(f)$ and is shown on the right of the figure as the output of the filter cascade. Here you will note that the cleanup filter is designed with a very narrow transition band. Its purpose is to eliminate the residue transition band frequencies left over from the half band filtering. In this example, where the sample rate has been reduced, the cleanup filter has the advantage of operating at the lower sample rate and therefore is simpler to design.

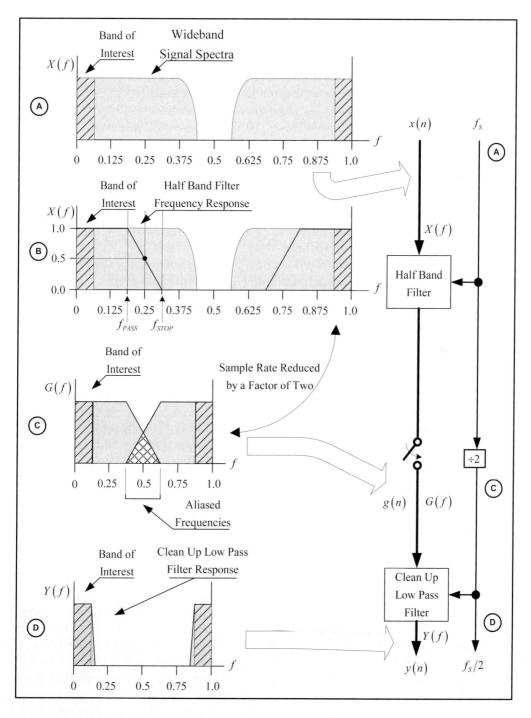

**Figure 6.33**   Using a half band filter to decimate the sample rate by $M = 2$

## 6.3 CASCADED INTEGRATOR COMB (CIC) FILTER

A half band filter or PPF are appropriate for use where the overall change in sample rate is small to moderate. For large rate changes, however, the number of half band filters in cascade can grow to be quite large and the complexity of a PPF can increase rapidly with an increase in sample rate change. For example, a decimation factor of 64 would require six half band filters in cascade, each decimating the sample rate by a factor of two, or five half band filters in cascade followed by a decimate by two cleanup filter. In this example, the cascade of half band filters rapidly becomes inefficient.

A more efficient class of filter that can be used for large sample rate changes is the CIC filter developed by E. B. Hogenauer in 1981 [1]. The CIC filter can be implemented either as an interpolation filter or as a decimation filter.

Just as its name suggests, a CIC filter is composed of what are called *integrator sections* and *comb sections* in cascade with one another. Configured as a decimation filter, the CIC is implemented by cascading $N$ integrator sections followed by $N$ comb sections. When configured as an interpolation filter, the CIC is implemented by $N$ comb sections in cascade, followed by $N$ integrator sections.

Functionally, each comb section is paired to an integrator section, and each comb/integrator pair is referred to as a *stage*. A CIC filter is usually defined by the number of stages $N$ that it contains. For example, an $N = 4$ stage CIC filter would have four comb sections in cascade and four integrator sections in cascade, for a total of $2N$, or eight sections.

The CIC filter has a unique characteristic that makes it attractive for applications where the sample rate change is large. That is, all the filter coefficients are equal to unity, so there are absolutely no multipliers required. This attribute substantially reduces the number of computations necessary to implement a large rate change filter, which translates into a significantly simpler design that requires considerably less hardware. In addition, since all the CIC filter coefficients are unity, the filter is symmetric and therefore has a linear phase response and a constant group delay.

Some example applications where the CIC can make an impact include zoom spectral transforms, narrow band signal extraction in digital receivers, preprocessing for a PPF used for signal channelization, and a host of other applications that require efficient processing to implement large rate changes.

We will begin the discussion of this unique class of filter by first listing the parameters that we use to both define the filter and to specify its performance. Then we will discuss the individual comb and integrator sections. Once this is complete, we will have a platform from which to discuss the architecture of the CIC filter and then derive its z-transform, impulse response, and frequency response. Once this is accomplished, we will design an example decimation filter and example interpolation filter and discuss the performance of each.

In keeping with the notation in Hogenauer's IEEE paper [1], we define the following filter parameters that are specific to CIC filters:

$R$    The sample rate change implemented by the filter

$M$   The comb filter differential delay; typically $M$ is set to 1 or 2

$N$   The number of stages in a CIC filter

$f_{S\uparrow}$   The high sampling rate $f_{S\uparrow} = Rf_{S\downarrow}$, always used by the integrator section

$f_{S\downarrow}$   The low sampling rate $f_{S\downarrow} = f_{S\uparrow}/R$, always used by the comb section

$f_{\uparrow}$   Frequencies referenced to the high sample rate

$f_{\downarrow}$   Frequencies referenced to the low sample rate

$f_{PASS}$   The pass band cutoff frequency of the input signal of interest

$H_C(\omega)$   The frequency response of the CIC comb filter stage

$H_I(\omega)$   The frequency response of the CIC integrator stage

$\theta(\omega)$   The phase response of the composite CIC filter

The CIC filter is a multirate filter, so it makes sense to view it as a device that operates at two separate clock rates. The integrator section of the CIC filter always operates at the high sample rate of $f_{S\uparrow} = Rf_{S\downarrow}$, and the comb filter section of a CIC filter always operates at the low sample rate of $f_{S\downarrow} = f_{S\uparrow}/R$.

## 6.3.1   CIC Comb Filter Stage

Two configurations of a CIC comb filter are illustrated in Figure 6.34. The configuration in part A of the figure illustrates a comb section for the case where the differential delay is set to $M = 1$.

This simple structure generates an output sequence given by the difference equation

$$y(n) = x(n) - x(n-1)$$

The configuration in part B illustrates a comb section for the case where the differential delay is set to $M = 2$. The output sequence generated by this structure is given by the difference equation

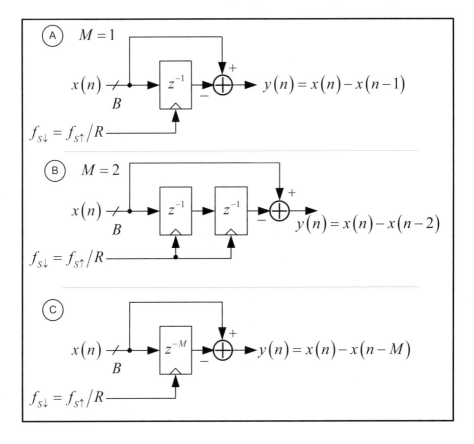

**Figure 6.34**  CIC filter comb building block

$$y(n) = x(n) - x(n-2)$$

The block diagram in part C illustrates the general form of the comb filter architecture where the parameter $M$ specifies the programmable comb filter differential delay. The output sequence generated by this structure is given by the difference equation

$$y(n) = x(n) - x(n-M)$$

**Equation 6.10**

Although $M$ could take on many possible values, the best CIC filter performance is generally obtained by limiting $M$ to be equal to either 1 or 2. In all cases, the bit width of the comb filter input $x(n)$ is given by $B$ bits. The corresponding z-transform is easily determined to be

$$Y(z) = \left(1 - z^{-M}\right) X(z)$$
$$H_c(z) = Y(z)/X(z)$$
$$H_c(z) = \left(1 - z^{-M}\right)$$

**Equation 6.11**

We can obtain the comb filter frequency response by evaluating its z-transform on the unit circle in the $z$-plane. We accomplish this by replacing the complex variable $z$ with the complex exponential $re^{j\omega}$ such that $z = re^{j\omega}$. Since the evaluation takes place on the unit circle, the magnitude $r$ equals unity and therefore $z = e^{j\omega}$. Making this substitution into Equation 6.11, we get

$$H_c(z)\big|_{z=e^{j\omega}} = \left(1 - e^{-j\omega M}\right) = \left(1 - e^{-j2\pi M f_\downarrow}\right)$$

**Equation 6.12**

The expression in Equation 6.12 represents the z-transform of the comb filter evaluated on the unit circle relative to the low sample rate $f_{S\downarrow}$.

The composite CIC filter response is composed of both the comb frequency response $H_C(z)$ and the integrator frequency response $H_I(z)$ in cascade. The composite frequency response is given by $H(z) = H_I(z)H_C(z)$ for a single stage filter. Each of these two frequency response components operate at different sample rates. When we discuss the stand-alone comb filter frequency response, we would like to do so relative to the low sample rate $f_{S\downarrow}$. Therefore we will use Equation 6.12. However, in order to discuss the composite CIC filter frequency response, we will need to reference the frequency response of the comb filter sections relative to the high sample rate of $f_{S\uparrow} = Rf_{S\downarrow}$ of the integrator. We can derive this high rate reference as follows. When we substitute the high rate frequency $f_\uparrow = Rf_\downarrow$ into Equation 6.12, we obtain

$$H_C(z)\big|_{z=e^{j\omega}} = \left(1 - e^{-j2\pi f_\downarrow RM}\right)$$

**Equation 6.13**

The expression in Equation 6.13 represents the z-transform of the comb filter evaluated on the unit circle relative to the high sample rate. From Equation 6.13, we can see that the corresponding z-transform of the comb filter referenced to the high sample rate is expressed as

$$H_C(z) = \left(1 - z^{-RM}\right)$$

**Equation 6.14**

### 6.3.2 CIC Integrator Stage

The block diagram of a CIC integrator stage is illustrated in Figure 6.35. The difference equation that describes the integrator stage is given by

$$y(n) = x(n) + y(n-1)$$

**Equation 6.15**

The corresponding z-transform is given by

$$Y(z) = X(z) + z^{-1}Y(z)$$
$$H_I(z) = Y(z)/X(z)$$
$$H_I(z) = \frac{1}{1-z^{-1}}$$

**Equation 6.16**

The integrator has unity feedback. In a CIC filter configured to operate as a decimation filter, this will result in register overflow in all integrator stages. This will not be a problem if two conditions are met [1]:

1. The filter is implemented using two's complement arithmetic that wraps around from the most positive to most negative number representations.
2. The range of numbers supported by the bit width of the filter is greater than the maximum value expected at the output of the composite CIC filter.

As was the case with the comb stage, we can evaluate the integrator frequency response by evaluating its z-transform on the unit circle. We do that by replacing all occurrences of $z$ with $e^{j\omega}$ so that

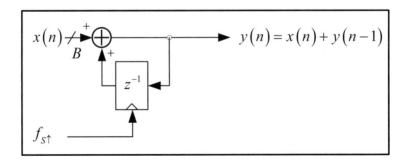

**Figure 6.35**  CIC filter integrator building block

$$H_I(z)\big|_{z=e^{j\omega}} = H_I\left(e^{j\omega}\right) = \frac{1}{1-e^{-j\omega}} = \frac{1}{1-e^{-j2\pi f_{\uparrow}}}$$

The integrator section of the CIC filter always operates at the high sample rate of $f_{s\uparrow}$.

### 6.3.3 Composite CIC Filter Z-Transform Relative to the High Rate

The composite CIC filter is implemented by a cascade of $N$ comb sections and $N$ integrator sections. The $N$ section cascade of comb filters referenced to the high sample rate is expressed as

$$\left[H_C(z)\right]^N = \left(1-z^{-RM}\right)^N$$

and the $N$ section cascade of integrators is given by

$$\left[H_I(z)\right]^N = \left[\frac{1}{\left(1-z^{-1}\right)}\right]^N$$

Because the comb sections and integrator sections are in cascade with one another, the z-transform of the composite CIC filter can then be expressed as

$$H(z) = \left[H_C(z)\right]^N \left[H_I(z)\right]^N = \left[\frac{\left(1-z^{-RM}\right)}{\left(1-z^{-1}\right)}\right]^N$$

**Equation 6.17**

The z-transform for a single CIC filter stage is given by

$$H(z) = \left[H_C(z)\right]^1 \left[H_I(z)\right]^1 = \left[\frac{\left(1-z^{-RM}\right)}{\left(1-z^{-1}\right)}\right]$$

which is recognized to be the closed form expression for the series

$$H(z) = \sum_{n=0}^{RM-1} z^{-n}$$

**Equation 6.18**

This can be easily proved by the following. If we expand Equation 6.18, we obtain

$$\sum_{n=0}^{RM-1} z^{-n} = 1 + z^{-1} + z^{-2} + z^{-3} + \cdots + z^{-(RM-1)}$$

**Equation 6.19**

If we multiply both sides of Equation 6.19 by the term $z^{-1}$, we get a second series given by

$$z^{-1}\sum_{n=0}^{RM-1} z^{-n} = z^{-1} + z^{-2} + z^{-3} + \cdots + z^{-(RM-1)} + z^{-RM}$$

**Equation 6.20**

If we subtract Equation 6.20 from Equation 6.19, we arrive at

$$\sum_{n=0}^{RM-1} z^{-n} - z^{-1}\sum_{n=0}^{RM-1} z^{-n} = \left\{1 + z^{-1} + z^{-2} + z^{-3} + \cdots + z^{-(RM-1)}\right\} - \left\{z^{-1} + z^{-2} + z^{-3} + \cdots + z^{-(RM-1)} + z^{-RM}\right\}$$

This equation collapses to

$$\sum_{n=0}^{RM-1} z^{-n}\left(1 - z^{-1}\right) = 1 - z^{-RM}$$

Dividing both sides by the term $\left(1 - z^{-1}\right)$ leaves us with

$$\sum_{n=0}^{RM-1} z^{-n} = \frac{1 - z^{-RM}}{1 - z^{-1}}$$

**Equation 6.21**

which is the proof we seek.

Equation 6.21 represents the z-transform for a single stage CIC filter. The cascade of $N$ such stages can then be expressed as Equation 6.17, which is repeated here for clarity:

$$H(z) = \left[\sum_{n=0}^{RM-1} z^{-n}\right]^N = \left[\frac{1 - z^{-RM}}{1 - z^{-1}}\right]^N$$

**Equation 6.22**

The CIC system function of Equation 6.22 clearly shows that the CIC filter is a cascade of $N$ uniform FIR filters. As we can deduce from Equation 6.19, each filter of the CIC cascade has $RM$ coefficients, all of which are unity.

### 6.3.4 Composite CIC Filter Z-Transform Relative to the Low Rate

The CIC filter design parameters $R$, $M$, and $N$ are chosen to provide satisfactory pass band specifications over the frequency band ranging from 0 Hz to $f_{PASS}$ Hz, referenced to the low sampling rate. The term $f_{PASS}$ is defined as the pass band cutoff frequency of the filter input signal of interest. In order to view how these parameters affect the CIC low rate frequency response, we will need to derive the low rate Z transform. We can derive the low sample rate z-transform by beginning with the low rate comb system function $H_C(z)$ of Equation 6.11 and the integrator system function $H_I(z)$ of Equation 6.16, which are repeated here for clarity:

$$H_C(z) = (1 - z^{-M}) \quad H_I(z) = \frac{1}{1 - z^{-1}}$$

The composite z-transform for a single stage of the CIC filter referenced to the low sampling rate is then given by

$$H(z) = H_C(z)H_I(z) = \left[ \frac{1 - z^{-M}}{1 - z^{-1}} \right]$$

The low rate z-transform for $N$ stages in cascade is then given by

$$H(z) = H_C(z)^N H_I(z)^N = \left[ \frac{1 - z^{-M}}{1 - z^{-1}} \right]^N$$

**Equation 6.23**

As we will soon see, Equation 6.23 provides us with a very useful design tool to view the low rate CIC spectral response from 0 Hz to the first response null.

### 6.3.5 CIC Decimation Filter Architecture

The functional block diagram of a CIC configured as a three-stage decimation filter is illustrated in part A of Figure 6.36. The input sequence $x(n)$ sampled at the high rate clock $f_{S\uparrow}$ is applied to the input of an $N$ section integrator. The data and clock output from the integrator section are decimated by a factor of $R$ and then fed to the input of an $N$ section comb filter. The output of the comb filter is the decimated output sequence $y(n)$ at the low rate clock $f_{S\downarrow} = f_{S\uparrow}/R$.

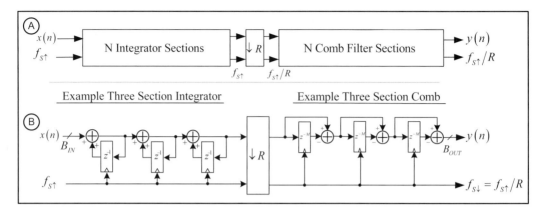

**Figure 6.36** The CIC configured as a three-stage decimation filter

The down sample operation represented by the symbol $\downarrow R$ is implemented by passing a single sample from the integrator section to the comb section every $R$ samples. That is, for every $R$ samples output by the integrator section, we input only one to the comb section and throw away the remaining $R-1$ samples.

Part B of the figure shows a functional example of the architecture for a three-stage CIC decimation filter. Each section of the filter, whether it be an integrator or comb, contains a register whose bit width is dependent upon the filter gain from section to section. Typically, all registers are of different bit widths. We will compute the bit widths of these registers later.

It is easily seen from the example CIC architecture that the only components other than the down sampler are registers and summers. The architecture is a simple pipeline of two basic structures. This attribute makes the CIC architecture extremely easy to implement in digital signal processing (DSP) devices such as FPGAs.

## 6.3.6  CIC Interpolation Filter Architecture

The functional representation of a CIC filter configured as an interpolator is illustrated in part A of Figure 6.37. The input sequence $x(n)$ sampled at the low rate clock $f_{S\downarrow} = f_{S\uparrow}/R$ is applied to the input of an $N$ section comb filter. The sample rate of the data and the frequency of the clock output from the comb are increased by a factor of $R$ and then fed to the input of an $N$ section integrator. The integrator output is the interpolated output sequence $y(n)$ at high rate clock $f_{S\uparrow}$. The sample rate increase in the data stream represented by the symbol $R\uparrow$ is achieved by inserting $R-1$ zeros between adjacent comb filter output samples.

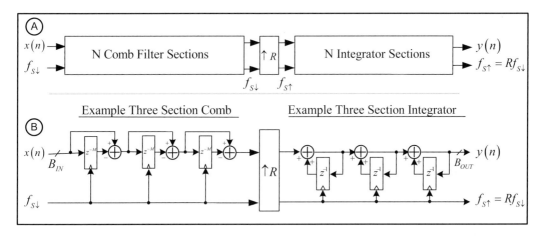

**Figure 6.37**  The CIC filter configured as a three-stage interpolation filter

    Part B of the figure shows a functional example of the architecture for a
three-stage CIC interpolation filter. As was the case with the decimation CIC
filter, the bit width of each register in the cascade will be different. We will
compute these widths later.

### 6.3.7  CIC Filter High Sample Rate Frequency Response

We derive the composite CIC high rate frequency response by evaluating the
CIC filter z-transform given in Equation 6.17 on the unit circle in the z-plane.
We begin with the high rate system function repeated here for clarity:

$$H(z) = \left[ \frac{\left(1 - z^{-RM}\right)}{\left(1 - z^{-1}\right)} \right]^{N}$$

**Equation 6.24**

    We evaluate Equation 6.24 on the unit circle in the z-plane by replacing
the complex variable $z$ with the complex exponential $e^{j\omega}$ to get

$$H(z)\Big|_{z=e^{j\omega}} = H\left(e^{j\omega}\right) = \left[ \frac{1 - e^{-j\omega RM}}{1 - e^{-j\omega}} \right]^{N}$$

We can factor the term $e^{-j\frac{\omega}{2}RM}$ from the numerator and factor the term $e^{-j\frac{\omega}{2}}$
from the denominator to produce

$$H\left(e^{j\omega}\right)=\left[\frac{e^{-j\frac{\omega}{2}RM}}{e^{-j\frac{\omega}{2}}}\left(\frac{e^{+j\frac{\omega}{2}RM}-e^{-j\frac{\omega}{2}RM}}{e^{+j\frac{\omega}{2}}-e^{-j\frac{\omega}{2}}}\right)\right]^{N}$$

**Equation 6.25**

Euler's identity for the sin function is given by $\left(e^{j\omega}-e^{-j\omega}\right)=2j\sin\left(\omega\right)$. We can use Euler's identity, collect terms, and collapse Equation 6.25 to get

$$H\left(e^{j\omega}\right)=\left[e^{-j\frac{\omega}{2}(RM-1)}\left(\frac{\sin\left(\dfrac{\omega}{2}RM\right)}{\sin\left(\dfrac{\omega}{2}\right)}\right)\right]^{N}$$

**Equation 6.26**

The term outside the parenthesis is recognized as being the system response phase term, and the quantity within the parenthesis is the system response amplitude term. We can express Equation 6.26 in terms of its magnitude and phase response as

$$H\left(e^{j\omega}\right)=\left|H\left(e^{j\omega}\right)\right|_{\uparrow}e^{-j\theta(\omega)}$$

**Equation 6.27**

where $\left|H\left(e^{j\omega}\right)\right|_{\uparrow}$ is the system high sample rate magnitude response and $\theta(\omega)$ is the system phase response. Both response terms are a function of $\omega$. The magnitude of high sample rate response $H\left(e^{j\omega}\right)$ is given by

$$\left|H\left(e^{j\omega}\right)\right|_{\uparrow}=\left|\frac{\sin\left(\dfrac{\omega}{2}RM\right)}{\sin\left(\dfrac{\omega}{2}\right)}\right|^{N}$$

**Equation 6.28**

We note from Equation 6.28 that

$$\left|H\left(e^{j\omega}\right)\right|_{\uparrow}=0$$

whenever

$$\sin\left(\frac{\omega}{2}RM\right) = 0 \quad \text{or when} \quad \frac{\omega}{2}RM = k\pi \quad \{\text{for } k = 1,2,3,\cdots\}$$

If we expand the argument of the sine, we see that

$$\frac{(2\pi f)RM}{2} = k\pi \quad \{\text{for } k = 1,2,3,\cdots\}$$

This tells us that the CIC high rate frequency response will experience nulls at the normalized frequencies

$$f = \frac{k}{RM} \quad \left\{\text{for } k = 1,2,3,\cdots,\left\lfloor\frac{R}{2}\right\rfloor\right\}$$

**Equation 6.29**

where the $\lfloor x \rfloor$ symbol is referred to as the floor(x) and is a function that returns the highest integer less than or equal to $x$. We express Equation 6.29 in terms of unnormalized frequencies as

$$f = \frac{kf_{s\uparrow}}{RM} \quad \left\{\text{for } k = 1,2,3,\cdots,\left\lfloor\frac{R}{2}\right\rfloor\right\}$$

**Equation 6.30**

There is one important case to consider: when $\omega = 0$ in Equation 6.28, we end up with the indeterminate form

$$\left|H\left(e^{j0}\right)\right|_{\uparrow} = \left|\frac{\sin(0)}{\sin(0)}\right|^{N} = \left|\frac{0}{0}\right|^{N}$$

We address this issue by using l'Hôpital's rule to differentiate both numerator and denominator and then evaluate the result at $\omega = 0$ to get

$$\left|H\left(e^{j\omega}\right)\right|_{\uparrow\omega=0} = \left|RM\frac{\cos\left(\frac{\omega}{2}RM\right)}{\cos\left(\frac{\omega}{2}\right)}\right|^{N}\Bigg|_{\omega=0} = \left[RM\left(\frac{1}{1}\right)\right]^{N} = (RM)^{N}$$

The magnitude at $\omega = 0$ turns out to be the maximum magnitude of $\left| H\left(e^{j\omega}\right)\right|_{\uparrow}$. Typically, when we plot Equation 6.28, we do so by plotting its magnitude relative to the maximum magnitude in decibels. This gives us the following relation:

$$\left| H\left(e^{j\omega}\right)\right|_{\uparrow db} = 20\text{Log}\left[\frac{\left| H\left(e^{j\omega}\right)\right|^{N}}{\left| H\left(e^{j\omega}\right)\right|_{MAX}^{N}}\right] = 20\text{Log}\left[\frac{\left| H\left(e^{j\omega}\right)\right|^{N}}{(RM)^{N}}\right] = 20\text{Log}\left[\left| \frac{1}{RM}\frac{\sin\left(\frac{\omega}{2}RM\right)}{\sin\left(\frac{\omega}{2}\right)}\right|^{N}\right]$$

**Equation 6.31**

When $\omega = 0$, Equation 6.31 reduces to

$$\left| H\left(e^{j0}\right)\right|_{\uparrow db} = 20\text{Log}\left[\left| \frac{1}{RM}RM\right|^{N}\right] = 20\text{Log}[1] = 0\text{ db}$$

The maximum value in decibels for this normalized function is therefore 0 db and occurs at $\omega = 0$. The high rate CIC frequency response for a differential delay $M = 1$ is illustrated in Figure 6.38. It is evident that the frequency response takes on a $\sin x / x$ "like" appearance. There are three different but equivalent calibrated frequency axes in the figure:

- Axis $A$ is calibrated in units of normalized low rate sampling frequency given by $f = \left(f_{\downarrow} / f_{s\downarrow}\right)$. This axis is included in the high rate frequency response plot only to add insight into the response of the composite filter. Here we see the low rate frequency response continuing on indefinitely with nulls at multiples of the low rate frequency.
- Axis $B$ is calibrated in units of high sampling rate frequency with nulls at $f_{\uparrow} = \frac{kf_{s\uparrow}}{RM}$.
- Axis $C$ is calibrated in units of high sampling rate frequency nulls at $f_{\uparrow} = \frac{k}{RM}$ where the frequency has been normalized to the high sampling rate $f_{s\uparrow}$.

The hatched area of Figure 6.38 represents an example narrow band signal of bandwidth $f_{PASS}$ located at the base band of a much wider band signal. This band of frequencies is usually referred to as the *band of interest* since this is the narrow band of input signal frequencies that we want to extract from a much wider signal bandwidth. In many industry applications, a wideband signal is

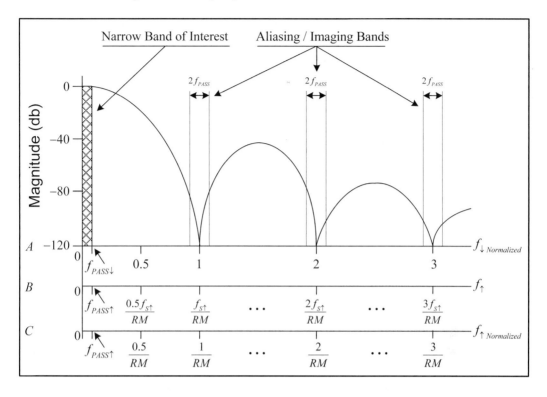

**Figure 6.38** CIC high sample rate frequency response $M = 1$

processed through a CIC decimation filter to substantially reduce the sample rate prior to extracting and processing the narrow band signal of interest.

When the sample rate of the signal is reduced or decimated by some factor $R$ there will be portions of the wideband input signal that will alias down to base band and interfere with the narrow band of interest. The aliasing bands shown in Figure 6.38 represent bands of frequencies that will fold down to base band and corrupt the signal band of interest when the sample rate is reduced. It is important to note how the nulls of the CIC frequency response fall directly in the center of these alias bands and substantially attenuate these frequencies. In Figure 6.38 we can see that for this particular CIC decimation filter, the frequencies that will alias down to the band of interest have been attenuated by more than 80 db by the response nulls.

When the sample rate of a narrow band signal is increased or interpolated by a factor $R$, the spectrum of the narrow band signal is replicated at multiples of the low sample rate. These replicated bands are referred to as *image bands*. Note that these bands, created by an interpolation filter, are identical in both position and bandwidth to the aliasing bands of a decimation filter. Expressed in a different manner, the bandwidth of the band of interest shown in the figure will be replicated in the image bands as a result of a sample rate increase. The reader should note how the nulls of the CIC frequency response substantially attenuate the frequencies in these image bands.

Stated another way, for CIC decimation filters, the frequency bands centered at each null and of width $2f_{PASS}$ will be folded back onto the pass band $-f_{PASS} \le f \le f_{PASS}$ when the sample rate is decreased by a factor of $R$, resulting in aliasing error. For CIC interpolation filters, these bands also represent the imaging bands that occur when the sample rate is increased by a factor of $R$. The bandwidth and placement of the aliasing/imaging bands are given by

$$B_{Alias/Image} = \frac{k}{RM} \pm f_{PASS} \text{ for } k = 1, 2, 3, \cdots$$

The frequency response of the CIC filter referenced to the high sample rate for the case where the comb differential delay is $M = 2$ is illustrated in Figure 6.39. We can see from the figure that the number of nulls at $k/RM$ have doubled from the case where $M = 1$. For the $M = 1$ case, the null occurs at frequencies $f_\uparrow = k/R$. In the $M = 2$ case, the nulls occur at frequencies $f_\uparrow = k/2R$. This illustrates that the differential delay $M$ can be used in a limited fashion to place the frequency response nulls. Note that the pass band roll-off is sharper for the $M = 2$ case, which can increase the attenuation at the band edge frequencies within the pass band of interest. This phenomenon is known as *pass band droop* and will be discussed in more detail later.

For the case where the CIC filter is used to increase the sample rate by a factor of $R$, the attenuation of the imaging bands is obvious from Figure 6.38. For the case where the CIC filter is used to decrease the sample rate by a factor of $R$, the attenuation of the aliasing bands is not as obvious. For this case, the frequency response is best viewed when referenced to the low sample rate, which is the recommended method when designing a CIC decimator [1].

For clarity, however, let's first use a high sample rate perspective to look at the aliasing that occurs when the sample rate is decreased by a factor of $R$. This is illustrated in Figure 6.40. For this example, we use axis $A$, which is calibrated in terms of the low sample rate. We can derive a great deal of information from this figure. First of all, we know that the original signal was sampled at the high rate of $f_{s\uparrow}$ and that it was decimated by a factor $R$. We also know that since the frequency response nulls occur at integer multiples of the decimated sample rate, the comb differential delay is set to $M = 1$. The normalized sample rate of the decimated signal is $f_{s\downarrow} = 1$, and the folding frequency is $f_\downarrow = 0.5$. We know all the frequencies in the aliasing bands will fold back or alias to base band and will corrupt the signal band of interest. The frequency bands defined by

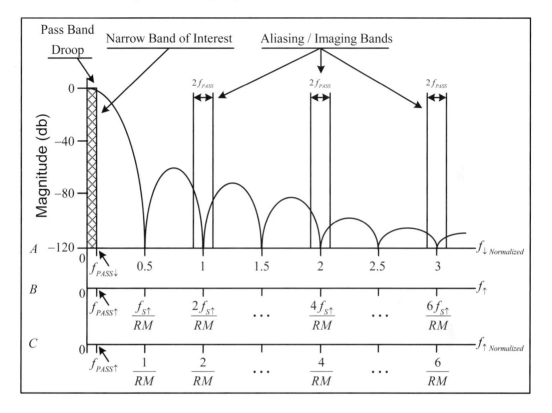

**Figure 6.39**  CIC high sample rate frequency response $M = 2$

$$\left(k - f_{PASS}\right) \leq f \leq \left(k + f_{PASS}\right) \quad \left\{\text{for } k = 1, 2, 3, \cdots, \lfloor R/2 \rfloor\right\}$$

will alias back onto the base band frequencies defined by $-f_{PASS} \leq f \leq +f_{PASS}$. The band that contributes the greatest amount of aliasing lies within the first aliasing band of

$$\left(1 - f_{PASS}\right) \leq f \leq \left(1 + f_{PASS}\right)$$

Therefore this is the only aliasing contribution illustrated in the figure. The frequencies from $0.5 \leq f \leq 1.0$ that fold back onto $0 \leq f \leq 0.5$ are shown. Figure 6.40 shows the amount of aliasing that might be expected for a particular CIC decimation filter design. As can be seen the aliasing due to alias band 1, for this example the CIC filter is less than −80 db. Contributions from other aliasing bands will be less.

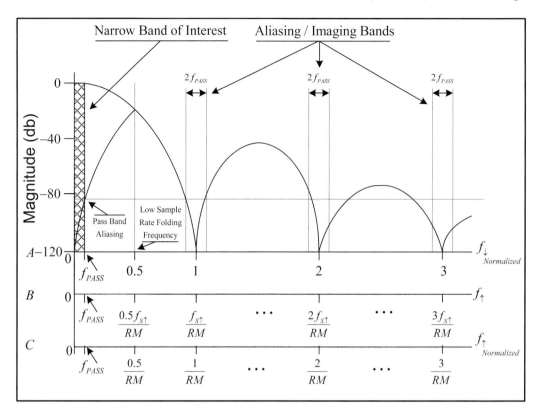

**Figure 6.40**   CIC decimation from a high sample rate perspective

### 6.3.8   CIC Filter Low Sample Rate Frequency Response

We derive the low rate frequency response magnitude $\left|H\left(e^{j\omega}\right)\right|_\downarrow$ by beginning with the high rate response $\left|H\left(e^{j\omega}\right)\right|_\uparrow$ in Equation 6.28. We know that the low sample rate is defined by $f/R$, or $\omega = 2\pi\left(f/R\right)$. Substitution of the low rate $\omega$ into Equation 6.28 gives us

$$\left|H\left(e^{j\omega}\right)\right|_\downarrow = \left|\frac{\sin\left(\pi M f\right)}{\sin\left(\pi f/R\right)}\right|^N$$

**Equation 6.32**

where the symbol $|\ |_\downarrow$ denotes the CIC low sample rate magnitude response. The CIC filter frequency response referenced to the low sample rate is illustrated in Figure 6.41.

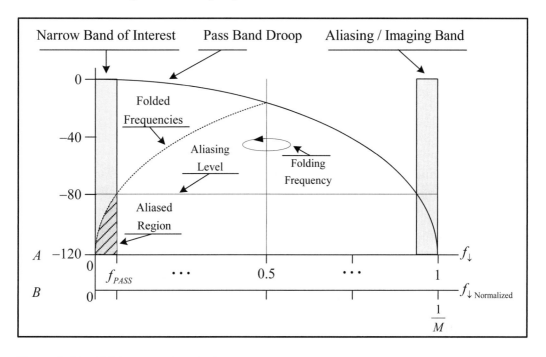

**Figure 6.41**   CIC low sample rate frequency response

Equation 6.32 is the "precise" CIC frequency response referenced to the low sample rate. We compute the null for the low rate response by observing that

$$\left|H\left(e^{j\omega}\right)\right|_{\downarrow} = 0 \ \text{ when } \ \sin\left(\pi M f\right) = 0 \ \text{ or when } \ \pi M f = \pi$$

This condition is satisfied when

$$f = \frac{1}{M}$$

**Equation 6.33**

The CIC filter is most efficiently used when implementing large sample rate changes. As the rate change $R$ becomes large, the argument of the sinusoid in the denominator term of Equation 6.32 becomes small, and the sin term can be approximated by its argument $\pi f / R$. Making this substitution, we arrive at

$$\left|H\left(e^{j\omega}\right)\right|_{\downarrow} \cong \left|\frac{\sin\left(\pi M f\right)}{\pi f / R}\right|^{N} \quad \left\{\text{For large } R\right\}$$

We can factor out $1/R$ from the denominator and multiply both numerator and denominator by $M$ to produce

$$\left| H\left(e^{j\omega}\right)\right|_{\downarrow} \cong \left| RM \frac{\sin\left(\pi M f\right)}{\pi M f}\right|^{N} \quad \{\text{For large } R\}$$

**Equation 6.34**

Equation 6.34 is the "approximate" frequency response for the CIC filter when the rate change $R$ is large. Hogenauer [1] states that the performance difference between the precise frequency response of Equation 6.32 and the approximate frequency response of Equation 6.34 for $0 \le f \le \dfrac{1}{M}$ is less than 1 db for the cases that employ the following parameter settings:

- $RM \ge 10$
- $1 \le N \le 7$
- $0 \le f_{\downarrow Normalized} \le 255/(256M)$

Thus there is little difference between the precise frequency response representation and the approximate frequency response. We can use the approximation expression in Equation 6.34 in most of the design applications we will ever run across.

We notice from Figure 6.41 that the band of frequencies that fold back and alias into the pass band of interest extend from the normalized frequency $\left(1 - f_{PASS}/f_{\downarrow}\right)$ to 1. The frequencies folded back from the first null of the CIC response make the largest contribution to the overall aliasing and therefore can be used to make a very close approximation to the total alias sum. We can calculate the amount of aliasing by using Equation 6.34 as follows:

$$\left| H\left(e^{j\omega}\right)\right|_{\downarrow db} \cong 20 \text{Log}_{10} \left| RM \frac{\sin\left(\pi M f\right)}{\pi M f}\right|^{N}$$

Since we are only interested in the aliasing relative to the normalized value of $\left| H\left(e^{j\omega}\right)\right|_{\downarrow}$, we need not consider the factor $RM$, which gives us

$$\left| H\left(e^{j\omega}\right)\right|_{\downarrow db} \cong 20 \text{Log}_{10} \left| \frac{\sin\left(\pi M f\right)}{\pi M f}\right|^{N} = 20 N \text{Log}_{10} \left| \frac{\sin\left(\pi M f\right)}{\pi M f}\right|$$

**Equation 6.35**

The maximum contribution to the overall aliasing level occurs at the normalized frequency $(1 - f_{PASS}/f_{\downarrow})$. We can substitute this value into Equation 6.35 to get

$$\left|H\left(e^{j\omega}\right)\right|_{\downarrow db}\ 20 \cdot N \cdot \mathrm{Log}_{10}\left|\frac{\sin\left(\pi M\left[1 - f_{PASS}/f_{\downarrow}\right]\right)}{\pi M\left[1 - f_{PASS}/f_{\downarrow}\right]}\right|$$

**Equation 6.36**

As an example, let's assume we implement a CIC filter that has one stage (i.e., $N = 1$) and has a differential delay $M = 1$ with a normalized pass bandwidth of $f_{PASS} = 1/32$. We could compute the amount of aliasing by using Equation 6.36:

$$\left|H\left(e^{j\omega}\right)\right|_{\downarrow db}\ = 20 \cdot N \cdot \mathrm{Log}_{10}\left|\frac{\sin\left(\pi\left[1 - 1/32\right]\right)}{\pi\left[1 - 1/32\right]}\right|$$

$$= 20 \cdot 1 \cdot \mathrm{Log}_{10}\left|\frac{\sin\left(\pi\left[31/32\right]\right)}{\pi\left[31/32\right]}\right|$$

$$= 29.8412\ db\ \cong\ 29.8\ db$$

**Table 6.1**   Pass Band Aliasing / Image Band Attenuation versus Number of Stages $N$

| $M$ | $f_{PASS}$ | Pass band aliasing attenuation / image band rejection as a function of the number of stages $N$ | | | | | |
|---|---|---|---|---|---|---|---|
| | | 1 | 2 | 3 | 4 | 5 | 6 |
| 1 | 1/128 | 42.1 | 84.2 | 126.2 | 168.3 | 210.4 | 252.5 |
| 1 | 1/64 | 36.0 | 72.0 | 108.0 | 144.0 | 180.0 | 215.9 |
| 1 | 1/32 | 29.8 | 59.7 | 89.5 | 119.4 | 149.2 | 179.0 |
| 1 | 1/16 | 23.6 | 47.2 | 70.7 | 94.3 | 117.9 | 141.5 |
| 1 | 1/8 | 17.1 | 34.3 | 51.4 | 68.5 | 85.6 | 102.8 |
| 1 | 1/4 | 10.5 | 20.9 | 31.4 | 41.8 | 52.3 | 62.7 |
| 2 | 1/256 | 48.1 | 96.3 | 144.4 | 192.5 | 240.7 | 288.8 |
| 2 | 1/128 | 42.1 | 84.2 | 126.2 | 168.3 | 210.4 | 252.5 |
| 2 | 1/64 | 36.0 | 72.0 | 108.0 | 144.6 | 180.0 | 216.0 |
| 2 | 1/32 | 29.8 | 59.8 | 89.6 | 119.5 | 149.4 | 179.3 |
| 2 | 1/16 | 23.6 | 47.5 | 71.2 | 95.0 | 118.7 | 142.5 |
| 2 | 1/8 | 17.8 | 35.6 | 53.4 | 71.3 | 89.1 | 106.29 |

It's a simple matter to multiply by other values of $N$ to get aliasing levels for multistage filters. For example, for an $N = 2$ stage filter, we get 59.68 db of aliasing attenuation; for a $N = 3$ stage filter, we get 89.52 db of aliasing attenuation; and so on. If we tabulate these computations for different values of $N$, $M$, and $f_{PASS}$, we end up with Table 6.1. This is an important table. We will use it in all of our CIC filter designs. The reader is encouraged to use Equation 6.36 to expand this table in an Excel spreadsheet to gain more resolution.

### 6.3.9  CIC Filter Phase Response

It was mentioned earlier that all coefficients in the CIC filter are unity and the impulse response is symmetric. Therefore the filter should have a linear phase response and a constant group delay. We saw in Chapter 5, "Finite Impulse Response Digital Filtering," that the phase response for a linear phase filter has the form

$$\theta(\omega) = -b\omega$$

**Equation 6.37**

The CIC high rate frequency response is given in Equation 6.26 and is repeated here for clarity:

$$H(e^{j\omega}) = \left[ e^{-j\frac{\omega}{2}(RM-1)} \left( \frac{\sin\left(\frac{\omega}{2}RM\right)}{\sin\left(\frac{\omega}{2}\right)} \right) \right]^N = \left| H(e^{j\omega}) \right| e^{-j\theta(\omega)}$$

For single stage CIC filter, we can write

$$H(e^{j\omega}) = e^{-j\frac{\omega}{2}(RM-1)} \left[ \frac{\sin\left(\frac{\omega}{2}RM\right)}{\sin\left(\frac{\omega}{2}\right)} \right] = \left| H(e^{j\omega}) \right| e^{-j\theta(\omega)}$$

The CIC high rate phase response has this form $\theta(\omega) = -b\omega$ and is given by

$$\theta(\omega) = -\frac{\omega}{2}(RM - 1)$$

where

$$b = \frac{1}{2}(RM-1)$$

If the phase response of a filter meets the requirement of Equation 6.37, then it will have a constant phase delay and it will exhibit a constant group delay as well. The phase response of the CIC meets this requirement. The CIC filter group delay is a constant and is given by

$$-\frac{d\theta}{d\omega} = -\frac{d}{d\omega}\left[-\frac{\omega}{2}(RM-1)\right] = \frac{(RM-1)}{2}$$

These results are also true for an $N$ stage CIC. It's a simple exercise to prove and is left to the reader for his or her own verification.

### 6.3.10   CIC Impulse Response

We have mentioned several times that the impulse response $h(n)$ of the CIC filter is composed of nothing but unity coefficients. This fact is obvious from the closed form transfer function given by Equation 6.22, which is repeated here for clarity:

$$H(z) = \left[\sum_{n=0}^{RM-1} z^{-n}\right]^N = \left[\frac{1-z^{-RM}}{1-z^{-1}}\right]^N$$

**Equation 6.38**

It is interesting and somewhat satisfying, however, to derive the same result by taking the inverse z-transform of the closed form expression for $H(z)$. For a single stage CIC filter, the z-transform given in Equation 6.38 becomes

$$H(z) = \frac{1-z^{-RM}}{1-z^{-1}}$$

**Equation 6.39**

We can determine the impulse response coefficients by computing the inverse z-transform. One method of doing this is performing the long division indicated by Equation 6.39. The result of the division is illustrated as follows:

$$
1-z^{-1} \overline{\smash{\big)}\ \begin{matrix} 1 \quad +z^{-1} \quad +z^{-2} \quad +z^{-3} \quad \cdots \quad +z^{-(RM-1)} \\ 1-z^{-RM} \end{matrix}}
$$

$$
\begin{aligned}
&1-z^{-1} \\
&\overline{\quad z^{-1}-z^{-RM}} \\
&\quad z^{-1}-z^{-2} \\
&\overline{\qquad\quad z^{-2}-z^{-RM}} \\
&\qquad\quad z^{-2}-z^{-3} \\
&\overline{\qquad\qquad\quad z^{-3}-z^{-RM}} \\
&\qquad\qquad\qquad \ddots \\
&\qquad\qquad\qquad\quad z^{-(RM-1)}-z^{-RM} \\
&\qquad\qquad\qquad\quad z^{-(RM-1)}-z^{-RM} \\
&\overline{\qquad\qquad\qquad\qquad\qquad 0}
\end{aligned}
$$

**Equation 6.40**

The coefficients of the polynomial in $z$ shown as the quotient in Equation 6.40 represent the coefficients of the filter impulse response $h(n)$. We can see from the quotient that $h(n) = 1$ $\{$for $n = 0, 1, 2, \cdots, RM\text{-}1\}$, and therefore all the coefficients are unity. The inverse z-transform shows that a single stage $(N = 1)$ CIC filter has $RM$ coefficients and that they are indeed all equal to 1. If we replace the implied unity coefficients in Equation 6.38 with the unity sequence $h(n)$, we arrive at

$$
H(z) = \sum_{n=0}^{RM-1} h(n)z^{-n} = 1 + z^{-1} + z^{-2} + z^{-3} + \cdots + z^{-(RM-1)}
$$

**Equation 6.41**

Of course, for an $N$ stage CIC, we will have $N(RM)$ total coefficients, which can be expressed as

$$
H(z) = \left[ \sum_{n=0}^{RM-1} h(n)z^{-n} \right]^N = \left[ 1 + z^{-1} + z^{-1} + z^{-1} + \cdots + z^{-(RM-1)} \right]^N
$$

This result clearly shows that the CIC filter is a cascade of $N$ uniform FIR filters, each of length $RM$ and all of which have unity coefficients.

### 6.3.11  CIC Filter Pole-Zero Placement

Design engineers usually have a deep-seeded desire to understand the filters they create by observing the pole-zero plots for those filters. A quick look at a pole-zero plot can show potential design flaws that could be showstoppers, such as filter instability. A careful study of the actual locations of poles and the zeros will provide critical information on filter performance. Let's begin this discussion by defining a few terms:

- $z$ is defined to be a complex variable.
- $z$ can be represented in the Cartesian coordinate system by $z = x \pm jy$.
- $z$ can be represented in the Polar coordinate system by $z = \rho e^{\pm j\theta}$.
- The magnitude of $z$ is given by $|z| = \rho = \sqrt{x^2 + y^2}$.
- The phase of $z$ is given by $\angle z = \theta = \tan^{-1}\left(\dfrac{y}{x}\right)$.

For the purposes of this discussion, the frequencies are referenced to the high sample rate. Let's begin by looking at the transfer function for a single stage CIC filter, given by

$$H(z) = \left[\frac{1 - z^{-RM}}{1 - z^{-1}}\right]$$

We can locate the poles of transfer function by examining the term in the denominator.

A pole occurs at the roots of the denominator. At frequencies near a pole, the magnitude of the transfer function approaches infinity and the system that owns the z-transform becomes unstable.

It is clear that the denominator term $1 - z^{-1} = 0$ for the case where $z = 1$. Therefore a pole exists on the unit circle at $z = 1$. This is represented by an X in part A of the pole-zero plot in Figure 6.42.

The numerator term $1 - z^{-RM}$ is a little bit more complicated. We can locate the zeros in the numerator term by setting it equal to zero and rearranging terms to produce

$$z^{RM} = 1$$

To solve for $z$, we need to compute the $RM$ roots of 1, such that

$$z = (1)^{\frac{1}{RM}}$$

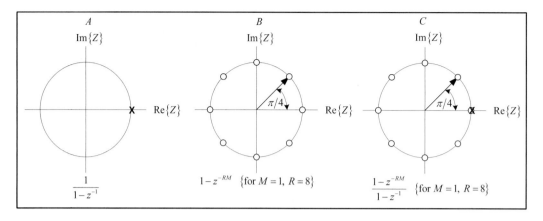

**Figure 6.42** CIC pole-zero placement

From Chapter 2, "Review of Complex Variables," we know that in polar notation, these roots are computed by

$$z = \rho^{\frac{1}{RM}} e^{j\left(\frac{\theta + 2\pi k}{RM}\right)} \quad \{\text{for } k = 0, 1, 2, \cdots, RM - 1\}$$

Expansion of the exponential term gives us

$$z = \rho^{\frac{1}{RM}} \left[ \cos\left(\frac{\theta + 2\pi k}{RM}\right) + j\sin\left(\frac{\theta + 2\pi k}{RM}\right) \right] \quad \{\text{for } k = 0, 1, 2, \cdots, RM - 1\}$$

**Equation 6.42**

We can simplify Equation 6.42 by noting that the polar magnitude $\rho$ for the principle root is

$$\rho = |z| = \sqrt{x^2 + y^2} = \sqrt{(1)^2 + (0)^2} = 1$$

and the phase $\theta$ for the principle root is computed to be

$$\theta = \tan^{-1}\left(\frac{y}{x}\right) = \tan^{-1}\left(\frac{0}{1}\right) = 0$$

Substituting the values for $\rho$ and $\theta$ into Equation 6.42, the expression for the $RM$ roots of 1 is

$$z = \left[ \cos\left( \frac{2\pi k}{RM} \right) + j\sin\left( \frac{2\pi k}{RM} \right) \right] \quad \{ \text{for } k = 0, 1, 2, \cdots, RM - 1 \}$$

If, for example, we were computing the zeros for a CIC filter with parameters $N = 1$, $M = 1$, and $R = 8$, as illustrated in part B of Figure 6.42, we would insert the values for $R$ and $M$ into this equation to get

$$z = \left[ \cos\left( \frac{2\pi k}{8} \right) + j\sin\left( \frac{2\pi k}{8} \right) \right] \quad \{ \text{for } k = 0, 1, 2, \cdots, 7 \}$$

or

$$z = \left[ \cos\left( \frac{\pi k}{4} \right) + j\sin\left( \frac{\pi k}{4} \right) \right] \quad \{ \text{for } k = 0, 1, 2, \cdots, 7 \}$$

We can easily see that for $k = 0$, $z = 1$, so there will be a zero located on the unit circle at $z = 1$. We know that the zeros will be uniformly spaced around the unit circle, so all we need to do is determine the phase angle for $k = 1$, place a zero on the unit circle at that angle, and then place the remainder of the zeros around the unit circle at multiples of that angle. In this example, the phase angle is easily seen to be $\pi/4$, so the second and successive zeros will occur at

$$|z| = 1, \quad \theta = \left( \frac{k\pi}{4} \right) \quad \{ \text{for } k = 1, 2, 3, \cdots, RM - 1 \}$$

**Equation 6.43**

The zeros for this example are shown in part B of Figure 6.42. The composite pole-zero diagram is illustrated in part C of Figure 6.42.

Note that the integrator pole at $z = 1$ is cancelled by a comb filter zero at $z = 1$. Ordinarily this might cause some concern on the part of a design engineer. This is because quantization error in the filter coefficients might alter the pole-zero placement just enough such that the pole at $z = 1$ may not be completely cancelled out by the zero at $z = 1$. This, of course, could lead to an unstable filter. However, this is not the case for the CIC filter. All coefficients are equal to unity; therefore there is no coefficient quantization error and the cancellation of the pole at $z = 1$ is perfect.

In this example, we will expect to see the occurrence of a frequency response null due to a response zero at multiple frequencies of $\pi/4$ radians, $f_s/8$ Hz, or 1/8 normalized.

The frequency response for this example is shown in Figure 6.43, which is the response of an actual filter. The design parameters for this filter are $M = 1$, $R = 8$, and $N = 4$. The high sample rate is 128 KHz, and the decimated sample rate is 16 KHz. The pass band edge is 2 KHz or $1/8 = 0.125$ normalized to the low sample rate. As can be seen from Figure 6.43, the frequency response nulls occur at multiples of $f_{S\uparrow}/8$, as predicted by Equation 6.43, where $f_{S\uparrow}$ is the high sample rate of 128 KHz. For the readers' convenience, the frequency axes for both the high and low rate are included for cross-reference purposes. It is interesting to note that the aliasing level shown in the figure is 68.5 db, which is exactly what's predicted in Table 6.1.

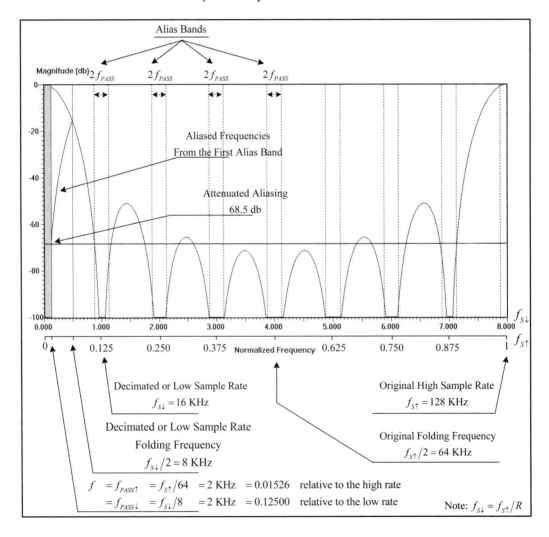

**Figure 6.43**    CIC decimate by eight filter response relative to the high sample rate

The poles and zeros for each stage of the CIC filter are identical. Therefore as the number of CIC stages increases the order of the poles and zeros also increases. An $N$th order CIC will have $N$th order poles and $N$th order zeros, all sitting on top of one another. Increasing the number of zeros has the effect of increasing the filter response roll-off and hence the attenuation at the frequencies near the zero. This roll-off, as a function of the number of stages $N$, is clearly shown in Figure 6.44 where the low rate frequency responses of eight separate CIC filters are superimposed on one another. The filter order $N$ is indicated by the symbol (N), and the response curve is displayed for each respective filter.

Note that as the filter order increases, the rate of the response roll-off increases, the width of the nulls increases, and the pass band aliasing decreases. The benefit of increasing the filter order is increased aliasing/image attenuation. The penalty paid for increasing the filter order is increased response droop that occurs at the pass band edge—an issue we will discuss later.

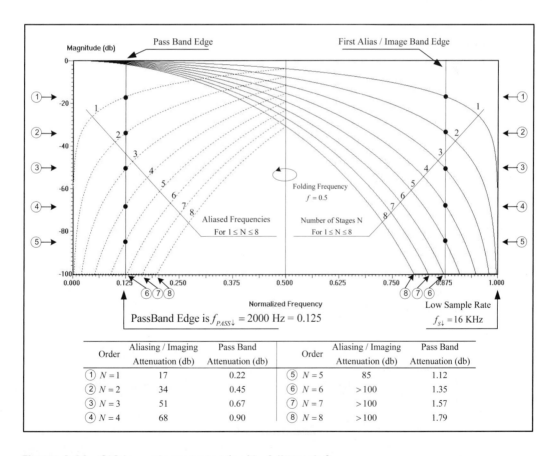

| Order | Aliasing / Imaging Attenuation (db) | Pass Band Attenuation (db) | Order | Aliasing / Imaging Attenuation (db) | Pass Band Attenuation (db) |
|---|---|---|---|---|---|
| ① $N = 1$ | 17 | 0.22 | ⑤ $N = 5$ | 85 | 1.12 |
| ② $N = 2$ | 34 | 0.45 | ⑥ $N = 6$ | > 100 | 1.35 |
| ③ $N = 3$ | 51 | 0.67 | ⑦ $N = 7$ | > 100 | 1.57 |
| ④ $N = 4$ | 68 | 0.90 | ⑧ $N = 8$ | > 100 | 1.79 |

**Figure 6.44**   CIC low rate response for $N = 1$ through 8

In the application depicted in Figure 6.44, each filter was a decimate by eight filter that reduced the sample rate by a factor of eight, from 128 KHz to 16 KHz. The pass band of interest is 2000 Hz or 0.125 normalized to the low sample rate. The comb differential delay is set to $M = 1$. The pass band attenuation, or droop, and alias attenuation measured at $f_{PASS}$ for the eight cases $(N = 1, 2, \cdots, 8)$ are shown in the table accompanying the figure.

We can see from the figure that a single stage CIC for these design parameters results in a pass band alias attenuation and image band attenuation of 17 db and a pass band edge droop of 0.22 db, while a second order filter produces a pass band alias attenuation and image band attenuation of 34 db and a pass band droop of 0.45 db, and so on, down to 85 db and 1.12 db for a fifth order filter. Additional stages produce aliasing/image attenuation greater than 100 db.

## 6.3.12  CIC Filter Gain

The DC gain of the CIC filter is a function of the rate change $R$, the comb differential delay $M$, and the number of stages $N$. The gain at the output of the final stage of a CIC decimation filter is

$$G = (RM)^N$$

**Equation 6.44**

and the gain at the output of the final stage of a CIC interpolation filter is

$$G = \frac{(RM)^N}{R}$$

**Equation 6.45**

For the purpose of computing the register growth in each section of a CIC interpolation filter, the gain of the $j$th section register is given by

$$
G_j = \left\{
\begin{array}{lll}
2^j & \text{for } j = 1, 2, 3, \cdots N & \text{Comb registers} \\[2mm]
\dfrac{2^{2N-j}(RM)^{j-N}}{R} & \text{for } j = N+1, N+2, \cdots, 2N & \text{Integrator registers}
\end{array}
\right\}
$$

**Equation 6.46**

### 6.3.13 CIC Decimation Filter Maximum Register Growth

The maximum bit width of the registers in a CIC decimation filter data path is a function of the filter parameters $R$, $M$, $N$, and the width of the input data register $B_{IN}$. If the bit width of the sequence input to the CIC filter is given by $B_{IN}$, then the maximum register width can be computed as the sum of the input data bit width plus the bit width of the overall register growth. This is just the input bit width summed with the $Log_2$ of the filter output gain, where the gain is determined by Equation 6.44. The maximum register bit width $B_{MAX}$ in a decimation filter occurs at the first integrator register (section $j = 1$), given by

$$B_{MAX} = \lceil N \, Log_2 (RM) + B_{IN} \rceil$$

**Equation 6.47**

where $\lceil x \rceil$ is the smallest integer greater than or equal to $x$. To make this clear, the register width is $B_{MAX}$ bits wide. All the registers following the first register must support the maximum accumulation produced by the first integrator section. Therefore $B_{MAX}$ represents the bit width for all the registers in each CIC integrator and comb section registers. $B_{MAX}$ is usually a fairly large number depending on the rate change $R$. It is shown in reference [1] that the bit width of the registers in sections that follow the first integrator section can be reduced significantly by truncating or rounding a number of least significant bits (LSBs) to streamline the decimation filter architecture. The number of retained register bits decreases monotonically from section to section, up to the final output register.

### 6.3.14 CIC Decimation Filter LSB Pruning

Hogenauer [1], provides a procedure that leads to an equation that is used to compute the number of LSBs $B_j$ in each of the $j$th section registers, which may be truncated or rounded without an increase in output quantization noise. This procedure has been aptly dubbed by many authors as *Hogenauer bit pruning*. The truncation of LSBs at each section register creates a source of quantization noise that flows from that register through to the filter output. The number of pruned LSBs is computed such that the noise source introduced by truncation or rounding at each section register is less than the total noise created by the final truncation to $B_{OUT}$ bits at the output of the filter. Unfortunately, the statistical mathematics associated with the pruning procedure is computationally intensive, prone to error, and will not be dealt with here. Instead we will utilize a set of precomputed tables that list the maximum number of possible truncated bits over a wide range of the most common CIC implementations. The tables that follow provide bit pruning data for CIC filters composed of 2, 3, 4, and 5 stages.

We'll introduce the bit pruning tables here and then demonstrate their uses later on. The reader is referred to the text by Uwe Meyer-Baese [4], which simplifies the bit pruning computation by providing a program on a companion CD-ROM that computes the number of truncated bits at each section register.

The maximum number of bits that can be pruned per section register for a two-stage CIC decimator is listed in Table 6.2 for 18 different values of the product $RM$. The notation $2N + 1$ refers to the truncation from the last comb section to the output register. The index $j$ identifies the CIC section register. Remember that for a two-stage CIC filter, there are four sections: two integrator sections followed by two comb sections. These sections are numbered $j = 1, 2, 3, 4$.

We will illustrate the use of Table 6.2 with a simple example. Suppose we were designing a two-stage CIC filter that input a sampled data stream that was $B_{in} = 16$ bits wide. Let's further suppose that the output register $B_{OUT}$ was also 16 bits wide and that the $RM$ product was 32. We can use Equation 6.47, repeated here for clarity, to compute the maximum width of each section register:

$$\begin{aligned} B_{MAX} &= \lceil N \, \mathrm{Log}_2(RM) + B_{IN} \rceil \\ &= \lceil 2 \, \mathrm{Log}_2(32) + 16 \rceil \\ &= 26 \end{aligned}$$

We can see from Table 6.2 that for an $RM$ product of 32, we can, as a maximum, prune 1 LSB from the section 1 integrator register $(j = 1)$, resulting in a register width of $B_{MAX} - 1$, or $26 - 1 = 25$ bits. We can prune up to a maximum of 6 bits from the second section register $(j = 2)$, resulting in a register width of $26 - 6 = 20$ bits. We can prune up to 7 LSBs from the third section register $(j = 3)$, giving us a register width of $26 - 7 = 19$ bits, and up to 8 LSBs from the fourth section register $(j = 4)$, making that register $26 - 8 = 18$ bits wide. Finally we can prune up to 10 LSBs from the output register $(j = 2N + 1)$ to produce a $26 - 10 = 16$ bit wide filter output. For this example, after bit pruning, the minimum width of each section register is given in Table 6.3. We should note that the width of the output register should be greater than or equal to $B_{MAX} - B_{2N+1}$, or $B_{OUT} \geq (B_{MAX} - B_{2N+1})$.

In case the reader is wondering, the pruning tables make the assumption that the minimum bit width of the output register $B_{OUT}$ is equal to the bit width of the input register $B_{IN}$. Why is this? The reason is that it is difficult to generate a one-size-fits-all table, and this restriction seems to satisfy the most common design criteria. That's why the number of bits pruned at the output register left us with $B_{OUT} = B_{MAX} - B_{j=2N+1} = 26 - 10 = 16$.

**Table 6.2**  Maximum Number of Pruned Bits for a Two-Stage CIC Decimator

| Number of stages $N$ | $RM$ | Max number of LSBs that can be discarded for $j = 1,2,3,4,2N+1$ | | | | |
|:---:|:---:|:---:|:---:|:---:|:---:|:---:|
| | | 1 | 2 | 3 | 4 | $2N+1$ |
| 2 | 8 | 0 | 3 | 3 | 4 | 6 |
| 2 | 16 | 1 | 4 | 5 | 6 | 8 |
| 2 | 24 | 2 | 6 | 7 | 8 | 10 |
| 2 | 32 | 1 | 6 | 7 | 8 | 10 |
| 2 | 40 | 2 | 6 | 8 | 9 | 11 |
| 2 | 48 | 2 | 7 | 9 | 10 | 12 |
| 2 | 56 | 2 | 7 | 9 | 10 | 12 |
| 2 | 64 | 2 | 7 | 9 | 10 | 12 |
| 2 | 72 | 3 | 8 | 10 | 11 | 13 |
| 2 | 80 | 2 | 8 | 10 | 11 | 13 |
| 2 | 88 | 2 | 8 | 10 | 11 | 13 |
| 2 | 96 | 3 | 9 | 11 | 12 | 14 |
| 2 | 104 | 3 | 9 | 11 | 12 | 14 |
| 2 | 112 | 3 | 9 | 11 | 12 | 14 |
| 2 | 128 | 2 | 9 | 11 | 12 | 14 |
| 2 | 256 | 3 | 10 | 13 | 14 | 16 |
| 2 | 512 | 3 | 12 | 15 | 16 | 18 |
| 2 | 1024 | 4 | 13 | 17 | 18 | 20 |

Suppose we design the same two-stage CIC filter but this time with an $RM$ product of 104 and an input register width of $B_{IN} = 11$ bits. We can compute the maximum register bit width to be

$$
\begin{aligned}
B_{MAX} &= \lceil N \, \mathrm{Log}_2(RM) + B_{IN} \rceil \\
&= \lceil 2 \, \mathrm{Log}_2(104) + 11 \rceil \\
&= 25
\end{aligned}
$$

From the table we see that the maximum number of truncated bits from each section is given by $B_J = 3,9,11,12,14$. The maximum LSB truncation at

**Table 6.3**   Bit Pruning Example

| Register section ($j$) | Maximum register width ($B_{MAX}$) | Maximum number of pruned LSBs ($B_j$) | Minimum section register width (*bits*) |
|---|---|---|---|
| 1 | 26 | 1 | 25 |
| 2 | 26 | 6 | 20 |
| 3 | 26 | 7 | 19 |
| 4 | 26 | 8 | 18 |
| $2N + 1$ (output register) | 26 | 10 | 16 |

the output register is 14. Therefore we compute the minimum output register width to be $B_{OUT} = 25 - 14 = 11$, which is equal to the width of the input register.

The maximum number of bits that can be pruned for a three-stage CIC decimator is listed in Table 6.4 for 18 different values of the product $RM$. As an example of Table 6.4, suppose we design a three-stage CIC filter with an $RM$ product of 88 and an input register width of $B_{IN} = 17$ bits. We compute the maximum register bit width to be

$$
\begin{aligned}
B_{MAX} &= \lceil N \, \mathrm{Log}_2\left(RM\right) + B_{IN} \rceil \\
&= \lceil 3 \, \mathrm{Log}_2\left(88\right) + 17 \rceil \\
&= 37
\end{aligned}
$$

From the table, we see that the maximum number of truncated bits from each section is given by $B_j = 2, 9, 14, 16, 17, 18, 20$. The maximum LSB truncation at the output register is 20. Therefore we compute the minimum output register width to be $B_{OUT} = 37 - 20 = 17$. So we see that the minimum bit width of the output register is equal to the bit width of the input register. We will learn more about these tables when we actually design a CIC filter for decimation.

The maximum number of bits that can be pruned for a four-stage CIC decimator is listed in Table 6.5 for 18 different values of the product $RM$. And, finally, the maximum number of bits that can be pruned for a five-stage CIC decimator is listed in Table 6.6 for 18 different values of the product $RM$.

The entries in the tables are sufficient for a wide range of CIC decimation designs. The entries correspond to $RM$ multiples of 8, ranging from 8 to 112. Additional entries correspond to an $RM$ product of 128, 256, 512,

**Table 6.4**  Maximum Number of Pruned Bits for a Three-Stage CIC Decimator

| Number of stages $N$ | $RM$ | Max number of LSBs that can be discarded for $j = 1,2,3,4,5,6,2N + 1$ | | | | | | |
|---|---|---|---|---|---|---|---|---|
| | | 1 | 2 | 3 | 4 | 5 | 6 | $2N + 1$ |
| 3 | 8 | 0 | 3 | 4 | 5 | 6 | 7 | 9 |
| 3 | 16 | 1 | 4 | 7 | 8 | 9 | 10 | 12 |
| 3 | 24 | 1 | 5 | 9 | 10 | 11 | 12 | 14 |
| 3 | 32 | 1 | 6 | 9 | 11 | 12 | 13 | 15 |
| 3 | 40 | 1 | 6 | 10 | 12 | 13 | 14 | 16 |
| 3 | 48 | 2 | 7 | 11 | 13 | 14 | 15 | 17 |
| 3 | 56 | 2 | 7 | 12 | 14 | 15 | 16 | 18 |
| 3 | 64 | 2 | 7 | 12 | 14 | 15 | 16 | 18 |
| 3 | 72 | 2 | 8 | 13 | 15 | 16 | 17 | 19 |
| 3 | 80 | 2 | 8 | 13 | 15 | 16 | 17 | 19 |
| 3 | 88 | 2 | 9 | 14 | 16 | 17 | 18 | 20 |
| 3 | 96 | 2 | 8 | 14 | 16 | 17 | 18 | 20 |
| 3 | 104 | 3 | 9 | 15 | 17 | 18 | 19 | 21 |
| 3 | 112 | 3 | 9 | 15 | 17 | 18 | 19 | 21 |
| 3 | 128 | 2 | 9 | 14 | 17 | 18 | 19 | 21 |
| 3 | 256 | 3 | 10 | 17 | 20 | 21 | 22 | 24 |
| 3 | 512 | 3 | 12 | 19 | 23 | 24 | 25 | 27 |
| 3 | 1024 | 4 | 13 | 22 | 26 | 27 | 28 | 30 |

and 1024. If the $RM$ product of a reader's CIC design corresponds to one of these table entries, then the maximum number of bits, as specified in the table, can be pruned at each section. If, however, the design happens to fall in between two row entries, then simply choose the previous row entry. The result will then contain a few bits more than optimum, but the number of bits pruned will still result in a significant hardware reduction.

For example, suppose a fourth order CIC design called for an RM product of 54. We look at the entries in Table 6.5 and see that it falls between the row entries for 48 and 56. We then choose to prune the number of bits specified for the $RM = 48$ case. This leaves a few extra bits in the design, but the filter performance is not affected.

**Table 6.5**  Maximum Number of Pruned Bits for a Four-Stage CIC Decimator

| Number of stages N | RM | Max number of LSBs that can be discarded for $j = 1,2,3,4,5,6,7,8,2N+1$ | | | | | | | | |
|---|---|---|---|---|---|---|---|---|---|---|
| | | 1 | 2 | 3 | 4 | 5 | 6 | 7 | 8 | 2N + 1 |
| 4 | 8 | 0 | 3 | 5 | 6 | 7 | 8 | 9 | 10 | 12 |
| 4 | 16 | 1 | 4 | 7 | 10 | 11 | 12 | 13 | 14 | 16 |
| 4 | 24 | 1 | 6 | 9 | 13 | 14 | 15 | 16 | 17 | 19 |
| 4 | 32 | 1 | 6 | 10 | 13 | 15 | 16 | 17 | 18 | 20 |
| 4 | 40 | 2 | 7 | 11 | 15 | 17 | 18 | 19 | 20 | 22 |
| 4 | 48 | 2 | 7 | 12 | 16 | 18 | 19 | 20 | 21 | 23 |
| 4 | 56 | 2 | 8 | 13 | 17 | 19 | 20 | 21 | 22 | 24 |
| 4 | 64 | 2 | 7 | 12 | 17 | 19 | 20 | 21 | 22 | 24 |
| 4 | 72 | 2 | 8 | 13 | 18 | 20 | 21 | 22 | 23 | 25 |
| 4 | 80 | 2 | 8 | 14 | 19 | 21 | 22 | 23 | 24 | 26 |
| 4 | 88 | 2 | 8 | 14 | 19 | 21 | 22 | 23 | 24 | 26 |
| 4 | 96 | 2 | 9 | 14 | 20 | 22 | 23 | 24 | 25 | 27 |
| 4 | 104 | 2 | 9 | 14 | 19 | 22 | 23 | 24 | 25 | 27 |
| 4 | 112 | 3 | 9 | 15 | 20 | 23 | 24 | 25 | 26 | 28 |
| 4 | 128 | 2 | 9 | 15 | 20 | 23 | 24 | 25 | 26 | 28 |
| 4 | 256 | 3 | 10 | 17 | 24 | 27 | 28 | 29 | 30 | 32 |
| 4 | 512 | 3 | 12 | 20 | 27 | 31 | 32 | 33 | 34 | 36 |
| 4 | 1024 | 4 | 13 | 22 | 31 | 35 | 36 | 37 | 38 | 40 |

As an example of Table 6.5, suppose we design a four-stage CIC filter with an *RM* product of 56 and an input register width of $B_{IN} = 7$ bits. We can compute the maximum register bit width to be

$$B_{MAX} = \lceil N \, Log_2 (RM) + B_{IN} \rceil$$
$$= \lceil 4 \, Log_2 (56) + 7 \rceil$$
$$= 31$$

From the table, we see that the maximum number of truncated bits from each section is given by $B_j = 2,8,13,17,19,20,21,22,24$. The maximum LSB truncation at the output register is 24. Therefore we compute the minimum

**Table 6.6**  Maximum Number of Pruned Bits for a Five-Stage CIC Decimator

| Number of stages N | RM | Max number of LSBs that can be discarded for $j = 1,2,3,4,5,6,7,8,9,10,2N+1$ | | | | | | | | | | |
|---|---|---|---|---|---|---|---|---|---|---|---|---|
| | | 1 | 2 | 3 | 4 | 5 | 6 | 7 | 8 | 9 | 10 | 2N+1 |
| 5 | 8 | 0 | 3 | 5 | 7 | 8 | 9 | 10 | 11 | 12 | 12 | 15 |
| 5 | 16 | 0 | 4 | 8 | 10 | 13 | 14 | 15 | 16 | 17 | 17 | 20 |
| 5 | 24 | 1 | 5 | 9 | 12 | 15 | 17 | 18 | 19 | 20 | 20 | 23 |
| 5 | 32 | 1 | 6 | 10 | 14 | 17 | 19 | 20 | 21 | 22 | 22 | 25 |
| 5 | 40 | 1 | 7 | 11 | 15 | 19 | 21 | 22 | 23 | 24 | 24 | 27 |
| 5 | 48 | 1 | 7 | 12 | 16 | 20 | 22 | 23 | 24 | 25 | 25 | 28 |
| 5 | 56 | 2 | 8 | 13 | 18 | 22 | 24 | 25 | 26 | 27 | 27 | 30 |
| 5 | 64 | 1 | 7 | 13 | 17 | 22 | 24 | 25 | 26 | 27 | 27 | 30 |
| 5 | 72 | 2 | 8 | 13 | 18 | 23 | 25 | 26 | 27 | 28 | 28 | 31 |
| 5 | 80 | 2 | 8 | 14 | 19 | 24 | 26 | 27 | 28 | 29 | 29 | 32 |
| 5 | 88 | 2 | 9 | 14 | 20 | 25 | 27 | 28 | 29 | 30 | 30 | 33 |
| 5 | 96 | 2 | 8 | 14 | 19 | 24 | 27 | 28 | 29 | 30 | 30 | 33 |
| 5 | 104 | 2 | 9 | 15 | 20 | 25 | 28 | 29 | 30 | 31 | 31 | 34 |
| 5 | 112 | 3 | 10 | 15 | 21 | 26 | 29 | 30 | 31 | 32 | 32 | 35 |
| 5 | 128 | 2 | 9 | 15 | 21 | 26 | 29 | 30 | 31 | 32 | 32 | 35 |
| 5 | 256 | 2 | 10 | 18 | 24 | 31 | 34 | 35 | 36 | 37 | 37 | 40 |
| 5 | 512 | 3 | 12 | 20 | 28 | 35 | 39 | 40 | 41 | 42 | 42 | 45 |
| 5 | 1024 | 3 | 13 | 23 | 31 | 40 | 44 | 45 | 46 | 47 | 47 | 50 |

output register width to be $B_{OUT} = 31 - 24 = 7$. Again we can see that the table has been designed so that the minimal bit width of the output register is equal to the bit width of the input register.

Suppose we design a five-stage CIC filter with an $RM$ product of 24 and an input register width of $B_{IN} = 9$ bits. We can compute the maximum register bit width as

$$
\begin{aligned}
B_{MAX} &= \lceil N \, \mathrm{Log}_2 (RM) + B_{IN} \rceil \\
&= \lceil 5 \, \mathrm{Log}_2 (24) + 9 \rceil \\
&= 32
\end{aligned}
$$

From the table, we see that the maximum number of truncated bits from each section is given by $B_J = 1, 5, 9, 12, 15, 17, 18, 19, 20, 20, 23$. The maximum number of LSB truncation at the output register is $23$. Therefore we compute the minimum output register width to be $B_{OUT} = 32 - 23 = 9$.

### 6.3.15  CIC Interpolation Filter Register Growth

The minimum width $W_j$ of the register in the $j$th section of a CIC interpolation filter is given by

$$W_j = \left\lceil B_{IN} + \text{Log}_2\left(G_j\right) \right\rceil \quad \text{for } j = 1, 2, 3, \cdots, 2N$$

**Equation 6.48**

where $G_j$ is the gain of the $j$th section and is given in Equation 6.46. The maximum register bit width of an interpolation filter section occurs at the last integrator register and is given by

$$B_{MAX} = W_{j=2N} = \left\lceil N \, \text{Log}_2\left(RM\right) - \text{Log}_2\left(R\right) + B_{IN} \right\rceil$$

**Equation 6.49**

where the symbol $\lceil x \rceil$ is the ceiling operator, a function that returns the smallest integer greater than or equal to $x$. We will demonstrate the use of these equations later on when we design a CIC interpolation filter.

### 6.3.16  CIC Pass Band Droop

The CIC magnitude response approximates a $\sin(x)/x$ roll-off at the edge of the desired signal pass band. This roll-off increases as the number of stages $N$ increases. Depending on the width of the pass band $f_{PASS}$, this roll-off can produce considerable attenuation at frequencies near the pass band edge.

For example, a pass band whose width relative to the low rate sample frequency of $f_{PASS}/f_{S\downarrow} = 1/128$ has a 0.0 db roll-off at the edge frequency for up to five filter stages. However, as the width of the pass band grows, the magnitude of the roll-off increases. A pass band edge of $f_{PASS}/f_{S\downarrow} = 1/4$, for example, has a roll-off of 0.91 db at band edge for a single stage filter and 1.82 db for a two-stage filter, increasing to 5.47 db for a six-stage filter. This can be a significant problem.

The droop can be easily calculated by substituting the frequency of the band edge into the CIC frequency response given by Equation 6.34. This equation normalized to unity gain is repeated here for clarity:

$$\left|H\left(e^{j\omega}\right)\right|_{\downarrow} = \left|\frac{\sin\left(\pi M f\right)}{\pi M f}\right|^{N}$$

Substitution of the normalized band edge frequency yields

$$\left|H\left(e^{j\omega}\right)\right|_{\downarrow} = \left|\frac{\sin\left[\pi M\left(f_{PASS}/f_{S\downarrow}\right)\right]}{\pi M\left(f_{PASS}/f_{S\downarrow}\right)}\right|^{N}$$

The magnitude of the pass band edge droop in decibels is then given by

$$\left|H\left(e^{j\omega}\right)\right|_{\downarrow db} = 20N\ \mathrm{Log}_{10}\left\{\left|\frac{\sin\left[\pi M\left(f_{PASS}/f_{S\downarrow}\right)\right]}{\pi M\left(f_{PASS}/f_{S\downarrow}\right)}\right|\right\}$$

**Equation 6.50**

For example, the expected droop in decibels for a four-stage CIC decimation filter with a pass bandwidth of $f_{PASS}/f_{S\downarrow} = 1/8$ and a comb differential delay of $M = 1$ is computed to be

$$\left|H\left(e^{j\omega}\right)\right|_{\downarrow db} = 20\cdot 4\ \mathrm{Log}_{10}\left\{\left|\frac{\sin\left[\pi\left(1/8\right)\right]}{\pi\left(1/8\right)}\right|\right\} = 80\ \mathrm{Log}_{10}\left(0.974495358\right) = -0.8976\ \mathrm{db} \cong -0.90\ \mathrm{db}$$

For the same filter, but with a comb differential delay of $M = 2$, the droop is computed as

$$\left|H\left(e^{j\omega}\right)\right|_{\downarrow db} = 20\cdot 4\ \mathrm{Log}_{10}\left\{\left|\frac{\sin\left[\pi\left(1/4\right)\right]}{\pi\left(1/4\right)}\right|\right\} = 80\ \mathrm{Log}_{10}\left(0.9003163\right) = -3.6484\ \mathrm{db} \cong -3.65\ \mathrm{db}$$

Increasing $M$ from 1 to 2 results in a significant increase in pass band roll-off. Care should be taken when choosing the value of $M$ in relation to the bandwidth of the pass band. A table of droop values in decibels at the pass band edge is generated using these equations for several values of $M\cdot\left(f_{PASS}/f_{S\downarrow}\right)$. The results are tabulated versus the number of stages $N$ in Table 6.7. This is an important table. We will use it in all of our CIC filter designs. If you foresee the design of a CIC filter in your near future, it would

**Table 6.7**   Band Edge Droop for $M\left(f_{PASS}/f_{s\downarrow}\right)$ versus the Number of Stages $N$

| $M \cdot f_{PASS}/f_{s\downarrow}$ | Pass band droop (db) at $f_{PASS}$ as a function of the number of stages $N$ | | | | | |
|---|---|---|---|---|---|---|
| | 1 | 2 | 3 | 4 | 5 | 6 |
| 1/128 | 0.00 | 0.00 | 0.00 | 0.00 | 0.00 | 0.01 |
| 1/64 | 0.00 | 0.01 | 0.01 | 0.01 | 0.02 | 0.02 |
| 1/32 | 0.01 | 0.03 | 0.04 | 0.06 | 0.07 | 0.08 |
| 1/16 | 0.06 | 0.11 | 0.17 | 0.22 | 0.28 | 0.34 |
| 1/8 | 0.22 | 0.45 | 0.67 | 0.90 | 1.12 | 1.35 |
| 1/4 | 0.91 | 1.82 | 2.74 | 3.65 | 4.56 | 5.47 |

be worthwhile to use Equation 6.50 to generate a much larger table using an Excel spreadsheet.

### 6.3.17   CIC Interpolation Filter Example Design

The design of a CIC interpolation filter is a bit simpler than the design of a decimation filter, so we will perform an example design of an interpolation filter first. As mentioned earlier, the CIC filter is best used in applications that require a large sample rate change. Let's assume that you have been tasked to design a filter that increases the sample rate of an input sequence with the following specifications:

1. The filter input bit width: $B_{IN} = 16$
2. The filter output bit width: $B_{OUT} = 16$
3. The image band attenuation: IBA $\geq 90$ db
4. The pass band edge roll-off: PBR $\leq 0.3$ db
5. The input sample rate: $f_{s\downarrow} = 32$ KHz
6. The interpolated output sample rate: $f_{s\uparrow} = 16.384$ MHz
7. The pass band edge: $f_{PASS} = 2$ KHz

We first compute at the sample rate change:

$$R = \frac{16.384 \text{ MHz}}{32 \text{ KHz}} = 512$$

This is a huge rate change. Implementing a conventional FIR filter with this large of sample rate change would require a significant amount of hardware resources. The number of coefficient multiplies would be large, which would translate into issues such as on-chip and on-board real estate, increased power consumption, increased design complexity, and increased test and integration time. So, being an astute design engineer, you immediately determine that the most economical filter you can build to satisfy this requirement is a CIC filter. Congratulations, you have made the right choice, and those who attend your first design review will have a high opinion of your intellect.

The next item of business is to select the comb filter differential delay to be $M = 1$. This is done primarily to reduce the filter main lobe roll-off attenuation and to ease the requirements of the follow on $x/\sin x$ compensation and cleanup filter, should one be necessary. The next thing we compute is the normalized low sample rate pass band edge. In this example we get

$$\frac{f_{PASS}}{f_{S\downarrow}} = \frac{2\,\text{KHz}}{32\,\text{KHz}} = \frac{1}{16} = 0.0625$$

We use the tables provided to determine the number of CIC stages necessary to perform the task. For the readers' convenience, two small tables (Tables 6.8 and 6.9) are shown that are excerpts from Table 6.1 and Table 6.7 (presented previously).

Table 6.8 shows three rows of Table 6.1, which tabulates the alias/image attenuation in decibels versus the number of filter stages $N$ for various values of the normalized pass band edge $f_{PASS}$. Table 6.9 shows three rows of Table 6.7, which tabulates the pass band roll-off or droop in decibels versus the number of filter stages $N$ for various values of the product $M \cdot (f_{PASS}/f_{S\downarrow})$.

From Table 6.8, we find the entry for the case where $M = 1$ and normalized $f_{PASS} = 1/16$ and then scroll across the columns until we find an acceptable image band attenuation. Our specification requires that the image band attenuation be greater than or equal to 90 db. Under the column for four stages, we see that the attenuation figure is 94.3 db, which exceeds the specification by more than 4 db.

Next, in Table 6.9 we find the row entry for the normalized $M \cdot f_{PASS} = 1/16$ and the column entry for $N = 4$ to find the pass band roll-off figure. The design specification defines the value to be $\leq 0.3$ db. From this table, we see that for a four-stage filter the pass band roll-off is 0.22 db, which exceeds the filter specification.

So now we know from the tables that a CIC filter with four stages exceeds the filter specifications we were given and therefore will be sufficient for our design. The next step in the design process is to compute the number of bits that are necessary to implement each of the $j = 1, 2, 3, \cdots, 2N$ comb and integrator section registers in the filter.

**Table 6.8**   Alias/Image Band Attenuation Table Excerpt

| | | Pass band aliasing / image band attenuation as a function of the number of stages $N$ | | | | | |
|---|---|---|---|---|---|---|---|
| $M$ | $f_{PASS}$ | 1 | 2 | 3 | 4 | 5 | 6 |
| 1 | 1/32 | 29.8 | 59.7 | 89.5 | 119.4 | 149.2 | 179.0 |
| 1 | 1/16 | 23.6 | 47.2 | 70.7 | 94.3 | 117.9 | 141.5 |
| 1 | 1/8 | 17.1 | 34.3 | 51.4 | 68.5 | 85.6 | 102.8 |

**Table 6.9**   Pass Band Roll-Off Table Excerpt

| | Pass band droop (db) at $f_{PASS}$ as a function of the number of stages $N$ | | | | | |
|---|---|---|---|---|---|---|
| $M \bullet f_{PASS}$ | 1 | 2 | 3 | 4 | 5 | 6 |
| 1/32 | 0.01 | 0.03 | 0.04 | 0.06 | 0.07 | 0.08 |
| 1/16 | 0.06 | 0.11 | 0.17 | 0.22 | 0.28 | 0.34 |
| 1/8 | 0.22 | 0.45 | 0.67 | 0.90 | 1.12 | 1.35 |

We accomplish this by first using Equation 6.46 to calculate the gain $G_j$ at each of the $j$th comb and integrator section registers and then by using Equation 6.48 to calculate the required register bit width $W_j$ at each of the $j$th registers for $j = 1, 2, 3, \cdots, 2N$. These two equations are repeated here for clarity:

$$G_j = \begin{cases} 2^j & \text{for } j = 1, 2, 3, \cdots N & \text{Comb registers} \\ \dfrac{2^{2N-j}(RM)^{j-N}}{R} & \text{for } j = N+1, N+2, \cdots, 2N & \text{Integrator registers} \end{cases}$$

$$W_j = \left\lceil B_{IN} + Log_2 \left( G_j \right) \right\rceil \text{ for } j = 1, 2, 3, \cdots, 2N$$

We can easily calculate the gain for the four comb sections as

$$G_{j=1,2,3,4} = 2, \ 4, \ 8, \text{ and } 16$$

respectively. Then we calculate the gain for the four integrator sections to be

**Table 6.10**  CIC Interpolation Filter Section Register Bit Widths

| Section register ($j$) | CIC section type | Gain $G_j$ | Register bit width $W_j$ |
|---|---|---|---|
| 1 | Comb | 2 | 17 |
| 2 | Comb | 4 | 18 |
| 3 | Comb | 8 | 19 |
| 4 | Comb | 16 | 20 |
| 5 | Integrator | 8 | 19* |
| 6 | Integrator | 2048 | 27 |
| 7 | Integrator | 524,288 | 35 |
| 8 | Integrator | 134,217,728 | 43 |

$$G_{j=5} = \qquad\qquad 8$$
$$G_{j=6} = \qquad\qquad 2048$$
$$G_{j=7} = \qquad 524{,}288$$
$$G_{j=8} = \quad 134,\,217{,}728$$

respectively. As an example, let's compute the gain for the eighth section as

$$G_8 = \frac{2^{8-8}\left(512\right)^4}{512} = 134{,}217{,}728$$

Substituting these gain values into the register width equation, we can compute the number of bits required for each of the internal integrator registers. For example, the bit width of the seventh register is computed by

$$W_7 = \left\lceil 16 + \mathrm{Log}_2\left(\frac{2\left(512\right)^3}{512}\right)\right\rceil = 35 \text{ bits}$$

The gains and the corresponding register widths for each CIC section are listed in Table 6.10.

The low rate frequency response for this interpolation filter example is illustrated in Figure 6.45. Note from the horizontal cursor that the image

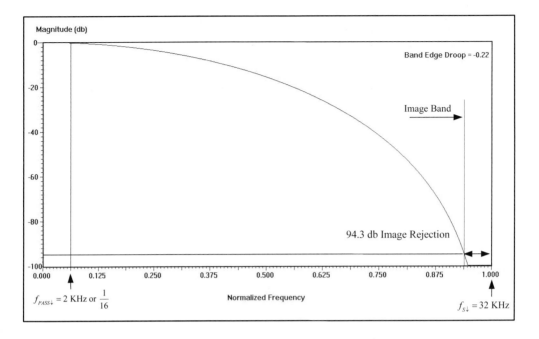

**Figure 6.45** Low rate frequency response of the CIC interpolation filter example

rejection is measured to be 94.3 db at the first image band, and the pass band droop is measured to be 0.22 db, just as the design tables predicted.

The architecture for this interpolation filter is illustrated in Figure 6.46. The first four registers in the filter are used to implement the comb sections, and the last four registers are used to implement the integrator sections in the filter. In between the comb and integrator sections is an up sampler that inserts 511 zeros in between sequential samples passed from the comb to the integrator sections. Note in the figure that the last comb section bit width is calculated to be 20 bits but the follow on integrator section is calculated to have only 19 bits. This anomaly is noted in Table 6.10 by an asterisk in the register width column. It turns out that this apparent inconsistency occurs when designing an interpolation filter with the value of $M = 1$. This gives rise to a special condition documented by Hogenauer in reference [1]: when $M = 1$, the width of the last comb section should be calculated by the following relation:

$$W_N = B_{IN} + N - 1$$

**Equation 6.51**

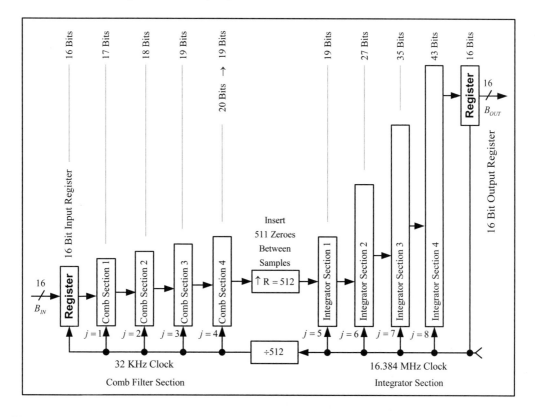

**Figure 6.46**   Example CIC interpolation filter design architecture

If this rule is followed, then the bit width of the last comb section becomes 19 bits instead of 20. This 1-bit difference in the last comb section is illustrated in Figure 6.46.

The number of register bits discarded at the output register is given by

$$B_{Discard} = W_{j=2N} - B_{out}$$

For this example, the number of discarded bits is computed to be

$$B_{Discard} = 43 - 16 = 27 \text{ bits}$$

That's all there is to it. The design of the interpolation filter is complete. As the reader can see, the interpolator architecture lends itself well to implementation on FPGAs and application-specific integrated circuits (ASICs). The comb and integrator sections within an FPGA or ASIC are built up by repetitive utilization of the same development software library components or through the use of predesigned and scalable FPGA cores. These library components and/or cores are usually designed and supplied by the FPGA chip manufacturer as part of their software development station tools.

## 6.3.18  CIC Decimation Filter Example Design

As mentioned previously, the CIC filter is best used in applications that require a large sample rate change. The definition of the term *large* is usually made clear by computing the number of coefficient multiplies required to implement a rate change filter using conventional FIR filters versus using a CIC in cascade with a smaller FIR cleanup filter. This comparison, which takes into account the difference in real estate usage, power consumption, expended resources, and time to delivery, always makes for a good bullet on a design review presentation slide. In addition, a review slide with this trade-off information gives your design decision a good dose of credibility.

The following discussion will walk you through the design of a CIC decimation filter. Let's assume that you have been tasked by your supervisor to design a decimation filter with the following specifications:

1. The filter input bit width: $B_{IN} = 16$
2. The filter output bit width: $B_{OUT} = 16$
3. The pass band aliasing attenuation: PBA $\geq 65$ db
4. The pass band roll-off: PBR $\leq 1$ db
5. The input sample rate: $f_{S\uparrow} = 4.096$ MHz
6. The decimated output sample rate: $f_{S\downarrow} = 128$ KHz
7. The pass band edge: $f_{PASS} = 16$ KHz

A quick calculation shows that the sample rate change $R = (4.096 \text{ MHz} / 128 \text{ KHz}) = 32$. Since this is a fairly large rate change, and since you are a competent engineer, you compare the computational requirements between a FIR filter and a CIC/FIR cascade implementation and decide to use a CIC filter followed by a FIR cleanup filter to perform the multirate filtering task. So far, so good; your first engineering decision is a correct one. So what's the next step? The first thing to do is to compute the normalized pass band edge, which is

$$f_{PASS} = (16 \text{ KHz} / 128 \text{ KHz}) = 1/8 = 0.125$$

Since we wish to minimize the pass band roll-off, the next thing to do is to set the comb differential delay to $M = 1$. Now comes the simple part. We use Table 6.1 and Table 6.7 to determine the number of CIC stages necessary to perform the task. Once again, excerpts from these two tables are shown here for convenience.

Table 6.11 shows three rows of the original Table 6.1 that list the alias / image band attenuation in decibels versus the number of filter stages for various values of the normalized pass band edge $f_{PASS}$.

Table 6.12 shows the corresponding three rows of the original Table 6.7 that lists the pass band droop in decibels versus the number of filter stages for various values of the normalized product $M \cdot f_{PASS}$. From Table 6.11, we find the entry for the case where $M = 1$ and $f_{PASS} = 1/8$ and then scroll across the columns until we find an acceptable pass band aliasing attenuation. Under the column for four stages, we see that the attenuation figure is 68.5 db, which beats the pass band attenuation specification we were given of $PBA \geq 65$ db.

Next, from Table 6.12 we find the row entry for $M \cdot f_{PASS} = 1/8$ and the column entry for $N = 4$ to find the pass band roll-off figure. From this table, we see that for a four-stage filter we should expect a pass band roll-off of 0.90 db, which beats the pass band roll-off specification we were given of $PBR \leq 1 \ db$.

So now we know that a CIC filter consisting of four stages will give us the performance necessary to exceed the filter specification given by your supervisor. So far, because of your design choices, your supervisor is still a happy camper.

**Table 6.11** Pass Band Attenuation Table

| | | Pass band aliasing / image band attenuation as a function of the number of stages $N$ | | | | | |
|---|---|---|---|---|---|---|---|
| $M$ | $f_{PASS}$ | 1 | 2 | 3 | 4 | 5 | 6 |
| 1 | 1/16 | 23.6 | 47.2 | 70.7 | 94.3 | 117.9 | 141.5 |
| 1 | 1/8 | 17.1 | 34.3 | 51.4 | 68.5 | 85.6 | 102.8 |
| 1 | 1/4 | 10.5 | 20.9 | 31.4 | 41.8 | 52.3 | 62.7 |

**Table 6.12** Pass Band Roll-Off Table

| | Pass band droop ($db$) at $f_{PASS}$ as a function of the number of stages $N$ | | | | | |
|---|---|---|---|---|---|---|
| $M \bullet f_{PASS}$ | 1 | 2 | 3 | 4 | 5 | 6 |
| 1/16 | 0.06 | 0.11 | 0.17 | 0.22 | 0.28 | 0.34 |
| 1/8 | 0.22 | 0.45 | 0.67 | 0.90 | 1.12 | 1.35 |
| 1/4 | 0.91 | 1.82 | 2.74 | 3.65 | 4.56 | 5.47 |

Now we need to do a little bit of math. We need to compute the most significant bit $B_{MAX}$, required of all the registers in each section of the filter. To accomplish this, we use Equation 6.47, repeated here for clarity:

$$
\begin{aligned}
B_{MAX} &= \left\lceil N \, \text{Log}_2 \left( RM \right) + B_{IN} \right\rceil \\
&= \left\lceil 4 \, \text{Log}_2 \left( 32 \cdot 1 \right) + 16 \right\rceil \\
&= 36
\end{aligned}
$$

At this point in the design, the architecture of the filter takes on the shape shown in Figure 6.47. Note that the input register is 16 bits wide, to support an input sequence bit width $B_{IN}$, and the output resister is also 16 bits wide to support the specified output sequence bit width $B_{OUT}$. The individual section registers at this point are all $B_{MAX} = 36$ bits wide. The first four section registers are used to implement the integrator section of the CIC filter. The last four section registers are used to implement the comb filter section of the CIC. In between the two filter sections is a down sampler that passes

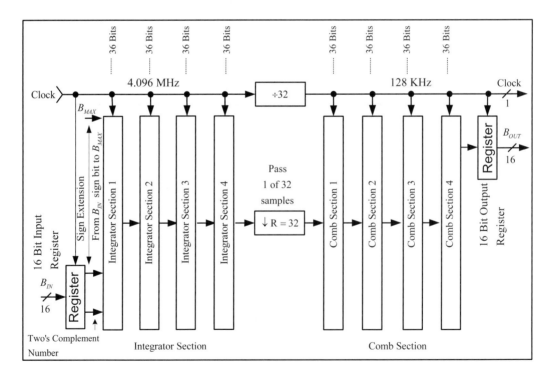

**Figure 6.47**   Four-stage decimate by 32 CIC filter without pruned LSBs

1 sample for every 32 between the integrator section and the comb section. The reader should take note that the data is encoded in two's complement format. We are adding the content of the 16-bit input register to the content of the first stage integrator that is 36 bits wide. Therefore two's complement number in the input register will have to be sign extended on the input to the first integrator register, all the way to $B_{MAX}$.

Now we have arrived at the point where we can truncate some of the LSBs in the section registers by following the procedure outlined in reference [1]. In other words, we can use the Hogenauer LSB pruning procedure to compute the number of truncated LSBs $B_j$ for the $j = 1, 2, 3, \cdots, 2N$ internal registers and the output register $j = (2N + 1)$. For the purposes of this design example, we will avoid the lengthy pruning computations. We will opt instead to determine the number of bits to prune for each section by using Table 6.5. We scroll down the rows of the table until we find the entry for $RM = 32$ and then move across the row and read off the number of bits we can truncate at each section of the filter. The resultant register widths for our design example are then tabulated in Table 6.13.

We can use the information in Table 6.13 to reduce the size of the registers within the CIC filter. The resulting CIC decimate by 32 filter architecture with LSB pruning is illustrated in Figure 6.48.

**Table 6.13**   Intermediate Register Truncated LSBs

| $j$ | Section register | Truncated LSBs $B_j$ | Register width |
|:---:|:---|:---:|:---:|
| 1 | Integrator register 1 | $B_1 = 1$ | $36 - 1 = 35$ |
| 2 | Integrator register 2 | $B_2 = 6$ | $36 - 6 = 30$ |
| 3 | Integrator register 3 | $B_3 = 10$ | $36 - 10 = 26$ |
| 4 | Integrator register 4 | $B_4 = 13$ | $36 - 13 = 23$ |
| 5 | Comb register 1 | $B_5 = 15$ | $36 - 15 = 21$ |
| 6 | Comb register 1 | $B_6 = 16$ | $36 - 16 = 20$ |
| 7 | Comb register 3 | $B_7 = 17$ | $36 - 17 = 19$ |
| 8 | Comb register 4 | $B_8 = 18$ | $36 - 18 = 18$ |
| 2N+1 | Output register | $B_{2N+1} = 20$ | $36 - 20 = 16$ |

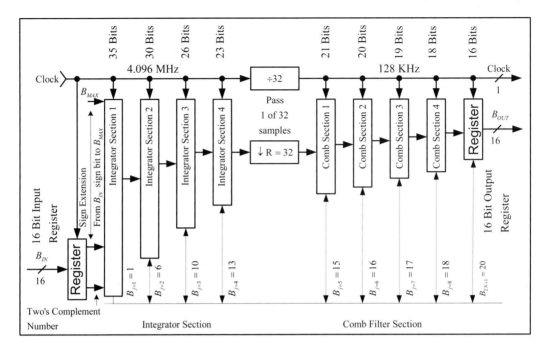

**Figure 6.48**   Four-stage decimate by 32 CIC filter with pruned LSBs

The low rate frequency response for the filter in this example is illustrated in Figure 6.49. The decimated sample rate is $f_{S\downarrow} = 128\,\text{KHz}$. The pass band edge is 16 KHz or 1/8 normalized. The measured pass band attenuation is 68.6 db, which, when the precision of the graphics is taken into account, is essentially equal to the 68.5 db specified in Table 6.11. The reader should also note in the figure that the pass band droop measured at the pass band edge is 0.9 db, exactly as specified in Table 6.12.

### 6.3.19   CIC Decimation Filter Graphical Summary

This is a good time to pause, collect our thoughts, and put all the previous CIC filter discussions into some sort of meaningful perspective. A long time ago, someone said that a picture is worth a thousand words. In these days of conservation, perhaps we could save a couple thousand words and a few sheets of paper if we invested some time developing a few pictures to help us summarize the operational characteristics of a CIC decimation filter.

Let's begin by looking at a generic base band signal spectrum, as illustrated in Figure 6.50. The frequency axis (1) in this figure is calibrated at the high sample rate in units of $2\pi k/16$ radians ranging from 0 to $2\pi$. The

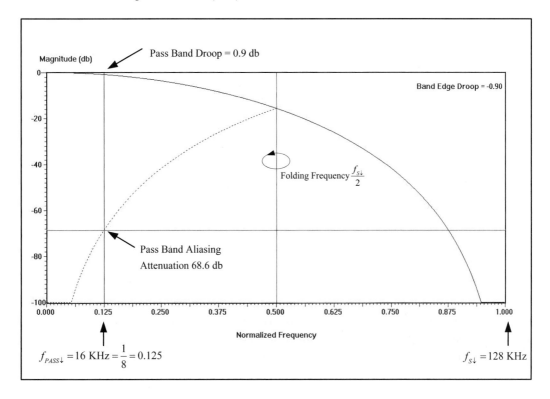

**Figure 6.49**   CIC decimate by 32 low rate frequency response

frequency axis (2) is calibrated at the high sample rate in units of $f_{S\uparrow}/16$ Hz ranging from 0 to $f_{S\uparrow}$.

The signal spectrum range is $\left[-8f_{S\uparrow}/16\right] \le f \le \left[+8f_{S\uparrow}/16\right]$. The standard discrete Fourier transform (DFT) display convention is to show the positive frequencies ranging from 0 Hz to $8f_{S\uparrow}/16$, and the negative frequencies ranging from $8f_{S\uparrow}/16$ to $f_{S\uparrow}$. This latter band of frequencies is typically thought of as being either the negative frequencies of the signal spectra on the other side of 0 Hz or as the negative frequencies associated with the image band centered at $f_{S\uparrow}$ Hz. This is the band of frequencies that will alias about the folding frequency of $f_{S\downarrow}/2$ when the sample rate is reduced.

It is our desire to process this signal through a set of three different CIC decimation filters for the three simple cases where the comb differential delay $M = 1$ and the sample rate change $R \downarrow$ is set to 2, 4, or 8. The only portion of the signal spectra that we are interested in lies within the frequency band $-f_{PASS} \le f \le f_{PASS}$, illustrated in Figure 6.50. We have no concern for the band

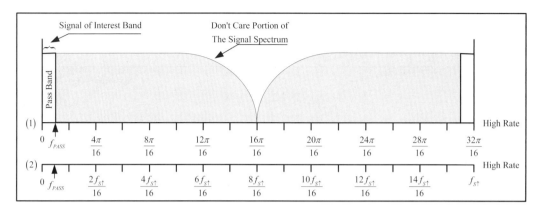

**Figure 6.50**   CIC decimation filter graphical summary base band spectrum

of frequencies in the input signal represented by the grayed region of the spectrum. This area is termed the "don't care" region of the input spectrum.

About the only application that immediately comes to mind where a CIC filter would be used for such small rate changes would be the case where the signal sample rate is so high as to make processing with a conventional multiplier based FIR unfeasible. Other than that, one would probably never use a CIC filter for such small rate changes, but our objective here is to graphically observe the physics of the CIC decimation filter operation with the aid of some very simple and easy-to-plot examples.

### 6.3.19.1   Case I: Decimation by Two

Let's begin this discussion by processing the signal whose spectrum is illustrated in Figure 6.50 with a CIC filter configured to decimate the sample rate by a factor of two. Decimation is accomplished by passing every other sample from the CIC filter integrator section to the filter comb section. The spectrum of the processed signal is shown in part A of Figure 6.51, after it has been filtered and decimated by a factor of two. The figure includes three different frequency axes: Axis 1 is calibrated at the high sample rate in units of $2\pi k/16$ radians, ranging from $0$ to $2\pi$. Axis 2 is calibrated at the high sample rate in units of $f_{S\uparrow}/16$ Hz, ranging from $0$ to $f_{S\uparrow}$. Axis 3 is calibrated at the low sample rate in units of $f_{S\downarrow}/8$ Hz, ranging from $0$ to $f_{S\downarrow}$.

Along the left side of the plot is the pole-zero diagram for the decimate by two CIC filter where $RM = 2$. Note that there are two zeros in the diagram. One zero occurs at $z = 0$ and the other at $z = \pi$ radians. Remember that the frequency response nulls occur for frequencies located at a zero. Comparing the pole-zero plot with the frequency response plot, we see that we do indeed get a frequency response null at $\pi$ radians relative to the high sample rate. We do not see a frequency response null at $f = 0$ Hz because the corresponding zero is cancelled by the integrator pole.

When the sample rate is reduced by a factor of two, the new sample rate is represented by $f_{S\downarrow}$. The only frequencies that we are interested in are the original pass band and any band of frequencies that will alias into the pass band after the sample rate reduction. The band of frequencies within the range of $f_{S\uparrow}/2 - f_{PASS} \leq f \leq f_{S\uparrow}/2 + f_{PASS}$ are the frequencies that will alias

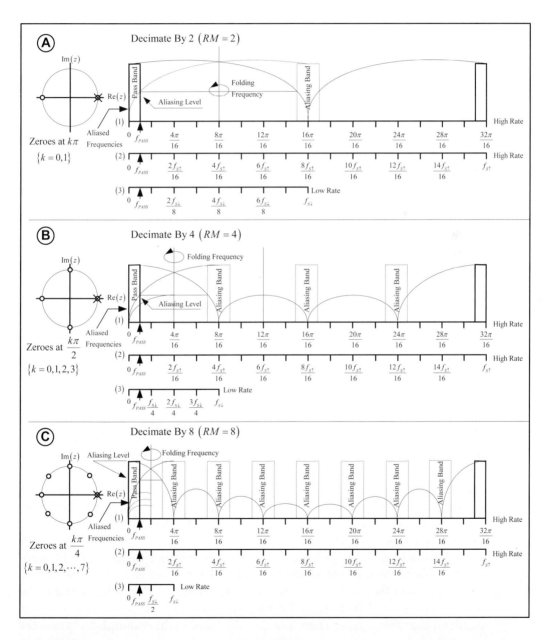

**Figure 6.51** CIC decimation filter graphical summary

down to base band when the sample rate is decimated by a factor of two. This band is centered at the new decimated sample rate of $f_{s\downarrow} = f_{s\uparrow}/2$ or, as shown in the figure, is centered at $\pi$ radians relative to the high sample rate.

If the sample rate reduction occurred with no filtering, then these frequencies would alias back unattenuated, completely corrupting the frequencies in our sacred pass band. It can be seen, however, that the frequency response of the CIC filter has a spectral null at $\pi$ radians, which is dead center in the aliasing band frequencies. In a properly designed CIC filter, these frequencies are heavily attenuated and the pass band signal degradation due to aliasing is minimal.

### 6.3.19.1 Case II: Decimation by Four

If we process the exact same input signal with a CIC filter configured to decimate by a factor of $RM = 4$, we end up with the frequency response plot shown in part B of Figure 6.51. Again the sample rate reduction by a factor of four is accomplished by passing only one sample for every four that pass between the input integrator section and the output comb section.

We see that frequency bands centered at $4kf_{s\uparrow}/16 = kf_{s\uparrow}/4$ $\{\text{for } k = 1,2,3\}$ or $k\pi/2$ $\{\text{for } k = 1,2,3\}$ relative to the high sample rate are the three bands that will alias into our pass band at the reduced sample rate.

The accompanying pole-zero plot shows the filter has four zeros spaced at intervals of $\pi/2$ radians. Therefore we should expect to see frequency response nulls at intervals of $\pi/2$ radians relative to the high sample rate, and we do. Again there is no null at 0 since that zero is cancelled by the integrator pole. We also see that the CIC filter response nulls are dead center within each of the alias bands, which severely attenuates these frequencies. The figure shows the magnitude of the aliased frequencies that extend into the pass band. The figure illustrates that the highest aliasing contribution is made by the first alias band. All other bands make contributions of a lesser degree.

### 6.3.19.3 Case III: Decimation by Eight

Part C of Figure 6.51 shows the resultant spectrum when we process the same input signal with a CIC filter configured to decimate by a factor of $RM = 8$. The decimation is accomplished by passing one sample for each eight samples passed between the filter integrator section and the filter comb section. Note here that there are now seven alias frequency bands that we should be concerned with. These bands are centered at multiples of $\pi/4$ relative to the high sample rate.

Also note that the pole-zero diagram shows that there are eight zeros evenly spaced around the unit circle. One of these zeros cancels the pole at $z = 0$, and the other seven are associated with filter response nulls that occur at multiples of $\pi/4$ relative to the high sample rate. The frequency response confirms that this is true. Therefore the CIC filter in this example provides the attenuation necessary to avoid serious degradation to the pass band frequencies caused by frequency components of all the aliasing bands.

In all three of these cases there remains a great deal of post filter residue frequencies that are not contained in the desired pass band. For this reason, in most if not all cases, the CIC filter is followed by a small FIR filter with a sharp pass band roll-off and deep stop band attenuation that operates at the low sample rate $f_{s\uparrow}/R$ to perform residual frequency cleanup duties. As you might expect, this filter is sometimes called a *cleanup filter*. We will discuss this CIC/FIR architecture a bit later.

### 6.3.20 CIC Interpolation Filter Graphical Summary

Next we graphically summarize the operation of a CIC interpolation filter in Figure 6.52, where we illustrate three cases where the filter is configured to increase the sample rate by a factor of 2, 4, and 8. In all three cases, the value of the comb differential delay is set to $M = 1$. Once again it should be pointed out that the odds are that no one would ever use a CIC filter for such a small sample rate increase. The objective here is to discuss examples that can easily be illustrated graphically.

Part A of the figure illustrates the spectrum of the original base band signal. It is represented by two rectangular blocks that are the positive and negative frequencies of the base band signal or, alternatively, the positive frequencies of the base band signal and the negative frequencies of the band of digital image frequencies centered at $f_{s\downarrow}$. Either interpretation is correct. In both cases, the two-sided bandwidth is $2f_{PASS}$.

As we go through the three cases, the reader should see that the CIC filter we are describing is identical to those we described in the decimation filter graphical summary. The only difference is that architecturally the comb sections and the integrator sections have been swapped. We can do this because the filter is a linear system.

The frequency response plots for all three cases (B, C, and D) have three frequency axes: Axis 1 is referenced to the low sample rate in units of $f_{s\downarrow}$ Hz. Axis 2 is referenced to the high sample rate in units of $k\pi$ radians. Axis 3 is referenced to the high sample rate in fractional units of $f_{s\uparrow}$.

#### 6.3.20.1 Case I: Interpolation by Two

In part B, we use the CIC filter to increase the sample rate of the signal by a factor of two. In other words, we set the CIC parameters $RM = 2$. This is the same setting we used for the CIC decimation filter that reduced the sample rate by a factor of two in the previous section. The actual sample rate increase is achieved by inserting a zero sample between adjacent samples as they are clocked from the comb section of the filter to the integrator section.

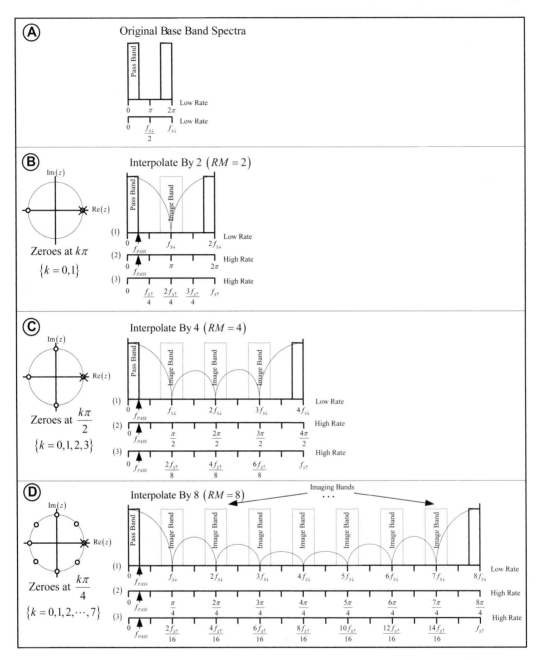

**Figure 6.52** CIC interpolation filter graphical summary

Along the left side the frequency response is the corresponding pole-zero diagram. Once again, it has two zeros. It has a zero at $z = 0$, which is cancelled by the corresponding integrator pole, and a zero at $z = \pi$. From this diagram, we expect to see a filter response null at $\pi$ radians, and we do.

When the sample rate is increased by a factor of two, the resultant frequency spectrum is widened by a factor of two. This new spectrum consists of what looks like the two original pass bands plus an image band centered at $f_{s\downarrow}$ when referenced to the low sample rate and centered at $\pi$ or $f_{s\uparrow}/2$ when referenced to the high sample rate. The reader will note that the filter response null occurs squarely in the center of the image band. The image band frequency components will be attenuated proportionally to the order of the filter.

A tool that the reader can use to visualize the interpolation process is to think of the digital spectrum as being replicated an infinite number of times. That is, the original two-sided signal spectra repeat at multiples of the original low sample rate. We look at the spectrum through a window whose width is equal to the sample rate. When we increase the sample rate by a factor of two, all we do is widen the window from which we can view more of this replicated spectrum (in this case twice as much). By doing so, we end up viewing two copies of the signal spectra. This visualization tool is easy to use and works for all integer increases in the sample rate.

### 6.3.20.2   Case II: Interpolation by Four

Part C of the figure illustrates the resultant signal spectrum when a CIC filter is used to increase the sample rate by a factor of four. This is accomplished by setting $RM = 4$ and inserting three zeros between adjacent samples as these samples were passed from the comb section to the integrator section.

The new sample rate is given by $f_{s\uparrow} = 4f_{s\downarrow}$. The spectral window has widened so that we can see four copies of the original base band spectrum—that is, the original and three image bands. The image bands are centered at multiples of the low sample rate $kf_{s\downarrow}$ $\{\text{for } k = 1, 2, 3\}$ or when referenced to the high sample rate at multiples of $k\pi/2$ $\{\text{for } k = 1, 2, 3\}$.

The corresponding pole-zero diagram shows that this filter has four zeros also located at multiples of $\pi/2$. Therefore we should expect to see filter response nulls at multiples of $\pi/2$ radians, and we do. These filter nulls correspond to the center frequencies of each of the three image bands. Therefore these bands are attenuated substantially.

### 6.3.20.3   Case III: Interpolation by Eight

Part D of the figure illustrates the resultant signal spectrum when a CIC filter is used to increase the sample rate by a factor of eight. We accomplished this feat simply by setting the filter parameters $RM = 8$ and inserting seven zeros between adjacent samples as these samples were passed from the comb section to the integrator section.

The window through which we view the digital spectrum is widened by a factor of eight. We can now view eight copies of the original spectrum, consisting of the original signal spectrum and seven image copies. When referenced to the low sample frequency, the image copies are centered at multiples of $kf_{s\downarrow}$ $\{$for $k = 1, 2, 3, 4, 5, 6, 7\}$. When referenced to the high sample rate, the image bands are centered at multiples of $k\pi/4$ $\{$for $k = 1, 2, 3, 4, 5, 6, 7\}$.

The associated pole-zero diagram shows there are eight zeros spaced at multiples of $\pi/4$ radians around the unit circle. From this diagram, we expect to see CIC filter response nulls at multiples of $\pi/4$ referenced to the high sample rate, and of course, we do. The nulls of the CIC filter fall precisely in the center of each of the image bands. The amount of attenuation for each of the image bands depends on the order $N$ of the CIC filter (i.e., the number of stages the filter has).

As was the case with the CIC decimation filters, the CIC interpolation filters are generally but not necessarily followed by a short FIR cleanup filter to get rid of the residue image bands. This cleanup filter, if used, does not necessarily require a sharp pass band roll-off, since the transition band can theoretically extend to the edge of the first image band. This is a good thing since this cleanup filter will be operating at the high sample rate and a relaxed transition band results in fewer filter multiplies. The cleanup filter, however, should have a deep stop band attenuation in order to knock down the image band residue frequencies.

## 6.3.21   CIC Cleanup Filter

Because of the $\sin(x)/x$ like frequency response of the CIC filter, some applications will require that a small FIR filter follow in cascade to clean up the frequency residue left behind by the CIC. We can think of the CIC as doing the bulk of the rate change computations and the cleanup filter as doing a small amount of the overall computations to tidy up the final output.

This is especially true for sample rate decimation applications. The cascaded FIR in a CIC/FIR decimation filter has the advantage of operating at the reduced sample rate so it can be implemented much more efficiently. The cleanup filter can also be designed to take on an $x/\sin x$ frequency response over the output pass band to compensate for the band edge droop cased by the CIC.

In interpolation or sample rate increase applications, a FIR filter in cascade with the CIC may not be necessary. If, for example, the CIC is designed so that residue image bands are below the specified filter stop band, then no cleanup is necessary. This is a good thing because an interpolation FIR filter would be tasked to operate at the higher sample rate, which can make the filter more difficult to design and build. We will discuss the need for a cleanup filter in the next two paragraphs.

### 6.3.21.1   CIC/FIR Decimation Filter Cascade

A CIC decimation filter is almost always followed in cascade with a cleanup FIR filter that is used to eliminate aliased frequencies beyond the desired pass band edge. The readers first concern might be, why bother with a CIC/FIR cascade? Why not dispense with the complexity of a two-filter cascade and just use a poly phase FIR for the entire filtering task? Using a PPF is definitely an option and is the correct solution when the sample rate change is relatively small. But remember that we are discussing the applications here that require large sample rate changes.

Now suppose that the sample rate change is large and further suppose that the band of interest occupies a very narrow band within a wideband signal. In order to extract this narrow band signal, the filter would require a sharp transition band. One of the drivers in determining the length of a filter impulse response is the width of its transition band. The narrower the transition band, the longer the impulse response. The number of filter multipliers increases rapidly as the transition band is narrowed. The number of phases of a PPF is equal to the decimation ratio. Large rate changes translate to a large number of filter phases and hence a greater complexity in the control hardware.

For large rate changes, it is best to dispense with the multipliers and use a CIC filter to perform the bulk of the work. The FIR used for cleanup duties will be operating at a much lower sample rate and therefore will require far less hardware to implement. Common CIC/FIR cascade techniques are as follows:

1. Use the CIC to reduce the sample rate of the input signal to the final value; then follow up with a narrow band FIR cleanup filter.
2. Use the CIC to reduce the sample rate of the input signal to twice the final value; then follow up with a narrow band FIR decimate by two cleanup filter.

What defines a small rate change versus a large rate change is totally determined by the application for which the filter is intended, the amount of real estate is available, the design power budget, the program design and test schedule, and of course, the cost involved. Usually the definition is based on a comparison between the number of computations and hardware inventory required by competing rate change architectures.

A conceptual block diagram of the CIC/FIR decimation filter cascade is illustrated in Figure 6.53. In part A of the figure, we see a block diagram of a CIC filter followed by a FIR cleanup filter. In parts B, C, D, and E of the figure, we see the spectrum of the input, intermediate, and output signal.

In part A, we see the block diagram of the CIC filter followed by a FIR cleanup filter. The input sequence $x(n)$ is clocked into the CIC filter at a rate of $f_{S\uparrow}$ samples per second. In part B, we see the conceptual spectral diagram of the wideband input sequence with a very narrow band of interest positioned at the base band. The frequency axis is calibrated in units of $f_{S\uparrow}/R$ from $0$ to $4f_{S\uparrow}/R$ Hz. The reader should be able to determine by inspection that $R = 4$ in this example.

In part C of the figure, we see the CIC filter frequency response superimposed on the input signal spectrum. We also see the location of the bands of frequencies that will alias back onto the desired narrow pass band after the sample rate is reduced.

In part D of the figure, we can see the conceptual signal spectrum at the output of the CIC filter. It is clear that some aliasing has occurred in the narrow base band signals, as expected. We can also see that all the residue frequencies beyond $f_{PASS}$ are aliased. It is apparent that at this point there is a need for a FIR cleanup filter to remove these aliased frequencies.

In part E of the figure, we see the decimated signal spectrum at the output of the cleanup filter. The cleanup FIR is designed to have narrow transition bands and deep stop band attenuation. The resultant output is the clean narrow band signal $y(n)$, sampled at the much lower rate of $f_{S\downarrow} = f_{S\uparrow}/R$.

This is a very simple design that can be the kernel for much more complex sample rate reduction applications. For example, suppose we added a digital tuner to the front end of this two-filter cascade. With this simple addition to the design, we now can tune to any narrow band within the wideband signal and down convert this narrow band to base band, where it is filtered, extracted, reduced in sample rate, and provided to some follow on digital signal processor.

Guess what, guys? The description of this application is nothing more than a digital radio. The follow on processor can be any number of things such as an AM/FM demodulator, a digital automatic gain control (dAGC), or perhaps some sort of long term storage to freeze and catalog captured signals of interest for later analysis. The possibilities are many. This is what makes a career as a design engineer in the field of signal processing so much fun.

### 6.3.21.2   CIC/FIR Interpolation Filter Cascade

Why use a CIC/FIR interpolation cascade as opposed to a single poly phase FIR filter? This is answered in the same way that we answered the CIC/FIR

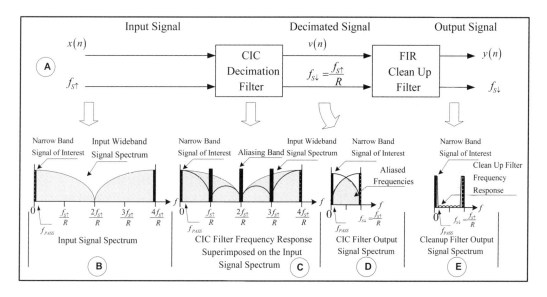

**Figure 6.53**   CIC/cleanup decimation filter cascade

decimation cascade question. The CIC/FIR cascade is a much more efficient choice for large sample rate change applications.

It should be noted that a cascade cleanup filter may not be necessary for many CIC interpolation filter designs. This will be true if the CIC is designed to attenuate the interpolated image bands so that they are below the specified filter minimum stop band. This is the preferred method and is the one that is usually implemented. With this said, to be thorough, we will discuss the CIC/FIR interpolation cascade because it provides us the opportunity to present a bigger conceptual picture to visualize the operation of a CIC interpolation filter.

The CIC/FIR interpolation filter cascade is illustrated in Figure 6.54. In part A of the figure, we can see the block diagram of the two-filter cascade. We can see that the input to the CIC interpolation filter is the signal $x(n)$, sampled at the rate of $f_{S\downarrow}$. We see the intermediate signal $v(n)$ output from the CIC, sampled at a rate of $f_{S\uparrow} = R f_{S\downarrow}$, and the FIR cleanup output signal $y(n)$, also sampled at a rate of $f_{S\uparrow} = R f_{S\downarrow}$.

In part B of the figure, we can see the narrow spectrum of the input signal. For this example, it is assumed that the input signal has been generated and band limited by some previous signal processing function.

In part C, we can see the frequency response of the CIC filter superimposed on the interpolated input signal spectrum. Note that $N-1$ image bands are created when the sample rate is increased by a factor of $N$. Also

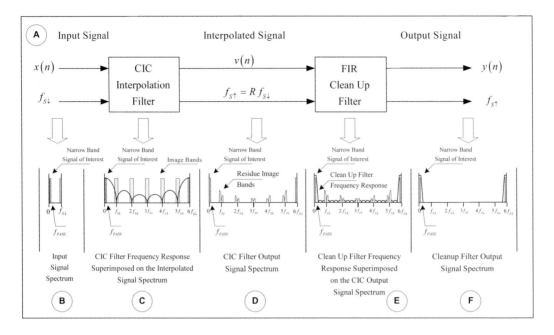

**Figure 6.54**  CIC/cleanup interpolation filter cascade

note also that the nulls of the CIC occur at the centers of each image band. The reader should be able to determine by inspection of the spectrum in part C that the sample rate is increased by a factor of six in this simple example.

In part D of the figure, we see the spectrum of the signal output from the CIC interpolation filter. The reader should pay special attention to the residue image band frequencies left behind by the CIC $\sin(x)/x$ like frequency response. If the CIC has been designed with enough stages so that these residue frequencies are below the specified filter stop band, then there is no need for a FIR cleanup filter in cascade.

Part E of the figure illustrates the frequency response of the FIR cleanup filter superimposed on the spectrum of the CIC output signal. If a FIR cleanup filter is necessary, the designer can take advantage of the fact that its transition band can be relatively wide since the first image band is centered at $f_{S\downarrow}$ or $f_{S\uparrow}/R$. This will make the design of the cleanup FIR a great deal simpler.

Part F of the figure shows the signal spectrum output by the cleanup filter. Notice that the image band residue has been eliminated.

If a FIR cleanup filter is required, another method to consider is to use the CIC to interpolate the sample rate two or four times less than the desired output rate. Then design the FIR cleanup filter as an interpolate by two or four PPF to complete the rate change and clean up the output signal spectrum. This method has the advantage of operating the cleanup filter at a much lower sample rate, which might be simpler to design.

So there you have it. In this chapter, we discussed several methods that can be used to change the sample rate of a discrete signal. We discussed poly phase architectures, half band architectures, and CIC filter architectures. All these architectures are quite common, and all are easily implemented in discrete hardware or in FPGAs.

## 6.4 REFERENCES

[1] Eugene B. Hogenauer. "An Economical Class of Digital Filters for Decimation and Interpolation." *IEEE Transactions of Acoustics, Speech, and Signal Processing*, vol. ASSP-29, no. 2 (April 1981).

[2] Ronald E. Crochiere and Lawrence R. Rabiner. *Multirate Digital Signal Processing*. Englewood Cliffs, NJ: Prentice Hall, 1983.

[3] Xilinx CIC Compiler v1.2 Product Specification DS613, April 26, 2008.

[4] Uwe Meyer-Baese. *Digital Signal Processing with Field Programmable Gate Arrays*. 3rd ed., with companion CD-ROM. Springer: 2007.

# CHAPTER SEVEN

# Complex to Real Conversion

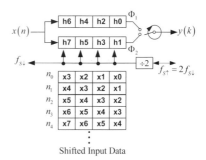

Shifted Input Data

There are many instances where signal processing hardware and/or software operate on real signals. One example is passing a real digitized signal $x_R(n)$ through a digital filter to shape the signal bandwidth. Another example is computing a histogram of a real discrete signal by collecting statistics on one or more of its parameters.

For the most part, when dealing with communication signals, it is either convenient or mandatory that a signal be processed in its complex form. That is, the signal $x_R(n)$ is first converted from its real form to its complex form $x_C(n) = a(n) + jb(n)$, where $a(n)$ is the real part of $x_C(n)$ and $b(n)$ is the imaginary part of $x_C(n)$.

A very good example of complex signal processing is seen in digital systems that employ a front-end tuner. These systems fall into a category that can be loosely described as "digital radio," in that an input wideband signal is tuned down or up in frequency and passed through a low pass or band pass filter to isolate a narrow band of interest. The mathematics of the tuning function converts the real input signal into a complex signal. The filtered narrow band signal is then processed in its complex form to implement whatever the particular application requires. After the intermediate processing is complete, the complex signal is generally converted back to real and provided as an output.

The process of tuning a signal is nothing more than multiplying the real digitized input signal $x(n)$ by a complex sinusoid $m(n)$ given by

$$m(n) = \cos\left(\frac{2\pi k}{N}n\right) \pm j\sin\left(\frac{2\pi k}{N}n\right)$$

**533**

to produce a product signal of the form

$$
\begin{aligned}
x(n)m(n) &= x(n)\left\{\cos\left(\frac{2\pi k}{N}n\right)\pm j\sin\left(\frac{2\pi k}{N}n\right)\right\} \\
&= x(n)\cos\left(\frac{2\pi k}{N}n\right)\pm jx(n)\sin\left(\frac{2\pi k}{N}n\right) \\
&= c(n)\pm jd(n)
\end{aligned}
$$

**Equation 7.1**

The quantity $k/N$ is the normalized frequency of the sinusoidal terms. This complex multiply in Equation 7.1 translates the spectrum of the input signal either up (plus sign) or down (minus sign) in frequency and in the process converts the real signal to a complex signal. The narrow band signal of interest is obtained by passing the translated signal $a(n)\pm jd(n)$ through a digital filter with impulse response $h(n)$. There are actually two identical filters, each with an impulse response $h(n)$. One filters the real part of the translated signal $c(n)$, and the other filters the imaginary part of the translated signal $d(n)$. The block diagram of a signal processing system that tunes, filters, processes, and then performs a complex to real conversion is illustrated in Figure 7.1.

The narrow band of frequencies that this processing function in Figure 7.1 isolates is a complex signal composed of the real and imaginary sequences $\hat{a}(n)$ and $\hat{b}(n)$, respectively. Typically all follow on processing on this signal is performed in the complex domain as well. The end result of the intermediate processing is a complex signal $a(n)\pm jb(n)$ that is applied to the input of the complex to real conversion logic. The conversion output is a real signal labeled $y(n)$. If necessary, this signal is applied to a D/A converter, a sample and hold (to eliminate converter code glitches), and a reconstruction filter, to produce an analog output.

The sections that follow discuss a procedure that is commonly used to perform this complex to real signal conversion. But first it is a good idea to acquaint ourselves with the big picture, and that involves understanding the architecture of a typical digital signal processing (DSP) system.

## 7.1   A TYPICAL DIGITAL SIGNAL PROCESSING (DSP) SYSTEM

A block diagram of a typical DSP system is illustrated in Figure 7.1. This typical system inputs an analog signal, band limits it with an antialiasing filter, digitizes it with an A/D converter, digitally processes the signal, and then

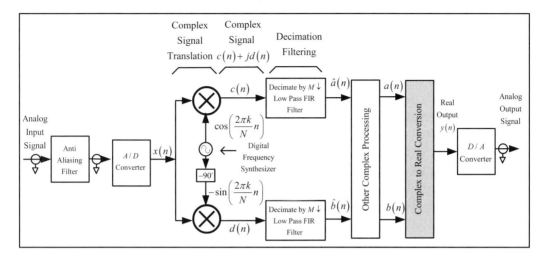

**Figure 7.1**   Block diagram of a typical signal processing system

converts it back to an analog signal on output. To be thorough, it makes sense to familiarize ourselves with this typical system by tracing a signal from input to output.

Using the digital radio analogy, the input signal might represent a stack of 12 frequency division multiplexed (FDM) channels. Each channel occupies 4 KHz of the stack bandwidth. The total signal bandwidth is then equal to $12 \times 4\,\text{KHz} = 48\,\text{KHz}$.

The spectrum of such an analog input signal is represented by the triangle diagram in Figure 7.2. Each triangle, numbered from 0 to 11, represents one independent channel that occupies a 4 KHz bandwidth.

The frequency axis in part A of the figure is calibrated in kilohertz with axis tick marks representing the center of each channel band. Part B of the figure is the equivalent discrete frequency axis calibrated in terms of normalized frequency $k/N$. This notation is typical of a discrete Fourier transform (DFT) and is discussed in Chapter 3, "Review of the Fourier Transform." Chapter 3 illustrates that a normalized frequency is just the frequency $f$ divided by the sample frequency $f_s$, or $f/f_s$. This can be expressed as a ratio of two numbers typically given by $k/N$. We can equate the two ratios such that $f/f_s = k/N$, where the variable $N$ is the number of points in the DFT and the variable $k$ is the discrete frequency index or bin number that ranges from $k = 0, 1, 2, \cdots, N-1$.

The objective in this example is to digitize the 48 KHz composite wideband analog signal, select one of the twelve 4-KHz narrow band channels by tuning or down converting that channel to base band, and then isolating it by

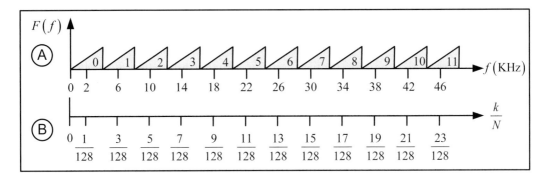

**Figure 7.2**  Example input signal spectrum

passing it through a 4-KHz low pass digital filter. As illustrated in Figure 7.1, the digitized input signal $x(n)$ is a real sequence and the process of tuning the signal down in frequency transforms $x(n)$ into a complex signal. Other digital processing may occur, depending on the application, and then the resultant signal is converted from complex back to real and then back to analog for input to an earphone or speaker. The complex to real processing part of this typical system is represented by the shaded block in Figure 7.1. From that figure, we can see that the complex to real processing block is only a small part of the overall processing system.

We begin processing by passing this signal through an antialiasing filter, which limits the bandwidth of the signal to prevent undersampling, which would result in aliasing. We choose our sample rate to be $f_s$ = 256 KHz. This results in approximately

$$(256K \text{ sample}/\text{second})/(48K \text{ cycle}/\text{second}) = 5.33 \text{ sample}/\text{cycle}$$

at band edge, which is well above the Nyquist rate. If we interpret part A of Figure 7.2 to be the positive frequencies of a 128-point DFT of the digitized input signal, then the axis in part B can be calibrated in terms of normalized frequency as $k/128$. We can convert any normalized frequency back to actual frequency through the simple relation

$$f = \frac{k}{N} f_s$$

As an example, converting the normalized frequency $k/N = 23/128$ back to frequency is a simple as

$$f = \frac{23}{128} 256 \text{ KHz} = 46 \text{ KHz}$$

which correlates with the value on frequency axis A in the figure. The digitized signal $x(n)$ is fed directly to the input of two quadrature multipliers. The output of the multiply operation is a complex signal that can be expressed as

$$x(n)e^{-j\left(\frac{2\pi k}{N}n\right)} = x(n)\cos\left(\frac{2\pi k}{N}n\right) - jx(n)\sin\left(\frac{2\pi k}{N}n\right)$$

The top multiplier forms the product

$$c(n) = x(n)\cos\left(\frac{2\pi k}{N}n\right)$$

which is the real part of the complex signal and the bottom multiplier forms the product

$$d(n) = -x(n)\sin\left(\frac{2\pi k}{N}n\right)$$

which is the imaginary or quadrature part of the complex signal.

The complex signal out of the multiply operation is a tuned version of the input. In this simple example, we limit the tuner frequencies to the 12 center frequencies given by the 12 values of $k/N$ shown in Figure 7.2. This is sufficient to individually tune the spectra of each individual channel in the 12-channel stack down in frequency to 0 Hz.

The reader should note that this discussion is not a tutorial on signal tuning. For a detailed analysis on that subject, the reader should consult Chapter 9, "Signal Tuning." Our objective here is to merely walk through the operations illustrated in the block diagram of Figure 7.1 and to accomplish this, we need to briefly mention the concept of signal tuning or, as it is sometimes called, signal translation.

Let's assume that we wish to isolate channel 4 from the channel stack for further processing. In this example, we can accomplish this by setting the value of $k/N$ to $9/128$. The complex tuner will translate (or down convert) the signal spectra by $f = (9/128)256$ KHz $= 18$ KHz so that the tuned signal spectra look like those illustrated in Figure 7.3.

Superimposed on Figure 7.3 is the conceptual response of a low pass filter used to extract the channel of interest. In this case, this is channel 4. This filtering function is performed by the two low pass filters illustrated in the block diagram of Figure 7.1. These two filters are identical. One filter is used

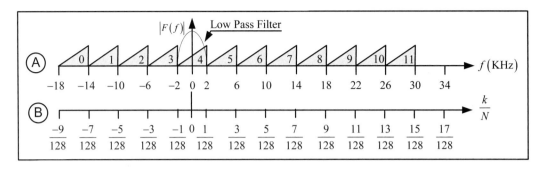

**Figure 7.3**  Tuned signal spectrum

to filter the real part of the signal $c(n)$ and the second filter is used to filter the imaginary part of the signal $d(n)$.

The magnitude of the filtered signal spectrum is illustrated in part A of Figure 7.4. The horizontal axis of part A is calibrated in frequency and the horizontal axis of part B is calibrated in terms of normalized frequency. Note that the tuning and filtering operations have isolated channel 4 of the 12-channel stack.

Since the complex bandwidth of the filtered signal is only 4 KHz or 1/12th of the 48 KHz bandwidth of the original digitized signal, we no longer need to operate at the high sample rate of 256 KHz. It makes sense to reduce the sample rate to the complex Nyquist sample rate. In the case of a complex signal, the Nyquist sample rate is equal to the signal bandwidth. Therefore the Nyquist sample rate for our 4-KHz wide complex signal is 4 KHz. Reducing the sample rate whenever possible makes sense because computations performed at the lower sample rate produce a much simpler and more efficient hardware design and/or real time software design.

A reduction in sample rate from 256 KHz to 4 KHz represents a sample rate decimation by a factor of 64. The reader should keep in mind that we are discussing a simple example. The determination of the actual decimated sample rate may be based on many factors other than achieving the minimal sample rate. Our motivation in selecting a complex sample rate of 4 KHz is purely to develop a system application baseline from which we can demonstrate the procedures for digital complex to real conversion.

As the block diagram in Figure 7.1 indicates, the sample rate decimation is performed by the two low pass filters. Both filters in this example are implemented as a decimate by $M_\downarrow = 64$ low pass poly phase filter (PPF) or a decimate by $M_\downarrow = 64$ cascaded integrator comb (CIC) filter, followed by a finite impulse response (FIR) cleanup filter. At the output of the filter, we now have a complex signal $x_C(n) = \hat{a}(n) + j\hat{b}(n)$ of bandwidth 4 KHz sampled at

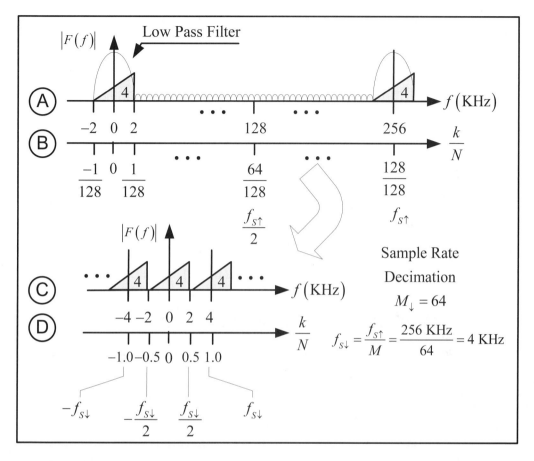

**Figure 7.4**   Filtered and decimated spectra magnitude

a rate of $f_{S\downarrow} = 4$ KHz. For a tutorial on PPFs and CIC filters that are used for sample rate reduction, see Chapter 6, "Multirate Finite Impulse Response Filter Design."

In part A of the figure, we see the combination of tuning and low pass filtering has isolated channel 4 from the 12-channel FDM stack. The signal at this point is complex. It is clear from the figure that this 4 KHz signal is way oversampled at 256 KHz. We have an opportunity here to improve the efficiency of our hardware design by reducing the sample rate. In this case, we choose to reduce or decimate the sample rate by a factor of 64 to 4 KHz. The decimated complex spectrum calibrated in frequency is illustrated in part C of the figure, and the same spectrum is calibrated in terms of normalized frequency in part D of the figure.

At this point, as implied in the block diagram of Figure 7.1, we can further process the signal to achieve whatever the particular application requires. Our objective may be to simply collect statistics on the recovered channel, or we may wish to determine if a signal of interest is present or absent, or we may want to digitally control the gain of the signal. The applications are limitless. At the end of this application-specific processing we will need to perform the complex to real conversion of the signal and output the real version of the signal to a D/A converter. In this simple example, the system output is the signal originally transmitted in channel 4 of the composite signal stack.

This brings us to the intent of this chapter, which is to determine the processing that goes on within the shaded block in Figure 7.1, labeled "Complex to Real Conversion." Our purpose from this point forward is to provide a detailed discussion of the procedure used to convert a complex digital signal such as the one we just discussed to a real digital signal.

## 7.2 CONVERSION OF A COMPLEX SIGNAL TO A REAL SIGNAL

We begin our discussion by expanding the shaded complex to real block in Figure 7.1 into a much more detailed block diagram that illustrates all the internal processing required to implement a complex to real conversion function. The expanded block diagram is illustrated in Figure 7.5.

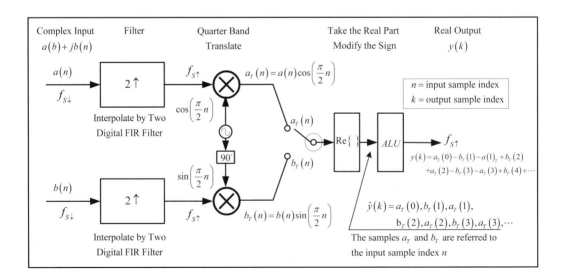

**Figure 7.5**   Block diagram of complex to real conversion logic

Perhaps the best way to approach this discussion is to paint the big picture first and then follow with the technical details that make the picture work. In this case, the big picture is the visualization of the signal spectra as the signal is sequentially processed by each block in the complex to real conversion block diagram of Figure 7.5. The technical details will then serve to illustrate how each block in Figure 7.5 is implemented. In a multirate system such as this, the notation $f_{S\downarrow}$ represents the lower sample rate prior to a sample rate increase and the higher or interpolated sample rate is represented by the notation $f_{S\uparrow}$.

### 7.2.1  Signal Spectra in the Complex to Real Conversion Process

The spectrum of the complex signal that we isolated in the previous section, just prior to the complex to real conversion, is illustrated in part A of Figure 7.6. Here we see the magnitude spectrum of the complex signal $x(n) = a(n) + jb(n)$ that represents the 4-KHz wide channel 4 sampled at the complex Nyquist rate of $f_{S\downarrow} = 4$ KHz. The spectrum of this digital signal repeats every $f_{S\downarrow}$ Hz. The first thing we need to do in this conversion process is to increase the signal sample rate of the complex signal such that as a minimum, the real version of this signal is sampled at the real Nyquist rate. In this example, the Nyquist rate for a real signal with a spectral bandwidth of 4 KHz is 8 Ksps. Therefore the real and imaginary parts of the complex input signal first pass through identical interpolate by two FIR filters, which increases the sample rate from $f_{S\downarrow} = 4$ KHz to $f_{S\uparrow} = 8$ KHz.

In addition, the filters implement a low pass filter characteristic to eliminate the image band created by the sample rate increase. Both of these processes occur together within the filters. It is simpler, however, if we can visualize these two processes as occurring sequentially. This process is illustrated in the spectral diagrams of Figure 7.6. In part B of the figure, we see the signal spectrum after a having its sample rate increased or interpolated by a factor of two to produce a new signal sampled at a rate of $f_{S\uparrow} = 2f_{S\downarrow}$. We can see from the diagram that when the sample rate was increased by a factor of two, an image band was introduced that is identical to the original base band spectra. This image band is illustrated by the shaded band in the figure.

In general, when the sample rate of a signal is increased by factor of $N$, there will be $N-1$ image bands created in the output signal spectrum. These image bands need to be removed.

The second function performed by the two low pass filters is to eliminate the image bands. This is illustrated in part C of the figure, where the filter is superimposed on the signal spectrum. It is clear that the interpolate by two multirate filters perform a critical role in the conversion process and are the technical heart of the entire design. Therefore they need to be designed carefully. The design techniques, discussed in Chapter 5, "Finite Impulse Response

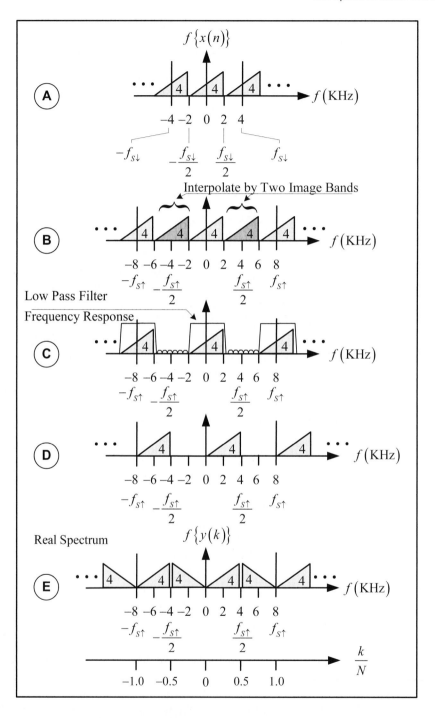

**Figure 7.6** Complex to real signal conversion spectra

Digital Filtering," and Chapter 6, "Multirate Finite Impulse Response Filter Design," can be used to produce an optimum filter for this application.

The next thing we need to do is to quarter band translate the interpolated complex signal spectra so that the spectral band is centered at $f_{s\uparrow}/4$, or 2 KHz in this example. The translated signal spectrum is illustrated in part D of Figure 7.6. The translation is performed by a complex multiplier at the output of the multirate filters, as shown in Figure 7.5. The tuned complex signal is translated up in frequency by $f_{s\uparrow}/4$ Hz.

The final piece of the conversion process is to simply take the real part of the complex signal and adjust the sign of the output samples. This process is indicated in the block diagram of Figure 7.5 by the blocks labeled $\mathrm{Re}\{\ \}$ and ALU. The spectrum of the real signal output by the complex to real converter is illustrated in part E of Figure 7.6.

The term *ALU* is an abbreviation for *arithmetic logic unit*. In days gone by, an ALU was a chip that performed a handful of different arithmetic functions like addition and subtraction and so forth. These days, the individual functions performed by an ALU are implemented in a line or two of VHSIC hardware description language (VHDL) code, but the term *ALU* still lives on because functionally it is a very descriptive term. If the reader ever hears someone mention an ALU, it's a good bet that it was uttered by a grizzled veteran of the early days of DSP.

In summary, when the real part extraction and sign adjustment process is complete, the spectrum of the real output signal appears as shown in part E of Figure 7.6. The reader who is getting his first exposure to a complex to real conversion theory is probably saying to himself, "This stuff is pretty darned complicated! It will take me a couple of months to figure out how to design one of these!"

Well, for the readers who fall into this category, there is some good news and some bad news. The bad news is, yes, this conversion algorithm does *appear* difficult. The good news is that it's not! As a matter of fact, the conversion algorithm is really very simple to implement. It turns out that through the magic of signal processing tricks of the trade, many of the logic blocks shown in the conversion block diagram of Figure 7.5 can be reduced to a level of simple data manipulation. We will delve into the technical details of this conversion process next.

### 7.2.2  Interpolation Filter for Real Signals

Since interpolation filters are the heart of a complex to real conversion process, let's discuss the design and operation of these filters first. The interpolate by two digital filter used for the complex to real conversion application can be easily designed using the methods described in Chapter 5, "Finite Impulse Response Digital Filtering," and Chapter 6, "Multirate Finite Impulse

Response Filter Design." In the interest of maintaining continuity within this chapter, we will take the time to review the theory of an interpolate by two filter.

The operation of an interpolate by two filter is very straightforward. Let's assume that we have a discrete real data sequence called $x(n)$. Each sample in the sequence is sequentially numbered $0, 1, 2, \cdots$. Although it is not important for this discussion, we will define each sample to be $B$ bits wide. The low rate sample clock is labeled $Clk(f_{s\downarrow})$, and its rate is $f_{s\downarrow}$ Hz. The input data sequence and its sample clock are illustrated in Figure 7.7.

We assume that the sequence $x(n)$ is the output of some preceding processor that we apply to the input of an interpolate by two multirate FIR filter illustrated in Figure 7.8.

The filter processes the input data sequence and outputs a new sequence $y(k)$ at twice the sample rate. We use the index $k$ for the output samples and the index $n$ for the input samples because of the different clock rates. The up sampled or interpolated output sequence $y(k)$ and associated clock $2xClk(f_{s\uparrow})$ are also shown in Figure 7.7.

Note that from an overall system point of view, the system clock generation and distribution network supplies the filter with both the high rate and low rate clocks. Normally the clock generation and distribution network is an independent design that synthesizes and delivers precisely timed and

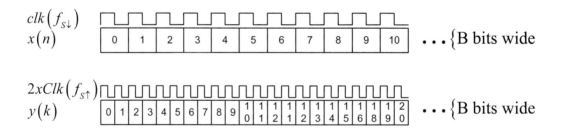

**Figure 7.7**  Data sequence and associated sample clock

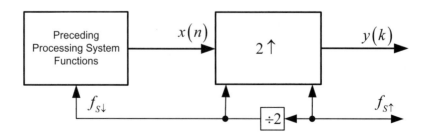

**Figure 7.8**  Interpolate by two filter

phase aligned clocks to all the subsystem components, such as this conversion component. The clock network is a nontrivial design, especially when dealing with clocks that range from the high megahertz to gigahertz. We will not deal with the issues associated with clock network design in this book. However, it should be noted that the system clock and all its derivatives represent the gold standard timing reference for that system, and therefore its design should be well thought out.

A brute force method for increasing the sample rate by a factor of two is to simply insert a zero valued sample every second sample in the input sequence, pass this sequence into the digital filter, and let the filter interpolate or smooth the value of the output samples. This insertion process would be done at twice the input clock rate, so the final sample rate is twice that of the input. The input data sequence for this brute force method would resemble Figure 7.9. The shaded time slots contain zero valued samples.

This method will work, but it is very inefficient. First of all, you would have to design the circuitry that would interleave data samples and zero valued samples at twice the clock rate. Second, you would have to design filters that operated at twice the clock rate. This might not sound like a big deal, but what if twice the clock rate was extremely high? A lot of processing must take place each clock period. As the clock period is shortened, there is less time to perform computations, and when the computational timing budget is reduced, things start to get tight. Third, why waste valuable computational resources? If the filter were of length $N$, for example, then approximately $N/2$ filter coefficients would be multiplying a zero valued data sample to produce a zero output product, which contributes nothing to the computational process. If this filter is implemented in hardware, it would be a tremendous waste of power consuming resources. If this filter is implemented in software—either on a state of the art digital signal processor chip or on a core processor in a field programmable gate array (FPGA)—then it would consume twice as many operations as necessary and would be an excellent way to slow down the entire system.

A design engineer's life is tough enough, and there is absolutely no need to make life tougher. In addition to these three reasons, it is also wise to remember that engineers are notorious for pulling technical "one-up-man-ships" on other engineers. You do not want to be giving a design review

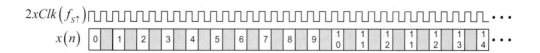

**Figure 7.9**   Increasing the sample rate by zero insertion

presentation on an inefficient system with these guys in the audience. They will find it very difficult to pass up an opportunity throw a few zingers in your direction. An experienced engineer will look at a design from all angles and will do his best to make it zinger-proof.

That said, although the interleaved zero sequence in Figure 7.9 is a sub-optimum solution, it is an excellent starting point from which to begin the discussion on the interpolation filter design. We will start with this worst case implementation, analyze it, and then use the information gleaned from that analysis to design a more efficient filter.

Suppose we had a 32-tap FIR filter and we wished to process the zero filled data sequence shown in Figure 7.9. We could build a filter out of a 32-stage, B-bit-wide shift register, 32 multipliers, and a huge summation tree. In doing so, we would end up with a filter architecture that looks like the one illustrated in Figure 7.10.

Operationally, the alternating zero data sequence $x(n)$ is shifted in at the high sample clock rate beginning with sample 0 and continuing on forever. A snapshot of the numbered samples in the shift register is shown in the figure. Each clock period, a new sample is shifted in, the oldest sample is shifted out, and all the samples in between are multiplied by the filter coefficient vector $h(n)$. The products are summed to produce a single output sample $y(k)$.

The first thing we notice with this architecture is that we have 32 multiply accumulate operations occurring at each tick of the high rate output sample clock. If the data rate was high enough, this may require 32 separate hardware multipliers. The second thing we notice about this architecture is that 16 of the 32 multipliers are multiplying a filter coefficient with a zero valued data sample and therefore contributing absolutely nothing to the overall summation. For this and many other reasons, the filter architecture in Figure 7.10 is not acceptable. We will investigate the operation of this filter a little more closely and see what can be done to eliminate the inefficiency of the zero multiplies. Before we begin, let's present a block diagram of a FIR digital filter and the short hand drawing notation that we use to represent

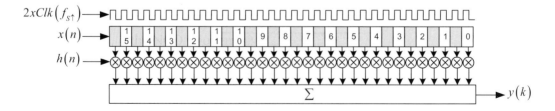

**Figure 7.10**   Thirty-two tap FIR filter architecture

these filters. The typical block diagram of a FIR filter that we usually see in engineering publications is illustrated in part A of Figure 7.11. In order to simplify the following discussion, we will use the equivalent FIR filter block diagram representation illustrated in part B of the figure.

Let's take a look at the timing diagram in Figure 7.12. For a moment, instead of visualizing the data sequence $x(n)$ shifting by a fixed filter coefficient vector $h(n)$, let's visualize an infinitely long and static data vector $x(n)$. At each sample clock instant, let's visualize that the filter coefficient vector $h(n)$ is shifted one sample to the right, relative to the data sequence. The dot product of the two is then computed to produce a single output sample $y(k)$. To make this simple, let's assume that the filter only has eight coefficients, numbered $h_0$ through $h_7$. We will also assume that the filter is processing an input sequence that has alternating zero filled samples. The timing diagram

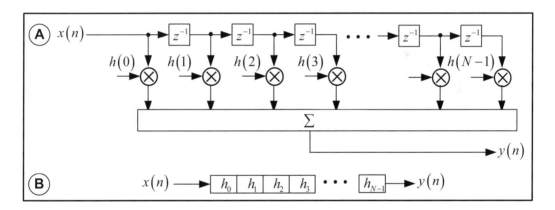

**Figure 7.11** Shorthand drawing notation for a digital FIR filter

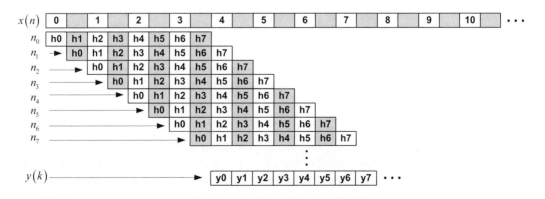

**Figure 7.12** Timing diagram: First eight filter output samples

showing the sliding filter coefficient vector $h(n)$ relative to the static data sequence $x(n)$ for the first eight sample instants $(n_0, n_2, \cdots, n_7)$ is illustrated in Figure 7.12.

Note that at the discrete sample time $n_0$ only the even numbered filter coefficients $(h_0, h_2, h_4, h_6)$ line up with the actual input data samples. The odd numbered coefficients $(h_1, h_3, h_5, h_7)$ line up with the zero valued samples. The dot product of the data and coefficient vectors produces output sample $y(0)$. It is clear there is no contribution to the output dot product $y(0)$ from any of the odd numbered filter coefficients.

At the next sample time $n_1$, the coefficient vector is shifted one data sample to the right, relative to the static data sequence. The data and coefficient vectors are now positioned so that only the odd numbered filter coefficients are lined up with actual data samples. The even numbered filter coefficients line up with the zero valued samples in the input data stream. The dot product of the data and coefficient vectors at this sample instant produces the output sample $y(1)$. The dot product computations at this sample instant did not utilize any of the even numbered filter coefficients.

The zero valued data samples and the unused filter coefficients at each sample instant in Figure 7.12 are shown as shaded blocks to highlight the fact that they do not make a contribution in the computation of a filter output sample. There is a pattern here that we can exploit. At the even sample time instants $(n_0, n_2, n_4, \cdots)$, the filter only uses the even numbered filter coefficients to produce an even numbered output sample $(y_0, y_2, y_4, \cdots)$. At the odd sample time instants $(n_1, n_3, n_5, \cdots)$, the filter only uses the odd coefficients to produce an odd numbered output sample $(y_1, y_3, y_5, \cdots)$. This pattern repeats forever. We can write the effective dot product for the first five filter outputs as

$$
\begin{aligned}
y(0) &= x(0)h(0) &&+x(1)h(2) &&+x(2)h(4) &&+x(3)h(6) \\
y(1) &= &&x(1)h(1) &&+x(2)h(3) &&+x(3)h(5) &&+x(4)h(7) \\
y(2) &= &&x(1)h(0) &&+x(2)h(2) &&+x(3)h(4) &&+x(4)h(6) \\
y(3) &= &&&&x(2)h(1) &&+x(3)h(3) &&+x(4)h(5) &&+x(5)h(7) \\
y(4) &= &&&&x(2)h(0) &&+x(3)h(2) &&+x(4)h(4) &&+x(5)h(6) \\
&\vdots &&&&\vdots
\end{aligned}
$$

**Equation 7.2**

We observe from Equation 7.2 that there exists a very symmetrical pattern that we can take advantage of in the hardware and/or software architecture used to implement this filter. It is clear that for sequential output sample instants, the filter output alternates between the dot product of the input data vector and the even filter coefficients and the dot product of the input data and the odd filter coefficients. Therefore we can split the coefficients into two separate vectors. One vector will contain all the even coefficients, and the other vector will contain all the odd coefficients. Using these two coefficient vectors, we can build a two-phase filter such as the one illustrated in Figure 7.13.

In this architecture, there are two shift registers, labeled $\Phi_1$ for phase 1 and $\Phi_2$ for phase 2. The phase 1 filter contains all the even filter coefficients, and the phase 2 filter contains all the odd filter coefficients. The same data sample is clocked into both phases at the same input sample instant by the low rate clock $f_{s\downarrow}$. Two dot products are formed: one for each phase. The output wiper arm rotates at twice the input sample rate and selects the output of the phase 1 filter followed by the output of the phase 2 filter. The rotation of the wiper arm is synchronized to the input data clock so that a new data sample is clocked in every time the wiper arm leaves the $\Phi_1$ position. The fact that the wiper arm is selecting two output samples for every input sample effectively increases the sample rate by a factor of two, which is our objective. Note that this method does not waste any computational resources on the multiplication of zero valued data.

The data content in each of the two shift registers is shown in the bottom of the figure for the first five input sample instants. The data is shifted down the filter delay line register at the input sample rate. The reader should note that an important attribute of this filter architecture is that all the filter computations are performed at the low clock rate.

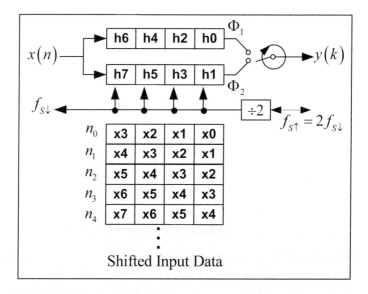

**Figure 7.13**  Real signal poly phase interpolate by two filter

Since this filter has more than one phase, it is called a *poly phase filter* (PPF). The wiper arm used by the filter is a descriptive symbol that clearly shows the order of sample selection. This wiper arm is actually nothing more than a simple two-to-one multiplexer.

Interpolation of the input signal at a low sample rate to obtain even higher output clock rates would be implemented in the same fashion. Increasing the sample rate by a factor of eight, for example, would split the filter coefficients into eight phases, and the wiper arm would sample the output of each of the eight phases for every input sample.

Once again, if we label the output of the two filter phases as $\Phi_1$ and $\Phi_2$, then the operational sequence for this interpolate by two filter would be to first initialize the filter by shifting in enough data samples to fill the filter registers and then sequentially process the data as outlined in the following sequence:

| Sample Instant | Operation | Filter Output |
|:---:|:---:|:---:|
| | Compute | $y(0) = \Phi_1$ |
| $n_0$ | Clock in next input sample | |
| | Compute | $y(1) = \Phi_2$ |
| | Compute | $y(2) = \Phi_1$ |
| $n_1$ | Clock in next input sample | |
| | Compute | $y(3) = \Phi_2$ |
| | Compute | $y(4) = \Phi_1$ |
| $n_2$ | Clock in next input sample | |
| | Compute | $y(5) = \Phi_2$ |
| | Compute | $y(6) = \Phi_1$ |
| $\vdots$ | $\vdots$ | $\vdots$ |

**Equation 7.3**

There is no smoke or mirrors. Building an interpolate by two filter is really this simple, folks.

## 7.2.3 Interpolation Filter for Complex Signals

Our discussion of the interpolation filter has so far has only dealt with real input signals. Our objective is to increase the sample rate of a complex signal by a factor of two. So how do we implement a complex interpolate by two filter? The answer is pretty simple. We use a filter identical to the one just discussed for real signals, but in the complex case, we use two of these filters. We use one filter to process the real part of the complex signal, and we use the identical second filter to process the imaginary part of the complex signal. The complex version of the interpolate by two filter is illustrated in Figure 7.14. Here we see two identical filters configured in the two-phase poly phase form. One filter is processing the real part of the complex signal and the second filter is processing the imaginary or quadrature part of the complex signal. The operation of each of the two filters that make up the complex filter is identical to that previously described for the real interpolate by two filter.

In the complex filters, the two phases are labeled $R_{\Phi 1}$ and $R_{\Phi 2}$ for the filter that operates on the real part of the complex input signal and $I_{\Phi 1}$ and $I_{\Phi 2}$ for the filter that operates on the imaginary part of the complex signal. As was the case with the interpolate by two filter that processed real data, the output data for each of the two filters that process the real part and imaginary parts of the input data will be computed as described by Equation 7.2 and by the sequence of events described by Equation 7.3. This means that the filter will initiate operation with the wiper arms connecting to phases $R_{\Phi 1}$ and $I_{\Phi 1}$ at the output of the filter.

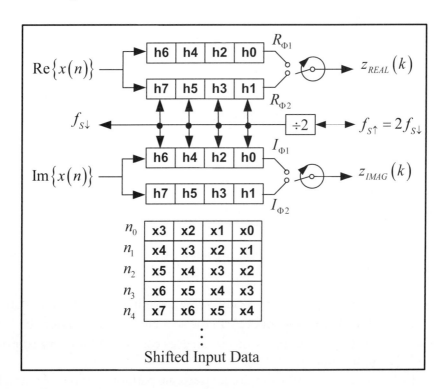

**Figure 7.14**   Complex interpolate by two filter

The first complex output will be taken from wiper post $R_{\Phi 1}$ and $I_{\Phi 1}$. Then a new complex input data sample will be shifted into the delay register, and the next complex output sample is taken from wiper post $R_{\Phi 2}$ and $I_{\Phi 2}$, followed by $R_{\Phi 1}$ and $I_{\Phi 1}$. Then a new complex input sample is clocked into the filter delay register, and the process repeats itself. If we define the complex filter output $z(k)$ as a complex variable so that

$$z(k) = z_{REAL}(k) + jz_{IMAG}(k) = R_\Phi + jI_\Phi$$

then the computational sequence begins by first filling the filter delay registers with input samples and from then on the computations would follow this sequence:

| Sample Instant | Operation | Filter Output |
|---|---|---|
|  | Compute | $z(0) = R_{\Phi 1} + jI_{\Phi 1}$ |
| $n_0$ | Clock in next input sample |  |
|  | Compute | $z(1) = R_{\Phi 2} + jI_{\Phi 2}$ |
|  | Compute | $z(2) = R_{\Phi 1} + jI_{\Phi 1}$ |
| $n_1$ | Clock in next input sample |  |
|  | Compute | $z(3) = R_{\Phi 2} + jI_{\Phi 2}$ |
|  | Compute | $z(4) = R_{\Phi 1} + jI_{\Phi 1}$ |
| $n_2$ | Clock in next input sample |  |
|  | Compute | $z(5) = R_{\Phi 2} + jI_{\Phi 2}$ |
|  | Compute | $z(6) = R_{\Phi 1} + jI_{\Phi 1}$ |
| $\vdots$ | $\vdots$ | $\vdots$ |

**Equation 7.4**

The important information the reader should glean from this discussion is summarized as follows:

1. Both filters in the complex processing separately implement Equation 7.2.
2. Together, both filters operate in the sequence given by Equation 7.4.
3. There are two complex output samples computed for every complex filter input sample.
4. The filter computations or dot products are performed at the low sample rate.
5. The wiper arm effectively increases the sample by a factor of two.

6. The index of the input data sequence $n$ increments at the low sample rate $f_{S\downarrow}$.

7. The index of the output data sequence $k$ increments at the high sample rate $f_{S\uparrow}$.

8. Each of the two filters is implemented as a two-phase PPF structure.

### 7.2.4   Quarter Band Translation

Our next objective is to translate the filtered and interpolated complex signal so that its spectrum is centered at $f_S/4$. This is accomplished by placing a simple quarter band translator on the output of the complex interpolation filter, as illustrated in Figure 7.15.

The quarter band translation process is nothing more than multiplying the output of the real interpolation filter by $\cos(\pi k/2)$ and multiplying the output of the imaginary interpolation filter by $\sin(\pi k/2)$.

The reader should keep in mind that we are actually multiplying two complex numbers. If the product of the two complex numbers is represented by $g(k) = g_{REAL}(k) + jg_{IMAG}(k)$, then the complex multiply is given by

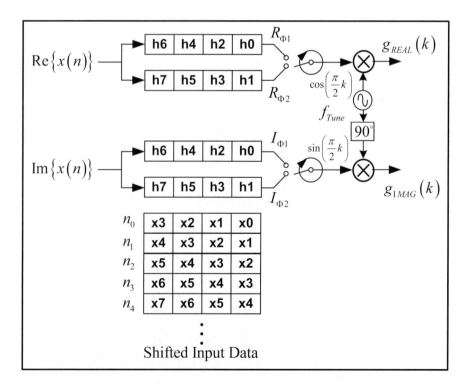

**Figure 7.15**   Complex interpolation filter followed by a quarter band translator

$$g(k) = g_{REAL}(k) + jg_{IMAG}(k) = (R_\Phi + jI_\Phi)\left[\cos\left(\frac{\pi}{2}k\right) + j\sin\left(\frac{\pi}{2}k\right)\right] \quad k = 1,2,3,\cdots$$

**Equation 7.5**

The reader should also remember that the complex quantity $j = \sqrt{-1}$ and its square $j^2 = -1$ are mathematical abstractions and therefore the multiply operations indicated in Equation 7.5, whether they be implemented in hardware or software, do not include these $j$ terms. Since inclusion of the abstract $j$ affects the sign of the product terms, we will have to account for these terms ourselves after the multiplications are complete. We will discuss this important point later. For now, let's concentrate on the format of the complex multiplication.

It is interesting to note that the complex sinusoid term in Equation 7.5 takes on a periodic sequence of only four distinct values as the index $k$ is incremented. This sequence is well known in the DSP world and is given here:

| Index | Complex sinusoid | Complex value |
|---|---|---|
| $k=0$ | $\cos\left(\frac{\pi}{2}0\right) + j\sin\left(\frac{\pi}{2}0\right)$ | $= (1 + j0)$ |
| $k=1$ | $\cos\left(\frac{\pi}{2}1\right) + j\sin\left(\frac{\pi}{2}1\right)$ | $= (0 + j1)$ |
| $k=2$ | $\cos\left(\frac{\pi}{2}2\right) + j\sin\left(\frac{\pi}{2}2\right)$ | $= (-1 + j0)$ |
| $k=3$ | $\cos\left(\frac{\pi}{2}3\right) + j\sin\left(\frac{\pi}{2}3\right)$ | $= (0 - j1)$ |
| $\vdots$ | $\vdots$ | $\vdots$ |

**Equation 7.6**

Thus the complex product $g(k) = g_{REAL}(k) + jg_{IMAG}(k)$ takes on the sequence given in Equation 7.7:

$$
\begin{aligned}
g(0) &= g_{REAL}(0) + jg_{IMAG}(0) = (R_{\Phi1} + jI_{\Phi1}) \times (+1 + j0)\\
g(1) &= g_{REAL}(1) + jg_{IMAG}(1) = (R_{\Phi2} + jI_{\Phi2}) \times (+0 + j1)\\
g(2) &= g_{REAL}(2) + jg_{IMAG}(2) = (R_{\Phi1} + jI_{\Phi1}) \times (-1 + j0)\\
g(3) &= g_{REAL}(3) + jg_{IMAG}(3) = (R_{\Phi2} + jI_{\Phi2}) \times (+0 - j1)\\
&\vdots
\end{aligned}
$$

**Equation 7.7**

Notice that we begin with the initial condition $k = 0$, which connects the wiper arms to the $\Phi_1$ wiper posts. Doing so allows us to implement the processing sequence given by Equation 7.2. Also notice that because the real and imaginary components of the complex sinusoid only take on values of 1 and 0, we really do not need to use multipliers in the quarter band translation process. When we design a complex to real converter in industry, we don't utilize any multipliers. The multipliers that are illustrated in the block diagrams of Figure 7.5 and Figure 7.15 are there strictly for functional clarity. The computed values for the complex output $g(k)$ in Equation 7.7 is given by

$$
\begin{aligned}
g(0) &= +R_{\Phi 1} & +jI_{\Phi 1} \\
g(1) &= -I_{\Phi 2} & +jR_{\Phi 2} \\
g(2) &= -R_{\Phi 1} & -jI_{\Phi 1} \\
g(3) &= +I_{\Phi 2} & -jR_{\Phi 2} \\
&\;\;\vdots & \vdots
\end{aligned}
$$

**Equation 7.8**

The sequence illustrated in Equation 7.8 repeats every four output samples. The signs in Equation 7.8 are correct. The reader must remember, however, that the sign of each of the terms in Equation 7.8 are the result of including the $j$ term in the complex multiplies. Since the quarter band translation multipliers in hardware or software do not take the $j$ term into account, the sign of the $g(k)$ terms out of the complex multiply operation will not be correct. We will deal with this issue a bit later.

The reader might be asking the question, "The two multiplier structure shown in Figure 7.15 is only performing a portion of a complex multiply, so how does that give us the outputs shown in Equation 7.8?" The answer is, it doesn't. However, since the complex sinusoid only takes on the four complex values shown in Equation 7.6, the partial multiply structure computes all the output values we need.

## 7.2.5 Computing the Real Part

The block diagram of Figure 7.16 shows the added function $\mathrm{Re}\{\ \}$, which computes the real part of the translated complex signal $g(k)$. This is the easy part of this conversion process. In order to convert the quarter band translated complex signal $g(k)$ to a real signal, all we need to do is take the real part of $g(k)$ and discard the imaginary part. If the signal output by the block labeled $\mathrm{Re}\{\ \}$ is named $m(k)$ then we can say $m(k) = \mathrm{Re}\{g(k)\}$. It's as simple as that. If the $j$ terms are taken into account, the sequence $m(k)$ is then given by

$$m(k=0) \quad = +R_{\Phi 1} = +R_{\Phi 1}(n=0)$$
$$m(k=1) \quad = -I_{\Phi 2} = -I_{\Phi 2}(n=1)$$
$$m(k=2) \quad = -R_{\Phi 1} = -R_{\Phi 1}(n=1)$$
$$m(k=3) \quad = +I_{\Phi 2} = +I_{\Phi 2}(n=2)$$

$$\left. \begin{array}{c} \text{where } k = \text{the output sample index} \\ n = \text{the input sample index} \end{array} \right\}$$

$$\vdots \qquad \vdots$$

**Equation 7.9**

All the $\text{Re}\{\ \}$ function is doing is alternately selecting the outputs of the real and imaginary ports of the two interpolation filters. If the reader examines the output sequence for $m(k)$, it clearly shows that the real part of the complex signal is a simple multiplex between the real part and the imaginary part of the complex interpolated and frequency translated signal.

In addition, if we take a close look at the sequence $m(k)$, we see something very interesting. We note that the real part of the sequence is constructed using only the phases $R_{\Phi 1}$ and $I_{\Phi 2}$. The contribution of filter phases $R_{\Phi 2}$ and $I_{\Phi 1}$ is not included in the real part of the sequence $m(k)$. This is exciting news because if we don't use these terms, why should we even bother computing

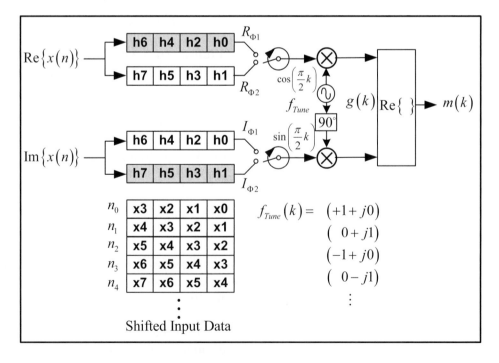

**Figure 7.16**  Taking the real part of the quarter band translated complex signal $g(k)$

them? If we take a look at Figure 7.16, we can see that filter delay lines for phases $R_{\Phi 1}$ and $I_{\Phi 2}$ are highlighted, indicating that they are providing the only filter outputs we need. The other two filter phases $R_{\Phi 2}$ and $I_{\Phi 1}$ are being computed, but the resulting dot products end up being discarded and are never used. We could save a lot of hardware and/or software resources if we eliminate the computational horsepower expended on the unused filter phase computations. We could eliminate a great deal more resources if we discard the unnecessary quarter band translation multipliers. In doing so, we can re-draw the complex to real conversion block diagram in its new and more efficient form, as illustrated in the block diagram of Figure 7.17. Here we see that the unused filter phases have been eliminated and the complex translation multipliers have been removed. Also note that since we only compute the real part of a complex number, the real function operator $\operatorname{Re}\{\ \}$ has been removed and replaced with a two-to-one multiplexer, illustrated in the figure as a rotating wiper arm. The resulting architecture is a great deal simpler than what we had before.

### 7.2.6 Sign Correction

Now that we have an efficient architecture and we have our desired output data sequence, we need to address the sign correction issue that we have mentioned several times. If we look at the block diagram of Figure 7.17, it is clear that the raw output sequence is a virgin sequence in that the sign of each term due to the complex $j$ operator has not yet been included. A simple method of handling the sign issue is with the inclusion some logic that is the

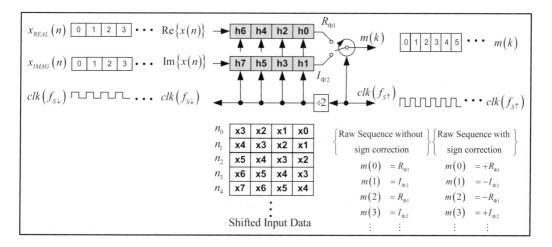

**Figure 7.17** Efficient complex to real architecture

equivalent of an ALU, as illustrated in the block diagram of Figure 7.18. The ALU operates in synchronism with the wiper arm.

The sequence $m(k)$ is input to the $A$ port of the ALU, and a ground or logic "zero" is input to the $B$ port of the ALU. The ALU is programmed to produce two possible outputs. It can produce the output $B + A$, which is the same thing as $0 + m(k) = m(k)$, or it can produce the output $B - A$, which is the same as $0 - m(k) = -m(k)$. Therefore if the ALU is programmed to perform the periodic sequence of operations

$$(B+A),\ (B-A),\ (B-A),\ (B+A),\ \cdots$$

in synchronism with the wiper arm, then the output sequence will take on the form

$$y(k) = +m(k),\ -m(k),\ -m(k),\ +m(k),\ \cdots\ = +R_{\Phi 1},\ -I_{\Phi 2},\ -R_{\Phi 1},\ +I_{\Phi 2},\ \cdots$$

This is the desired output sequence with the proper signs. The output sequence $y(k)$ is the real signal obtained by the complex to real conversion of the complex input sequence $x(n)$.

The inclusion of an ALU to perform the sign conversion logic is purely for clarity of discussion. The designer will more than likely implement his or her circuit in an FPGA of some sort, and therefore the ALU function will reduce to nothing more than a simple VHDL statement.

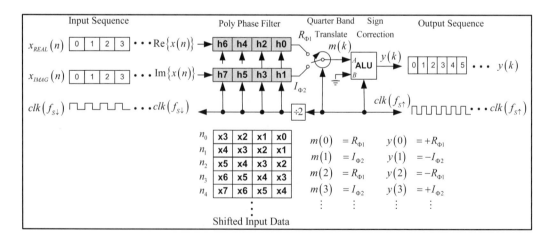

**Figure 7.18**  Final complex to real conversion architecture

### 7.2.7  **Functional Design Summary**

To review our previous steps, we began our discussion using a functionally accurate but very inefficient conversion architecture as a starting point, and then we walked through a step-by-step development to obtain a very efficient and simple structure that can be used to perform complex to real conversion. The final version of the complex to real conversion architecture is illustrated in Figure 7.18.

It is sometimes beneficial to use the more inefficient but functionally correct representation of a complex to real converter that is illustrated in Figure 7.19. This representation is useful in a high-level slideshow presentation at design reviews because the individual processing functions are clearly shown and because this functional representation makes for an easy slide to speak to. Other uses for this type of representation might be in formal proposal write-ups in response to a customer-generated request for quote (RFQ). The reader should also note that if the sample rate of the input signal is already at the Nyquist rate for real signals, then the need for the interpolate by two complex filter vanishes. In this case, the absence of the interpolation filters really simplifies the complex to real conversion logic design.

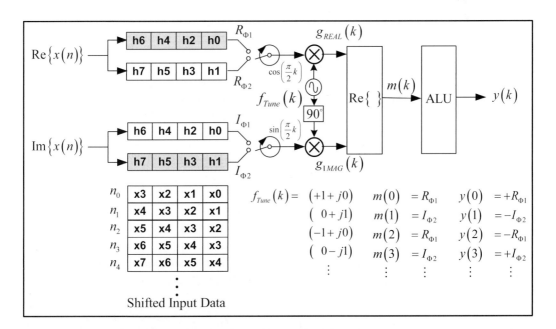

**Figure 7.19**   Complex signal poly phase interpolate by two architecture

## 7.3  COMPLEX TO REAL SIMULATION RESULTS

In the preceding sections, we learned how to design a complex to real conversion circuit. Now is a good time to reinforce this knowledge with a computer simulation that inputs and processes an actual signal. The simulation we discuss here implements the exact complex to real architecture illustrated in Figure 7.18. The reader is urged to compare the simulation results given in this paragraph with the technical discussion of the previous paragraphs.

### 7.3.1  Interpolation Filter Simulation

We will begin by graphically illustrating the complex to real conversion process, and then we will demonstrate the same process through the use of a computer simulation that operates on an actual signal. Our starting point will be the spectral interpretation of the complex to real conversion interpolation filter, as illustrated in Figure 7.20. Part A of the figure shows the two-sided frequency response of a complex interpolate by two digital filter. If we define the filter impulse response as a sequence labeled $h(n)$, then the frequency response of the filter is represented by the Fourier transform of this impulse response sequence given by

$$H\left(\frac{k}{N}\right) = f\{h(n)\}$$

Note that the frequency axis is calibrated in terms of normalized frequency $k/N$ and that the filter pass band extends from $-f_S/4$ to $+f_S/4$ or $-0.25$ to $+0.25$. We now introduce a complex sinusoid $f(n)$ given by

$$f(n) = e^{j\left(\frac{2\pi k_0}{N}n\right)} = \cos\left(\frac{2\pi k_0}{N}n\right) + j\sin\left(\frac{2\pi k_0}{N}n\right)$$

**Equation 7.10**

The normalized frequency of this complex sinusoid is set to $k_0/N = 1/4 = 0.25$, which is one quarter of the sample rate $f_S$. If we substitute this value back into Equation 7.10, we arrive at

$$f(n) = \cos\left(\frac{\pi}{2}n\right) + j\sin\left(\frac{\pi}{2}n\right)$$

**Equation 7.11**

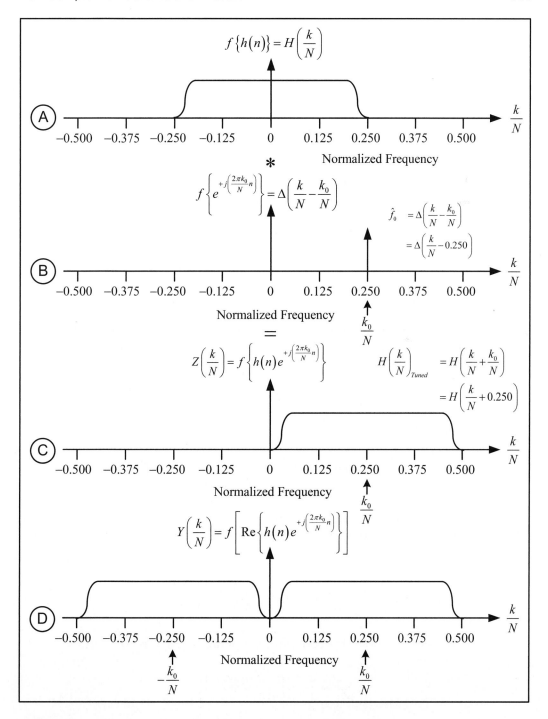

**Figure 7.20**  Complex to real conversion spectral interpretation

As the sample index increments, $n = 0, 1, 2, 3, \cdots$, the complex sinusoid in Equation 7.11 traces out the periodic four-sample sequence given by

Index     Complex sinusoid     Complex value

$$n = 0 \quad \cos\left(\frac{\pi}{2}0\right) + j\sin\left(\frac{\pi}{2}0\right) \quad = (1 + j0)$$

$$n = 1 \quad \cos\left(\frac{\pi}{2}1\right) + j\sin\left(\frac{\pi}{2}1\right) \quad = (0 + j1)$$

$$n = 2 \quad \cos\left(\frac{\pi}{2}2\right) + j\sin\left(\frac{\pi}{2}2\right) \quad = (-1 + j0)$$

$$n = 3 \quad \cos\left(\frac{\pi}{2}3\right) + j\sin\left(\frac{\pi}{2}3\right) \quad = (0 - j1)$$

$$\vdots \qquad\qquad \vdots \qquad\qquad \vdots$$

This is the exact same sequence that we derived for the quarter band translation logic in Equation 7.6. Since the sinusoid is complex, its spectrum across the band ranging from $-f_s/2$ to $+f_s/2$ consists of a single impulse at $k/N = 0.25$. The spectrum of this complex sinusoid is computed by taking the DFT of $f(n)$, which gives us

$$f\{f(n)\} = f\left\{e^{j\left(\frac{2\pi k_0}{N}n\right)}\right\} = F\left(\frac{k}{N}\right) = \Delta\left(\frac{k}{N} - \frac{k_0}{N}\right) = \Delta\left(\frac{k}{N} - 0.25\right)$$

The line spectrum of this complex sinusoid is shown in part B of Figure 7.20.

Both the complex interpolation filter impulse response $h(n)$ and the complex sinusoid $f(n)$ are discrete time series. Multiplication of these two time series results in the convolution of their respective spectra. If we define this time series product as the sequence $z(n)$, then we can write the Fourier transform pair as

$$z(n) = h(n)f(n) \overset{f}{\leftrightarrow} H\left(\frac{k}{N}\right) * F\left(\frac{k}{N}\right)$$

The symbol $*$ denotes the convolution operation. We can now write

$$Z\left(\frac{k}{N}\right) = f\left\{h(n)e^{+j\left(\frac{2\pi k_0}{N}n\right)}\right\}$$

Convolution of the filter frequency response with the complex sinusoid centered at $k/N = 0.25$ translates the center frequency of the filter response from 0 Hz to the frequency of the sinusoid $k_0/N = 0.25$. This is illustrated in part C of Figure 7.20. This is the same process that was performed by the quarter band translation logic in Figure 7.15.

If the filter response $H(k/N)$ was originally centered at the normalized frequency $k_m/N = 0$ Hz, then we can compute the translated frequency response as

$$h(n)e^{+j\left(\frac{2\pi k_0}{N}n\right)} \overset{f}{\leftrightarrow} Z\left(\frac{k}{N}\right) = H\left(\frac{k}{N} - \frac{k_0}{N} - \frac{k_m}{N}\right)$$

$$= H\left(\frac{k}{N} - 0.250 - 0.00\right)$$

$$= H\left(\frac{k}{N} - 0.250\right)$$

Now we see that the complex low pass interpolation filter frequency response has been quarter band translated so that it is centered on the normalized frequency $k_0/N = 0.25$, or $f_s/4$. Don't forget that this translated frequency response is that of a complex sequence. This fact is easy to recognize, since the spectrum has no negative mirror image frequency.

Now when we take the real part of the translated sequence $z(n)$, we end up with the final complex to real conversion logic output $y(n) = \text{Re}\{z(n)\}$. In the frequency domain, this is given by

$$Y\left(\frac{k}{N}\right) = f\left\{\text{Re}\left[h(n)e^{+j\left(\frac{2\pi k_0}{N}n\right)}\right]\right\} = f\left\{h(n)\cos\left(\frac{2\pi k_0}{N}n\right)\right\} = \frac{1}{2}\left\{H\left(\frac{k}{N} - \frac{k_0}{N}\right) + H\left(\frac{k}{N} + \frac{k_0}{N}\right)\right\}$$

$$= \frac{1}{2}\left\{H\left(\frac{k}{N} - 0.25\right) + H\left(\frac{k}{N} + 0.25\right)\right\}$$

When we take the real part of the complex sequence, the spectrum of the filter response is modified as illustrated in part D of Figure 7.20. The real signal spectrum includes the negative frequencies that are a mirror image of the positive frequencies.

In part D of the figure, we have drawn the negative frequencies as ranging from $-f_s/2$ to 0 or $-0.5$ to 0. This is the equivalent of a normal DFT output, which usually displays the negative frequencies from $f_s/2$ to $f_s$ or normalized frequency of 0.5 to 1. Our computer simulations will use the latter presentation.

We will validate the previous discussion by first designing an actual interpolation filter and then using this filter in the computer simulation. In the simulation, we will monitor via software test points the filter frequency response, the quarter band translated response, and the response following the extraction of the real part.

The first item of business is to use the Parks-McClellan algorithm to design an optimum FIR filter relative to the high sample rate $f_{s\uparrow}$. The listing in Table 7.1 is the output of the PM algorithm that tabulates the double-precision floating point and 16-bit integer coefficients of a 46-tap low pass filter that we will use in the complex to real simulation. For an optimum filter response, we could use the floating point coefficients. Here we opt to simulate an integer processor and use the integer coefficients to implement the polyphase interpolate by two filter. By doing so, we lose some performance, particularly in the depth of the filter stop band.

This filter was designed using the procedures outlined in Chapter 5, "Finite Impulse Response Digital Filtering." We note from Table 7.1 that the filter pass band extends from the normalized frequency 0 to 0.16. The stop band extends from 0.25 to 0.5. The transition bandwidth is 0.09. Other filters with a sharper roll-off and narrower transition band can be easily designed at the expense of using a higher number of coefficients. The stop band for this filter is –77 db when using the double-precision floating point coefficients. For this particular filter, the use of the 16-bit integer coefficients reduces the stop band attenuation by 6 db to –71 db. The frequency response for this 46-tap filter is illustrated in Figure 7.21.

We implement the polyphase interpolate by two filter by splitting the filter coefficients in half. We assign all the even coefficients $h(0), h(2), h(4), \cdots, h(44)$ to the phase $R_{\Phi 1}$ filter, and we assign all the odd coefficients $h(1), h(3), h(5), \cdots, h(45)$ to the phase $I_{\Phi 2}$ filter, as illustrated in Figure 7.18.

This filter is a good starting point for a complex to real conversion design. The reader may find it beneficial to experiment by using additional coefficients to narrow the transition band and increase the stop band attenuation. As always, the system specifications should drive the filter design parameters.

The reader should remember that, at this point, we have a complex filter that inputs a complex signal and outputs a complex signal. When the data output by the filter is quarter band translated, we can visualize the pass band of the filter as being translated as well. The translated frequency response illustrated in Figure 7.22 clearly shows that the filter is indeed complex because there are no mirrored negative frequencies.

**Table 7.1**   Forty-Six Tap Simulation Complex to Real Interpolation Filter Coefficients

Finite Impulse Response FIR
Linear Phase Digital Filter Design
Remez Exchange Algorithm
Band Pass Filter
Filter Length = 46

***** Impulse Response *****

| | | | | | |
|---|---|---|---|---|---|
| H(0) | = | 0.00001044 | = H(45) | = | 0 |
| H(1) | = | 0.00030543 | = H(44) | = | 10 |
| H(2) | = | 0.00026549 | = H(43) | = | 9 |
| H(3) | = | -0.00056252 | = H(42) | = | -17 |
| H(4) | = | -0.00122561 | = H(41) | = | -39 |
| H(5) | = | 0.00013065 | = H(40) | = | 4 |
| H(6) | = | 0.00269161 | = H(39) | = | 88 |
| H(7) | = | 0.00209116 | = H(38) | = | 69 |
| H(8) | = | -0.00326351 | = H(37) | = | -106 |
| H(9) | = | -0.00647199 | = H(36) | = | -211 |
| H(10) | = | 0.00035919 | = H(35) | = | 12 |
| H(11) | = | 0.01118153 | = H(34) | = | 366 |
| H(12) | = | 0.00832817 | = H(33) | = | 273 |
| H(13) | = | -0.01145878 | = H(32) | = | -374 |
| H(14) | = | -0.02228997 | = H(31) | = | -729 |
| H(15) | = | 0.00065039 | = H(30) | = | 21 |
| H(16) | = | 0.03592747 | = H(29) | = | 1177 |
| H(17) | = | 0.02754549 | = H(28) | = | 903 |
| H(18) | = | -0.03729601 | = H(27) | = | -1221 |
| H(19) | = | -0.07990643 | = H(26) | = | -2617 |
| H(20) | = | 0.00085972 | = H(25) | = | 28 |
| H(21) | = | 0.19930561 | = H(24) | = | 6531 |
| H(22) | = | 0.37289983 | = H(23) | = | 12219 |

| | Band 1 | Band 2 |
|---|---|---|
| Lower Band Edge | 0.0000000 | 0.2500000 |
| Upper Band Edge | 0.1600000 | 0.5000000 |
| Desired Value | 1.0000000 | 0.0000000 |
| Weighting | 1.0000000 | 1.2000000 |
| Deviation | 0.0001547 | 0.0001289 |
| Deviation(db) | 0.0013437 | -77.7932655 |
| Sum of Floating Point Coefficients = | | 1.0001546 |

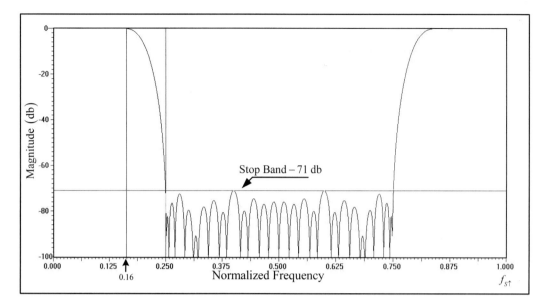

**Figure 7.21**   Forty-six tap simulated complex interpolation filter composite frequency response

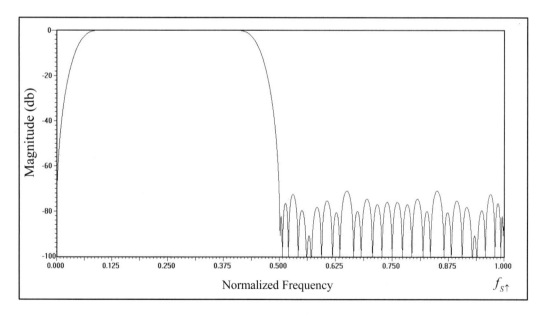

**Figure 7.22**   Software simulation of a quarter band translated complex interpolation filter

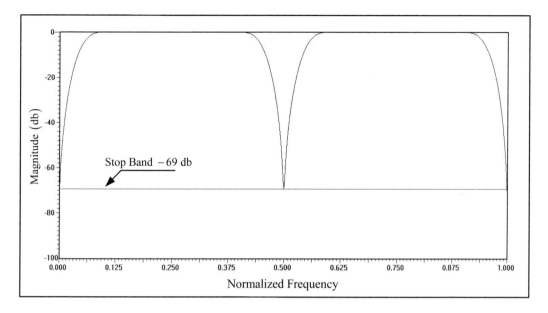

**Figure 7.23**  Software simulation interpolation filter after conversion to a real filter

Finally we take the real part of the translated sequence. This is accomplished by discarding the imaginary part of the complex time series sequence after frequency translation, retaining the real part and adjusting the sign sequence to reflect the inclusion of the $j$ terms in the complex translation multiply. This is identical to the process discussed in section 7.2.6.

After the real part is taken and the sign sequence is modified, we end up with the modified frequency response shown in Figure 7.23. We can think of this spectral plot as the translated frequency response of the low pass interpolation filter, or we can think of this as the data output by the low pass filter being translated in frequency. Either way is correct. Once again, we can tell by the inclusion of the negative frequency mirror extending from the normalized frequency 0.5 to 1.0 that this is the spectrum of a real filter impulse response or, alternatively, the spectrum of a real base band signal translated to a quarter of the sample rate.

### 7.3.2  Data Simulation

Now that we have discussed complex to real conversion from the interpolation filter frequency response point of view, it's time embark on a similar discussion, this time from the input data point of view.

Let's begin by assuming a normalized complex sinusoid of frequency $f_0 = k_0/N$ is the input signal to a complex to real conversion logic circuit. The spectrum of this sinusoid is the single impulse illustrated in part A of Figure 7.24. As shown in the figure, the spectrum of a digital signal is periodic in $f_{s\downarrow}$ Hz.

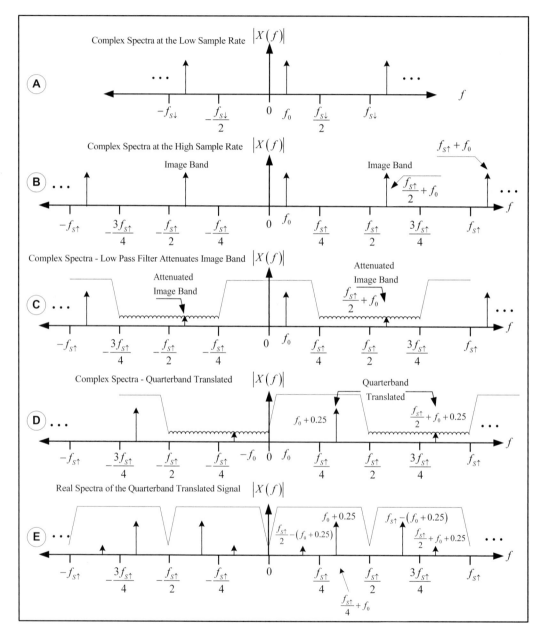

**Figure 7.24** Example of the conversion of a complex sinusoid

At this point, we are only concerned with the portion of the spectrum between 0 and $f_{s\downarrow}$ or, alternatively, the portion of the spectrum between $-f_{s\downarrow}/2$ and $+f_{s\downarrow}/2$, which consists of a single impulse at frequency $f_0$.

To make things simpler, we would like to visualize the complex to real conversion as being a sequence of independent events consisting of the following:

1. The increase of the complex sample rate by a factor of two
2. The low pass filtering of the interpolated complex signal to eliminate the resulting image bands
3. The quarter band translation of the complex signal to center its spectrum to $f_s/4$
4. The taking of the real part of the complex quarter band translated signal

The visualization of this sequence is illustrated in parts B through E in Figure 7.24.

In part B of the figure, we see the spectrum of the input signal after the sample rate has been increased by a factor of two. Since the spectrum of a digital signal is repetitive, there isn't much difference between the spectrum of a low rate and high rate signal. It is important to note, however, that the sample rate has doubled and is given by the notation $f_{s\uparrow}$. We interpret the impulse at $f_{s\uparrow}/2 + f_0$ to be an "image band" that is a by-product of the sample rate increase. If, for example, we had increased the sample rate by a factor of three, we would have seen two image bands between 0 Hz and $f_{s\uparrow}$ Hz. In general, if we increase the sample rate by a factor of $L$, we would end up with $L-1$ image bands.

The purpose of the low pass digital filter in the complex to real conversion process is to remove these image bands. In part C of the figure, we see the filter response superimposed on the input signal spectra. The stop band of the low pass filter response attenuates the image band to a negligible level. What we have left is a complex signal that is sampled at twice the rate of the original input signal with attenuated image bands that are below the stop band of the interpolation filter.

Next, we quarter band translate the complex signal to center the signal spectrum at $f_{s\uparrow}/4$. This is illustrated in part D of the figure. For illustrative purposes, we also show the translated filter pass band. Note that we see not only the spectrum of the original sinusoid but also the attenuated signal of the image sinusoid created when we increased the sample rate by a factor of two. The attenuated image is below the stop band of the interpolation filter.

Finally, we take the real part of the interpolated complex sequence and include the sign operations imposed on our computations by the $j$ operator and end up with a real signal whose spectrum is illustrated in part E of Figure 7.24. Once again, to help in the visualization of this process, we show the overlaid frequency response of the interpolation filters. And now that the theory has been graphically illustrated, it is time to validate it with a computer simulation on actual input data. Our computer simulation generates a complex sinusoid for use as an input signal to the simulator's complex to real logic circuit. The spectrum of this input signal is illustrated in Figure 7.25. The frequency of the input sinusoid is $f_0 = f_{s\downarrow}/8$ or a normalized frequency of 0.125. Once again, the reader can identify this signal as complex because there are no negative mirror frequencies. If we take the exact same sinusoid and do nothing else to it except sample it at twice the rate, we would end up with the plot representation shown in Figure 7.26.

We need to stop here for a moment and explain the difference between the spectral plots in Figure 7.25 and Figure 7.26. The plot width of the $x$-axis is fixed at 1024 points for all plots. Figure 7.25 is a spectral plot for a complex sinusoid at the low sample rate. The frequency of the complex sinusoid is $f_0 = f_{s\downarrow}/8$ or 0.125. Figure 7.26 is a spectral plot for the same complex sinusoid but relative to the high sample rate $f_{s\uparrow} = 2f_{s\downarrow}$. Therefore the frequency of the complex sinusoid relative to the higher sample rate is

$$\frac{f_{s\downarrow}}{8} = \frac{2f_{s\downarrow}}{16} = \frac{f_{s\uparrow}}{16} = 0.0625$$

Because the length of the $x$-axis is fixed, a cursory glance at the plots (without taking into account the difference in sample rates) and the resultant frequency scaling might be a source of confusion.

The output of the complex to real converter simulation for our complex input sinusoid is illustrated in Figure 7.27. Here we see the two-sided spectrum of a real signal that has been translated by the normalized frequency of $f_{s\uparrow}/4 = 0.25$. The real signal has two frequency impulses centered at $f = 0.3125$ and $f = 0.6875$. If the DFT were plotted as a two-sided spectrum, the negative frequencies would have been centered at a normalized frequency of $f = -0.3125$. We also see the filtered remainder of the attenuated image frequency that was created when the sample rate was increased by a factor of two.

The horizontal cursor of the complex to real simulator measures the magnitude of the attenuated image frequency as 78.4 db down. We can correlate this to the magnitude of the interpolation filter stop band by slewing the simulator cursors to the location of the original pretranslated image frequency, as illustrated in Figure 7.28. The pretranslated image frequency was located at

$$\frac{f_{s\uparrow}}{2} + \frac{f_{s\uparrow}}{16} = 0.5000 + 0.0625 = 0.5625$$

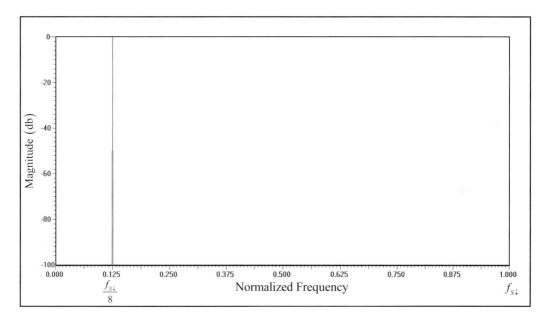

**Figure 7.25**  Spectrum of a complex sinusoid at the low sample rate

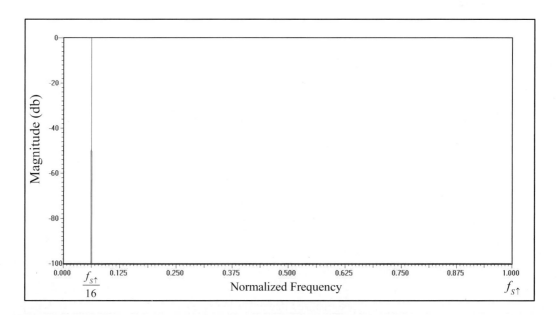

**Figure 7.26**  Spectrum of a complex sinusoid at the high sample rate

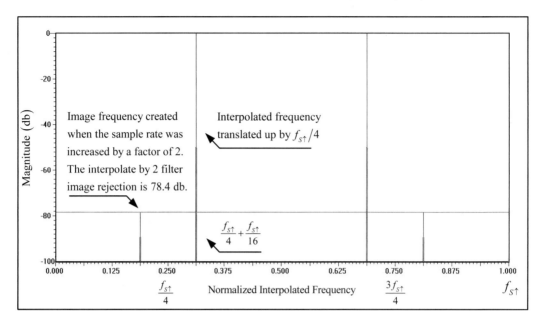

**Figure 7.27**   Output of the complex to real conversion logic

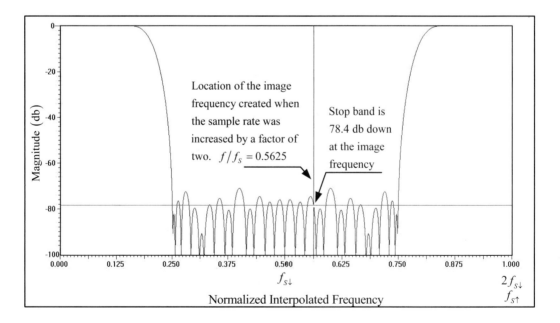

**Figure 7.28**   Interpolation filter image frequency rejection

The vertical cursor in the simulation plot is placed at 0.5625, and the horizontal cursor is slewed down to the intersection of the vertical cursor and the magnitude of the filter frequency response.

From the horizontal cursor, we can see that the magnitude of the interpolation filter stop band at that point is equal to –78.4 db, which is identical to the magnitude of the residual image in Figure 7.27. As long as the image frequency is below the specified filter stop band, the rejection specifications are satisfied and downstream processing will not be adversely affected.

The design of a complex to real signal processing function is straightforward. The heart of the design is the interpolation filter. Care should be taken to design this filter to meet or exceed the overall system specifications. Using the Parks-McClellan optimum FIR filter design procedures discussed in Chapter 5, "Finite Impulse Response Digital Filtering," will simplify this process significantly. The reader should note that if the sample rate of the original complex input signal is already at twice that of its Nyquist rate, then the interpolation filter will not be a necessary component in the overall complex to real conversion algorithm.

The tutorial given in this chapter is by no means the only solution to a complex to real conversion application. Harris [1] provides a very interesting treatment of this application by converting a complex 6-KHz telephony voice channel to an 8-KHz real voice signal using a four-phase interpolation filter and a translator (or *heterodyne*, as he calls it), followed by a three-phase decimation filter. In short, he interpolates the sample rate by a factor of 4 to 24 KHz, translates the signal, and then decimates the signal by a factor of three down to 8 KHz. The point to be made is that the procedure outlined in this chapter can be modified to suit many different processing applications.

## 7.4  REFERENCE

[1] Frederic J. Harris. "Multirate Filters." In *Handbook of Digital Signal Processing Engineering Applications*. Edited by Douglas F. Elliot. San Diego, CA: Academic Press, 1987.

# CHAPTER EIGHT

# Digital Frequency Synthesis

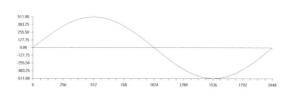

## 8.1 NUMERICALLY CONTROLLED OSCILLATOR (NCO)

There are numerous applications in the world of digital signal processing (DSP) that utilize a numerically controlled oscillator (NCO). An NCO is a programmable oscillator that outputs a digital sinusoid at some user specified frequency and phase. The sinusoid can be fixed at some programmed frequency, or it can be swept or hopped over a band of frequencies. The sinusoid can have a constant phase, or it can be programmed to have multiple or switched phases. It can be a simple or a complex device, depending on the requirements of the application in which the NCO is used. A typical application utilizes the NCO to produce a programmable sinusoid to tune band pass signals down to base band for filtering and postprocessing, similar to the local oscillator in an analog AM radio receiver.

The most basic NCO consists of three components: a phase register, a phase accumulator, and a sine look up table. The trick to designing an NCO is to understand how to implement each of these three components in order to guarantee that the device meets or exceeds the application specifications. We will start our discussion by looking at the three components of a basic NCO, and then we will expand on that to produce a device that can be readily used in most applications.

### 8.1.1 NCO Phase Register

The NCO phase register is nothing more than a simple storage device that is utilized to hold the NCO phase word. If the NCO is implemented in hardware, then the storage device is a simple shift register. If the NCO is implemented in software, then the storage device is a simple variable.

In our initial discussions, the phase register will be $M$ bits wide, where the value of $M$ is chosen to satisfy several of the overall NCO design performance specifications. In later discussions, we will expand the width of the register to $M + N$ bits wide, where value of $N$ is chosen to increase the precision of the NCO output frequency.

For now, however, let us briefly visualize the NCO as being nothing more than the simple mathematical expression given by $y(n) = \sin(\Phi n)$, where

> $y(n)$ = the steady stream of NCO sinusoidal output samples
>
> $\sin(\Phi n)$ = the waveform generator that produces samples of a sinusoidal signal
>
> $n$ = the sample index that takes on sequential values of $0, 1, 2, 3, ...., \infty$ at some distinct periodic rate, such as the system clock rate
>
> $\Phi$ = the phase word that defines the frequency of the generated sine wave

In this purely mathematical example, the phase word $\Phi$ is a constant. It is the incremental phase of the waveform generator. $\Phi$ is multiplied by the periodic sample index $n$ to produce an incrementally increasing argument of the sine function. The increasing argument over time produces the sequential output samples of the generated sinusoidal waveform.

For the same sequence $n$, different values of $\Phi$ will produce sinusoidal output sequences of different frequency. If the phase word $\Phi$ is small, the resultant sinusoidal sequence will have a lower frequency. If the phase word $\Phi$ is large, the resultant sinusoidal sequence will have a higher frequency.

In a typical NCO design, the value of the phase word is programmable, and it is used to define the frequency of the sinusoid generated by the NCO. If the phase word is held constant, the NCO will produce a constant output tone. However, there is no reason to assume that the phase word needs to be held constant. The value of $\Phi$ can change over time in some specified manor peculiar to the NCO application. For example, if the phase word switches between two distinct values in response to a serial bit stream of ones and zeros, the NCO could conceivably be utilized to generate a frequency shift keyed (FSK) communications system. Another example would be that if the value of the phase word were incremented in an orderly fashion between two defined limits, then the NCO would output a swept frequency that could be useful in bench test applications where the frequency response of units under test are analyzed and verified. In summary, the NCO phase register is nothing more than a device that stores a programmable numerical value we call $\Phi$, which is used to determine the NCO output frequency.

## 8.1.2  NCO Phase Accumulator

The NCO phase accumulator consists of an adder, or arithmetic logic unit (ALU), and an output register. A block diagram for a phase accumulator is illustrated in Figure 8.1. For purposes of this initial discussion, the phase register that we just discussed provides an input to this simple accumulator and is defined to be $M$ bits wide. The accumulator is also $M$ bits wide. We will see later that the value of M will play a significant role in determining the overall performance and tuning precision of the NCO.

As illustrated in Figure 8.1, the NCO phase register is loaded from some external source such as a microprocessor bus. The phase register holds an $M$ bit wide binary number that is called the NCO *phase word* and is represented by the variable $\Phi$. For our immediate purposes, the phase value $\Phi$ can be considered a programmable value that is set and held constant for the duration of the NCO operation. Later we will see that this is not an operational requirement. The registered value of $\Phi$ is applied to the B input port of an $M$ bit adder, where it is summed with the value input to the A input port of the same adder. The A input is simply the feedback of the registered adder output. The resultant sum appears at the output port C and is clocked into the output register by the sample clock $F_s$. The combination of the adder and output register form an accumulator that continuously adds, in a modulo $2^M$ fashion, the value of $\Phi$ to the running sum, and it does so at the sample clock rate. That is, the running sum is incremented by the value stored in the phase register on every sample clock. The running sum output at port C can be expressed as

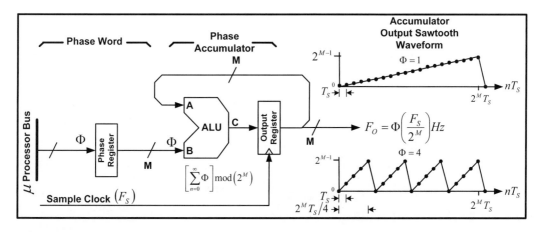

**Figure 8.1**  Block diagram of an NCO phase accumulator

$$C = \left[\sum_{n=0}^{\infty} \Phi\right] \mathrm{mod}\left(2^M\right)$$

It is easy to see that if $\Phi$ were set to the value of one, the running sum would be incremented by one on each sample clock and the output register would hold that value until the next clock. In this simple example, the accumulation sequence would be periodic and would appear as

$$\cdots 0, 1, 2, 3, 4, \cdots \left(2^M - 1\right), 0, 1, 2, 3, 4, \cdots \left(2^M - 1\right) \cdots$$

As this sequence shows, the phase accumulator is nothing more than a simple modulo $2^M$ counter that counts in increments of $\Phi$.

The dots on the accumulator waveforms shown in Figure 8.1 are intended to graphically represent accumulator output samples. As illustrated in Figure 8.1, the accumulation stored in the output register would, over time, look like a digital saw tooth waveform with samples spaced every

$$\frac{1}{F_S} = T_S \text{ seconds}$$

The period of the saw tooth waveform $\left(T_{SAW\ TOOTH}\right)$ for this example would be equal to

$$T_{SAW\ TOOTH} = \frac{2^M T_S}{\Phi} \text{ seconds}$$

For this example, if $\Phi = 1$, $M = 10$, and $F_S = 1.024$ MHz, then successive samples would be spaced $0.9765625\ \mu$ seconds apart and the saw tooth waveform would have a period of

$$T_{SAW\ TOOTH} = \frac{2^M T_S}{\Phi} = \frac{2^{10}\left(0.9765625\right)}{1} = 1.0 \text{ mseconds}$$

Therefore the frequency of the accumulator output saw tooth is given by

$$F_{SAW\ TOOTH} = \frac{1}{T_{SAW\ TOOTH}} = 1 \text{ KHz}$$

Let's take a look at another example. If the value of $\Phi$ were increased from 1 to 4, then the accumulation sequence would be

$$\cdots 0,4,8,12,\cdots,0,4,8,12,\cdots \ \left\{ \mathrm{modulo}\left(2^M\right) \right\}$$

The accumulation produces a saw tooth waveform that skips three out of every four possible accumulation results to produce a saw tooth waveform with a period that is four times shorter than in the case where $\Phi = 1$, which means the saw tooth frequency is four times greater. The sample clock frequency is still the same, and the output accumulation samples are still spaced

$$\frac{1}{F_S} = T_S \text{ seconds or } 0.9765625 \ \mu \text{ seconds apart}$$

Therefore the period of the saw tooth waveform is

$$T_{\text{SAW TOOTH}} = \frac{2^M T_S}{\Phi} = \frac{2^{10}\left(0.9765625\right)}{4} = 0.25 \text{ mseconds}$$

and the output saw tooth frequency is 4 KHz. The saw tooth waveforms for these two examples are illustrated in Figure 8.1.

So we can see that for the same sample clock $F_S$, the value we choose for $\Phi$ controls the frequency of the accumulator output saw tooth waveform. Since the accumulator is $M$ bits wide, there are a maximum of $2^M$ samples per saw tooth cycle when the value of the phase word is set to unity. It's easily seen that, for the case where $\Phi$ is set to unity, the saw tooth frequency can be computed to be $\left(F_S/2^M\right)$ Hz. For this simple example, this value also represents the lowest or fundamental frequency saw tooth that can be obtained out of the accumulator. Therefore the frequency resolution of this NCO is equal to its fundamental frequency. We will see later that there exists a simple method to significantly increase the NCO frequency resolution.

Setting $\Phi$ to zero would result in an infinite period that is commensurate with an output frequency of 0 Hz or DC. As the phase word $\Phi$ increases in value, the running sum will accumulate at a faster rate resulting in an increase in output saw tooth frequency. The actual saw tooth frequency out of the accumulator is given by the expression

$$F_O = \Phi\left(\frac{F_S}{2^M}\right) \text{ Hz}$$

**Equation 8.1**

where

$F_0$ is the frequency of the saw tooth waveform output from the phase accumulator,

$2^M$ is the maximum number of accumulations possible per saw tooth period,

$F_S$ is the accumulator sample clock rate, and

$\Phi$ is the phase word that is repetitively accumulated at the sample rate.

The lowest frequency that can be produced by the NCO accumulator is DC, and this obviously occurs when the phase word is equal to zero. The fundamental frequency of the NCO accumulator occurs when the phase word is set to its minimum value other than zero. For the NCO in the previous example, the minimum value of the phase word was unity and therefore the fundamental frequency for that particular NCO accumulator is given by

$$\frac{F_S}{2^M} \text{ Hz}$$

The highest frequency saw tooth that can be produced by the accumulator occurs when the phase word is set to its maximum value of $2^{M-1}$. If we substitute this maximum value for $\Phi$ in Equation 8.1, then we can see that the highest frequency we can achieve out of the phase accumulator is

$$F_O = \Phi\left(\frac{F_S}{2^M}\right) = 2^{M-1}\left(\frac{F_S}{2^M}\right) = \frac{F_S}{2} \text{ Hz}$$

**Equation 8.2**

Equation 8.2 shows that the highest frequency obtainable out of the phase accumulator is equal to half the sample rate. Therefore we can state that as the phase word $\Phi$ increases through its range of values

$$0 \le \Phi \le 2^{M-1}$$

the accumulator output frequency $F_O$ will increase through its frequency range according to

$$0 \le F_O \le F_S/2 \text{ Hz}$$

**Equation 8.3**

It should be noted here that since the bit width of the phase register is $M$ bits, the maximum value this register can hold is equal to $(2^M - 1)$. This is almost twice as large as the upper limit imposed on the value of $\Phi$, which is $2^{M-1}$. Values of $\Phi$ larger than $2^{M-1}$ will violate the Nyquist sample rate limit. In this case, the output sinusoid will be under sampled, it will be aliased, and the NCO will produce erroneous output frequencies. This is a significant point, and it will be discussed in further detail in the next section.

### 8.1.3   NCO Sine Look Up Table

We can complete the design of the basic NCO by adding a sine look up table. For NCOs implemented in hardware, look up tables are realized in memory, such as static RAM, ROM, PROM, flash memory, or core memory, should the NCO be implemented in a field programmable gate array (FPGA). For NCOs implemented in software, look up tables are realized by large arrays typically implemented in heap memory. The type of memory used is application dependent and is determined by the size of the table, the speed of the memory, and the power consumption of the memory. The look up table for most designs will contain one complete period of a sine waveform. The block diagram of a complete NCO is illustrated in Figure 8.2.

We can see from the block diagram that the output of the phase accumulator is applied directly to the address port of the look up table memory. The digital saw tooth waveform produced by the phase accumulator serves as a linear address ramp to the look up table memory. From this point forward, the phrases "phase accumulator saw tooth waveform" and "look up table address ramp" will be considered one in the same and will be used

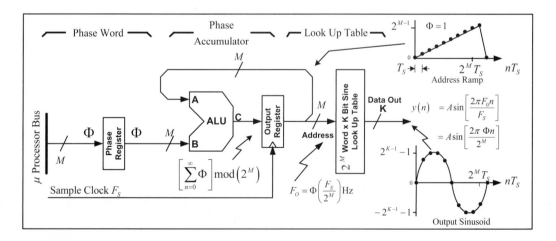

**Figure 8.2**   Block diagram of a complete NCO

interchangeably. The bit width of the address ramp determines the number of words in the look up table memory. The $M$-bit-wide address ramp fixes the number of words in the look up memory to be $2^M$.

One complete cycle of a sine wave is stored in the look up table. The table is repetitively addressed modulo $2^M$ by the phase accumulator address ramp. The output of the table will be a continuous digital sinusoidal waveform whose period is identical the period of the accumulator address ramp. This is illustrated in Figure 8.2. If the period of the address ramp and the sinusoid waveform are identical, then so are their frequencies. Therefore the frequency of the look up table sinusoid can be represented by the same Equation 8.1 as was used for representing the accumulator saw tooth output frequency. This equation is repeated here for convenience:

$$F_O = \Phi\left(\frac{F_S}{2^M}\right) \text{ Hz}$$

**Equation 8.4**

where

$F_0$ is the NCO sinusoidal output frequency,
$2^M$ is the size of the sine look up table,
$F_S$ is the accumulator sample clock rate, and
$\Phi$ is the phase word that is repetitively accumulated at the sample rate.

In the analog world, the value of the time variable $t$ can be considered continuous. The value of an analog sinusoid at any time $t$ is represented by an equation of the form

$$y(t) = A\sin(\omega_0 t) = A\sin(2\pi f_0 t)$$

In the digital domain, however, the time variable is quantized and takes on values that are periodic increments of the sample period $T_S$. The value of a digital sinusoidal waveform sample is defined at periodic intervals of the sample period or $nT_S$, where the index $n$ takes on discrete values of $n = 0, 1, 2, 3, \cdots, \infty$. Therefore the sinusoid output from the look up table can be expressed by the equation

$$y(n) = A\sin(\omega_0 nT_S) = A\sin(2\pi F_0 nT_S)$$

If we substitute $1/F_S$ for $T_S$, we end up with the following:

$$y(n) = A \sin\left[\frac{2\pi F_0 n}{F_S}\right]$$

**Equation 8.5**

Equation 8.5 clearly shows that the argument of the NCO sinusoid can be expressed as a ratio of the output frequency $F_0$ to the sample frequency $F_S$.

If we substitute the NCO output frequency $\Phi\left(\dfrac{F_S}{2^M}\right)$ for $F_0$ in Equation 8.5, we arrive at

$$y(n) = A \sin\left(\frac{2\pi}{2^M}\frac{\Phi}{} n\right)$$

**Equation 8.6**

Equation 8.6 clearly shows that the argument of the NCO sinusoid can also be expressed as a ratio of the programmable phase word $\Phi$ to size of the look up table $2^M$. Both Equation 8.5 and Equation 8.6 are useful in describing the sinusoidal waveform output from the look up table.

It should be noted for a $K$-bit-wide look up table that contains two's complement sinusoidal samples that the maximum amplitude of the NCO generated sinusoid is given by $A = 2^{K-1} - 1$. This will be discussed in detail later when we actually design a look up table.

Now let's take a look at the mechanics of the NCO operation. If the phase word $\Phi$ is set to unity then every possible point in the address ramp will be generated and every address of the look up table will be accessed. One cycle of the sinusoidal output waveform will contain a sample from every location in the look up table memory.

If the phase word $\Phi$ is doubled to a value of 2, then the address ramp will increment by two every sample clock cycle, and every other location in the look up table will be accessed. This means the address ramp will cycle through the look up table address space twice as fast, producing two cycles of the output waveform in the same time it took to produce one cycle when $\Phi$ was set to unity. Incrementing $\Phi$ from 1 to 2 caused the NCO output frequency $F_O$ to double. However, each cycle of the $\Phi = 2$ sinusoidal waveform contains only half of the possible look up table samples.

Similarly, if $\Phi$ is set to a value of 4, then every fourth address in the look up table will be accessed and the frequency of the output sinusoid will be four times faster. For this case, there will be one fourth the number of samples per cycle of the output waveform. So as $\Phi$ increases, the frequency of the output sinusoid increases. This will continue until $\Phi$ reaches its maximum value of $2^{M-1}$. At this point, the look up table output sinusoid will be represented by just two samples per cycle.

At first glance, two samples do not seem to be enough to represent a cycle of a sinusoidal waveform. In principle, however, the minimal number of samples necessary to reconstruct an analog sinusoid from a digital sinusoidal waveform is two samples per cycle. This phenomenon is discussed thoroughly in just about every textbook written on the subject of DSP, and it is called the *Nyquist sampling theorem*. The Nyquist theorem states that the minimum number of digital samples necessary to reconstruct an analog sinusoid from a digital sinusoid is two samples per cycle. The designer should make sure to thoroughly investigate the spectrum of the NCO output across its entire tuning range to verify the purity of the sinusoidal signal. Remember that even though the Nyquist theorem says that two samples per cycle are sufficient to reconstruct an analog sinusoid from a digital sinusoidal signal, it does not provide any guarantees as to the purity of that reconstructed sinusoid. For example, it does not address issues like harmonics and spurs in the reconstructed sinusoidal spectrum due to nonlinearities in the sinusoidal sample generation process. When generating frequencies near the upper limit of $F_S/2$, it may be necessary to adjust the phase offset of the ramp address so as to utilize look up table samples whose amplitudes are at or near the peaks of the stored waveform. We will discuss this in greater detail in the section that deals with NCO phase offset.

Part of the design effort should include a thorough software bit-level simulation of the NCO before the design is actually built to verify proper NCO performance. The time to correct or modify the design is *before* it gets built, not after. A good software simulation is easy to write and it's worth its weight in gold because it allows the designer to see exactly how his design will perform before it's set in concrete.

### 8.1.4  Construction of a Sine Look Up Table

Equation 8.4 defines the frequency produced by an NCO and is repeated here for clarity:

$$F_O = \Phi\left(\frac{F_S}{2^M}\right) \text{ Hz}$$

In the simple example of the NCO phase accumulator that we have discussed, the output frequency $F_O$ is an integer multiple of the fundamental frequency given by the term

$$\left(\frac{F_S}{2^M}\right)$$

when the phase word is set to unity. We are still concerned with understanding the basics of a simple NCO, so for now, the integer multiplier is the phase word $\Phi$. In later paragraphs, we will discuss modifications to $\Phi$ that will vastly improve the capabilities and performance of the NCO.

When an NCO is designed, there are certain design points that are fixed based upon the requirements of the application in which the NCO will be asked to perform. A good percentage of the time, the designer is constrained to work with a sampling frequency $F_S$ that has been dictated by outside requirements. If this is the case, it makes sense to choose the value for $M$ based on obtaining an efficient fundamental frequency. Additionally, the size of the look up table will be determined by the value of $2^M$. So the value of $M$ also plays an important role in the type, size, speed, and power consumption of the memory used to implement the look up table.

The quantization of the stored samples depends on the number of bits that are necessary to adequately represent the amplitude of the output sine wave. Care must be taken when samples of the stored sine wave are computed, in order to minimize quantization error and nonlinearities that can cause unwanted harmonics or spurs in the NCO output signal spectrum.

The most common numerical representation for the stored look up table samples is two's complement format. In Figure 8.2 we see that the width of the look up table output is $K$ bits. Since the samples are encoded in two's complement format, at least one bit (the most significant bit of the $K$ bits) must be used to represent the sign of the sample. This leaves a maximum of $K-1$ bits to represent the sample magnitude. The maximum amplitude range $R$ that can be represented by $K-1$ bits in two's complement representation is

$$-\left(2^{K-1}\right) \le R \le +\left(2^{K-1}-1\right)$$

Note that two's complement representation has an inherent DC offset in the negative direction due to the fact that the range of negative values is one greater than the range of positive values. For this reason, designers typically do not allow the use of the most negative two's complement value and limit the range of the sinusoid samples to a maximum range of $\pm\left(2^{K-1}-1\right)$.

For example, assume the width of the look up table $K$ is specified as 10 bits. If we reserve the most significant bit for the sign, then we have 9 bits left to represent magnitude. Since the two's complement number representation is bipolar, the maximum range of the $K$ bit number is limited to $-512 \le R \le +511$, but we would compute the look up table samples to be within the range of $-511 \le R \le +511$ to avoid the problems incurred by a negative DC offset in the NCO waveform. The look up table samples can be generated as follows:

$$y(n) = \left(2^{K-1} - 1\right)\left[\sin\left(\frac{2\pi}{2^M}\right)n\right] \quad \text{for} \quad 0 \le n \le 2^M - 1$$

**Equation 8.7**

where

$n$           is the sample index,

$y(n)$        is the $n$th sample of the computed sinusoid,

$\left(2^{K-1} - 1\right)$   is the maximum amplitude of the computed sinusoid, and

$\left[\dfrac{2\pi}{2^M}\right]$        is the sinusoid phase increment in radians.

Since the sine function ranges between +1 and −1 inclusive, $y(n)$ by definition is a floating point number, ranging between $\pm\left(2^{k-1} - 1\right)$. The designer must address the method used to convert the floating point numbers produced by Equation 8.7 to the integer numbers utilized by the look up table. For simple designs, truncation or rounding to eliminate the fractional part of $y(n)$ may be sufficient. If not, then other, more refined methods may be necessary to minimize the stored waveform nonlinearities. A powerful tool to use in the determination of the purity of the look up table sinusoid is the discrete Fourier transform (DFT). The DFT can be used to compute the spectrum of the sinusoid and expose any spurs and harmonics that may be present in the synthesized NCO signal. Using the DFT as a performance checker allows the method of generation of the stored samples to be modified until optimum performance is achieved. Purchased devices that utilize an NCO always characterize the performance of the NCO in terms of its spectral response. We will discuss this issue in greater detail in later sections.

Choosing a value for $M$ and $K$ will size the look up table. Generally speaking, the larger the table, the better the NCO performance. This is because a larger table will accommodate the storage of a single sinusoid period represented by more samples and will increase the NCO frequency resolution. A wider table will increase the number of bits used to represent the amplitude of the sinusoid and will therefore reduce amplitude quantization noise.

Let's use what we have learned so far to design a simple NCO, using specifications that would be fairly common in industry. Keep in mind that better performing NCOs are just as easily designed and will be dealt with in later paragraphs. For now, however, we are only interested in demonstrating the design concepts.

Let's assume for our design that the sample rate $F_s$ is fixed by program requirements to be 16.384 MHz, a common off-the-shelf oscillator frequency offered by many manufacturers. Let's also assume that the designer wants to be able to tune the NCO in increments of 2 KHz. For the simple NCOs we have discussed so far, this would mean that the fundamental frequency of the NCO will be set at 2 KHz.

Let's assume that the data path in the overall system is 12 bits wide. Therefore, for the NCO to be compatible with the rest of the system, its output should be 12 bits wide as well. This sets the value for the output bit width $K$ to be 12. Therefore the range of numbers $R$ that we will use to represent the look up table samples is

$$-\left(2^{K-1}-1\right) \le R \le +\left(2^{K-1}-1\right)$$

$$-2047 \le R \le +2047$$

Since we have decided that the NCO fundamental frequency will be 2 KHz, we can then rearrange the terms in Equation 8.4 and calculate the value of $M$:

$$F_O = \Phi\left(\frac{F_s}{2^M}\right)$$

$$2^M = \Phi\left(\frac{F_s}{F_O}\right)$$

$$\text{Log}_{10}\left(2^M\right) = \text{Log}_{10}\left(\Phi\left(\frac{F_s}{F_O}\right)\right)$$

$$M = \frac{\text{Log}_{10}\left(\Phi\frac{F_s}{F_O}\right)}{\text{Log}_{10}\left(2\right)} = \frac{\text{Log}_{10}\left(\dfrac{16.384 \times 10^6}{2 \times 10^3}\right)}{\text{Log}_{10}\left(2\right)} = 13$$

The value of $M$ is set to 13 bits, so the size of the look up table will be $2^M = 8192$ words long by 12 bits wide. Plugging these values for $M$ and $K$ into Equation 8.7, we can generate all 8192 samples that will be stored in the look up table:

$$y(n) = 2047\left[\sin\left(\frac{2\pi}{2^{13}}\right)n\right] \quad \text{for } 0 \le n \le 8191$$

**Equation 8.8**

It is interesting to view just a few of the samples computed by Equation 8.8. Eight samples of the total 8192 samples stored in the look up table generated by Equation 8.8 are illustrated in Table 8.1. Table entries are selected for samples where the argument of the sine function takes on values that are multiples of $\pi/4$. For purposes of illustration, the numerical representation of the sinusoid samples is formatted as integer decimal (base 10), hexadecimal (base 16), and binary (base 2). The hexadecimal and binary representations are in two's complement format.

All 8192 samples in the look up table are plotted in Figure 8.3 through Figure 8.7. The plot of the stored waveform represents the fundamental frequency of the NCO when $\Phi$ is set to its minimum value of unity. When $\Phi$ is set to unity, the period of this waveform is

$$T_S = \left(\frac{2^M}{F_S}\right) = \left(\frac{2^{13}\,\text{sample}}{16.384 \times 10^6 \,\dfrac{\text{sample}}{\text{seconds}}}\right) = 0.5m \text{ seconds}$$

and the output fundamental frequency is 2 KHz. As long as $\Phi$ remains constant, this frequency sinusoid will be output repetitively from the look up table.

The annotation at the bottom of each plot in Figure 8.3 through Figure 8.7 provides information on the plot and signal parameters:

> M.N = The programmable phase word that is M + N bits wide
> Phi = The value of the phase word in hexadecimal format
> K = The bit width of the sin look up table

**Table 8.1**   Example Samples Stored in the Look Up Table

| $n$ | $(2\pi/2^{13})n$ | $y(n)$ (decimal) | $y(n)$ (hexadecimal) | $y(n)$ (binary) |
|---|---|---|---|---|
| 0 | 0 | 0 | 0x000 | 0000 0000 0000 |
| 1024 | $\pi/4$ | 1447 | 0x5A7 | 0101 1010 0111 |
| 2048 | $2\pi/4$ | 2047 | 0x7FF | 0111 1111 1111 |
| 3072 | $3\pi/4$ | 1447 | 0x5A7 | 0101 1010 0111 |
| 4096 | $4\pi/4$ | 0 | 0x000 | 0000 0000 0000 |
| 5120 | $5\pi/4$ | −1447 | 0xA59 | 1010 0101 1001 |
| 6144 | $6\pi/4$ | −2047 | 0x801 | 1000 0000 0001 |
| 7168 | $7\pi/4$ | −1447 | 0xA59 | 1010 0101 1001 |

Fs = The sample frequency at which the phase accumulator is clocked

Fo = The value of the NCO output frequency

Po = The phase offset of the NCO output sinusoid

Len = The length of the plotted waveform

The plot of the NCO output waveform when $\Phi$ is set to a value of 2 is illustrated in Figure 8.4. We can easily see that the complete address space of the sine look up table has been accessed twice, producing two complete cycles of sine wave in the same 0.5 $m$ second time span. The NCO output frequency has doubled to 4 KHz. In this example, the output sinusoid is composed of every second sample stored in the look up table, and each cycle of the sinusoid is represented by 4096 samples.

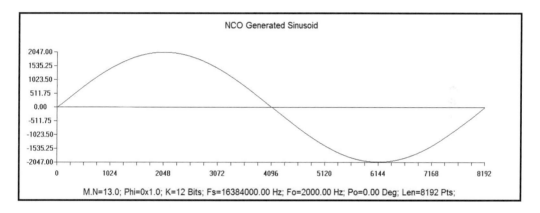

**Figure 8.3**  Plot of the NCO output with $\Phi = 1$

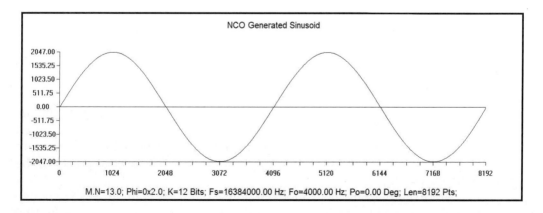

**Figure 8.4**  Plot of the NCO output with $\Phi = 2$

The plot of the NCO output waveform when $\Phi$ is set to a value of 4 is illustrated in Figure 8.5. The output frequency has increased to 8 KHz. In this example, the output sinusoid utilizes every fourth sample stored in the look up table and each cycle of the sinusoid is represented by 2048 samples.

The plot of the stored waveform when $\Phi$ is set to a value of 5 is illustrated in Figure 8.6. The output frequency has increased to 10 KHz. In this example, the output sinusoid utilizes every fifth sample stored in the look up table. There are 8192 samples in the sine look up table. The number of look up table samples is not evenly divisible by five. On average, each cycle of the sinusoid is represented by $8192/5 = 1638.4$ samples. This simply means that the set of samples accessed from the table that represents one cycle of the output waveform are not repetitive for successive cycles. That is, the sample sequence does not repeat itself for five complete cycles of the sinusoidal waveform or five complete cycles through the sine look up table. In this example, the sample sequence for a single period of the generated sinusoid does not repeat for 8192 look up table samples. The modulo $2^M$ phase accumulation and therefore the modulo $2^M$ address ramp tends to slowly walk through the table sample space over time. Each cycle of the five cycle sinusoidal waveform begins on a different phase. This is referred to later in the text as an *optimal frequency* because there is no periodicity of the table quantized sample error over the entire period of the five cycle waveform. We will see the advantage of this later when we discuss the ways to maximize the spurious free dynamic range (SFDR) of the NCO.

Finally, the plot of the stored waveform when $\Phi$ is set to a value of 32 is illustrated in Figure 8.7. The output frequency has increased to 64 KHz. In this example, the output sinusoid utilizes every 30-second sample stored in the look up table, and each cycle of the sinusoid is represented by 256 samples.

We can continue with this sequence until $\Phi$ reaches its maximum value of $2^{M-1}$ and $F_O$ reaches its maximum value of $F_S/2 = 8.192$ MHz. At this limit, a single cycle of the output waveform is represented by only $\left(2^M/2^{M-1}\right) = (16384/8192) = 2$ samples, which of course is the Nyquist limit.

The plots of the NCO output frequencies illustrated in Figure 8.3 through Figure 8.7 are useful to observe, and they certainly convey some basic engineering information, but the fact is that time domain waveforms such as these seldom convey a great deal of information regarding the actual performance of a system such as an NCO. For that kind of detailed engineering data, we will need to use the DFT to convert the time domain waveforms into the frequency domain and look at the spectral purity of the synthesized waveform.

For example, let's take a look at an NCO that is configured to generate a 34 KHz sinusoid. Specifically, we will take a look at two cases: In the first case, we will analyze the performance of a look up table that was generated

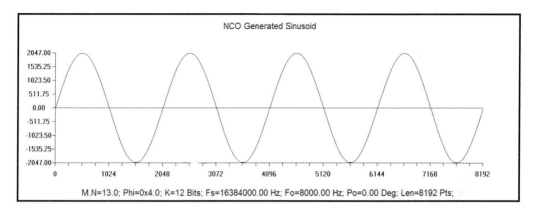

**Figure 8.5**    Plot of the NCO output with Φ = 4

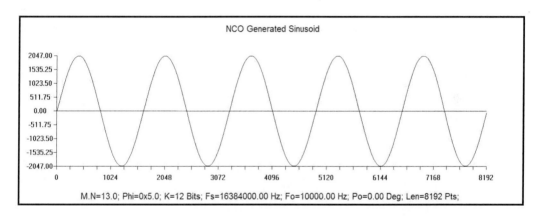

**Figure 8.6**    Plot of the NCO output with Φ = 5

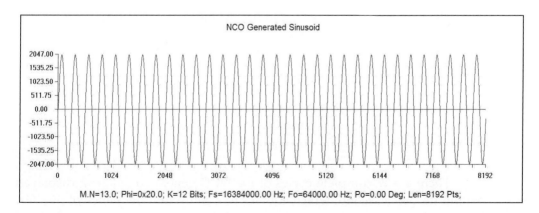

**Figure 8.7**    Plot of the NCO output with Φ = 32

by storing truncated sinusoid samples. By *truncated* we mean that when generating the sample content of the look up table, the fractional part of the sample value generated by Equation 8.7 is simply deleted, leaving only the integer part. In the second case, we will analyze the performance of the same look up table that was generated by storing rounded sinusoid samples. That is, the fractional part of the samples generated by Equation 8.7 is rounded up or down to form the nearest integer. For both cases, the time domain waveforms of the NCO sinusoid appear to be identical, but the spectral plots of the two cases show quite a different story.

### 8.1.4.1   Case I: Analysis of Truncated Look Up Table Samples

Let's take a look at the case where the samples stored in the look up table were computed in floating point arithmetic and then truncated to remove the fractional part of each sample value. The NCO parameters used in this example are listed at the bottom of the following figures and are repeated here for clarity. Two of the parameters listed at the bottom of the figure have not yet been defined and will be discussed in later sections. The parameter $M.N$ is the width of the enhanced phase word and for this example is equal to $M$. The parameter $P_O$ is the measure of phase offset and for this example is set to zero:

- Sample rate: $F_S = 4.096$ MHz
- Address width: $M = 14$
- Look up table length: $2^M = 16,384$ samples
- Look up table bit width: $K = 16$ bits
- NCO phase word: $\Phi = 0x88$ hex $= 136$ decimal
- NCO output frequency: $F_O = 34$ KHz
- Number of points in the DFT analysis: $Len = 2048$

The time domain representation of the NCO generated sinusoid is illustrated in Figure 8.8. As you can see from the figure, the sample rate is 4.096 MHz and the number of samples displayed is 2048. The time length of the $x$-axis in this plot represents

$$\frac{2048 \text{ samples}}{4.096 \times 10^6 \dfrac{\text{samples}}{\text{second}}} = 500 \ \mu \text{ second}$$

There are 17 complete cycles of the NCO generated sinusoid displayed within the 500 $\mu$ second, which equates to a sinusoidal frequency of 34 KHz. The time domain waveform looks great, but it does not provide much detail into the purity of the waveform.

The spectral representation of the sinusoid in Figure 8.8 was computed with 2048-point DFT and is shown in Figure 8.9. The frequency values on the horizontal axis are shown in their conventional format in that they are normalized to the sample rate of 4.096 MHz. For example, the point on the axis labeled 0.5000 represents half the sample rate $Fs/2$, or 2.048 MHz; the point labeled 0.2500 represents $F_s/4$, or 1.024 MHz; the point labeled 0.0625 represents $F_s/16$, or 256 KHz; and so on. For clarity, the setup parameters of the NCO were chosen so that the NCO synthesized output frequency fell into bin 17 of the 2048-bin DFT. This was done to prevent DFT spectral leakage. The leakage is an artifact of the DFT itself and isn't a part of the actual signal. Leakage makes for a degraded plot and obscures the data we are trying to investigate. To keep from generating

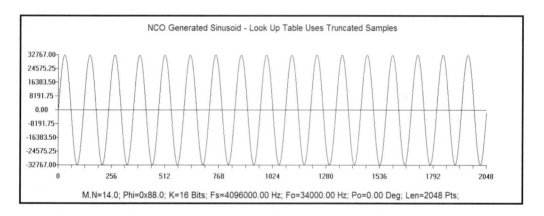

**Figure 8.8** NCO generated sinusoid using 16-bit truncated look up table

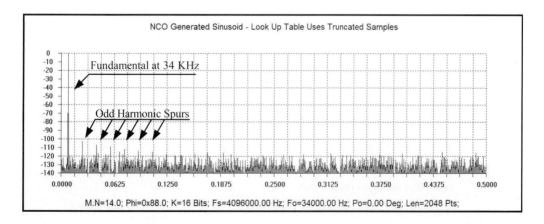

**Figure 8.9** Spectral plot of sinusoid using a 16-bit truncated look up table

this leakage noise for analysis purposes, we generate a sinusoid that falls into exactly one DFT bin. Since this test sinusoid fell squarely into a DFT bin, we did not have to preprocess the time domain data with a window prior to performing the DFT. DFT spectral leakage is discussed in depth in Chapter 3, "Review of the Fourier Transform."

If we label the NCO output signal as $y(n)$ and it's DFT as $Y(k)$, then we can plot the normalized waveform power with respect to frequency by the relation

$$P(k) = 10\text{Log}_{10}\left(\frac{|Y(k)|^2}{|Y(k)_{Max}|^2}\right)$$

where

$$|Y(k)|^2 = Y(k)_{REAL}^2 + Y(k)_{IMAG}^2$$

and where $|Y(k)_{Max}|^2$ is the maximum sample magnitude in the spectrum of $Y(k)$. Therefore the maximum value of $P(k)$ will be equal to $10\text{Log}_{10}(1) = 0$ db. All other values of $P(k)$ will be relative to the max value and will be less than 0 db.

Although it is difficult to determine from the plot, the 0 db spectral line in Figure 8.9 is located in bin 17. The spectrum of a periodic sequence consists of the fundamental frequency and its associated harmonics. It's the harmonics that we are interested in observing here. The frequency represented by this spectral line can be computed using the relation

$$F_O = n\left(\frac{F_S}{L}\right)$$

**Equation 8.9**

where

$F_O$ = the NCO frequency
$n$ = the DFT bin number occupied by the NCO frequency
$F_S$ = the sample frequency
$L$ = the number of points (bins) in the DFT

The frequency of the spectral line in Figure 8.9 is thus given by

$$F_O = 17\left(\frac{4.096 \times 10^6}{2048}\right) = 34 \text{ KHz}$$

It is comforting to know that both the time domain plot and the spectral domain plot agree that the NCO generated sinusoid frequency is equal to 34 KHz. It is interesting to observe the frequency domain plot of Figure 8.9 and see the four spectral spurs that occur, with the first one being about −103 db down from the fundamental. As listed in Table 8.2, these spurs are the third, fifth, seventh, and ninth harmonics of the fundamental frequency.

When the look up table is accessed repetitively by the periodic $M$ bit address ramp, the discontinuities in the table due to sample truncation errors repeat at a regular rate. Spurs are generated because this repetition concentrates the error energy into a single frequency. The difference measured in decibels between the peak of the fundamental and the peak of the highest spur is called the *spur free dynamic range*, or SFDR. This is an important measurement in most systems, whether they are analog or digital. In this example, the SFDR is measured as 103 db.

The SFDR in a digital system such as an NCO is proportional to the number of bits in the digital sinusoidal sample. In an ideal system, the worst case signal to noise ratio (SNR) is calculated by

$$\text{SNR} = (6.02n + 1.76)\,\text{db}$$

**Equation 8.10**

where $n$ is the sample width in bits. Equation 8.10 says that we should expect close to 6 db of SNR per bit of quantization. Using this equation, the worst case SNR for a 16-bit NCO in the previous example is $6.02(16) + 1.76 \text{ db} = 98 \text{ db}$. The problem with Equation 8.10 is that it is more applicable to the performance

**Table 8.2**  Harmonic of the 34 KHz NCO Sinusoid

| Spectral line | DFT bin | Frequency (KHz) | Log magnitude (db) |
|---|---|---|---|
| Fundamental | 17 | 34 | 0 |
| Third harmonic | 51 | 102 | −103 |
| Fifth harmonic | 85 | 170 | −107 |
| Seventh harmonic | 119 | 238 | −109 |
| Ninth harmonic | 153 | 306 | −112 |

analysis of an A/D converter, which must deal with undesirable issues such as spurious noise, clipping, and signal distortion. A digital system such as an NCO does not have spurious noise, clipping, or distortion influencing its spectral performance, so its SNR is usually much better than that predicted by Equation 8.10. We can, however, use the SNR equation to predict the worst case SNR for our NCO. That equation predicted 98-db SNR, and our simulation measured 117 db of SNR and 103 db of SFDR for this particular NCO.

It is interesting to plot the truncation error for the sinusoid in Figure 8.8. When we generate a truncated sample from a floating point sample, we are simply lopping off the fractional part of the sample. That is, if the time series $y(n)$ represents the series of floating point samples from which we derive the series of truncated samples $y_T(n)$, then the truncation error for that string of samples can be represented by

$$\varepsilon_T(n) = y(n) - y_T(n)$$

Thus the difference between the magnitude of a floating point sinusoid sample and the magnitude of the truncated sinusoid sample stored in look up table memory never exceeds unity. The range of truncation error $\varepsilon_T(n)$ is given by $-1 < \varepsilon_T < +1$. The plot of the truncation error time series for the 34 KHz waveform is illustrated in Figure 8.10. Note that while the error lies between $\pm 1$, there is a clear correlation between the positive error and the positive portion of the sinusoid waveform and between the negative error and the negative portion of the sinusoid waveform.

The plot of the truncation error clearly takes on the appearance of a noisy square wave whose frequency is equal to the original sinusoid frequency. We would expect that this truncation error would be the source of the spectral spurs we saw in Figure 8.9. To see if this is true, all we need to do is generate a spectral plot of the truncation error time series and compare the spurs in the two spectral plots. The spectrum of the truncation error is illustrated in Figure 8.11. In that plot, we can clearly see the exact same spurs that were present in the truncated sinusoidal signal spectra.

The spurs that are so evident in Figure 8.9 are the result of the truncation of the look up table samples. Let's verify this with a quick simulation. We can generate a noiseless 34 KHz square wave of amplitude $\pm 1$ that approximates the truncation square wave in Figure 8.10. The only difference between the two waveforms is the absence of noise on the square wave peaks in the simulated waveform. The simulated truncation error is illustrated in Figure 8.12. The associated spectral plot is illustrated in Figure 8.13.

On initial inspection, we can see that the spectral lines of the error waveform line up in the same frequency bins as the spurs in the spectral plot of the truncated waveform. Since the square wave is at the same frequency as the

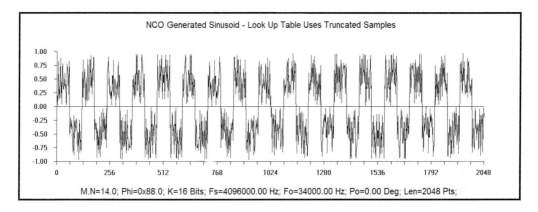

**Figure 8.10** Time domain plot of the 34 KHz sinusoid truncation error

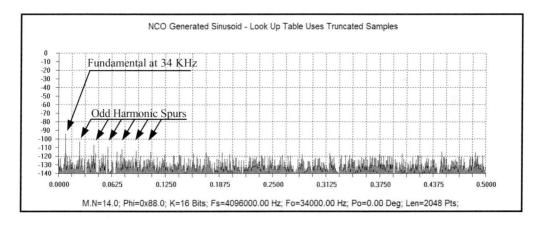

**Figure 8.11** Spectral plot of the truncation error time series

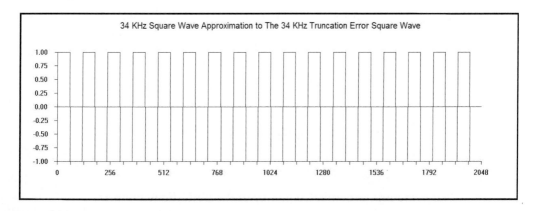

**Figure 8.12** Simulated truncation error square wave

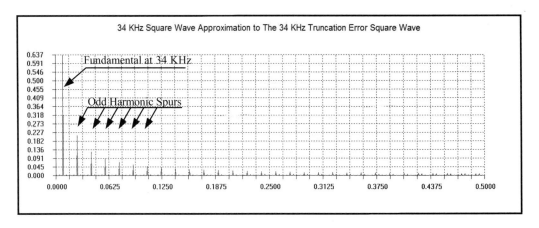

**Figure 8.13**  Simulated truncation error square wave spectrum

NCO-generated sinusoid, we would expect to see the fundamental frequency in bin 17, and we do. The other spectral lines are the odd harmonics of the square wave itself, and they are measured to be in bins 51, 85, 119, and 153, respectively. We saw in Chapter 3, "Review of the Fourier Transform," that we can decompose a periodic waveform into a DC component, the sum of its fundamental frequency, and all harmonics of that frequency, via the Fourier series. Any periodic signal $g(t)$ can be represented as a series of sinusoidal components such that

$$g(t) = a_0 + \sum_{n=1}^{\infty} a_n \cos(n\omega_0 t) + \sum_{n=1}^{\infty} b_n \sin(n\omega_0 t)$$

If we use the procedures discussed in Chapter 3 to break down the square wave in Figure 8.12 into its component frequencies, we discover that the coefficients $a_0$ and $b_n$ are all equal to zero, so we end up with

$$g(t) = \sum_{n=1}^{\infty} a_n \cos(n\omega_0 t)$$

**Equation 8.11**

where

$\omega_0$ = the fundamental frequency of the periodic waveform

$a_n$ = the amplitude of the fundamental frequency $(n = 1)$ and all harmonic frequencies $(n > 1)$

$n$ = the summation index of the fundamental and all harmonic frequencies

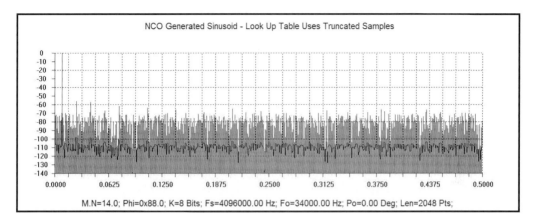

**Figure 8.14**   Spectral plot of sinusoid using 8-bit truncated look up table

Using the procedures discussed in Chapter 3, we can solve for $a_n$ to get

$$a_n = \frac{2}{\pi}\left[\frac{1}{n}\sin\left(\frac{n\pi}{2}\right)\right]$$

Substituting this expression for $a_n$ back into Equation 8.11, we arrive at

$$g(t) = \frac{2}{\pi}\sum_{n=1}^{\infty}\left[\frac{1}{n}\sin\left(\frac{n\pi}{2}\right)\right]\cos\left(n\omega_0 t\right)$$

We expand the equation for $g(t)$ to include the first five terms of the Fourier series to end up with

$$g(t) = 0.637\cos\left(\omega_0 t\right) + 0.212\cos\left(3\omega_0 t\right) + 0.127\cos\left(5\omega_0 t\right) + 0.091\cos\left(7\omega_0 t\right) + 0.070\cos\left(9\omega_0 t\right) + \cdots$$

**Equation 8.12**

Equation 8.12 clearly shows the magnitude of the fundamental and first four harmonic components of the 34 KHz square wave. Comparing these computed results with the actual magnitude of the fundamental and first four harmonics in the plot of Figure 8.13 reveals an identical match.

For comparison purposes, Figure 8.14 illustrates the effects of using smaller word width samples in the look up table. In this figure, we see the spectral plot of the same 34 KHz sine wave, but in this case the look up table sample quantization has been reduced from $K = 16$ bits to $K = 8$ bits. As you would expect, the spurs for 8-bit quantization are identical to those shown for the 16-bit quantization case. This is because the spurs are the product of the same 34 KHz truncation error square wave. Therefore they will be largely independent of the NCO bit width $K$.

Another important thing to note is the significant increase in the overall noise floor of the spectrum. For the $K = 16$ bit case we had an SFDR of 103 db, whereas the SFDR for the $K = 8$ bit case is measured to be 58 db—a significant reduction.

### 8.1.4.2   Case II: Analysis of Rounded Look Up Table Samples

In the previous paragraph, we were concerned with the performance of an NCO-generated sinusoid utilizing a look up table composed of truncated samples. Now let's turn our attention to the case where the NCO look up table is composed of rounded samples. The same 34 KHz NCO generated sinusoid is illustrated in Figure 8.15, only this time it has been generated with an NCO look up table that contains samples that have been rounded to remove the fractional part. Visual comparison between this figure and the plot of the sinusoid produced by using a truncated sample look up table shown in Figure 8.8 shows little if any difference. Both waveforms appear to be the same. A look at the spectrum for the rounded sample sinusoid, however, clearly shows a noticeable reduction in the spur levels. The spectral plot for this sinusoid is shown in Figure 8.16. For this particular example, the harmonics are equal to or below the noise floor and are not observable. The plot of the rounded sample error is illustrated in Figure 8.17. As we should expect, the magnitude of the round off error $\varepsilon_R(n)$ falls within the range given by $-0.5 < \varepsilon_R(n) \le +0.5$.

Even though the NCO generated the same 34 KHz sinusoid, there doesn't appear to be any apparent periodicity in the time domain plot of the rounding error. On first glance, one might think that the rounding error for all the samples of a sinusoidal cycle should be identical to the rounding errors for

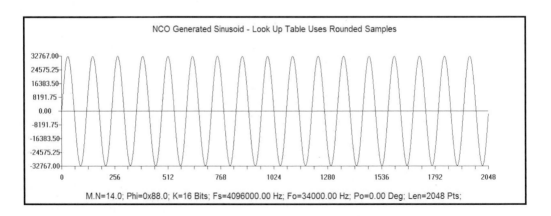

**Figure 8.15**   NCO generated sinusoid using a 16-bit rounded look up table

successive cycles. And if that is the case, then we should see some sort of periodic error waveform in the time domain. For this particular case, this is not true. Why do you suppose this is? In order to generate a 34 KHz sinusoid, the NCO phase word $\Phi$ for this example was set to 136. The look up table contains 16,384 samples. $\Phi$ is not an integer submultiple of 16,384:

$$\frac{16,384}{136} = 120.470588235$$

The NCO phase accumulator will continuously sum the value 136 modulo 16,384 to produce a look up table address. Since $\Phi$ is not an integer

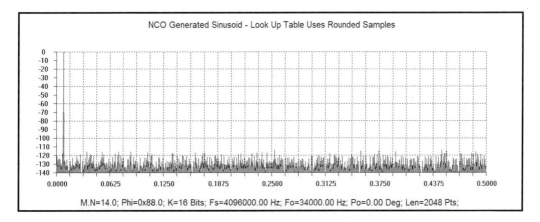

**Figure 8.16** Spectral plot of sinusoid using 16-bit rounded look up table samples

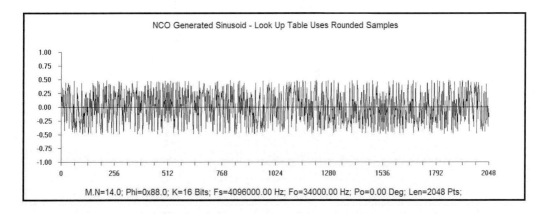

**Figure 8.17** Time domain plot of the 34 KHz sinusoid rounding error

submultiple of the look up table length, the address will tend to walk through the look up table. Therefore a sequence of successive cycles of the NCO output sinusoid will begin on a different phase from the preceding cycle. When this happens, the nonlinearities in the look up table waveform will not be periodic for that sequence. As the reader might expect, there are different sequence lengths for different frequencies. That is, there are different sequence lengths for different values of $\Phi$. We will discuss this in greater detail in the analysis tips that follow.

Since the sequence of samples that make up each cycle of the displayed sinusoid waveform are different, the sequence of rounding errors for each successive cycle will be different as well. Therefore we see no periodicity of error and the DFT doesn't either. The spectrum of the rounding error waveform is illustrated in Figure 8.18. For the case where the look up table was composed of truncated samples, the same argument applies, but in that case we cannot escape the fact that the truncation error produced a 34 KHz square wave that added its own Fourier components to the signal.

### 8.1.4.3  Tips for Generating NCO Test Sinusoids for DFT Analysis

**Analysis tip 1**. Many times it's convenient when characterizing the performance of systems like an NCO to generate tones that fall directly within a DFT frequency bin. What this allows us to do is to generate a tone that has an integer number of cycles within the number of points processed by the DFT. By doing this, we eliminate the time domain discontinuity in the waveform at the DFT boundaries and therefore eliminate the associated spectral leakage that these discontinuities cause. This allows us to analyze the output of the system without having to deal with all the spectral clutter that can occur due to these DFT artifacts.

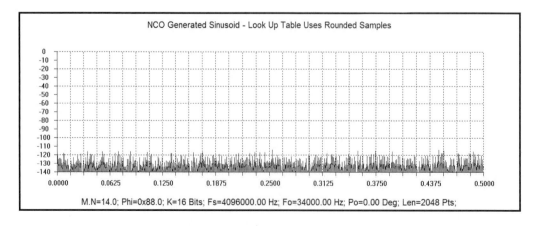

**Figure 8.18**  Spectral plot of the rounded error time series

Setting the NCO phase word $\Phi$ to produce exact DFT bin frequencies is straightforward. We know that the NCO output frequency is given by Equation 8.1:

$$F_0 = \Phi\left(\frac{F_S}{2^M}\right)$$

We can identify the frequency corresponding to any DFT bin through the use of Equation 8.9:

$$F_0 = n\left(\frac{F_S}{L}\right)$$

We can set these two equations equal to one another to produce the relationship:

$$F_0 = n\left(\frac{F_S}{L}\right) = \Phi\left(\frac{F_S}{2^M}\right)$$

We can then solve for $\Phi$ by rearranging terms to produce

$$\Phi = n\left(\frac{2^M}{L}\right)$$

**Equation 8.13**

where

$\Phi$ = the NCO phase word
$n$ = the desired DFT bin number of the NCO generated sinusoid
$M$ = the power of two size of the look up table memory
$L$ = the number of points in the DFT

Equation 8.13 allows us to determine the correct value for $\Phi$ to cause the NCO to produce a sinusoid waveform that will fall within the $n$th DFT bin. This equation is extremely useful for generating sinusoids with a spectral representation that is not cluttered by DFT artifacts.

**Analysis tip 2.** When it is desirable to generate test frequencies with minimal spurs and DFT artifacts, one should consider using NCO frequencies that are referred to as *optimum frequencies* or *sweet frequencies* [3]. If the sampling frequency is $F_s$ and the length of the DFT is $L$, then the frequency resolution of the DFT is given by $F_s/L$. From Equation 8.9 we know that the NCO frequency represented by the $n$th bin of the DFT is given by $F_0 = n(F_s/L)$. If the length of the DFT $L$ is a power of two and if the value of $n$ is odd, representing an odd DFT bin, then the resulting frequency $F_0$ is referred to as an *optimum frequency*. The resulting NCO phase word $\Phi$ will be such that it will cause successive cycles of the sinusoid generated by the NCO to begin on a different phase across the span of samples equal to the length of the DFT. For these cases, the sine wave sequence out of the NCO will repeat itself every $L$ points, which happens to be the length of the DFT. From a DFT perspective, this eliminates the periodicity of look up table quantization discontinuities and therefore lowers the visibility of the spurs inherent in the table design. These optimum DFT frequencies are the result of maximal length sample sequences.

Let's look at an example. Suppose we have an NCO whose sample clock rate is $F_s = 1.024$ MHz, with a look up table that is $2^M = 2^{14} = 16,384$ samples in length. Further suppose that we wished to analyze the output of the NCO using an $L = 2048$ point DFT. We can verify the maximal length sequences output by the NCO by sequentially generating a sinusoid for each of the DFT bins from 0 to 1024, which would correspond to NCO frequencies ranging from 0 Hz to $F_s/2$ Hz. We can accomplish this by computing the phase words for such an analysis by using Equation 8.13, which is repeated here for clarity:

$$\Phi = n\left(\frac{2^M}{L}\right)$$

We would compute the phase words for each of the 1024 sinusoids by incrementing the value of $n$ by 1, through a range of

$$0 \leq n \leq \frac{L}{2} \text{ or } 0 \leq n \leq \frac{2048}{2}$$

This would fix the value of the NCO phase word $\Phi$ to be within the range of

$$0 \leq \Phi \leq 2^{M-1} \text{ or } 0 \leq \Phi \leq 2^{13} \text{ or } 0 \leq \Phi \leq 8192$$

which in turn would cause the NCO to generate output frequencies within the range of

$$0 \leq F_O \leq \frac{F_s}{2} \text{ or } 0 \leq F_O \leq 5.12 \text{ MHz}$$

For each sinusoidal sequence output by the NCO, we could count the number of samples per each sequence repetition. For the case outlined here, we would find 512 maximal length sequences, each of which is 2048 samples in length. These sequences would correspond to the odd bin frequencies of the DFT (i.e., bins $n = 1, 3, 5, 7, \cdots 1023$). We would find repetitive sequences of varying lengths for the even bin frequencies, none of which would be the maximal length of $L = 2048$.

As an example, the rounding error sequence plot of Figure 8.17 is repeated in Figure 8.19. This plot was for a 34 KHz sinusoid that fell into DFT bin number 17, an odd bin. This sequence is periodic in 2048 samples. The rounding error sequence of Figure 8.20 shows a plot for a 36 KHz sinusoid that fell into DFT bin number 18, an even bin. A quick inspection of this figure will reveal that there are two periods of error sequence, each 1024 samples in length, and therefore this is not a maximal length sequence.

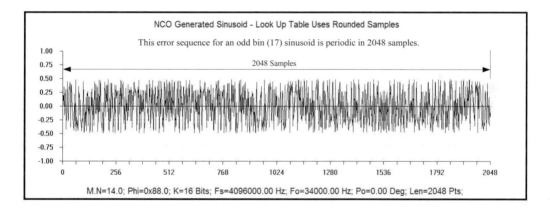

**Figure 8.19** Rounding error sequence for an odd bin sinusoid ($n = 17$)

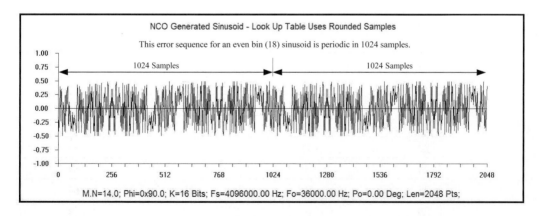

**Figure 8.20** Rounding error sequence for an even bin sinusoid ($n = 18$)

As a short example, let us look at the NCO output sequence lengths versus bin numbers for the case where the NCO look up table length $2^M$ is 128 samples and the DFT length $L$ is 64 samples. The sequence lengths for all the bins are tabulated in Table 8.3.

You will note that the sequence length for all the odd bins is maximal and equal to 64, the same length as the DFT. The sequence lengths for all the even bins vary in length but never equal the maximal length. It is interesting to note that the sequence length for bin 0 is 1. This is true because bin 0 is DC and its sequence repeats itself every sample.

A second example is shown in Table 8.4 for the case where the NCO look up table length $2^M$ is 16,384 samples and the DFT length $L$ is 128 samples. The DFT length was kept short only to produce a smaller table. As you can clearly see, every odd bin corresponds to a maximal length sequence of 128 samples while the even bins correspond to a varying length sequence, none of which is maximal.

It is clear from these two tables that the optimum or sweet frequencies for DFT analysis are the ones that occupy the odd bins. If you should ever be generating a data sheet to tout your product, and if that data sheet includes a

**Table 8.3**   Example NCO Sequence Length $M = 7$, $L = 64$

| Even bin number | Sequence length | Odd bin number | Sequence length |
| :---: | :---: | :---: | :---: |
| 0 | 1 | 1 | 64 |
| 2 | 32 | 3 | 64 |
| 4 | 16 | 5 | 64 |
| 6 | 32 | 7 | 64 |
| 8 | 8 | 9 | 64 |
| 10 | 32 | 11 | 64 |
| 12 | 16 | 13 | 64 |
| 14 | 32 | 15 | 64 |
| 16 | 4 | 17 | 64 |
| 18 | 32 | 19 | 64 |
| 20 | 16 | 21 | 64 |
| 22 | 32 | 23 | 64 |
| 24 | 8 | 25 | 64 |
| 26 | 32 | 27 | 64 |
| 28 | 16 | 29 | 64 |
| 30 | 32 | 31 | 64 |

**Table 8.4**   Example NCO Sequence Length $M = 14$, $L = 128$

| Even bin number | Sequence length | Odd bin number | Sequence length |
|:---:|:---:|:---:|:---:|
| 0 | 1 | 1 | 128 |
| 2 | 64 | 3 | 128 |
| 4 | 32 | 5 | 128 |
| 6 | 64 | 7 | 128 |
| 8 | 16 | 9 | 128 |
| 10 | 64 | 11 | 128 |
| 12 | 32 | 13 | 128 |
| 14 | 64 | 15 | 128 |
| 16 | 8 | 17 | 128 |
| 18 | 64 | 19 | 128 |
| 20 | 32 | 21 | 128 |
| 22 | 64 | 23 | 128 |
| 24 | 16 | 25 | 128 |
| 26 | 64 | 27 | 128 |
| 28 | 32 | 29 | 128 |
| 30 | 64 | 31 | 128 |
| 32 | 4 | 33 | 128 |
| 34 | 64 | 35 | 128 |
| 36 | 32 | 37 | 128 |
| 38 | 64 | 39 | 128 |
| 40 | 16 | 41 | 128 |
| 42 | 64 | 43 | 128 |
| 44 | 32 | 45 | 128 |
| 46 | 64 | 47 | 128 |
| 48 | 8 | 49 | 128 |
| 50 | 64 | 51 | 128 |
| 52 | 32 | 53 | 128 |
| 54 | 64 | 55 | 128 |
| 56 | 16 | 57 | 128 |
| 58 | 64 | 59 | 128 |
| 60 | 32 | 61 | 128 |
| 62 | 64 | 63 | 128 |

spectral plot to demonstrate system SNR or SFDR, be sure to maximize the perceived performance of your system by using an odd bin sinusoid. On a separate note, if you happen to be reading a data sheet, pay attention to the frequencies the manufacturer chose to demonstrate system SFDR and SNR performance.

## 8.2  ENHANCED NCO PHASE ACCUMULATOR

In the previous sections, we studied the basic architecture of an NCO. In this basic form, the phase register was $M$ bits wide and contained an integer $\Phi$, which could take on values from 0 to $2^{M-1}$. Remember that the synthesized frequency produced by the NCO was given by Equation 8.4 and is repeated here in Equation 8.14 for convenience:

$$F_0 = \Phi\left(\frac{F_S}{2^M}\right) \text{Hz}$$

**Equation 8.14**

where

> $F_0$ is the NCO output frequency,
> $2^M$ is the size of the memory used for the sine look up table,
> $F_S$ is the accumulator sample clock rate, and
> $\Phi$ is the phase word that is repetitively accumulated at the sample rate.

Also remember that the fundamental frequency out of the basic NCO, the lowest possible frequency other than DC, was achieved when the value of $\Phi$ was set to unity. All other possible NCO output frequencies were simple integer multiples of the fundamental frequency and that integer multiplier was $\Phi$.

Limiting $\Phi$ to be an integer value seriously limits the frequency resolution of the NCO. From an operational point of view, this coarse tuning limitation would prevent the NCO from performing more exact processing functions, like tuning a band pass signal to base band within some small specified number of hertz accuracy.

One could increase the NCO frequency resolution by making the sine look up table larger, which would lower the fundamental frequency. An increase in memory size, however, could rapidly become unwieldy, since doubling the look up table size would only halve the fundamental frequency, which doesn't provide much return on memory investment. In addition, larger memory brings with it several design problems that need to be addressed such as additional power consumption, slower speed, and increased real estate.

A simpler and much more efficient method to increase the NCO frequency resolution without increasing the size of the look up table is to change the format of $\Phi$ from an integer representation to a fractional representation. The term *fractional representation* only means that we will be adding an imaginary binary point in the phase word. Instead of the phase word being $M$ bits wide, it will be $M + N$ bits wide where $M$ represents the integer part of $\Phi$ and $N$ represents the fractional part of $\Phi$. The $M$ bits are the $M$ most significant bits of the phase word $\Phi$. They are the bits located to the left of the binary point and are the same bits we have been talking about in the previous sections. The $N$ bits are the least significant bits (LSBs) of the phase word $\Phi$ and are located to the right of the binary point. The notation for this format is $M.N$, where there is a binary point located between $M$ and $N$. The format of the fractional representation for $\Phi$ is illustrated in Figure 8.21.

It is important to note that for this enhanced version of an NCO, the accumulator sums all $M + N$ bits, but as illustrated in Figure 8.22, only the $M$ most significant bits out of the accumulator are utilized as the address ramp to the look up table. Think about it for a minute. If $\Phi$ is formatted as $M.N$, where $M = 5$ bits and $N = 1$ bit, then $M.N = 5.1$, and the smallest binary number that can be represented in this format other than zero is $M.N = 00000.1$. In this case, the single bit in the $N$ bit field takes on the value of $2^{-1} = 1/2$. So it will take two periods of the sample clock before the accumulation carries from the most significant bit of the $N$ field into the least significant bit of the 5-bit $M$ field that represents the integer portion of $\Phi$. That is, it takes two sample clocks of accumulation for the $2^{-1}$ bit to increment the $2^0$ bit in the phase word. Since only the $M$ bit field of the accumulation is used as the look up table address ramp, it will take two sample clocks to increment the table address by 1. For this case, $\Phi_{Min} = 1/2$ and the NCO fundamental output frequency is half that for the case where the phase word only consisted of $M$ bits and had a minimum $\Phi$ of unity. This change is reflected

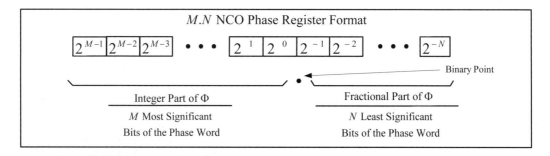

**Figure 8.21** Fractional format representation

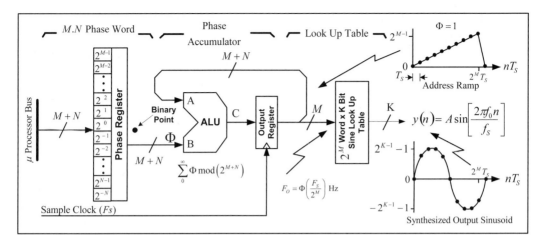

**Figure 8.22**   Block diagram of an enhanced NCO

in Figure 8.22. The initial sequence for the first eight sample clocks of the accumulator output shown in Figure 8.22 is listed in Table 8.5 for the case where $M.N = 5.1$ and the phase word $\Phi$ is set to 00000.1. It is clear from this table that the address ramp increments every second accumulation cycle.

Using the fractional definition for $\Phi$, we can define its range to be

$$2^{-N} \leq \Phi \leq 2^{M-1}$$

We can substitute the range for $\Phi$ into Equation 8.1, repeated here for clarity to compute the frequency range of the NCO:

$$F_O = \Phi\left(\frac{F_S}{2^M}\right)$$

$$2^{-N}\left(\frac{F_S}{2^M}\right) \leq F_O \leq 2^{M-1}\left(\frac{F_S}{2^M}\right)$$

or

$$\left(\frac{F_S}{2^{M+N}}\right) \leq F_O \leq \left(\frac{F_S}{2}\right)$$

Let's look at a series of simple examples to illustrate the concept of a fractional phase word. The example NCOs shown in Figure 8.23 are intended

**Table 8.5**   Fractional Phase Accumulator Output Sequence

| Sample clock cycle | Accumulator output $(M.N = 5.1)$ | Look up table address ramp $(M = 5)$ |
|:---:|:---:|:---:|
| 1 | 00000.0 | 00000 |
| 2 | 00000.1 | 00000 |
| 3 | 00001.0 | 00001 |
| 4 | 00001.1 | 00001 |
| 5 | 00010.0 | 00010 |
| 6 | 00010.1 | 00010 |
| 7 | 00011.0 | 00011 |
| 8 | 00011.1 | 00011 |

to illustrate to the reader the implementation of and the benefits of using a fractional phase word. For all the examples, imagine a base NCO architecture that operates with a sample clock whose frequency is $F_S = 1024$ Hz and has a 32-word by 8-bit look up table. Since the look up table is 32 words in length, the value of $M$ is fixed at 5. The architecture for all these examples is illustrated in Figure 8.23 A through D.

The first example shown in Figure 8.23, part A, shows the phase word of the base NCO is formatted as $M.N = 5.0$. The minimum value of the phase word is unity. The fundamental frequency for this NCO where $\Phi_{Min} = 1$ is determined by

$$F_O = \Phi\left(\frac{F_S}{2^M}\right) = (1)\left(\frac{F_S}{2^5}\right) = (1)\left(\frac{1024}{32}\right) = 32 \text{ Hz}$$

The second example shown in Figure 8.23, part B, shows the same base NCO but with an added bit to the phase word such that $\Phi$ is formatted as $M.N = 5.1$. Therefore $\Phi_{Min} = 2^{-1} = 1/2$ and the fundamental frequency for this NCO is given by

$$F_O = \Phi\left(\frac{F_S}{2^M}\right) = \left(\frac{1}{2}\right)\left(\frac{F_S}{2^5}\right) = \left(\frac{1}{2}\right)\left(\frac{1024}{32}\right) = 16 \text{ Hz}$$

It is interesting to note that we have increased the NCO frequency resolution by a factor of two simply by adding a single bit to the width of the

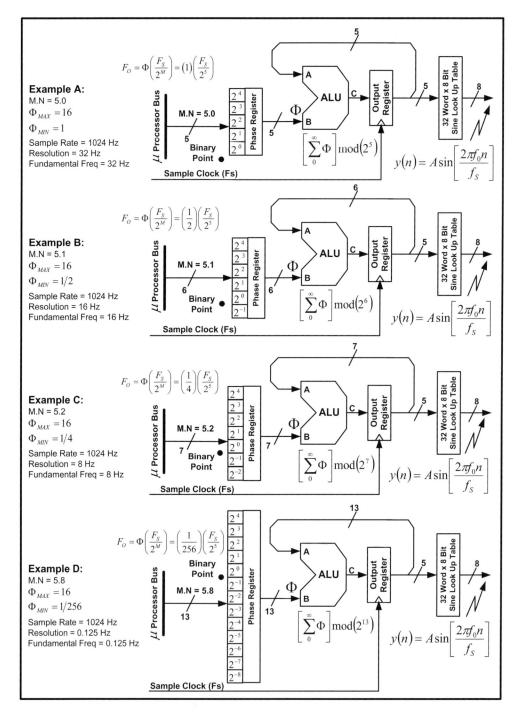

**Figure 8.23**   Example fractional point phase accumulator NCOs

phase accumulator, and we did not have to change the size of the sinusoidal look up table.

For the same base NCO, if we add two bits to the phase word such that $\Phi$ is now formatted as $M.N = 5.2$, then $\Phi_{Min} = 2^{-2} = 1/4$ and the fundamental frequency for this NCO illustrated in Figure 8.23, part C, is given by

$$F_O = \Phi\left(\frac{F_S}{2^M}\right) = \left(\frac{1}{4}\right)\left(\frac{F_S}{2^5}\right) = \left(\frac{1}{4}\right)\left(\frac{1024}{32}\right) = 8 \text{ Hz}$$

As can be seen, by adding two bits to the phase accumulator, we have quadrupled the frequency resolution of this simple NCO.

In the last example, for the same base NCO we increase the size of the phase word to 13 bits and format it as $M.N = 5.8$. In this case, $\Phi_{Min} = 2^{-8} = 1/256$ and the fundamental frequency for this NCO illustrated in Figure 8.23, part D, is given by

$$F_O = \Phi\left(\frac{F_S}{2^M}\right) = \left(\frac{1}{256}\right)\left(\frac{F_S}{2^5}\right) = \left(\frac{1}{256}\right)\left(\frac{1024}{32}\right) = 0.125 \text{ Hz}$$

Now we begin to see some exciting results. Our simple NCO with a $M.N = 5.8$ phase word now has a subhertz frequency resolution of 0.125 Hz. Think what this means. If we design this NCO carefully, taking into account and minimizing the effects of look up table sample quantization and minimizing spectral spurs and harmonics, and if we carefully choose other prime components of our NCO, like the sample clock oscillator, then we will be able to realize a pretty good NCO that has the capability to tune in 1/8th Hz steps. There is no reason why the frequency resolution for this NCO can't be even greater. The rule of thumb here is the greater the frequency resolution of the NCO, the greater care that must be taken in the design of the NCO.

## 8.3 NCO SYNTHESIZED OUTPUT FREQUENCY ERROR

The importance of the NCO sample clock oscillator was mentioned previously as being critical to the overall performance of the NCO. As a matter of fact, the choice of a digital oscillator to serve as the NCO clock source is probably the single most important part of an NCO design. Why do you suppose this is true? The answer is that the stability and accuracy of the clock source determines the stability and accuracy of the frequency synthesized by the NCO. We will see why this is the case in this section.

We have shown that the ideal NCO synthesized output frequency is given by

$$F_O = \Phi\left(\frac{F_S}{2^M}\right) \quad \left\{\text{for } 2^{-N} \leq \Phi \leq 2^{M-1}\right\}$$

**Equation 8.15**

where

$F_O$ is the NCO output frequency,
$2^M$ is the size of the memory used for the sine look up table,
$M$ is the width of the integer part of the NCO phase word,
$N$ is the width of the fractional part of the NCO phase word,
$F_S$ is the accumulator sample clock rate, and
$\Phi$ is the $M.N$ phase word that is repetitively accumulated at the sample rate.

Equation 8.15 is truly accurate only for the unique case where a perfect sample clock oscillator is utilized to generate the sample frequency $F_S$. In practice, no oscillator is perfect. All oscillators have a specified range of frequency drift due to variations in temperature and component aging. When you purchase an oscillator, you will be getting a device that outputs a clock frequency that is specified to be within some finite band of frequencies that encapsulate the desired center frequency. That band is usually specified in terms of parts per million (ppm). A manufacturer will specify that their device is guaranteed to output a square wave with a specified duty cycle at some center frequency $F_S \pm (xx)\text{ppm}$ over a specified temperature range for a specified length of time. The designer should also be well aware of the duty cycle specification. If your design, for whatever reason, uses both edges of the system clock, then it is important to ensure that the clock has a 50% duty cycle. When possible, it is always a good idea to purchase an oscillator at twice the desired frequency and use it to clock a divide by two D flip-flop. The flip-flop output will be a 50% duty cycle square wave and can be used as the system clock.

There are many manufacturers that sell clock oscillators. The performance specifications for these devices can vary over a wide range depending on the quality of the device. Important considerations in choosing the right oscillator certainly include price, availability, and lead time, but the designer should pay special attention to the NCO application in which the clock oscillator will be used. In a great many applications, the synthesized frequency specifications will be constrained enough to warrant a significant amount of design time spent on vender and part selection. In general, the tighter the specification, the more expensive the oscillator, and the greater the lead time between the order and receipt of the device.

Many oscillators, especially the ones with the tighter specifications, are not off-the-shelf purchases and must be manufactured, tested, and burned in prior to delivery. This of course takes time. Delivery of these devices often takes several months after receipt of order, so it is essential that the designer know and understand the performance specifications of the clock oscillator and how these specifications will affect the overall performance of the NCO. You definitely do not want to discover during the test phase of your design that the clock oscillator you selected is degrading your NCO's performance. By then, it's probably too late to order a higher performance device without having to add months to the program schedule. Imagine the effect that would have on your performance review!

In almost all but the simplest of cases, the designer must take into account the error in the NCO synthesized output frequency. This is especially true if the synthesizer is being used to generate a precise frequency such as a reference tone or if it's being used to tune an input signal or serving as a stable carrier in a frequency modulation (FM) or phase shift keyed (PSK) application. If, for example, the synthesized frequency output by the NCO is offset by several hundred hertz, then the reference tone will be in error, the tuned signal will suffer an unwanted frequency offset, and the modulated signal application will suffer from carrier frequency error.

There are several sources of error to be accounted for when generating a synthesized frequency. By far the biggest source of error is the stability of the sample clock frequency $F_S$. Remember that the sample clock controls the rate at which the phase accumulator produces its output address ramp and the frequency of that address ramp fixes the sinusoidal frequency out of the sine look up table. For the same phase word $\Phi$, a sample clock that is slower than the specified $F_S$ will produce a lower NCO output frequency, while a sample clock that is faster than the specified $F_S$ will produce a higher NCO output frequency. This means the clock oscillator that generates the sample clock used to clock the synthesizer phase accumulator is indeed a pivotal piece of the overall synthesizer design. A loosely specified sample clock oscillator could wreak havoc on your design.

Let's revisit the waveform of an ideal phase accumulator address ramp, as illustrated in Figure 8.24, for the simple case where the phase word $\Phi$ is set to $\Phi_{Min}$ and the output frequency $F_O$ is the NCO fundamental frequency. In this example, the NCO accumulator is operating with a perfect sample clock oscillator that produces a sample clock whose frequency is $F_S \pm 0$ ppm and has zero drift over time and temperature. As we can see, the phase accumulator continuously adds the phase word to the accumulated sum at the sample clock rate $F_S$ and produces a periodic output whose numeric amplitude resembles a saw tooth waveform. When this saw tooth is applied as the ramp address to the sine look up table, the table outputs a sinusoid whose period is identical to the period of the saw tooth address. The resulting frequency of the synthesized sinusoid is identical to the frequency of the accumulator saw tooth output.

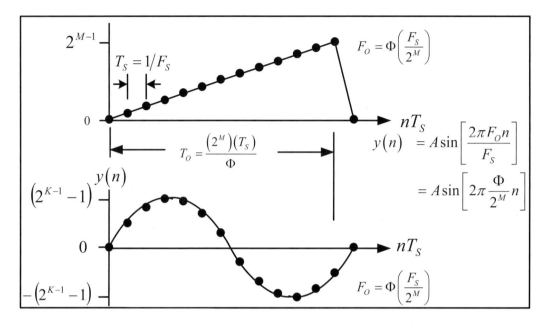

**Figure 8.24**    Ideal phase accumulator address ramp

Now let's take a look at an exaggerated waveform of a phase accumulator address ramp, as illustrated in Figure 8.25 for the simple case where the phase word $\Phi$ is set to $\Phi_{Min}$. In this example, the sample clock oscillator frequency is specified as $F_S \pm (xx)$ ppm. The sample clock frequency will lie somewhere within a frequency band of $F_S \pm F_\Delta$, where $F_\Delta$ is the (xx) ppm frequency error. It is very important to note that the term $F_S \pm F_\Delta$ is not intended to represent two separate frequencies that are added together but rather two numbers that are summed together to represent a single frequency. The two interpretations are quite different, and the former interpretation is erroneous for this discussion.

It is clear that a phase accumulator utilizing the clock generated by this nonideal oscillator will produce an address ramp whose period will lie somewhere within the range

$$\left(\frac{2^M}{\Phi}\right)(T_{S-}) \le T_{Ramp} \le \left(\frac{2^M}{\Phi}\right)(T_{S+})$$

where the term $T_S$ represents the ideal sample clock period, the term $T_{S+}$ represents a lengthened sample clock period due to a slower sample clock, and the term $T_{S-}$ represents an shortened sample clock period due to a faster sample clock.

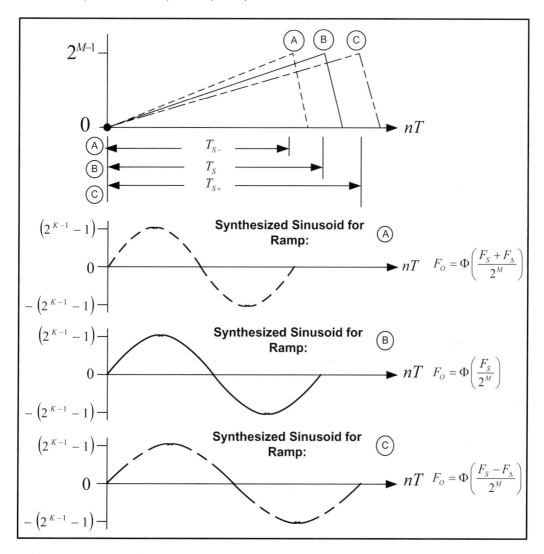

**Figure 8.25**   Actual phase accumulator address ramp due to sample clock offset

The synthesized sinusoid produced by the NCO will experience the same range of periods as the accumulator ramp. Figure 8.25 illustrates the ramp address periods and subsequent NCO output sinusoid periods for the case where the sample clock frequency is ideal and for the cases where the sample clock frequency equals its maximum drift limits.

The ideal situation occurs when the frequency of the clock oscillator is exactly equal to the specified frequency—that is, $F_s \pm 0$ ppm. This case is illustrated by waveform B in the figure. This situation rarely happens, so we have to make sure the design is robust enough to handle the extreme limit cases, as illustrated by waveforms A and C. Waveform A represents the case where the frequency of the clock oscillator is higher than the ideal specified frequency by some amount $+F_\Delta$. Waveform C represents the case where the frequency of the clock oscillator is lower than the ideal specified frequency by some amount $-F_\Delta$. We can see the effect that a clock oscillator can have on the NCO output sinusoid.

Now let's compute the NCO frequency offset error that can be expected as a result of a frequency offset in the sample clock $F_S$. We will model the frequency of the sample clock as a sum of two components, consisting of the ideal clock frequency $F_S$ and the maximum possible offset to the clock frequency $\pm F_\Delta$. This offset is specified by the manufacturer as the *sample clock oscillator drift*. Drift is mainly due to component aging over time and changes in operational temperature. The resulting NCO output synthesized frequency due to these two components is designated by $\widehat{F_0}$.

Substituting $(F_S \pm F_\Delta)$ for $F_S$ in Equation 8.15 produces Equation 8.16, which takes into account the contribution of the sample clock offset:

$$\widehat{F_0} = \Phi\left(\frac{F_S \pm F_\Delta}{2^M}\right) = \Phi\left(\frac{F_S}{2^M}\right) \pm \Phi\left(\frac{F_\Delta}{2^M}\right)$$

**Equation 8.16**

The second term in Equation 8.16

$$\Phi\left(\frac{F_\Delta}{2^M}\right)$$

represents the maximum NCO synthesized frequency offset error due to the maximum oscillator drift specified by the manufacturer. Rearranging terms in Equation 8.15, we can solve for $\Phi$, which gives us

$$\Phi = F_0\left(\frac{2^M}{F_S}\right)$$

**Equation 8.17**

Substituting Equation 8.17 into Equation 8.16 and eliminating terms, we see that

$$\hat{F}_0 = F_0 \left( \frac{2^M}{F_S} \right) \left( \frac{F_S}{2^M} \right) \pm F_0 \left( \frac{2^M}{F_S} \right) \left( \frac{F_\Delta}{2^M} \right) = F_0 \pm F_0 \left( \frac{F_\Delta}{F_S} \right)$$

or

$$\hat{F}_0 = F_0 \left( 1 \pm \frac{F_\Delta}{F_S} \right)$$

**Equation 8.18**

Equation 8.18 clearly shows that the resultant NCO output frequency consists of the sum of the ideal frequency $F_0$ and an error frequency that is proportional to $F_0$. The error term is the ratio of the maximum drift frequency range $F_\Delta$ to the ideal sample frequency $F_S$. This error ratio is a fraction that is multiplied by the ideal NCO output frequency $F_0$. Therefore we can see that as the NCO output frequency $F_0$ increases, the bandwidth of possible frequency error also increases.

We have seen from previous discussion that the NCO synthesized output frequency $F_0$ can range from 0 Hz to half the sample frequency so its limits are clearly defined by

$$0 \leq F_0 \leq \frac{F_S}{2}$$

**Equation 8.19**

By substituting the end points of the output frequency range, as defined by Equation 8.19 for $F_0$ into Equation 8.18, we can determine the bounds on the error bandwidth. When the NCO is programmed to produce an output frequency of 0 Hz, then $F_0$ is zero and the output error due to the sample clock oscillator is zero as well. This is because

$$\hat{F}_0 = F_0 \left( 1 \pm \frac{F_\Delta}{F_S} \right) = 0 \left( 1 \pm \frac{F_\Delta}{F_S} \right) = 0 \text{ Hz}$$

When the NCO is programmed to produce its highest possible output frequency of $F_S/2$ Hz, then the value of $\hat{F}_0$ is given by

$$\hat{F}_0 = F_0 \left( 1 \pm \frac{F_\Delta}{F_S} \right) \quad = \quad \frac{F_S}{2} \left( 1 \pm \frac{F_\Delta}{F_S} \right) \quad = \quad \frac{F_S}{2} \pm \left( \frac{F_\Delta}{2} \right)$$

**Equation 8.20**

So we see that the bandwidth of possible synthesizer output frequency error is zero when the NCO output is programmed to be 0 Hz, and it increases linearly to a maximum of $\pm F_\Delta/2$ Hz when the synthesizer output is set to its maximum frequency of $F_S/2$ Hz. Maximum values for all the offset frequency errors that lie between these two limit points can be found simply by substituting the desired value of $F_0$ into Equation 8.18. For example, when the NCO phase word is chosen to produce an output frequency $F_0$ equal to $F_S/4$ Hz, then the maximum expected output frequency error will be less than or equal to $\pm F_\Delta/4$ Hz.

The NCO frequency error bounds are illustrated in Figure 8.26. In the diagram, the ideal NCO output frequency is plotted along the horizontal axis. The maximum amount of frequency error is plotted relative to the vertical axis. The error of the NCO output frequency will lie somewhere in the hatched area of the triangle. The NCO error when using an ideal oscillator that exhibits absolutely no frequency offset would reside somewhere along the horizontal frequency axis of the error plot, depending on the value of $F_0$. The NCO error when using a real clock oscillator will reside somewhere within the hatched area or, in a worst case scenario, on the bounding triangle lines, depending on the value of $F_0$ and the oscillator drift. If the value for the ideal NCO output frequency $F_0$ is plotted as a dot on the horizontal axis, the frequency error will lie somewhere on a vertical line that passes through that dot.

This doesn't mean that operationally you will ever see the maximum possible frequency error. Equation 8.18 simply provides the bounds for the error versus the ideal output frequency. This is important because it tells us that the error in the NCO output frequency is bounded and it varies linearly from 0 Hz to a maximum possible $\pm F_\Delta/2$ Hz as the frequency of the

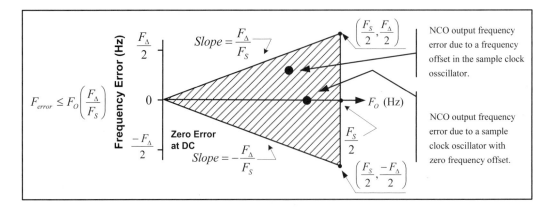

**Figure 8.26**   NCO output frequency error bounds

synthesized output is tuned from 0 Hz to $F_s/2$ Hz (i.e., the maximum error limit increases as the synthesizer output frequency increases). Let's stop here and take a look at two examples.

**Example 1**. Suppose you have an application where the sample clock $F_s$ is fixed at 16.384 MHz. You purchase a clock oscillator from an established vendor that is specified to have a clock frequency of 16.384 MHz with a maximum clock drift of $\pm 25$ ppm over a temperature range of –40°C to 85°C. What should you expect your output synthesized frequency error bound to be? You can compute the drift frequency range as

$$F_{\Delta} = \pm\left(\frac{25}{10^6}\right)\left(16.384 \times 10^6\right) = \pm 409.6 \text{ Hz}$$

So you can see that the clock frequency, as specified by the manufacturer, has a maximum potential drift across the –40°C to 85°C temperature range of $\pm 409.6$ Hz about the desired center frequency. To compute the effect this has on the error of your NCO synthesized output frequency, simply substitute this maximum drift number into Equation 8.20 to get

$$\hat{F}_0 = \frac{F_s}{2} \pm \left(\frac{F_{\Delta}}{2}\right)$$

$$\hat{F}_0 = \frac{16.384 \times 10^6 \text{ Hz}}{2} \pm \left(\frac{409.6 \text{ Hz}}{2}\right) = 8.192 \text{ MHz} \pm 204.8 \text{ Hz}$$

Now you know that when your NCO is programmed to produce its maximum output frequency of 8.192 MHz, the maximum frequency error your design should ever see is $\pm 204.8$ Hz. You can then point out with authority at your design review that the absolute maximum limits of synthesized frequency error will be 0 Hz at DC and will increase linearly to a maximum possible offset due to the potential drift of the sample clock oscillator of $\pm 204.8$ Hz when the NCO synthesizer is set to output is highest frequency of $F_s/2 = 8.192$ MHz. If this maximum potential frequency error will not work for your application, then you will need to revisit the vendor's product catalog and find a clock oscillator with tighter drift specifications.

**Example 2**. Let's say that the clock oscillator drift in the previous example was too large and would not work for the application for which it was intended. Further searching produces a vendor who will provide a custom-built clock oscillator module with a specified drift due to aging over time and a temperature of 0.25 ppm guaranteed for a period of five years at an operating temperature of $20°C \pm 4°C$, with a 10-month lead time. If the temperature range and aging time limits for this clock oscillator are acceptable, then the frequency synthesizer you design will be a great deal more expensive but will have a far better error performance than the one in the previous example. Let's run the numbers and see just how much better the performance would be. For the same oscillator center frequency of 16.384 MHz, the maximum drift you should expect for this clock oscillator is

$$F_\Delta = \pm \left( \frac{0.25}{10^6} \right) (16.384 \times 10^6) = \pm 4.096 \text{ Hz}$$

This a hundred times better than the design in example 1. The corresponding synthesizer output frequency will have a maximum potential error when tuned from DC to half the sample rate of $\pm 2.048$ Hz. This example may be overkill for some applications, but it clearly shows the importance the clock oscillator has on the overall performance of the frequency synthesizer.

## 8.4  ADDING A PROGRAMMABLE PHASE OFFSET TO THE NCO OUTPUT

One example where a programmable phase offset is useful is the case where the NCO is used in an application where a complex signal is tuned or translated in frequency. In this case, the NCO would have two identical look up tables, each capable of producing an output sinusoid of the same frequency with an independent phase offset represented by

$$\sin(\omega_0 t + \Theta_1) \text{ and } \sin(\omega_0 t + \Theta_2)$$

If we choose the value of $\Theta_1$ to be 90° ($\pi/2$ radians) and $\Theta_2$ to be 0°, then we would have an NCO that will produce a complex sinusoidal output commonly utilized for quadrature tuning of a signal in a digital receiver, as represented by

$$\sin\left(\omega_0 t + \pi/2\right) \pm j \sin\left(\omega_0 t\right) = \cos\left(\omega_0 t\right) \pm j \sin\left(\omega_0 t\right)$$

**Equation 8.21**

A second example of the benefits of adding phase offset would be the real time modification or modulation of the phase term $\Theta$ as a mechanism to transmit information, such as might be contained in a serial digital bit stream. For this case, $\Theta$ is not a constant and would vary over time so the NCO would produce an output of the form $\sin\left(\omega_0 t + \Theta(t)\right)$. For example, information might be transmitted if the time varying phase offset $\Theta(t)$ term takes on a sequence of $\pm\pi/2$ in response to a serial digital bit stream of ones and zeros.

### 8.4.1   NCO Phase Offset

The NCOs we have discussed so far have all had zero phase offset in the output sinusoid. That is, the output sinusoid sequences all began with a phase offset of $0°$. The waveform produced by the NCO was of the form $y(t) = \sin(\omega_0 t)$. In many applications, it may be desirable to add a phase offset to the NCO output. The NCO would then produce a sinusoid of the form

$$y(t) = \sin(\omega_0 t + \Theta)$$

**Equation 8.22**

where the phase offset is either a fixed value given by $\Theta$ or a time varying value given by $\Theta(t)$. In both cases, the phase offset can take on a range of $0 \le \Theta < 2\pi$. We can express Equation 8.22 in terms relating to the NCO as follows:

$$y(t) = \sin\left(\omega_0 t + \Theta\right)$$

$$y(t) = \sin\left(2\pi f_0 t + \Theta\right)$$

In the analog world, the variable $t$ is continuous, but in the digital world, $y(t)$ takes on values at discrete increments of time equivalent to periodic increments in the sample period $nT_S$. To describe this equation in the digital domain, the variable $t$ can be replaced by $nT_S$ to produce a discrete time mathematical representation given by

$$y(nT_S) = \sin\left(2\pi f_0 nT_S + \Theta\right)$$

**Equation 8.23**

We know from previous paragraphs that the frequency produced by an NCO is given by

$$f_0 = \Phi\left(\frac{F_S}{2^M}\right)$$

**Equation 8.24**

If we substitute Equation 8.24 into Equation 8.23, we get the expression

$$y(nT_s) = \sin\left[2\pi F_s\left(\frac{\Phi}{2^M}\right)nT_s + \Theta\right]$$

After canceling terms, we arrive at

$$y(nT_s) = \sin\left[2\pi\left(\frac{\Phi}{2^M}\right)n + \Theta\right]$$

Since it is implicit that $y(nT_s)$ is defined only at the sample instants, we can drop the $nT_s$ notation and replace it with the digital sample index $n$, resulting in Equation 8.25. This equation enables us to mathematically express the NCO output in terms of the NCO variables:

$$y(n) = \sin\left[2\pi\left(\frac{\Phi}{2^M}\right)n + \Theta\right]$$

**Equation 8.25**

Since the sine look up table contains $2^M$ samples, the phase offset resolution or the smallest increment of phase is given by $2\pi/2^M$ radians or $360/2^M$ degrees. Computation of the two phase offset values can be achieved through the use of either Equation 8.26 or Equation 8.27:

$$\Theta = \frac{2\pi}{2^M}\Omega \text{ radians for } 0 \leq \Omega < 2^M - 1$$

**Equation 8.26: Phase offset in radians**

$$\Theta = \frac{360°}{2^M}\Omega \text{ degrees for } 0 \leq \Omega < 2^M - 1$$

**Equation 8.27: Phase offset in degrees**

where

$\Theta$ = the desired phase in radians or degrees
$M$ = the power of two size of the look up memory
$\Omega$ = the $M$-bit-wide programmable phase offset word expressed in radians or degrees

Substitution of either Equation 8.26 or Equation 8.27 into Equation 8.25 will give us a complete mathematical representation of the NCO output signal with phase offset for the case where the sinusoid argument is expressed in radians or degrees:

$$y(n) = \sin\left[ 2\pi\left(\frac{\Phi}{2^M}\right)n + \left(\frac{2\pi}{2^M}\right)\Omega \right]$$

**Equation 8.28: NCO output expression in radians**

$$y(n) = \sin\left[ 2\pi\left(\frac{\Phi}{2^M}\right)n + \left(\frac{360°}{2^M}\right)\Omega \right]$$

**Equation 8.29: NCO output expression in degrees**

### 8.4.2   Implementing an NCO Phase Offset

The NCO phase offset is achieved by summing the offset value with the NCO ramp address, as illustrated in the block diagram in Figure 8.27. As can be seen from the block diagram, implementation of the phase offset can be as

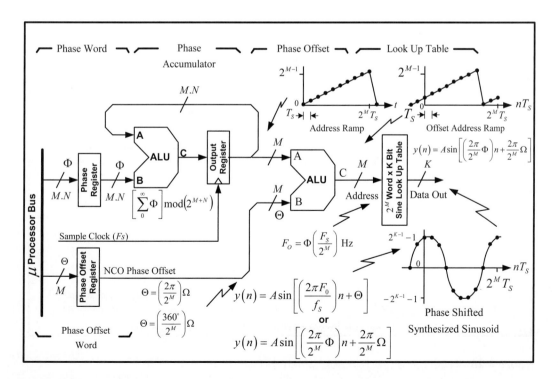

**Figure 8.27**   Block diagram of an NCO with phase offset

simple as loading the value for $\Theta$ from an external bus, similar to the loading of the NCO phase word $\Phi$. This is a convenient method for loading a constant phase offset. Applications that require real time phase offset changes may have need of a separate circuit that modifies the phase offset in response to some stimulus like a serial bit stream.

### 8.4.3  Phase Offset Examples

Examples of phase offset are illustrated in Figure 8.28 through Figure 8.32. One complete cycle of a 500-Hz sine wave is shown in each figure, each with a different offset phase. The phase offset values are 0°, 22.5°, 45°, 90°, and 135°. The synthesized sinusoid in all five examples was generated with an NCO operating at a sample rate of $F_S = 1.024$ MHz, $M.N = 12.0$ for a 4096 sample look up table, and $\Phi = 2$. Therefore the synthesized frequency is computed as

$$f_O = \Phi\left(\frac{F_S}{2^M}\right) = \Phi\left(\frac{F_S}{2^{12}}\right) = 2\left(\frac{1.024 \times 10^6 \frac{\text{sample}}{\text{second}}}{4096 \text{ sample}}\right) = 500 \text{ Hz}$$

If we load the NCO parameters into Equation 8.29, we arrive at the sinusoidal time series for all five phase offset examples.

$$y(n) = \sin\left[\left(\frac{2\pi}{2^M}\Phi\right)n + \left(\frac{360}{2^M}\right)\Omega\right] = \sin\left[\left(\frac{4\pi}{2^{12}}\right)n + \left(\frac{360}{2^{12}}\right)\Omega\right] \left\{\text{for } \begin{array}{l} n = 0, 1, 2, 3, \cdots \\ \Omega = 0, 256, 512, 1024, 1536 \end{array}\right\}$$

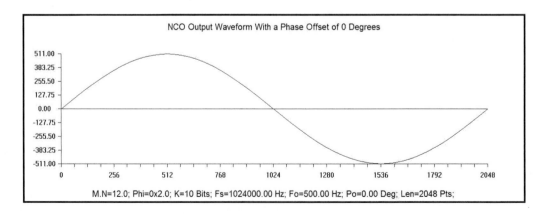

$$\text{M.N=12.0; Phi=0x2.0; K=10 Bits; Fs=1024000.00 Hz; Fo=500.00 Hz; Po=0.00 Deg; Len=2048 Pts;}$$

**Figure 8.28**  NCO generated sinusoid with 0° phase offset

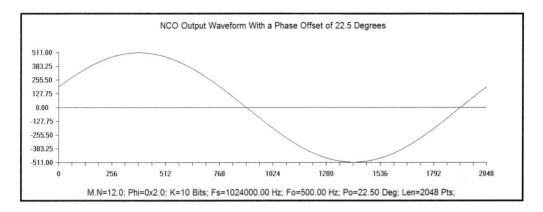

**Figure 8.29**   NCO generated sinusoid with 22.5° phase offset

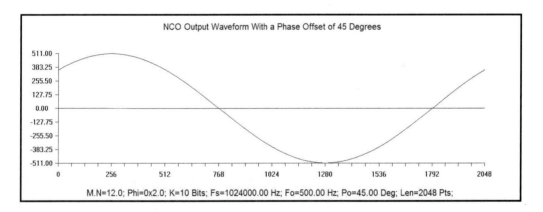

**Figure 8.30**   NCO generated sinusoid with 45° phase offset

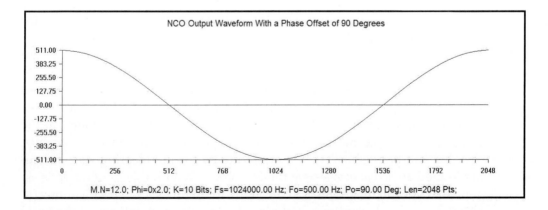

**Figure 8.31**   NCO generated sinusoid with 90° phase offset

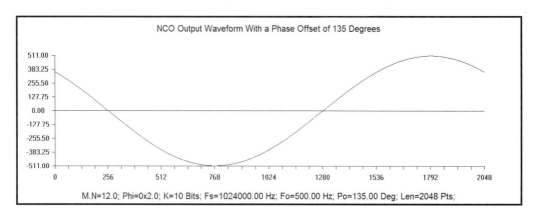

M.N=12.0; Phi=0x2.0; K=10 Bits; Fs=1024000.00 Hz; Fo=500.00 Hz; Po=135.00 Deg; Len=2048 Pts;

**Figure 8.32**  NCO generated sinusoid with 135° phase offset

## 8.5  DESIGN OF AN INDUSTRY-GRADE NCO

Now that we have a working knowledge of digital NCOs and how they are designed, let's design an NCO suitable for use in real-world applications. With minor modifications to be compliant with specific program requirements, this NCO would be useable in most real-world designs. We will design an industry grade NCO in this section and enhance its performance later on when we introduce the concept of phase dithering.

Typically a company will bid a program based on a set of high-level requirements and specifications published by the prospective customer and provided to each competing contractor as a data source on which to base their bid. This is usually termed a *request for proposal* (RFP) or a *request for quote* (RFQ). The winning contractor usually uses the proposal period and the first few months after contract award to break down the high-level specifications provided by the customer into a set of detailed performance specifications for each of the program design components. It is up to the component design engineer to implement these specifications into his or her particular design and to verify compliance at each program design review. At some point in the program, after the software is written and the hardware is designed and built, the designer is going to have to demonstrate the performance of his or her creation and demonstrate to management and the customer that the design performance meets or exceeds all specifications and requirements. This dog and pony show, as it's referred to by the engineering staff, is usually termed an *acceptance test* by the customer. So with this in mind, let's assume that your manager just gave you the following set of specifications for an NCO and has tasked you to design, build, test, and deliver it. The NCO specifications are as follows:

- Project sample clock $F_S = 32.768$ MHz $\pm 10$ ppm
- NCO tuning resolution = 500 Hz
- NCO tuning range = 0 Hz to 16.384 MHz
- NCO maximum tuning error $\leq \pm 180$ Hz at 16.384 MHz
- Synthesized sine wave quantization = $K \geq 11$ bits
- Quantization signal to noise ration > 70 db
- Spur free dynamic range > 70 db
- Phase offset resolution $\leq 0.2$ degrees
- The NCO should at a minimum be capable of tuning in steps of 500 Hz through a range of 0 Hz to 16.384 MHz.

Now that you have the specifications, you can sit down at your desk and hammer out this design before lunchtime. Let's begin. Remember that the basic equation that describes the operation of the NCO is

$$F_O = \Phi \left( \frac{F_S}{2^M} \right) \quad \left\{ \text{for} \quad 2^{-N} \leq \Phi \leq 2^{M-1} \right\}$$

**Equation 8.30**

where

$F_0$ is the NCO output frequency,
$2^M$ is the size of the memory used for the sine look up table,
$M$ is the integer part of the NCO phase word,
$N$ is the fractional part of the NCO phase word,
$F_S$ is the accumulator sample clock rate, and
$\Phi$ is the phase word that is repetitively accumulated at the sample rate.

Also remember that the equation for computing the samples in the look up table is

$$y(n) = \left( 2^{K-1} - 1 \right) \left( \sin \left[ \frac{2\pi}{2^M} n \right] \right) \text{ for } 0 \leq n \leq 2^M - 1$$

**Equation 8.31**

where

$n$ is the running sample index,

$y(n)$ is the $n$th sample of the computed sinusoid,

$K$ is the bit width of the sine look up table,

$\left(2^{K-1}-1\right)$ is the maximum amplitude of the computed sinusoid, and

$\left[2\pi/2^{M}\right]$ is the sinusoid phase increment in radians.

In addition, you should jot down the equation for the incremental NCO phase offset as

$$\Theta = \frac{360°}{2^{M}}\,\Omega \text{ degrees for } 0 \le \Omega < 2^{M}-1$$

So now that you have all your tools, it's time to design an NCO.

## 8.5.1 Determine the Bit Width (*K*) of the Sine Look Up Table Samples

Let's begin by examining the signal to noise ratio (SNR) specification. The SNR of a digitized analog signal is estimated by

$$\text{SNR} = (6.02B + 1.76)\text{db}$$

**Equation 8.32**

where $B$ is the number of bits in the quantized sample. Equation 8.32 represents a rule of thumb for estimating the quantized SNR for the case where an analog signal is sampled by an A/D converter. The derivation of this equation is treated thoroughly in just about every DSP and modern communication textbook. The derivation takes into account the probability of signal distortion due to clipping of the analog signal at the input to the A/D. Statistical assumptions in the derivation minimize the probability of signal clipping by reducing the amplitude of the analog input to the A/D. This in turn reduces the maximum possible quantized SNR. Since we are going to mathematically compute our look up table sinusoid under more ideal conditions than sampling with an A/D converter, the SNR degradation due to signal clipping will not be a factor and we should expect a better SNR than Equation 8.32 would suggest. Therefore we can rightly assume that we can use Equation 8.32 to estimate the number of bits needed to represent the samples of our look up table sinusoid, and in doing so, we can expect a better SNR figure than what that equation predicts.

The 70 db SNR specification we need to be compliant with should allow us to determine the minimum number of bits necessary to represent the signal amplitude of the look up table sinusoid. Solving for $B$ in Equation 8.32, we get

$$B = \frac{(\text{SNR} - 1.76)}{6.02} = \frac{(70 - 1.76)}{6.02} = 11.335 \text{ bits} \approx 12 \text{ bits}$$

The computed result is rounded up to the next highest number of bits to give us 12 bits of quantization. If 12 bits is equal to the number of bits utilized in the system for downstream computations, then this is a good initial number for the designer to use. If the number of bits utilized in the system for downstream computations is larger than 12 bits, then the designer can increase the size of his look up table to accommodate larger bit widths and enjoy an increase in SNR at the same time. If for some reason the number of bits utilized in the system for downstream computations is less than 12 bits, then there could be a major problem in the overall system design, in that it doesn't make sense to expend the extra hardware in the NCO to maintain a larger SNR figure than can be maintained by follow on hardware. If this is the case, then it is time to have a meeting with all the program system engineers, design engineers, and program manager, where the collective assembly of heads can address the problem and come up with a solution.

Since our calculation gives us 12 bits and meets the program specifications, and the follow on hardware is also using 12 bits, we are good to go and can assign a value of $K = 12$ in Equation 8.31 when we compute the look up table samples.

## 8.5.2  Determine the Size (*M*) in the Look Up Table

Now we want to determine the number of samples that need be stored in the look up table. As a first estimate, let's use the ratio of the NCO sample rate to the NCO look up table size to get us in the ball park of the 500 Hz tuning resolution requirement, all the time keeping in mind that we would like to keep the size of the look up table to a reasonable value. Let's start by assuming the phase word of the NCO is unity. Then

$$F_0 = \Phi\left(\frac{F_S}{2^M}\right) = 1 \cdot \left(\frac{F_S}{2^M}\right)$$

$$2^M = \frac{F_S}{F_0} = \frac{32.768 \times 10^6 \text{ Hz}}{500 \text{ Hz}} = 65{,}536 \text{ words}$$

**Equation 8.33**

A 65,000-word by 12-bit look up table is fairly large. If we choose to go with this large of a table, then the bit width of the address ramp is computed to be

$$M = \frac{\text{Log}_{10}\left(65,536\right)}{\text{Log}_{10} 2} = 16 \text{ bits}$$

Although it can be implemented either in discrete static ram (SRAM) devices, in core memory in an FPGA, or in heap memory for a software NCO, it is still a bit unwieldy for our design. So let's reduce the table size by a factor of eight to more reasonable size of 8192 words, which sets the value of $M = 13$. We can now compute the initial design fundamental frequency for the case where $\Phi$ is unity:

$$F_O = \Phi\left(\frac{F_S}{2^M}\right) = 1 \cdot \left(\frac{32.768 \times 10^6 \text{ Hz}}{2^{13}}\right) = 4000 \text{ Hz}$$

If we decide to proceed with $M = 13$ then the size of the look up table is completely defined to be 8192 samples by 12 bits. Our tuning frequency resolution however is way out of spec. No problem—we will deal with the tuning resolution discrepancy of 4000 Hz versus 500 Hz later when we design the $M.N$ phase word $\Phi$.

Now that we know the values for $M$ and $K$, we can use Equation 8.31 to compute the look up table samples:

$$y(n) = \left(2^{K-1} - 1\right)\left(\sin\left[\frac{2\pi}{2^M}\right]n\right) \text{ for } 0 \leq n \leq 2^M - 1$$

$$y(n) = \left(2^{11} - 1\right)\left(\sin\left[\frac{2\pi}{2^{13}}\right]n\right) = (2047)\left(\sin\left[\frac{2\pi}{8192}\right]n\right) \text{ for } 0 \leq n \leq 8191$$

Remember that $y(n)$ is a floating point number and needs to be converted to an integer prior to storage in the look up table. Special attention should be paid to the method of conversion to minimize quantization noise. For the purpose of this design, we will round the floating point number to the nearest whole digit. After the look up table is computed, the designer should use a computer simulation to sweep the frequency of the NCO sinusoid through its range of 0 Hz to $F_S/2$ Hz and perform a spectral analysis of the swept sinusoid to ensure that the SNR across the band meets or exceeds design specifications. Special attention should be paid to the SNR and spurious free dynamic range (SFDR) of the NCO across the entire frequency range. If the computer analysis of the look up table shows that the NCO performance exceeds program specifications, then the table is good to go. Otherwise the table may have to be tweaked by increasing the values for $K$ and possibly $M$.

### 8.5.3   Determine the Size and Format (*M.N*) of the Phase Word

Now that we know the value of $M$, we need to compute the value of $N$ so we can determine the size of the phase word $\Phi$ that is formatted as $M.N$ bits wide. We choose the value of $N$ based on the specified frequency resolution of 500 Hz. To do this, we can rearrange the terms in the NCO Equation 8.30, replace $F_O$ with the fundamental frequency of 500 Hz, and solve for $\Phi_{Min}$:

$$F_O = \Phi_{Min}\left(\frac{F_S}{2^M}\right)$$

$$\Phi_{Min} = F_O\left(\frac{2^M}{F_S}\right) = \left(500\,\frac{\text{cycle}}{\text{second}}\right)\left(\frac{8192\,\dfrac{\text{sample}}{\text{cycle}}}{32.768x10^6\,\dfrac{\text{sample}}{\text{second}}}\right) = 0.125 = 2^{-3}$$

**Equation 8.34**

Equation 8.34 says that $\Phi_{Min}$ must be set to a value of 0.125 or $2^{-3}$ or 1/8 in order to achieve a fundamental frequency out of the NCO of 500 Hz. This sets the minimum value of $N$ to 3 bits. The format of the NCO phase word will be $M.N = 13.3$. Fortunately this is a very convenient value since the composite phase word will be 16 bits.

Sixteen bits is a very common bit width, so it makes the parts selection for building the NCO that much easier, assuming of course that the NCO is being built out of discrete parts. If this NCO is being built in a gate array or an application-specific integrated circuit (ASIC) or in software, the importance of standard bit widths is significantly reduced. In either case, the designer is free to increase the value of $N$ to further increase the NCO frequency resolution and lower the NCO fundamental frequency even further. For an $M.N$ phase accumulator, the fundamental frequency for an NCO is given by

$$\Phi_{Min}\left(\frac{F_S}{2^M}\right) = \left(\frac{1}{2^N}\right)\left(\frac{F_S}{2^M}\right) = \frac{F_S}{2^{M+N}}$$

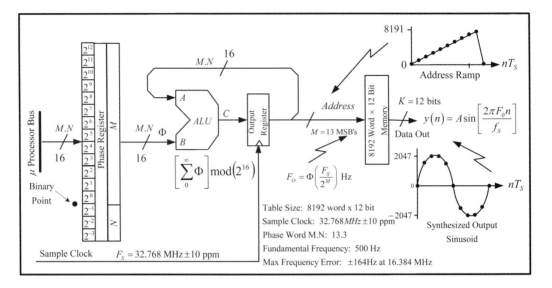

**Figure 8.33**   NCO design block diagram

In this sample design, the fundamental frequency is computed to be

$$\frac{F_S}{2^{M+N}} = \frac{32.768 \times 10^6 \text{ Hz}}{2^{13+3}} = 500 \text{ Hz}$$

which meets our design specification. If we had chosen the value of $N$ to be 12, the fundamental frequency of the NCO would be 0.9766 Hz. Setting the value of $N$ to a greater number of bits than required does not affect the capability of this NCO to tune in increments of 500 Hz; it only increases the tuning resolution. The $M.N$ binary format for the phase word can be visualized to be 0000000000000.000. The phase word in this design example consists of 16 bits and is represented in hexadecimal format as $0x0000$. Setting the phase word to any arbitrary value through its range of $0x0000$ to $0x8000$ allows the NCO we just designed to tune from 0 Hz to 16.384 MHz in 500 Hz steps. The block diagram of this NCO at this point in the design is illustrated in Figure 8.33. Table 8.6 lists the NCO output frequency in response to the individual setting of each bit of the phase word.

## 8.5.4   Determine the NCO Tuning Accuracy

The specification required that the tuning error of our NCO be no greater than 180 Hz at half the sample rate of 16.384 MHz. It is important to note that the frequency error and hence the tuning accuracy of the NCO is not a

**Table 8.6**   NCO Phase Word Bit Value versus Output Frequency

| Accumulator phase word (bit number) | Phase word binary point | Phase word bit value (power of 2) | NCO phase word value (decimal) | NCO phase word value (Hex) | NCO output frequency (Hz) | NCO output sinusoid (sample/cycle) |
|---|---|---|---|---|---|---|
| 15 | | 12 | 4096.000 | 0x8000 | 16384000 | 2 |
| 14 | | 11 | 2048.000 | 0x4000 | 8192000 | 4 |
| 13 | | 10 | 1024.000 | 0x2000 | 4096000 | 8 |
| 12 | | 9 | 512.000 | 0x1000 | 2048000 | 16 |
| 11 | | 8 | 256.000 | 0x0800 | 1024000 | 32 |
| 10 | | 7 | 128.000 | 0x0400 | 512000 | 64 |
| 9 | | 6 | 64.000 | 0x0200 | 256000 | 128 |
| 8 | | 5 | 32.000 | 0x0100 | 128000 | 256 |
| 7 | | 4 | 16.000 | 0x0080 | 64000 | 512 |
| 6 | | 3 | 8.000 | 0x0040 | 32000 | 1024 |
| 5 | | 2 | 4.000 | 0x0020 | 16000 | 2048 |
| 4 | Binary | 1 | 2.000 | 0x0010 | 8000 | 4096 |
| 3 | point • | 0 | 1.000 | 0x0008 | 4000 | 8192 |
| 2 | | -1 | 0.500 | 0x0004 | 2000 | 16384 |
| 1 | | -2 | 0.250 | 0x0002 | 1000 | 32768 |
| 0 | | -3 | 0.125 | 0x0001 | 500 | 65536 |

function of any design implementation of the NCO itself. The NCO output frequency error is dependent on and bounded by the stability of the sample clock oscillator. Recall that the oscillator frequency is the sum of its desired output frequency and its offset or error frequency and is given by

$$\widehat{F}_O = F_O \left( 1 \pm \frac{F_\Delta}{F_S} \right)$$

where

$\widehat{F}_O$ = the actual NCO output frequency
$F_O$ = the desired NCO output frequency

$\left(F_\Delta / F_S\right)$ = the normalized NCO frequency error

$F_\Delta = F_S \cdot (xx)$ ppm = the calculated oscillator frequency deviation in hertz, based on the manufacturer's drift specifications

Working backward, we know from the specifications that when $F_O = F_S/2$, the NCO frequency error is specified to be ≤ 180 Hz, so we can write

$$F_O\left(\frac{F_\Delta}{F_S}\right) = \frac{F_S}{2}\left(\frac{F_\Delta}{F_S}\right) = \left(\frac{F_\Delta}{2}\right) \le 180 \text{ Hz therefore } F_\Delta \le 360 \text{ Hz}$$

We know that for this error specification $F_\Delta = F_S \times (xx)$ ppm, so

$$xx(\text{ppm}) \le \frac{360 \text{ Hz}}{F_S} = \frac{360 \text{ Hz}}{32.768 \times 10^6 \text{ Hz}} \le 10.986 x 10^{-6} \text{ or } 10.98 \text{ ppm}$$

Now all we have to do is verify that the system oscillator is specified to fall in the frequency range over the time and temperature of 32.768 MHz ± 10.98 ppm. For our design, the system sample clock oscillator is specified by the manufacturer to operate at a frequency of 32.768 MHz ± 10 ppm over time and temperature. This means the maximum NCO frequency error will meet and exceed the ±180 Hz tuning accuracy specification. By how much you may ask? The actual performance specification calculated at 10 ppm is ± 164 Hz at the maximum NCO output frequency of half the sample rate. So we beat the specification by 16 Hz. The accuracy of ±164 Hz sounds like a pretty hefty frequency error, especially when the tuning resolution is specified to be 500 Hz. Remember that ±164 Hz is the maximum possible frequency error that can be expected when the NCO is operating at its absolute maximum frequency limit of 16.384 MHz. To put this error in better perspective, consider two issues:

- *Issue 1.* It is reasonable to assume that the NCO will never be configured to operate at or even near its maximum output frequency. Rarely if ever do we actually operate digital systems at or near the Nyquist sample rate. A good rule of thumb is to limit the maximum NCO frequency to $F_S/2.5$ Hz, which in this case is $32.768 \text{ MHz}/2.5 = 13.1072$ MHz. This means we should expect a maximum theoretical frequency error of $13.1072 \text{ MHz}(164 \text{ Hz}/16.384 \text{ MHz}) = 131.2$ Hz.
- *Issue 2.* The frequency error or the slope of the maximum expected error curve is

$$\frac{\pm F_\Delta}{F_S} = \frac{\pm 328\ \text{Hz}}{32.768 x 10^6\ \text{Hz}} = \frac{\pm 10.0\ \mu\ \text{Hz}}{\text{Hz}}$$

This means that we would expect an envelope of maximum possible frequency error of $\pm 10.0\ \mu\ \text{Hz/Hz}$ for every hertz the NCO is tuned. When the NCO is programmed to produce a 500 Hz output tone, the maximum frequency error envelope we should ever expect to see is

$$500\ \text{Hz}\left(\frac{\pm 10.0\ \mu\ \text{Hz}}{\text{Hz}}\right) = \pm 0.0050\ \text{Hz or } \pm 5.0\ \text{m Hz}$$

Oscillators that operate at 32.768 MHz with a 10 ppm drift specification due to temperature and component aging are not too difficult to find. If not, most oscillator manufacturers will be more than happy to build a custom oscillator for you that will meet or significantly beat this specification. On the down side, the cost will be high, and the lead time will be very long. If you have to go this route, remember that not only will the manufacturer have to design, package, and test the custom oscillator, he will also have to burn it in over an extended period of time. Therefore it is imperative, from a program schedule point of view, that the system clock oscillators are made a high program design priority and that they are ordered as close to day one of the project as possible.

If the standard off the shelf oscillators won't do the job, there are other routes available to the overworked and harried design engineer. At the time of this writing, one can purchase more capable types of oscillators such as temperature compensated voltage controlled crystal oscillators (TCVCXOs) that range in frequency to well above 160 MHz. These are precision oscillators that have an internal temperature compensation circuit that corrects for drift caused by changes in temperature to specifications as tight as $\pm 2.5$ ppm and drift due to aging as tight as $\pm 2.0$ ppm. These devices can operate in the 0°C to 70°C range. They can be purchased in dual inline packages (DIP) or surface mount (SMT) packages and can operate from 2.8, 3.0, 3.3, or 5.0 volts. On the down side, they do consume more power, and they are expensive, even when purchased in quantity.

### 8.5.5 Adding a Programmable Phase Offset

Adding a phase offset capability to this NCO is as simple as adding a 13-bit phase offset register and a second ALU that accumulates modulo $2^{13}$. The smallest increment of phase offset obtainable with this design is when $\Omega = 1$:

$$\Theta = \frac{360°}{2^M}\Omega = \frac{360°}{2^{13}}\Omega = \frac{360°}{8192}\cdot 1 \cong 0.044°$$

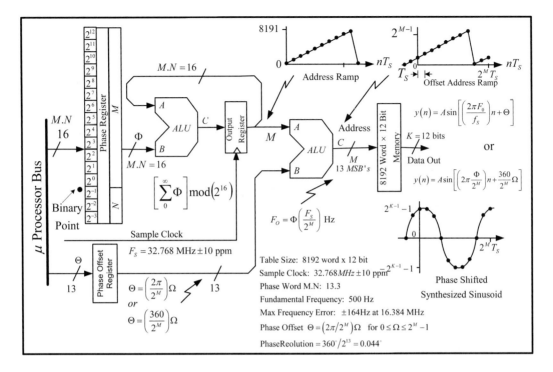

**Figure 8.34** NCO block diagram with phase offset

This beats the specification of 0.2 degrees by a factor greater than four. So our design is complete. The finished NCO design is illustrated in Figure 8.34. So you spent 15 minutes on the design, it is complete, and you barely got your pencil warm.

### 8.5.6 NCO Design Performance

A representative sinusoid generated by the newly designed NCO is illustrated in Figure 8.35. The sinusoid is a tone at $720,000$ Hz. For a sample rate of $F_S = 32.768$ MHz, the 2048 points in the plot span $62.5\ \mu$ seconds and there are 45 complete cycles in the plot. The plot of the look up table sample rounding error is illustrated in Figure 8.36. The NCO frequency falls directly into the DFT bin number 45. Therefore the sequence processed by the 2048 bin DFT is a maximal length sequence and no periodicity of rounding error is present. The spectral plot of the NCO output is shown in Figure 8.37. The highest spurs in the noise floor are 92 db down from the spectral peak, which exceeds the design specification of 70 db. Although not shown, the highest spurs for the same NCO utilizing truncated samples was 82 db down from the spectral peak, which is still within spec but is a significantly poorer

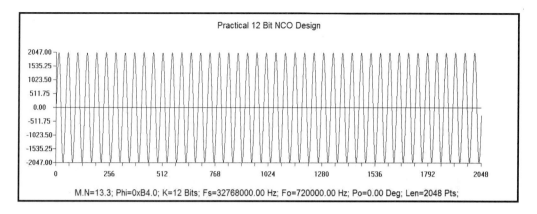

**Figure 8.35**   NCO sinusoidal output

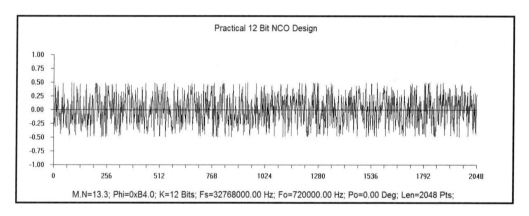

**Figure 8.36**   NCO rounding error

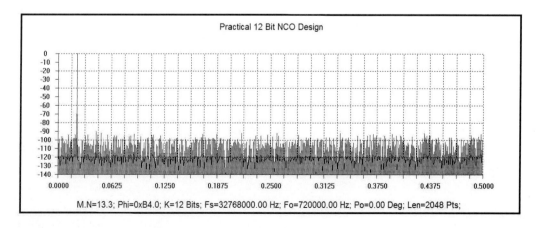

**Figure 8.37**   NCO output spectra

performance than that demonstrated for the rounded samples. The exact same NCO sinusoidal output, but with a 180° phase offset, is illustrated in Figure 8.38.

For demonstration purposes, the spectral plot for the same NCO but with 16-bit sample quantization is illustrated in Figure 8.39. For this case, the highest spur was 117 db below the spectral peak, a 25 db gain over the 12-bit quantization case. This 25 db gain should be expected, since the added 4 bits of quantization allows Equation 8.32 to predict an SNR increase of $SNR = (6.02B + 1.76)\,db$ or $(6.02(4) + 1.76)\,db = 25.84$ db. This comparison between the 12-bit NCO and the 16-bit NCO shows an impressive increase in performance. The price paid is a look up table that is four times larger.

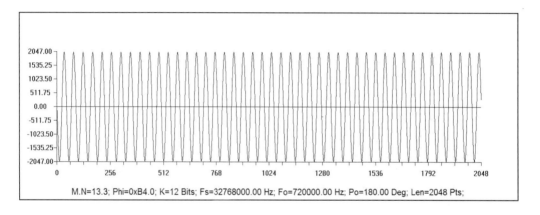

M.N=13.3; Phi=0xB4.0; K=12 Bits; Fs=32768000.00 Hz; Fo=720000.00 Hz; Po=180.00 Deg; Len=2048 Pts;

**Figure 8.38**   NCO sinusoidal output with 180° phase offset

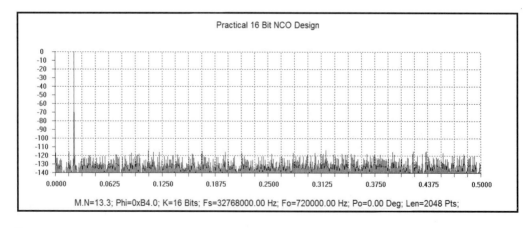

M.N=13.3; Phi=0xB4.0; K=16 Bits; Fs=32768000.00 Hz; Fo=720000.00 Hz; Po=0.00 Deg; Len=2048 Pts;

**Figure 8.39**   NCO output spectra-bit width increased from 12 to 16 bits

## 8.6   **NCO PHASE DITHER**

In the discussion on generating the samples of a sinusoid used in the sine look up table, we mentioned the effects that truncation and rounding have on the purity of the output NCO sinusoidal waveform. Both methods produce discontinuities in the signal due to quantization of the sample values and due to the distortion caused by the conversion from floating point to integer.

When the look up table is being addressed continuously, these discontinuities are periodic and all the energy associated with these discontinuities is concentrated into unwanted low level frequencies, sometimes referred to as *spurs*. We have seen an example of these spurs in Figure 8.9 that illustrated the spectrum of a 34 KHz sinusoid generated by an NCO with a sine look up table that was designed using truncated samples.

It was clear from that discussion that truncation of the fractional part of the table sample produced much larger spurs than the rounding of the fractional part. The rounding method, however, is not necessarily going to produce spur free output signals. Spurs are a fact of life that the designer has to live with. Although the spurs cannot be completely eliminated, they can be significantly attenuated.

One method of reducing the periodic energy that results in unwanted spurs is called *phase dithering*. In this method, all we do is add pseudo random digital noise to the LSBs of the address ramp. The random noise is nothing more than a random number generated by a simple maximal length pseudo random sequence generator, implemented by a shift register identical to those described by Dixon [4]. The generator is constructed from an $N$ stage by a 1-bit-wide shift register with feedback from various stages. The selection of feedback stages determines the sequence length. There are special sets of feedback stages that will produce a maximal length code. For example, an 8-bit register with the proper feedback can generate a sequence that does not repeat itself for 255 consecutive bits. In general, an $N$ bit register can be configured with the proper feedback to generate a $2^N - 1$ bit maximal length length sequence. These configurations are termed *maximal length sequence generators*. The selection of the feedback stages is critical to the generation of a maximal length random sequence. Dixon [4] provides tables that list all the combinations of register length and feedback stages that produce maximal length codes.

Phase dither sums the $R$ LSBs of the address ramp with the $R$ bits of the pseudo random sequence. This essentially adds a small random phase jitter to the look up table address. This randomness disrupts the periodic occurrence of look up table discontinuities and spreads the associated spur energy across the output signal spectrum. Although the dither tends to spread the energy of the low level spurs, it does not substantially affect the energy in the much larger fundamental signal. Dither is a widely used method to increase the SFDR of digital NCOs. A good design will include a dither option that can be enabled or disabled remotely via the microcontroller command bus. An example of an $R = 7$ bit maximal length pseudo random sequence generator is illustrated in Figure 8.40. In this example, the nonrepeatable code length is 127 bits.

The generator all zero state occurs when all bits of the pseudo random word are zero. This condition is not allowed. If all the bits in the generator are equal to zero, the generator will remain in this state until reset. The maximal length pseudo random number (PRN) sequence generator must be designed to detect the all zero state and reset itself to a known seed value if this condition should ever occur.

If the phase offset $\Theta$ is constant, then we can separately represent the phase dither by the term $D(n)$. In doing so, we can now write the equation for the NCO output as

$$y(n) = A\sin\left[\left(2\pi \frac{\Phi}{2^M}\right)n + \Theta + D(n)\right]$$

**Equation 8.35**

The sine look up table is $2^M$ samples long. As presented earlier, the constant phase offset in degrees is determined by

$$\Theta = \left(360°/2^M\right)\Omega$$

where $\Omega$ is the desired phase offset in units of look up table samples. The length of the PRN shift register is $R$ bits long. The maximum value of the phase dither term can be represented as

$$D(n)_{MAX} = \left(360°/2^M\right)\left(2^R - 1\right)$$

The composite phase term can be represented by

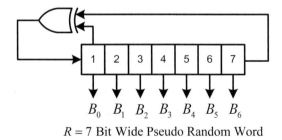

$R = 7$ Bit Wide Pseudo Random Word

**Figure 8.40**   Seven-bit maximal-length PRN sequence generator

$$\Theta + D(n)_{MAX} = \left(360°/2^M\right)\left(\Omega + D(n)_{MAX}\right) = \left(360°/2^M\right)\left(\Omega + \left[2^R - 1\right]\right)$$

Since the dither term $D(n)$ changes with each tick of the sample clock, we can represent the instantaneous phase terms as $\Theta \pm D(n)$ and in doing so arrive at Equation 8.35. For a sine look up table of length 8192, the maximum value of the phase dither for a $R$ bit register is tabulated in Table 8.7.

Figure 8.41 illustrates the block diagram of a complete NCO with phase offset and phase dither capability. The figure shows the phase dither being summed with the LSBs of the look up table address ramp.

**Table 8.7** Dither Variance

| Register length $R$ | Range | Max dither (degrees) | Max dither (radians) |
|:---:|:---:|:---:|:---:|
| 5 | 31 | 1.36 | 0.0237 |
| 6 | 63 | 2.77 | 0.0482 |
| 7 | 127 | 5.58 | 0.0974 |
| 8 | 255 | 11.21 | 0.1956 |

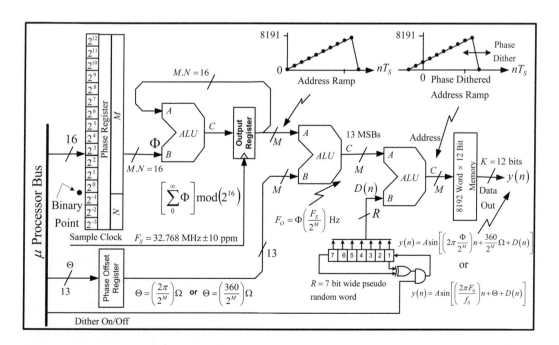

**Figure 8.41** NCO block diagram with phase dither

The phase dither option can significantly improve the SFDR specifications for a given NCO. The bit width of the dither generator was set at 7 in this example only for demonstration purposes. The actual length of the generator will depend on the application and should be derived based on the design specifications.

Although not shown in the figure, the reader should note that the shift register pseudo random code generator needs to detect the condition where all the bits in the register are zero. This detection can be as simple as an $R$ input $OR$ gate that recognizes the all zero state and synchronously resets the register to some known seed value on the next sample clock.

Once the NCO design is complete, it should be thoroughly simulated via a program written by the design engineer to verify proper operation. Investing a day to write, test, and debug a simple software simulator to verify the proper performance of an NCO function will pay big dividends after the circuit is built and works the first time it is turned on.

## 8.7 REFERENCES

[1] Intersil Technical Brief. "Measuring Spurious Free Dynamic Range in a D/A Converter," TB326, January 1995.

[2] C. C. Bissel and D. A. Chapman. *Digital Signal Transmission*. Cambridge, MA: Cambridge University Press, 1992.

[3] Tsui James. *Digital Techniques for Wideband Receivers*. 2nd ed. Boston, MA: Artech House, 2001.

[4] Robert Dixon. *Spread Spectrum Systems*. 2nd ed. New York: John Wiley and Sons, 1984.

# CHAPTER NINE

# Signal Tuning

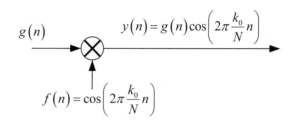

It is oftentimes necessary when processing a signal to move it from one region of the frequency spectrum to another region. This is especially true when processing communications signals, where a band limited signal centered at frequency $f_1$ is tuned to another center frequency $f_2$ in order to simplify downstream processing.

There are several synonymous terms that are commonly used to describe a signal whose spectrum has been moved from one center frequency to another. A signal whose spectrum has been moved is referred to as having been *tuned*, or it may be referred to as having been *translated*. In many cases when a signal's spectrum is tuned from some higher center frequency to a lower center frequency, it is referred to as having been *down converted*. Similarly if a signal's spectrum is tuned from a lower center frequency to some higher center frequency, it is referred to as having been *up converted*. For example, if a band limited signal is translated from a center frequency of 20 MHz down to 0 Hz, we can simply say that the signal was down converted to DC. There is nothing wrong with saying the signal was tuned to DC or translated to DC either. We will use all four descriptions—tuned, translated, down converted, and up converted—in this book. It is also common terminology to say that a signal has been down converted or tuned to *base band* when its center frequency has been translated from some other center frequency to 0 Hz or DC. We will use the term *base band* frequently in this book to indicate a band of frequencies centered at DC.

A good example of signal tuning is in the generation of a frequency division multiplex (FDM) signal. In such a signal, many band limited base band signals that are independent of one another are translated up in frequency so that they occupy adjacent frequency bands called *channels*. These translated signals are then summed together to form a single signal composed of several adjacent and independent channels. This is done so that multiple independent signals can simultaneously utilize the same transmission medium, like a single copper wire. As an example, a 12-channel FDM stack is referred to as a *group* in the telephone industry and allows for the transmission of 12 independent telephone channels on a single conductor. As another example, a 60-channel FDM stack is composed of five groups and allows for the transmission of 60 independent telephone channels on a single conductor. The 60-channel multiplex is referred to as a *super group*. There are several multiplex protocols above the super group level, all designed to maximize the use of a single transmission channel.

Another good example has to do with antennas that transmit and receive communications over free space [1]. Antennas require that their dimensions be proportional to the wavelength of the signal being transmitted. The formula for the wavelength is $\lambda = c/f$, where $\lambda$ is the signal wavelength, $c$ is the speed of light ($3 \times 10^8$ meters per second), and $f$ is the signal frequency. We can see that a low frequency signal such as a 10 KHz audio tone would require an antenna with dimensions on the order of 30,000 meters. An antenna of this size would be a bit cumbersome. To make the antenna realizable, it is first necessary to translate the signal up in frequency prior to transmission. If, for example, the signal were translated to 1 GHz, the antenna dimensions would be on the order of 0.3 meters, a much more manageable size.

An example that is more suited to digital signal processing (DSP) is the down conversion of a band limited signal to base band, where it can be filtered and efficiently processed at a much lower sample rate. This technique is analogous to a digital radio.

We will devote the first half of this chapter to the development of concepts and procedures for continuous time or analog tuning. This will give us a firm understanding of the concept of signal translation. In the second half of this chapter, we will develop the same concepts and procedures for discrete time or digital tuning. Even if the reader is only interested in the study of digital tuning, it is recommended that he or she read the section on analog tuning as well. This is because most of the basic tuning concepts are formulated in the study of analog tuning and then built upon in the study of digital tuning.

It is the objective of this chapter to illustrate the methods used to translate the spectrum of a signal both up and down in frequency. Our discussions will require working with signals in both the time domain and the frequency domain. Thus it makes sense to spend some time reviewing the properties of the Fourier transform that are specific to the tuning application.

## 9.1   CONTINUOUS TIME (ANALOG) FOURIER TRANSFORM

Our intent here is to discuss the use and properties of the Fourier transform that are directly associated with the translation of narrow band signals. If the reader needs to refresh his or her memory on the concepts, mathematics, and properties of the Fourier transform, it would be beneficial to read Chapter 3, "Review of the Fourier Transform."

All Fourier transform pairs mentioned in this chapter are derived in Chapter 3. There have been many great books written on the subject of the Fourier transform, some of which are listed in this chapter's references. The reader is referred to these works if a more in-depth analysis of the Fourier transform is desired.

We begin by studying the unit impulse both in the time and frequency domain. The concept of an impulse is important. As we will see time and time again, a good understanding of the time domain and frequency domain properties of the impulse function is absolutely critical to understanding the concepts of signal translation.

A function of the variable $t$ in the time domain and a function of the variable $f$ in the frequency domain are said to be a Fourier transform pair if one function can be mathematically converted to the other function via the Fourier transform. In our study, we will utilize the symbol $\overset{f}{\leftrightarrow}$ to denote the relationship between a Fourier transform pair. The notation $Y(f) = f\{y(t)\}$ is often used to indicate that $Y(f)$ is the forward Fourier transform of $y(t)$, and the notation $y(t) = f^{-1}\{Y(f)\}$ is used to indicate that $y(t)$ is the inverse Fourier transform of $Y(f)$.

The definition of both the forward and inverse Fourier transform is fairly straightforward. The forward Fourier transform computes the frequency domain representation $Y(f)$ of a time domain function $y(t)$. This computation is achieved via the expression

$$Y(f) = \int_{-\infty}^{\infty} y(t)\, e^{-j(2\pi f)t}\, dt$$

**Equation 9.1**

The inverse Fourier transform computes the time domain representation $y(t)$ from its frequency domain representation $Y(f)$ via the expression

$$y(t) = \int_{-\infty}^{\infty} Y(f)\, e^{+j(2\pi f)t}\, df$$

**Equation 9.2**

If both $y(t)$ and $Y(f)$ are such that Equation 9.1 and Equation 9.2 are valid, then we can state that $y(t)$ and $Y(f)$ are a Fourier transform pair, and we can indicate this is so by the expression

$$y(t) \overset{f}{\leftrightarrow} Y(f)$$

If we choose to use transform functions expressed in terms of radian frequency $\omega$ rather than in terms of frequency $f$, then Equation 9.1 and Equation 9.2 can be rewritten as

$$Y(\omega) = \int_{-\infty}^{\infty} y(t)\, e^{-j\omega t}\, dt$$

**Equation 9.3**

$$y(t) = \frac{1}{2\pi} \int_{-\infty}^{\infty} Y(\omega)\, e^{+j\omega t}\, d\omega$$

**Equation 9.4**

Equation 9.4 is derived from Equation 9.2 by replacing $2\pi f$ with $\omega$ and noting that $df = (1/2\pi)d\omega$. The use of either set of equations is determined by the reader's preference and by the application in which they are used. The remainder of this chapter will utilize the transform pair expressed in terms of the frequency variable $f$, as shown in Equation 9.1 and Equation 9.2.

## 9.1.1  The Time Domain Unit Impulse

We will define the continuous time domain unit impulse function as

$$\delta(t) \begin{cases} = 1 & \text{for } t = 0 \\ = 0 & \text{for } t \neq 0 \end{cases}$$

The Fourier transform of the unit impulse function is given by

$$f\{\delta(t)\} = \int_{-\infty}^{\infty} \delta(t) e^{-j(2\pi f)t}\, dt \;=\; \delta(0) e^{-j0} = 1$$

**Equation 9.5**

Equation 9.5 leads to the definition of the Fourier transform pair:

$$\delta(t) \overset{f}{\leftrightarrow} 1$$

**Equation 9.6**

The unit impulse shifted in time by $t_0$ is defined as

$$\delta(t - t_0) \left\{ \begin{array}{ll} = 1 & \text{for } t = t_0 \\ = 0 & \text{for } t \neq t_0 \end{array} \right\}$$

The Fourier transform of the time shifted unit impulse function is

$$f\{\delta(t - t_0)\} = \int_{-\infty}^{\infty} \delta(t - t_0) e^{-j(2\pi f)t} \, dt \; = \; e^{-j(2\pi f)t_0}$$

**Equation 9.7**

Equation 9.7 leads to the definition of the Fourier transform pair:

$$\delta(t - t_0) \overset{f}{\leftrightarrow} e^{-j(2\pi f)t_0}$$

**Equation 9.8**

Similarly

$$f\{\delta(t + t_0)\} = \int_{-\infty}^{\infty} \delta(t + t_0) e^{-j(2\pi f)t} \, dt \; = \; e^{+j(2\pi f)t_0}$$

**Equation 9.9**

Equation 9.9 leads to the definition of the Fourier transform pair:

$$\delta(t + t_0) \overset{f}{\leftrightarrow} e^{+j(2\pi f)t_0}$$

**Equation 9.10**

It is evident from Equation 9.5, Equation 9.7, and Equation 9.9 that the spectrum of the unit impulse has a uniform magnitude over the entire frequency range. The phase is linear and is proportional to the time shift $t_0$.

### 9.1.2   The Frequency Domain Unit Impulse

The frequency domain unit impulse response is defined to be

$$\delta(f) \ = \left\{ \begin{array}{l} = 1 \text{ for } f = 0 \\ = 0 \text{ for } f \neq 0 \end{array} \right\}$$

The inverse Fourier transform of the frequency impulse function is

$$f^{-1}\{\delta(f)\} = \int_{-\infty}^{\infty} \delta(f) e^{j(2\pi f)t} df = 1$$

This leads us to the Fourier transform pair:

$$\delta(f) \overset{f}{\leftrightarrow} 1$$

**Equation 9.11**

We define a unit impulse response shifted in frequency by $f_0$ as $\delta(f - f_0)$. The inverse Fourier transform of $\delta(f - f_0)$ by

$$f^{-1}\{\delta(f - f_0)\} = \int_{-\infty}^{\infty} \delta(f - f_0) e^{j2\pi ft} df = e^{j(2\pi f_0)t}$$

**Equation 9.12**

Therefore the Fourier transform pair for the shifted frequency impulse is given by

$$e^{j(2\pi f_0)t} \overset{f}{\leftrightarrow} \delta(f - f_0)$$

**Equation 9.13**

The spectrum of $e^{j(2\pi f_0)t}$ is illustrated in Figure 9.1.

By changing the quantity $(f - f_0)$ in Equation 9.12 to $(f + f_0)$, we can similarly conclude that Equation 9.14 defines the Fourier transform pair:

$$e^{-j(2\pi f_0)t} \overset{f}{\leftrightarrow} \delta(f + f_0)$$

**Equation 9.14**

The spectrum of $e^{-j(2\pi f_0)t}$ is illustrated in Figure 9.2.

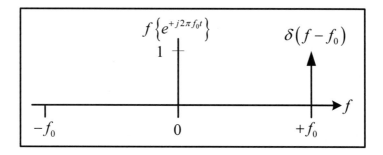

**Figure 9.1**   Positive exponential impulse spectrum

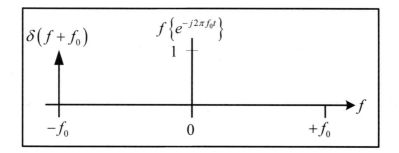

**Figure 9.2**   Negative exponential impulse spectrum

### 9.1.3   The Cosine Function cos(2πf₀t)

The Fourier transform of $\cos(2\pi f_0 t)$ is defined by

$$f\left\{\cos(2\pi f_0 t)\right\} = \int_{-\infty}^{+\infty} \cos(2\pi f_0 t)e^{-j(2\pi f)t}\,dt = \frac{1}{2}\left\{\delta(f-f_0)+\delta(f+f_0)\right\}$$

**Equation 9.15**

Equation 9.15 leads to the definition of the Fourier transform pair:

$$\cos(2\pi f_0 t) \overset{f}{\leftrightarrow} \frac{1}{2}\left\{\delta(f-f_0)+\delta(f+f_0)\right\}$$

**Equation 9.16**

The spectrum for $\cos(2\pi f_0 t)$ is illustrated in Figure 9.3.

### 9.1.4   The Sine Function sin(2πf₀t)

The Fourier transform of $\sin(2\pi f_0 t)$ is defined by

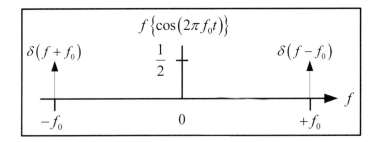

**Figure 9.3**   Line spectrum of $\cos(2\pi f_0 t)$

$$f\{\sin(2\pi f_0 t)\} = \int_{-\infty}^{+\infty} \sin(2\pi f_0 t) e^{-j(2\pi f)t}\, dt = \frac{1}{2j}\{\delta(f-f_0)-\delta(f+f_0)\}$$

**Equation 9.17**

Equation 9.17 leads to the definition of the Fourier transform pair:

$$\sin(2\pi f_0 t) \overset{f}{\leftrightarrow} \frac{1}{2j}\{\delta(f-f_0)-\delta(f+f_0)\}$$

**Equation 9.18**

If we multiply the right side of Equation 9.18 by the quantity $j/j$, we can define the alternate Fourier transform pair:

$$\sin(2\pi f_0 t) \overset{f}{\leftrightarrow} \frac{-j}{2}\{\delta(f-f_0)-\delta(f+f_0)\}$$

**Equation 9.19**

The spectrum of $\sin(2\pi f_0 t)$ as defined by Equation 9.18 is illustrated in Figure 9.4. The spectrum of the alternate form of $\sin(2\pi f_0 t)$ as defined by Equation 9.19 is illustrated in Figure 9.5. It is important that the reader understand that Figure 9.4 and Figure 9.5 are plots of spectral amplitude and not spectral magnitude. If the plots were spectral magnitude, then both impulses would appear as positive and the $j$ terms would be missing.

### 9.1.5   The Exponential Function $e^{j(2\pi f_0)t}$

We have already discussed the Fourier transform of an exponential function. Because the exponential plays a huge role in signal tuning, we will use this

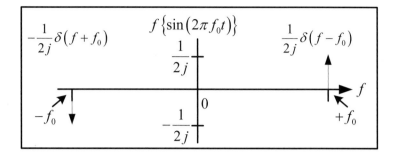

**Figure 9.4** Line spectrum of $\sin(2\pi f_0 t)$

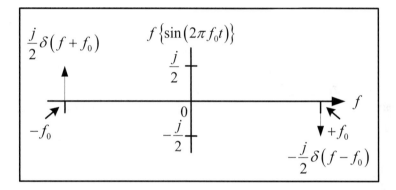

**Figure 9.5** Alternate line spectrum of $\sin(2\pi f_0 t)$

section to graphically illustrate the derivation of this transform. The spectrum of the exponential function $f(t) = e^{j(2\pi f_0)t}$ can be computed first by using Euler's formula to expand $e^{j2\pi f_0 t}$ and then by using the transform identities we derived in Equation 9.13 and Equation 9.14. Expansion of $e^{j2\pi f_0 t}$ results in

$$
\begin{aligned}
e^{+j(2\pi f_0)t} \quad &= \cos\left(2\pi f_0 t\right) + j\sin\left(2\pi f_0 t\right) \\[6pt]
&= \left[\frac{e^{j2\pi f_0 t} + e^{-j2\pi f_0 t}}{2}\right] + j\left[\frac{e^{j2\pi f_0 t} - e^{-j2\pi f_0 t}}{2j}\right] \\[6pt]
&= \frac{e^{j2\pi f_0 t}}{2} + \frac{e^{-j2\pi f_0 t}}{2} + \frac{e^{j2\pi f_0 t}}{2} - \frac{e^{-j2\pi f_0 t}}{2}
\end{aligned}
$$

**Equation 9.20**

Now we can compute the Fourier transform for each of the four terms in Equation 9.20. The Fourier transform in Equation 9.21 produces four frequency-shifted unit impulse terms that are labeled 1, 2, 3, and 4. These impulses are plotted in Figure 9.6. The impulse terms 1 and 3 are plotted off-set from $+f_0$ only to show that there are two terms of identical amplitude that sum together. The impulse terms 2 and 4 at $-f_0$ are $180°$ out of phase and cancel one another.

$$f\{e^{+j2\pi f_0 t}\} = \frac{1}{2}\int_{-\infty}^{\infty} e^{j2\pi f_0 t} e^{-j2\pi ft} \quad +\frac{1}{2}\int_{-\infty}^{\infty} e^{-j2\pi f_0 t} e^{-j2\pi ft} \quad +\frac{1}{2}\int_{-\infty}^{\infty} e^{j2\pi f_0 t} e^{-j2\pi ft} \quad -\frac{1}{2}\int_{-\infty}^{\infty} e^{-j2\pi f_0 t} e^{-j2\pi ft}$$

$$= \frac{1}{2}\int_{-\infty}^{\infty} e^{-j2\pi(f-f_0)t} \quad +\frac{1}{2}\int_{-\infty}^{\infty} e^{-j2\pi(f+f_0)t} \quad +\frac{1}{2}\int_{-\infty}^{\infty} e^{-j2\pi(f-f_0)t} \quad -\frac{1}{2}\int_{-\infty}^{\infty} e^{-j2\pi(f+f_0)t}$$

$$= \underbrace{\frac{1}{2}\delta(f-f_0)}_{①} \quad \underbrace{+\frac{1}{2}\delta(f+f_0)}_{②} \quad \underbrace{+\frac{1}{2}\delta(f-f_0)}_{③} \quad \underbrace{-\frac{1}{2}\delta(f+f_0)}_{④}$$

**Equation 9.21**

The plot in Figure 9.7 shows the resultant spectrum when all four terms of Equation 9.21 are summed. This allows us to define the Fourier transform pair:

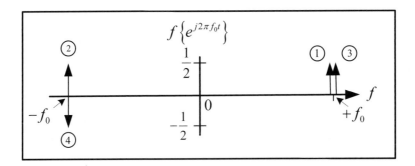

**Figure 9.6**  Positive exponential spectral terms

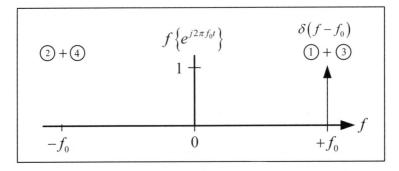

**Figure 9.7**  Positive exponential spectrum

$$e^{j(2\pi f_0)t} \overset{f}{\leftrightarrow} \delta\left(f - f_0\right)$$

**Equation 9.22**

This is identical to the transform pair that we originally described in Equation 9.13.

### 9.1.6 The Exponential Function $e^{-j(2\pi f_0)t}$

In a similar manner, the spectrum of the exponential function $f(t) = e^{-j2\pi f_0 t}$ can be computed first by using Euler's formula to expand $e^{-j2\pi f_0 t}$ and then by using the same transform identities we derived in Equation 9.13 and Equation 9.14. Expansion of $e^{-j2\pi f_0 t}$ results in

$$
\begin{aligned}
e^{-j(2\pi f_0)t} &= \cos\left(2\pi f_0 t\right) - j\sin\left(2\pi f_0 t\right) \\
&= \left[\frac{e^{j2\pi f_0 t} + e^{-j2\pi f_0 t}}{2}\right] - j\left[\frac{e^{j2\pi f_0 t} - e^{-j2\pi f_0 t}}{2j}\right] \\
&= \frac{e^{j2\pi f_0 t}}{2} + \frac{e^{-j2\pi f_0 t}}{2} - \frac{e^{j2\pi f_0 t}}{2} + \frac{e^{-j2\pi f_0 t}}{2}
\end{aligned}
$$

**Equation 9.23**

We can compute the Fourier transform on the four terms of Equation 9.23 to arrive at Equation 9.24. This equation contains four frequency shifted unit impulse terms that are labeled 1, 2, 3, and 4. These impulses are plotted in Figure 9.8.

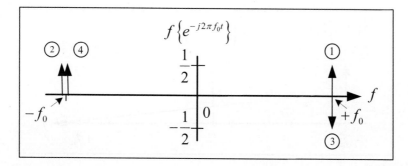

**Figure 9.8**  Negative exponential spectral terms

$$f\{e^{-j2\pi f_0 t}\} \quad = \frac{1}{2}\int_{-\infty}^{\infty} e^{j2\pi f_0 t} e^{-j2\pi f t} \quad +\frac{1}{2}\int_{-\infty}^{\infty} e^{-j2\pi f_0 t} e^{-j2\pi f t} \quad -\frac{1}{2}\int_{-\infty}^{\infty} e^{j2\pi f_0 t} e^{-j2\pi f t} \quad +\frac{1}{2}\int_{-\infty}^{\infty} e^{-j2\pi f_0 t} e^{-j2\pi f t}$$

$$= \frac{1}{2}\int_{-\infty}^{\infty} e^{-j2\pi(f-f_0)t} \quad +\frac{1}{2}\int_{-\infty}^{\infty} e^{-j2\pi(f+f_0)t} \quad -\frac{1}{2}\int_{-\infty}^{\infty} e^{-j2\pi(f-f_0)t} \quad +\frac{1}{2}\int_{-\infty}^{\infty} e^{-j2\pi(f+f_0)t}$$

$$= \underbrace{\frac{1}{2}\delta(f-f_0)}_{①} \quad \underbrace{+\frac{1}{2}\delta(f+f_0)}_{②} \quad \underbrace{-\frac{1}{2}\delta(f-f_0)}_{③} \quad \underbrace{+\frac{1}{2}\delta(f+f_0)}_{④}$$

**Equation 9.24**

The impulse terms 2 and 4 are plotted offset from $-f_0$ only to show that there are two terms of identical amplitude that sum together. The impulse terms 1 and 3 at $+f_0$ are 180° out of phase and cancel one another. The plot in Figure 9.9 illustrates the resultant spectrum for the negative exponential when all four terms of Equation 9.24 are summed. This allows us to define the Fourier transform pair:

$$e^{-j(2\pi f_0)t} \overset{f}{\leftrightarrow} \delta(f+f_0)$$

**Equation 9.25**

This is identical to the transform pair we originally described in Equation 9.14.

We spent a fair amount of time illustrating the summing and cancellation of frequency impulses because it is important and it will be a major factor when we discuss complex frequency translation later in this chapter. We will frequently use the identities given in Equation 9.13 and Equation 9.14. The reader should be able to visualize the cancellation property that is associated with complex exponentials because it makes our lives a whole lot easier when we work with communication signals and need to translate them up and down in frequency.

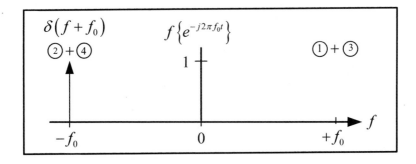

**Figure 9.9**  Negative exponential spectrum

## 9.1.7 Frequency Translation

To simplify our discussion on frequency translation, we need to develop a few additional tools in the form of Fourier transform pairs. For the first pair, let us begin by defining the base band function $f(t)$ to be a band limited signal with a continuous frequency spectrum denoted by $F(f)$. Furthermore we define the function $y(t) = f(t)\cos(2\pi f_0 t)$ with a frequency spectrum represented by $Y(f)$. The Fourier transform of $y(t)$ that was derived in Chapter 3 can be written as

$$f\{y(t)\} = \int_{-\infty}^{\infty} f(t)\cos(2\pi f_0 t)e^{-j2\pi ft}\, dt = \frac{1}{2}\{F(f-f_0) + F(f+f_0)\}$$

**Equation 9.26**

The transform pair for this signal product is then given by

$$f(t)\cos(2\pi f_0 t) \overset{f}{\leftrightarrow} \frac{1}{2}\{F(f-f_0) + F(f+f_0)\}$$

**Equation 9.27**

This is important. Equation 9.27 tells us that if we multiply a band limited base band signal $f(t)$ by a cosine wave at frequency $f_0$ we can expect to see the spectrum of $f(t)$ translated to both $+f_0$ and $-f_0$, and both of these translations will be half the amplitude of the original. We will explore the graphical development of this in the next section.

The Fourier transform for $f(t)\sin(2\pi f_0 t)$ was also derived in Chapter 3 and can be written as

$$f\{y(t)\} = \int_{-\infty}^{\infty} f(t)\sin(2\pi f_0 t)e^{-j2\pi ft}\, dt = \frac{1}{2j}\{F(f-f_0) - F(f+f_0)\}$$

**Equation 9.28**

The transform pair for this signal product is then given by

$$f(t)\sin(2\pi f_0 t) \overset{f}{\leftrightarrow} \frac{1}{2j}\{F(f-f_0) - F(f+f_0)\}$$

**Equation 9.29**

We also know that the Fourier transform of $f(t)$ is given by

$$F(f) = \int_{-\infty}^{\infty} f(t) e^{-j2\pi ft} \, dt$$

If we change the variable $f$ to $f - f_0$, we get

$$F(f - f_0) = \int_{-\infty}^{\infty} f(t) e^{-j2\pi(f-f_0)t} \, dt$$

We can rearrange terms to produce

$$F(f - f_0) = \int_{-\infty}^{\infty} \left[ f(t) e^{+j2\pi f_0 t} \right] e^{-j2\pi ft} \, dt$$

The right-hand side of the equation is just the Fourier transform of $f(t) e^{+j2\pi f_0 t}$, giving us the Fourier transform:

$$f(t) e^{+j(2\pi f_0)t} \overset{f}{\longleftrightarrow} F(f - f_0)$$

**Equation 9.30**

The dual to this pair can be computed in a similar fashion and is given here:

$$f(t) e^{-j(2\pi f_0)t} \overset{f}{\longleftrightarrow} F(f + f_0)$$

**Equation 9.31**

The Fourier transform pairs in Equation 9.30 and Equation 9.31 are extremely important. The reader should store these two identities in the top of his or her memory stack for quick retrieval.

The next Fourier transform pair to be defined here involves the concept of convolution. Convolution is one of the most important concepts in signal processing. It is covered in every single communications and DSP textbook. It is assumed the reader has some background in communication theory and/ or DSP and is familiar with the topic of convolution. Thus we will not deal with the basics here. We will only deal with the concept of convolution as it applies to the topic of frequency translation. If the reader wishes to study the subject of convolution further, all the texts listed in this chapter's references provide an excellent treatment of the subject.

Convolution is a big player in the translation of signal spectra. Frequency translation involves the multiplication of two-time domain sequences such as a band limited signal $x(t)$ and a sinusoidal signal $\sin(2\pi f_0 t)$. It turns out that multiplication of two signals in the time domain results in the convolution of their spectra in the frequency domain. If, for example, we have two time domain signals $x(t)$ and $h(t)$ whose Fourier transforms are given by $f\{x(t)\} = X(f)$ and $f\{h(t)\} = H(f)$, then we can state that

$$x(t)h(t) \overset{f}{\leftrightarrow} X(f) * H(f)$$

**Equation 9.32**

The symbol $*$ is used to denote the convolution operation. Equation 9.32 is a Fourier transform pair. Similarly, convolution in the time domain results in multiplication in the frequency domain. This defines another Fourier transform pair:

$$x(t) * h(t) \overset{f}{\leftrightarrow} X(f)H(f)$$

**Equation 9.33**

We can easily provide proof of the Fourier transform pair given in Equation 9.32 by the following. Let's define the convolution of $X(f)$ and $H(f)$ to be in the standard form given by

$$X(f) * H(f) = \int_{-\infty}^{\infty} X(u)H(f-u)\,du$$

**Equation 9.34**

The inverse Fourier transform of Equation 9.34 can be written as

$$f^{-1}\{X(f) * H(f)\} = \int_{-\infty}^{\infty}\left[\int_{-\infty}^{\infty} X(u)H(f-u)\,du\right]e^{j2\pi ft}\,df$$

We can rearrange terms to get

$$f^{-1}\{X(f) * H(f)\} = \int_{-\infty}^{\infty} X(u)\left[\int_{-\infty}^{\infty} H(f-u)e^{j2\pi ft}\,df\right]du$$

**Equation 9.35**

We recognize that the term within the brackets is just the inverse Fourier transform of the frequency-shifted function $H(f-u)$. We already know from the Fourier transform pair given in Equation 9.30 that

$$h(t)e^{j(2\pi u)t} \overset{f}{\leftrightarrow} H(f-u)$$

so we can substitute this expression into Equation 9.35 to get

$$f^{-1}\left\{X(f)*H(f)\right\} = h(t)\left[\int_{-\infty}^{\infty}X(u)\,e^{j(2\pi u)t}du\right]$$

If we replace $u$ with $f$ as the variable of integration, then $du = df$, and we can easily see that the quantity in the brackets is just the inverse Fourier transform of $x(t)$. Thus

$$f^{-1}\left\{X(f)*H(f)\right\} = h(t)\left[\int_{-\infty}^{\infty}X(f)\,e^{j(2\pi f)t}df\right] = x(t)h(t)$$

We end up with the Fourier transform pair:

$$x(t)h(t)\overset{f}{\leftrightarrow}X(f)*H(f)$$

**Equation 9.36**

In Equation 9.33, the proof of the Fourier transform pair

$$x(t)*h(t)\overset{f}{\leftrightarrow}X(f)H(f)$$

is achieved in a similar manner. To be thorough, we will present the proof here. Let's say we have a convolution function

$$y(t) = x(t)*h(t) = \int_{-\infty}^{\infty}x(\tau)h(t-\tau)d\tau$$

We compute the Fourier transform $Y(f) = f\left\{y(t)\right\}$ by

$$Y(f) = \int_{-\infty}^{\infty}y(t)\,e^{-j(2\pi f)t}dt = \int_{-\infty}^{\infty}\left[\int_{-\infty}^{\infty}x(\tau)h(t-\tau)d\tau\right]e^{-j(2\pi f)t}dt$$

We can change the order of integration so that

$$Y(f) = \int_{-\infty}^{\infty}x(\tau)\left[\int_{-\infty}^{\infty}h(t-\tau)\,e^{-j(2\pi f)t}dt\right]d\tau$$

Now let's define the variable $p = t - \tau$. This means that $t = p + \tau$ and $dt = dp$. If we make these substitutions in the previous equation and collect terms, we get

$$
\begin{aligned}
Y(f) &= \int_{-\infty}^{\infty} x(\tau) \left[ \int_{-\infty}^{\infty} h(p) e^{-j2\pi f(p+\tau)} dp \right] d\tau \\
&= \left[ \int_{-\infty}^{\infty} x(\tau) e^{-j2\pi f\tau} d\tau \right] \left[ \int_{-\infty}^{\infty} h(p) e^{-j2\pi fp} dp \right] \\
&= \left[ \int_{-\infty}^{\infty} x(\tau) e^{-j2\pi f\tau} d\tau \right] H(f) \\
&= X(f) H(f)
\end{aligned}
$$

We end up with the Fourier transform pair:

$$
x(t) * h(t) \overset{f}{\leftrightarrow} X(f) H(f)
$$

## 9.1.8 Real Frequency Translation

Frequency translation of a signal by a sine or cosine function is oftentimes referred to as real frequency translation simply because the translating signal is a real valued function.

In sections 9.1.3 and 9.1.4 we introduced the spectra of the real signals commonly used for frequency translation—that is, $\cos(2\pi f_0 t)$ and $\sin(2\pi f_0 t)$. In this section, we will observe the frequency domain translations resulting from a time domain multiplication such as $y(t) = g(t) f(t)$, where $g(t)$ is defined to be a narrow band signal, and $f(t)$ is defined to be a sinusoidal tuning signal.

### 9.1.8.1 Translation of a Real Signal by cos(2πf₀t)

Suppose we have a band limited message signal $g(t)$ that is multiplied by a translating signal $f(t) = \cos(2\pi f_0 t)$. We end up with a product signal $y(t) = g(t) f(t) = g(t) \cos(2\pi f_0 t)$, as illustrated in Figure 9.10.

If the message signal $g(t)$ has a spectrum defined by $G(f)$, and if the translating signal $f(t)$ has a spectrum defined by $F(f)$, then the multiplication of $g(t)$ and $f(t)$ will result in the convolution of their spectra $G(f)$ and $F(f)$. That is,

$$
y(t) = g(t) f(t) \overset{f}{\leftrightarrow} Y(f) = G(f) * F(f)
$$

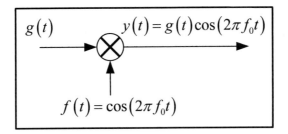

**Figure 9.10**  Translation of a real signal by cos(2πf₀t)

Let's begin by looking at the case where $g(t)$ is a simple cosine wave-form at frequency $f_m$ such that $g(t) = \cos(2\pi f_m t)$. We know from previous study that the spectrum of $f(t)$ is given by

$$F(f) = \frac{1}{2}\{\delta(f - f_0) + \delta(f + f_0)\}$$

Similarly we know that the spectrum of $g(t)$ is given by

$$G(f) = \frac{1}{2}\{G(f - f_m) + G(f + f_m)\} = \frac{1}{2}\{\delta(f - f_m) + \delta(f + f_m)\}$$

The convolution of $G(f)$ and $F(f)$ can be written as

$$G(f) * F(f) = \left[\frac{1}{2}\{\delta(f - f_m) + \delta(f + f_m)\}\right] * \left[\frac{1}{2}\delta(f + f_0) + \frac{1}{2}\delta(f - f_0)\right]$$

**Equation 9.37**

Using the distribution property, we can split Equation 9.37 into two terms to get

$$G(f) * F(f) = \left[\frac{1}{2}\{\delta(f - f_m) + \delta(f + f_m)\}\right] * \left[\frac{1}{2}\delta(f + f_0)\right] + \left[\frac{1}{2}\{\delta(f - f_m) + \delta(f + f_m)\}\right] * \left[\frac{1}{2}\delta(f - f_0)\right]$$

**Equation 9.38**

Convolving each of the two terms in Equation 9.38, we arrive at

$$G(f) * F(f) = \frac{1}{4}\left\{\delta(f + f_0 - f_m) + \delta(f + f_0 + f_m)\right\} + \frac{1}{4}\left\{\delta(f - f_0 - f_m) + \delta(f - f_0 + f_m)\right\}$$

**Equation 9.39**

The process by which we arrived at Equation 9.39 is illustrated graphically in Figure 9.11. The line spectrum of the message signal $g(t) = \cos(2\pi f_m t)$ is illustrated in part A of the figure. The line spectrum of the translating or tuning signal $f(t) = \cos(2\pi f_0 t)$ is illustrated in part B of the figure. The result of the convolution of the two is illustrated in part C of the figure.

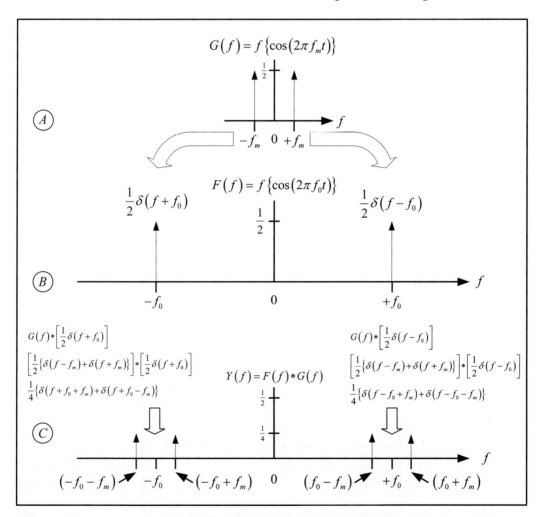

**Figure 9.11**   Translation of message signal $\cos(2\pi f_m t)$ by $\cos(2\pi f_0 t)$

We can clearly see that the line spectra of the message signal have been translated both up and down in frequency such that its original center frequency of 0 Hz has been moved to $\pm f_0$ Hz. The end result is identical to the message signal being convolved with the two cosine frequency impulses. Note that the amplitude of the translated $G(f)$ line spectra was scaled from 1/2 to 1/4, as we would expect.

Now lets's take a look at the case where $g(t)$ is a finite bandwidth signal with a continuous spectrum such that that illustrated in Figure 9.12. We multiply $g(t)$ by the same signal $f(t) = \cos(2\pi f_0 t)$ as before to translate the spectrum of $g(t)$ to $\pm f_0$ Hz. The Fourier transform pair defined by Equation 9.27 tells us that when we perform the multiplication, we should expect to see

$$g(t)\cos(2\pi f_0 t) \overset{f}{\leftrightarrow} \frac{1}{2}\{G(f - f_0) + G(f + f_0)\}$$

**Equation 9.40**

The frequency domain convolution of the carrier signal $f(t) = \cos(2\pi f_0 t)$ and the message signal $g(t)$ are illustrated graphically in Figure 9.13. The spectrum of the message signal $g(t)$ is shown in part A of the figure. The line spectrum of the translating or tuning signal $f(t) = \cos(2\pi f_0 t)$ is shown in part B of the figure. The convolution of the two signal spectra is illustrated in part C of the figure.

As we can see, the resultant spectrum agrees with Equation 9.40 in that the message signal has been translated both up and down in frequency from

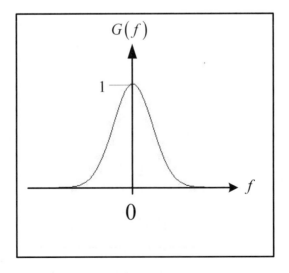

**Figure 9.12** Finite bandwidth message signal spectrum

a center frequency of $0$ Hz to a new center frequency of $\pm f_0$ Hz, and the amplitude has been scaled by $1/2$.

This type of frequency translation is sometimes referred to as *linear modulation* [3], in which the amplitude of the carrier $f_0$ in our example varies in a one-to-one correspondence to the amplitude of the message signal $g(t)$. This type of frequency translation is also referred to as *double sideband amplitude modulation* (DSB-AM). A DSB-AM signal results from the multiplication of the message signal by a carrier. The portion of the message signal $\frac{1}{2}G(f - f_0)$ that is above the carrier $+f_0$ in Figure 9.13 is called the *upper sideband*, and

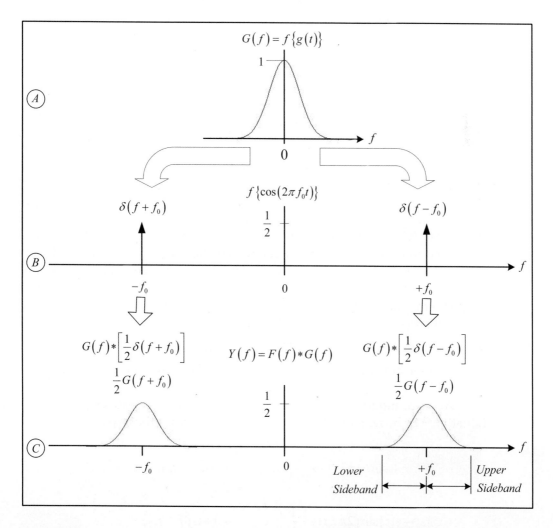

**Figure 9.13** Translation of a band limited signal with a continuous spectrum

the portion of the message signal $\frac{1}{2}G(f-f_0)$ that is below the carrier $+f_0$ is called the *lower sideband*.

### 9.1.8.2 Translation of a Real Signal by sin(2πf₀t)

Now let's take a look at the spectral translation that results when a band limited message signal $g(t)$ with a spectrum as shown in Figure 9.12 is multiplied by the translation or tuning signal $\sin(2\pi f_0 t)$. The block diagram of the multiplication of the two signals is illustrated in Figure 9.14. We could derive all the mathematics that describe the spectral translation, but we've already done that and all the information we need is found in Equation 9.18 and Equation 9.29, which are repeated here for clarity:

$$\sin(2\pi f_0 t) \overset{f}{\leftrightarrow} \frac{1}{2j}\{\delta(f-f_0)-\delta(f+f_0)\}$$

$$g(t)\sin(2\pi f_0 t) \overset{f}{\leftrightarrow} \frac{1}{2j}\{G(f-f_0)-G(f+f_0)\}$$

Figure 9.15 illustrates the frequency domain convolution of the line spectra of the tuning signal $\sin(2\pi f_0 t)$ and the narrow band spectrum of the message signal $g(t)$. The narrow band spectrum of the message signal $G(f)$ is shown in part A of the figure. The line spectrum of the tuning signal $f\{\sin(2\pi f_0 t)\}$ is shown in part B of the figure. The convolution of the two is illustrated in part C of the figure. The reader should keep in mind that this is a plot of amplitudes, not magnitudes. This is why we see negative spectral amplitudes and values like $1/2j$ for $Y(f)$.

The magnitude plot of $Y(f)$ is shown in Figure 9.16. This is the more common representation of translated signal spectra.

### 9.1.8.3 Translation of a Real Signal by –sin(2πf₀t)

Now let's see how the spectrum of a signal $g(t)$ is translated when it is multiplied by $-\sin(2\pi f_0 t)$, as illustrated in Figure 9.17.

We can modify the Fourier transform pairs of Equation 9.18 and Equation 9.29 by negating the sine function to produce the two new pairs:

$$-\sin(2\pi f_0 t) \overset{f}{\leftrightarrow} \frac{1}{2j}\{-\delta(f-f_0)+\delta(f+f_0)\}$$

$$-g(t)\sin(2\pi f_0 t) \overset{f}{\leftrightarrow} \frac{1}{2j}\{-G(f-f_0)+G(f+f_0)\}$$

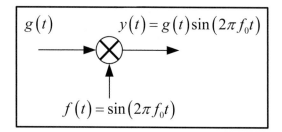

**Figure 9.14**   Translation of a real signal by sin($2\pi f_0 t$)

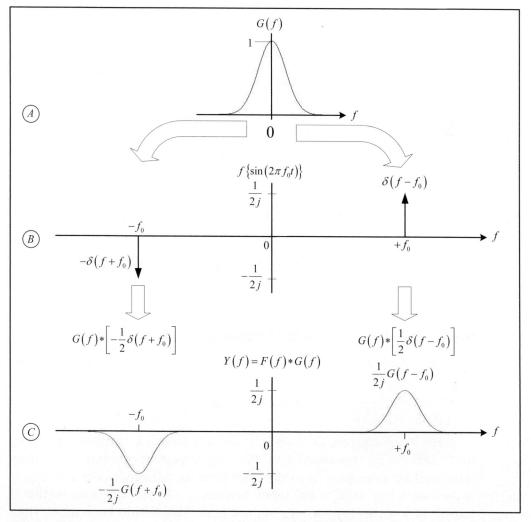

**Figure 9.15**   Spectral amplitude of $G(f)$ translated by sin($2\pi f_0 t$)

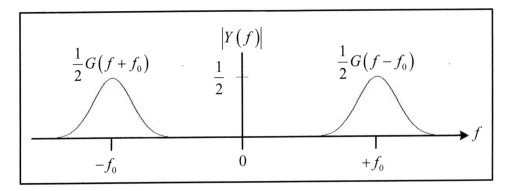

**Figure 9.16**  Spectral magnitude of $G(f)$ translated by $\sin(2\pi f_0 t)$

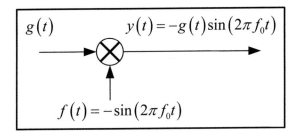

**Figure 9.17**  Translation of a real signal by $-\sin(2\pi f_0 t)$

These equations are graphically represented in Figure 9.18. We can see from the figure that when we change the sign of the tuning signal, we end up inverting the translated spectral amplitudes.

## 9.1.9  Complex Frequency Translation

A complex tuning signal can be represented by a simple exponential, such as

$$e^{\pm j(2\pi f_0)t} = \cos(2\pi f_0 t) \pm j\sin(2\pi f_0 t)$$

For many reasons, DSP designs typically utilize a complex tuning signal to perform spectral translation. One very important reason is that a signal translated by a complex tuner does not produce mirrored spectra at $\pm f_0$, as is the case when using a real tuner. Instead, it will produce a signal that is translated to either $+f_0$ or $-f_0$, but not both. This should be intuitive since the spectra of a complex exponential is a single frequency impulse at either

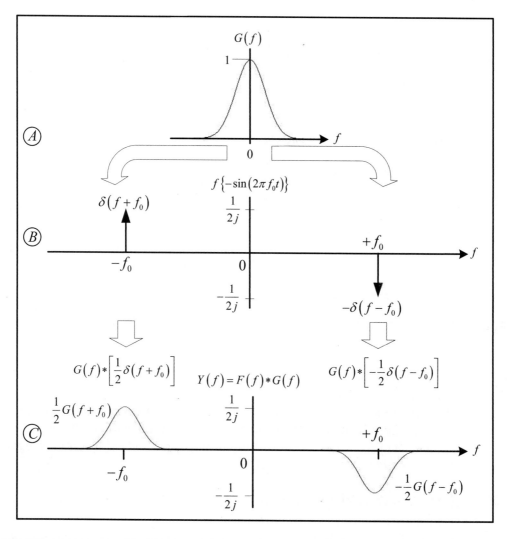

**Figure 9.18** Spectral amplitude of $G(f)$ translated by $-\sin(2\pi f_0 t)$

$+f_0$ or $-f_0$, as defined in Equation 9.13 and Equation 9.14. This means that the spectra of a base band signal will be convolved with a single frequency domain impulse as opposed to two.

### 9.1.9.1 Translation of a Real Signal by $e^{j(2\pi f_0)t}$

Let's begin by multiplying the real input message signal $g(t) = \cos(2\pi f_m t)$ by the complex exponential tuning signal $f(t) = e^{j(2\pi f_0)t}$. The tuner that performs the multiplication of the two signals is illustrated in Figure 9.19.

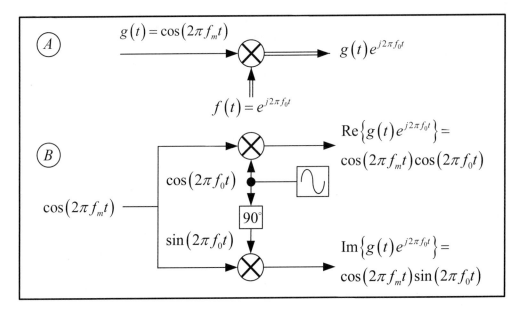

**Figure 9.19** Translation of a real signal by $e^{j2\pi f_0 t}$

Part A of the figure is a graphic shorthand interpretation of the complex tuner. The single line with an arrow is typically used to indicate a real signal, while the double line with an arrow typically indicates a complex signal. Part B of the figure is the expanded block diagram of the complex tuner. As can be seen, the product signal out of the multiplier is indeed a complex signal expressed as

$$y(t) = g(t)e^{j(2\pi f_0)t} = \cos(2\pi f_m t)e^{j(2\pi f_0)t} = \cos(2\pi f_m t)\cos(2\pi f_0 t) + j\cos(2\pi f_m t)\sin(2\pi f_0 t)$$

If implemented in hardware, the product signal would require two signal paths: one for the real part of the product and another for the imaginary part of the product. The sinusoid drawn in a box represents a simple tunable tone oscillator used to generate the tuning signal. The oscillator output is used as the real part of the complex tuning signal. The $90°$ phase shift creates the imaginary or quadrature component of the complex tuning signal. If we expand the exponential term, the complex product signal becomes

$$y(t) = \cos(2\pi f_m t)\left[\cos(2\pi f_0 t) + j\sin(2\pi f_0 t)\right]$$

The Fourier transform of $y(t)$ is then computed by

$$Y(f) = \int_{-\infty}^{\infty} \left[ g(t)e^{j(2\pi f_0)t} \right] e^{-j(2\pi f)t} dt$$

$$= \int_{-\infty}^{\infty} \left[ \cos(2\pi f_m t)e^{j(2\pi f_0)t} \right] e^{-j(2\pi f)t} dt$$

$$= \int_{-\infty}^{\infty} \left[ \cos(2\pi f_m t) \right] e^{-j2\pi(f-f_0)t} dt$$

$$= \int_{-\infty}^{\infty} \left[ \frac{e^{j(2\pi f_m)t} + e^{-j(2\pi f_m)t}}{2} \right] e^{-j2\pi(f-f_0)t} dt$$

Separation of terms produces the sum of two transforms:

$$Y(f) = \frac{1}{2}\int_{-\infty}^{\infty} e^{j(2\pi f_m)t} \, e^{-j2\pi(f-f_0)t} dt \;+\; \frac{1}{2}\int_{-\infty}^{\infty} e^{-j(2\pi f_m)t} \, e^{-j2\pi(f-f_0)t} dt$$

$$= \frac{1}{2}\int_{-\infty}^{\infty} e^{-j2\pi(f-f_0-f_m)t} \;+\; \frac{1}{2}\int_{-\infty}^{\infty} e^{-j2\pi(f-f_0+f_m)t} \, dt$$

Through the use of previously defined transform pairs, we arrive at the translated spectral representation of $y(t) = \cos(2\pi f_m t)e^{j2\pi f_0 t}$:

$$Y(f) = \frac{1}{2}\left\{ \delta(f - f_0 - f_m) + \delta(f - f_0 + f_m) \right\}$$

**Equation 9.41**

The spectrum of the real input message signal $g(t) = \cos(2\pi f_m t)$ is shown in Figure 9.20. As we would expect, the spectrum consists of two frequency impulses at $\pm f_m$.

The spectrum of the tuning signal $f(t) = e^{j2\pi f_0 t}$ is shown in Figure 9.21. As we would expect from our previous derivations, the spectrum of the tuning

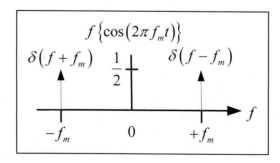

**Figure 9.20** Spectrum of the input message signal $\cos(2\pi f_m t)$

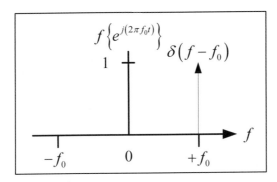

**Figure 9.21**   Spectrum of the tuning signal $e^{j2\pi f_0 t}$

signal consists of a single frequency impulse centered at $f_0$. Multiplication of the base band message signal by the tuning frequency results in the convolution of the base band spectra with the impulse spectra of the tuning frequency. The spectrum of the translated product signal is illustrated in Figure 9.22. We should also note that, unlike a real tuner, the product signal is complex and the spectrum at the output of a complex tuner does not have a mirror image centered $-f_0$.

The same results are true if we are dealing with an input signal message $g(t)$ that has a finite continuous bandwidth. This is illustrated in Figure 9.23. The spectrum of the band limited input signal $g(t)$ is shown in part A of the figure. The spectrum of the complex tuning signal $f(t) = e^{j(2\pi f_0)t}$ is illustrated in part B of the figure. The translated spectrum resulting from the multiplication of $f(t)$ and $g(t)$ is shown in part C of the figure. It is clear from the figure that the spectra of the two signals have been convolved with one another. Note that there is no mirror image band at $-f_0$.

### 9.1.9.2   Translation of a Real Signal by $e^{-j(2\pi f_0)t}$

Now we will look at the results of a real, band limited input message signal $g(t) = \cos(2\pi f_m t)$ that is multiplied by the complex exponential tuning signal $f(t) = e^{-j(2\pi f_0)t}$. The multiplication of the two signals is illustrated in Figure 9.24. Part A of the figure is a graphic shorthand interpretation of the complex multiplication. Once again, standard convention has the single line arrow indicating a real signal while the double line arrow indicates a complex signal. Part B of the figure is the expanded block diagram of the complex multiplication. The product signal out of the multiplier is

$$y(t) = g(t)e^{-j(2\pi f_0)t} = \cos(2\pi f_m t)e^{-j(2\pi f_0)t}$$

If we expand the exponential term, the complex product signal becomes

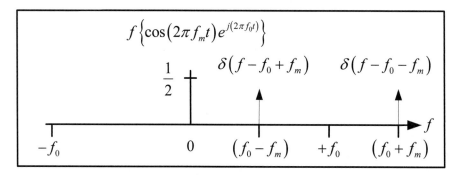

**Figure 9.22**   Spectrum $\cos(2\pi f_m t)$ translated to $+f_0$

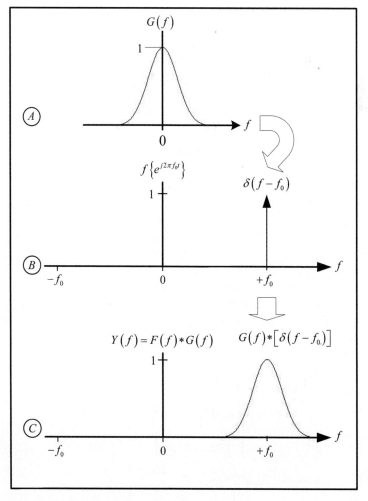

**Figure 9.23**   Translation of a band limited signal by $e^{j(2\pi f_0)t}$

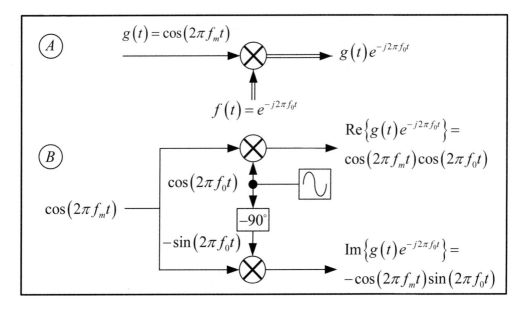

**Figure 9.24** Translation of a real signal by $e^{-j2\pi f_0 t}$

$$y(t) = \cos(2\pi f_m t)\left[\cos(2\pi f_0 t) - j\sin(2\pi f_0 t)\right]$$

With the exception of the conceptual $j$ term, the complex product is identical to that illustrated in the figure. The product signal out of the multiplier is indeed a complex signal. If implemented in hardware, the product signal would require two signal paths: one for the real part of the product and another for the imaginary part of the product. Multiplication in the time domain is the equivalent of convolution in the frequency domain. We could fall back on our previously derived identities to compute the Fourier transform of the product signal, but this is such an important concept we will invest the time in deriving the mathematics that describe this complex tuning process.

The Fourier transform of $y(t)$ is computed by

$$
\begin{aligned}
Y(f) &= \int_{-\infty}^{\infty}\left[g(t)e^{-j(2\pi f_0)t}\right]e^{-j(2\pi f)t}\,dt \\
&= \int_{-\infty}^{\infty}\left[\cos(2\pi f_m t)e^{-j(2\pi f_0)t}\right]e^{-j(2\pi f)t}\,dt \\
&= \int_{-\infty}^{\infty}\left[\cos(2\pi f_m t)\right]e^{-j2\pi(f+f_0)t}\,dt \\
&= \int_{-\infty}^{\infty}\left[\frac{e^{j(2\pi f_m)t}+e^{-j(2\pi f_m)t}}{2}\right]e^{-j2\pi(f+f_0)t}\,dt
\end{aligned}
$$

Separation of terms produces the sum of two transforms:

$$Y(f) = \frac{1}{2}\int_{-\infty}^{\infty} e^{j(2\pi f_m)t} \, e^{-j2\pi(f+f_0)t} dt \quad + \frac{1}{2}\int_{-\infty}^{\infty} e^{-j(2\pi f_m)t} \, e^{-j2\pi(f+f_0)t} dt$$

$$= \frac{1}{2}\int_{-\infty}^{\infty} e^{-j2\pi(f+f_0-f_m)t} dt \quad + \frac{1}{2}\int_{-\infty}^{\infty} e^{-j2\pi(f+f_0+f_m)t} dt$$

We can substitute previously defined transform pairs for each of the two terms and arrive at the translated spectral representation of $y(t) = \cos(2\pi f_m t)e^{-j2\pi f_0 t}$:

$$Y(f) = \frac{1}{2}\left\{ \delta(f+f_0-f_m) + \delta(f+f_0+f_m) \right\}$$

**Equation 9.42**

The spectrum of the real input signal $g(t) = \cos(2\pi f_m t)$ was shown in Figure 9.20 and consists of two frequency impulses at $\pm f_m$. The spectrum of the tuning signal $f(t) = e^{-j2\pi f_0 t}$ is shown in Figure 9.25.

The spectrum of the tuned or translated signal is shown in Figure 9.26. Note that when multiplied by a complex exponential, the original signal spectrum is merely translated so that it becomes centered on frequency of the tuning signal. Since the tuning signal is complex, there is no mirror image at $+f_0$, as was the case when the tuning signal was real. It's easily seen that the spectrum of the input signal has been convolved with the center frequency of the complex tuning signal.

The same results are obtained if the input signal has a continuous band limited spectrum, as illustrated in Figure 9.27. The spectrum of the narrow band message signal is illustrated in part A of Figure 9.27. The spectrum of the tuning signal is illustrated in part B, and the tuned or translated signal is illustrated in part C.

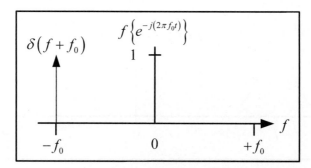

**Figure 9.25**   Spectrum of the tuning signal $e^{-j2\pi f_0 t}$

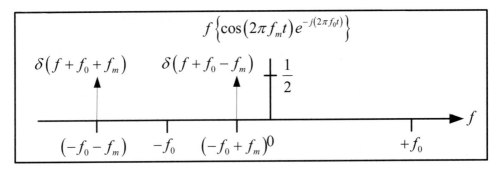

**Figure 9.26**  Spectrum $\cos(2\pi f_m t)$ translated to $-f_0$

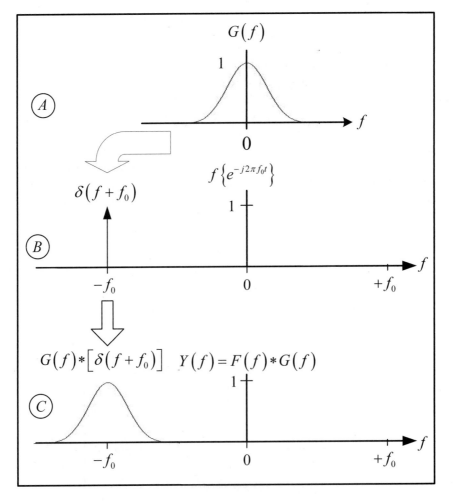

**Figure 9.27**  Translation of a band limited signal by $e^{-j2\pi f_0 t}$

This is a pretty handy result. We now know how to move a signal up and down in frequency. If a real signal is applied to a tuner, the signal spectra can be translated in frequency and filtered by an intermediate frequency filter to extract a specific narrow band signal for further processing. This is analogous to a digital AM radio where the wideband signal is the entire AM spectrum and the narrow bands output from the tuner filter combination are the individual AM radio station channels. In a "digital-radio-like" processor, the intermediate frequency (IF) would be 0 Hz or DC and the intermediate frequency filter would more than likely be a decimate by $M$ low pass filter.

### 9.1.9.3 Translation of a Complex Signal by $e^{j(2\pi f_0)t}$

In this section, we will illustrate the frequency translation of a complex input message signal $g(t) = e^{j(2\pi f_m)t}$ by multiplying it with a complex tuning signal $f(t) = e^{j(2\pi f_0)t}$. The block diagram of this complex multiply is illustrated in Figure 9.28.

Part A of the figure is the shorthand drawing notation of a complex tuner, and part B of the figure is the expanded block diagram of the tuner that performs a complex by complex multiply.

The complex input signal in this example is a simple complex exponential given by

$$g(t) = e^{j(2\pi f_m)t} = \cos(2\pi f_m t) + j\sin(2\pi f_m t)$$

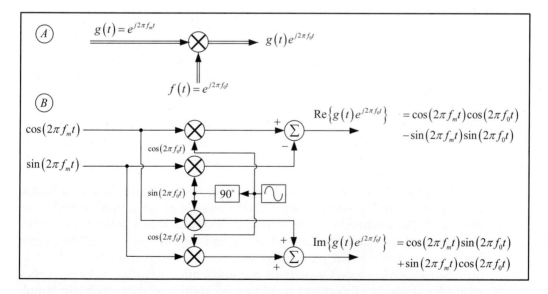

**Figure 9.28** Block diagram of translation of a complex signal by $e^{j2\pi f_0 t}$

The input signal is multiplied by a complex tuning exponential given by

$$f(t) = e^{j(2\pi f_0)t} = \cos(2\pi f_0 t) + j\sin(2\pi f_0 t)$$

The tuned output or product signal can be mathematically represented by

$$y(t) = g(t)f(t) = e^{j(2\pi f_m)t}e^{j(2\pi f_0)t}$$

The Fourier transform of $Y(f) \overset{f}{\longleftrightarrow} y(t)$ is computed to be

$$Y(f) = \int_{-\infty}^{\infty}\left[g(t)e^{j(2\pi f_0)t}\right]e^{-j(2\pi f)t}\,dt$$

$$= \int_{-\infty}^{\infty}\left[e^{j(2\pi f_m)t}\,e^{j(2\pi f_0)t}\right]e^{-j(2\pi f)t}\,dt$$

$$= \int_{-\infty}^{\infty}\left[e^{j2\pi(f_0+f_m)t}\right]e^{-j(2\pi f)t}\,dt$$

$$= \int_{-\infty}^{\infty}\left[e^{-j2\pi(f-f_0-f_m)t}\right]dt$$

Using the transform pair identities that we previously derived, we arrive at

$$Y(f) = \delta(f - f_0 - f_m)$$

The spectrum of $g(t)$, $f(t)$, and $y(t)$ are shown in Figure 9.29. In part A of the figure, we see the line spectra of the input message signal $g(t)$. Notice that $G(f)$ is a one-sided spectrum consisting of a single frequency impulse at $f_m$. One of the advantages of processing complex signals is the fact that they have a one-sided spectrum.

Part B of the figure shows the line spectra of the tuning signal $f(t)$, which consists of a single frequency impulse at $f_0$. Part C of the figure shows that the input message signal spectra have been convolved with the tuning signal spectra, resulting in a frequency translation of the input from $f_m$ to $f_0 + f_m$.

Let's stop here for a moment and imagine in our minds a system that extracts and then processes narrow band signals at some intermediate frequency $f_0$. All we need to do is translate the input signal either up or down in frequency, so that it is centered at $f_0$, and extract the desired frequency band with a fixed band pass filter. We then pass the filtered band off to downstream processors. Or perhaps our application requires us to frequency multiplex several independent band limited signals so that we might simultaneously transmit them all to some remote location on a single copper wire.

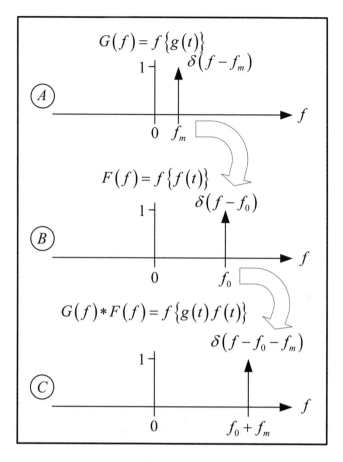

**Figure 9.29**   Translation of a complex signal by $e^{j2\pi f_0 t}$

This process is called *frequency division multiplexing,* or FDM for short. In this application, we might have a hundred or so independent base band signals, such as 4 KHz bandwidth telephone channels, that need multiplexed onto a single transmission medium. In this application, we could have a hundred multiply circuits of the type shown in Figure 9.28, with tuning frequencies spaced at 4 KHz intervals. The translated signals would be summed together and then transmitted on a single wire, with each signal occupying its own 4 KHz band. The reader might scoff at using 100 complex tuners to multiplex 100 independent signals. Consider the alternative of using 100 separate wires to transmit the same information. The trade off of using 1 or 100 wires in terms of expense, manpower, and environmental impact is significant, and this only takes into consideration a mere 100 telephone circuits. Multiply this by millions of telephones, and compare it to the miniscule on-chip real estate required for the necessary multiplex tuners, and then compute the trade off.

FDM, however, is old technology and pales in comparison to the more modern digital time division multiplexing (TDM) technology that can multiplex thousands of independent telephone channels on a single fiber-optic cable. Demultiplexing a particular channel of an FDM signal at the destination only requires that the multiplexed signal be down converted or tuned to base band and then low pass filtered and sent off a follow on processor such as a demodulator.

### 9.1.9.4 Translation of a Complex Signal by $e^{-j(2\pi f_0)t}$

Now let's take a look at the case where the same input signal $g(t)$ is multiplied by a negative exponential waveform, as illustrated in Figure 9.30. The compact or shorthand drawing notation of the complex tuner is shown in part A of the figure, and the expanded block diagram of the complex tuner is shown in part B of the figure. Note that the block diagram of Figure 9.30 is identical to that block diagram in Figure 9.28, with the exception of the sign changes on the input to the two summers. The input message signal in this example is once again a simple complex exponential given by

$$g(t) = e^{j(2\pi f_m)t} = \cos(2\pi f_m t) + j\sin(2\pi f_m t)$$

The input signal is multiplied by the tuning signal, which is another complex exponential given by

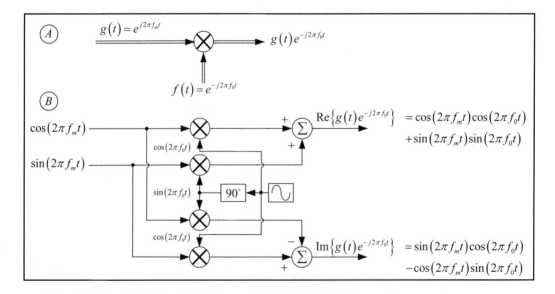

**Figure 9.30** Block diagram of translation of a complex signal by $e^{-j2\pi f_0 t}$

$$f(t) = e^{-j(2\pi f_0)t} = \cos(2\pi f_0 t) - j\sin(2\pi f_0 t)$$

The translated output signal can be mathematically represented by

$$y(t) = g(t)f(t) = e^{j(2\pi f_m)t}e^{-j(2\pi f_0)t}$$

The Fourier transform of $y(t)$ is computed to be

$$Y(f) = \int_{-\infty}^{\infty} \left[ g(t)e^{-j(2\pi f_0)t} \right] e^{-j(2\pi f)t} \, dt$$

$$= \int_{-\infty}^{\infty} \left[ e^{j(2\pi f_m)t} \; e^{-j(2\pi f_0)t} \right] e^{-j(2\pi f)t} \, dt$$

$$= \int_{-\infty}^{\infty} \left[ e^{-j2\pi(f_0 - f_m)t} \right] e^{-j(2\pi f)t} \, dt$$

$$= \int_{-\infty}^{\infty} \left[ e^{-j2\pi(f + f_0 - f_m)t} \right] dt$$

Using the transform pair identities that we previously derived, we arrive at

$$Y(f) = \delta(f + f_0 - f_m)$$

The spectrum of $g(t)$, $f(t)$, and $y(t)$ are shown in Figure 9.31. The one-sided line spectrum of the input signal $g(t)$ is shown in part A. The line spectrum of the tuning signal $f(t)$ is shown in part B. Part C of the figure shows that the input signal spectra have been convolved with the tuning signal spectra, resulting in a frequency down conversion from $f_m$ down to $-f_0 + f_m$.

In both the previous examples, the message frequency was translated by $f_0$ Hz. The positive exponential tuning signal translated the message signal by $+f_0$ Hz, and the negative exponential tuning signal translated the message signal by $-f_0$ Hz. The whole operation is very straightforward.

## 9.1.10   Complex Conjugate Symmetry

Let's multiply the narrow band message signal $2g(t)$ by a real tuning signal $\cos(2\pi f_0 t)$ to produce a translated signal that we call $y(t)$. That is,

$$y(t) = 2g(t)\cos(2\pi f_0 t)$$

Since we are dealing with a real tuning signal, we will expect to see the narrow band spectrum of $g(t)$ translated to two new center frequencies of $\pm f_0$. That is, we would expect the modulated or tuned spectrum to be

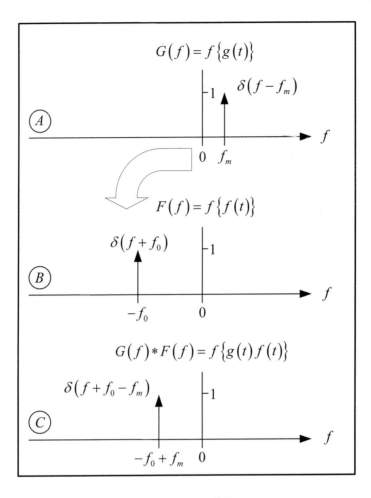

**Figure 9.31**  Translation of a complex signal by $e^{-j2\pi f_0 t}$

$$2g(t)\cos(2\pi f_0 t) \overset{f}{\leftrightarrow} 2\left[\frac{1}{2}\{G(f-f_0) + G(f+f_0)\}\right] = \{G(f-f_0) + G(f+f_0)\}$$

The time domain multiplication results in a frequency domain convolution to produce the spectrum illustrated in Figure 9.32. We expand the output product in the time domain to get

$$
\begin{aligned}
y(t) &= 2g(t)\cos(2\pi f_0 t) \\
&= g(t)\left[e^{j2\pi f_0 t} + e^{-j2\pi f_0 t}\right] \\
&= g(t)e^{j2\pi f_0 t} + g(t)e^{-j2\pi f_0 t}
\end{aligned}
$$

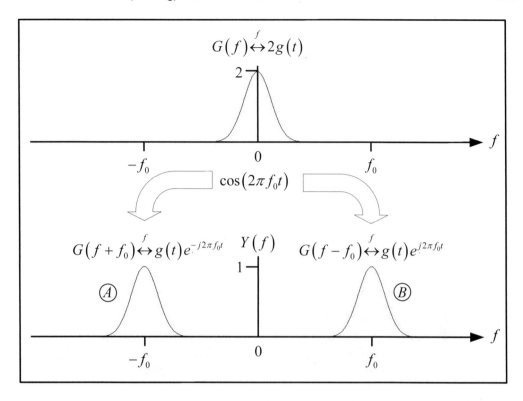

**Figure 9.32** Conjugate symmetric spectrum

Further expansion using Euler's identity produces

$$y(t) = g(t)\left[\cos(2\pi f_0 t) + j\sin(2\pi f_0 t)\right] + g(t)\left[\cos(2\pi f_0 t) - j\sin(2\pi f_0 t)\right]$$

**Equation 9.43**

The two terms on the right-hand side of Equation 9.43 are complex conjugates of one another. Each term corresponds to one side of the two-sided spectrum illustrated in Figure 9.32. It is important to understand that each of the two components of $y(t)$ can be reconstructed from the other. Therefore, after a frequency translation of this type, it is only necessary for a system to utilize either one of the two translated components.

For example, if we wished to demodulate $y(t)$, we only need to multiply by the complex signal $e^{-j2\pi f_0 t}$ to get

$$\begin{aligned} y(t)e^{-j2\pi f_0 t} &= \left[g(t)e^{j2\pi f_0 t} + g(t)e^{-j2\pi f_0 t}\right]e^{-j2\pi f_0 t} \\ &= g(t) + g(t)e^{-j2\pi 2 f_0 t} \end{aligned}$$

Taking the Fourier transform of both sides of this equation shows the translated signal spectrum to be

$$f\left\{y(t)e^{-j2\pi f_0 t}\right\} = G(f) + G(f + 2f_0)$$

This would down convert the spectrum $Y(f)$ such that spectrum component $B$ of Figure 9.32 would be translated down in frequency by $f_0$ Hz and centered at 0 Hz, while the spectrum component $A$ would be translated down in frequency by $f_0$ Hz and centered at $-2f_0$ Hz. The down converted signal would then be passed through a low pass filter. The filter would pass the down converted spectral component $B$ and reject the down converted spectral component $A$. This down conversion and filter process reduces the signal bandwidth without signal degradation. This process is illustrated in section 9.1.11.

## 9.1.11 Translation Example

At this point, a graphical example might be useful to illustrate the properties of signal tuning. Figure 9.33 shows a simple artist's conception of an FDM signal $g(t)$, composed of $N$ real channels. Each channel carries a signal independent of all others and is identified by its unique spectra. The figure depicts these channels as occupying equal bandwidth slots but centered at constant increments of center frequency. The artist's conception of independent signals with unique spectra is illustrated in the figure by a collection of imaginative but simple spectral representations.

This example defines the composite FDM signal $g(t)$ to be a real signal. Why? Because the spectra are two sided. It is our task to input $g(t)$ to into a box of hardware that we design and extract a particular channel for

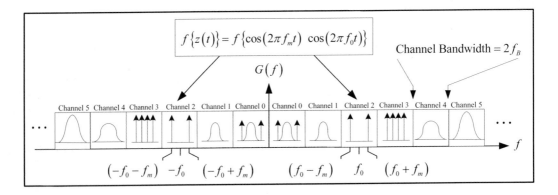

**Figure 9.33** Spectrum of an FDM signal

downstream processing. In this case, our objective is to extract and process the signal positioned in channel 2. We will define the signal that occupies the second channel slot to be $z(t) = \cos(2\pi f_m t)\cos(2\pi f_0 t)$. For ease of demonstration, $f\{z(t)\}$ is a simple line spectra centered at frequency $f_0$, but it could just as easily be a band limited continuous spectrum such as that which occupies channel 4. The first thing we need to do is down convert $G(f)$ by $f_0$ Hz such that the channel of interest (in this case, channel 2) is centered at base band or 0 Hz. This means we multiply $g(t)$ by the complex tuner signal $e^{-j2\pi f_0 t}$. This process is illustrated in Figure 9.34. Part A of the figure shows the spectrum of the original input signal $g(t)$. Part B of the figure shows the same spectrum translated down in frequency by $f_0$ Hz. We name the translated signal $y(t)$. Note that channel 2 is now a base band signal centered at 0 Hz.

We can describe this process mathematically by the following. The signal that occupies channel 2 is real and is given by

$$z(t) = \cos(2\pi f_m t)\cos(2\pi f_0 t)$$

If we multiply by the complex tuning signal $f(t) = e^{-j2\pi f_0 t}$, we get

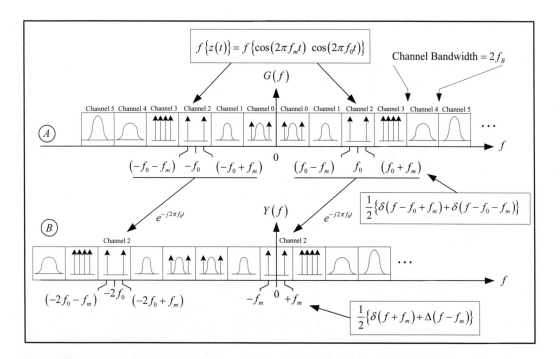

**Figure 9.34**   Down conversion of an FDM signal

$$z(t)e^{-j(2\pi f_0)t} = \left[ \cos(2\pi f_m t)\cos(2\pi f_0 t) \right] e^{-j(2\pi f_0)t}$$

**Equation 9.44**

We can expand Equation 9.44 using the Euler identity to get

$$z(t)e^{-j(2\pi f_0)t} = \left[ \frac{e^{j(2\pi f_m)t} + e^{-j(2\pi f_m)t}}{2} \right]\left[ \frac{e^{j(2\pi f_0)t} + e^{-j(2\pi f_0)t}}{2} \right] e^{-j(2\pi f_0)t}$$

We can separate terms to get

$$z(t)e^{-j(2\pi f_0)t} = \left[ \frac{e^{j2\pi(f_0+f_m)t}}{4} + \frac{e^{-j2\pi(f_0-f_m)t}}{4} + \frac{e^{j2\pi(f_0-f_m)t}}{4} + \frac{e^{-j2\pi(f_0+f_m)t}}{4} \right] e^{-j(2\pi f_0)t}$$

Multiplication by the tuning exponential $e^{-j(2\pi f_0)t}$ produces the final time domain equation:

$$z(t)e^{-j(2\pi f_0)t} = \left[ \left( \frac{e^{j(2\pi f_m)t}}{4} + \frac{e^{-j(2\pi f_m)t}}{4} \right) + \left( \frac{e^{-j2\pi(2f_0-f_m)t}}{4} + \frac{e^{-j2\pi(2f_0+f_m)t}}{4} \right) \right]$$

**Equation 9.45**

For convenience, let's name the two terms on the right-hand side $q_1(t)$ and $q_2(t)$, such that

$$z(t)e^{-j(2\pi f_0)t} = \left[ \left( \frac{e^{j(2\pi f_m)t}}{4} + \frac{e^{-j(2\pi f_m)t}}{4} \right) + \left( \frac{e^{-j2\pi(2f_0-f_m)t}}{4} + \frac{e^{-j2\pi(2f_0+f_m)t}}{4} \right) \right] = q_1(t) + q_2(t)$$

**Equation 9.46**

Using the Fourier transform pair identities, we can easily convert Equation 9.46 to the frequency domain. The resulting frequency domain equation is

$$f\{z(t)e^{-j(2\pi f_0)t}\} = Z(f) = \frac{1}{4}\{\delta(f - f_m) + \delta(f + f_m)\} + \frac{1}{4}\{\delta(f + 2f_0 - f_m) + \delta(f + 2f_0 + f_m)\}$$

**Equation 9.47**

We can name the two terms on the right-hand side of Equation 9.47 $Q_1(f)$ and $Q_2(f)$, respectively, such that

$$Z(f) = \frac{1}{4}\left\{\delta(f - f_m) + \delta(f + f_m)\right\} + \frac{1}{4}\left\{\delta(f + 2f_0 - f_m) + \delta(f + 2f_0 + f_m)\right\} = Q_1(f) + Q_2(f)$$

**Equation 9.48**

We recognize that the translated spectrum defined by $Z(f)$ is the spectrum of the second channel of the translated wideband spectrum $Y(f)$. That is, multiplying the time domain signal $g(t)$ by the complex exponential $e^{-j(2\pi f_0)t}$ has down converted the entire wideband spectrum $G(f)$ by $f_0$ Hz. By definition, the time domain signal that occupied channel two represented by $z(t)$ has also been down converted by $f_0$ Hz, such that its two-sided spectrum $Z(f)$ is now centered at 0 Hz and $-2f_0$ Hz, defined by $Q_1(f)$ and $Q_2(f)$, respectively, in Equation 9.48. All we need to do now to extract the signal that occupies channel 2 is to low pass filter $y(t)$.

A block diagram of the signal translation and extraction process is illustrated in Figure 9.35. In part B of the figure, the wideband input signal $g(t)$ is multiplied by the tuning signal $f(t) = e^{-j(2\pi f_0)t}$ to produce a complex product signal $y(t) = g(t)e^{-j(2\pi f_0)t}$. As illustrated at the bottom of Figure 9.35, the signal $y(t)$ is the spectrally down converted version of $g(t)$. The spectrum has been down converted by $f_0$ Hz such that the portion of the spectrum originally occupied by channel 2 and centered at $f_0$ Hz is now centered at 0 Hz.

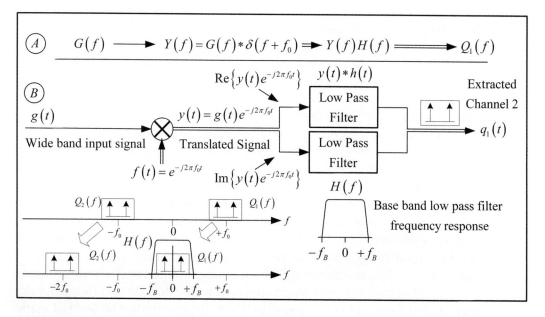

**Figure 9.35**   FDM signal extraction block diagram

The portion of the original spectrum occupied by channel 2 and centered at $-f_0$ Hz is now centered at $-2f_0$ Hz.

Now all that is necessary to extract channel 2 is to low pass filter the complex translated signal. The intermediate signal $y(t)$ is then convolved with the impulse response of the low pass filter $h(t)$ to extract channel 2. It does this by passing the frequency band represented by $Q_1(f)$ and rejecting all other frequencies. Note that both the real part and the imaginary part of the complex signal are each filtered by identical low pass filters with bandwidth equal to $2f_B$.

The down converted and filtered output spectrum is defined by $Q_1(f)$. We can see from Figure 9.35 that the frequency components of $Q_2(f)$ are outside the band of the low pass filter and are eliminated. We interpret this process in the frequency domain in part A of the figure. The input signal spectra $G(f)$ is convolved with the spectra of the tuning signal $F(f) = \delta(f + f_0)$ to produce an intermediate signal spectra $Y(f) = G(f) * \delta(f + f_0)$. $Y(f)$ is then multiplied by the frequency response of the low pass filter $H(f)$ to produce the output

$$Q_1(f) = Y(f)H(f) = \{G(f) * \Delta(f + f_0)\} H(f) = \frac{1}{4}\{\delta(f - f_m) + \delta(f + f_m)\}$$

We note that $q_1(t)$ is complex and therefore does not contain the negative or mirror spectrum. If downstream processing dictates, this signal could be converted to a real signal at the output of the filter.

If so we could use the techniques described in Chapter 7, "Complex to Real Conversion," to perform this task. If, however, the sample rate of the output signal $q_1(t)$ were high enough, then the two low pass filters defined by $H(f)$ could also be utilized to decimate the output sample rate to the Nyquist rate for real signals. From there, the conversion from complex to real is a simple multiplex and sign manipulation operation. The process outlined here is just as easily implemented if the input message signal represented by $g(t)$ were complex with a one-sided spectrum.

The powerful Fourier transform pairs that we derived earlier make an engineer's life a great deal simpler when working the mathematics of a problem that requires flipping back and forth between the time domain and frequency domain. Expressing the time domain signal as a collection of exponentials significantly eases the pain of mathematical derivation. The reader can prove this by investing some time working on some frequency translation problems that require converting between the time domain and frequency domain, using sines and cosines as opposed to exponentials. The level of mathematical complexity increases considerably.

## 9.2 DISCRETE TIME (DIGITAL) FOURIER TRANSFORM

All the discussions in the previous sections have dealt with the translation of continuous time analog signals. In the second half of the chapter, we will be discussing the tuning of digital signals. All discussions in the following text are analogous to the previous discussions. The following dialogue, however, will require an understanding of the mathematics of digital frequency and the discrete Fourier transform (DFT). If the reader needs to refresh his or her memory on these two subject areas, it is suggested that he or she read the discussion on the concepts in Chapter 1, "Review of Digital Frequency," and the discussion in Chapter 3, "Review of the Fourier Transform." In addition, the reader can obtain a more in-depth treatment of these subjects from the references given at the end of this chapter.

In order to set the stage for the treatment of digital signal tuning, it is appropriate to briefly review the properties of both the DFT and the notion of discrete frequency, which are specific to this discussion. A function of the variable $n$ in the discrete time domain and a function of the variable $k/N$ in the discrete frequency domain are said to be a discrete transform pair if each function can be mathematically computed from the other via the DFT or its inverse. The term $k/N$ is a fraction that represents normalized discrete frequency.

In our study, we will utilize the symbol $\overset{f}{\leftrightarrow}$ to denote the relationship between a discrete transform pair, such as

$$h(n) \overset{f}{\leftrightarrow} H(k/N)$$

The notation $H(k/N) = f\{h(n)\}$ is often used to indicate that $H(k/N)$ is the forward DFT of $h(n)$, and the notation $h(n) = f^{-1}\{H(k/N)\}$ is used to indicate that $h(n)$ is the inverse discrete Fourier transform (IDFT) of $H(k/N)$. The definitions of both the forward DFT and IDFT are discussed in detail in Chapter 3 and are repeated here only for clarity. The DFT of $h(n)$ is computed by the relation

$$H\left(\frac{k}{N}\right) = \sum_{n=0}^{N-1} h(n) \, e^{-j\left(2\pi\frac{k}{N}\right)n} \quad \text{for} \left\{ \begin{array}{l} 0 \leq n \leq N-1 \\ 0 \leq k \leq N-1 \end{array} \right\}$$

**Equation 9.49**

The IDFT of $H(k/N)$ is computed by the relation

$$h(n) = \frac{1}{N} \sum_{k=0}^{N-1} H\left(\frac{k}{N}\right) e^{j\left(2\pi\frac{k}{N}\right)n} \quad \text{for} \left\{ \begin{array}{l} 0 \leq n \leq N-1 \\ 0 \leq k \leq N-1 \end{array} \right\}$$

**Equation 9.50**

We recall from Chapter 1 that a digital radian frequency is described by

$$\omega_k = \left(2\pi k\right)\frac{f_s}{N}$$

**Equation 9.51**

The term $N$ is the number of equidistant samples located on the unit circle where $\omega_k$ is evaluated. The angular distance between adjacent samples is fixed at $2\pi/N$. Since one revolution around the unit circle is equivalent to one cycle in frequency, $N$ can also be thought of as the maximum number of samples per period of a single cycle of a digital sinusoid. The term $k$ is the integer frequency domain index. The digital frequency is simply

$$f_k = \frac{\omega_k}{2\pi}$$

**Equation 9.52**

The frequency index term $k$ ranges from $0 \le k \le N-1$. The normalized digital radian frequency is given by

$$\omega_k = \frac{2\pi k}{N}$$

and the normalized frequency is given by

$$f_k = \frac{k}{N}$$

The normalized digital radian frequency ranges from $0 \le \omega_k < 2\pi$ in increments of $2\pi/N$, and the corresponding normalized digital frequency ranges from $0 \le f_k < 1$ in increments of $1/N$. The actual or unnormalized digital frequency is simply

$$f = f_k f_s = \frac{k}{N} f_s$$

where $f_s$ is the sample rate of the digital system. The frequency terms $\omega_k$ and $f_k$, defined in Equation 9.51 and Equation 9.52, are single valued. The value of the frequency is determined by the frequency index $k$. For example, suppose we are working with a digital system that operates at a sample rate of $f_s = 128$ KHz. Further suppose that we have divided the unit circle into

$N = 32$ equal arc segments and that we assign $k = 8$. The normalized radian frequency for these particular parameters is given by

$$\omega_k \frac{2\pi}{N} k = \frac{2\pi}{32} 8 = \frac{\pi}{2}$$

The normalized frequency for these same parameters is

$$f_k = \frac{k}{N} = \frac{8}{32} = 0.25$$

The unnormalized radian frequency is then

$$\omega = \omega_k f_s = \frac{\pi}{2}(128 \text{ KHz}) = (64K)\pi \frac{\text{radians}}{\text{second}}$$

The unnormalized frequency is then

$$f = f_k f_s = (0.25)128 \text{ KHz} = 32 \text{ KHz} = f_s/4 \text{ Hz}$$

The results of this example are accurate, but once again they are just numbers that indicate a particular value that we can write down on paper. In digital hardware design, we need more than a value. We need to repetitively produce all the samples for an entire period of a particular frequency. In order for $\omega_k$ and $f_k$ to be cyclic over some fixed period of time, they need to be indexed and evaluated sequentially at some subset of $N$, equidistant points around the unit circle. This is accomplished by including a sample index $n$ that increments at the sample rate. Including the sample index gives us the standard implementation representation of a digital frequency:

$$\omega_k = \left(\frac{2\pi k}{N}\right)n \text{ and } f_k = \left(\frac{k}{N}\right)n$$

**Equation 9.53**

The term $k/N$ is the normalized frequency, and the term $n$ is the time domain sample index that ranges from $0 \leq n \leq \infty$ and increments at the sample rate $f_s$. You can think of $\omega_k$ and $f_k$ as the tips of a vector extending from the origin to the unit circle, rotating around the unit circle in angular chunks of $k/N$, at a rate equal to the sample rate $f_s$. Both $\omega_k$ and $f_k$ are evaluated at each radian chunk and in doing so take on values that are sinusoidal over time.

The subscript $k$ in the radian frequency term $\omega_k$ and the frequency term $f_k$ denotes that $\omega$ and $f$ are digital quantities. In this section it is understood that we are dealing exclusively with digital frequencies, so for simplicity the subscript $k$ will be dropped from future notation. For purposes of simplicity, we will also use the terms *Fourier transform* and *DFT* interchangeably when discussing digital signals.

In the treatment of analog or continuous time signal tuning we used the notation $\delta(t)$ and $\delta(f)$ to represent a time domain and a frequency domain impulse. In dealing with the subject of discrete time signal tuning, we will use the notation $\Delta(n)$ to represent the discrete time domain impulse and we will use the notation $\Delta(k)$ or $\Delta(k/N)$ to represent the discrete frequency domain impulse. In these representations, $n$ is the discrete time domain sample index that increments at the sample rate, and $k$ is the discrete frequency domain sample index that can take on any value within the range of $0 \le k \le (N-1)$. In our discussions on digital or discrete time tuning, we will work with normalized frequency notation, or $f = k/N$.

## 9.2.1   Discrete Time Domain Unit Impulse Response

We will define the discrete time domain unit impulse function as

$$\Delta(n) = \left\{ \begin{array}{ll} 1 & \text{for } n = 0 \\ 0 & \text{for } n \ne 0 \end{array} \right\}$$

**Equation 9.54**

The Fourier transform of the unit time domain impulse function is given by

$$f\{\Delta(n)\} = \sum_{n=0}^{N-1} \Delta(n)\, e^{-j\left(2\pi\frac{k}{N}\right)n} = \Delta(0)e^{-j0} = 1$$

**Equation 9.55**

We can see from Equation 9.55 that the Fourier transform of a discrete unit impulse centered at $n = 0$ has unity magnitude and 0 phase in the frequency domain. Equation 9.55 leads to the definition of the discrete transform pair

$$\Delta(n) \overset{f}{\longleftrightarrow} 1$$

**Equation 9.56**

The discrete unit impulse shifted in time by $m$ samples is defined as

$$\Delta(n-m) = \left\{ \begin{array}{ll} 1 & \text{for } n = m \\ 0 & \text{for } n \neq m \end{array} \right\}$$

The Fourier transform of a discrete unit impulse shifted by $m$ samples is derived as follows:

$$f\{\Delta(n-m)\} = \sum_{n=0}^{N-1} \Delta(n-m) \, e^{-j\left(2\pi\frac{k}{N}\right)n}$$

**Equation 9.57**

If we let the variable $p = n - m$, then $n = p + m$. This means that when $n = 0$, $p = -m$, and when $n = N - 1$, $p = N - 1 - m$. If we substitute this change of variables into Equation 9.57, we get

$$f\{\Delta(p)\} = \sum_{p=-m}^{N-1-m} \Delta(p) \, e^{-j\left(2\pi\frac{k}{N}\right)(p+m)}$$

**Equation 9.58**

The summation in Equation 9.58 is periodic in $N$ samples. The summation indices represent the same period as the summation indices in Equation 9.57. The DFT thinks that the signal it is processing is periodic in $N$ samples, so any set of $N$ sequential samples is equivalent. Therefore we can drop the $-m$ in the summation index without modifying the result. We end up with

$$f\{\Delta(p)\} = \sum_{p=0}^{N-1} \Delta(p) \, e^{-j\left(2\pi\frac{k}{N}\right)(p+m)}$$

We can move the terms not associated with index $p$ out of the summation to get

$$f\{\Delta(p)\} = e^{-j\left(2\pi\frac{k}{N}\right)m} \sum_{p=0}^{N-1} \Delta(p) \, e^{-j\left(2\pi\frac{k}{N}\right)p}$$

Now since $\Delta(p)$ equals unity at $p = 0$ and is zero elsewhere, and since $e^{-j\left(2\pi\frac{k}{N}\right)p}$ equals unity at $p = 0$, the summation reduces to

$$f\{\Delta(p)\} = e^{-j\left(2\pi\frac{k}{N}\right)m}$$

Since $p = n - m$, we replace the variable $p$ to get

$$f\{\Delta(n-m)\} = e^{-j\left(2\pi\frac{k}{N}\right)m}$$

**Equation 9.59**

Equation 9.59 leads to the definition of the discrete transform pair:

$$\Delta(n-m) \overset{f}{\leftrightarrow} e^{-j\left(2\pi\frac{k}{N}\right)m}$$

**Equation 9.60**

Similarly,

$$f\{\Delta(n+m)\} = \sum_{n=0}^{N-1} \Delta(n+m)\, e^{-j\left(2\pi\frac{k}{N}\right)n} = e^{+j\left(2\pi\frac{k}{N}\right)m}$$

**Equation 9.61**

Equation 9.61 leads to the definition of the discrete transform pair:

$$\Delta(n+m) \overset{f}{\leftrightarrow} e^{+j\left(2\pi\frac{k}{N}\right)m}$$

**Equation 9.62**

As was the case for continuous time, it is evident from Equation 9.61 and Equation 9.62 that the spectrum of the discrete unit impulse has a uniform magnitude over the entire frequency range. The phase is linear and since it is evaluated on the unit circle it is given by

$$\theta(k) = \left(2\pi\frac{k}{N}\right)m\ Mod(N)$$

The phase is proportional to the value of the time shift $m$. It is instructional to see what the phase looks like for several different values of the sample shift $m$. Figure 9.36, Figure 9.37, and Figure 9.38 show the phase response for an impulse evaluated at 64 equally spaced points on the unit circle that is delayed in time by $m = 1$, 2, and 3 samples, respectively. It is interesting to note that if sample shift index $m = 0$, as would be the case for the time domain function $\Delta(n)$, then the value of the phase would compute to $\theta = 0$, as predicted.

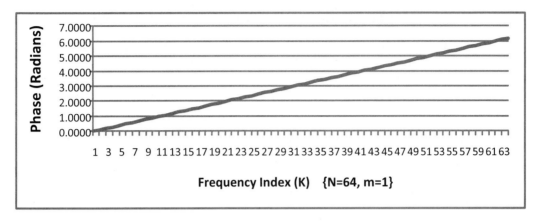

**Figure 9.36**  Impulse phase response *m* = 1

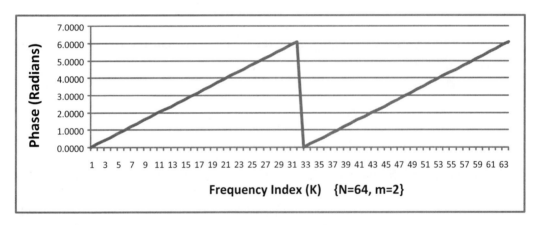

**Figure 9.37**  Impulse phase response for *m* = 2

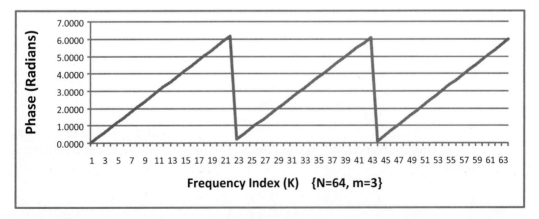

**Figure 9.38**  Impulse phase response for *m* = 3

## 9.2.2 Discrete Frequency Domain Unit Impulse Response

Before we begin, it is important to refresh our memory regarding digital frequency and normalized digital frequency. In the digital world just about everything is governed by the system sample rate $f_s$. The theoretically highest frequency that can be achieved in a digital system is half the sample rate or $f_s/2$. This is called the *Nyquist rate*. Nyquist stated that reconstruction of an analog signal from a digital signal requires that the highest frequency in the digital signal have at least two samples per cycle. Stated another way, the highest frequency that is obtainable in a digital system is equal to half the sample rate. It is common in the DSP world to represent a discrete frequency as being normalized to the sample rate. For example, given the digital frequency $f_K$, DSP engineers typically refer to this frequency as being a fraction of the sample frequency or $f_K/f_s$. For example if the digital frequency $f_K = 25$ KHz and the system sample rate was $f_s = 125$ KHz, then an engineer will usually refer to the frequency as $0.2$.

If we obey the Nyquist rule, the range of normalized digital frequencies is confined to

$$0 \le \frac{f_k}{f_s} \le 0.5$$

We can define a digital frequency simply by writing $f_K = \cos\left(2\pi \frac{k}{N} n\right)$. The fundamental frequency for this expression is determined by evaluating it at the sample rate at $N$ equally spaced points on the unit circle. Each revolution of the unit circle represents a single cycle of a discrete frequency. With exception of DC, the smallest or fundamental frequency obtainable is given by

$$f_s/N$$

If we normalize this to the sample rate $f_s$, then the fundamental normalized frequency becomes

$$1/N$$

The normalized frequency has no units. The range of normalized frequencies is given by

$$0 \le \frac{k}{N} \le 0.5$$

The frequency index $k$ can take on the discrete integer values given by $\{0,1,2,3,\cdots,N/2\}$.

We can convert back to the unnormalized frequency representation simply by multiplying the normalized value by the sample rate $f_s$. For example, suppose $N = 32$, $k = 8$, and $f_s = 256$ KHz. The normalized frequency would be $k/N = 8/32 = 1/4 = 0.25$. The unnormalized frequency is computed by

$$f_k = \left(\frac{k}{N}\right) f_s = \left(\frac{8}{32}\right) 256 \text{ KHz} = 64 \text{ KHz}$$

In the limit, if we set $k = N/2$, then

$$f_k = \left(\frac{k}{N}\right) f_s = \left(\frac{1}{2}\right) f_s = \frac{f_s}{2}$$

which is the Nyquist rate and therefore the highest possible frequency we can represent digitally.

The important thing to remember is that for the rest of this chapter we are dealing with normalized frequency expressed as $k/N$.

DSP engineers usually evaluate a DFT over the full frequency range of $0 \le f < f_s$, which in the normalized frequency notation is $0 \le k/N \le (N-1)/N$, which means the frequency index $k$ has the range $0 \le k \le N - 1$.

We interpret the frequencies computed on the unit circle from $\pi$ to $2\pi$ radians as having a frequency index $k$ ranging from

$$\frac{N}{2} < k \le N - 1$$

This corresponds to the frequencies

$$\frac{f_s}{2} < f < \frac{(N-1)}{N} f_s$$

These frequencies are the mirror image of the frequencies computed from 0 to $\pi$ radians and are usually interpreted as being negative frequencies. If plotted, the negative frequencies would extend on a plot, from $-f_s/2$ to 0 Hz.

Getting back to the DFT, we can rewrite the expression for the DFT so that its notation is more suitable for working with normalized frequencies. We do this simply by moving the quantity $N$ out from under the $2\pi$ term and place it directly under the $k$ term to get

$$H\left(\frac{k}{N}\right) = \sum_{n=0}^{N-1} h(n)\, e^{-j(2\pi)\left(\frac{k}{N}\right)n} \quad \text{for} \left\{ \begin{array}{c} 0 \le n \le N-1 \\ 0 \le k \le N-1 \end{array} \right\}$$

The IDFT is given by

$$h(n) = \sum_{k=0}^{N-1} H\left(\frac{k}{N}\right) e^{j(2\pi)\left(\frac{k}{N}\right)n} \quad \text{for} \left\{ \begin{array}{c} 0 \le n \le N-1 \\ 0 \le k \le N-1 \end{array} \right\}$$

Some applications prefer to scale the IDFT by $1/N$, which results in the expression

$$h(n) = \frac{1}{N}\sum_{k=0}^{N-1} H\left(\frac{k}{N}\right) e^{j(2\pi)\left(\frac{k}{N}\right)n} \quad \text{for} \left\{ \begin{array}{c} 0 \le n \le N-1 \\ 0 \le k \le N-1 \end{array} \right\}$$

Now both equations are written in terms of normalized frequency $k/N$. Now, if we specify a frequency index $k_0$ we are referring to the index for the normalized frequency of $k_0/N$. This is important when visualizing a discrete frequency plot. Keeping this concept foremost in our minds, we will begin our discussion with the definition of the frequency domain unit impulse function

$$\Delta\left(\frac{k}{N}\right) = \left\{ \begin{array}{l} 1 \text{ for } k = 0 \\ 0 \text{ for } k \ne 0 \end{array} \right\}$$

This is easily seen by taking the IDFT of the impulse to arrive at

$$f^{-1}\left\{\Delta\left(\frac{k}{N}\right)\right\} = \sum_{k=0}^{N-1} \Delta\left(\frac{k}{N}\right) e^{+j(2\pi)\left(\frac{k}{N}\right)n} = \Delta(0)e^{-j0} = 1$$

This gives us the discrete transform pair:

$$1 \overset{f}{\leftrightarrow} \Delta\left(\frac{k}{N}\right)$$

**Equation 9.63**

The frequency domain impulse that is shifted in frequency by an index $k_0$ is expressed as $Y(k) = \Delta(k - k_0)$. Since we wish to work in terms of normalized frequency, this expression becomes

$$Y\left(\frac{k}{N}\right) = \Delta\left(\frac{k}{N} - \frac{k_0}{N}\right)$$

Let's calculate the inverse Fourier transform of $Y\left(\frac{k}{N}\right)$ by

$$y(n) = \sum_{k=0}^{N-1} Y\left(\frac{k}{N}\right) e^{j(2\pi)\left(\frac{k}{N}\right)n}$$

$$= \sum_{k=0}^{N-1} \Delta\left(\frac{k}{N} - \frac{k_0}{N}\right) e^{j(2\pi)\left(\frac{k}{N}\right)n}$$

**Equation 9.64**

If we let the variable

$$\frac{p}{N} = \frac{k}{N} - \frac{k_0}{N}$$

then

$$\frac{k}{N} = \frac{p}{N} + \frac{k_0}{N}$$

Also, when the summation index $k = 0$,

$$\frac{p}{N} = -\frac{k_0}{N} \rightarrow p = -k_0$$

and when the summation index $k = N - 1$,

$$\frac{p}{N} = \frac{N-1}{N} - \frac{k_0}{N} \rightarrow p = N - 1 - k_0$$

If we substitute this change of variables back into Equation 9.64, we end up with

$$y(n) = \sum_{p=-k_0}^{N-1-k_0} \Delta\left(\frac{p}{N}\right) e^{j(2\pi)\left(\frac{p}{N} + \frac{k_0}{N}\right)n}$$

**Equation 9.65**

As we mentioned before, the summation in Equation 9.65 is periodic in $N$ samples. The summation indices represent the same period as the summation indices in Equation 9.64. Because the DFT believes the $N$ sample function, it is processing is periodic we can drop the $-k_0$ term in the indices without modifying the result. We end up with the expression

$$y(n) = \sum_{p=0}^{N-1} \Delta\left(\frac{p}{N}\right) e^{j(2\pi)\left(\frac{p}{N} + \frac{k_0}{N}\right)n}$$

**Equation 9.66**

We can move the terms not associated with index $p$ outside of the summation to get

$$y(n) = e^{j(2\pi)\left(\frac{k_0}{N}\right)n} \sum_{p=0}^{N-1} \Delta\left(\frac{p}{N}\right) e^{j(2\pi)\left(\frac{p}{N}\right)n}$$

**Equation 9.67**

Since the impulse function $\Delta(p/N) = 1$ for $p = 0$ and is zero elsewhere, and since the term

$$e^{j(2\pi)\left(\frac{p}{N}\right)n} = 1 \text{ for } p = 0$$

the expression in Equation 9.67 reduces to

$$y(n) = e^{j(2\pi)\left(\frac{k_0}{N}\right)n}$$

**Equation 9.68**

Thus from Equation 9.64 and Equation 9.68, we can conclude that

$$e^{j(2\pi)\left(\frac{k_0}{N}\right)n} \overset{f}{\leftrightarrow} \Delta\left(\frac{k}{N} - \frac{k_0}{N}\right)$$

**Equation 9.69**

is a discrete transform pair expressed in terms of normalized frequency. The discrete spectrum of $e^{j(2\pi)(k_0/N)n}$ is illustrated in Figure 9.39. Note that the discrete frequency axis is scaled in terms of normalized frequency.

By changing the quantity $\left(\dfrac{k}{N} - \dfrac{k_0}{N}\right)$ in Equation 9.69 to $\left(\dfrac{k}{N} + \dfrac{k_0}{N}\right)$, we can similarly conclude that Equation 9.70 defines the Fourier transform pair expressed in terms of normalized frequency

$$e^{-j(2\pi)\left(\frac{k_0}{N}\right)n} \overset{f}{\longleftrightarrow} \Delta\left(\frac{k}{N} + \frac{k_0}{N}\right)$$

**Equation 9.70**

The spectrum of the negative exponential $e^{-j(2\pi)(k_0/N)n}$ is illustrated in Figure 9.40.

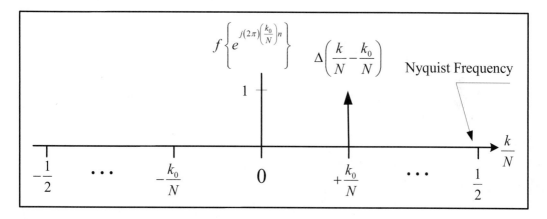

**Figure 9.39** Spectrum of a positive exponential

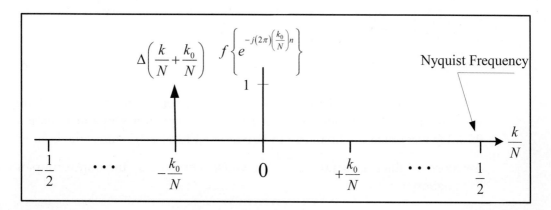

**Figure 9.40** Spectrum of a negative exponential

The discrete transform pairs that we have derived here for the unit impulse are extremely important to the understanding of discrete signal translation. The reader should be familiar with these transform pairs.

### 9.2.3  Discrete Cosine Function cos(2πk/N)n

The spectrum of the function $x(n) = \cos\left(2\pi \frac{k_0}{N} n\right)$ can be computed using a combination of Euler's identity and the Fourier transform identities that we derived in Equation 9.69 and Equation 9.70. We start by writing the equation for the Fourier transform of $x(n)$ using the notation for normalized frequencies:

$$X\left(\frac{k}{N}\right) = \sum_{n=0}^{N-1} x(n) e^{-j(2\pi)\left(\frac{k}{N}\right)n} \qquad \text{for } \left\{ \begin{array}{l} 0 \le k \le N-1 \\ 0 \le n \le N-1 \end{array} \right\}$$

$$= \sum_{n=0}^{N-1} \left[ \cos\left(2\pi \frac{k_0}{N} n\right) \right] e^{-j(2\pi)\left(\frac{k}{N}\right)n}$$

We use Euler's identity to expand the cosine function into its exponential components:

$$X\left(\frac{k}{N}\right) = \sum_{n=0}^{N-1} \left[ \frac{e^{j(2\pi)\left(\frac{k_0}{N}\right)n} + e^{-j(2\pi)\left(\frac{k_0}{N}\right)n}}{2} \right] e^{-j(2\pi)\left(\frac{k}{N}\right)n}$$

Now we split the summation into two separate terms to get

$$X\left(\frac{k}{N}\right) = \frac{1}{2} \sum_{n=0}^{N-1} \left[ e^{j(2\pi)\left(\frac{k_0}{N}\right)n} \right] e^{-j(2\pi)\left(\frac{k}{N}\right)n} + \frac{1}{2} \sum_{n=0}^{N-1} \left[ e^{-j(2\pi)\left(\frac{k_0}{N}\right)n} \right] e^{-j(2\pi)\left(\frac{k}{N}\right)n}$$

**Equation 9.71**

The reader should be able to complete the computation by inspection with the aid of the impulse identities given in Equation 9.69 and Equation 9.70. We recognize the first term in Equation 9.71 to be $1/2$ times the Fourier transform of the exponential $e^{j(2\pi)\left(\frac{k_0}{N}\right)n}$, which is defined by the transform pair in Equation 9.69 as

$$e^{j(2\pi)\left(\frac{k_0}{N}\right)n} \overset{f}{\leftrightarrow} \Delta\left(\frac{k}{N} - \frac{k_0}{N}\right)$$

We also recognize that the second term in Equation 9.71 to be $1/2$ times the Fourier transform of the exponential $e^{-j(2\pi)\left(\frac{k_0}{N}\right)n}$, which is defined by the transform pair in Equation 9.70 to be

$$e^{-j(2\pi)\left(\frac{k_0}{N}\right)n} \overset{f}{\leftrightarrow} \Delta\left(\frac{k}{N}+\frac{k_0}{N}\right)$$

Therefore we can write the final result by inspection as

$$X\left(\frac{k}{N}\right) = \frac{1}{2}\left\{\Delta\left(\frac{k}{N}-\frac{k_0}{N}\right)+\Delta\left(\frac{k}{N}+\frac{k_0}{N}\right)\right\}$$

This is one way of using the impulse identities. A second, more common method is to get Equation 9.71 into more recognizable form and then determine the result by inspection. We begin by performing the multiplications in Equation 9.71 to get

$$X\left(\frac{k}{N}\right) = \frac{1}{2}\sum_{n=0}^{N-1}\left[e^{-j(2\pi)\left(\frac{k}{N}-\frac{k_0}{N}\right)n}\right] + \frac{1}{2}\sum_{n=0}^{N-1}\left[e^{-j(2\pi)\left(\frac{k}{N}+\frac{k_0}{N}\right)n}\right]$$

**Equation 9.72**

This is a more standard form, and it is easy to recognize that

$$X\left(\frac{k}{N}\right) = \frac{1}{2}\left\{\Delta\left(\frac{k}{N}-\frac{k_0}{N}\right)+\Delta\left(\frac{k}{N}+\frac{k_0}{N}\right)\right\}$$

As a result of these calculations, we define Equation 9.73 to be a discrete transform pair expressed in normalized frequency notation:

$$\cos\left(2\pi\frac{k_0}{N}n\right) \overset{f}{\leftrightarrow} \frac{1}{2}\left\{\Delta\left(\frac{k}{N}-\frac{k_0}{N}\right)+\Delta\left(\frac{k}{N}+\frac{k_0}{N}\right)\right\}$$

**Equation 9.73**

The line spectrum for $x(n) = \cos\left(2\pi\frac{k_0}{N}n\right)$ is illustrated in Figure 9.41.

Now is a pretty good time for an example. Suppose we had a cosine waveform given by

$$x(n) = \cos\left(2\pi\frac{k_0}{N}n\right)$$

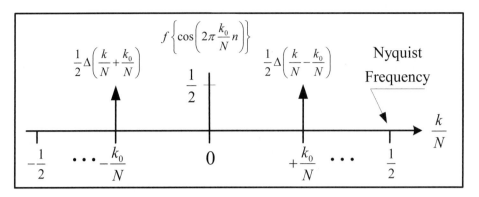

**Figure 9.41**  Line spectrum of cos(($2\pi k_0/N$)n)

Further suppose that the cosine was generated on a unit circle that was divided into $N = 64$ equally spaced points and that the frequency index $k_0 = 16$. The cosine waveform is expressed as

$$x(n) = \cos\left(2\pi\left(\frac{16}{64}\right)n\right)$$

This means that the unit circle has 64 evenly spaced points at which we can evaluate the cosine waveform, and we are going to evaluate it at every sixteenth point. Therefore each cycle of the cosine waveform will consist of 4 points, and there will be a total of 16 complete cycles every 64 ticks of the sample clock. The alert reader will recognize that the cosine frequency is $f_s/4$. This cosine waveform is illustrated in Figure 9.42. Note that each sample is indicated by a black dot. There are 4 samples per cycle, and we have drawn 16 complete cycles in the span of 64 samples.

Now let's compute the spectrum of this waveform by taking its DFT. We could choose any arbitrary length for the DFT, but some choices are much better than others. For this example, we will choose a convenient length of $N = 64$. Why is this convenient? It's convenient because 64 samples encapsulate exactly 16 cycles of the sinusoidal waveform and therefore present the DFT input with an integer number of cycles of the discrete time waveform. Since the DFT believes the waveform it is operating on is periodic, this choice eliminates any discontinuities that would occur if the waveform had a partial cycle at the end of the sequence. We compute the DFT and plot the results, as illustrated in Figure 9.43.

The spectral plot in Figure 9.43 is drawn against two horizontal axes. The first axis is calibrated in units of normalized frequency $k/N$. The second axis is calibrated in frequency $f$. On the normalized frequency axis, we can see that the line spectra occur at

$$\frac{k}{N} = \pm\frac{16}{64} = \pm\frac{1}{4} = 0 \pm .25$$

On the frequency axis we see that the line spectra occur at a frequency of

$$\pm\frac{f_s}{4}$$

The highest frequency we could have computed with the DFT is

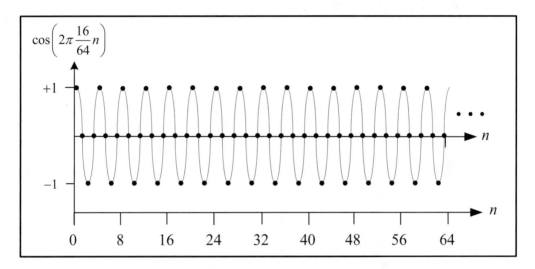

**Figure 9.42** Example cosine waveform time series

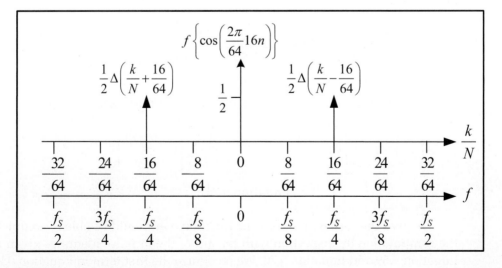

**Figure 9.43** Example cosine waveform spectrum

$$k/N = 32/64 = 0.5$$

or

$$f = f_s/2$$

The reader is invited to work through the math to verify these results.

### 9.2.4 Discrete Sine Function sin(2πk/N)n

The Fourier transform of the sine function is very similar to that of the cosine function. The spectrum of the function $x(n) = \sin\left(2\pi \dfrac{k_0}{N} n\right)$ can be computed using a combination of Euler's identity and the Fourier transform identities that we derived in Equation 9.69 and Equation 9.70. We start by writing the equation for the Fourier transform of $x(n)$ using the notation for normalized frequencies:

$$x\left(\frac{k}{N}\right) = \sum_{n=0}^{N-1} x(n)\, e^{-j(2\pi)\left(\frac{k}{N}\right)n} \qquad \text{for} \left\{\begin{array}{c} 0 \le k \le N-1 \\ 0 \le n \le N-1 \end{array}\right\}$$

$$= \sum_{n=0}^{N-1}\left[\sin\left(2\pi \frac{k_0}{N} n\right)\right] e^{-j(2\pi)\left(\frac{k}{N}\right)n}$$

We use Euler's identity to expand the sine function into its exponential components to arrive at

$$x\left(\frac{k}{N}\right) = \sum_{n=0}^{N-1}\left[\frac{e^{j(2\pi)\left(\frac{k_0}{N}\right)n} - e^{-j(2\pi)\left(\frac{k_0}{N}\right)n}}{2j}\right] e^{-j(2\pi)\left(\frac{k}{N}\right)n}$$

Now we split the summation into two separate terms to get

$$x\left(\frac{k}{N}\right) = \frac{1}{2j}\sum_{n=0}^{N-1}\left[e^{j(2\pi)\left(\frac{k_0}{N}\right)n}\right] e^{-j(2\pi)\left(\frac{k}{N}\right)n} - \frac{1}{2j}\sum_{n=0}^{N-1}\left[e^{-j(2\pi)\left(\frac{k_0}{N}\right)n}\right] e^{-j(2\pi)\left(\frac{k}{N}\right)n}$$

**Equation 9.74**

As was the case with the cosine function, we should be able to complete the computation by inspection with the aid of the impulse identities given in Equation 9.69 and Equation 9.70. We recognize the first term in Equation 9.74

to be $1/2j$ times the Fourier transform of the exponential $e^{j(2\pi)\left(\frac{k_0}{N}\right)n}$, which is defined by Equation 9.69 as

$$e^{j(2\pi)\left(\frac{k_0}{N}\right)n} \overset{f}{\leftrightarrow} \Delta\left(\frac{k}{N} - \frac{k_0}{N}\right)$$

We also recognize that the second term in Equation 9.74 is $-1/2j$ times the Fourier transform of the exponential $e^{-j(2\pi)\left(\frac{k_0}{N}\right)n}$, which is defined by Equation 9.70 to be

$$e^{-j\left(\frac{2\pi}{N}\right)k_0 n} \overset{f}{\leftrightarrow} \Delta\left(\frac{k}{N} + \frac{k_0}{N}\right)$$

Therefore we can write the final result by inspection to be

$$X\left(\frac{k}{N}\right) = \frac{1}{2j}\left\{\Delta\left(\frac{k}{N} - \frac{k_0}{N}\right) - \Delta\left(\frac{k}{N} + \frac{k_0}{N}\right)\right\}$$

As a result of these calculations, we define Equation 9.75 as a discrete transform pair expressed in normalized frequency notation:

$$\sin\left(2\pi\frac{k_0}{N}n\right) \overset{f}{\leftrightarrow} \frac{1}{2j}\left\{\Delta\left(\frac{k}{N} - \frac{k_0}{N}\right) - \Delta\left(\frac{k}{N} + \frac{k_0}{N}\right)\right\}$$

**Equation 9.75**

The line spectrum of $\sin\left(2\pi\frac{k_0}{N}n\right)$ is illustrated in Figure 9.44. The reader should take note that this figure is an amplitude plot and not a magnitude plot.

This is one method of using the impulse identities. A more common method is to get Equation 9.74 into a more recognizable form and then determine the result by inspection. We begin by performing the multiplications on Equation 9.74 to get

$$x\left(\frac{k}{N}\right) = \frac{1}{2j}\sum_{n=0}^{N-1}\left[e^{-j(2\pi)\left(\frac{k}{N} - \frac{k_0}{N}\right)n}\right] - \frac{1}{2j}\sum_{n=0}^{N-1}\left[e^{-j(2\pi)\left(\frac{k}{N} + \frac{k_0}{N}\right)n}\right]$$

**Equation 9.76**

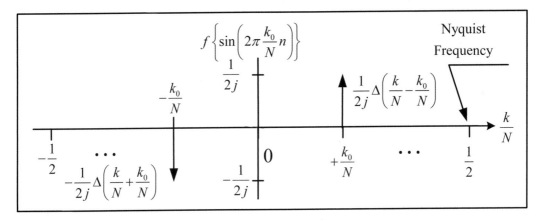

**Figure 9.44**   Line spectrum of $\sin((2\pi k_0/N)n)$

The expression in Equation 9.76 is a more standard form and it is easy to recognize that

$$X\left(\frac{k}{N}\right) = \frac{1}{2j}\left\{\Delta\left(\frac{k}{N}-\frac{k_0}{N}\right)-\Delta\left(\frac{k}{N}+\frac{k_0}{N}\right)\right\}$$

We can modify the expression for the Fourier transform pair in Equation 9.75 by multiplying the right-hand side by $j/j$ to arrive at an alternate transform pair:

$$\sin\left(2\pi\frac{k_0}{N}n\right) = \frac{-j}{2}\left\{\Delta\left(\frac{k}{N}-\frac{k_0}{N}\right)-\Delta\left(\frac{k}{N}+\frac{k_0}{N}\right)\right\}$$

**Equation 9.77**

Equation 9.77 is illustrated in Figure 9.45.

Let's illustrate all this with an example. Suppose we had a sine waveform given by

$$x(n) = \sin\left(2\pi\frac{k_0}{N}n\right)$$

Assume that the sine wave was generated on a unit circle that was divided into $N = 64$ equally spaced points and that the frequency index $k_0 = 16$. Therefore our sine waveform is given by

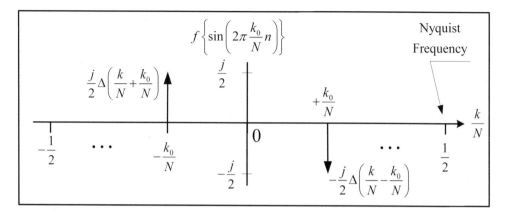

**Figure 9.45**   Alternate line spectrum of sin$((2\pi k_0/N)n)$

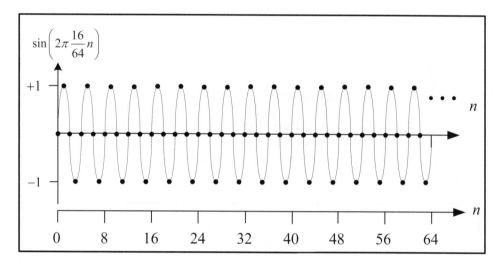

**Figure 9.46**   Example sine waveform time series

$$x(n) = \sin\left(2\pi\left(\frac{16}{64}\right)n\right)$$

This means that the unit circle has 64 evenly spaced points, at which we can evaluate the sine waveform, and we are going to evaluate it at every sixteenth point. Therefore each cycle of the sine waveform will consist of 4 points, and there will be a total of 16 complete cycles every 64 ticks of the sample clock. This sine waveform is illustrated in Figure 9.46. Note that each sample is indicated by a black dot. There are 4 samples per cycle, and we have drawn 16 complete cycles in the span of 64 samples.

Now let's compute the spectrum of this waveform by taking its DFT. For this example, we will choose a DFT length of $N = 64$. We compute the DFT and plot the results as illustrated in Figure 9.47. The spectral plot is drawn against two horizontal axes. The first axis is calibrated in units of normalized frequency $k/N$. The second axis is calibrated in units of frequency $f$.

On the normalized frequency axis, we can see that the line spectra occur at

$$\frac{k}{N} = \pm\frac{16}{64} = \pm\frac{1}{4} = \pm0.25$$

On the frequency axis, we see that the line spectra occur at a frequency of

$$\pm\frac{f_s}{4}$$

The highest frequency we could have computed with the DFT is

$$64 \text{ or } f = f_s/2$$

### 9.2.5  Interpreting the DFT

In regard to the previous discussions on the frequency domain representation of the discrete sine and cosine functions, there is one real-world issue concerning the DFT that needs to be addressed.

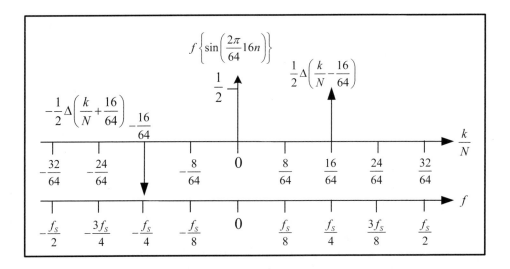

**Figure 9.47**  Example sine waveform spectrum

There is a property of the DFT that often causes confusion in the interpretation of the data that it computes. The analog or continuous Fourier transform (CFT) is a "pencil-and-paper" computation that has infinite frequency precision and infinite frequency range and therefore can produce precise results. The DFT, on the other hand, is a computer-based computation that operates on a limited amount of data and produces results only at $N$ discrete frequencies and therefore does not have infinite precision or infinite range.

A common misconception among first-time DFT users is that if they wish to compute the frequency response of a discrete time signal $x(n)$, then they would simply compute the DFT of $x(n)$ and get the discrete spectral representation of that signal. Although the previous statement is true, it is not completely true because it does not take into account a common DFT data interpretation issue that will be addressed here.

The equation that defines a DFT is repeated here for clarity:

$$X\left(\frac{k}{N}\right) = \sum_{n=0}^{N-1} x(n)\, e^{-j\left(2\pi\frac{k}{N}\right)n} \quad \text{for} \quad \left\{ \begin{array}{l} 0 \le k \le N-1 \\ 0 \le n \le N-1 \end{array} \right\}$$

We can see from the equation that the sequence $x(n)$, of length $N$ samples, is multiplied by the $N$ sample complex exponential:

$$e^{-j\left(2\pi\frac{k}{N}\right)n} = \left[ \cos\left(2\pi\frac{k}{N}\right)n - j\sin\left(2\pi\frac{k}{N}\right)n \right] \quad \left\{ \begin{array}{l} n = 0,1,2,3,\cdots N-1 \\ k = 0,1,2,3,\cdots N-1 \end{array} \right\}$$

This $N$ point multiplication takes place $N$ separate times, each time with an incremented value of the frequency index $k$. You can think of this as $N$ vector multiplies, and all vectors are $N$ points long. Each vector multiply sums the product of the $N$ sample input vector to a complex sinusoidal vector of normalized frequency $k/N$. The frequency index $k$ points to the $k$th vector multiply and increments through the vectors $0 \le k < N-1$.

The result or sum of the $k$th vector product is stored in what can be thought of as a DFT frequency bin labeled $X(k/N)$. This process can be likened to a correlation of the discrete time signal $x(n)$ with $N$ distinct complex sinusoids that are spaced at frequency intervals of $2\pi/N$ radians. The sum of each correlation is referred to as the correlation score. The correlation score can be a fairly large number or a fairly small number, depending on how closely the input signal matches any of the $N$ different DFT sinusoids.

The alert reader, and by that I mean the reader who hasn't fallen asleep yet, probably noticed that the sine and cosine waveforms discussed in the previous sections had a frequency that was exactly equal to one of the $N$ possible frequencies generated by the DFT. That is, the radian frequency of the discrete sine and cosine waves were given as

$$\omega_0 = \left( 2\pi \frac{k_0}{N} \right) n$$

where $k_0$ is a fixed integer that falls somewhere within the range of $k = 0, 1, 2, 3, \cdots N/2$ for real signals and $k = 0, 1, 2, 3, \cdots N - 1$ for complex signals. The finite number of integer radian frequencies at which the DFT is evaluated is given by

$$\omega_{DFT} = \left( 2\pi \frac{k}{N} \right) n \quad \text{for } k = 0, 1, 2, 3, \cdots N - 1$$

The collection of frequencies referred to as $\omega_{DFT}$ are called *analysis frequencies*. This is because all the frequencies in the input waveform are correlated or analyzed with this set of $N$ frequencies. This means that $\omega_0$ for each of the sine and cosines we have discussed has been exactly equal to one of the $N$ possible values of $\omega_{DFT}$. This is an optimum condition that certainly exists, but since most signals contain many frequencies, this is not typical in the everyday engineering environment.

You can think of the DFT as producing $N$ distinct correlation scores that are stored into $N$ equal width frequency bins. The DFT essentially correlates the input time domain waveform against the $N$ discrete complex frequencies of $\omega_{DFT}$. Each of these $N$ correlations will produce a correlation score that represents the degree of similarity between the frequency components that make up the input signal and the $k$th frequency of $\omega_{DFT}$.

Let's assume that the input signal $x(n)$ is a pure tone with a frequency $k/N$ that is identical to one of the $N$ possible discrete DFT frequencies. In this case, the DFT will produce a correlation score close to zero for all $N$ analysis frequencies except the $k$th analysis frequency, which corresponds to the one that matches the frequency of the input. The $k$th score will produce a near perfect correlation, and all the computed energy between these two frequencies will reside in the $k$th DFT bin. A plot of the DFT output will show that all the correlation energy is concentrated as a single spectral line in that bin. This will result in the type of plots we have shown for the discrete sine and cosine waves in Figure 9.41 through Figure 9.45.

What happens if we compute the DFT of a single sinusoidal input whose frequency does not exactly equal one of the $N$ possible discrete DFT frequencies? This is a great question, and the remainder of this section is devoted to the answer. Let's take the case where the DFT is evaluated at $N$ discrete frequency points and the input waveform sequence is a sinusoid whose frequency does not fall exactly on a DFT bin center.

That is suppose $k_0/N$ is a normalized bin center frequency and that $k_\Delta/N$ is some fractional offset such that the normalized frequency

$(k_0 + k_\Delta)/N$ does not fall exactly on a DFT bin center. We can represent this frequency by

$$x(n) = \cos\left(\frac{2\pi(k_0 + k_\Delta)}{N}\right)n \quad \text{for } \{0 \le n \le N-1\}$$

Thus there will not be an integer number of complete periods of $x(n)$ within the $N$ samples of the DFT. For example, if $N = 32$, $k_0 = 3$, and $k_\Delta = 0$, then as the index $n$ increments from $0$ to $(N-1) = 31$, the cosine waveform will have traced out exactly three complete cycles from $0$ to $2\pi$. If we increase the offset so that $k_\Delta = 0.1$, then $k_0 + k_\Delta = 3.1$ and for the same range of the index $n$ the cosine waveform will trace out approximately 3.1 cycles. The extra fraction of a cycle is a truncation of a complete period.

The point is, there will be a partial cycle of the input waveform within the $N$ samples. The DFT believes the input signal is periodic in $N$ samples, so from the DFT point of view the partial cycle will create a periodic discontinuity. To see what happens, let's compute the DFT of $x(n)$ in the following steps:

$$X\left(\frac{k}{N}\right) = \sum_{n=0}^{N-1}\left[\cos\left(2\pi\frac{(k_0 + k_\Delta)}{N}\right)n\right]e^{-j(2\pi)\left(\frac{k}{N}\right)n} \quad \text{for } \left\{\begin{array}{l} 0 \le n \le N-1 \\ 0 \le k \le N-1 \end{array}\right\}$$

The first thing we always do is use Euler's identity to expand the cosine function. After doing so, we arrive at

$$X\left(\frac{k}{N}\right) = \sum_{n=0}^{N-1}\left[\frac{e^{j(2\pi)\left(\frac{k_0+k_\Delta}{N}\right)n} + e^{-j(2\pi)\left(\frac{k_0+k_\Delta}{N}\right)n}}{2}\right]e^{-j(2\pi)\left(\frac{k}{N}\right)n}$$

The second thing we always do is to perform the multiplication of exponentials and then split the summation into two terms to get

$$X\left(\frac{k}{N}\right) = \frac{1}{2}\sum_{n=0}^{N-1}\left[e^{-j\left(\frac{2\pi}{N}\right)(k-[k_0+k_\Delta])n}\right] + \frac{1}{2}\sum_{n=0}^{N-1}\left[e^{-j\left(\frac{2\pi}{N}\right)(k+[k_0+k_\Delta])n}\right]$$

**Equation 9.78**

The third thing we always do is to use the identities given by Equation 9.69 and Equation 9.70 to write the equation for the line spectrum:

$$X\left(\frac{k}{N}\right)=\frac{1}{2}\left\{\Delta\left(\frac{k}{N}-\frac{k_0+k_\Delta}{N}\right)+\Delta\left(\frac{k}{N}+\frac{k_0+k_\Delta}{N}\right)\right\}$$

**Equation 9.79**

Now let's interpret Equation 9.79 graphically by looking at some line spectra plots. Figure 9.48 shows the line spectrum of a cosine wave whose frequency is identical to one of the $N$ discrete DFT frequencies. In this case $k_\Delta = 0$, and the normalized cosine frequency is given by $f = k_0/N$. Because the frequency of the cosine wave falls directly in a DFT frequency bin, we have a nonzero correlation and all the cosine energy is confined to a pair of impulses located at $f = (k/N \pm k_0/N)$. All other correlation results are zero for the cases where $k \neq k_0$.

Figure 9.49 illustrates the case of the same cosine wave whose frequency is offset by $k_\Delta$ and does not identically match any of the $N$ possible discrete DFT frequencies. In this case, the normalized frequency is $f = (k_0 + k_\Delta)/N$. In this figure, we have drawn the cosine line spectra with dotted lines to indicate where spectral lines would be located on the DFT output spectral plot if the DFT had infinite precision. As we can clearly see, the cosine frequency impulses $f = (k \pm k_0)/N$ fall between two adjacent DFT frequency bins. This means that when the sinusoid is correlated against the $N$ discrete DFT frequencies, there will be no exact correlation and therefore the energy associated with this waveform will not fall in to a single DFT bin. Instead we will get a partial or nonzero correlation between the frequency of the input cosine waveform and several adjacent DFT frequencies. This will result in partial correlation scores spread across several adjacent bins. The resulting spectral plot looks very similar to the artist's concept shown in Figure 9.50.

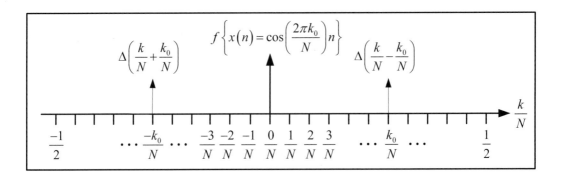

**Figure 9.48**   Sinusoid falls directly in a DFT bin

We see from this figure that the partial or nonzero correlations between the $N$ DFT frequencies and the input cosine frequency have spread the waveform energy across several adjacent DFT bins. This phenomenon is referred to as *spectral leakage* and is discussed in some detail in Chapter 3, "Review of the Fourier Transform." In that chapter, we also discuss methods and procedures that can be used to mitigate the effects of leakage. Chapter 3 also discusses this property in the discrete time domain by realizing that the DFT considers all input signals that it processes to be periodic in $N$ samples. The leakage problem graphically demonstrated in Figure 9.50 is due simply to the fact that the $N$ sample period of the input cosine waveform

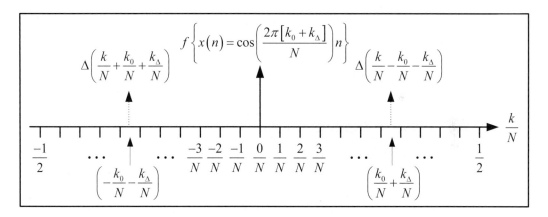

**Figure 9.49**   Sinusoid falls outside a DFT bin

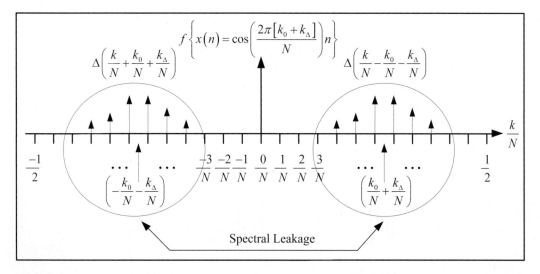

**Figure 9.50**   Spectral leakage

$$x(n) = \cos\left(2\pi \frac{k_0 + k_\Delta}{N}\right)n \quad \text{for } \{0 \le n \le N - 1\}$$

was not an integer submultiple of the $N$ sample DFT. The partial period of the cosine wave introduced a discontinuity in the waveform that the DFT assumes to be periodic. This sharp discontinuity introduced a whole lot of extra frequencies. The DFT dutifully computed all these frequencies, and we see the results in the spectral plot.

The reader should understand that the spectral leakage is not caused by the input cosine waveform. It is caused by the fact that the input waveform does not have an integer number of complete cycles in the $N$ sample space of the DFT. There is nothing wrong with the cosine waveform. It is still a pure tone. The problem exists only in the spectral computation and display.

This is one of the drawbacks when working in an arena where we do not have the luxury of infinite precision. The reader should be aware of this DFT property when analyzing the frequency response of any digital system.

## 9.2.6 Discrete Exponential Function $e^{j(2\pi k/N)n}$

The Fourier transform pair of the exponential function $x(n) = e^{j(2\pi k_0/N)n}$ given in Equation 9.69 is repeated here for clarity:

$$e^{j(2\pi)\left(\frac{k_0}{N}\right)n} \overset{f}{\leftrightarrow} \Delta\left(\frac{k}{N} - \frac{k_0}{N}\right)$$

We could leave it here and be done, but it is instructive to take the long computational path and achieve the same results, only this time using a more visual graphics approach. The spectrum of the exponential function $x(n) = e^{j(2\pi k_0/N)n}$ can be computed first by using Euler's identity to expand $e^{j(2\pi k_0/N)n}$ and then by using the transform identities for frequency domain impulses that we derived in Equation 9.69 and Equation 9.70. Expansion of $x(n) = e^{j(2\pi k_0/N)n}$ using Euler's identity gives us

$$x(n) = \cos\left(2\pi \frac{k_0}{N}\right)n + j\sin\cos\left(2\pi \frac{k_0}{N}\right)n$$

$$x(n) = \left[\frac{e^{j(2\pi)\frac{k_0}{N}n} + e^{-j(2\pi)\frac{k_0}{N}n}}{2}\right] + j\left[\frac{e^{j(2\pi)\frac{k_0}{N}n} - e^{-j(2\pi)\frac{k_0}{N}n}}{2j}\right]$$

Now we can separate the equation into its four terms to get

$$x(n) = \frac{1}{2}e^{j(2\pi)\left(\frac{k_0}{N}\right)n} + \frac{1}{2}e^{-j(2\pi)\left(\frac{k_0}{N}\right)n} + \frac{1}{2}e^{j(2\pi)\left(\frac{k_0}{N}\right)n} - \frac{1}{2}e^{-j(2\pi)\left(\frac{k_0}{N}\right)n}$$

**Equation 9.80**

Let's compute the Fourier transform on all the terms in Equation 9.80. We start with

$$X\left(\frac{k}{N}\right) = \sum_{n=0}^{N-1}\left[\frac{1}{2}e^{j(2\pi)\left(\frac{k_0}{N}\right)n} + \frac{1}{2}e^{-j(2\pi)\left(\frac{k_0}{N}\right)n} + \frac{1}{2}e^{j(2\pi)\left(\frac{k_0}{N}\right)n} - \frac{1}{2}e^{-j(2\pi)\left(\frac{k_0}{N}\right)n}\right]e^{-j\left(2\pi\left(\frac{k}{N}\right)n\right)}$$

**Equation 9.81**

Now we can use the frequency domain impulse identities in Equation 9.69 and Equation 9.70 to compute the Fourier transform for each term of Equation 9.81. The resulting transform is

$$X\left(\frac{k}{N}\right) = \underbrace{\frac{1}{2}\Delta\left(\frac{k}{N}-\frac{k_0}{N}\right)}_{A} + \underbrace{\frac{1}{2}\Delta\left(\frac{k}{N}+\frac{k_0}{N}\right)}_{B} + \underbrace{\frac{1}{2}\Delta\left(\frac{k}{N}-\frac{k_0}{N}\right)}_{C} - \underbrace{\frac{1}{2}\Delta\left(\frac{k}{N}+\frac{k_0}{N}\right)}_{D}$$

**Equation 9.82**

The Fourier transform in Equation 9.82 contains four frequency-shifted unit impulse terms that are labeled $A$, $B$, $C$, and $D$. All four of these impulses are plotted in Figure 9.51. The impulse terms $A$ and $C$ are plotted offset from their normalized frequency $+k_0/N$ only to show that there are two terms of identical amplitude that sum together. The impulse terms B and D plotted at the normalized frequency $-k_0/N$ are opposite in phase and cancel one another. The plot in Figure 9.52 illustrates the resultant spectrum when all four terms of Equation 9.82 are summed.

It is important to note that the line spectrum for the time domain function $e^{+j(2\pi)(k_0/N)n}$ is a single impulse located at the normalized frequency $k_0/N$. This is the graphical derivation of the Fourier transform pair:

$$e^{+j(2\pi)\left(\frac{k_0}{N}\right)n} \overset{f}{\leftrightarrow} \Delta\left(\frac{k}{N}-\frac{k_0}{N}\right) \quad \{\text{Where } k/N \text{ is normalized frequency}\}$$

**Equation 9.83**

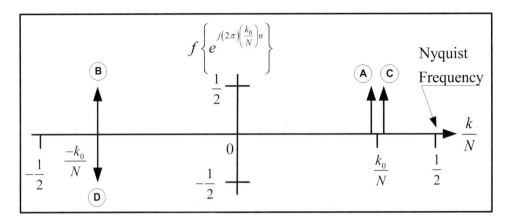

**Figure 9.51**  Cancellation of line spectra for $e^{+j(2\pi k_0/N)n}$

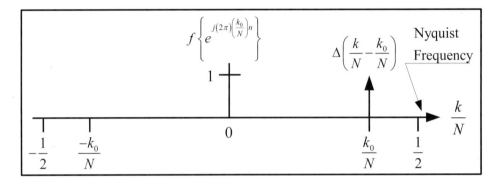

**Figure 9.52**  Line spectrum of $e^{+j(2\pi k_0/N)n}$

### 9.2.7  Discrete Exponential Function $e^{-j(2\pi/N)nk}$

The spectrum of the exponential function $x(n) = e^{-j(2\pi k_0/N)n}$ can be computed first by using Euler's identity to expand $e^{-j(2\pi k/N)n}$ and then by using the transform identities for frequency domain impulses that we derived in Equation 9.69 and Equation 9.70. Expansion of $x(n) = e^{-j(2\pi k_0/N)n}$ using Euler's identity gives us

$$x(n) = \cos\left(2\pi \frac{k_0}{N}\right)n - j\sin\cos\left(2\pi \frac{k_0}{N}\right)n$$

$$x(n) = \left[ \frac{e^{j(2\pi)\left(\frac{k_0}{N}\right)n} + e^{-j(2\pi)\left(\frac{k_0}{N}\right)n}}{2} \right] - j\left[ \frac{e^{j(2\pi)\left(\frac{k_0}{N}\right)n} - e^{-j(2\pi)\left(\frac{k_0}{N}\right)n}}{2j} \right]$$

Now we can separate the equation into its four terms to get

$$x(n) = \frac{1}{2}e^{j(2\pi)\left(\frac{k_0}{N}\right)n} + \frac{1}{2}e^{-j(2\pi)\left(\frac{k_0}{N}\right)n} - \frac{1}{2}e^{j(2\pi)\left(\frac{k_0}{N}\right)n} + \frac{1}{2}e^{-j(2\pi)\left(\frac{k_0}{N}\right)n}$$

**Equation 9.84**

We adopt the same strategy as before, by taking the longer but more instructional computational path that will allow us to graphically illustrate the whole process. We begin by computing the Fourier transform of Equation 9.84 such that

$$X\left(\frac{k}{N}\right) = \sum_{n=0}^{N-1}\left[\frac{1}{2}e^{j(2\pi)\left(\frac{k_0}{N}\right)n} + \frac{1}{2}e^{-j(2\pi)\left(\frac{k_0}{N}\right)n} - \frac{1}{2}e^{j(2\pi)\left(\frac{k_0}{N}\right)n} + \frac{1}{2}e^{-j(2\pi)\left(\frac{k_0}{N}\right)n}\right]e^{-j\left(2\pi\left(\frac{k}{N}\right)n\right)}$$

Now we can use the frequency domain impulse identities in Equation 9.69 and Equation 9.70 to compute the Fourier transform of Equation 9.84. The resulting transform is

$$X\left(\frac{k}{N}\right) = \underbrace{\frac{1}{2}\Delta\left(\frac{k}{N} - \frac{k_0}{N}\right)}_{A} + \underbrace{\frac{1}{2}\Delta\left(\frac{k}{N} + \frac{k_0}{N}\right)}_{B} - \underbrace{\frac{1}{2}\Delta\left(\frac{k}{N} - \frac{k_0}{N}\right)}_{C} + \underbrace{\frac{1}{2}\Delta\left(\frac{k}{N} + \frac{k_0}{N}\right)}_{D}$$

**Equation 9.85**

The Fourier transform in Equation 9.85 contains four frequency-shifted unit impulse terms that are labeled A, B, C, and D. All four of these impulses are plotted in Figure 9.53. The impulse terms B and D are plotted offset from their normalized frequency $-k_0/N$ only to show that there are two terms of identical amplitude that sum together. The impulse terms A and C plotted at the normalized frequency $+k_0/N$ are opposites and cancel one another. The plot in Figure 9.54 illustrates the resultant spectrum when all four terms of Equation 9.85 are summed.

It is important to note that the line spectrum for the time domain function $e^{-j(2\pi/N)nk}$ is a single impulse located at the normalized frequency $-k_0/N$. This is the graphical derivation of the Fourier transform pair:

$$e^{-j(2\pi/N)k_0 n} \overset{f}{\leftrightarrow} \Delta\left(\frac{k}{N} + \frac{k_0}{N}\right) \quad \{\text{Where } k/N \text{ is normalized frequency}\}$$

**Equation 9.86**

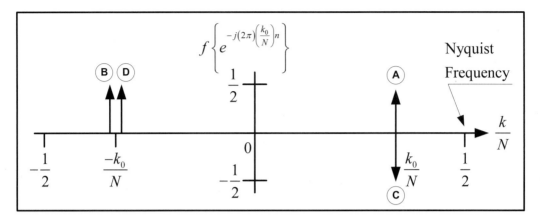

**Figure 9.53**  Cancellation of line spectra for $e^{-j(2\pi k_0/N)n}$

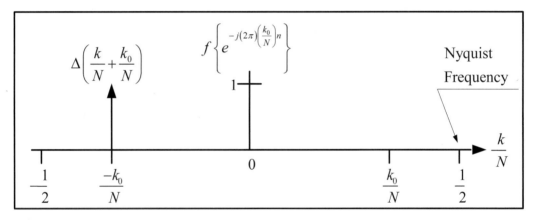

**Figure 9.54**  Line spectrum of $e^{-j(2\pi k_0/N)n}$

### 9.2.8  Discrete Frequency Identities

Before we get into the discussion on digital frequency translation, we need to develop a few additional tools in the form of Fourier transform pairs. Let's begin be defining the function $f(n)$ as a band limited signal with a discrete frequency spectrum denoted by $F(k/N)$. Furthermore we define the function

$$y(n) = f(n)\cos\left(2\pi \frac{k_0}{N} n\right)$$

as a discrete frequency spectrum represented by $Y(k/N)$. The Fourier transform of $y(n)$ is

$$f\{y(n)\} = Y\left(\frac{k}{N}\right) = \sum_{n=0}^{N-1}\left[f(n)\cos\left(2\pi\frac{k_0}{N}n\right)\right]e^{-j\left(2\pi\frac{k}{N}\right)n}$$

Expansion of the cosine function gives us

$$Y\left(\frac{k}{N}\right) = \sum_{n=0}^{N-1}f(n)\left[\frac{e^{j2\pi\left(\frac{k_0}{N}\right)n} + e^{-j2\pi\left(\frac{k_0}{N}\right)n}}{2}\right]e^{-j2\pi\left(\frac{k}{N}\right)n}$$

We can rearrange terms to get

$$Y\left(\frac{k}{N}\right) = \frac{1}{2}\sum_{n=0}^{N-1}f(n)\left[e^{-j2\pi\left(\frac{k}{N}-\frac{k_0}{N}\right)n}\right] + \frac{1}{2}\sum_{n=0}^{N-1}f(n)\left[e^{-j2\pi\left(\frac{k}{N}+\frac{k_0}{N}\right)n}\right]$$

**Equation 9.87**

We know that the Fourier transform of $f(n)$ expressed in terms of normalized frequency is

$$F\left(\frac{k}{N}\right) = \sum_{n=0}^{N-1}f(n)e^{-j\left(2\pi\frac{k}{N}\right)n}$$

If we change the variable

$$\frac{k}{N}$$

to

$$\frac{k}{N} - \frac{k_0}{N}$$

we get

$$F\left(\frac{k}{N} - \frac{k_0}{N}\right) = \sum_{n=0}^{N-1}f(n)e^{-j2\pi\left(\frac{k}{N}-\frac{k_0}{N}\right)n}$$

**Equation 9.88**

Similarly, if we change the variable

$$\frac{k}{N}$$

to

$$\frac{k}{N}+\frac{k_0}{N}$$

we get

$$F\left(\frac{k}{N}+\frac{k_0}{N}\right)=\sum_{n=0}^{N-1}f(n)e^{-j2\pi\left(\frac{k}{N}+\frac{k_0}{N}\right)n}$$

**Equation 9.89**

Substituting Equation 9.88 and Equation 9.89 back into Equation 9.87, we arrive at

$$Y\left(\frac{k}{N}\right)=\frac{1}{2}\left\{F\left(\frac{k}{N}-\frac{k_0}{N}\right)+F\left(\frac{k}{N}+\frac{k_0}{N}\right)\right\}$$

This gives us the valuable Fourier transform pair, defined by

$$f(n)\cos\left(\frac{2\pi}{N}k_0 n\right)\overset{f}{\leftrightarrow}\frac{1}{2}\left\{F\left(\frac{k}{N}-\frac{k_0}{N}\right)+F\left(\frac{k}{N}+\frac{k_0}{N}\right)\right\}$$

**Equation 9.90**

We can derive the Fourier transform pair for

$$y(n)=f(n)\sin\left(2\pi\frac{k_0}{N}n\right)$$

in a similar manner to produce the Fourier transform pair:

$$f(n)\sin\left(\frac{2\pi}{N}k_0 n\right)\overset{f}{\leftrightarrow}\frac{1}{2j}\left\{F\left(\frac{k}{N}-\frac{k_0}{N}\right)-F\left(\frac{k}{N}+\frac{k_0}{N}\right)\right\}$$

**Equation 9.91**

We can easily derive another valuable transform pair by reworking Equation 9.88, which is repeated here for clarity:

$$F\left(\frac{k}{N}-\frac{k_0}{N}\right)=\sum_{n=0}^{N-1}f(n)e^{-j2\pi\left(\frac{k}{N}-\frac{k_0}{N}\right)n}$$

We can rearrange terms to produce

$$F\left(\frac{k}{N}-\frac{k_0}{N}\right)=\sum_{n=0}^{N-1}\left[f(n)e^{j2\pi\left(\frac{k_0}{N}\right)n}\right]e^{-j2\pi\left(\frac{k}{N}\right)n}$$

The right-hand side of the equation is just the Fourier transform of $f(n)e^{j2\pi\left(\frac{k_0}{N}\right)n}$ giving us the Fourier transform pair:

$$f(n)e^{j2\pi\left(\frac{k_0}{N}\right)n}\overset{f}{\leftrightarrow}F\left(\frac{k}{N}-\frac{k_0}{N}\right)$$

**Equation 9.92**

The dual to this pair can be computed in a similar fashion:

$$f(n)e^{-j(2\pi)\left(\frac{k_0}{N}\right)n}\overset{f}{\leftrightarrow}F\left(\frac{k}{N}+\frac{k_0}{N}\right)$$

**Equation 9.93**

The next Fourier transform pair to be defined here involves the concept of convolution. As mentioned previously, convolution is one of the most important concepts in signal processing. Convolution plays a huge part in the translation of signal spectra. Frequency translation involves the multiplication of two time domain sequences such as a band limited sequence $x(n)$ and a sinusoidal sequence such as $\sin\left(2\pi\frac{k_0}{N}n\right)$. It turns out that multiplication of two sequences in the time domain results in the convolution the spectra of these two sequences in the frequency domain. If, for example, we have two discrete time domain series $x(n)$ and $h(n)$ whose Fourier transforms are given by $f\{x(n)\}=X\left(\frac{k}{N}\right)$ and $f\{h(n)\}=H\left(\frac{k}{N}\right)$, then we can state that

$$x(n)h(n)\overset{f}{\leftrightarrow}\frac{1}{N}X\left(\frac{k}{N}\right)*H\left(\frac{k}{N}\right)$$

**Equation 9.94**

is a Fourier transform pair. The symbol $*$ is used to denote the convolution operation, and the argument $k/N$ is a fraction representing normalized discrete frequency. Similarly, convolution in the discrete time domain is the equivalent to multiplication in the frequency domain, which defines another Fourier transform pair:

$$x(n)*h(n)\overset{f}{\leftrightarrow}X\left(\frac{k}{N}\right)H\left(\frac{k}{N}\right)$$

**Equation 9.95**

We can easily provide proof of the Fourier transform pair given in Equation 9.94 by the following. Let's define $X\left(\dfrac{k}{N}\right)$ and $H\left(\dfrac{k}{N}\right)$ as functions of normalized frequency. We can write the convolution of these two functions in conventional form as the expression

$$X\left(\frac{k}{N}\right)*H\left(\frac{k}{N}\right)=\sum_{p=0}^{N-1}X\left(\frac{p}{N}\right)H\left(\frac{k}{N}-\frac{p}{N}\right)$$

**Equation 9.96**

We know from previous discussion that the term $X(p/N)$ is the Fourier transform of the discrete time series $x(n)$ and is given by

$$X\left(\frac{p}{N}\right)=\sum_{n=0}^{N-1}x(n)\,e^{-j2\pi\left(\frac{p}{N}\right)n}$$

**Equation 9.97**

We also know from the Fourier transform pair in Equation 9.88 that

$$H\left(\frac{k}{N}-\frac{p}{N}\right)=\sum_{n=0}^{N-1}h(n)e^{-j2\pi\left(\frac{k}{N}-\frac{p}{N}\right)n}$$

**Equation 9.98**

We can substitute Equation 9.97 and Equation 9.98 back into Equation 9.96 and then manipulate terms to get

$$X\left(\frac{k}{N}\right) * H\left(\frac{k}{N}\right) = \sum_{p=0}^{N-1}\left[\sum_{n=0}^{N-1} x(n)\, e^{-j2\pi\left(\frac{p}{N}\right)n}\right]\left[\sum_{n=0}^{N-1} h(n)\, e^{-j2\pi\left(\frac{k}{N}-\frac{p}{N}\right)n}\right]$$

$$= \sum_{p=0}^{N-1}\left[\sum_{n=0}^{N-1} x(n)\, e^{-j2\pi\left(\frac{p}{N}\right)n}\right]\left[\sum_{n=0}^{N-1} h(n)\, e^{j2\pi\left(\frac{p}{N}\right)n}\, e^{-j2\pi\left(\frac{k}{N}\right)n}\right]$$

Now we can rearrange this expression to separately collect the summation over all the $p$ terms and the summation over all the $n$ terms. In doing so, we wind up with the expression

$$X\left(\frac{k}{N}\right) * H\left(\frac{k}{N}\right) = \sum_{p=0}^{N-1}\left[e^{-j2\pi\left(\frac{p}{N}\right)n}\, e^{+j2\pi\left(\frac{p}{N}\right)n}\right]\left[\sum_{n=0}^{N-1}[x(n)h(n)]\, e^{-j2\pi\left(\frac{k}{N}\right)n}\right]$$

The expression $e^{-j2\pi\left(\frac{p}{N}\right)n}\, e^{+j2\pi\left(\frac{p}{N}\right)n}$ collapses to unity and the summation of the $p$ terms equals $N$, which results in

$$X\left(\frac{k}{N}\right) * H\left(\frac{k}{N}\right) = N\sum_{n=0}^{N-1} x(n)h(n)e^{-j2\pi\left(\frac{k}{N}\right)n}$$

**Equation 9.99**

We recognize Equation 9.99 to be the Fourier transform of the product $x(n)h(n)$. Therefore

$$X\left(\frac{k}{N}\right) * H\left(\frac{k}{N}\right) = N f\{x(n)h(n)\}$$

Now we can write the Fourier transform pair:

$$x(n)h(n) \overset{f}{\leftrightarrow} \frac{1}{N}X\left(\frac{k}{N}\right) * H\left(\frac{k}{N}\right)$$

The proof for the Fourier transform pair in Equation 9.95 is achieved in a similar manner. Let's begin with the assumption that we have a function $y(n)$ that is the convolution of two other functions $x(n)$ and $h(n)$. The convolution is represented in the following expression:

$$y(n) = x(n) * h(n) = \sum_{p=0}^{N-1} x(p) h(n-p)$$

**Equation 9.100**

We can compute the DFT of $y(n)$ by

$$f\{y(n)\} = \sum_{n=0}^{N-1} y(n)\, e^{-j2\pi\left(\frac{k}{N}\right)n} \quad \text{for } 0 \le k < N-1$$

**Equation 9.101**

If we substitute Equation 9.100 into Equation 9.101, we arrive at

$$f\{y(n)\} = \sum_{n=0}^{N-1}\left[\sum_{p=0}^{N-1} x(p) h(n-p)\right] e^{-j2\pi\left(\frac{k}{N}\right)n}$$

**Equation 9.102**

If we define a new variable $r = n - p$, then when $n = 0$, $p = -r$ and when $n = N-1$, $p = N-1-r$. We also see that $n = r+p$. Similar arguments can be made for the summation variable $p$. If we make these substitutions in to Equation 9.102, we end up with

$$f\{y(n)\} = \sum_{p=-r}^{(N-1)-r}\left[\sum_{r=-p}^{(N-1)-p} x(p) h(r)\right] e^{-j2\pi\left(\frac{k}{N}\right)(r+p)}$$

The DFT assumes that the waveform it is processing is periodic in $N$ samples, so the new limits on the summation produce the same results as if the limits ranged from $r = 0$ to $r = N-1$ and from $p = 0$ to $p = N-1$. Making this change gives us

$$f\{y(n)\} = \sum_{p=0}^{N-1}\left[\sum_{r=0}^{N-1} x(p) h(r)\right] e^{-j2\pi\left(\frac{k}{N}\right)(r+p)}$$

After rearranging terms, we get

$$f\{y(n)\} = \left[\sum_{p=0}^{N-1} x(p) e^{-j2\pi\left(\frac{k}{N}\right)p}\right]\left[\sum_{r=0}^{N-1} h(r) e^{-j2\pi\left(\frac{k}{N}\right)r}\right]$$

We recognize the two terms on the right side of the equation as being the DFTs of $x(n)$ and $h(n)$. Therefore we can state that

$$f\{y(n)\} = f\{x(n) * h(n)\} = X\left(\frac{k}{N}\right)H\left(\frac{k}{N}\right)$$

This gives us the Fourier transform pair we seek:

$$x(n) * h(n) \overset{f}{\leftrightarrow} X\left(\frac{k}{N}\right)H\left(\frac{k}{N}\right)$$

We spent the time here deriving the poof for the time and frequency domain convolution transform pairs because they are important in the study of frequency translation. We will rely heavily on the frequency convolution identity in the following discussions on discrete frequency translation.

### 9.2.9 Discrete Real Frequency Translation

Frequency translation of a signal by a sine or cosine function is oftentimes referred to as *real frequency translation*. In sections 9.2.3 and 9.2.4, we introduced the spectra of the real discrete signals commonly used for frequency translation—that is, $\cos(2\pi k_0 n/N)$ and $\sin(2\pi k_0 n/N)$. In this section, we will observe the frequency domain translations that result from a time domain multiplication, such as $y(n) = g(n)f(n)$, where $f(n)$ is a discrete sinusoidal function and $g(n)$ is a discrete message signal.

We can observe these translations by remembering that multiplication of two sequences in the discrete time domain results in the convolution of the spectra of these same two sequences in the frequency domain. That is, if

$$y(n) \overset{f}{\leftrightarrow} Y(k)$$

and

$$f(n) \overset{f}{\leftrightarrow} F(k)$$

are Fourier transform pairs, then

$$y(n) = g(n)f(n) \overset{f}{\leftrightarrow} Y(k) = \sum_{i=-\infty}^{i=+\infty} G(i)F(k-i)$$

### 9.2.9.1 *Translation of a Discrete Real Signal by cos ($2\pi k_0 n/N$)*

Suppose we have a band limited message signal $g(n)$ that is multiplied with a translating signal $f(n) = \cos(2\pi k_0 n/N)$. We end up with a product signal $y(n) = g(n)\cos(2\pi k_0 n/N)$, as illustrated in Figure 9.55.

If the message signal $g(n)$ has a spectrum defined by $G(k/N)$ and the translating signal $f(n)$ has a spectrum defined by $F(k/N)$, then the multiplication of $g(n)$ and $f(n)$ will result in the convolution of their spectra. That is,

$$y(n) = g(n)f(n) \overset{f}{\leftrightarrow} Y(k/N) = G(k/N) * F(k/N)$$

We will begin by looking at the case where the message signal $g(n)$ is a simple cosine waveform at the normalized frequency $k_m/N$ such that $g(n) = \cos(2\pi k_m n/N)$. We know from previous study that the spectrum of $f(n)$ is given by

$$F\left(\frac{k}{N}\right) = \frac{1}{2}\left\{\Delta\left(\frac{k}{N} - \frac{k_0}{N}\right) + \Delta\left(\frac{k}{N} + \frac{k_0}{N}\right)\right\}$$

Similarly, the spectrum of $g(n)$ is given by

$$G\left(\frac{k}{N}\right) = \frac{1}{2}\left\{\Delta\left(\frac{k}{N} - \frac{k_m}{N}\right) + \Delta\left(\frac{k}{N} + \frac{k_m}{N}\right)\right\}$$

The convolution of $G\left(\dfrac{k}{N}\right)$ and $F\left(\dfrac{k}{N}\right)$ can be written as

$$Y\left(\frac{k}{N}\right) = \left[G\left(\frac{k}{N}\right) * F\left(\frac{k}{N}\right)\right] = \left[\frac{1}{2}\left\{\Delta\left(\frac{k}{N} - \frac{k_m}{N}\right) + \Delta\left(\frac{k}{N} + \frac{k_m}{N}\right)\right\}\right] * \left[\frac{1}{2}\left\{\Delta\left(\frac{k}{N} - \frac{k_0}{N}\right) + \Delta\left(\frac{k}{N} + \frac{k_0}{N}\right)\right\}\right]$$

**Equation 9.103**

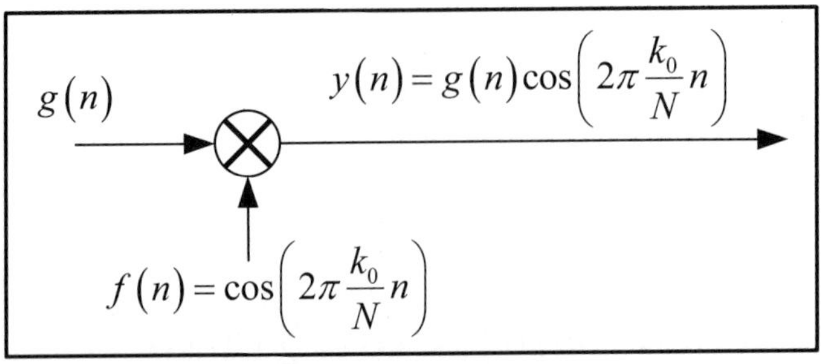

**Figure 9.55**  Translation of a real signal by cos($2\pi k_0/N$)

We can split Equation 9.103 into two terms to get

$$Y\left(\frac{k}{N}\right) = \left[\frac{1}{2}\left\{\Delta\left(\frac{k}{N} - \frac{k_m}{N}\right) + \Delta\left(\frac{k}{N} + \frac{k_m}{N}\right)\right\}\right] * \left[\frac{1}{2}\Delta\left(\frac{k}{N} - \frac{k_0}{N}\right)\right] + \left[\frac{1}{2}\left\{\Delta\left(\frac{k}{N} - \frac{k_m}{N}\right) + \Delta\left(\frac{k}{N} + \frac{k_m}{N}\right)\right\}\right] * \left[\frac{1}{2}\Delta\left(\frac{k}{N} + \frac{k_0}{N}\right)\right]$$

**Equation 9.104**

Convolving each of the two terms in Equation 9.104, we arrive at

$$Y\left(\frac{k}{N}\right) = \frac{1}{4}\left\{\Delta\left(\frac{k}{N} - \frac{k_0}{N_0} - \frac{k_m}{N}\right) + \Delta\left(\frac{k}{N} - \frac{k_0}{N} + \frac{k_m}{N}\right)\right\} + \frac{1}{4}\left\{\Delta\left(\frac{k}{N} + \frac{k_0}{N} - \frac{k_m}{N}\right) + \Delta\left(\frac{k}{N} + \frac{k_0}{N} + \frac{k_m}{N}\right)\right\}$$

**Equation 9.105**

The process by which we arrived at Equation 9.105 is illustrated graphically in Figure 9.56. The line spectrum of the message signal $g(n) = \cos(2\pi k_m n/N)$ is illustrated in part A of the figure. The line spectrum of the translating or tuning signal $f(n) = \cos(2\pi k_0 n/N)$ is illustrated in part B of the figure. The result of the convolution of the two is illustrated in part C of the figure.

We can clearly see from the figure that the line spectrum of the message signal has been translated both up and down in frequency such that its original center frequency of 0 Hz has been moved to $\pm k_0/N$ Hz. The end result is identical to the message signal being convolved with the two cosine frequency impulses. Also note that the amplitude of the translated $G(k/N)$ line spectra were scaled from $1/2$ to $1/4$, as we would expect.

Now let's take a look at the case where $g(n)$ is a finite bandwidth signal with a spectrum such as that illustrated in part A of Figure 9.57. We multiply the narrow band signal $g(n)$ by the same signal $f(n) = \cos(2\pi k_0 n/N)$ as before to translate the spectrum of $g(n)$ to $\pm k_0/N$ Hz. The Fourier transform pair defined in Equation 9.90 tells us that when we perform the multiplication, we should expect to see the spectral translation given by

$$g(n)\cos\left(\frac{2\pi}{N}k_0 n\right) \overset{f}{\leftrightarrow} \frac{1}{2}\left\{G\left(\frac{k}{N} - \frac{k_0}{N}\right) + G\left(\frac{k}{N} + \frac{k_0}{N}\right)\right\}$$

**Equation 9.106**

The convolution of the translating signal $f(n) = \cos(2\pi k_0 n/N)$ and the band limited message signal $g(n)$ is illustrated graphically in Figure 9.57. The spectrum of the message signal $g(n)$ is shown in part A of the figure. The line spectrum of the translating, or tuning signal $f(n)$, is shown in part B of the figure. The convolution of the two signals is illustrated in part C of the figure.

As we can see, the resultant spectrum agrees with Equation 9.106 in that the message signal has been translated both up and down in frequency from a center frequency of $0\,\text{Hz}$ to a new center frequency of $\pm k_0/N\,\text{Hz}$, and the amplitude has been scaled by $1/2$. This type of frequency translation is sometimes referred to as *linear modulation* [3], in which the amplitude of the carrier, $\cos(2\pi k_0 n/N)$, varies in a one-to-one correspondence to the amplitude of the message signal, $g(n)$. This is the same double sideband (DSB) modulation previously discussed for analog signals. The portion of the message signal

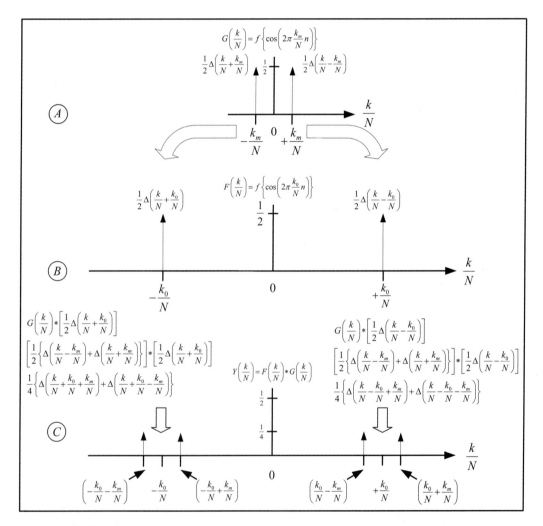

**Figure 9.56** Translation of a cosine waveform at discrete frequency $k_m/N$

$\frac{1}{2}G\left(\frac{k}{N}-\frac{k_0}{N}\right)$ that is above the carrier $k_0/N$ in Figure 9.57 is called the *upper sideband*, and the portion of the message signal $\frac{1}{2}G\left(\frac{k}{N}-\frac{k_0}{N}\right)$ that is below the carrier is called the *lower sideband*.

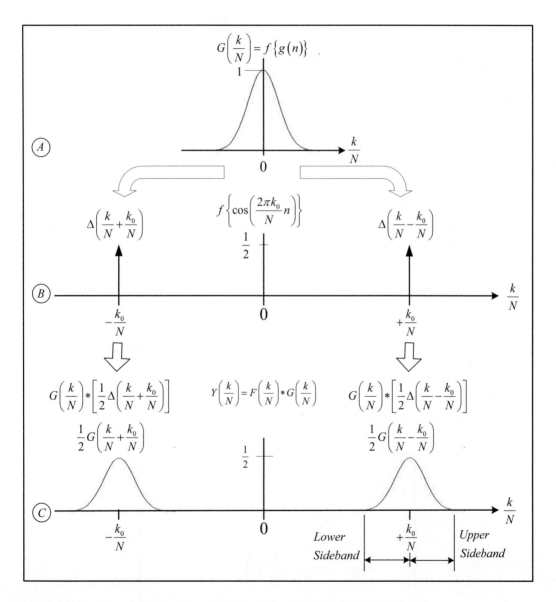

**Figure 9.57**   Translation of a band limited signal by $\cos(2\pi k_0/N)$

### 9.2.9.2 *Translation of a Discrete Real Signal by sin(2πk₀n/N)*

We will now look at the spectral translation that results when a band limited input message signal $g(n)$ is multiplied by the translation or tuning signal $f(n) = \sin(2\pi k_0 n/N)$. The block diagram of the multiplication of the two signals is illustrated in Figure 9.58. The discrete time domain product signal is given by $y(n) = g(n) f(n)$. The spectrum of the product signal is determined by the convolution of the two input signal spectra $Y(k/N) = G(k/N) * F(k/N)$.

    We could derive all the mathematics that describe the spectral translation, but we've already done that. All the information we need is found in Equation 9.75 and Equation 9.91, which are repeated here for clarity:

$$\sin\left(2\pi \frac{k_0}{N} n\right) \overset{f}{\leftrightarrow} \frac{1}{2j}\{\Delta(k - k_0) - \Delta(k + k_0)\}$$

$$f(n)\sin\left(2\pi \frac{k_0}{N} n\right) \overset{f}{\leftrightarrow} \frac{1}{2j}\left\{F\left(\frac{k}{N} - \frac{k_0}{N}\right) - F\left(\frac{k}{N} + \frac{k_0}{N}\right)\right\}$$

    Figure 9.59 illustrates the frequency domain convolution of the line spectra of the tuning signal $f(n) = \sin(2\pi k_0 n/N)$ and the narrow band spectrum of the message signal $g(n)$. The narrow band spectrum of the message signal $G(k/N)$ is shown in part A of the figure.

    The line spectrum of the tuning signal $f\{\sin(2\pi k_0 n/N)\}$ is shown in part B of the figure and is determined from Equation 9.75. The convolution of the two signal spectra is illustrated in part C of the figure and is determined from Equation 9.91. The reader should keep in mind that this is a plot of amplitudes, not magnitudes. This is why we see negative amplitudes and imaginary values like $\pm 1/2j$ for $Y(k/N)$. The magnitude plot of $Y(k/N)$ is shown in Figure 9.60. This is the more common representation of translated signal spectra.

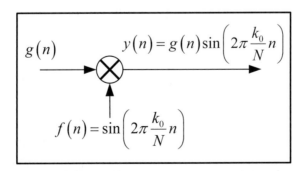

**Figure 9.58** Translation of a real signal by sin(2πk₀/N)

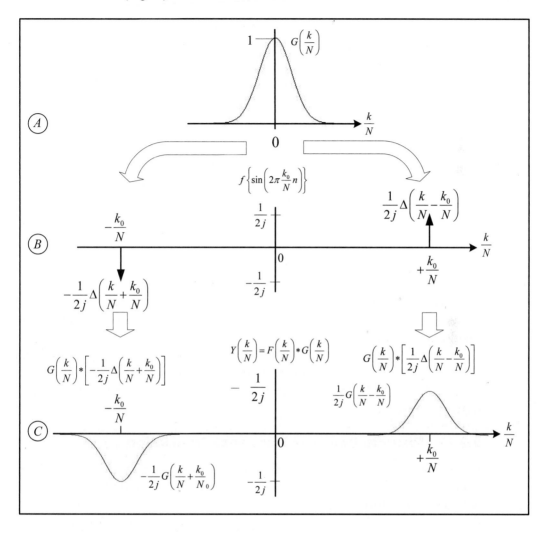

**Figure 9.59**   Spectrum of a narrow band signal translated by $\sin(2\pi n k_0/N)$

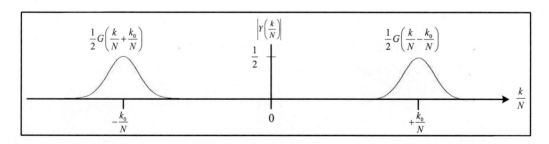

**Figure 9.60**   Spectral magnitude of translated narrow band signal

### 9.2.9.3   Translation of a Discrete Real Signal by –sin(2πk₀n/N)

In the interest of being thorough, let's see how the spectrum of a narrow band signal $g(n)$ is translated when it is multiplied by a discrete signal $f(n) = -\sin(2\pi n k_0/N)$, as illustrated in Figure 9.61.

We can modify the Fourier transform pairs of Equation 9.75 and Equation 9.91 by negating the sine function to produce the two new pairs graphically represented in Figure 9.62:

$$-\sin\left(2\pi \frac{k_0}{N}n\right) \overset{f}{\leftrightarrow} \frac{1}{2j}\left\{-\Delta(k-k_0) + \Delta(k+k_0)\right\}$$

$$-g(n)\sin\left(2\pi \frac{k_0}{N}n\right) \overset{f}{\leftrightarrow} \frac{1}{2j}\left[-G\left(\frac{k}{N}-\frac{k_0}{N}\right) + G\left(\frac{k}{N}+\frac{k_0}{N}\right)\right]$$

Part A of the figure shows the spectrum of the narrow band message $g(n)$. Part B of the figure shows the spectrum of the tuning signal $f(n) = -\sin(2\pi n k_0/N)$. Part C of the figure shows the convolution of the message spectrum and the tuning spectrum.

We can see that the spectrum of the original message signal $G(k/N)$ has been halved in amplitude and translated to the normalized frequencies $\pm k_0/N$. This behavior is what we expect, because it was defined in Equation 9.91.

## 9.2.10   Discrete Complex Frequency Translation

In the world of digital hardware, the translation of a digital narrow band signal up or down in frequency is almost always done by multiplying that signal with a complex tuning signal, such as

$$e^{+j(2\pi k_0 n/N)} \quad \text{or} \quad e^{-j(2\pi k_0 n/N)}$$

In the following sections we will demonstrate how this is done.

### 9.2.10.1   Translation of a Discrete Real Signal by e^{+j(2πk₀n/N)}

If a discrete time signal $x(n)$ is multiplied by the exponential

$$e^{j2\pi\left(\frac{k_0}{N}\right)n}$$

then the spectrum of $x(n)$ is translated in frequency by some constant $k_0/N$. This is given by the expression

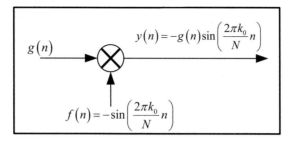

**Figure 9.61**   Translation of a real signal by $-\sin(2\pi k_0 n/N)$

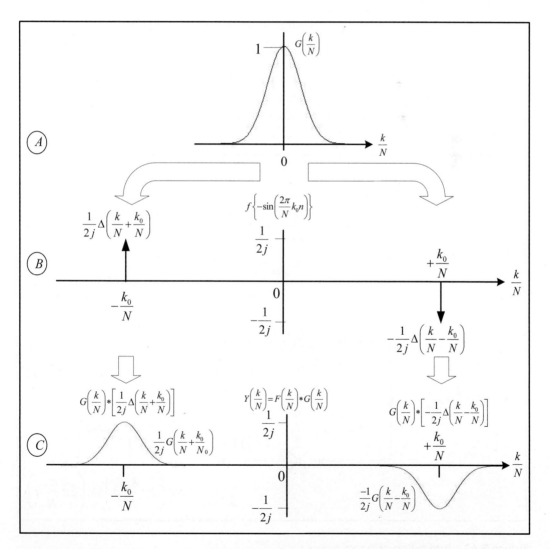

**Figure 9.62**   Spectrum of a narrow band signal translated by $-\sin(2\pi n k_0/N)$

$$x(n)e^{j2\pi\left(\frac{k_0}{N}\right)n} \overset{f}{\leftrightarrow} X\left(\frac{k}{N} - \frac{k_0}{N}\right) \quad \left\{\text{Where } \frac{k}{N} \text{ is normalized frequency}\right\}$$

### Equation 9.107

This expression is defined in Equation 9.92 as a Fourier transform pair. This is a very powerful statement. It provides us with a means to move the spectrum of any discrete narrow band signal up in frequency by a simple complex multiplication operation.

We will illustrate this by multiplying the real input message signal $g(n) = \cos(2\pi k_m n/N)$ by the complex exponential tuning signal $f(n) = e^{j(2\pi k_0 n/N)}$. The block diagram of the tuner that performs the complex multiplication of the two signals is illustrated in Figure 9.63. Part A of the figure is a graphic shorthand interpretation of the complex tuner. The single line arrow is typically used to indicate a real signal, while the double line arrow typically indicates a complex signal. Part B of the figure is the expanded block diagram of

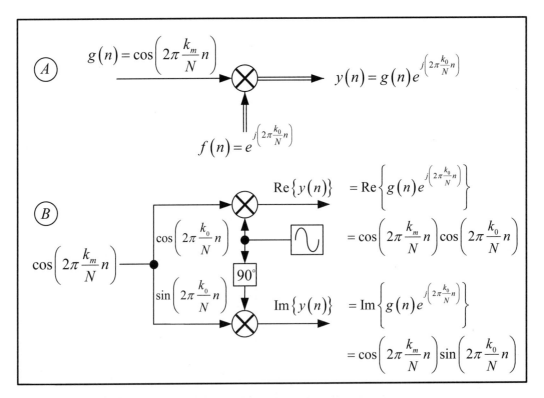

**Figure 9.63** Translation of a real signal by $e^{+j(2\pi k_0 n/N)}$

the complex tuner. As can be seen, the product signal out of the multiplier is indeed a complex signal expressed as

$$y(n) = g(n)e^{j\left(2\pi\frac{k_0}{N}n\right)} = \left[\cos\left(2\pi\frac{k_m}{N}n\right)\right]e^{j\left(2\pi\frac{k_0}{N}n\right)}$$

If implemented in hardware, the product signal would require two signal paths: one for the real part of the product and another for the imaginary or quadrature part of the product. The sinusoid drawn in a box represents a simple tunable tone oscillator used to generate the tuning signal. The oscillator output is used as the real part of the complex tuning signal. The 90° phase shifter creates the imaginary or quadrature component of the complex tuning signal. If we expand the exponential term, the complex product signal becomes

$$y(n) = \cos\left(2\pi\frac{k_m}{N}n\right)\left[\cos\left(2\pi\frac{k_0}{N}n\right) + j\sin\left(2\pi\frac{k_0}{N}n\right)\right]$$

The Fourier transform of $y(n)$ is computed by

$$Y\left(\frac{k}{N}\right) = \sum_{n=0}^{N-1}\left[g(n)e^{j\left(2\pi\frac{k_0}{N}n\right)}\right]e^{-j\left(2\pi\frac{k}{N}n\right)}$$

$$= \sum_{n=0}^{N-1}\left[\cos\left(2\pi\frac{k_m}{N}n\right)e^{j\left(2\pi\frac{k_0}{N}n\right)}\right]e^{-j\left(2\pi\frac{k}{N}n\right)}$$

$$= \sum_{n=0}^{N-1}\left[\cos\left(2\pi\frac{k_m}{N}n\right)e^{-j2\pi\left(\frac{k}{N}-\frac{k_0}{N}\right)n}\right]$$

$$= \sum_{n=0}^{N-1}\left[\frac{e^{j\left(2\pi\frac{k_m}{N}n\right)} + e^{-j\left(2\pi\frac{k_m}{N}n\right)}}{2}\right]e^{-j\left(2\pi\left(\frac{k}{N}-\frac{k_0}{N}\right)n\right)}$$

Separation of terms produces the sum of two transforms

$$Y\left(\frac{k}{N}\right) = \frac{1}{2}\sum_{n=0}^{N-1}e^{j\left(2\pi\frac{k_m}{N}n\right)}e^{-j\left(2\pi\left(\frac{k}{N}-\frac{k_0}{N}\right)n\right)} + \frac{1}{2}\sum_{n=0}^{N-1}e^{-j\left(2\pi\frac{k_m}{N}n\right)}e^{-j\left(2\pi\left(\frac{k}{N}-\frac{k_0}{N}\right)n\right)}$$

$$= \frac{1}{2}\sum_{n=0}^{N-1}e^{-j2\pi\left(\frac{k}{N}-\frac{k_0}{N}-\frac{k_m}{N}\right)n} + \frac{1}{2}\sum_{n=0}^{N-1}e^{-j2\pi\left(\frac{k}{N}-\frac{k_0}{N}+\frac{k_m}{N}\right)n}$$

Through the use of previously defined transform pairs, we arrive at the translated spectral representation expressed as

$$Y\left(\frac{k}{N}\right) = \frac{1}{2}\left\{\Delta\left(\frac{k}{N} - \frac{k_0}{N} - \frac{k_m}{N}\right) + \Delta\left(\frac{k}{N} - \frac{k_0}{N} + \frac{k_m}{N}\right)\right\}$$

**Equation 9.108**

The spectrum of the real input signal $g(n) = \cos(2\pi k_m n/N)$ is shown in Figure 9.64. As we would expect, the spectrum consists of two frequency impulses at $\pm k_m/N$. The spectrum of the tuning signal $f(n) = e^{j2\pi\frac{k_0}{N}n}$ is shown in Figure 9.65. As we would expect from our previous discussions, the spectra of the complex tuning signal consist of a single frequency impulse centered at $k_0/N$. Multiplication of the base band message signal by the tuning frequency results in the convolution of the base band spectra with the impulse spectra of the tuning frequency. The translated product signal is illustrated in Figure 9.66.

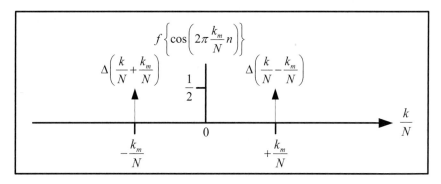

**Figure 9.64** Spectrum of the input message signal $\cos(2\pi k_m n/N)$

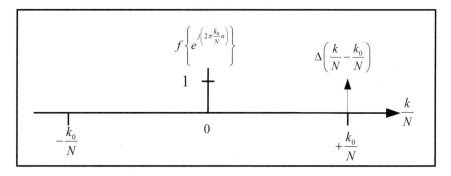

**Figure 9.65** Spectrum of the tuning signal $e^{j(2\pi k_0 n/N)}$

It is extremely important to note that, unlike a real tuner, the product signal for this tuner is complex and the spectrum at the output of a complex tuner does not have a mirror image centered $-k_0/N$. The output spectrum is one sided.

Let's quickly verify this result. Suppose that $G(k/N)$ is the two-sided spectrum of a sinusoid with spectral lines at $\pm k_m/N$. Further suppose that $F(k/N)$ is the spectrum of a complex tuning signal with a single spectral line at $k_0/N$. Let's assign DFT bin values of $k_m = \pm 1$ and $k_0 = 3$. We also drop the normalized frequency notation for a moment in favor of using the DFT bin numbers as the sequence index and perform the standard convolution of these two signals using the equation

$$Y(m) = \sum_{k=-2}^{2} G(k)F(m-k) \quad \text{for } 0 \le m \le 4$$

In doing so, we compute the first five convolutional sums of $Y(m)$:

$$
\begin{aligned}
Y(0) &= G(-2)F(2) \ +G(-1)F(1) & +G(0)F(0) \ +G(1)F(-1) & +G(2)F(-2) &= 0 \\
Y(1) &= G(-2)F(3) \ +G(-1)F(2) & +G(0)F(1) \ +G(1)F(0) & +G(2)F(-1) &= 0 \\
Y(2) &= G(-2)F(4) +\big[G(-1)F(3)\big] & +G(0)F(2) \ +G(1)F(1) & +G(2)F(0) &= 1 \\
Y(3) &= G(-2)F(5) \ +G(-1)F(4) & +G(0)F(3) \ +G(1)F(2) & +G(2)F(1) &= 0 \\
Y(4) &= G(-2)F(6) \ +G(-1)F(5) & +G(0)F(4) \ +\big[G(1)F(3)\big] & +G(2)F(2) &= 1
\end{aligned}
$$

We can see from the bracketed terms in the convolutional sums that $k_0 - k_m = 2$ and $k_0 + k_m = 4$, so $Y(m/N)$ would have translated spectral lines located at $2/N$ and $4/N$.

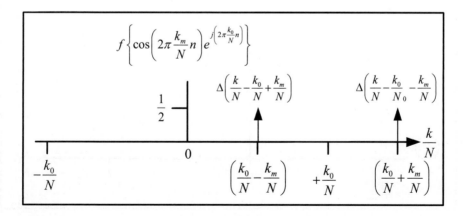

**Figure 9.66**   Spectrum $\cos(2\pi k_m n/N)$ translated to $k_0/N$

The results we obtained with the simple cosine message waveform are identical to the results we would obtain when dealing with an input narrow band message signal $g(n)$. This case is illustrated in Figure 9.67.

The spectrum of the band limited input signal $g(n)$ is shown in part A of Figure 9.67. The spectrum of the tuning signal $f(n) = e^{j(2\pi k_0 n/N)}$ is illustrated in part B of the figure. The translated spectrum resulting from the multiplication of $f(n)$ and $g(n)$ is shown in part C of the figure.

It is clear from the figure that the spectra of the two signals have been convolved with one another. The reader should note that the signal output by the tuner is a complex time series. It should also be noted that there is no mirror image spectral band at $-k_0/N$. At some point in the processing system

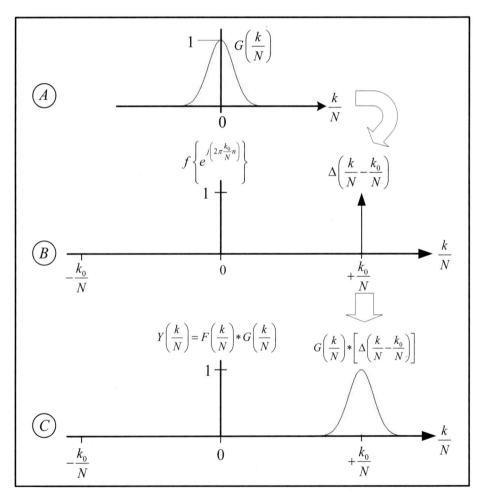

**Figure 9.67**  Translation of a band limited signal by $e^{j(2\pi k_0 n/N)}$

downstream from the tuner, the complex signal will more than likely be converted back to a real signal. The topic of complex to real signal conversion is discussed in detail in Chapter 7, "Complex to Real Conversion."

### 9.2.10.2   Translation of a Discrete Real Signal by $e^{-j(2\pi k_0 n/N)}$

Now we will look at the results of a real message signal $g(n) = \cos(2\pi k_m n/N)$ that is multiplied by the complex exponential tuning signal given by

$$f(n) = e^{-j\frac{2\pi k_0 n}{N}}$$

The block diagram depicting the multiplication of the two signals is illustrated in Figure 9.68. Part A of the figure is a shorthand interpretation of the complex multiplication. Once again standard convention has the single line arrow indicating a real signal while the double line arrow indicates a complex signal. Part B of the figure is the expanded block diagram of the complex multiplication. As can be seen, the product signal out of the multiplier is indeed a complex signal. If implemented in hardware, the product signal would require two signal paths: one for the real part of the product and another for the imaginary part of the product.

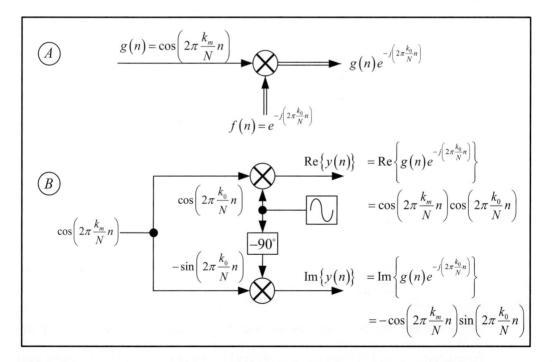

**Figure 9.68**   Translation of a real signal by $e^{-j(2\pi k_0 n/N)}$

Multiplication in the time domain is the equivalent of convolution in the frequency domain. We could explain the frequency translation by a negative exponential simply by using the previously derived identities, but this is such an important concept we will invest some time in the mathematics that describe this process. The product signal out of the multiplier is

$$y(n) = g(n)e^{-j\left(2\pi\frac{k_0}{N}n\right)} = \cos\left(2\pi\frac{k_m}{N}n\right)e^{-j\left(2\pi\frac{k_0}{N}n\right)}$$

If we expand the exponential term, the complex product signal becomes

$$y(n) = \left\{\cos(2\pi k_m n/N)\right\}\left\{\cos(2\pi k_0 n/N) - j\sin(2\pi k_0 n/N)\right\}$$

The Fourier transform of $y(n)$ is computed by

$$Y\left(\frac{k}{N}\right) = \sum_{n=0}^{N-1}\left[g(n)e^{-j2\pi\frac{k_0}{N}n}\right]e^{-j2\pi\frac{k}{N}n}$$

$$= \sum_{n=0}^{N-1}\left[\cos\left(2\pi\frac{k_m}{N}n\right)e^{-j2\pi\frac{k_0}{N}n}\right]e^{-j2\pi\frac{k}{N}n}$$

$$= \sum_{n=0}^{N-1}\cos\left(2\pi\frac{k_m}{N}n\right)e^{-j(2\pi)\left(\frac{k}{N}+\frac{k_0}{N}\right)n}$$

$$= \sum_{n=0}^{N-1}\left[\frac{e^{+j2\pi\left(\frac{k_m}{N}\right)n} + e^{-j2\pi\left(\frac{k_m}{N}\right)n}}{2}\right]e^{-j(2\pi)\left(\frac{k}{N}+\frac{k_0}{N}\right)n}$$

Separation of terms produces the sum of two transforms:

$$Y\left(\frac{k}{N}\right) = \frac{1}{2}\sum_{n=0}^{N-1}e^{+j2\pi\left(\frac{k_m}{N}\right)n}e^{-j(2\pi)\left(\frac{k}{N}+\frac{k_0}{N}\right)n} + \frac{1}{2}\sum_{n=0}^{N-1}e^{-j2\pi\left(\frac{k_m}{N}\right)n}e^{-j(2\pi)\left(\frac{k}{N}+\frac{k_0}{N}\right)n}$$

$$= \frac{1}{2}\sum_{n=0}^{N-1}e^{-j(2\pi)\left(\frac{k}{N}+\frac{k_0}{N}-\frac{k_m}{N}\right)n} + \frac{1}{2}\sum_{n=0}^{N-1}e^{-j(2\pi)\left(\frac{k}{N}+\frac{k_0}{N}+\frac{k_m}{N}\right)n}$$

Through the use of previously defined transform pairs, we arrive at the translated spectral representation of $y(n) = \cos(2\pi k_m n/N)e^{-j\left(2\pi\frac{k_0}{N}n\right)}$ expressed as follows:

$$Y\left(\frac{k}{N}\right) = \frac{1}{2}\left\{\Delta\left(\frac{k}{N} + \frac{k_0}{N} - \frac{k_m}{N}\right) + \Delta\left(\frac{k}{N} + \frac{k_0}{N} + \frac{k_m}{N}\right)\right\}$$

### Equation 9.109

The spectrum of the real input signal $g(n) = \cos(2\pi k_m n/N)$ was illustrated in Figure 9.64 and consists of two frequency impulses at $\pm k_m/N$. The spectrum of the tuning signal $f(n) = e^{-j2\pi\frac{k_0}{N}n}$ is shown in Figure 9.69.

The spectrum of the tuned or translated signal is shown in Figure 9.70. Note that, when multiplied by a complex exponential, the original signal spectrum is merely translated in frequency to the center frequency of the tuning signal. Since the tuning signal is complex, there is no spectral mirror

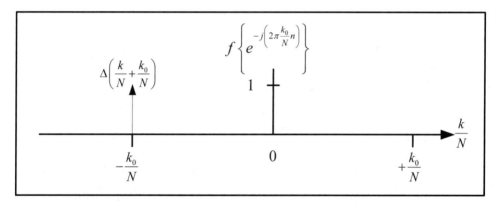

**Figure 9.69** Spectrum of the tuning signal $e^{-j(2\pi k_0 n/N)}$

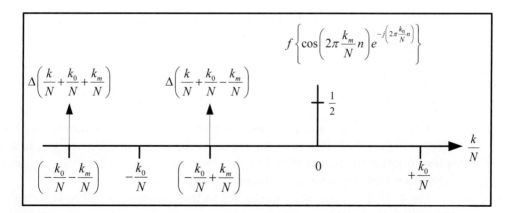

**Figure 9.70** Spectrum $\cos(2\pi k_m n/N)$ translated to $-k_0/N$

image at $+k_0/N$, as happens when the tuning signal is real. It's easily seen that the spectrum of the input signal has been convolved with the center frequency of the complex tuning signal.

This is a pretty handy result. We now know from this section and the previous section how to move the spectrum of a signal up and down in frequency. The results we obtained in this section are the direct result of the Fourier transform pair

$$x(n)e^{-j\left(\frac{2\pi n k_0}{N}\right)} \overset{f}{\leftrightarrow} X\left(\frac{k}{N}+\frac{k_0}{N}\right) \quad \left\{\text{Where } \frac{k}{N},\frac{k_0}{N} \text{ are normalized frequencies}\right\}$$

**Equation 9.110**

If a signal is applied to a tuner, the signal spectra can be translated in frequency and then filtered by an IF filter to extract a specific narrow band signal for further processing. This is analogous to a digital AM radio, where the wideband signal is the entire AM spectrum and the narrow bands output from the tuner filter combination are the individual AM radio station channels. In the world of DSP, the IF filter is generally a low pass filter. Typically the spectral band associated with a signal of interest is down converted to base band and passed through a low pass filter; the sample rate is decimated, and then it is passed on to follow on circuits for processing at the lower sample rate.

The same results are obtained if the input signal has a narrow band spectrum as illustrated in Figure 9.71. The spectrum of the narrow band message signal is illustrated in part A of the figure. The line spectrum of the tuning signal is illustrated in part B and the tuned or translated signal is illustrated in part C.

### 9.2.10.3  Translation of a Discrete Complex Signal by $e^{j(2\pi k_0 n/N)}$

In this section we will illustrate the frequency translation of a complex input message signal $g(n)=e^{j\left(2\pi\frac{k_m}{N}n\right)}$ by multiplying it with a complex tuning signal $f(n)=e^{j\left(2\pi\frac{k_0}{N}n\right)}$.

The block diagram of the complex multiply hardware structure is illustrated in Figure 9.72. Part A of the figure is the shorthand drawing notation for the complex tuner, and part B of the figure is the expanded block diagram of the tuner that performs a complex-by-complex multiply.

The complex input message signal in this example is a simple complex exponential given by

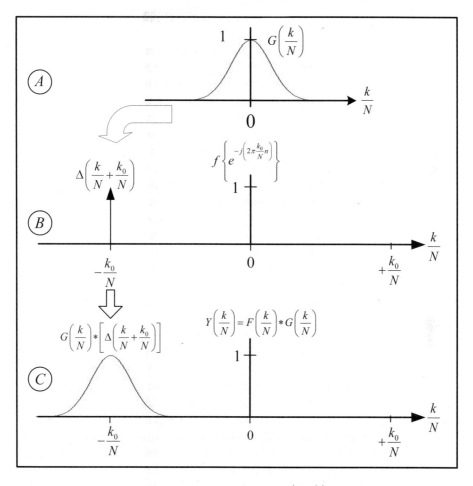

**Figure 9.71**   Translation of a band limited signal by $e^{-j(2\pi k_0 n/N)}$

$$g(n) = e^{j\left(2\pi \frac{k_m}{N} n\right)} = \cos\left(2\pi \frac{k_m}{N} n\right) + j\sin\left(2\pi \frac{k_m}{N} n\right)$$

The input signal is multiplied by another complex exponential given by

$$f(n) = e^{j\left(2\pi \frac{k_0}{N} n\right)} = \cos\left(2\pi \frac{k_0}{N} n\right) + j\sin\left(2\pi \frac{k_0}{N} n\right)$$

The tuned output or product signal can be mathematically represented by

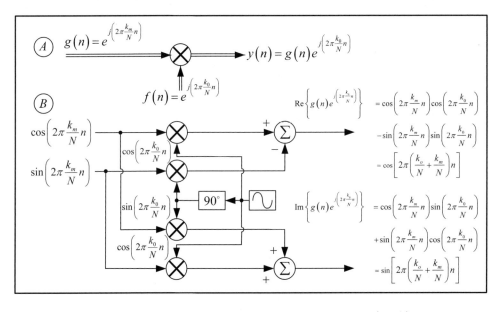

**Figure 9.72** Block diagram of translation of a complex signal by $e^{j(2\pi k_0 n/N)}$

$$y(n) \; = \; g(n)f(n) \; = \; e^{j\left(2\pi \frac{k_m}{N}n\right)} e^{j\left(2\pi \frac{k_0}{N}n\right)}$$

The Fourier transform of $Y\left(\dfrac{k}{N}\right) \overset{f}{\leftrightarrow} y(n)$ is computed to be

$$Y\left(\frac{k}{N}\right) \; = \sum_{n=0}^{N-1}\left[g(n)f(n)\right]e^{-j\left(2\pi \frac{k}{N}n\right)}$$

$$= \sum_{n=0}^{N-1}\left[e^{j\left(2\pi \frac{k_m}{N}n\right)} e^{j\left(2\pi \frac{k_0}{N}n\right)}\right]e^{-j\left(2\pi \frac{k}{N}n\right)}$$

$$= \sum_{n=0}^{N-1}\left[e^{j2\pi\left(\frac{k_0}{N}+\frac{k_m}{N}\right)n}\right]e^{-j\left(2\pi \frac{k}{N}n\right)}$$

$$= \sum_{n=0}^{N-1}\left[e^{-j2\pi\left(\frac{k}{N}-\frac{k_0}{N}-\frac{k_m}{N}\right)n}\right]$$

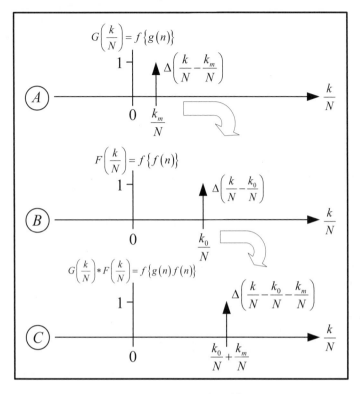

**Figure 9.73**   Translation of a complex signal by $e^{j(2\pi k_0 n/N)}$

Using the transform pair identities that we previously derived, we arrive at

$$Y\left(\frac{k}{N}\right) = \Delta\left(\frac{k}{N} - \frac{k_0}{N} - \frac{k_m}{N}\right)$$

The spectrum of $g(n)$, $f(n)$, and $y(n)$ are shown in Figure 9.73. In part A of the figure, we see the line spectra of the input message signal $g(n)$. Notice that $G(k/N)$ is a one-sided spectrum consisting of a single frequency impulse at $k_m/N$. Part B of the figure shows the line spectra of the tuning signal $f(n)$, which consist of a single frequency impulse at $k_0/N$. Part C of the figure shows that the input message signal spectra have been convolved with the tuning signal spectra, resulting in a frequency translation of the input from

$$\frac{k_m}{N} \rightarrow \frac{k_0}{N} + \frac{k_m}{N}$$

### 9.2.10.4 *Translation of a Discrete Complex Signal by* $e^{-j(2\pi k_0 n/N)}$

Now let's take a look at the case where the same input signal $g(n)$ is multiplied by a negative exponential waveform, as illustrated in the hardware block diagram in Figure 9.74. The compact or shorthand drawing notation is shown in part A of the figure, and the expanded block diagram of the complex tuner is shown in part B of the figure. Note that the block diagram of Figure 9.74 is identical to that block diagram in Figure 9.72, with the exception of the sign changes on the input of the two summers.

The input message signal in this example is once again a simple complex exponential given by

$$g(n) = e^{j\left(2\pi \frac{k_m}{N} n\right)} = \cos\left(2\pi \frac{k_m}{N} n\right) + j\sin\left(2\pi \frac{k_m}{N} n\right)$$

The input signal is multiplied by the tuning signal, which is another complex exponential given by

$$f(n) = e^{-j\left(2\pi \frac{k_0}{N} n\right)} = \cos\left(2\pi \frac{k_0}{N} n\right) - j\sin\left(2\pi \frac{k_0}{N} n\right)$$

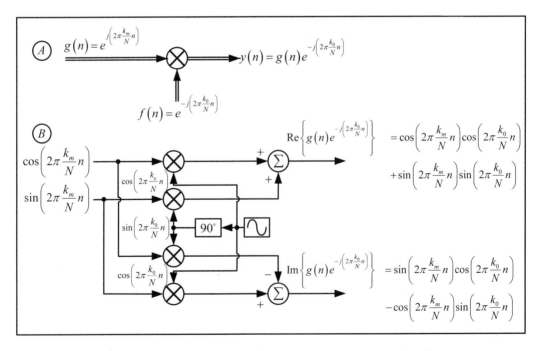

**Figure 9.74** Block diagram of translation of a complex signal by $e^{-j(2\pi k_0 n/N)}$

The translated output signal can be mathematically represented by

$$y(n) = g(n)f(n) = e^{j\left(2\pi\frac{k_m}{N}n\right)}e^{-j\left(2\pi\frac{k_0}{N}n\right)}$$

The Fourier transform pair $Y\left(\frac{k}{N}\right) \overset{f}{\leftrightarrow} y(n)$ is computed as

$$Y\left(\frac{k}{N}\right) = \sum_{n=0}^{N-1}\left[g(n)e^{-j\left(2\pi\frac{k_0}{N}n\right)}\right]e^{-j\left(2\pi\frac{k}{N}n\right)} = \sum_{n=0}^{N-1}\left[e^{j\left(2\pi\frac{k_m}{N}n\right)}e^{-j\left(2\pi\frac{k_0}{N}n\right)}\right]e^{-j\left(2\pi\frac{k}{N}n\right)}$$

$$= \sum_{n=0}^{N-1}\left[e^{-j2\pi\left(\frac{k_0}{N}-\frac{k_m}{N}\right)n}\right]e^{-j\left(2\pi\frac{k}{N}n\right)} = \sum_{n=0}^{N-1}\left[e^{-j2\pi\left(\frac{k}{N}+\frac{k_0}{N}-\frac{k_m}{N}\right)n}\right]$$

Using the transform pair identities that we previously derived, we arrive at

$$Y\left(\frac{k}{N}\right) = \Delta\left(\frac{k}{N}+\frac{k_0}{N}-\frac{k_m}{N}\right)$$

The spectrum of $g(n)$, $f(n)$, and $y(n)$ are shown in Figure 9.75. The one-sided line spectrum of the input signal $g(n)$ is shown in part A. The line spectrum of the tuning signal $f(n)$ is shown in part B. Part C of the figure shows that the input signal spectra have been convolved with the tuning signal spectra, resulting in a frequency down conversion from

$$\frac{k_m}{N} \rightarrow \frac{k_m}{N}-\frac{k_0}{N}$$

All things considered, discrete signal tuning is very straightforward and easy to compute. The one thing that needs to be kept in mind at all times is the frequency limitations imposed by Nyquist.

### 9.2.10.5 General Depiction of Complex Signal Translation

Let's now describe a more general complex translation of a complex signal. Suppose we have a narrow band complex message signal $x(n)$ whose spectrum is centered at the normalized frequency $k_m/N = 3/8 = 0.375$, as illustrated in Figure 9.76, part A. It is our objective to translate or down convert this signal so that it is centered at $k_{Tuned}/N = 1/8 = 0.125$.

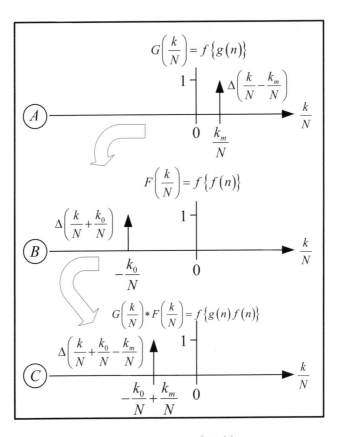

**Figure 9.75** Translation of a complex signal by $e^{-j(2\pi k_0 n/N)}$

We can see that the translation involves shifting the signal spectrum down in frequency by an amount equal to $f = 0.375 - 0.125 = 0.250$. All we need to do to accomplish this spectral translation is to multiply the time domain version of the signal by the complex tuning signal

$$f_0 = e^{-j\left(2\pi \frac{k_0}{N} n\right)} \quad \text{where} \quad \frac{k_0}{N} = 0.250$$

The line spectrum for this complex tuning signal is illustrated in part B of the figure. In the time domain, the translated signal would be represented by the product of the two complex signal sequences given by

$$y(n) = x(n) e^{-j\left(2\pi \frac{k_0}{N} n\right)}$$

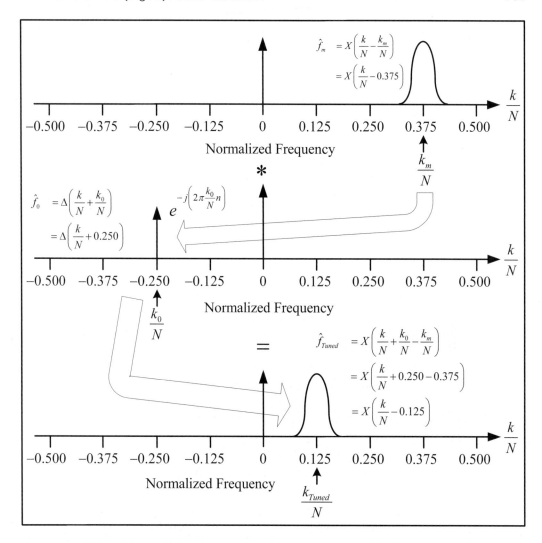

**Figure 9.76**   General depiction of complex signal translation

In the frequency domain, the translated signal would be represented by the convolution of the Fourier transforms of the two signals and is given by

$$Y\left(\frac{k}{N}\right) = f\{x(n)\} * f\left\{e^{-j\left(2\pi\frac{k_0}{N}n\right)}\right\}$$

The Fourier transform of the complex signal $x(n)$ is given by

$$f\{x(n)\} = X\left(\frac{k}{N} - \frac{k_m}{N}\right)$$

The Fourier transform of the complex tuning signal is given by

$$f\left\{e^{\left[-j\left(2\pi\frac{k_0}{N}n\right)\right]}\right\} = \Delta\left(\frac{k}{N} + \frac{k_0}{N}\right)$$

The convolution of the two transforms is given by

$$Y\left(\frac{k}{N}\right) = X\left(\frac{k}{N} - \frac{k_m}{N}\right) * \Delta\left(\frac{k}{N} + \frac{k_0}{N}\right) = X\left(\frac{k}{N} + \frac{k_0}{N} - \frac{k_m}{N}\right)$$

$$= X\left(\frac{k}{N} + 0.250 - 0.375\right) = X\left(\frac{k}{N} - 0.125\right)$$

as shown in part C of the figure. This result is exactly what we intended, and it points out the simplicity of working with complex digital signals.

Let's repeat this example, only this time we will take a quick look at the convolution equation. For ease of demonstration, we will replace both the narrow band message signal $X(k)$ and the tuning signal $H(k)$ with a unity amplitude impulse. The convolution equation is given by

$$Y\left(\frac{k}{N}\right) = \sum_{m=0}^{N-1} X\left(\frac{m}{N}\right) H\left(\frac{k}{N} - \frac{m}{N}\right)$$

If we replace the normalized frequency notation $k/N$ on the frequency axis with the frequency index $k$, and if we convolve the first five samples, we end up with the more conventional version of the convolution equation:

$$Y(k) = \sum_{m=0}^{4} X(m) H(k - m) \quad \text{for } 0 \le k \le 4$$

The convolution of these two signal spectra is illustrated in Figure 9.77. Here we have depicted a message signal $X(k)$ as being a frequency impulse of unity amplitude at $k_m/N = 0.375$, or $k = 3$. We have also depicted the tuning signal $H(k)$ as being a frequency impulse at $k_o/N = -0.250$, or $k = -2$. That is,

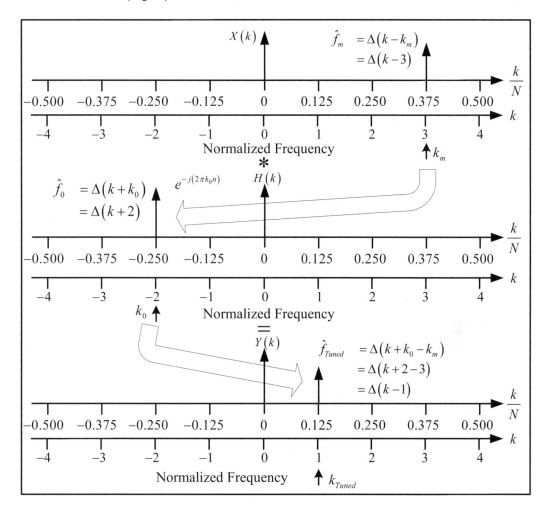

**Figure 9.77** Example frequency domain convolution

$$X(k) = \begin{cases} 1 \text{ for } k = 3 \\ 0 \text{ for } k \neq 3 \end{cases} \qquad H(k) = \begin{cases} 1 \text{ for } k = -2 \\ 0 \text{ for } k \neq -2 \end{cases}$$

We can write the convolution equation operation in its entirety for this particular example, as illustrated in Equation 9.111. It is clear from both the equation and the figure that the tuning frequency at $k_0/N = -0.250$, or $k = -2$, has translated the message signal at $k_m/N = 0.375$, or $k = 3$, down in frequency to $k_{Tuned}/N = 0.125$, or $k = 1$.

$$
\begin{aligned}
Y(0) &= X(0)H(0) &&+X(1)H(-1) &&+X(2)H(-2) &&+X(3)H(-3) &&+X(4)H(-4) &&= 0\\
Y(1) &= X(0)H(1) &&+X(1)H(0) &&+X(2)H(-1) &&+\big[X(3)H(-2)\big] &&+X(4)H(-3) &&= 1\\
Y(2) &= X(0)H(2) &&+X(1)H(1) &&+X(2)H(0) &&+X(3)H(-1) &&+X(4)H(-2) &&= 0\\
Y(3) &= X(0)H(3) &&+X(1)H(2) &&+X(2)H(1) &&+X(3)H(0) &&+X(4)H(-1) &&= 0\\
Y(4) &= X(0)H(4) &&+X(1)H(3) &&+X(2)H(2) &&+X(3)H(1) &&+X(4)H(0) &&= 0
\end{aligned}
$$

**Equation 9.111**

The bracketed term in the convolutional sum equations is equal to unity, making the output signal $Y(k) = Y(1) = 1$ or, in normalized frequency terms, $Y(k/N) = Y(1/8) = Y(0.125) = 1$ and all other $Y(k/N) = 0$.

For the interested reader, the best graphical interpretation of convolution ever given by an author can be found in Cadzow's book [8], pages 96 to 100.

## 9.3 USEFUL EQUATIONS

This section presents a number of useful equations.

### 9.3.1 Continuous Time Expressions

Fourier transform as a function of frequency:

$$
Y(f) = \int_{-\infty}^{\infty} y(t) e^{-j2\pi ft} dt
$$

**Equation 9.1**

Inverse Fourier transform as a function of frequency:

$$
y(t) = \int_{-\infty}^{\infty} Y(f) e^{+j2\pi ft} df
$$

**Equation 9.2**

Fourier transform as a function of radian frequency:

$$
Y(\omega) = \int_{-\infty}^{\infty} y(t) e^{-j\omega t} dt
$$

**Equation 9.3**

Inverse Fourier transform as a function of radian frequency:

$$y(t) = \frac{1}{2\pi} \int_{-\infty}^{\infty} Y(\omega) \, e^{+j\omega t} \, d\omega$$

**Equation 9.4**

Spectrum of the $\delta$ function:

$$f\{\delta(t)\} = \int_{-\infty}^{\infty} \delta(t) e^{-j2\pi ft} \, dt \;=\; e^{-j0} = 1$$

**Equation 9.5**

Spectrum of the time shifted $\delta(t - t_0)$ function:

$$f\{\delta(t - t_0)\} = \int_{-\infty}^{\infty} \delta(t - t_0) e^{-j2\pi ft} \, dt \;=\; e^{-j2\pi f t_0}$$

**Equation 9.7**

Spectrum of the time shifted $\delta(t + t_0)$ function:

$$f\{\delta(t + t_0)\} = \int_{-\infty}^{\infty} \delta(t + t_0) e^{-j2\pi ft} \, dt \;=\; e^{+j2\pi f t_0}$$

**Equation 9.9**

## 9.3.2   Fourier Transform Pairs

$$\delta(t - t_0) \overset{f}{\leftrightarrow} e^{-j2\pi f t_0}$$

**Equation 9.8**

$$\delta(t + t_0) \overset{f}{\leftrightarrow} e^{+j2\pi f t_0}$$

**Equation 9.10**

$$e^{j(2\pi f_0 t)} \overset{f}{\leftrightarrow} \delta(f - f_0)$$

**Equation 9.13**

$$e^{-j(2\pi f_0 t)} \overset{f}{\leftrightarrow} \delta\left(f + f_0\right)$$

**Equation 9.14**

$$\cos\left(2\pi f_0 t\right) \overset{f}{\leftrightarrow} \left\{\delta\left(f - f_0\right) + \delta\left(f + f_0\right)\right\}$$

**Equation 9.16**

$$\sin\left(2\pi f_0 t\right) \overset{f}{\leftrightarrow} \frac{1}{2j}\left\{\delta\left(f - f_0\right) - \delta\left(f + f_0\right)\right\}$$

**Equation 9.18**

$$\sin\left(2\pi f_0 t\right) \overset{f}{\leftrightarrow} \frac{j}{2}\left\{-\delta\left(\omega - \omega_0\right) + \delta\left(\omega + \omega_0\right)\right\}$$

**Equation 9.19**

$$f(t)\cos\left(2\pi f_0 t\right) \overset{f}{\leftrightarrow} \frac{1}{2}\left\{F\left(f - f_0\right) + F\left(f + f_0\right)\right\}$$

**Equation 9.27**

$$f(t)\sin\left(2\pi f_0 t\right) \overset{f}{\leftrightarrow} \frac{1}{2j}\left\{F\left(f - f_0\right) - F\left(f + f_0\right)\right\}$$

**Equation 9.29**

$$f(t)e^{+j2\pi f_0 t} \overset{f}{\leftrightarrow} F\left(f - f_0\right)$$

**Equation 9.30**

$$f(t)e^{-j2\pi f_0 t} \overset{f}{\leftrightarrow} F\left(f + f_0\right)$$

**Equation 9.31**

$$x(t)h(t) \overset{f}{\leftrightarrow} X(f) * H(f)$$

**Equation 9.32**

$$x(t) * h(t) \overset{f}{\leftrightarrow} X(f)H(f)$$

**Equation 9.33**

### 9.3.3 Discrete Time Expressions

Forward DFT:

$$H\left(\frac{k}{N}\right) = \sum_{n=0}^{N-1} h(n)\, e^{-j\left(\frac{2\pi}{N}\right)nk} \quad \text{for} \left\{ \begin{array}{c} 0 \le n \le N \\ 0 \le k \le N \end{array} \right\}$$

**Equation 9.49**

IDFT:

$$h(n) = \frac{1}{N}\sum_{k=0}^{N-1} H\left(\frac{k}{N}\right) e^{j\left(\frac{2\pi}{N}\right)nk} \quad \text{for} \left\{ \begin{array}{c} 0 \le n \le N \\ 0 \le k \le N \end{array} \right\}$$

**Equation 9.50**

Fourier transform of a discrete impulse $\Delta$ function:

$$f\{\Delta(n)\} = \sum_{n=0}^{N-1} \Delta(n)\, e^{-j\left(\frac{2\pi}{N}\right)nk} = e^{-j0} = 1$$

**Equation 9.55**

Fourier transform of discrete unit impulse shifted by $m$ samples:

$$f\{\Delta(n-m)\} = e^{-j\left(\frac{2\pi}{N}\right)mk}$$

**Equation 9.59**

### 9.3.4  Discrete Fourier Transform Pairs

$$\Delta(n-m) \overset{f}{\leftrightarrow} e^{-j\left(\frac{2\pi}{N}\right)mk}$$

**Equation 9.59**

$$\Delta(n+m) \overset{f}{\leftrightarrow} e^{+j\left(\frac{2\pi}{N}\right)mk}$$

**Equation 9.62**

$$e^{j(2\pi)\left(\frac{k_0}{N}\right)n} \overset{f}{\leftrightarrow} \Delta\left(\frac{k}{N} - \frac{k_0}{N}\right)$$

**Equation 9.69**

$$e^{-j(2\pi)\left(\frac{k_0}{N}\right)n} \overset{f}{\leftrightarrow} \Delta\left(\frac{k}{N} + \frac{k_0}{N}\right)$$

**Equation 9.70**

$$\cos\left(\frac{2\pi k_0}{N}n\right) \overset{f}{\leftrightarrow} \frac{1}{2}\left\{\Delta\left(\frac{k}{N} - \frac{k_0}{N}\right) + \Delta\left(\frac{k}{N} + \frac{k_0}{N}\right)\right\}$$

**Equation 9.73**

$$\sin\left(\frac{2\pi k_0}{N}n\right) \overset{f}{\leftrightarrow} \frac{1}{2j}\left\{\Delta\left(\frac{k}{N} - \frac{k_0}{N}\right) - \Delta\left(\frac{k}{N} + \frac{k_0}{N}\right)\right\}$$

**Equation 9.75**

$$\sin\left(\frac{2\pi k_0}{N}n\right) \overset{f}{\leftrightarrow} \frac{-j}{2}\left\{\Delta\left(\frac{k}{N} - \frac{k_0}{N}\right) - \Delta\left(\frac{k}{N} + \frac{k_0}{N}\right)\right\}$$

**Equation 9.77**

$$f(n)\cos\left(\frac{2\pi}{N}k_0n\right) \overset{f}{\leftrightarrow} \frac{1}{2}\left\{F\left(\frac{k}{N}-\frac{k_0}{N}\right) + F\left(\frac{k}{N}+\frac{k_0}{N}\right)\right\}$$

**Equation 9.90**

$$f(n)\sin\left(\frac{2\pi}{N}k_0n\right) \overset{f}{\leftrightarrow} \frac{1}{2j}\left\{F\left(\frac{k}{N}-\frac{k_0}{N}\right) - F\left(\frac{k}{N}+\frac{k_0}{N}\right)\right\}$$

**Equation 9.91**

$$x(n)h(n) \overset{f}{\leftrightarrow} \frac{1}{N}X\left(\frac{k}{N}\right) * H\left(\frac{k}{N}\right)$$

**Equation 9.94**

$$x(n)*h(n) \overset{f}{\leftrightarrow} X\left(\frac{k}{N}\right)H\left(\frac{k}{N}\right)$$

**Equation 9.95**

## 9.4 REFERENCES

[1] Herbert Taub and Donald L. Schilling. *Principles of Communication Systems*. San Francisco: McGraw-Hill, 1971.

[2] Ron Bracewell. *The Fourier Transform and Its Applications*. San Francisco: McGraw-Hill, 1965.

[3] R. E. Ziemer and W. H. Tranter. *Principles of Communications Systems, Modulation, and Noise*. Boston: Houghton Mifflin Company, 1976.

[4] Ferrel G. Stremler. *Introduction to Communication Systems*. 2nd ed. Workingham, UK: Addison-Wesley, 1982.

[5] Athanasios Papoulis. *The Fourier Integral and Its Applications*. San Francisco: McGraw-Hill, 1962.

[6] E. Oran Brigham. *The Fast Fourier Transform*. Englewood Cliffs, NJ: Prentice Hall, 1974.

[7] Roger L. Freeman. *Telecommunication Transmission Handbook*. 3rd ed. New York: John Wiley and Sons, 1991.

[8] James A. Cadzow. *Discrete Time Systems*. Englewood Cliffs, NJ: Prentice Hall, 1973.

# CHAPTER TEN

# Elastic Store Memory

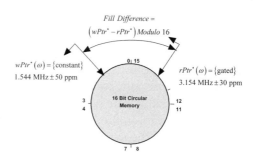

$$\textit{Fill Difference} =$$
$$\left(wPtr^{*} - rPtr^{*}\right)\textit{Modulo } 16$$

$wPtr^{*}(\omega) = \{\text{constant}\}$
$1.544 \text{ MHz} \pm 50 \text{ ppm}$

$rPtr^{*}(\omega) = \{\text{gated}\}$
$3.154 \text{ MHz} \pm 30 \text{ ppm}$

0 | 15

16 Bit Circular
Memory

3
4

12
11

7 | 8

Most designers during their career have designed interfaces between two or more data processing systems that utilized synchronous data streams. The term *synchronous*, as used in this chapter, refers to the case where the different data streams are synchronous or time aligned to the same system clock. When dealing with synchronous interfaces, the design is usually relatively straightforward.

There are occasions, however, when a designer must interface two or more processing systems or data streams where the data rates are asynchronous to one another. For purposes of this chapter, the term *asynchronous* refers to the case where the each data stream is time aligned to its own clock, generated by an independent clock oscillator. The frequency and phase of each clocked data stream are similar but not necessarily identical. Each clock oscillator's output frequency uniquely varies over time and temperature. In many cases, these clocks may differ by as much as a few thousand hertz. In this chapter, we will refer to these asynchronous data streams as *tributaries*.

There are a great many applications where two or more asynchronous data streams or tributaries are multiplexed together to form a single data stream at a higher output clock rate. The output clock is usually generated by an independent oscillator as well, which makes the multiplex operation even more difficult because this just adds one more asynchronous clock to the mix. The reader should stop and think for a moment about how a design like this could be implemented without experiencing a single bit error. The only method that I know is with the use of an elastic store memory.

The function of the elastic store memory is to serve as an elastic interface buffer that inputs a tributary at one clock rate and outputs it at another clock rate. The output clock and the output bit stream are synchronous to the follow on processing system. Therefore the elastic store memory can be viewed as a bit stream rubber band that absorbs differences between two or more asynchronous bit streams. Several elastic store memories can be used to multiplex and synchronize several input asynchronous tributaries to a common output clock.

## 10.1  EXAMPLE APPLICATION OF AN ELASTIC STORE MEMORY

An excellent application of an elastic store memory is in the everyday world of digital telephony. Therefore we will use the digital telephony application as a means to illustrate how an elastic store memory works.

In this chapter, the terms *channel*, *telephone channel*, *call*, *voice channel*, and *circuit* all mean the same thing and will be used interchangeably. A channel is nothing more than a connection between two telephones that is used to carry out a conversation between two parties.

An example of asynchronous multiplexing is seen in the transmission of telephone channels from one point to some other remote point. The difficulty in the transmission of many asynchronous telephone channels is that not all calls originate from the same location, nor is the destination of every call the same. Every telephone cannot have a dedicated copper wire, fiber-optic cable, radio channel, or other transmission medium to every other telephone in the world. If this were the case, the skies would be black with transmission lines, the ground would be full of fiber optic cables, and the mountains would be covered with radio towers. In the interest of efficiency, it is desirable to collect calls initiated in one locality at one call center and then multiplex as many of these call channels as possible onto a single medium, such as a copper wire or an optical fiber, for transmission to some other remote call center for distribution.

The problem telephone engineers face is the multiplexing of hundreds of asynchronous digital telephone channels, each associated with their own clock, to produce a single bit stream at a higher clock rate where all the channels are synchronous to a single independent transmission clock.

Channels are multiplexed in a hierarchical scheme that consists of several levels of protocol. Higher levels can support a greater number of channels than the lower levels. Tributaries at each level are constructed by multiplexing several asynchronous tributaries from the next lower level. Higher level tributaries enable a greater number of channels to be transmitted, which more efficiently utilizes the transmission medium.

Designing an interface that inputs and multiplexes several asynchronous tributaries (such as those used in the transmission of digital telephone channels) and outputs a single data stream that is synchronous to a higher frequency clock is a nontrivial engineering problem. This is especially true the first time an engineer attempts to do it. As a matter of fact, the design issues may initially seem complex enough to make one feel the interface will never work reliably. If the reader has just gotten involved with a project of this type, there is no need to panic and look for another job just yet. The entire concept of asynchronous multiplexing can be easily implemented through the use of an elastic store memory.

The remainder of this chapter deals with the design and implementation of an elastic store memory. In order for us to make this design representative of relevant technology, we will design this elastic store memory to specifically handle the multiplexing of real-world asynchronous telephone tributaries. Even though we are specifically dealing with telephone signal formats in this chapter, the elastic store memory we discuss here can be used for just about any asynchronous application. Before we begin, it is necessary to first build a solid conceptual foundation for this elastic memory. Therefore we will first need to briefly discuss the format and protocol of the telephone circuit multiplex hierarchy. The first thing we need to know is that in the telephone industry, digital telephony is usually referred to as *pulse code modulation*, or PCM.

## 10.2  PCM MULTIPLEXING HIERARCHY

There are four basic protocols in the world of telephony:

- North American Digital Hierarchy
- European or CEPT Digital Hierarchy
- Japanese Digital Hierarchy
- Synchronous Optical Network (SONET) Digital Hierarchy

There are many excellent texts that discuss in detail all the tributary formats for each of the four protocols listed. Many of these texts are listed in this chapter's references. It is not the intention of this book to repeat any of the fine works in this area, especially since it would require a complete book to do so. It will suffice only to mention the existence of these different protocols.

For the purposes of discussing the design and operation of an elastic store memory, we will need to develop some input tributary bit streams. Rather than create fake bit streams for this discussion, it makes more sense to design a real elastic store memory that inputs and outputs common, everyday, real-world bit streams. This will go a long way to verify the utility of this type of memory.

We will utilize two bit stream formats as defined by the North American Digital Hierarchy protocol. Specifically, we will design an elastic store memory for use in multiplexing two asynchronous DS-1 input tributaries to a single clock synchronous DS-1C output tributary. In order to accomplish this, we must first give the reader a brief overview of the North American hierarchy, protocol, and tributary formats.

A brief summary of the North American Digital Hierarchy and the call-carrying capacity for each level is given in Table 10.1. There are several "DS" levels in the North American Digital Hierarchy. The term DS is an abbreviation for *digital signal* or *data service level*. As tabulated in Table 10.1, a DS-0 is a single digitized telephone channel. It is digitized to a 64 Kbps rate. Twenty-Four of these DS-0s are collected and multiplexed together to form a single bit stream called a DS-1 that is clocked at a rate of 1.544 Mbps $\pm$ 50 ppm. The term *ppm* stands for parts per million. Therefore the 1.544 Mbps data rate of a DS-1 multiplex protocol would have a clock variance of $\pm$77.2 Hz.

Two DS-1s carrying up to 24 independent telephone channels each can be multiplexed to the 48 channel DS-1C hierarchical level, sometimes referred to as an intermediate level. The DS-1C multiplex inputs two 1.544 Mbps $\pm$ 50 ppm asynchronous bit streams and outputs a single 3.152 Mbps $\pm$ 30 ppm synchronous bit stream. We will design the elastic store memory that helps perform this DS-1C multiplex in this chapter.

Alternatively, four DS-1s carrying up to 24 independent telephone channels each can be multiplexed into the next hierarchical level, called a 96-channel DS-2. This process continues up to the DS-4 hierarchical level, which can carry up to 4032 independent telephone channels. Each DS level has its own format and protocol. For our purposes, we will only consider the DS-1C multiplex. Our discussion for this multiplex level will suffice to demonstrate an elastic store design that, if designed properly, can be utilized at any level of the hierarchy. The same concepts and procedures used for this multiplex can be used for just about any other multiplex application that deals with asynchronous tributaries.

**Table 10.1**  North American Digital Hierarchy

| Level | Number of tributaries | Number of voice circuits | Output tributary bit rate (bps) |
|-------|-----------------------|--------------------------|----------------------------------|
| DS-0 | N/A | 1 | 64 Kbps |
| DS-1 | 24 DS-0s | 24 | 1.544 Mbps $\pm$ 50 ppm |
| DS-1C | 2 DS-1s | 48 | 3.152 Mbps $\pm$ 30 ppm |
| DS-2 | 4 DS-1s | 96 | 6.312 Mbps $\pm$ 30 ppm |
| DS-3 | 7 DS-2s | 672 | 44.735 Mbps $\pm$ 20 ppm |
| DS-4 | 6 DS-3s | 4032 | 274.176 Mbps $\pm$ 10 ppm |

As illustrated in Figure 10.1, 24 individual 64 Kbps call channels (DS-0s) are multiplexed to the DS-1 level. Then two asynchronous DS-1 data streams are multiplexed to the DS-1C level. Each DS-1 multiplexer uses its own 1.544 MHz ± 50 ppm system clock, which means the two DS-1 bit streams are clock asynchronous with respect to one another, with a maximum clock frequency difference between the two bit streams of $2 \times 77.2$ Hz = 144.4 Hz or 144.4 bit/sec.

An elastic store memory must be employed by the DS-1C multiplex to service each DS-1 input to enable the generation of an output 48-channel multiplexed signal that is clock synchronous to the 3.152 MHz ± 30 ppm DS-1C system clock.

Let's begin by briefly discussing the bit-level format and protocol of a DS-1 and DS-1C tributary. Once we become familiar with and understand the characteristics of these tributaries, we can formulate a design for the DS-1C elastic store.

### 10.2.1  North American Digital Hierarchy DS-1 Frame Format

Figure 10.2 illustrates the bit-level format for the North American Digital Hierarchy DS-1 bit stream. There are 24 channels, and each channel sample is

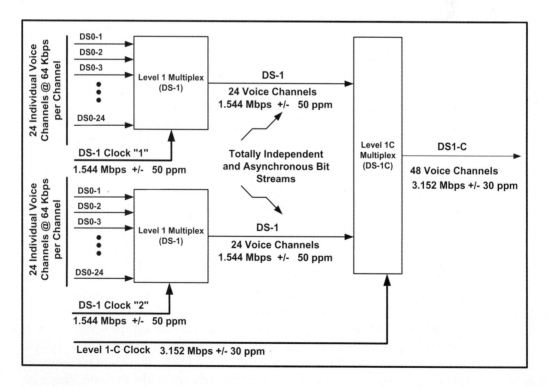

**Figure 10.1**  Functional DS-0 to DS-1C multiplex

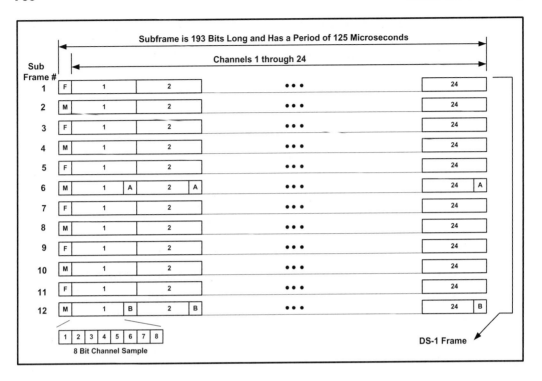

**Figure 10.2**  North American Digital Hierarchy DS-1 frame format

quantized to 8 bits. The bit stream is divided into a bunch of bits called a *frame*. The DS-1 frame is composed of 12 subframes. Each subframe is 193 bits long and is transmitted every 125 $\mu$ seconds, which equates to a subframe rate of 8 KHz. The resulting ideal bit rate is given by 193 bits/125 $\mu$ seconds = 1.544 Mbps.

When designing the DS-1C multiplex elastic store memory, we could actually care less about the input tributary bit stream format or properties other than the bit rate of the data stream itself. Our only concern is to input the two independent DS-1 tributaries and multiplex them into a single output tributary at a higher clock rate. The DS-1C multiplexer doesn't know or care about the content of the tributaries. It is only concerned with the fact that it must combine and reclock the two tributaries to a common output clock. However, since we have gone this far, it would be useful from an informational point of view to briefly define the bit fields in the DS-1 frame format.

Each of the 12 subframes contains 24 voice channel samples. Each sample is digitized at an 8 KHz rate and is quantized to 8 bits, resulting in a channel bit rate of 64 Kbps. As illustrated in Figure 10.2, a frame alignment bit sequence is formed by inserting *F* bits at the beginning of each odd subframe. This 6-bit frame alignment sequence is a fixed 101010 pattern. A multiframe alignment bit sequence is formed by inserting an $M$ bit at the beginning of each even sub-frame. The 6-bit multiframe alignment pattern is a fixed 001110 sequence.

The least significant bit for each of the 24 channel samples in the sixth subframe is replaced with a signaling bit sequence labeled $A$. The least significant bit for each channel sample in the twelfth subframe is replaced with a signaling bit sequence labeled $B$. The $A$ and $B$ bit sequences form two signaling subchannels, each with a bit rate of

$$\left(\frac{24 \text{ bits}}{12 \text{ subframe}}\right)\left(\frac{\text{subframe}}{125 \ \mu \text{ seconds}}\right) = 16 \text{ kbps}$$

When we design the DS-1C elastic store, we will treat the input tributaries as simple serial bit streams. The only specification that is important for our elastic store design is the clock rates of the input bit streams. The DS-1 clock rate is specified by the International Telephone and Telegraph Consultative Committee (CCITT) (since 1992 known as the International Telecommunications Union [ITU-T]) based in Geneva, Switzerland, to be 1.544 Mbps $\pm$ 50 ppm. This means that the allowable DS-1 bit stream clock rate is 1.544 MHz $\pm$ 77.2 Hz.

### 10.2.2  North American Digital Hierarchy DS-1C Frame Format

The DS-1C multiplex inputs two tributaries at a bit rate of 1.544 MHz $\pm$ 50 ppm each and multiplexes them into an output stream whose bit rate is 3.152 MHz $\pm$ 30 ppm. Figure 10.3 illustrates the bit-level format for the North American Digital Hierarchy DS-1C bit stream. Since we are going to be forming this output stream in our design, the format of this frame is important.

The DS-1C frame is 1272 bits in length and is composed of four subframes. Each subframe is 318 bits long. Each of the four subframes contains six microframes consisting of 53 bits each. Every microframe begins with an overhead bit. The remaining 52 bits in a microframe are formed by alternating bits from two input DS-1 tributaries, labeled T1 and T2.

The bit sequence M0, M1, M1, Z is a multiframe alignment word and takes on the sequence of 011Z. The Z bit is an alarm signaling bit and takes on the values of 1 for a normal condition and 0 for an alarm condition. The frame alignment word is formed by the bit sequence F0,F1,F0,F1,F0,F1,F0,F1 and is a fixed pattern equal to 01010101.

Bits labeled C1 form a 3-bit field in subframes 1 and 3 and are stuff control bits for tributary T1. They indicate if tributary T1 contains a stuff or information bit. Bits C2 form a 3-bit field in subframes 2 and 4 and are stuff control bits for tributary T2. These stuff bits indicate if tributary T2 contains a stuff or information bit. A stuff control bit sequence of 000 indicates that the stuff

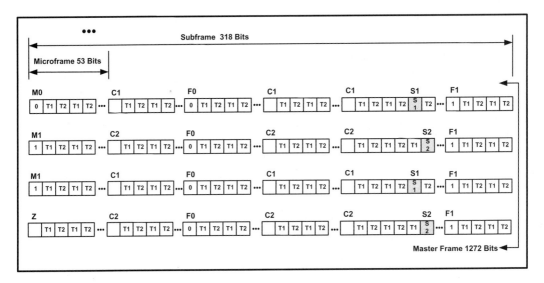

**Figure 10.3**   North American Digital Hierarchy DS-1C frame format (2)

bit location contains an information bit and a sequence of 111 indicates if the stuff bit location contains a stuff or garbage bit.

The stuff bit location for each tributary is in the third tributary bit position in the fifth microframe. The stuff bit locations are labeled S1 and S2 for tributary T1 and T2, respectively.

The stuff bits allow the DS-1C multiplexer to adjust the average output bit rate of an input tributary to accommodate input data streams whose bit rate fluctuates due to clock oscillator frequency variations. There are two stuff bits per frame allocated for each of the two input tributaries. The DS-1C multiplexer can, on a frame-by-frame basis, load these stuff bit positions with 0, 1, or 2 input tributary bits. This allows the multiplexer to modify the output bit rate so that on average the bit rate is identical to that of the input tributaries. When the stuff bit contains a tributary bit, it is said to contain an *information bit*. When the stuff bit contains a fill or garbage bit, which is the same as a noninformation bit, it is said to contain a stuff bit. As we will see shortly, the insertion or noninsertion of information bits into the stuff bit positions provides a data throttle mechanism to increase or decrease the average output bit rate such that it is sufficient to compensate for the wide variance in input data rates.

## 10.3  DS-1C MULTIPLEXER DESIGN OVERVIEW

As mentioned in the previous paragraphs, we will design a DS-1C multiplexer that utilizes an elastic store memory to perform the multiplex between the asynchronous DS-1 input and DS-1C bit output streams. Our elastic store design will allow for the multiplexing of two 24-channel input tributaries that are clock asynchronous to one another into a single 48-channel clock synchronous output tributary.

The concepts introduced in this design will be identical for any level of multiplex. The major difference between levels is in the speed at which the elastic store operates. At the lower hierarchical levels, a simple FIFO (first in first out) memory could be used to implement the elastic store. However, at the higher levels, using a simple FIFO or even a conventional SRAM (static random access memory) to implement an elastic store would be cumbersome at best and more than likely would not be capable of operating at the higher clock rates.

For this reason, even though our elastic store design will be operating at the lower clock rates, it will be fully capable of operating at the high bit rates of the DS-4 multiplex and higher. For example, if this elastic store design is implemented on a high-speed application-specific integrated circuit (ASIC), it could easily operate at 600 Mbps or more.

The design will utilize simple functional devices such as flip-flops that can operate at very high speed when implemented in modern logic such as field programmable gate arrays (FPGAs) or ASICs. This way, the reader can become familiar with the high-speed design, implementation concepts, and architecture and will be able to use this design with a few parameter changes to handle most applications. The maximum data rate of an elastic store is limited only by the speed of the digital logic selected for the design implementation.

### 10.3.1   Elastic Store Memory Design Conceptual Overview

The heart of the multiplex structure is the memory that implements the interface between two or more asynchronous data streams. In this case the memory is an elastic store that buffers input data bits at a fairly constant low clock rate and then outputs these bits as necessary at a higher clock rate.

The memory must be designed such that once an input data bit stream is applied, it will never reach an underflow or overflow condition, where data is interrupted or lost because the memory is either empty or filled to capacity.

The input tributary clock is utilized to write data bits into the elastic store memory. The output tributary clock is utilized to read data bits from the elastic store memory and to build the output data frame such as the DS-1C illustrated in Figure 10.3. In addition to reading information bits from the elastic store, the multiplexer must also use the output clock to insert overhead bits from other sources such as frame alignment bits, stuff control bits, alarm bits, and the stuff bits necessary to build an output frame. The reader should understand that the high rate output clock is not synchronous with respect to the low rate input clock. However, on average, the number of information bits read from the elastic store at the high rate must equal the number of bits written to the elastic store at the low rate.

At the beginning of each output subframe period, the elastic store is checked to determine how many bits it has accumulated. If the accumulation is greater than some predefined threshold indicating that the input data is being written to memory faster than it is being read out, an extra information bit is read out during the subframe period. This extra bit will occupy the stuff bit position in the subframe. In this case, the three stuff control bits will be set to 000 to alert the downstream processing equipment that the stuff bit location contains a valid tributary or information bit.

If the accumulation of bits in the elastic store memory is less than the same predefined threshold, indicating that the input data is being written to memory slower than it is being read out, an extra information bit will not be read and the stuff bit position is loaded with a logic 1 or with some predetermined garbage bit. The three stuff control bits will be set to 111 to inform the downstream processing equipment that the stuff bit contains nothing and should be ignored.

## 10.3.2   Elastic Store Memory Input and Output Data Rates

In our design, we will use one elastic store memory for each of the two input tributaries. The elastic store input tributary data rate is 1.544 Mbps $\pm$ 50 ppm or 1.544Mbps $\pm$ 77.2 bps. The elastic store output data clock rate is specified to be 3.152 Mbps $\pm$ 30 ppm or 3.152 Mbps $\pm$ 94.56 bps.

It will make the description of the elastic store operational mechanics much simpler if we consider the DS-1C frame to consist of two half frames. The first half frame consists of subframe 1 and subframe 2. The second half frame consists of subframe 3 and subframe 4. If we look at Figure 10.3, we can see that the first half frame of a DS-1C frame contains 636 bits, consisting of 12 overhead bits, 311 information bits, and a stuff bit for tributary 1. It also contains 311 information bits and a stuff bit for tributary 2.

Each of the two input tributaries has a single stuff opportunity bit every half frame or every 636 bits. Each half frame contains 311 information bits and a stuff bit from each of the two tributaries. If the stuff bit location for a particular tributary contains information, then the half frame will contain 312 information bits for that tributary. If the stuff bit contains a "stuff bit" (i.e., no information), the half frame will contain only 311 information bits for that tributary.

We are interested in determining if the DS-1C bit clock that reads both tributary elastic store memories and also inserts the 12 overhead bits in each DS-1C half frame is sufficient to keep the elastic store memories from overflowing or underflowing. If the DS-1C output clock is too slow, then the

**Table 10.2** Elastic Store Input versus Output Data Rates

| Elastic store input bit rate (MHz) | Elastic store output bit rate (MHz) | | | Tributary bits |
|---|---|---|---|---|
| | Minimum DS-1C clock frequency 3.152 MHz – 30 ppm or 3.15190544 MHz | Ideal DS-1C clock frequency 3.152 MHz | Maximum DS-1C clock frequency 3.152 MHz + 30 ppm or 3.15209456 MHz | Tributary information bits read every two subframes or 636 bits |
| | Elastic store read bit rate (MHz) | | | |
| Input DS-1 tributary minimum bit clock frequency $f_{IN}$ 1.544 MHz – 50 ppm or 1.5439228 MHz | 1.541262 MHz or 3.15190544 MHz scaled by 311/636 | 1.541308 MHz or 3.152 MHz scaled by 311/636 | 1.541354 MHz or 3.15209456 MHz scaled by 311/636 | Stuff frame: The stuff bit position contains no information bit. Therefore, 311 information bits are read every two subframes. The number of data bits stored in elastic store memory will slowly increase. |
| Input DS-1 tributary maximum bit clock frequency $f_{IN}$ 1.544 MHz + 50 ppm or 1.5440772 MHz | 1.546218 MHz or 3.15190544 MHz scaled by 312/636 | 1.546264 MHz or 3.152 MHz scaled by 312/636 | 1.546310 MHz or 3.15209456 MHz scaled by 312/636 | Info frame: The stuff bit position contains an information bit. Therefore, 312 information bits are read every two subframes. The number of data bits stored in elastic store memory will slowly decrease. |

tributary information bits will accumulate in elastic store memory until no unused memory exists and an overflow condition occurs causing the loss of data. If the DS-1C output clock is too fast, then the tributary information bits in elastic store memory will slowly decrease until no information bits are stored in memory and an underflow condition occurs causing the insertion of meaningless data bits in the output data stream.

We will handle the difference between the input and output clock rates by "data throttling." That is, we will use the stuff bit as a means to increase and decrease the output information bit rate for each tributary. This will allow us to on average match input tributary bit rates. The number of bits held in elastic store will increase and decrease over time, but the average fill level will be constant.

Using this scheme, neither an overflow nor an underflow condition can occur due to the asynchronous nature of the input and output clocks. This can be verified by inspection of Table 10.2.

The first column of Table 10.2 lists the minimum and maximum allowable frequencies of the DS-1 tributary input clock. The 1.544 Mbps ± 50 ppm input tributary clock can vary from a minimum frequency of 1.5439228 MHz to a maximum frequency of 1.5440772 MHz. This means that the elastic store memory will be constantly filling with tributary information bits at some constant clock rate with a maximum variance of ±77.2 bps.

The output DS-1C clock also has a defined range of frequencies as well. From left to right, the columns of Table 10.2 list the minimum allowable frequency, the ideal frequency, and the maximum allowable frequency of the DS-1C tributary output clock. The 3.152 MHz ± 30 ppm output tributary clock can vary from a minimum frequency of 3.15190544 MHz to a maximum frequency of 3.15209456 MHz. This represents a range of ±94.56 bps.

The input and output clocks can vary in frequency through their specified allowable ranges due to many causes such as oscillator aging and the affects of ambient temperature. The elastic store must be guaranteed to work across all the combined ranges of all the two input tributary clock frequencies and the output clock frequency. This is where the tributary stuff bit comes in to play. The tributary bit stream written into an elastic store memory can be thought of as being constant. We can use the tributary stuff bit to throttle the rate at which we read the bit stream from the elastic store memory. We can on average slow down the read rate by not reading a tributary information bit during the output frame stuff bit times. We can also on average increase the read rate by reading an extra information bit from memory during the output frame stuff bit times

If we count the total number of bits in a DS-1C half frame, we come up with 636 bits. If we count the total number of information bits in a half frame contributed by a single tributary (1 or 2), we come up with 312 bits if the stuff bit contains an information bit, and 311 bits if the stuff bit contains no information bit. Stated another way, each DS-1C half frame will contain either 311 or 312 information bits read from elastic store for each of the two input tributaries. This means that the effective elastic store output bit rate for a given tributary is the output clock scaled by a factor of 311/636 for the case where an extra information bit is not read and 312/636 for the case where an extra information bit is read. In our design, we will utilize two identical elastic store memories. One memory will write and read tributary 1 data stream bits, and the second memory will write and read tributary 2 data stream bits.

Looking at Table 10.2 again, we see the far right-hand column lists the 311 and 312 information bit cases. Keep in mind that the DS-1C multiplexer is forming a frame of data by using the output clock at appropriate times to strobe out tributary 1 and tributary 2 information bits from two independent elastic store memories and to strobe out at the appropriate times the framing, stuff control, and other overhead bits. While the multiplexer is off doing overhead tasks or is off reading the content of one of the tributary elastic store memories, the data in the other elastic store are accumulating. At the start of each DS-1C half frame, a decision is made by the multiplexer to either read an extra information bit or not. This decision is made by observing the number of bits stored in elastic store. If the number is above some prespecified threshold, then an extra bit is read out during the half frame stuff bit time. If the number of bits in memory is below some threshold, then the multiplexer does not read out an extra information bit. The effective elastic store output bit read rate for both cases is tabulated in the shaded area of Table 10.2.

For the case where 311 information bits are read out during the course of a half frame, we can see that the output read clock rate is slower than the lowest allowable input write clock rate for all three columns in the table. We can see from the table that the minimum allowable value for the write clock is 1.543 MHz, and the minimum read clock is 1.541 MHz. So for the case were only 311 information bits are read from elastic store memory during an output half frame, the elastic store write clock is faster than the read clock and the fill level of the elastic store memory will slowly increase.

For the case where 312 information bits are read out during the course of a half frame, we can see that the read clock rate is faster than the fastest allowable input clock rate across all three columns of the table. From the table, we can see that the maximum allowable write clock rate is 1.544077.2 MHz, and the maximum read clock rate is on the order of 1.546 MHz. For the case where 312 information bits are read from elastic store during a half frame, it is clear that the read clock is faster than the write clock and the fill level of the elastic store memory slowly decreases.

The multiplexer makes the decision to read or not to read an extra information bit at the start of every half frame. It will on occasion read data bits out of memory at a slower rate than the bits are being written into memory. On other occasions, it will read data bits out of memory faster than the rate at which the bits are being written in. On average, however, it will read data bits at a rate that is identical to the input bit rate. In doing so it can keep the number of bits stored in elastic store memory at some average value such that the data bit content of the elastic store memory never comes close to being empty or full. The amount of data bits in the elastic store memory will undulate up and down over time, analogous to a person's chest expanding and contracting while inhaling and exhaling. One might view the fill level of the elastic store memory as being analogous to breathing.

The data tabulated in Table 10.2 clearly shows that for the allowable input and output tributary clock frequencies, the elastic store will operate as intended and no data bits will be lost. Now that the number crunching is done and we are satisfied the elastic store will work for this application, the next step is to actually design the elastic store hardware. Our design will depend heavily on the results obtained from bit-level simulations. These simulations provide us with a window to observe the real time operation of the elastic store internals. This will allow us to completely understand exactly how the elastic store works before we actually build the hardware and or write the software. This extra effort will allow us to build something that will work the first time it's tested.

## 10.4  DESIGN OF THE ELASTIC STORE MEMORY

For our purposes, we will design the elastic store with the assumption that it will be implemented in a high-speed FPGA or ASIC. In the interest of speed and simplicity, we will only utilize the very basic hardware components or macros that should be resident in any ASIC or FPGA device development library. If speed is not a consideration and if there are overriding reasons not to use programmable devices or foundry devices, then there is absolutely no reason why this design could not be implemented in discrete components as well.

### 10.4.1  Size of the Elastic Store Memory

The physical size of the elastic store memory is important. If it is too large, it wastes on chip resources, it will probably be slower, and it will consume more power than necessary. If it is too small, it runs the risk of underflowing or overflowing due the undulating bit accumulation levels caused by the variances of the input and output bit rates.

The following two cases illustrate two different approaches to implementing the elastic store memory. The first case is an inefficient brute force implementation intended for illustration purposes only. The second case is a high-speed efficient implementation that would be actually designed and built. In order to better understand the efficient method, it will be helpful to first understand the shortcomings of the inefficient method.

### 10.4.2 Case I: An Inefficient Brute Force Memory Implementation

Serial bit stream data is continually being written into the elastic store memory at the input tributary clock rate. At the start of each output tributary half frame, a decision is made as to whether stuffing is to occur or not occur, based on the number of bits stored in memory.

Remember that our DS-1C multiplexer design will utilize two identical elastic store memories: one for each input tributary. A DS-1C subframe will contain either 311 or 312 information bits from each of the two input tributaries. If the stuff bit for an individual tributary contains an information bit, then 312 information bits will be read from elastic store memory during the half frame period. If the stuff bit does not contain an information bit, then 311 information bits must be read from elastic store memory during the half frame period. Tributary bits are read from elastic store memory at the output or DS-1C clock rate.

A brute force method of implementing an elastic store memory would be to build a memory capable of holding two half frames of an input tributary or 624 bits, as illustrated in Figure 10.4. The stuff / no stuff decision

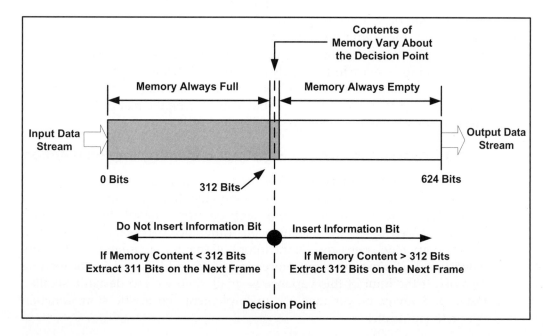

**Figure 10.4** Inefficient implementation of an elastic store memory

would be made at the start of each half frame by monitoring the fullness of this memory.

If at the start of a half frame the number of stored bits exceeded 312, then 312 bits would be read from memory and the following output frame would contain an information bit in the stuff bit position. For this case, the overall elastic store bit accumulation would be reduced on average by a fraction of a bit over the period of a half frame.

If at the start of a half frame the number of stored bits did not exceed 312, then the following frame would not include an information bit in the stuff bit position, and only 311 bits would be read from elastic store memory. In this case, the overall elastic store bit accumulation would grow on average by a fraction of a bit over the period of a half frame.

This implementation would work but it would be inefficient because on average the first 312 memory locations would always be nearly filled, and the remaining locations would on average remain empty. As shown in Figure 10.4 the only memory activity of note occurs in the address space immediately surrounding the decision point. This is where the variance of the memory bit fill oscillates about the 312-bit threshold, and for proper operation, this variance is the only information the DS-1C multiplexer needs to know. The remainder of elastic store memory for this case is wasted by storing either way too many information bits or storing nothing. This large memory ends up consuming a fair amount of chip resources, consumes more power than necessary, and is a more difficult design to implement.

A much more efficient and realizable elastic store implementation would be one that only stores the accumulated information bits that make up the variance address space immediately around the decision point. If the variance is small, then almost all of this 624-bit memory can be eliminated, not only saving on chip resources and power but also offering the real prospect of dramatically increasing the memory speed.

### 10.4.3  Case II: An Efficient High-Speed Memory Implementation

Since the rate that tributary data is written to memory and the rate that tributary data is read from memory is on average identical, it is intuitive that the elastic store memory only needs to be large enough to handle the memory fill variance caused by (1) the instantaneous difference between input and output tributary clock rates and (2) the information bit accumulation that occurs when the multiplexer is away processing overhead bits and is not reading memory. If the limit of this variance is small, then we can design a smaller, faster, and lower power memory to implement the elastic store function within the multiplexer.

In the next section we will use a simple computer simulation to quantify the size of our elastic store memory, but for now let's begin with the assumption that the elastic store memory will experience worst case a ±4 bit variance about the decision point. Therefore a 16-bit memory with a decision point set at 8 bits would be more than sufficient, since it would always have a 4-bit underflow and 4-bit overflow guard band. In this case, the elastic store can be modeled by the block diagram of Figure 10.5. Here we show the elastic store memory to be 16 bits in length with the theoretical decision point sitting at 8 bits. If our storage variance is ±4 bits, then the memory should slowly fill and deplete over time between the 4-bit and 12-fill levels. If this version of the elastic store memory proves to be correct, then it is easy to see that this is indeed the preferred structure since it is efficient, faster, and more easily designed. In order to determine the actual size of the elastic store memory, we will write a software simulator and compute statistics on the ebb and flow of data bit accumulation. This simulation is discussed next.

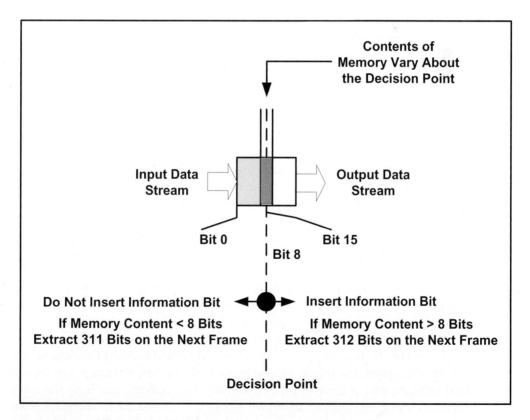

**Figure 10.5** Efficient implementation of an elastic store memory

## 10.4.4  Elastic Store Memory Simulation Results

The only method I feel comfortable with in determining the operational capability of any digital circuit, circuit card, or system I design is writing a simple computer program that performs a bit-level simulation of the design. This method almost always guarantees the success of the design before it's actually built.

Since I am a creature of habit, I wrote a simple computer program that simulates the operation of the elastic store memory. It simulates the DS-1C multiplexer elastic store input tributary low rate data clock, the output tributary high rate data clock, and it computes statistics on the elastic store memory minimum, maximum, and average fill. It also keeps tabs on the total number of bits written to and read from memory.

Operationally, an elastic store memory is designed to fill to a predefined initial fill level at system start-up before any memory reads can be performed. Typically this initial fill level is identical to the value used for the decision point. It is very interesting when looking at the program output statistics to observe how the number of bits stored in memory undulates about the decision point or memory threshold when the memory is operating in the steady state condition.

An even more interesting observation occurs when we set the initial fill to either a low number or a high number while setting the memory decision point somewhere midrange. In the first case, we can see the data bits in memory slowly grow over time until the decision point is reached and steady state operation is achieved. In the second case, we can watch the number of data bits in memory slowly decay until the number of bits reaches the decision point and steady state operation begins.

In the next section, we will present some computer-generated data that will illustrate both the initial or transient start-up behavior and the steady state operational characteristics observed for this elastic store memory, and we will utilize these data to determine the actual size of memory necessary to implement the design.

In order to simulate the data read from elastic store memory, it is critical to read the data bits at the appropriate times corresponding to the actual bit times of the output DS-1C frames. Therefore, for this simulation to be accurate, it needs to not only simulate the operation of the two elastic store memories in the DS-1C multiplexer but also simulate the entire multiplexer. So, for your reading pleasure, following the presentation of the data, we will look at some critical parts of the computer program. This will include how the asynchronous input and output tributary clocks are simulated in software, how the framing of output information bits is implemented, and how the insertion of overhead bits is implemented. The simulation program sounds complicated, but you will be surprised to discover just how simple this program really is. The program only required about 20 minutes to write and test, but it supplied an enormous amount of information concerning the validity of the design. The small effort devoted to writing this or any bit-level simulation usually pays a huge return on investment when validating a design.

### 10.4.5   Elastic Store Steady State Operational Statistics

The simulation statistics for the elastic store that we will design are shown in the tables that follow. Each table lists the clock rates of the input and output tributary bit streams, and each table lists the elastic store initial fill and threshold values. The initial fill value is the number of input bits that are allowed to accumulate in memory on start-up before any output bits are read. The threshold value is the elastic store decision point. If the number of unread bits in elastic store is greater than threshold, an extra information bit will be read during the half frame. If the number of unread bits in elastic store is less than the threshold value, an extra information bit will not be read during the half frame. This is called *throttling the data* and it maintains the memory fill level at or near the decision point.

We will explain the tabulation of simulation results by referencing Table 10.3. The data in each row of the table is printed out at the end of a complete DS-1C frame. The frame number is listed in the far left column. There are 24 microframes per frame, and the second column prints out the last microframe processed before a row of data is printed. This is done to verify that the data are printed out at the right time in the frame. The number of bits per frame is counted as the bits are processed and should always equal 1272 for each frame. This value is printed out as another means of verification that the statistics are collected and printed out at the proper time.

We will refer to the elastic store as "ESTO." The real information is contained in the next three columns, labeled "ESTO Fill," "ESTO Min," and "ESTO Max." The numbers for ESTO Fill represent the number of residual or unread data bits left in the elastic store memory at the end of a DS-1C frame. The ESTO Min and ESTO Max numbers represent the minimum and maximum number of bits that were stored in the elastic store memory since the beginning of operation (i.e., system start-up). This will allow us to determine the maximum value of bit variance about the threshold value, and it will alert us if an underflow or overflow has occurred at any point in time since the beginning of operation.

The next column, labeled "STUFF BITS USED," tabulates the total number of times a stuff bit in a frame contained a tributary information bit. And, finally, the last two columns tabulate the number of bits written into and read from elastic store memory up to the end of the current frame. This helps us keep track of all the bits and will aid us in doing some mental bit bookkeeping to verify the elastic store is working correctly.

**Table 10.3**   Elastic Store Steady State Simulation, Case I

```
Case 1
DS_1  Input  Tributary Clock = 1.544000 MHz
DS_1C Output Tributary Clock = 3.152000 MHz
Elastic Store Initial Fill   = 6 bits
Elastic Store Threshold      = 6 bits
```

| Frame # | Micro Frame # | Bits Per Frame | ESTO Fill | ESTO Min | ESTO Max | Stuff Bits Used | DS_1 Input Bits | DS-1C Output Bits |
|---|---|---|---|---|---|---|---|---|
| 39986 | 24 | 1272 | 6 | 4 | 7 | 43436 | 24914728 | 24914728 |
| 39987 | 24 | 1272 | 6 | 4 | 7 | 43437 | 24915351 | 24915351 |
| 39988 | 24 | 1272 | 6 | 4 | 7 | 43438 | 24915974 | 24915974 |
| 39989 | 24 | 1272 | 6 | 4 | 7 | 43439 | 24916597 | 24916597 |
| 39990 | 24 | 1272 | 6 | 4 | 7 | 43440 | 24917220 | 24917220 |
| 39991 | 24 | 1272 | 6 | 4 | 7 | 43441 | 24917843 | 24917843 |
| 39992 | 24 | 1272 | 7 | 4 | 7 | 43442 | 24918467 | 24918466 |
| 39993 | 24 | 1272 | 7 | 4 | 7 | 43443 | 24919090 | 24919089 |
| 39994 | 24 | 1272 | 7 | 4 | 7 | 43444 | 24919713 | 24919712 |
| 39995 | 24 | 1272 | 7 | 4 | 7 | 43445 | 24920336 | 24920335 |
| 39996 | 24 | 1272 | 7 | 4 | 7 | 43446 | 24920959 | 24920958 |
| 39997 | 24 | 1272 | 7 | 4 | 7 | 43447 | 24921582 | 24921581 |
| 39998 | 24 | 1272 | 6 | 4 | 7 | 43449 | 24922205 | 24922205 |
| 39999 | 24 | 1272 | 6 | 4 | 7 | 43450 | 24922828 | 24922828 |
| 40000 | 24 | 1272 | 6 | 4 | 7 | 43451 | 24923451 | 24923451 |

The following three tables (Table 10.1, Table 10.2, and Table 10.3) tabulate data for the cases where the input and output bit rates are at their ideal specified values and at their extreme variance values. All three simulations set the initial fill value to 6 bits and the threshold value to 6 bits. All three simulations process 40,000 frames of tributary data and print out the accumulated elastic store statistics during the last 15 frames prior to and including the forty-thousandth frame.

### 10.4.5.1   Case I: Ideal Tributary Input Clock / Output Clock

Table 10.3 tabulates the statistics for the case where the DS-1 input tributary bit rate is equal to its specified ideal value of 1.544 MHz and the DS-1C output tributary bit rate is also equal to its specified ideal value of 3.152 MHz.

By the time the forty-thousandth frame has been processed, all the initial transients have died down and the elastic store is essentially operating in a steady state condition. It is clear that the elastic store memory is stable and is operating as planned with the elastic store fill slowly varying between 6 and 7 bits, as specified by the threshold value.

The maximum number of bits stored in the elastic store memory over the entire 40,000 frames never exceeded 7, as indicated in the ESTO MAX column, while the minimum number of bits after the initial fill occurred was never less than 4, as indicated in the ESTO MIN column.

The simulation shows that when the two bit clocks are at their ideal values, the multiplexer uses at least one of the two stuff bits per frame to carry an information bit. It is interesting to note that when the elastic store fill increases from 6 bits to 7 bits, the second stuff bit in a frame is eventually used to carry an extra information bit, effectively increasing the output data bit rate, and the fill level drops from 7 back to 6 bits. These data are tabulated in the column labeled "Stuff Bits Used," which accumulates the number of stuff bits used to carry information bits.

At the end of 40,000 frames, the total number of bits written into elastic store memory is tabulated to be 24,923,451, and the total number of bits read from elastic store memory is also 24,923,451. This correlates with the fact that the elastic store fill level was equal to the threshold level of 6 bits, indicating that no extra or unaccounted bits are hanging around in memory. You may ask why the 6 bits still in elastic store are not included in the tabulation. The answer is that the simulator tabulation doesn't begin until the elastic store reaches the initial fill level, which in this case is 6 bits.

The data shows that over 40,000 frames, 43,451 information bits are read out of elastic store memory during a stuff bit time. This means that 43,451 − 40000 = 3,451 frames, required that two information bits be inserted into the two available stuff bit positions during a frame period. This clearly shows a throttling up of the output data rate on occasion to prevent the buildup of stored data bits beyond the threshold level. I find design statistics like these to be incredibly interesting. These statistics also give designers a warm and happy feeling, in that they not only quantify design parameters, like memory size and threshold levels, but also predict in a pretty convincing fashion that the proposed design will operate as desired.

### 10.4.5.2 Case II: Min Tributary Input Clock / Max Output Clock

Table 10.4 tabulates the data for the case where the input DS-1 tributary bit rate is at its lowest permissible extreme value of 1.544 MHz − 50 ppm and the output DS-1C tributary bit rate is at its highest permissible extreme value of 3.152 MHz + 30 ppm. This is one of two worst-case scenarios in terms of the relative difference between the two data stream bit rates. In this simulation, the elastic store initial fill and the elastic store threshold value are both set to 6 bits.

In this simulation, we observe similar statistical behavior to that of the previous example. Once again, the simulation is stopped after 40,000 frames when the elastic store is operating in its steady state condition. The average number of bits in the elastic store memory is slowly varying between

**Table 10.4** Elastic Store Steady State Simulation, Case II

```
Case 2
DS_1  Input  Tributary Clock = 1.543923 MHz
DS_1C Output Tributary Clock = 3.152095 MHz
Elastic Store Initial Fill   = 6 bits
Elastic Store Threshold      = 6 bits
```

| Frame # | Micro Frame # | Bits Per Frame | ESTO Fill | ESTO Min | ESTO Max | Stuff Bits Used | DS_1 Input Bits | DS-1C Output Bits |
|---|---|---|---|---|---|---|---|---|
| 39986 | 24 | 1272 | 7 | 4 | 7 | 41442 | 24912735 | 24912734 |
| 39987 | 24 | 1272 | 7 | 4 | 7 | 41443 | 24913358 | 24913357 |
| 39988 | 24 | 1272 | 6 | 4 | 7 | 41445 | 24913981 | 24913981 |
| 39989 | 24 | 1272 | 6 | 4 | 7 | 41446 | 24914604 | 24914604 |
| 39990 | 24 | 1272 | 6 | 4 | 7 | 41447 | 24915227 | 24915227 |
| 39991 | 24 | 1272 | 6 | 4 | 7 | 41448 | 24915850 | 24915850 |
| 39992 | 24 | 1272 | 6 | 4 | 7 | 41449 | 24916473 | 24916473 |
| 39993 | 24 | 1272 | 6 | 4 | 7 | 41450 | 24917096 | 24917096 |
| 39994 | 24 | 1272 | 6 | 4 | 7 | 41451 | 24917719 | 24917719 |
| 39995 | 24 | 1272 | 6 | 4 | 7 | 41452 | 24918342 | 24918342 |
| 39996 | 24 | 1272 | 6 | 4 | 7 | 41453 | 24918965 | 24918965 |
| 39997 | 24 | 1272 | 6 | 4 | 7 | 41454 | 24919588 | 24919588 |
| 39998 | 24 | 1272 | 6 | 4 | 7 | 41455 | 24920211 | 24920211 |
| 39999 | 24 | 1272 | 6 | 4 | 7 | 41456 | 24920834 | 24920834 |
| 40000 | 24 | 1272 | 6 | 4 | 7 | 41457 | 24921457 | 24921457 |

6 and 7 bits, as we would expect, and on average the number of bits written is equal to the number of bits read from elastic store memory.

Note for this case, after 40,000 frames there were 41,457 stuff bit locations used to carry information bits. This means that there were 1457 occurrences when two stuff bits in a single frame were utilized. In case II, the elastic store used 1994 fewer stuff bit locations to store information bits than it did in case I. This should be expected, since the output clock is at its maximum frequency and the input clock is at its minimum frequency, so the elastic store throttled down its output information bit rate.

### 10.4.5.3  Case III: Max Tributary Input Clock / Min Output Clock

Table 10.5 illustrates the tabulated data for the last steady state case. In this simulation the input DS-1 tributary clock is at its maximum permissible variance of 1.544 MHz + 50 ppm, and the output DS-1C tributary clock is at its minimum permissible variance of 3.152 MHz − 30 ppm.

Once again, after 40,000 frames, it is clear that the elastic store memory is operating in a stable manner. We do see a little more jitter in the elastic

**Table 10.5**   Elastic Store Steady State Simulation, Case III

Case 3
DS_1  Input   Tributary Clock = 1.544077 MHz
DS_1C Output Tributary Clock = 3.151905 MHz
Elastic Store Initial Fill   = 6 bits
Elastic Store Threshold      = 6 bits

| Frame # | Micro Frame # | Bits Per Frame | ESTO Fill | ESTO Min | ESTO Max | Stuff Bits Used | DS_1 Input Bits | DS-1C Output Bits |
|---|---|---|---|---|---|---|---|---|
| 39986 | 24 | 1272 | 6 | 4 | 7 | 45429 | 24916721 | 24916721 |
| 39987 | 24 | 1272 | 6 | 4 | 7 | 45430 | 24917344 | 24917344 |
| 39988 | 24 | 1272 | 7 | 4 | 7 | 45431 | 24917968 | 24917967 |
| 39989 | 24 | 1272 | 7 | 4 | 7 | 45432 | 24918591 | 24918590 |
| 39990 | 24 | 1272 | 7 | 4 | 7 | 45433 | 24919214 | 24919213 |
| 39991 | 24 | 1272 | 7 | 4 | 7 | 45434 | 24919837 | 24919836 |
| 39992 | 24 | 1272 | 6 | 4 | 7 | 45436 | 24920460 | 24920460 |
| 39993 | 24 | 1272 | 6 | 4 | 7 | 45437 | 24921083 | 24921083 |
| 39994 | 24 | 1272 | 6 | 4 | 7 | 45438 | 24921706 | 24921706 |
| 39995 | 24 | 1272 | 7 | 4 | 7 | 45439 | 24922330 | 24922329 |
| 39996 | 24 | 1272 | 7 | 4 | 7 | 45440 | 24922953 | 24922952 |
| 39997 | 24 | 1272 | 7 | 4 | 7 | 45441 | 24923576 | 24923575 |
| 39998 | 24 | 1272 | 7 | 4 | 7 | 45442 | 24924199 | 24924198 |
| 39999 | 24 | 1272 | 6 | 4 | 7 | 45444 | 24924822 | 24924822 |
| 40000 | 24 | 1272 | 6 | 4 | 7 | 45445 | 24925445 | 24925445 |

store fill level for this case than in the previous two, but the minimum and maximum bit store levels have not changed.

It is interesting to note that the number of stuff bit locations that contained information bits for this case was 45,445, which is 1994 greater than case I. This is to be expected, since the input bit clock is at its maximum value and the output bit clock is at its minimum value. We would expect to see bits written into the elastic store at a faster rate, and since the output clock is slower, we would expect to see more stuff bit utilization.

Only three of the many steady state simulation cases are shown here, but they clearly demonstrate that the elastic store steady state performance is as intended. The simulation shows that only 7 bits of memory are utilized, and the fill level undulates between 6 and 7 bits.

From this data, we may conclude that a 7- or 8-bit memory would be sufficient to implement this elastic store design. The number could be even smaller if the threshold were set lower than 6 bits. This may be true, but every design should have some headroom built in. In this case, it would make perfect sense to build a 12- or 16-bit elastic store and use the threshold setting to place the fill level near the memory midpoint. Before we get excited and jump right into the design, it would be wise to further simulate the operation of the elastic store and observe its transient characteristics right after start-up. After all, we don't want any surprises coming our way after this design in cemented in hardware.

## 10.4.6 Elastic Store Transient Statistics

Before we can be completely sure that this elastic store memory will operate correctly, we need to examine the transient properties of this memory during the start-up phase. In order to feel comfortable about our design, let's run the simulation again and observe the behavior of the elastic store during the first few frames after start-up.

### 10.4.6.1 Case IV: Elastic Store Memory Decay Transient Response

Table 10.6 tabulates the data for the case where the input DS-1 tributary bit rate is at its lowest permissible extreme value of 1.544 MHz – 50 ppm, and the output DS-1C tributary is at its highest permissible extreme value of 3.152 MHz + 30 ppm.

In this case we are interested observing the transient behavior of the elastic store when the initial fill level is set to 16 and the threshold is set to 6 bits. In this case, the elastic store will fill to 16 bits after start-up, switch on, and then we would expect its fill level to slow down or decay to its threshold operating point of 6 bits. We will observe the elastic store behavior for the first 36 frames at start-up, beginning after the elastic store reaches a fill level of 16 bits. The simulation data for case IV clearly shows that the elastic store initially reaches a maximum of 16 bits after start-up before any bits are read from memory. After bit reading was activated, the elastic store algorithm caused the multiplexer to utilize both stuff bits to carry information bits each frame, slowly reducing the fill level until it reaches the specified threshold level.

From the tabulated data, we clearly see the ESTO Fill level beginning at 15 bits in frame one and decaying to its steady state of 6-bit and 7-bit undulation. The decay time takes 8 frames to reach a fill of 7 bits and a total of 14 frames to reach the specified threshold level of 6 bits. At approximately 403 $\mu$ seconds per frame, it takes 3.2 m seconds to decay from 16 bits to 7 bits, and a total of 5.6 m seconds to decay down to the 6-bit threshold level. After this 5.6 m second settling time, the elastic store stabilizes and begins to operate in its steady state condition.

The alert reader might ask the question, "If the initial fill was set to 16 bits, why did the simulation show an initial ESTO fill of only 15 bits?" The answer is that the simulation only printed out the statistical data at the end of each frame. The simulation ESTO Max does show that the maximum number of bits in the elastic store was 16. It also shows that two stuff bits were used to

**Table 10.6**   Elastic Store Transient Simulation, Case IV

```
Case 4
DS_1  Input  Tributary Clock = 1.543923 MHz

DS_1C Output Tributary Clock = 3.152095 MHz
Elastic Store Initial Fill   = 16 bits
Elastic Store Threshold      = 6 bits
```

| Frame # | Micro Frame # | Bits Per Frame | ESTO Fill | ESTO Min | ESTO Max | Stuff Bits Used | DS_1 Input Bits | DS-1C Output Bits |
|---|---|---|---|---|---|---|---|---|
| 1 | 24 | 1272 | 15 | 14 | 16 | 2 | 623 | 624 |
| 2 | 24 | 1272 | 14 | 13 | 16 | 4 | 1246 | 1248 |
| 3 | 24 | 1272 | 13 | 12 | 16 | 6 | 1869 | 1872 |
| 4 | 24 | 1272 | 12 | 11 | 16 | 8 | 2492 | 2496 |
| 5 | 24 | 1272 | 11 | 10 | 16 | 10 | 3115 | 3120 |
| 6 | 24 | 1272 | 10 | 9 | 16 | 12 | 3738 | 3744 |
| 7 | 24 | 1272 | 9 | 8 | 16 | 14 | 4361 | 4368 |
| 8 | 24 | 1272 | 8 | 7 | 16 | 16 | 4984 | 4992 |
| 9 | 24 | 1272 | 7 | 6 | 16 | 18 | 5607 | 5616 |
| 10 | 24 | 1272 | 7 | 5 | 16 | 19 | 6230 | 6239 |
| 11 | 24 | 1272 | 7 | 5 | 16 | 20 | 6853 | 6862 |
| 12 | 24 | 1272 | 7 | 5 | 16 | 21 | 7476 | 7485 |
| 13 | 24 | 1272 | 7 | 5 | 16 | 22 | 8099 | 8108 |
| 14 | 24 | 1272 | 7 | 5 | 16 | 23 | 8722 | 8731 |
| 15 | 24 | 1272 | 6 | 5 | 16 | 25 | 9345 | 9355 |
| 16 | 24 | 1272 | 6 | 5 | 16 | 26 | 9968 | 9978 |
| 17 | 24 | 1272 | 6 | 5 | 16 | 27 | 10591 | 10601 |
| 18 | 24 | 1272 | 6 | 5 | 16 | 28 | 11214 | 11224 |
| 19 | 24 | 1272 | 6 | 5 | 16 | 29 | 11837 | 11847 |
| 20 | 24 | 1272 | 6 | 5 | 16 | 30 | 12460 | 12470 |
| 21 | 24 | 1272 | 6 | 5 | 16 | 31 | 13083 | 13093 |
| 22 | 24 | 1272 | 6 | 5 | 16 | 32 | 13706 | 13716 |
| 23 | 24 | 1272 | 6 | 5 | 16 | 33 | 14329 | 14339 |
| 24 | 24 | 1272 | 6 | 5 | 16 | 34 | 14952 | 14962 |
| 25 | 24 | 1272 | 6 | 5 | 16 | 35 | 15575 | 15585 |
| 26 | 24 | 1272 | 6 | 5 | 16 | 36 | 16198 | 16208 |
| 27 | 24 | 1272 | 6 | 5 | 16 | 37 | 16821 | 16831 |
| 28 | 24 | 1272 | 7 | 5 | 16 | 38 | 17445 | 17454 |
| 29 | 24 | 1272 | 7 | 5 | 16 | 39 | 18068 | 18077 |
| 30 | 24 | 1272 | 7 | 5 | 16 | 40 | 18691 | 18700 |
| 31 | 24 | 1272 | 7 | 5 | 16 | 41 | 19314 | 19323 |
| 32 | 24 | 1272 | 7 | 5 | 16 | 42 | 19937 | 19946 |
| 33 | 24 | 1272 | 7 | 5 | 16 | 43 | 20560 | 20569 |
| 34 | 24 | 1272 | 7 | 5 | 16 | 44 | 21183 | 21192 |
| 35 | 24 | 1272 | 7 | 5 | 16 | 45 | 21806 | 21815 |
| 36 | 24 | 1272 | 7 | 5 | 16 | 46 | 22429 | 22438 |

carry information in the first frame. This correlates with the simulation ESTO Min, which shows that the bit level dropped to 14 bits sometime during the first frame. The number of bits written to memory at the end of 36 frames is 22,429, plus the 16-bit initial fill, or 22,445 bits. The total bits read from memory at the end of 36 frames is 22,438 bits.

The 9-bit difference between the 22,429 bits written to memory and the 22,438 bits read is because we initially wrote 16 bits to memory as the initial fill value. Remember the initial fill is not counted in the write statistics. The residue number of bits in memory at the end of the thirty-sixth frame is 7 bits. That means the simulation read $16 - 7 = 9$ bits more than it thought it wrote in. So the bit bookkeeping balances, and we are satisfied that all bits are accounted for. Hey, this is exciting stuff! It is very much like watching your hardware design working for the first time during the test.

### 10.4.6.2  Case V: Elastic Store Memory Attack Transient Response

Table 10.7 tabulates the statistical data for the case where the input DS-1 tributary clock is at its maximum permissible variance of 1.544 MHz + 50 ppm and the output DS-1C clock is at its minimum permissible variance of 3.152 MHz – 30 ppm.

In this simulation, the elastic store threshold is set to high value of 16 bits and the initial fill value is set to a low value of 4 bits. In this simulation, we choose to observe the elastic store start-up behavior for the first 36 frames of output data. We would expect to see the memory fill level slowly increase or attack from the initial fill value of 4 up to the threshold value of 16, and we do. The fill transient lasts 11 frames or 4.4 m seconds before the elastic store reaches a condition of steady state operation.

At the conclusion of 36 frames, we wrote 22,432 bits to elastic store, and we read 22,420 bits, for a difference of 12 bits. Let's do some bit bookkeeping so we can explain this difference. We set the initial fill value to be 4 bits at which time we began reading the bits from memory. The threshold, however, was set to 16 bits, so the bit content of memory had to grow by 12 bits to reach the prescribed threshold. It took 11 frames for this growth to occur, as evidenced by the fact that no information bits were read out during the stuff bit times for these frames. So in order to reach the elastic store threshold value, we ended up writing 12 bits more to elastic store memory than we read out. The bit books balance again!

Pay attention to the wide range between the ESTO Min and ESTO Max values. The minimum fill was 2 bits and the maximum fill was 17 bits, a 15-bit difference. This shows the inherent problems with setting the initial fill and threshold values so far apart. A better strategy is to set the two values as close as possible to one another, based on the results of the simulation software.

This is exciting news. With these simulations, we have shown that the operation of the elastic store is orderly and predictable during both the

**Table 10.7**   Elastic Store Transient Simulation, Case V

```
Case 5
DS_1  Input  Tributary Clock = 1.544077 MHz
DS_1C Output Tributary Clock = 3.151905 MHz
Elastic Store Initial Fill   = 4 bits
Elastic Store Threshold      = 16 bits
```

| Frame # | Micro Frame # | Bits Per Frame | ESTO Fill | ESTO Min | ESTO Max | Stuff Bits Used | DS_1 Input Bits | DS-1C Output Bits |
|---|---|---|---|---|---|---|---|---|
| 1 | 24 | 1272 | 5 | 2 | 5 | 0 | 623 | 622 |
| 2 | 24 | 1272 | 6 | 2 | 6 | 0 | 1246 | 1244 |
| 3 | 24 | 1272 | 7 | 2 | 7 | 0 | 1869 | 1866 |
| 4 | 24 | 1272 | 8 | 2 | 8 | 0 | 2492 | 2488 |
| 5 | 24 | 1272 | 9 | 2 | 9 | 0 | 3115 | 3110 |
| 6 | 24 | 1272 | 10 | 2 | 11 | 0 | 3738 | 3732 |
| 7 | 24 | 1272 | 11 | 2 | 12 | 0 | 4361 | 4354 |
| 8 | 24 | 1272 | 13 | 2 | 13 | 0 | 4985 | 4976 |
| 9 | 24 | 1272 | 14 | 2 | 14 | 0 | 5608 | 5598 |
| 10 | 24 | 1272 | 15 | 2 | 15 | 0 | 6231 | 6220 |
| 11 | 24 | 1272 | 16 | 2 | 16 | 0 | 6854 | 6842 |
| 12 | 24 | 1272 | 16 | 2 | 17 | 1 | 7477 | 7465 |
| 13 | 24 | 1272 | 16 | 2 | 17 | 2 | 8100 | 8088 |
| 14 | 24 | 1272 | 16 | 2 | 17 | 3 | 8723 | 8711 |
| 15 | 24 | 1272 | 17 | 2 | 17 | 4 | 9347 | 9334 |
| 16 | 24 | 1272 | 17 | 2 | 17 | 5 | 9970 | 9957 |
| 17 | 24 | 1272 | 17 | 2 | 17 | 6 | 10593 | 10580 |
| 18 | 24 | 1272 | 17 | 2 | 17 | 7 | 11216 | 11203 |
| 19 | 24 | 1272 | 16 | 2 | 17 | 9 | 11839 | 11827 |
| 20 | 24 | 1272 | 16 | 2 | 17 | 10 | 12462 | 12450 |
| 21 | 24 | 1272 | 16 | 2 | 17 | 11 | 13085 | 13073 |
| 22 | 24 | 1272 | 16 | 2 | 17 | 12 | 13708 | 13696 |
| 23 | 24 | 1272 | 17 | 2 | 17 | 13 | 14332 | 14319 |
| 24 | 24 | 1272 | 17 | 2 | 17 | 14 | 14955 | 14942 |
| 25 | 24 | 1272 | 17 | 2 | 17 | 15 | 15578 | 15565 |
| 26 | 24 | 1272 | 17 | 2 | 17 | 16 | 16201 | 16188 |
| 27 | 24 | 1272 | 16 | 2 | 17 | 18 | 16824 | 16812 |
| 28 | 24 | 1272 | 16 | 2 | 17 | 19 | 17447 | 17435 |
| 29 | 24 | 1272 | 16 | 2 | 17 | 20 | 18070 | 18058 |
| 30 | 24 | 1272 | 17 | 2 | 17 | 21 | 18694 | 18681 |
| 31 | 24 | 1272 | 17 | 2 | 17 | 22 | 19317 | 19304 |
| 32 | 24 | 1272 | 17 | 2 | 17 | 23 | 19940 | 19927 |
| 33 | 24 | 1272 | 17 | 2 | 17 | 24 | 20563 | 20550 |
| 34 | 24 | 1272 | 16 | 2 | 17 | 26 | 21186 | 21174 |
| 35 | 24 | 1272 | 16 | 2 | 17 | 27 | 21809 | 21797 |
| 36 | 24 | 1272 | 16 | 2 | 17 | 28 | 22432 | 22420 |

transient and the steady state phases of operation. All that remains now is to design the hardware to implement this amazing algorithm, all the while feeling assured that the design will work correctly the first time and won't require hours of debugging, redesign, rework, and refabrication after the circuit is initially built.

### 10.4.6.3  Examining the Simulation Program

This section is supplied for those who may be interested in knowing how the quick-and-dirty simulation program works and how some of the code was written. If you are in a hurry to see the hardware realization of this elastic store algorithm, then this section can be skipped.

The simulation code snippets shown in this paragraph are written in C. The key element in the software simulation is in the generation of the two asynchronous clocks. This simulation had to generate both the DS-1 bit clock with a frequency range of 1.544 MHz $\pm 50$ ppm and the DS-1C bit clock with a frequency range of 3.152 MHz $\pm 30$ ppm. How is this done, you may ask? Well, it's pretty simple really. The following is a list of steps that one can use to generate two asynchronous software clocks. If more than two clocks are necessary, the same procedure can be followed for any number of additional clocks.

1. Select the highest frequency clock as the reference clock. In this case, it is the 3.152 MHz $\pm 30$ ppm DS-1C output clock.
2. Construct a C "**for**" loop with an index hClk (high clock) that represents the leading edge of the reference clock. Every time the hClk index is incremented, all the processing done on the leading edge of the reference clock is implemented in the "**for**" loop. For example,

```
for(hClk = 0;  hClk < NumberOfClocks; hClk ++) {
  … implement all processing that occurs on the leading
    edge of the reference clock.
}                                    // end of loop on hClk
```

3. Divide the frequency of the lower frequency clock lClk (low clock) by the frequency of the reference clock to create a fractional ratio of the two clocks. For example,

```
double hClk = 3152000;      // higher frequency clock
double lClk = 1544000;      // lower frequency clock
double dRatio = lClk/hClk;  // should be equal to
                            // 0.489847715736
```

4. Keep a running accumulation on lClk. Every time hClk is incremented by 1 in the **for** loop, increment lClk by the fractional value of dRatio. The lClk accumulation represents the phase of lClk relative to the reference clock. The value of lClk will incrementally increase by the fraction dRatio and will eventually become greater than 1. You can think of it as the phase lClk rolling over modulo 360°. When the lClk accumulation is greater than 1, an lClk period has elapsed and by definition an lClk leading edge has occurred.

5. Whenever the lClk phase accumulates to greater than 1, implement all the processing that is done on the leading edge of lClk, then subtract 1 from the accumulation. The resulting fractional residue in the lClk accumulator represents the fractional period or phase of lClk that remains. Later accumulations build on this residue phase. For example,

```
for(hClk = 0; hClk < NumberOfClocks; hClk++){
     [an hClk leading edge has occurred so implement all
     processing that takes place on the leading edge of
     hClk]
     lClk += dRatio;
     if(lClk ≥ 1) {
          [an lClk leading edge has occurred so implement
          all processing that takes place on the leading
          edge of lClk]
          lClk -= 1;       // decrement lClk phase by 360º
          }                // end of if statement on lClk
}                          // end of loop in hClk
```

The pseudocode listed here will keep track of the reference clock and the phase of the slower clock relative to the reference clock.

The processing of data frames is also very simple. Most frames will be repetitive. This repetitive feature lends itself well to a **C switch** statement. If each subframe within a frame is different, then the bit processing for each subframe can be handled by a separate **case** statement. Similarly, if two or more subframes are identical, they all can be bundled together and handled by a single **case** statement. The following pseudocode shows an example of the processing of a frame with eight subframes. Subframes 1, 2, 3, 4, and 5 are unique, and subframes 6, 7, and 8 are identical.

```
for (subframe =1; subframe < 9; subframe++) {
  switch (subframe) {
    case subframe1: {process bits in subframe 1}break;
    case subframe2: {process bits in subframe 2}break;
    case subframe3: {process bits in subframe 3}break;
```

```
    case subframe4: {process bits in subframe 4}break;
    case subframe5: {process bits in subframe 5}break;
    case subframe6:
    case subframe7:
    case subframe8: {process bits in subframes6,7,8}break;
    default:     break;
    }                          // end of switch on subframe
}                              // end of for loop on subframe
```

As an example, let's assume that the signal being simulated consists of repetitive frames and that each frame is 1024 bits in length. Assume that each frame consists of four subframes, each of which is 256 bits in length. Let's also assume that subframe 1 begins with four frame alignment bits followed by 252 information bits; subframes 2 and 3 contain 256 information bits; and subframe 4 begins with a single stuff control bit followed by a single stuff bit and then 254 information bits. An example of such a frame structure is illustrated in Figure 10.6. Let's further assume that the clock oscillators are perfect and produce two frequencies of 3.152 MHz for the output clock and 1.544 MHz for the input clock. The reader is cautioned that these clock frequencies were chosen purely as an example only, and since we are specifically interested in demonstrating the simulation code, they do not reflect any realistic clock pairs for this frame format.

The pseudolisting of this code is illustrated in Listing 10.1. Although not shown in this example, it is a simple matter to add code to implement nonperfect clocks that have some variance about the ideal frequency. All one has to do is add the variance to the desired clock and recompute the ratio of the two.

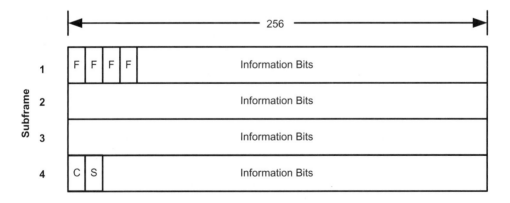

**Figure 10.6**  Example data transmission frame

## Listing 10.1  Pseudo Code for an Example Bit Stream Multiplex

```
double    hClk = 3152000;       //high rate clock
double    lClk = 1544000;       //low rate clock
double    dRatio = lClk/hClk;   //should equal 0.489847715736
ULONG     NumFrames = 40,000;   //number of frames to process
ULONG     NumSFrame = 4;        //number of subframes in frame
ULONG     NumClocks = 256;      //number of clocks per subframe
ULONG     iFrame, iSFrame;      //frame and subframe index
            ⋮
for(iFrame =1; iFrame < NumFrames + 1; iFrame ++){
   for(iSFrame =1; iSFrame < NumSFrame + 1; iSFrame ++){
   for(hClk=1; hClk < NumClocks+1; hClk ++) {
      switch(iSFrame) {
      case 1:
        if(hClk <5){read output frame alignment bit;}
        else        {read information bit from elastic store;}
        break;

      case 2:
      case 3:{read information bit from elastic store;}
        break;

      case 4:
        if(hClk == 1) {read stuff control bit; }
        if(hClk == 2) {read stuff or information bit; }
        if(hClk  > 2) {read information bit;}
        break;

      default:
      break;
      }                         // end of switch on iSFrame

      lClk += dRatio;           // increment phase of lClk
      if(lClk >=1) {
         {write an information bit into elastic store}
         lClk-= 1;              // dec lClk phase by 360 degrees
      }                         // end of if on lClk
   }                            // end of for loop on hClk
   }                            // end of for loop on iSFrame
}                               // end of for loop on iFrame
```

## 10.5  HARDWARE IMPLEMENTATION OF THE ELASTIC STORE MEMORY

It doesn't matter if the elastic store is implemented out of discrete parts, implemented out of ASIC development library macros, or constructed via VHSIC hardware description language (VHDL) design language. It makes sense to utilize the simplest components, macros, or constructs as possible in order to ensure the finished hardware design operates as fast as possible. Elastic store memories operating at the higher multiplex levels of any telephone transmission hierarchy need to be extremely fast. Therefore the design of our elastic store will reflect these concerns.

One of the simplest, smallest, and fastest memory devices available to the designer is the common everyday D flip-flop. These devices are easily implemented via VHDL and are a staple in any ASIC development library. Since we do not need a large storage capacity, we will utilize the D flip-flop to implement our elastic store bit memory.

Our simulation results show that we could implement the elastic store memory with as little as 8 bits of storage. However, it is always more comfortable when designs of this type have additional operational guard band. Therefore we will implement a 16-bit elastic store memory that requires no additional effort on our part and will give us more than a 50% guard band or, as some might say, design margin.

The actual design is very simple. The block diagram for the elastic store memory is illustrated in Figure 10.7. The memory itself is composed of 16 D flip-flops, which gives the elastic store a maximum storage of 16 bits. The input data bit stream is applied to the D input of each flip-flop. Each flip-flop is clocked with the DS-1 data stream bit clock. The output of a 4-to-16 decoder serves as the clock enable to each individual flip-flop. A modulo 16 counter provides the input to the decoder. The counter is also clocked by the DS-1 input bit stream clock. Using the decoder outputs as individual clock enables, as opposed to utilizing them as individual clocks, is a safer design because all flip-flops are clocked with the same clock, which simplifies timing variance and, more importantly, eliminates the risk of decoder output glitches that might false clock one or more of the flip-flops.

The data arrives at the DS-1 clock rate and is sequentially clocked into D flip-flop 0 through 15. Since the counter counts modulo 16, the next bit in the input sequence will be clocked into flip-flop 0 and the process is repeated. The string of flip-flops and clock decoders can be viewed as a 16-bit circular memory that is constantly overwriting the oldest bit in memory with the newest input bit. There is no complicated memory address to fool around with, which eliminates those pesky address access times and makes this an extremely fast architecture.

The output of each flip-flop is fed to the input of a 16-to-1 multiplexer. On the leading edge of the output DS-1C output clock, the multiplexer will

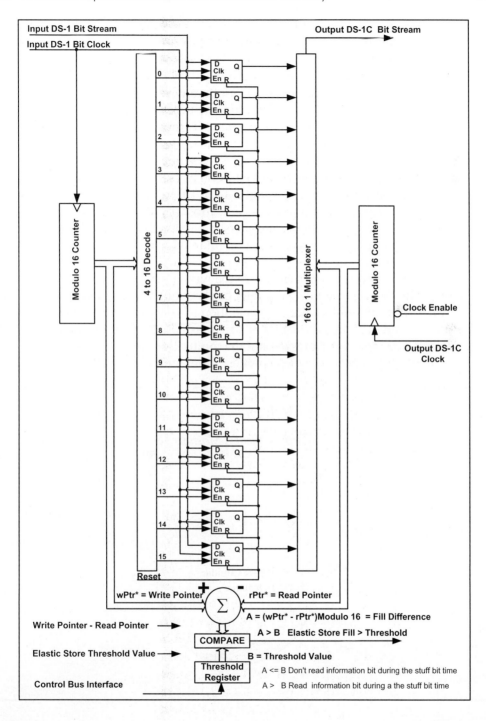

**Figure 10.7**   Sixteen-bit elastic store block diagram

sequentially select one flip-flop as the elastic store output. As was the case with the input decoder, the multiplexer output is determined by the address provided by the output of a modulo 16 counter. The memory can be thought of as being read in a circular fashion from flip-flop 0 through 15 and then back to 0, and this process repeats indefinitely.

The output counter is continuously clocked by the output DS-1C bit stream clock. Whenever an information bit is to be read from the elastic store memory, the multiplexer will issue a clock enable signal to the counter that will advance its count by one and access the next bit in memory.

You can think of the write operation as being a process where the input data bits are written at a constant rate in the circular memory, as illustrated in Figure 10.8. The write address produced by the modulo 16 counter can be thought of as a write pointer labeled *wPtr* ∗ that is continuously moving around a 16-bit circle at a constant angular velocity ($\omega$), defined by the input data stream clock rate.

The read operation is similar to the write operation. The read address produced by the read modulo 16 counter can be thought of as a read pointer labeled *rPtr* ∗. You can think of the read operation as being a process where the output data bits are read in a semijerky or gated fashion. By that, I mean that data bits are read out at a constant rate until the multiplexer has to go off and process an overhead bit such as a frame alignment bit or stuff control bit. During this time,

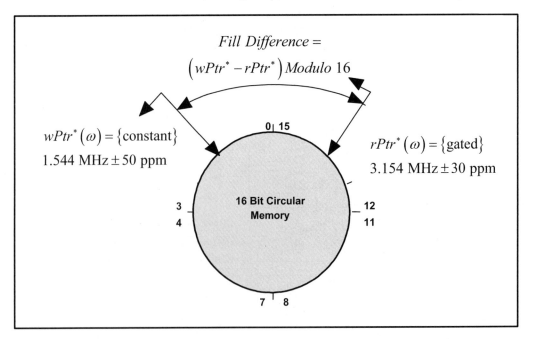

**Figure 10.8**   Conceptual elastic store circular memory

the reading of data bits is suspended until the next valid output information bit time. So the output read process is best modeled by a read pointer that advances around the circular memory in a start and stop or jerky fashion.

You can visualize the elastic store write and read process as two pointers chasing one another around a 16-bit memory circle. The write pointer is proceeding around the circle at a constant rate and the read pointer is proceeding around the same circle at a jerky or gated rate.

At this point we define the term *fill difference* as the number of unread bits in memory. It is the difference between the value of the write pointer and the value of the read pointer. This is an important number, and the elastic store needs to continuously compute and keep track of it.

The fill difference between the write pointer and the read pointer will increase during those times when the multiplexer is off servicing overhead bits and no information bits are read from memory. However, since the read pointer is clocked at a faster rate than the write pointer, the fill difference will decrease during the times when the multiplexer is reading bits from the elastic store memory. So it is expected that the fill difference will undulate about the programmed threshold value. This process is illustrated in Figure 10.8.

The computed fill difference is simply the modulo 16 result of the write pointer minus the read pointer. The elastic store memory computes the fill difference by subtracting the read pointer from the write pointer $wPtr*-rPtr*$ using two's complement arithmetic. The computation ignores any carry out bits and treats the result as a simple unsigned binary number. This method results in a modulo 16 subtraction and eliminates any boundary problems such as would occur when the write pointer and read pointer are on opposite sides of the address boundary, 0 and 15.

This fill difference is compared against the programmable threshold value residing in a holding register, and the resultant inequality is read by the multiplexer to determine if an extra information bit should be read from the elastic store memory during the next half frame. This is illustrated in the block diagram of Figure 10.7.

If the fill difference is greater than the threshold, the multiplexer will be instructed to read an extra information bit from elastic store memory and place that bit into the half frame stuff bit location. The multiplexer control will also place a sequence of 000 in the stuff control bit locations for the associated input tributary to inform follow on processors that the stuff bit position contains an information bit.

If the fill difference is less than the threshold, the multiplexer will be instructed not to read an extra information bit from elastic store memory, and the stuff bit location for that half frame will be filled with a stuff bit, which is a logic 1. The multiplexer control will also place a sequence of 111 in the stuff control bit locations for the associated input tributary.

For larger elastic store memories, the wider bit compare operation may be more time consuming than desired, which would result in slower operation. For larger memories, the speed of the elastic store operation can be increased somewhat by replacing the comparator with a second adder. The adder would be used to subtract the fill difference $(wPtr * - rPtr *)$ value from the threshold value. This produces an arithmetic relation of the form:

$$\text{Result} = \text{Threshold } - (wPtr * - rPtr *)$$

In this case, the sign bit of the result can be used as the stuff indicator. If the sign bit is negative, indicating the memory fill is larger than the threshold value, then an extra information bit should be read from elastic store and inserted into the frame stuff bit location. If the sign bit of the result is positive, indicating that the memory fill is less than the threshold value, then an extra information bit is not read from elastic store. Let's take a quick look at a couple of example computations using the comparator architecture.

**Example 1**. Figure 10.9 illustrates an instantaneous snapshot of a 16-bit elastic store circular memory in operation. In this case, the write pointer $(wPtr *)$ is moving around the circular memory at a constant rate of 1.544 MHz ± 50 ppm, or more correctly, the write pointer is writing serial data at a rate of 1.544 Mbps ± 77.2 bps. The instantaneous snapshot in Figure 10.9 shows the $wPtr *$ pointing to bit location 5 in memory. The read pointer $(rPtr *)$ is moving around the circular memory at a gated rate that averages 3.154 MHz ± 30 ppm when it is actively reading memory. In other words, when active, the read pointer is reading a serial bit stream from memory at an average rate of 3.154 Mbps ± 94.62 bps. The snapshot of Figure 10.9 shows the $rPtr *$ pointing to bit location 2 in the 16-bit circular memory. Bits 6 through 15 and bits 0 through 1 have already been read and are waiting to be written over.

By inspection, the fill difference is $5 - 2 = 3$. In this example, let's assume the threshold is set to a value of 6. When we do a subtraction $(wPtr * - rPtr *)$, we will perform the subtraction using two's complement arithmetic, ignore any carry bits, and treat the resultant sum as an unsigned binary number. The mathematical operation is given by

$$
\begin{array}{llll}
wPtr * & = 5 & = 0101 & \\
rPtr * & = 2 & = 0010 & \left.\right\} \text{Setup} \\
-rPtr * & = -2 & = 1110 &
\end{array}
$$

$$
\begin{array}{llll}
wPtr * & = 5 & = 0101 & \\
-rPtr * & = -2 & = 1110 & \left.\right\} \text{Computation} \\
wPtr * - rPtr * & = 3 & = \cancel{1}\ 0011 &
\end{array}
$$

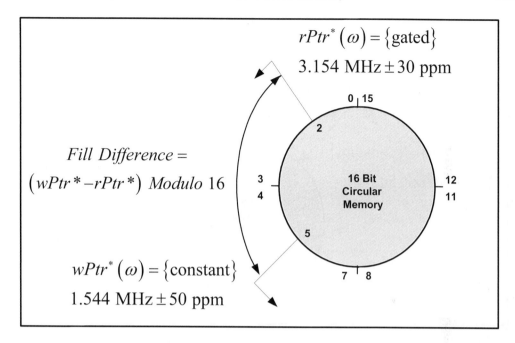

$$rPtr^*(\omega) = \{\text{gated}\}$$

$$3.154 \text{ MHz} \pm 30 \text{ ppm}$$

*Fill Difference =*

$$(wPtr * -rPtr *) \; Modulo \; 16$$

**16 Bit Circular Memory**

$$wPtr^*(\omega) = \{\text{constant}\}$$

$$1.544 \text{ MHz} \pm 50 \text{ ppm}$$

**Figure 10.9**   Elastic store fill computations, example 1

Since we are dealing with modulo 16 arithmetic, we toss out the carry bit and end up with

$$wPtr * -rPtr * \;\; = \;\; 3 \;\; = \;\; 0011$$

The resultant value is 3, which is what we would expect. The fill value is less than the threshold of 6, so no extra information bit will be read during the next stuff bit time.

**Example 2.** Figure 10.10 illustrates another instantaneous snapshot of an elastic store in operation. In this example, the write pointer $(wPtr *)$ is pointing to bit 2 and the read pointer $(rPtr *)$ is pointing to bit 13 in circular memory. Bits 3 through 12 have already been read and are waiting to be written over. By inspection, we see that the fill difference is 5. Let's do the calculations and see if we get the same answer. In this example, we will assume the threshold is set to a value of 4. When we do a subtraction $(wPtr * -rPtr *)$, we will again use two's complement modulo 16 arithmetic, ignore any carry bits, and treat the resultant sum as an unsigned binary number.

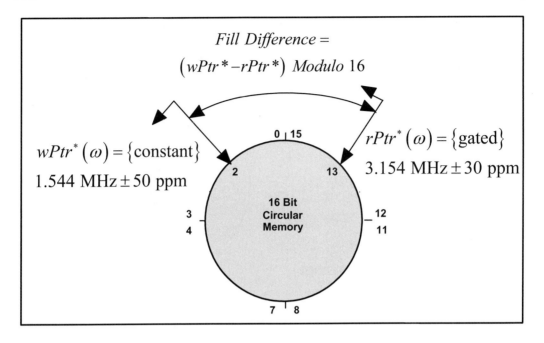

**Figure 10.10**   Elastic store fill computation, example 2

$$
\begin{array}{lll}
wPtr^* & = 2 & = 0010 \\
rPtr^* & = 13 & = 1101 \\
-rPtr^* & = -13 & = 0011
\end{array}
\Bigg\} \text{Setup}
$$

$$
\begin{array}{lll}
wPtr^* & = 2 & = 0010 \\
-rPtr^* & = -13 & = 0011 \\
wPtr^* - rPtr^* & = 5 & = \cancel{0}\ 0101
\end{array}
\Bigg\} \text{Computation}
$$

The resultant value of 5 is what we expected, and it is greater than the threshold of 4, so an extra information bit will be read during the next stuff bit time.

The alert reader might say, "Hey, wait a minute! 2 – 13 should equal –11, not +5! What's going on here?" Well, remember we are dealing with 4-bit two's complement modulo 16 arithmetic. We are ignoring the carry bit and treating the result as an unsigned number.

The 4-bit number field in two's complement representation can range from a negative value of 8 to a positive value of 7. Therefore the number –11 cannot be represented. We would have needed a 5-bit field ranging from –32 to +31 to be able to represent a value of –11. This would have been fine if we were using a 32-bit elastic store, but we aren't. We are using a 16-bit elastic store. The modulo 16 arithmetic we are using takes care of these nasty discrepancies and lets us do computations in a circular fashion.

**Example 3.** Let's do another quick example. Suppose the write pointer was pointing to bit 8 and the read pointer was pointing to bit 1 and the threshold value was set to 6:

$$
\left.
\begin{array}{rclcl}
wPtr^{*} & = & 8 & = & 1000 \\
rPtr^{*} & = & 1 & = & 0001 \\
-rPtr^{*} & = & -1 & = & 1111
\end{array}
\right\} \text{Setup}
$$

$$
\left.
\begin{array}{rclcl}
wPtr^{*} & = & 8 & = & 1000 \\
-rPtr^{*} & = & -1 & = & 1111 \\
wPtr^{*}-rPtr^{*} & = & 7 & = & \cancel{1}\ 0111
\end{array}
\right\} \text{Computation}
$$

The fill difference is 7, as we expected, and the threshold is 6, so the elastic store will instruct the multiplexer to read an extra information bit from elastic store memory during the next stuff bit time.

**Example 4.** Here's one more example, this time using extreme pointer values that will help cement the concept of pointer math. Consider the case illustrated in Figure 10.11. For this extreme case, the read pointer points to bit 13 in circular memory, and the write pointer for some reason has advanced to the point where it is one bit behind the read pointer at bit 12. This condition suggests that the write pointer is about to begin overwriting bits stored in memory before they have been read. This would lead to memory overflow. Let's assume that this limiting case has occurred because the threshold value has been mistakenly set to a large number such as 14.

The pointer computations show that the fill difference is 15 bits, which agrees with Figure 10.11. Since the threshold is set to 14, the elastic store memory will instruct the multiplexer to read an extra bit during the current half frame. This case illustrates the danger of setting the elastic store threshold too close to the memory limits. Operationally, this case would more than likely result in intermittent data errors due to random memory overflows. The reader can generate as many of these examples as necessary to verify this method of pointer arithmetic works for all possible combinations of pointer values.

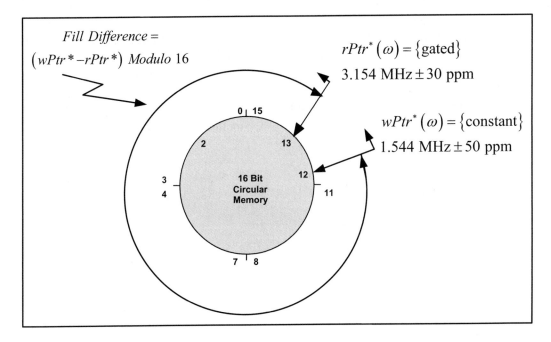

$$Fill\ Difference =$$
$$\left(wPtr* - rPtr*\right)\ Modulo\ 16$$

$$rPtr^*\left(\omega\right) = \left\{gated\right\}$$
$$3.154\ \text{MHz} \pm 30\ \text{ppm}$$

$$wPtr^*\left(\omega\right) = \left\{constant\right\}$$
$$1.544\ \text{MHz} \pm 50\ \text{ppm}$$

**Figure 10.11**   Elastic store fill computation, example 4

$$
\begin{array}{llll}
wPtr^* & = 12 & = 1100 \\
rPtr^* & = 13 & = 1101 \\
-rPtr^* & = -13 & = 0011
\end{array}\ \right\}\text{Setup}
$$

$$
\begin{array}{llll}
wPtr^* & = 12 & = 1100 \\
-rPtr^* & = -13 & = 0011 \\
wPtr^* - rPtr^* & = 15 & = 0\ 1111
\end{array}\ \right\}\text{Computation}
$$

To give you some level of confidence in the elastic store memory design that we just discussed, it should be noted that for a completely different application, I designed an almost identical elastic store to the one just discussed and it was successfully implemented in a gallium arsenide gate array.

The design consisted of four completely independent 16-bit elastic store memories, each of which input an asynchronous 145 Mbit/second serial bit stream. The elastic store memories allowed the four asynchronous input bit streams to be synchronously multiplexed into a single 600 MBit/second serial output stream. This was done way back in 1997. To put this design in perspective, the output bit clock had a period of 1.667 nanoseconds. The memory read accesses and pointer computations performed during each output clock cycle had to be completed in 1.667 billionths of a second. To me, this is pretty impressive, but even this level of performance is considered slow by today's standards.

At these speeds, clocks no longer look like textbook square waves. The transmission line capacitance causes the fast edge rate clocks to look like sine waves. As you might expect, the data driver source current for the clock and data bits, the transmission lines for the clock and data bits, and clock and data bit loads all had to be computed precisely and the resulting circuit had to be extensively simulated. Simulations were run overnight and the resulting gigabytes of stored simulation data were evaluated the next morning. This is a classic example of the need for bit-level simulations. Without simulation, this gallium arsenide chip would never have had a chance of working right out of the foundry. The chip did work correctly the first time, and the company did not suffer any schedule slips. In addition, the company did not have to pay for repeated fabrication costs or fees to the foundry for each three month refabrication cycle to incrementally fix design bugs. More importantly, I was able to keep my job and move on to the next project.

The reader should keep in mind that the elastic store memory is interfacing two asynchronous bit clocks and their associated bit streams. As such, this design can be considered an asynchronous design. In particular, it should be understood that because of the asynchronous relationship between the read and write clocks, the period of the fill difference measurement will not be a constant value. As the two clocks vary with respect to one another, the length of the fill difference period will undulate or breathe over time. It is entirely possible that on random occasions the computation of whether or not the fill difference is above or below threshold may violate digital timing rules such as setup and hold times. This could cause the elastic store to read a bit during a stuff bit time when it is not necessary, or to not read a bit when it is necessary. This is not a problem because, for the random occasions that this issue can occur, the memory will absorb the bit in question and will adjust its operation on the next computation cycle. This is a good reason for the designer to give him or herself extra headroom when determining the overall size of the elastic store memory. In this sense, the elastic store circuit is fault tolerant in that it compensates for possible computational glitches and the serial bit stream it is processing will suffer no bit errors. This design has been proven in the most extreme timing environments and it has always worked flawlessly.

## 10.6   OVERALL DS-1C MULTIPLEXER DESIGN BLOCK DIAGRAM

Now that we have designed an elastic store memory, let's complete the discussion by utilizing it in a multiplexer designed to multiplex two asynchronous DS-1 input tributaries to a single DS-1C output tributary. The block diagram of a multiplexer architecture that will perform this function is illustrated in Figure 10.12.

On the left side of the block diagram, we see that there are several processes going on, and all are referenced to the DS-1C output clock. The DS-1C output clock can vary anywhere within a 3.152 Mbps ± 30 ppm range. On the right side of the diagram, we see a simple multiplexer and a small amount of control logic that is used to formulate the output DS-1C bit stream.

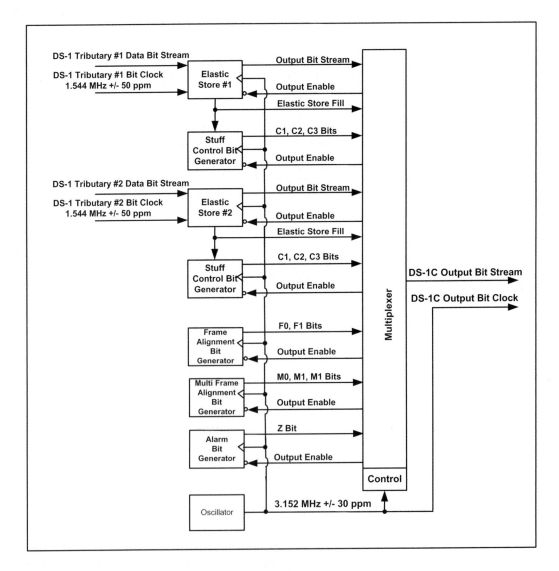

**Figure 10.12** DS-1C multiplexer block diagram

Note that in the block diagram there are two identical but completely independent elastic store memories, one processing the input bit stream for DS-1 tributary 1 and the second processing the input bit stream for DS-1 tributary 2. The two tributaries are completely independent of one another and their individual serial bit stream rates can independently vary anywhere within a 1.544 Mbps $\pm$ 50 ppm range.

Both elastic store memories accept output clock enables from the multiplexer and provide serial data and memory fill status to the multiplexer. Both elastic stores provide a fill status bit to their respective stuff bit control generator, which is nothing more than a flip-flop that is set or cleared at the beginning of each half frame.

The fill status flip-flop is cleared when the elastic store memory fill level is greater than the programmed threshold and is set when the elastic store fill level is less than the programmed threshold. So during the stuff control bit times, the multiplexer simply selects the output of the flip-flop and uses it to fill all the C1,C1,C1 or the C2,C2,C2 stuff control bits for tributary 1 or tributary 2. The sequence will be either 000 or 111, depending if the stuff bit location contains an information bit or a stuff bit.

The frame alignment bit generator is nothing more than a two bit circular shift register that, when enabled, clocks out a single bit of the frame alignment sequence of F0, F1, F0, F1,.... The multiframe alignment bit generator operates exactly the same as the frame alignment word generator, except that, when enabled, it outputs a bit sequence of M0, M1, M1,....

The alarm bit generator is a simple flip-flop as well. It is cleared if an alarm condition exists and is set if an alarm condition does not exist. The multiplexer simply selects the flip-flop output during the Z bit time in the output DS-1C frame. The whole process is clocked by an oscillator that is specified to provide an output of 3.152 MHz $\pm$ 30 ppm.

The bottom line is the design is straightforward: it's simple, but since it appears complex and difficult, it will impress your peers and dazzle management when it works correctly the first time.

## 10.7  ADDITIONAL INFORMATION

This chapter covers the basics for the design of an elastic store memory. To illustrate the operation of an elastic store memory, we utilized it in an application that reclocked asynchronous low rate serial input bit streams prior to multiplexing them into a higher clock rate synchronous serial output bit stream.

What about the reverse operation? Suppose we wanted to input the high-level DS-1C multiplex bit stream, demultiplex, and then output the two original low level asynchronous DS-1 bit streams at their original bit clock rates. This type of demultiplex operation takes some thought. Think about it. For this application, we would have to not only demultiplex the two DS-1 serial bit streams from the DS-1C bit stream but also synthesize independent bit clocks for each demultiplexed tributary that matched their original premultiplex clocks.

Again, we would use an elastic store memory for this demultiplex operation. The only difference is the elastic store would be slightly modified, and it would have a bit of extra logic associated with it to synthesize the two DS-1 bit stream clocks. Both synthesized DS-1 bit stream clocks would be independent of one another and would be synthesized to be on average identical to the original clocks used by the original input tributaries. In this application, the size of the elastic store directly determines the precision of the synthesized clock frequency. Not only that, the size of the elastic store directly determines the time constant of the synthesized clock frequency response to a change in tributary bit rate.

This certainly sounds like a tough nut to crack. Depending upon how it is designed, it can be very tough! If designed correctly, however, this version of an elastic store based tributary clock synthesizer can turn out to be a fairly straightforward design.

How does one synthesize a tributary clock from a demultiplexed tributary bit stream with no prior information about the original tributary clock frequency? If, for example, we want to demultiplex the two DS-1 bit streams from higher level the DS-1C bit stream, how would we know exactly what the frequency of the original bit clocks are? The answer is we don't know. The only information we have is the rate at which the DS-1 tributary bits are popping out of the higher order multiplex. Somehow from that information we need to synthesize two asynchronous DS-1 tributary clocks whose time varying frequency is somewhere within the range of 1.544 MHz ± 50 ppm.

In other words, after the demultiplex operation, the only information we have are the data bits themselves. If our synthesized tributary clock is too slow, we will end up dropping demultiplexed bits. If our synthesized clock is too fast, we will end up intermittently transmitting useless bits. This problem sure does sound difficult, if not impossible. We can come up with a solution, however, and that solution is to use an elastic store-based data locked loop (DLL). An elastic store memory for this type of application is discussed in detail Chapter 11, "Digital Data Locked Loops." If you think this is an advertisement for Chapter 11, you are correct. There is a great deal of useful engineering information for meaningful engineering applications packed neatly into that chapter.

## 10.8   REFERENCES

[1]  John C. Bellamy. *Digital Telephony*. New York: John Wiley and Sons, 1982.

[2]  Robert G. Winch. *Telecommunication Transmission Systems*. New York: McGraw-Hill, 1993.

[3]  David R. Smith. *Digital Transmission Systems*. 2nd ed. New York: Chapman and Hall, 1993.

[4]  Bell Labratories. *Transmission Systems for Communications*. Bell Laboratories, 1982.

[5]  Roger L. Freeman. *Telecommunication Transmission Handbook*. 3rd ed. New York: John Wiley and Sons, 1991.

[6]  Daniel Minoli. *Enterprise Networking Fractional T1 to Sonet, Frame Relay to BISDN*. Norwood, MA: Artech House, 1993.

[7]  Frank F. E. Owen. *PCM and Digital Transmission Systems*. New York: McGraw-Hill, 1982.

# CHAPTER ELEVEN

# Digital Data Locked Loops

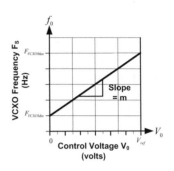

In the previous chapter we saw how elastic store memories are utilized to reclock asynchronous input tributary bit streams prior to being multiplexed into a synchronous output tributary. We utilized two levels of telephone multiplex signals to demonstrate the use of the elastic store memory. Specifically we used elastic store memories to multiplex two asynchronous DS-1 bit streams into a single DS-1C bit stream. Each of the input bit streams were associated with their own independent bit clock and were asynchronous to one another. Once the lower level bit streams are multiplexed into a higher level bit stream, all clock information associated with the lower level streams is essentially lost. The problem we have now is, how can we reverse this multiplex (i.e., how can we demultiplex the two DS-1 streams and synthesize a bit clock for each stream that is on average identical to its original clock)? The DS-1/DS-1C example is only one of an infinite number of possible examples. The same question can be asked of any demultiplex processing where the multiplexed tributaries were originally asynchronous to one another.

The answer to these questions is to utilize a digital data locked loop (DLL). The DLL is fairly simple device that uses an elastic store memory to synthesize a bit stream clock and then synchronizes the demultiplexed bit stream with that clock. All this takes place with no prior knowledge of the original clock frequency.

DLLs are suited for many applications. In order to maintain continuity within this book, we will describe and design a DLL that can be used to demultiplex the DS-1C tributary that we discussed in detail in Chapter 10, "Elastic Store Memory." There is no reason, however, to restrict the usage of a DLL to only telephony applications. The DLL we describe in this chapter can be considered a base model that with a few modifications can be used for many other applications as well.

## 11.1   DIGITAL DATA LOCKED DESIGN

To help us better understand the design of the DLL, we need to have an overall picture of the type of signals and the functional path of the signals we will be processing. For this reason we will utilize the DLL in a simple bit stream demultiplexer to synthesize a bit clock and resynchronize the recovered bit stream.

The functional blocks of a demultiplexer are illustrated in Figure 11.1. This book is only concerned with the shaded blocks in this figure. These blocks are the ones that utilize the DLL to synthesize and resync the recovered bit streams.

It is not the intention of this book to discuss all the other processing that goes on within a demultiplexer, but we will need to briefly describe the format of the demultiplexed signals that serve as inputs to the DLL. For this reason, we will briefly explain the end-to-end signal flow. The tributary demultiplex block receives the high-level multiplex input bit stream and then demultiplexes the bit streams associated with each tributary. At the output of the tributary demultiplex, the bit stream is accompanied by a gated clock that is used to indicate the existence of a valid tributary bit. You can envision the gated clock as the high rate input clock with missing teeth.

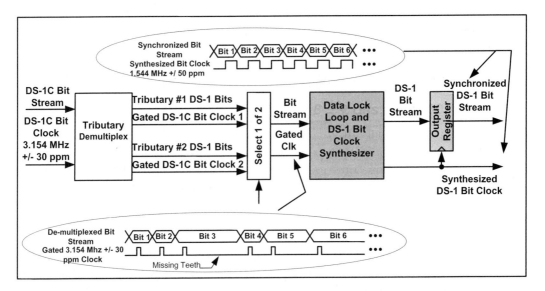

**Figure 11.1**   Simplified demultiplexer block diagram

As shown in Figure 11.1, the clock teeth are present whenever a new information bit is recovered by the demultiplexer. Clock teeth are missing whenever the demultiplexer is off processing an overhead bit or when it is off processing a bit from another embedded tributary, and the corresponding bit periods shrink and expand accordingly. Clearly this is not the desired format and timing of a recovered bit stream that we would like to hand off to any external processors. Instead, we would rather synthesize a valid 50% duty cycle bit clock and then synchronize the recovered bit stream to this clock.

We will use this chapter to design a DLL architecture that is relevant to a real-world digital signal processing (DSP) application. Since the reader is already familiar with the real-world DS-1 and DS-1C signals from the previous chapter, we will use these signals as the input and output of our DLL based demultiplexer.

The input to the demultiplexer is a DS-1C, which carries two DS-1 tributaries. The tributary demultiplex block in Figure 11.1 outputs the gated clock version of both tributaries. The block diagram shows that we select only one tributary for further processing. This will be sufficient for our discussion and development of the DLL architecture. Enhancing this design to process multiple bit streams is straightforward.

The selected bit stream and associated gated clock are fed to the DLL block, where the loop synthesizes a bit clock from the input bit stream and uses this clock to strobe the bit stream out of the DLL. The time aligned and newly formatted bit stream and 50% duty cycle bit clock output from the DLL circuit are illustrated in Figure 11.1.

The reader should remember that we have no idea what the frequency of the original tributary bit clock was. All we know is that it must be something within ±77.2 Hz of the 1.544 MHz center frequency, and even then the original clock may have been drifting over time between the two limits. The clock that our DLL synthesizes must on average match the frequency of the original clock, and it should track its drift over time.

### 11.1.1 Digital Data Locked Elastic Store

For those who read the previous chapter on elastic store memories, this section will contain some similar material. However, since the DLL application is significantly different from the bit stream multiplexer application of the preceding chapter, the elastic store memory we use here will be implemented quite a bit differently.

The block diagram of the modified elastic store for the DLL application is illustrated in Figure 11.2. For this application, we are using a 1024-by-1 bit, two-port memory. The logic behind selecting a memory 1024 bits in length will be made clear in later sections. It is enough to say here that we need a 10-bit address space in order to synthesize an accurate and stable DS-1 bit stream

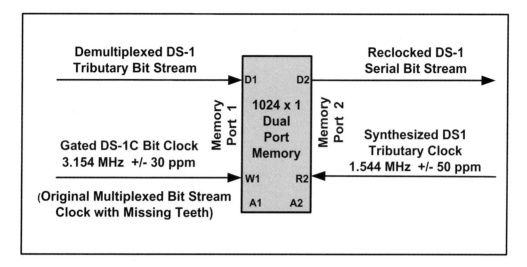

**Figure 11.2**   DLL elastic store memory

clock. We will delve deeper into the reasoning behind this as we proceed with the design.

Figure 11.2 shows a single 1024-by-1 bit, two-port memory. The two-port memory might be a discrete device manufactured by one of the many IC houses. It might also be a macro obtained from a field programmable gate array (FPGA) development software library, or it could have been designed by you, the engineer, using a hardware development language such as VHDL. (VHDL is an acronym for VHSIC hardware description language. VHSIC is an acronym for very high speed integrated circuits.)

No matter what the source, there are a great many methods that can be used to implement a dual port memory of this size. For our purposes, we have chosen to implement a two-port memory using the VHDL design language. This allows us the freedom to design our own architecture and define how the device operates. In this case we have designed a clock synchronous two-port memory. The term *synchronous* means that a data bit applied to the input data port D1 will be stored into memory at the address applied to port A1 on the leading edge of the clock presented at port W1. Similarly, a bit stored at the address applied to port A2 will be read from memory on the leading edge of the clock applied to port R2. The bit read from memory will appear at the output port D2.

The operation of our custom two-port memory isn't too much different from common standalone devices available from most memory manufacturers. However since we custom designed the memory ourselves using VHDL inside a programmable logic device, it fits our application perfectly and therefore provides us with our most efficient implementation. In addition, it allows us the flexibility to quickly and easily modify the size of the two-port memory. The importance of this will be shown in later sections.

Operationally, the recovered DS-1 bit stream is applied to the data pin D1 of memory port 1. The gated DS1-C bit clock is applied to the synchronous clock pin W1 of memory port 1. The bit applied to pin D1 will be written in to memory at the address applied to port A1 on the low to high transition of the gated clock. The reclocked DS-1 bit stream is output from the pin D2 of memory port 2. The synthesized DS-1 tributary clock is applied to the synchronous memory read port R2. Data stored at the address applied to port A2 will appear on pin D2 on the leading edge of the synthesized clock. For now the synthesized clock applied to memory port 2 pin R2 seems to mysteriously appear from out of the blue. We will spend much of this chapter discussing the derivation of this clock.

### 11.1.2  Digital Data Locked Elastic Store Write and Read Pointers

In Figure 11.3 we have added a modulo 1024 counter on both sides of the two-port memory, as indicated by the shaded boxes. These are write and read counters and are clocked with the write clock and synthesized read clock, respectively. The output of the counters are applied to the memory address ports and serve as the write port and read port memory address sources.

We can see from the figure that each time a recovered bit is clocked into memory port D1, the port 1 address is advanced by 1 and it points to the location in memory where the next recovered bit will be stored. The address counter is modulo 1024, so the writing of recovered bits progresses around a 1024-bit circle where the oldest bit stored is written over by the newest bit to arrive. The write address is referred to as a *write pointer*.

Similarly, each time a synthesized read clock strobes a bit from memory, the read address is incremented by one and it points to the address of the next bit waiting to be read. The read address counter is also modulo 1024, so the reading of data bits progresses around the same 1024-bit circle. The read address is referred to as a *read pointer*. We can envision the read pointer continuously chasing the write pointer around the circular memory.

The 10-bit write-and-read addresses also serve as 10-bit pointers that are used to calculate the fill difference in the memory. The fill difference is the difference between the write pointer $(wPtr\,^*)$ and the read pointer $(rPtr\,^*)$ and is a measure of how many unread bits are stored in memory. There is no hard and fast rule as to the number of bits needed for each pointer. The criteria for selecting the pointer bit width are based on the desired frequency precision, frequency jitter, and response time of the synthesized DLL read clock. As we will see later, the performance of the synthesized bit clock frequency diminishes as the bit width of the pointers decreases.

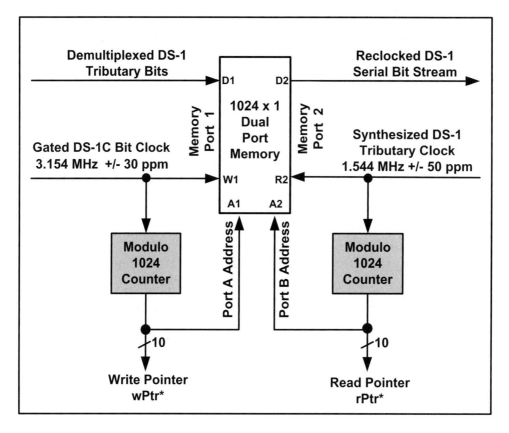

**Figure 11.3**  DLL write and read pointers

Now that we have created a write pointer and a read pointer, we need to compute the number of unread bits in memory. The number of residue bits is computed by subtracting the read pointer from the write pointer. The result is termed the *fill difference* and given by $wPtr^* - rPtr^*$.

This computation is illustrated in Figure 11.4. The adder shown as a shaded block in the figure is 10 bits wide and performs the modulo 1024 pointer subtraction.

To briefly summarize, the two 10-bit modulo 1024 counters provide the port addresses to the dual port memory, and they also provide address pointers that are used to compute the number of unread bits stored in the memory. The number of unread bits is termed the *fill difference* in the memory.

The reader should note that since the write pointer $wPtr^*$ and the read pointer $rPtr^*$ are asynchronous to one another, the period of the fill difference measurement will not be a fixed constant. As a matter of fact, the period of the fill difference measurement will breathe as the two pointers are chasing

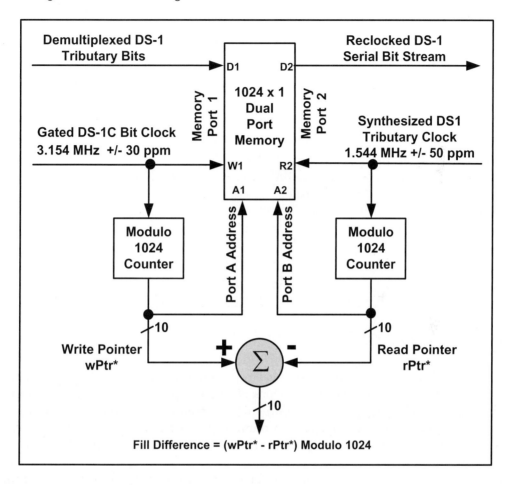

**Figure 11.4**   DLL fill difference calculation

one another around the circular memory address space. This should be of no consequence for most applications.

The graphical description of the write and read pointers and the difference between them is illustrated in Figure 11.5. The figure shows a 1024-bit circular memory with a write pointer $(wPtr^*)$ and a read pointer $(rPtr^*)$ chasing one another around the circle, with each pointer moving at its own rate. The write and read pointers can be visualized as moving around the circular memory with angular velocities determined by their instantaneous clock rate. Thus we can state

$$wPtr^*(\omega) = \{\text{gated}\}\ 3.154\ \text{MHz} \pm 30\ \text{ppm}$$

$$rPtr^*(\omega) = \{\text{synthesized}\}\ 1.544\ \text{MHz} \pm 50\ \text{ppm}$$

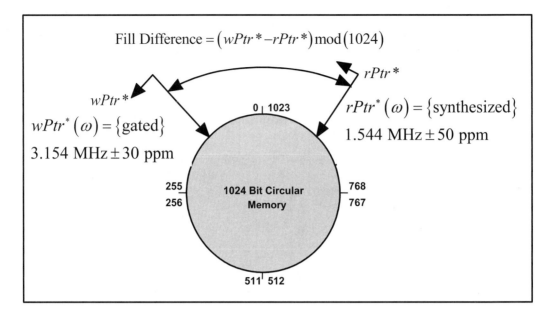

$$\text{Fill Difference} = \left(wPtr^* - rPtr^*\right)\bmod\left(1024\right)$$

$rPtr^*$

$wPtr^*$

$$wPtr^*\left(\omega\right) = \left\{\text{gated}\right\}$$

$$3.154\ \text{MHz} \pm 30\ \text{ppm}$$

$$rPtr^*\left(\omega\right) = \left\{\text{synthesized}\right\}$$

$$1.544\ \text{MHz} \pm 50\ \text{ppm}$$

0 | 1023

255
256

**1024 Bit Circular Memory**

768
767

511 | 512

**Figure 11.5**  DLL modulo 1024-bit circular memory

The angular velocity of the write pointer is expressed as the gated frequency of the base DS-1C bit clock. The write pointer clock is just the input 3.154 MHz ± 30 ppm clock with missing teeth. The gated clock teeth correspond to the occurrence of a demultiplexed tributary bit. Teeth that correspond to the occurrence of overhead bits or to the bits from the second tributary are removed.

So the write clock does not have a 50% duty cycle, and as illustrated in Figure 11.1, the period of each data bit is stretched whenever the demultiplexer is processing overhead bits or bits from the other tributary bit stream. Although the write pointer clock is not pretty, in a pure sense its clock tooth frequency will be equal to the original bit stream clock and will take on a value within the 1.544 MHz ± 50 ppm range of the original bit stream. Believe it or not, many demultiplex designs today treat the recovered but nonuniform bit stream and the random missing tooth clock as an acceptable design architecture. As we will see, we can use the DLL to make a significant performance improvement to this type of design.

The read pointer is the synthesized DS-1 output tributary clock, and it will be a 50% duty cycle clock with an angular velocity within the 1.544 MHz ± 50 ppm range of frequency specified by the International Telephone and Telegraph Consultative Committee (CCITT), known since 1992 as the International Telecommunications Union (ITU), based in Geneva, Switzerland.

When we compute the value for the fill difference, we use circular math or in this case modulo 1024 two's complement arithmetic. Let's do some pointer arithmetic so we can get a feel for how the modulo 1024 arithmetic works. When performing modulo 1024 pointer arithmetic, we will adhere to the following three rules:

- We will perform the $wPtr*-rPtr*$ subtraction using two's complement arithmetic.
- We will ignore any carry bits.
- We will treat the result as a normal unsigned binary number.

**Example 1.** An instantaneous snapshot in time of an operational elastic store memory write and read pointer is illustrated in Figure 11.6. In this figure, we can see that the write pointer is pointing to memory location 489 and the read pointer is pointing to location 241. The fill difference between them should be 248 bits. This means there are 248 unread bits in the two-port memory. Let's do the modulo 1024 computation and verify these numbers:

$$
\begin{array}{llllll}
wPtr* & = & 489 & = & 01\ 1110\ 1001 & \\
rPtr* & = & 241 & = & 00\ 1111\ 0001 & \left.\right\} \text{Setup} \\
-rPtr* & = & -241 & = & 11\ 0000\ 1111 & \\
\end{array}
$$

$$
\begin{array}{llllll}
wPtr* & = & 489 & = & 01\ 1110\ 1001 & \\
-rPtr* & = & -241 & = & 11\ 0000\ 1111 & \left.\right\} \text{Computation} \\
wPtr*-rPtr* & = & 248 & = & 1\ 00\ 1111\ 1000 & \\
\end{array}
$$

We ignore the carry bit and treat the result as an unsigned binary number. The fill difference is 248, as we expected, so the modulo 1024 math did the job.

**Example 2.** Let's work another example. This time we will choose a tougher problem. Since we are doing circular math, let's see what happens mathematically when the memory address 0 lies between the write and read pointers. An instantaneous snapshot of the elastic store memory in operation is illustrated in Figure 11.7. In this example, the write pointer points to memory location 127, and the read pointer points to memory location 890. The two pointers now are bracketing memory location 0. By inspection, we can see that the fill difference is equal to 261 bits. So let's do the modulo 1024 math and see if we come up with the same result:

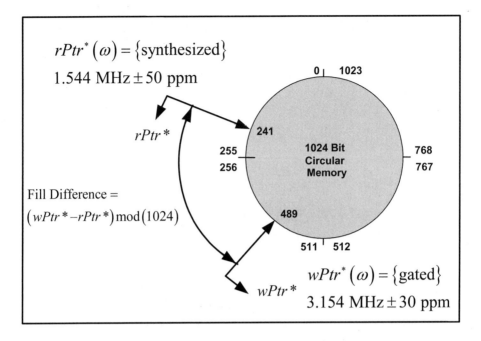

**Figure 11.6**  Pointer computation, example 1

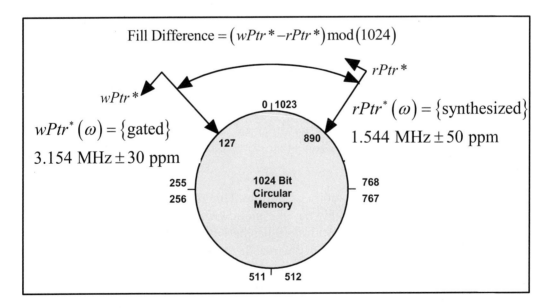

**Figure 11.7**  Pointer computation, example 2

$$
\left.
\begin{array}{llll}
wPtr^* & = 127 & = & 00\ 0111\ 1111 \\
rPtr^* & = 890 & = & 11\ 0111\ 1010 \\
-rPtr^* & = -890 & = & 00\ 1000\ 0110
\end{array}
\right\} \text{Setup}
$$

$$
\left.
\begin{array}{llll}
wPtr^* & = 127 & = & 00\ 0111\ 1111 \\
-rPtr^* & = -890 & = & 00\ 1000\ 0110 \\
wPtr^* - rPtr^* & = 261 & = & \cancel{0}\ 01\ 0000\ 0101
\end{array}
\right\} \text{Computation}
$$

There is no carry bit to ignore, and our pointer math yields the value of 261, so once again we are happy. The reader is invited to work through several examples until convinced that the modulo 1024 math produces correct results for all possible pointer locations.

### 11.1.3   Digital Data Locked Error Signal

The anxious reader might say, "Big deal—we have a fill difference. What can we do with it?" Good question! The answer may surprise you.

The DLL is essentially a feedback control loop. Proper operation of a control circuit requires that an error signal be computed and then fed back to the circuit as a second input. The purpose of the error signal is to iteratively drive the control circuit to some desired solution.

In our DLL design, the solution we seek is a synthesized bit clock that mirrors and tracks the frequency of the original tributary bit clock. Guess what? The elastic store fill difference we just discussed is the error signal that we will use for our DLL. The fill difference is a measure of how many unread bits we have in memory. It is also an instantaneous measure of just how close the write pointer and read pointer are to one another. As the two pointers chase one another around, the circular memory the fill difference will vary. It will undulate as one pointer advances and recedes relative to the other. The fill difference measurements taken over time will provide pointer tracking information that monitors this undulation. This tracking information is our error signal.

For example, assume the fill difference is slowly increasing, indicating that the write pointer is moving around the circular memory faster than the read pointer. This is interpreted by the DLL that the synthesized read clock is too slow and that it needs to increase the read clock frequency. If, for example, the fill difference is slowly decreasing, the DLL interprets this to mean that the read pointer frequency is faster than the write pointer frequency and therefore the synthesized read clock frequency needs to be reduced. Pretty simple, isn't it? Then the next question should be, "How can we convert this error information to a synthesized clock frequency?" The answer is found in the following section.

### 11.1.4  Digital Data Locked Analog Error Voltage

The main thrust of the design of the DLL is to synthesize the output bit stream clock so that on average it is equal in frequency to the gated input clock and thus equal in frequency to the original tributary bit clock. Even if the original bit stream clock was suffering from a serious frequency drift, as long as it remained within the ITU-specified frequency range, our synthesized clock should track it correctly.

Now let's put our error signal to work. For the DLL we are designing, we will apply our error signal to the input of a multiplying D/A converter (MDAC) [4]. This addition to our evolving DLL architecture is shown as the shaded block in Figure 11.8.

A MDAC is a simple device that multiplies a voltage reference $V_{REF}$ by an input digital word to produce an analog output voltage $V_0$. The MDAC is slow by modern day standards because it utilizes a fairly old R-2R ladder D/A architecture [2]. An MDAC was chosen for this architecture simply because it lends itself well to circuit description and helps us succinctly illustrate the derivation of a feedback control voltage.

It should be repeated right up front that the MDAC is inherently slow. For designs that need to operate at blazing speed, the MDAC will definitely be the long pole in the tent. For high-speed DLL applications, the MDAC would be replaced with circuit composed of a high-speed D/A and an analog multiplier built using a couple of operational amplifiers [1].

We can see from the figure that the digital error signal is applied to one input to the MDAC and a fixed voltage reference is applied to the other input. The digital error signal $\left(wPtr^* - rPtr^*\right)$ multiplies the reference voltage $V_{REF}$ to produce an output voltage $V_0$.

To better understand the derivation of the feedback error voltage $V_0$, let's take a look at how an MDAC works. Most MDACs can operate in one of two modes: a unipolar mode or a bipolar mode. A 10-bit MDAC operating in the bipolar mode will generate an analog output voltage according to the input/output relationship illustrated in Table 11.1. This table shows that there are $2^{10}$ or 1024 specific values of output voltage $V_0$. The MDAC output voltage ranges from a low of $-V_{REF}\left(512/512\right)$ volts through 0 volts to a high value of $+V_{REF}\left(511/512\right)$ volts. In this mode, the MDAC is initially tuned by setting the digital input word to its center value of 10 0000 0000 and then adjusting the MDAC bias network until the output voltage is 0 volts.

A 10-bit MDAC, operating in the unipolar mode will generate an analog output voltage according to Table 11.2. From this table, we see that the MDAC output ranges from low value of 0 volts to an almost full scale value of $V_{REF}\left(1023/1024\right)$ volts. For a 10-bit MDAC, we can see that the weight of a least significant bit (LSB) is equal to $V_{REF}\left(1/1024\right)$ volts. In this mode, the

MDAC is initially tuned by alternately setting the digital input word to all 0s and adjusting the MDAC bias network for an output of 0 volts and then setting the digital word to all 1s and adjusting the output voltage until it is equal to a value of $V_{REF}\left(1023/1024\right)$ volts.

For our DLL design we will opt to use the MDAC unipolar mode. We will apply the 10-bit digital fill difference signal directly to the input port of the MDAC, and we should expect to see the same MDAC input/output relationship as indicated in Table 11.2.

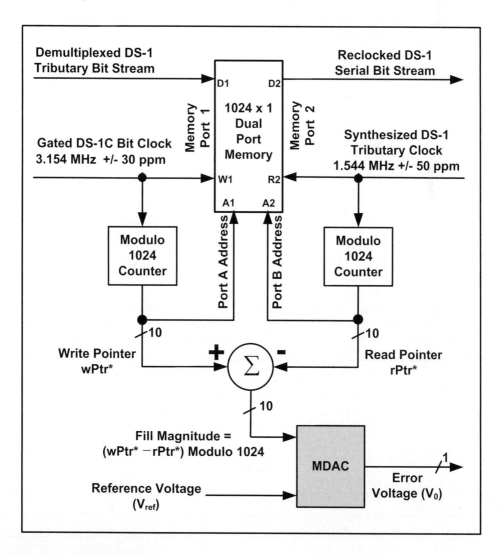

**Figure 11.8** DLL error voltage generation

So far we have derived our DLL digital error signal and we have defined a method for converting this to an analog error voltage. In the next paragraph will use this analog error voltage to control the synthesized frequency of a digital clock oscillator.

**Table 11.1**  MDAC Bipolar Configuration

| Digital input word | Analog output voltage |
| --- | --- |
| 11 1111 1111 | $V_{ref}(511/512)$ |
| 11 1111 1110 | $V_{ref}(510/512)$ |
| • | • |
| • | • |
| 10 0000 0001 | $V_{ref}(1/512)$ |
| 10 0000 0000 | 0 |
| 01 1111 1111 | $-V_{ref}(1/512)$ |
| • | • |
| • | • |
| 00 0000 0001 | $-V_{ref}(511/512)$ |
| 00 0000 0000 | $-V_{ref}(512/512)$ |

**Table 11.2**  MDAC Unipolar Configuration

| Digital input word | Analog output voltage |
| --- | --- |
| 11 1111 1111 | $V_{ref}(1023/1024)$ |
| 11 1111 1110 | $V_{ref}(1022/1024)$ |
| 11 1111 1101 | $V_{ref}(1021/1024)$ |
| • | • |
| • | • |
| 10 0000 0000 | $V_{ref}(512/1024)$ |
| • | • |
| • | • |
| 00 0000 0001 | $V_{ref}(1/1024)$ |
| 00 0000 0000 | 0 |

### 11.1.5 Digital Data Locked Clock Frequency Synthesis

The shaded blocks in Figure 11.9 show the addition of a voltage controlled crystal oscillator (VCXO) [3] and an output register to our DLL block diagram. The VCXO is a simple device that is capable of outputting a square wave with a variable frequency proportional to its input voltage. We will use this square wave as our variable frequency synthesized clock.

As illustrated in Figure 11.9, the MDAC output voltage $V_0$ is applied to the input of the VCXO. The VCXO synthesizes an output square wave or clock, whose frequency is determined by the value of $V_0$. The synthesized clock is utilized to strobe the modulo 1024 read counter that produces the elastic store read pointer $rPtr^*$. It is also utilized as the synchronous read strobe applied to pin R2 of port 2 of the dual port elastic store memory. A data bit stored at the address pointed to by the read pointer will be read from

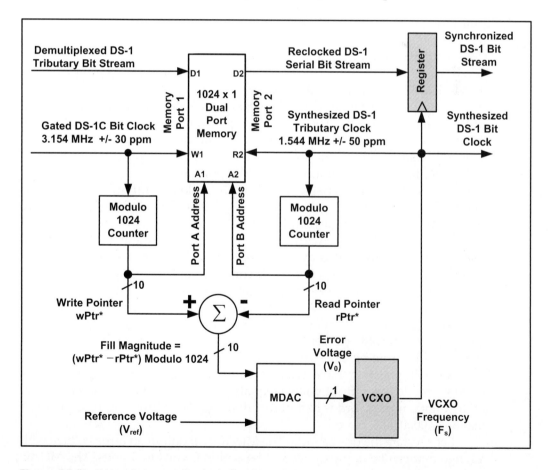

**Figure 11.9** DLL reference clock synthesis

memory on the rising edge of the synthesized clock. The bit will be strobed into the output register on the rising edge of the following clock. The output register is used to resync or time align the data bit read from memory with the output clock.

So we have come full circle, so to speak. The difference between the elastic store write and read pointers is computed to generate a digital error signal. The error signal is converted to an analog voltage that drives a VCXO, which in turn generates the read pointer clock. So far so good, but before we move on, we still need to better understand the operation of the VCXO and how it mates with the MDAC.

The synthesized clock frequency produced by the VCXO is determined by the value of an analog voltage at its input. The analog voltage is the error voltage from the output of the MDAC. When you purchase a VCXO from a manufacturer, you specify the center frequency and what is called the *pull range*. The pull range is generally specified in parts per million, or ppm. The pull range is the amount of frequency deviation a VCXO can exhibit about its center frequency. There are many standard frequencies and pull range options available for off the shelf VCXO devices.

We wish to synthesize a bit clock whose allowable frequency is 1.544 MHz ± 50 ppm. Therefore, for our design, we will choose from a vendor catalog a VCXO with a standard center frequency of 1.544 MHz with a pull range of ±100 ppm. The ±100 ppm pull range of the device will easily bracket the maximum ±50 ppm variance of the clock we wish to synthesize. This will also give us some design "headroom" to minimize the effects of VCXO nonlinearity commonly seen at the pull range limits. We will discuss this nonlinearity characteristic later. Our DLL, along with its VCXO, will be designed to run from a supply voltage of 3.3 volts. We will define the input to the MDAC voltage reference pin $(V_{REF})$ as 3.0 volts. Therefore the MDAC will produce an analog voltage with a swing between a minimum of 0 volts and a maximum of 3 volts, scaled by $1023/1024$, or 2.9971 volts. Keep in mind that these voltage selections are not set in concrete, and the designer is certainly free to select other values based on pertinent design constraints.

We would like the VCXO to generate its center frequency when the analog error signal is in the center of its 3-volt range, or 1.5 volts. This way we have 1.5 volts of control on either side of the center to increase or decrease the synthesized output bit clock frequency.

Referring back to Table 11.2, for an MDAC configured to operate in its unipolar mode, the output voltage equals $V_{REF}/2$ or 1.5 volts when the value of the digital input word is half of its full range. This turns out to be 512 decimal or 10 0000 0000 binary.

Let's restate the design facts. The MDAC will output an error voltage with a range that can be approximated to be within 0 volts to 3 volts. The MDAC will output a center voltage of 1.5 volts when its digital input is 512 decimal.

The VCXO will output a center frequency of 1.544 MHz when its input control voltage is 1.5 volts, which corresponds to an MDAC digital input of 512, and the value of 512 is the fill difference of the elastic store memory. When the input control voltage is at its minimum of 0 volts, the VCXO will theoretically output its minimum frequency of 1.544 MHz – 100 ppm, and when the control voltage is at its maximum 3.0 volts, it will theoretically output is maximum frequency of 1.544 MHz + 100 ppm. The word *theoretically* is used here because the output frequency versus control voltage curve for a VCXO is not perfectly linear, especially at each end of the curve. This is one reason why we selected a catalog VCXO with the rather large ±100 ppm variance—so we could extend the ends of the linear portion of the curve further from our expected operational range of ±50 ppm. Our design will never be asked to track in these extreme regions, so we will not worry too much about operating in the nonlinear areas.

Table 11.3 tabulates data that provides us with a better picture of how the ideal DLL frequency synthesizer would work. The first column of the table lists all possible values of the 10-bit binary error signal from 0 to 1023. This is the binary fill difference signal that is applied to the input of the MDAC. The second column lists the fractional values of the MDAC reference voltage $V_{REF}$ that produce the output error voltage $V_0$. The fractional multiplier is the fill difference or MDAC input divided by 1024. The third column lists the ideal values of the MDAC output error voltage $V_0$, which serves as the VCXO input control voltage. The fourth column lists the linear frequency offset from the VCXO center frequency for each value of the 10-bit digital error signal. And the fifth column lists the same frequency offset (or deviation) in parts per million (ppm).

The shaded rows in the table illustrate the range of frequency offset values that are allowed for our synthesized 1.544 MHz ± 50 ppm clock, which equate to a clock rate of 1.544 MHz ± 77.2 Hz. The synthesizer will only be asked to operate within this shaded area. There are 255 entries in the table above the shaded rows and 256 entries below the shaded rows. These unshaded areas of the table represent the extremes of the VCXO control voltage to synthesized output frequency curve, and it is these portions of the curve that will contain the nonlinear end points. This will provide more than enough headroom to shield us from the nonlinear areas of the VCXO response curve.

The anxious reader is probably asking, "So how do we initially establish the DLL operating point of 512 bits?" This is a good question, and the answer is easy. When the DLL is initially turned on, reading bits from the elastic store is disabled until half of the elastic store memory has been filled. In this case, that would be 512 bits. Once this initialization is complete, the reading of bits from the elastic store will be enabled. After initiation, the fill difference or digital error signal will be equal to 512 decimal or 10 0000 0000 binary, which

**Table 11.3**  VCXO Control Voltage versus Output Frequency Offset

| MDAC digital input word ($wPtr^* - rPtr^*$) | MDAC analog fractional output voltage ($V_0$) | VCXO control voltage (volts) | VCXO output frequency offset (Hz) | VCXO output frequency offset (ppm) |
|---|---|---|---|---|
| 11 1111 1111 | $V_{ref}(1023/1024)$ | 2.9971 | +154.09 | +100 |
| 11 1111 1110 | $V_{ref}(1022/1024)$ | 2.9941 | +153.79 | • |
| 11 1111 1101 | $V_{ref}(1021/1024)$ | 2.9912 | +153.49 | • |
| • | • | • | • | • |
| 11 0000 0000 | $V_{ref}(768/1024)$ | 2.25 | +77.20 | +50 |
| • | • | • | • | • |
| • | • | • | • | • |
| • | • | • | • | • |
| 10 0000 0000 | $V_{ref}(512/1024)$ | 1.50 | 0 | 0 |
| • | • | • | • | • |
| • | • | • | • | • |
| • | • | • | • | • |
| 01 0000 0000 | $V_{ref}(256/1024)$ | 0.75 | −77.20 | −50 |
| • | • | • | • | • |
| 00 0000 0010 | $V_{ref}(2/1024)$ | 0.0059 | −153.79 | • |
| 00 0000 0001 | $V_{ref}(1/1024)$ | 0.0020 | −154.09 | • |
| 00 0000 0000 | 0 | 0.0000 | −154.40 | −100 |

causes the MDAC to output its center voltage of $V_0/2 = 1.5$ volts, which in turn forces the VCXO synthesized bit stream clock to its center frequency value of 1.544 MHz.

At this point, if the frequency of the input bit stream write clock is greater than the frequency of the synthesized read clock, the elastic store write pointer will begin to advance relative to the elastic store read pointer and the fill difference will increase. Since the fill difference is the digital error term, the analog error voltage will increase as well, which will cause the VCXO output synthesized frequency to increase.

When the frequency of the synthesized read clock catches up and is equal to the original bit stream clock, the fill difference will remain constant, causing the error signal to remain constant, which in turn keeps the frequency of the synthesized bit clock locked to the frequency of the input clock. In this steady state condition, the average of the fill difference is constant and the read pointer will be chasing the write pointer around the circular 1024-bit circular memory at some constant average rate.

If, after initialization, the frequency of the input bit stream clock is less than the synthesized bit clock, then the elastic store write pointer position will recede relative to the elastic store read pointer and the fill difference will decrease. The decrease in the digital error signal will cause a similar decrease in the analog error voltage, which in turn causes a decrease in the VCXO synthesized bit clock frequency. Eventually the two clocks will be on average equal in frequency, the error signal will stabilize, and the synthesized frequency will remain locked to the input clock frequency. In this steady state condition, the ideal fill difference is constant and the read pointer continues to chase the write pointer around the elastic store circular memory.

### 11.1.6  Expected Digital Data Locked Operational Characteristics

In the real world nothing is perfect and the previously given scenario doesn't really happen—that is, not exactly. Here's why. The DLL is continuously iterating in an attempt to stabilize the magnitude of the error signal. The computations the DLL makes are not infinitely precise. Because the synthesized clock is quantized to 1024 possible discrete frequencies, it has finite precision and may never be exactly equal to the input bit clock frequency. It is easy to see that there will almost always be some small error between the true solution and the loop estimate of the solution. Typically the true error will be within ±1 LSB of the computed digital error signal. So we are already able to see one reason for making the two-port memory fairly large. That is, the larger the memory, the more precise the digital error signal will be, and the more resolution we get from the VCXO synthesized frequency.

Corrections will be continuously made by the loop to adjust the synthesized clock so that on average the frequencies of the two clocks are identical. That is, the actual frequency of the synthesized clock will slowly vary about the ideal frequency by a small amount. The magnitude of this variance will be the frequency equivalent of ±1 error signal LSB. Over time the differences between the two clocks will cause the number of bits in the elastic store memory to grow or recede as the corrected synthesized bit clock frequency swings between some fraction of a hertz below the input bit clock frequency and some fraction of a hertz above the input bit clock frequency. The undulating frequency difference, although small, is a prime source of synthesized clock jitter. So we see a second reason for making the two-port memory fairly large. An increased precision in synthesized frequency results in a smaller magnitude frequency error, which in turn manifests itself in a smaller frequency jitter.

The magnitude of synthesized clock jitter introduced by the tracking loop is limited by the word width of the digital error signal and hence the size of the elastic store memory. If, for example, the size of the elastic store memory were increased fourfold from 1024 bits to 4096 bits, the width of the digital error signal would increase from 10 to 12 bits. The voltage resolution of a 12-bit MDAC would increase fourfold. The resolution of an MDAC LSB would now be 1/4096, as opposed to the 1/1024, for the 10-bit error signal. The frequency resolution of the VCXO would therefore increase by a factor of four and the jitter would decrease.

This is a critical point, so let's restate what we just said. The loop is continuously making corrections in a vain attempt to reach a perfect solution and find an identical match between two frequencies. But the loop is limited by the number of discrete frequencies it can synthesize. The frequency solutions it finds will more than likely be either a fraction of a hertz greater or a fraction of a hertz less than the optimal solution. In a steady state condition, the fill difference in the elastic store memory slowly increases and decreases as the loop constantly iterates back and forth between the two closest solutions it can compute. A finer frequency resolution would reduce error between the optimum solution and the loop estimate of the solution. The residual error would take longer to accumulate, causing the growth or reduction of accumulated elastic store bits to occur over a longer period of time, which would result in a longer clock jitter period. So as the size of elastic store memory increases, so does the resolution of the synthesized frequency, which results in a decrease in clock jitter in both frequency and magnitude.

In the engineering world, however, nothing is without trade-off. The bad news is that increased resolution, which buys us a decrease in clock jitter, also causes the loop to become more lethargic, in that it cannot respond to instantaneous changes in input frequency as rapidly. In other words, the loop adapts to instantaneous input frequency changes more slowly as the precision of the loop error signal increases. The loop time constant is determined in part by the width in bits of the digital error signal. The DLL designer's determination of the loop time constant is a trade-off between the speed of adaption and steady state clock jitter. In some circles of engineering, the term *adaption* is referred to as the loop *attack rate*.

For example, a smaller digital error signal word width will result in a larger loop time constant. A larger time constant will allow the loop to adapt more quickly, but because the smaller width error signal has less frequency resolution, it will produce a higher amount of synthesized clock jitter on output. On the other hand, a digital error signal with a larger word width will result in a smaller loop time constant. The smaller time constant will increase the loop frequency resolution and reduce the amount of synthesized clock jitter, but it will also cause the loop to adapt more slowly.

As an example, if for some reason the original bit stream clock oscillator experienced a jump in frequency, the input bit stream would reflect that and the synthesized clock would track that change as well and produce (as closely as possible) a bit clock of equal frequency on output. If the original tributary bit clock instantaneously jumped from one extreme to another, such as from $F_0 - 50$ ppm to $F_0 + 50$ ppm, the frequency of the input bit stream would have increased as well. The elastic store will begin to fill at a faster rate, increasing the fill difference, which will cause an increase in the synthesized read clock frequency. Since the loop has a time constant, the laws of physics emphatically state that it cannot react instantaneously to sudden perturbations in the input bit clock. If one were to plot the change in the synthesized frequency over time in response to a frequency step in the input bit clock, it would take on the shape of a typical exponential, like the exaggerated plot shown in Figure 11.10. In this figure, the magnitude of the input bit clock frequency is represented by the solid line, and the magnitude of the synthesized bit clock frequency is represented by the dashed line. We see that at time $T_1$ the input frequency is steady at some value $F_1$ Hz. The synthesized bit clock frequency is shown in an exaggerated fashion as oscillating about the desired frequency of $F_1$ Hz with a magnitude equal to the frequency equivalent of ±1 LSB of error signal. At time $T_2$ the input frequency instantaneously steps up in value from $F_1$ to $F_2$ Hz. When this step occurs, we see that the loop response is to exponentially increase its synthesized frequency until it is within ±1 error signal LSB of the true frequency solution. The synthesized clock frequency will slowly oscillate about the optimum solution as the accumulated digital error alternately grows and decays by an LSB.

Figure 11.11 illustrates the exaggerated case where the word width of the digital error signal is reduced by one bit from that of Figure 11.10. Comparing

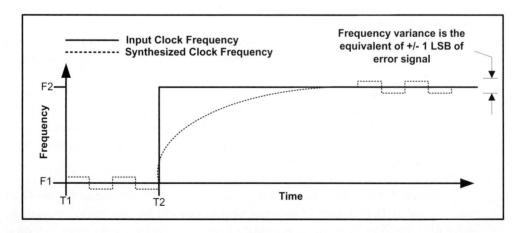

**Figure 11.10**   Example DLL frequency response: Long time constant

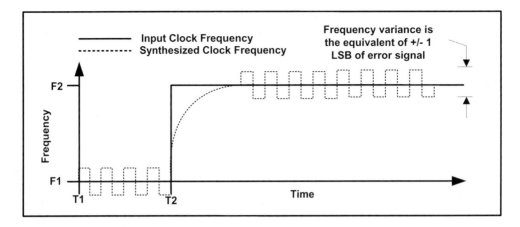

**Figure 11.11**   Example DLL frequency response: Short time constant

the two figures, we can immediately recognize three dramatic changes in loop response:

- The reduction in error signal bit width by 1 bit reduces the loop frequency resolution by half. Therefore the size of the discrete frequency steps in the synthesized output frequency is doubled. The loop time constant is doubled and, as shown in the figure, the loop responds to a step in input frequency significantly faster.
- The reduction in loop frequency resolution increases the difference between adjacent discrete VCXO synthesized output frequencies. Therefore the loop estimate of the frequency solution is not as precise. The steady state condition shows an increase in the magnitude of the synthesized clock frequency error.
- The digital error accumulates and then decays as it continually over corrects in its search for the perfect solution. The loop will continually jump between the two adjacent discrete frequency values that enclose the true solution. Since the difference between these values is greater, the loop will overcorrect by larger frequency estimates. The digital error will accumulate and decay at a faster rate. The higher frequency of loop correction will be reflected in an increased magnitude of synthesized clock frequency jitter.

The reader should remember that the clock jitter illustrated in Figure 11.10 and Figure 11.11 is drawn in an exaggerated fashion for illustration purposes. In reality the jitter magnitude is confined to ±1 LSB of the error signal, which can be very small. For instance, in our previous example, we happened to choose a VCXO with a frequency range of ±100 ppm and an elastic store memory 1024 bits in length. The frequency deviation of a single LSB of error signal is calculated to be 200 ppm/1024 = 0.195 ppm. If our VCXO center frequency is 1.544 MHz, then the frequency deviation due to one LSB of error signal would be $1.544\,\text{MHz}\left(0.195/10^{-6}\right) = 0.3011$ Hz. This is a small number. Because the tracking loop utilizes an elastic store memory, it has a bit storage reservoir that will fill or drain as needed to absorb input bit rate changes during the period of time the loop needs to exponentially acquire and track the input clock frequency.

Later we will look at the bit-level simulation of the DLL we just designed. We will use this simulation to compare the performance of our 10-bit loop to that of identical DLLs with error signals of different bit widths.

## 11.2  DIGITAL DATA LOCKED STEADY STATE BEHAVIOR

Now that we have discussed the functional operation of the DLL, let's take a more in-depth look by examining the equations that actually define the loop operation. We shall begin by referring to the DLL block diagram of Figure 11.9. In this figure, we can see that the two-port memory fill difference is computed by subtracting the value of the read pointer $(rPtr\,{}^{*})$ from the value of the write pointer $(wPtr\,{}^{*})$. The fill difference is the digital error signal that is fed to the MDAC to generate an analog error voltage $V_O$. For the purpose of mathematically describing the loop behavior, we will refer to the fill difference as the digital error signal $(\varepsilon_D)$, and we will represent this error signal by the equation

$$\varepsilon_D = \left[wPtr\,{}^{*} - rPtr\,{}^{*}\right]\text{mod}\left(2^B\right)$$

**Equation 11.1: General loop digital error signal**

where $B$ is the bit width of the digital error signal and $\text{mod}\left(2^B\right)$ allows us to compute the pointer difference in a circular fashion. In our design $B = 10$ bits, so $\text{mod}\left(2^B\right) = \text{mod}\left(2^{10}\right) = \text{mod}\left(1024\right)$ and $\varepsilon_D$ can be computed via the equation

$$\varepsilon_D = \left[wPtr\,{}^{*} - rPtr\,{}^{*}\right]\text{mod}\left(1024\right)$$

**Equation 11.2: Design specific loop digital error signal**

The error signal $\varepsilon_D$ is applied to the input of the MDAC to produce an output voltage $V_O$. The linear equation that describes this process is

$$V_O = \left(\frac{V_{REF}}{2^B}\right)\varepsilon_D$$

### Equation 11.3: MDAC voltage to error relationship

where $V_{REF}$ is the reference voltage applied to the input of the MDAC.

The VCXO generates a synthesized frequency $f_S$ proportional to the error voltage $V_O$ applied to its input. We can derive an ideal linear expression that represents this process by remembering the algebraic equation $y = mx + b$ that represents a straight line. In this equation $m$ represents the slope of the line and the quantity $b$ represents the y-axis intercept of the line. In this case, our independent variable is voltage, and the dependent variable is frequency. So the equation we will derive will take on the form of $f_S = mV_O + f$.

The VCXO has a center frequency and a maximum frequency variance about that center frequency. The highest frequency the VCXO can generate is termed $VCXO_{Max}$ and the lowest frequency the VCXO can generate is termed $VCXO_{Min}$. By inspecting the graph in Figure 11.12, we can see that for an ideal VCXO, the slope of the curve is given by

$$m = \left[\frac{F_{VCXOMax} - F_{VCXOMin}}{V_{REF}}\right]$$

The value for the frequency axis intercept is seen as $f = F_{VCXOMin}$. So the output frequency versus input voltage response for an ideal VCXO can be expressed as

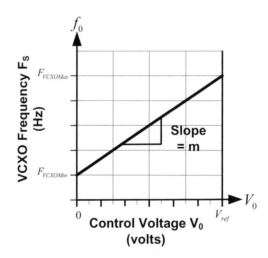

**Figure 11.12**   Linear VCXO output frequency versus voltage

$$f_S = \left[ \frac{F_{VCXOMax} - F_{VCXOMin}}{V_{REF}} \right] V_O + F_{VCXOMin} \quad \text{Hz}$$

**Equation 11.4: Ideal VCXO voltage versus frequency response**

If we substitute Equation 11.3 for $V_O$ in Equation 11.4, we will produce an equation that describes the VCXO output frequency response as a function of the digital error signal $\varepsilon_D$ :

$$f_S = \left[ \frac{F_{VCXOMax} - F_{VCXOMin}}{2^B} \right] \varepsilon_D + F_{VCXOMin} \quad \text{Hz}$$

**Equation 11.5: Ideal VCXO frequency versus digital error response**

Equation 11.5 provides us the means to compute the steady state VCXO output frequency as a function of the steady state digital error signal. However, since it does not contain an independent time variable $t$, it cannot be utilized to describe the loop transient response over time due to a change in the input tributary bit clock frequency. We will address this issue later.

Now let's take a closer look at the response curve for our ideal VCXO. This ideal response curve is defined by Equation 11.4 and is illustrated in Figure 11.13. The figure illustrates the output frequency versus input voltage response curve for the VCXO in our design example. Remember we chose a vendor-supplied VCXO with a specified center frequency of 1.544000 MHz and a pull range of ±100 ppm with an input control voltage that ranged from 0 to 3 volts.

The vertical axis on the left of the graph represents the offset from the center frequency in parts per million (ppm). This is also termed the VCXO *pull range*, and for the VCXO we have selected, this pull range is ±100 ppm. The vertical axis on the right of the plot represents the synthesized frequency range that we will ask the VCXO to operate over. This range is 1.544 MHz ± 50 ppm. This bounds the operating region of the VCXO to be somewhere within the bold black line portion of the curve. The horizontal axis represents the VCXO control voltage $V_O$, which for our design ranges from 0 volts to 3 volts.

The VCXO center frequency is specified to be 1.544 MHz when the input error voltage $V_O$ is half its maximum value or 1.5 volts. As discussed previously, we have designed the loop such that when the digital error signal $\varepsilon_D$ is equal to 512, the 10-bit MDAC error voltage $V_O$ will be equal to 1.5 volts and the output synthesized frequency will be 1.544 MHz. This set of conditions represents the zero error state of the loop. At the zero error state, the

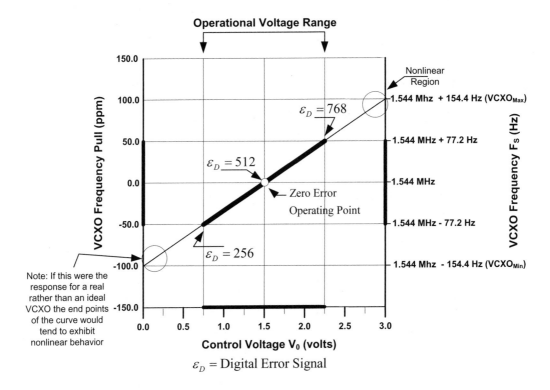

**Figure 11.13**   Ideal VCXO frequency versus voltage curve

VCXO operating point will be located at the midpoint of the curve, as shown in Figure 11.13.

Any deviation in the frequency of the input bit clock will cause the VCXO operating point to slide up or down this curve between its ±50 ppm boundaries. As we can see from the left side of the graph, the end points of the operating curve correspond to the input data clock limit frequencies of 1.544 MHz ± 50 ppm.

If we substitute values from Figure 11.13 into Equation 11.5, we arrive at

$$f_s = \left[ \frac{1544154.4 - 1543845.6}{1024} \right] \varepsilon_D + 1543845.6 \text{ Hz}$$

$$= \left[ \frac{308.8}{1024} \right] \varepsilon_D + 1543845.6 \text{ Hz}$$

$$= [0.3015625] \varepsilon_D + 1543845.6 \text{ Hz}$$

**Equation 11.6: Design example VCXO frequency versus error signal relationship**

Equation 11.6 illustrates very clearly the discrete frequencies that this 10-bit DLL can synthesize. The step size between adjacent realizable synthesized frequencies is 0.3015625 Hz across a range of 308.8 Hz. Our application will only utilize the center 154.4 Hz of this range, or 72.2 Hz on either side of the zero error operating point. This gives us plenty of guard band between our operating range and the nonlinear response of the VCXO.

It is also easy to see from this equation the role that the address size of the dual port memory plays in the precision of the synthesized frequencies. In our design example, the address size is 10 bits. Now, for example, if the size were increased to 12 bits, the synthesized frequency resolution would increase fourfold, from 0.3015625 Hz to 0.075390625 Hz.

We can rearrange Equation 11.5 to solve for $\varepsilon_D$ and determine the value of the digital error signal at the extreme operating points along the curve:

$$\varepsilon_D = \left(f_S - F_{VCXOMin}\right) \left[ \frac{2^B}{F_{VCXOMax} - F_{VCXOMin}} \right]$$

For example, when $f_O$ is equal to the minimal output frequency of 1.544 MHz − 50 ppm or 1,543,922.8 Hz, the value for $\varepsilon_D$ will be

$$\varepsilon_D = \left(1,543,922.8 - 1,543,845.6\right) \left[ \frac{1024}{308.8} \right] = 256$$

This is verified by the operating curve in Figure 11.13. The curve shows that the error signal will equal 256 when the synthesized clock is locked to the input bit clock and the frequency of the input clock is equal to its minimum specified value of 1.544 MHz − 50 ppm. Similarly, the error signal will equal 768 when the synthesized clock is locked to the input bit clock and the frequency of the input clock is equal to its maximum specified value of 1.544 Hz + 50 ppm. The expected operational error voltage range will be between 0.75 volts and 2.25 volts.

We are very fortunate in this particular design example in that the extreme digital error values are exact integer values. This means that the loop will be able to synthesize these extreme frequencies exactly, and therefore, at these limits, the loop should track the input frequency without error. This same behavior will occur at 512 discrete points along the operational portion of the VCXO response curve. For any points in between, the loop will not be able to synthesize an exact frequency, and it will wander between the two closest adjacent discrete frequencies.

This is a very interesting and informative graph. A plot similar to this is given in most if not all data sheets for VCXOs. The trick is to design the loop circuitry so the zero error voltage falls in the center of the curve. That way you can achieve a bipolar frequency response, given by $f_s \pm xxx$ ppm.

The VCXO described here is termed an ideal VCXO because the response curve is shown as being perfectly linear. Real VCXOs are not this well behaved. The response curve for real VCXOs are close but not perfectly linear over their specified input voltage range. The response curves tend to flatten out and become nonlinear at the end points of the specified operational range.

We chose a VCXO with an operational range twice as large as what we needed, simply to avoid these nonlinear segments at each end of the response curve. As a matter of fact, with this design, we really don't give a hoot about the VCXO response curve beyond the $\pm 50$ ppm range since we will never ask the VCXO to operate in these regions. By choosing a VCXO with the pull range of $\pm 100$ ppm, we designed in a significant 100% performance guard band into our DLL. This means we are going to experience far fewer "circuit funnies" during program testing and when verifying operational performance specifications across temperatures. In summary, it should be clear from the VCXO response curve in Figure 11.13 that our design operating range avoids the regions of the curve where these types of devices typically exhibit nonlinear behavior. This graphically illustrates the reasoning behind selecting a VCXO that supports twice the frequency deviation that our DLL design required.

To be thorough, we need to restate that the linear equation that describes the VCXO input voltage versus the output frequency response curve is a close approximation to but not the exact duplicate of the real response curve. The designer should expect to experience some small degree of synthesized frequency error due to the small error between the real versus the ideal response curves. The device data sheets should include a measure of the maximum nonlinearity figure over the specified operating temperature range. This information should allow the designer to put a bound on expected error.

## 11.3   DIGITAL DATA LOCKED TRANSIENT BEHAVIOR

In the previous section, we derived the equations that described the steady state behavior of the DLL. These equations gave us a method of determining what the synthesized output frequency of the DLL is as a function of either the digital error signal $\varepsilon_D$ or the VCXO input error voltage $V_O$. The phrase *steady state behavior* is meant to describe the condition where the DLL synthesized output bit clock frequency is locked to the input bit clock frequency and there is little if any frequency difference between the two.

The ever-questioning reader may ask, "How can we mathematically describe the DLL synthesized frequency response over time due to a change in the input bit clock frequency?" In other words, what is the mathematical expression that defines the DLL transient response to perturbations in the input bit clock frequency? We'll tackle the answer to that question next. In this section, we will derive an equation that describes the DLL transient response, and we will derive the time constant of the DLL, which is defined, as one might expect, in terms the designers' choice of specific circuit parameters. For purposes of this derivation, and to make the equation that describes the DLL transient response clearer, all constants will be represented by uppercase letters and all variables will be represented by lowercase letters.

In our studies of DLL transient behavior, we can make a logical assumption that the input bit clock frequency is a constant, for all time, with the exception of time $t = 0$, where it changes from one constant value to another constant value. That is, the input bit clock experiences a frequency step at $t = 0$ that initiates a DLL response in its synthesized frequency. The DLL synthesized transient frequency response is considered a function of time.

We will denote the input bit clock frequency as $F_I$, to indicate it will be treated as a constant at all times other than $t = 0$, and we will denote the synthesized bit clock frequency as $f_S$ or $f_S(t)$ to indicate that it is a variable and a function of time.

To paint a bigger picture, we will make the assumption that the input clock frequency can instantaneously change at any time from one value to another, and our derivation will treat that time of occurrence as $t = 0$. The synthesized clock frequency, however, will only change when the DLL is responding to a frequency change in the input bit clock. So we can view the DLL as a digital system with the frequency of the input bit clock as the system stimulus and the frequency of the synthesized bit clock as the system response.

The synthesized DLL clock frequency can only change when the accumulated digital error signal $\varepsilon_D = [wPtr * -rPtr *] \mod (1024)$ changes by a value of $\pm 1$ LSB or greater. After all, it's the digital error signal that generates the error voltage that in turn controls the VCXO output frequency. At this point, we need to ask ourselves, "How fast does the frequency of the synthesized DLL bit clock change in response to a change in the frequency of the input bit clock?"

To answer that question, we need to determine how much time it takes for the digital error signal to accumulate $\pm 1$ LSB of error. This is important because the growth or decay of the error signal depends on the difference in frequency between the input bit clock $F_I$ and the synthesized bit clock $f_S$. Remember that the two-port memory write pointer $wPtr *$ and the read pointer $rPtr *$ are chasing each other around the circular memory and the difference between the two is the digital error signal. A change in the error signal by $\pm 1$ LSB means that the fill difference has changed by one bit and the two pointers are chasing one another at slightly different rates. The accumulation or decay of the digital error signal occurs at a rate defined by the difference in the two clock frequencies, expressed as $F_I - f_S$ bit/sec. For example, if $F_I > f_S$ by 2 Hz, then the error accumulation would grow at a rate of 2 bits per second. If on the other hand, if $F_I < f_S$ by 2 Hz, then the error accumulation would decay at a rate of 2 bits per second.

It is easily seen that the rate of change of the error signal $\varepsilon_D$, and therefore the rate of change of the synthesized output bit clock frequency $f_S$, is dependent on the value of the input clock frequency and the value of the current synthesized frequency, or $(F_1 - f_S)$. More importantly, it is critical to note that the rate of change in the synthesized clock frequency is proportional to the current value of the synthesized clock frequency. This can be described by the relation

$$\frac{df_S}{dt} = k\left(F_I - f_S\right)$$

**Equation 11.7**

where $df_S/dt$ is the derivative of the function $f_S(t)$ with respect to time and can be read as the change in the value of $f_S(t)$ during a small change in time $t$, and where $k$ is some unknown proportionality constant. $F_I$ is the value of the input bit clock frequency and is considered a constant value both before and after time $t = 0$. This consideration is only necessary so we can easily derive an equation that predicts the response of the DLL to a *single* change in the frequency of the input bit clock. We can dismiss this restriction later. Moving along, Equation 11.7 tells us that the rate of change of the DLL synthesized frequency $f_S$ with respect to time is dependent on the current value of $f_S$. Now let's rearrange the terms to get

$$\frac{df_S}{dt} = -k\left(f_S - F_I\right)$$

**Equation 11.8**

If we divide both sides of the equation by the term $\left(f_S - F_I\right)$, and if we multiply both sides of the equation by $dt$, then we can get this equation into a standard form:

$$\frac{df_S}{\left(f_S - F_I\right)} = -kdt$$

**Equation 11.9**

Now let's integrate both sides of the equation to eliminate the differential:

$$\int \frac{df_s}{\left(f_s - F_I\right)} = -k\int dt$$

**Equation 11.10**

The term *standard form*, mentioned previously, refers to a very interesting and commonly used integral calculus identity that states if $f(t)$ is a differentiable function and if $f'(t)$ is its derivative, then

$$\int \frac{f'(t)}{f(t)}dt \;=\; \ln|f(t)|+C \;=\; \left\{ \begin{array}{ll} \ln\left(f(t)\right)+C & \text{if } f(t)>0 \\ \ln\left(-f(t)\right)+C & \text{if } f(t)<0 \end{array} \right\}$$

**Equation 11.11**

where $C$ is the constant of integration. Since $F_I$ is a constant, integrating both sides of Equation 11.10 gives us

$$\int \frac{df_s}{\left(f_s - F_I\right)} = -k\int dt$$

$$\ln\left|f_s - F_I\right| = -kt + C$$

If we keep in mind the sign bookkeeping of Equation 11.11, we can write

$$\ln\left(f_s - F_I\right) = -kt + C$$

We can simplify this equation further through the use of the relationship $e^{\ln(x)} = x$, resulting in

$$\left(f_s - F_I\right) = e^{-kt+C} = e^C e^{-kt}$$

The term $e^C$ is the constant of integration, which we can represent by the constant $A$, rearranging terms to get

$$f_s = F_I + Ae^{-kt}$$

**Equation 11.12**

Remember that we said $f_s$ and $f_s(t)$ represent the same variable. Therefore we also can say that

$$f_s(t) = F_I + Ae^{-kt}$$

**Equation 11.13**

We are almost there. We still need to determine the value for the constants $A$ and $k$. We can determine the value of $A$ simply by evaluating Equation 11.13 at time $t = 0$. We make the assumption that the DLL response is not instantaneous and that at $t = 0$ the DLL is in a steady state condition. That is, all transients have settled and the frequency synthesized by the DLL is constant, which enables us to write $f_s(0) = F_S(0)$. Also, at time $t = 0$, we can say that the input frequency $F_I$ steps up or down by some amount and is constant from that point forward. Furthermore, at time $t = 0$ the term $e^{-k0} = 1$. Inserting these values into Equation 11.13 gives us

$$f_s(t) = F_I + Ae^{-kt} \rightarrow F_S(0) = F_I + A$$

Now we can solve for $A$ to produce

$$A = F_S(0) - F_I$$

Substituting this value for $A$ back into Equation 11.13 leaves us with

$$f_s(t) = F_I + \left(F_S(0) - F_I\right)e^{-kt}$$

**Equation 11.14: DLL transient response**

Equation 11.14 describes the synthesized output frequency transient response of the DLL. With some extra manipulation, we can write this equation in an equivalent form given by

$$f_s(t) = F_S(0) + \left(F_I - F_S(0)\right)\left(1 - e^{-kt}\right)$$

**Equation 11.15: DLL transient response equivalent form**

Since we are dealing with a digital system, we need to convert the continuous time variable $t$ to the discrete time variable $nT$, where $n = 0, 1, 2, 3, 4, \ldots$ is the digital sample index and $T$ is the period of the input bit clock frequency.

The two versions of the discrete time transient response equation for the DLL are given by

$$f_S(nT) = F_I + (F_S(0) - F_I)e^{-knT}$$

and

$$f_S(nT) = F_S(0) + (F_I - F_S(0))(1 - e^{-knT})$$

**Equation 11.16: DLL discrete time transient response**

Now all we need to do to completely specify this equation is to compute the value of $k$. The value of $k$ is the DLL synthesis circuit time constant and is directly related to and determined by a set of fixed circuit parameters. We can determine the value of $k$ using one of two different methods. These two methods are discussed in the sections that follow.

## 11.3.1   Method 1: Determining the Value of *k*

The first method is to empirically determine the value of $k$ by measuring $F_I$, $F_S(0)$, and $f_S(nT)$, generated by an operational DLL at a few specific values of $nT$. We also have a second option, and that is to calculate the values from a bit-level simulation of the DLL. Either way, once these values are obtained we can insert them in to Equation 11.14 and calculate the value of $k$. We can rearrange the terms in Equation 11.14 so that

$$e^{-knT} = \left[ \frac{f_S(nT) - F_I}{F_S(0) - F_I} \right]$$

$$k = -\frac{1}{nT} \ln\left[ \frac{f_S(nT) - F_I}{F_S(0) - F_I} \right]$$

**Equation 11.17 Solution for *k***

We know either by measurement or calculation the values of $F_I$, $F_S(0)$, and $f_S(nT)$ for specific values of $nT$. Therefore we can substitute these values into Equation 11.17 and then solve for the constant $k$. The only way to get these values is to either take measurements on a DLL that has already been built or to use values produced by a bit-level software simulation of the DLL. The empirical or experimental solution to solving for the value of $k$ using a software simulation is extremely beneficial during the DLL design phase.

Solving for $k$ allows for the testing and verification of the overall DLL design prior to committing it to hardware. This is another great reason why all complex digital circuits should be bit-level simulated in software prior to their implementation.

Think of all the headaches and busted budgets that can be eliminated by investing the time to simulate and verify the proper operation of a circuit early in the design phase of a project. The alternative is to test after the circuit is built, and this method usually ends up costing big dollars for the iterative redesign, iterative refabrication of circuit cards, iterative retest, and of course, all the associated support-group touch labor that will be necessary. A good rule of thumb is that a single redline on an engineering drawing has the potential to keep 20 support people employed for a month, and that's expensive. Once the circuit is built, it is a good idea to perform these measurements to verify that the measured circuit value of $k$ is correct and that it agrees the value of $k$ obtained in method 2, discussed in the next section.

### 11.3.2  Method 2: Determining the Value of k

There is a second method for determining the value of $k$, which can be done during the design phase. Since this method gives the designer precious information early on in the design, this is the recommended approach. Remember that $k$ is the circuit transient response time constant and therefore must be a function of a few critical circuit parameters. It turns out that we have computed the value of $k$ before when we derived the steady state VCXO output in Equation 11.5. It was kept a secret back then, but it can be brought out of the closet now!

The VCXO steady state synthesized frequency is given in Equation 11.5 and is repeated here for clarity:

$$ f_S = \left[ \frac{F_{VCXOMax} - F_{VCXOMin}}{2^B} \right] \varepsilon_D + F_{VCXOMin} \ \text{Hz} $$

This is a linear equation that defines the frequency response of the VCXO that is illustrated in Figure 11.13. Even though we defined this equation to be the steady state response of the DLL, if we know the value of the error signal $\varepsilon_D$ at any instant of time, then we can compute the corresponding value of the DLL synthesized frequency. This is true even during the time of the DLL transient response. In other words, if we knew the history of the error signal for all time, we could use Equation 11.5 to compute the complete transient response of the synthesized DLL frequency $f_S$.

We can also say that a change in $f_S$ with respect to time $t$, due to a change in the error signal $\varepsilon_D(t)$, can be represented by $df_S/dt$ or

$$\frac{df_s}{dt} = \left[\frac{F_{VCXOMax} - F_{VCXOMin}}{2^B}\right] \varepsilon_D(t) \ \frac{\text{Hz}}{\text{second}}$$

**Equation 11.18**

The $F_{VCXOMin}$ term is a constant, and therefore its derivative with respect to time is zero. Equation 11.18 can be interpreted as follows: as the error function changes with time, the synthesized frequency changes with time as well. As we have seen, the error term at any given instant of time is given by $\varepsilon_D = (wPtr^* - rPtr^*)$ bits. The history of the error term track over all time can be written as

$$\varepsilon_D(t) = (wPtr^*(t) - rPtr^*(t)) \ \frac{\text{bits}}{\text{second}}$$

The track of the write pointer and the read pointer is nothing more than the frequency of the write and read clock. Therefore we can write the equivalent statement

$$\varepsilon_D(t) = (F_I - f_s(t)) \ \frac{\text{bits}}{\text{second}} \ = \ (F_I - f_s) \ \frac{\text{bits}}{\text{second}}$$

If we substitute this expression for $\varepsilon_D(t)$ back into Equation 11.18, we end up with

$$\frac{df_s}{dt} = \left\{\left[\frac{F_{VCXOMax} - F_{VCXOMin}}{2^B}\right]\frac{\text{Hz}}{\text{bit}}\right\}\left\{(F_I - f_s) \ \frac{\text{bits}}{\text{second}}\right\} = \left[\frac{F_{VCXOMax} - F_{VCXOMin}}{2^B}\right](F_I - f_s) \ \frac{\text{Hz}}{\text{second}}$$

**Equation 11.19**

We also know from Equation 11.8 that the change in $f_s$ with respect to time is given by

$$\frac{df_s}{dt} = -k(f_s - F_I)\frac{\text{bits}}{\text{second}}$$

If we equate the two expressions for $df_s/dt$, we get

$$\left[\frac{F_{VCXOMax} - F_{VCXOMin}}{2^B}\right](F_I - f_s)\left(\frac{\text{Hz}}{\text{second}}\right) = -k(f_s - F_I)\left(\frac{\text{bits}}{\text{second}}\right)$$

If we eliminate and rearrange terms, we determine the value for the DLL time constant:

$$k = \left[ \frac{F_{VCXOMax} - F_{VCXOMin}}{2^B} \right] \frac{\text{Hz}}{\text{bit}}$$

**Equation 11.20**

If we compare Equation 11.20 to Equation 11.5, we see that the time constant $k$ is just the slope of the linear operational curve of the VCXO.

This is a pretty neat result in that it clearly shows that the value of the DLL time constant is inversely proportional to the bit width of the digital error signal $(2^B)$, which is the depth of the two-port memory that serves as the DLL elastic store. Here we see that as the width of the digital error signal $(2^B)$ gets larger, the value of the DLL time constant $(k)$ gets smaller, the DLL transient response becomes longer, the synthesized frequency resolution increases, and the DLL becomes more sluggish in its response to changes in the input data stream bit clock. If the digital error signal $(2^B)$ becomes smaller, then just the opposite is true. In this case, the time constant gets bigger, the transient response gets shorter, the frequency resolution decreases, and the DLL response to a change in input frequency gets faster.

It always gives me a good feeling when the units of an equation make sense. Here the rate of change of the synthesized frequency $f_s$ with respect to time has the units of Hz/second, which makes sense to me. As a matter of fact, it makes a great deal more sense than if we had ended up with units like Hz/radian or something even more ridiculous like Hz/°C. My very first engineering problems professor constantly drilled his harried young students to always check the units of an equation. It is always a good indicator of whether or not the equation you just derived is correct or not.

Getting back to our DLL design, the synthesized frequency resolution of our 10-bit machine was determined to be

$$k = \left[ \frac{F_{VCXOMax} - F_{VCXOMin}}{2^B} \right] = \frac{308.8 \text{ Hz}}{1024 \text{ bits}} = 0.3016 \frac{\text{Hz}}{\text{bits}}$$

This says that a change in the digital error signal of ±1 LSB causes the DLL synthesized output frequency to change by ±0.3016 Hz. We can now complete Equation 11.16, which describes the transient behavior of our particular 10-bit DLL design by replacing the constant $k$ with the value 0.3016. We end up with

$$f_s(nT) = F_I + (F_s(0) - F_I)e^{-0.3016nT}$$

and

$$f_S(nT) = F_S(0) + (F_I - F_S(0))(1 - e^{-0.3016nT})$$

If we had opted in our design to use a two-port memory that was twice as large so the value of $2^B$ was 2048 as opposed to 1024, then the DLL frequency resolution would be twice as precise since $k$ would now be

$$k = \frac{308.8 \text{ Hz}}{2048 \text{ bits}} = 0.1508 \frac{\text{Hz}}{\text{bits}}$$

The value of $k$ is not only the frequency resolution of the DLL, it is also the time constant of the DLL transient response. So designers will need to make some sort of trade-off study to determine the optimum value of $k$ for the application in which the DLL will be used. The DLL frequency resolution and time constant $k$ are tabulated in Table 11.4 for several values of the digital error signal bit width $B$. A comparison is made between the calculated value of $k$ and the measured value of $k$ obtained from the bit-level simulator discussed in the next section.

Remember that $2^B$ is also the depth of the two-port memory used for the elastic store. It is clear from the table that for low values of $B$, the frequency resolution of the DLL in Hz/bit is fairly coarse, and for larger values of $B$, the frequency resolution becomes quite good. Note the second column from the right that shows the value of computed $k$ minus measured $k$. The two values are quite close. The percent error is for each case is tabulated in the last column.

The larger values of $B$ still equate to reasonable memory sizes. As an example, for $B = 12$, the two-port memory size is only 4096 by 1 bits. A memory

**Table 11.4** DLL Time Constant versus Error Signal Bit Width

| Digital error bit width (B) | Two port elastic store memory depth (bits) | Computed DLL time constant k (Hz/bit) | Measured DLL time constant k (Hz/bit) | R = Computed k – Measured k | % error = $\frac{100 R}{\text{measured } k}$ |
|---|---|---|---|---|---|
| 8 | 256 | 1.20625000 | 1.2166 | 0.0104 | −0.858 |
| 9 | 512 | 0.60312500 | 0.6052 | 0.0021 | −0.344 |
| 10 | 1024 | 0.30156250 | 0.3018 | 0.0002 | −0.079 |
| 11 | 2048 | 0.15078125 | 0.1509 | 0.0001 | −0.079 |
| 12 | 4096 | 0.07539063 | 0.0755 | 0.0001 | −0.145 |

of this size can be designed using VHDL or can be simply inserted from a development system library and easily implemented on any of a host of programmable logic arrays.

The important trade-offs to be understood from this discussion are as follows: First, as the value of $k$ gets smaller,

- The frequency resolution of the loop increases, which reduces the magnitude and frequency of the synthesized clock jitter.
- The circuit time constant is smaller, so the transient response becomes longer, which slows the attack time of the loop. That is, the loop responds more slowly to changes in the input bit clock frequency. You can think of this as a low pass filter that tends to smooth the high-frequency perturbations of its input, and as $k$ gets smaller, the low pass filter bandwidth gets narrower.

Second, as the value of $k$ gets larger,

- The frequency resolution of the loop decreases, which increases the magnitude and frequency of the synthesized clock jitter.
- The circuit time constant is larger, so the transient response becomes shorter, which speeds the attack time of the loop. That is, the loop responds more quickly to changes in the input bit clock frequency. You can think of this as a low pass filter whose bandwidth gets wider as $k$ gets larger.

So there is a design trade-off to be made here in terms of frequency resolution versus response time. A designer will have to determine, based on these two opposing circuit characteristics, what the optimum value for $k$ might be for a given application. The reader should understand that in most cases the oscillators used to clock data are usually well behaved. The clock frequency will slowly drift over time and temperature, but rarely if ever will they exhibit large step changes in frequency. If large step changes do occur, it is probably time to replace the oscillator. Since the variance in the write clock frequency is usually small, the loop transient response time due to perturbations in the input clock may not be a major factor in the design. That is, when performing design trade-offs, the loop frequency resolution and the loop frequency jitter concerns will more than likely trump the concerns for the loop transient response time.

We will demonstrate the performance of the DLL versus the value of $k$ and the size of the dual port memory in the next section, when we investigate some bit-level software simulations on our DLL design.

## 11.4  DATA LOCKED LOOP BIT-LEVEL SIMULATION

Equation 11.16, which is repeated here for clarity, says that we should expect the DLL transient bit clock frequency response to take on the shape of an exponential in response to a step in the input bit clock frequency

$$f_S(nT) = F_I + \left(F_S(0) - F_I\right)e^{-knT}$$

We will see in this section that this is true.

### 11.4.1  Description of the Software Simulator Screen

Before we begin, let's spend some time getting familiar with the author's DLL bit-level software simulator, which runs in Microsoft Windows. The parameters for each simulation are entered via the dialog box illustrated in Figure 11.14.

The dialog box contains three control groups:

- MDAC parameters group
  - *Reference voltage* $V_{REF}$. This allows the user to assign values for the MDAC input reference voltage. In our design, this is set to 3.0 volts.
  - *Resolution*. This allows the user to assign value for $B$, which is the width in bits of the digital error signal $\varepsilon_D$. It also sets the size of the elastic store memory to $2^B$. In our design, this is set to 10 bits.
- VCXO parameters group (note there are several parameters here but only the two we are interested in are listed)
  - *Center frequency* $F_S$. This is the center frequency of the VCXO. In our design, the value of this parameter is 1.544 MHz.
  - *Frequency deviation* $F_D$. This allows the user to assign the pull range of the VCXO. In our design, the value of this parameter is ±100 ppm.
- Input tributary parameters group
  - *Center frequency* $F_I$. This is the center frequency of the input tributary bit clock and is set to 1.544 MHz.
  - *Center frequency variance*. This allows the user to assign a value to the variance of the ideal center frequency. In our design, this value is set to 0 Hz.
  - *Center frequency max variance*. This allows the user to specify the maximum tributary frequency variance. In our design, this value is ±50 ppm.

Figure 11.15 illustrates a blank simulator plot screen. On the left of the screen are two vertical axes. The one on the far left represents the range of the

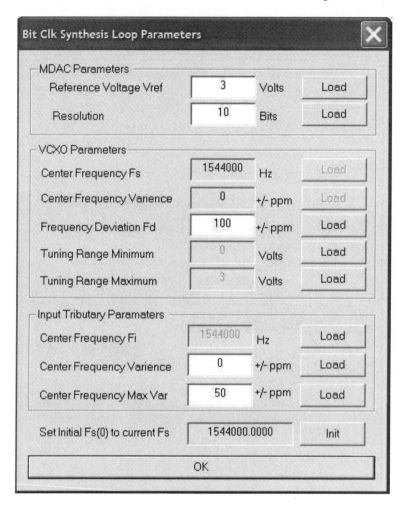

**Figure 11.14**  Software simulation parameters dialog box

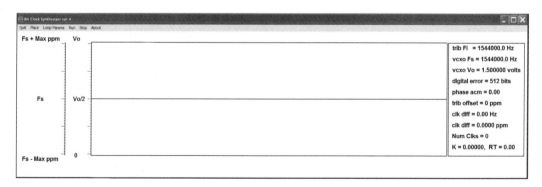

**Figure 11.15**  Software simulation plot screen

VCXO synthesized frequency with limits at $F_s \pm$ Max ppm. The VCXO center frequency on this vertical axis is labeled $F_s$. In this simulation, $F_s = 1.544$ MHz. The horizontal time axis passes through this center frequency point.

The second vertical axis represents the range of the VCXO error voltage $V_O$ with limits of 0 to $V_O$ volts. In our design, the max value for error voltage is 3 volts. The horizontal time axis passes through the midpoint of this range at $V_O/2 = 1.5$ volts. For our design, the maximum value of $f_s$ that we can plot is 1.544 MHz + 50 ppm or 1.544 MHz + 77.2 Hz, and the minimum value we can plot is 1.544 MHz – 50 ppm or 1.544 MHz – 77.2 Hz. The voltage range extends from 0 to 3 volts. In our 10-bit design, the maximum value of $V_O$ we can plot is 2.25 volts and the minimum is 0.75 volts.

The plot of the DLL transient response is read left to right on the screen, just like any normal graph. However, the plot data is entered on the right-hand side of the plot window and is scrolled to the left as additional plot data are entered. The plot continues to scroll across the plot window from right to left until the user stops the simulation. For this reason, a snapshot of a plot in progress may show the plot being skewed to the right of the vertical axis, but this is only because the plot has been stopped and has not completely scrolled across the full length of the plot window. The reader should visualize the point corresponding to the time $t = 0$ as scrolling from the right side of the plot to the left side and beyond. The points on the left of the plot correspond to the oldest in time and the points on the right of the plot correspond to the newest in time.

The minor and major tick marks on the time axis are scrolled from right to left along with the plot data. The distance between minor tick marks on the time axis is the equivalent of a million input bit clock periods. The distance between major tick marks on the time axis is the equivalent of 10 million input bit clock periods.

On the right of the plot screen are several variables that are constantly monitored and updated as the plot data is computed and as the plot scrolls across the plot window. These variables represent critical DLL circuit parameters. They are continuously being computed and displayed in real time as the data are plotted. This allows the user to view the behavior of the parameters with respect to time. These variables are defined from top to bottom in Table 11.5.

The simulation plots illustrate the synthesized frequency and the error voltage of the DLL over time in response to a step in the input bit clock frequency. In order to distinguish the two on a black and white plot, the synthesized frequency is shown as a thick line and the error voltage is shown as a thin line. With that said, let's take a look at a few simulations so we can graphically see the loop response in action. It is important to note that the software simulator does not plot data produced by the transient response equations we derived in the previous section. The simulator is a bit-level simulator that simulates the

**Table 11.5**   DLL Simulator Parameter Definitions

| Parameter | Definition |
|---|---|
| Trib $F_1$ | The frequency of the input tributary bit clock $F_1$ |
| VCXO $f_s$ | The frequency of the synthesized bit clock $f_s$ |
| VCXO $V_0$ | The loop error voltage output by the MDAC and input to the VCXO |
| Digital error | The digital error signal $\varepsilon_D = (wPtr^* - rPtr^*)$ expressed in bits |
| Phase accum | This is the accumulation of phase difference between the input bit clock and the synthesized bit clock. It is a monitor of the simulators synthesis of the two software clocks and is not important for our purposes here. |
| Trib offset | The frequency offset in ppm of the input bit clock from center frequency |
| Clk diff | The difference between the frequency of the input bit clock and the output synthesized bit clock $(F_1 - f_s)$, measured in Hz and in ppm |
| Num clks | The count of input bit clock cycles since $t = 0$ |
| $k$ | The measured value of the DLL circuit time constant. It is continuously measured during the 10% to 90% rise or fall times of the transient response. It should be noted that this measurement becomes less accurate as the rise/fall time of the transient response becomes shorter. This is due to the fact that there are far fewer quantized points to measure. |
| RT | The measure of the ratio of $[f_s - F_s(0)]/[F_1 - F_s(0)]$, which ranges from 0 to 1 and provides an indication of how the synthesized frequency is tracking the input frequency |

actual hardware architecture and plots the response of the DLL hardware circuit over time. We will expect to see the simulation results agree with the results independently obtained from our previously derived equations.

### 11.4.2   DLL Simulation Case I => B = 10 bits, k = 0.3016 Hz/bit

#### 11.4.2.1   Input Bit Clock Frequency Swings ±50 ppm

We spent a great amount of time designing a DLL with a 1024-bit elastic store memory and a 10-bit digital error signal, so we will observe the simulation results of that circuit first.

The simulation plot in Figure 11.16 shows the DLL synthesized frequency response to a step in frequency by the input bit clock of +50 ppm or +77.2 Hz above the center frequency, followed some time later by another frequency step of −100 ppm down to −77.2 Hz below center frequency.

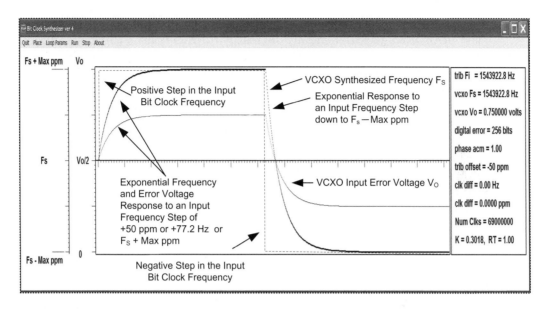

**Figure 11.16**   B = 10 frequency deviation = ±50 ppm

The distance between minor tick marks on the time axis is the equivalent of $10^6$ input bit clock periods. The distance between major tick marks is the equivalent of $10 \times 10^6$ input bit clock periods. The thick line is the plot of the synthesized VCXO clock frequency $f_s(t)$, and the thin line is the value of the loop error voltage $V_O(t)$. The values of the DLL circuit parameters listed to the right of the plot are frozen at the time the plot is stopped and represent the latest parameter measurements. These measurements correspond to the last plotted point, which is the last point on the right-hand side of the plot window.

The positive and negative frequency step of the input bit clock is shown as a dashed line. To avoid clutter on the following plots, this dashed step will not be shown, but the reader can rightfully assume that a step has occurred whenever an exponential response of $f_s(t)$ and $V_O(t)$ is initiated.

So what information can we glean from this graphical simulation? Well, we can immediately see from this plot that the loop response is indeed exponential, just as our derived transient response of Equation 11.16 predicted.

We can also see that after the input bit clock frequency instantaneously experiences a step of +77.2 Hz or +50 ppm, the DLL exponentially tracks and eventually locks to the input clock frequency. We can also see that when the input bit clock frequency experiences a negative step such that it is below the center frequency by 77.2 Hz, the DLL exponentially tracks and locks to the new input clock frequency. This time, the negative step was $-2 \times 77.2\ \text{Hz} = -154.4\ \text{Hz}$, which corresponds to a negative step of $-100$ ppm.

At time $t = 0$, the loop error voltage is set to its center value of 1.5 volts. The error voltage exponentially increases to 2.25 volts in response to the input frequency step of +77.2 Hz or +50 ppm and then exponentially decays to 0.75 volts following the negative tributary clock frequency step of –154.4 Hz or –100 ppm down to 1.544 MHz – 77.2 Hz.

It is interesting to note that the plot of both the synthesized bit clock frequency $f_S(t)$ curve and the VCXO control voltage $V_O(t)$ curve pass through the horizontal axis at the same point. This is as it should be, since the time axis is positioned at both the loop center frequency of 1.544 MHz and at the midpoint of the control voltage of 1.5 volts. Remember that when a 1.5 volt error signal is applied to the VCXO, it will synthesize the loop center frequency. So it is good to know that the hardware bit-level simulation behaves correctly in this regard.

We can also see from the parameter measurement list on the right-hand side of the plot that the value of the loop time constant was measured to be $k = 0.3018$ Hz/bit. As shown in Table 11.4, this value for $k$ is well within 0.1% of the calculated value of $k = 0.3016$ Hz/bit. The value of $k$ is continuously computed during the 10% to 90% rise/fall time of the response waveform. The accuracy of the computation for $k$ depends on the number of points in the rise or fall time of the curve. A shorter transient response curve will produce less accurate measurements.

The parameter measurement list also shows that when the simulation was stopped, the frequency of the input bit clock (trib offset) is equal to the center frequency –50 ppm and the input bit clock frequency is 1,543,922.8 Hz, both of which are expected. In addition we can see from the parameter list that the digital error signal $\varepsilon_D$ is equal to 256. The ideal VCXO frequency versus voltage curve of Figure 11.13 says that when the value of the error signal reaches 256, the VCXO output frequency should be 1.544 MHz – 50 ppm or 1,543,922.8 Hz and that the VCXO input error voltage should be 0.75 volts. A quick check of the plot measurement list shows that the synthesized frequency $F_S$ does indeed equal 1,543,922.8 Hz and the error voltage $V_O$ does indeed equal 0.75 volts.

The parameter list also shows that the difference between the input bit clock and the synthesized bit clock at this particular frequency is clk diff = 0 Hz. This tells us that the loop was able to exactly synthesize this particular frequency. This will only be true when the input frequency exactly equals one of the 1024 discrete frequencies that can be synthesized by this 10-bit DLL.

And finally we do not pay any attention to the "num clks" value in the plot parameter list. That is the number of clocks that have occurred since $t = 0$. This number is only meaningful during the loop transient response computations. Since the loop is already reached the steady state condition, this number is meaningless.

Now that you are familiar with the format of the simulation plot, let's view some additional plots to see how the DLL responds to input bit clock frequency steps of different magnitudes.

### 11.4.2.2  Input Bit Clock Frequency Deviation of +40 ppm

When the DLL was locked to the input bit clock frequency with a deviation of ±50 ppm, we saw that the DLL was able to synthesize a clock frequency that perfectly matched the frequency of the input bit clock. The DLL was able to do this because the digital error signal for those two extreme frequencies just happened to be integer values of 768 and 256. What happens when our DLL cannot synthesize a bit clock frequency that exactly matches the frequency of the input bit clock? Let's take a look at the case where the deviation of the input bit clock frequency is +40 ppm or +61.76 Hz. The digital error signal that would drive the DLL to synthesize this frequency is calculated by

$$\varepsilon_D = \left(f_O - F_{VCXOMin}\right)\left[\frac{2^B}{F_{VCXOMax} - F_{VCXOMin}}\right]$$

or

$$\varepsilon_D = \left(1544061.76 - 1543845.6\right)\left[\frac{1}{0.3016}\right] = 716.7109$$

Since $\varepsilon_D$ must be an integer, the closest error values that the loop can compute are the two that bracket 716.7109, which are $\varepsilon_D = 716$ and $\varepsilon_D = 717$. And that is exactly what the DLL will try to do. It will generate an error signal of $\varepsilon_D = 716$ and produce an output frequency $f_{S1}$ that is slightly less than the frequency of the input bit clock $F_I$. Slowly over time the small positive difference between the input bit clock frequency and the synthesized bit clock frequency $\left(F_I - f_{S1}\right)$ will accumulate until the error signal eventually increases by 1 bit to $\varepsilon_D = 717$. The extra error bit will cause the DLL to synthesize a higher output frequency $f_{S2}$ that is 0.3016 Hz greater than $f_{S1}$ and is a slightly greater than the frequency of the input bit clock $F_I$. Slowly over time the small negative difference between the input bit clock frequency and the synthesized bit clock frequency $\left(F_I - f_{S2}\right)$ will cause the error accumulation to decay until the error signal eventually decreases by 1 bit back to $\varepsilon_D = 716$. This process will repeat itself forever.

The switching back and forth between two synthesized frequencies that closely bracket the target frequency gives rise to synthesized clock jitter. The magnitude and the frequency of the jitter are dependent on the frequency resolution of the loop. The smaller the time constant $k$, the better the resolution and the smaller the jitter, both in magnitude and frequency. The jitter magnitude is smaller because the difference between the synthesized and target frequency is smaller, and the jitter frequency is smaller because it takes longer for the digital error to accumulate or decay by 1 LSB. The opposite is true if the value of $k$ gets larger. In this case, the resolution decreases and the magnitude and frequency of the synthesized frequency jitter increases.

Let's verify this by looking at the simulator plot for our design for the case where $B$ equals 10 and the calculated value for our loop time constant $k = 0.3016$. The plot of the synthesized frequency is illustrated in Figure 11.17. The parameter measurements on the right side of the plot show a post transient response clock difference of $(F_I - f_{S1})$ equal to –0.06 Hz. We can also see that the value for the digital error is $\varepsilon_D = 717$. The frequency resolution for our $B = 10$ bit design is good enough that we can barely see any jitter in the plot of the output frequency. In the plot window, an arrow points to a box that encloses the portion of the plot where the synthesized frequency is actually 0.06 Hz greater than the target input bit clock frequency.

Now take a look at Figure 11.18. Here the error has decayed to the point where the digital error signal $\varepsilon_D = 716$. Now the difference between the input bit clock frequency and the synthesized clock frequency $(F_I - f_{S2})$ is equal to 0.24 Hz. Once again there is an arrow in the plot window that points to the box that encloses the portion of the plot where this 0.24 Hz discrepancy occurs. This oscillation between $F_I - 0.24 \text{ Hz} \leq f_S \leq F_I + 0.06 \text{ Hz}$ will continue for forever and is the jitter about the center of the desired synthesized

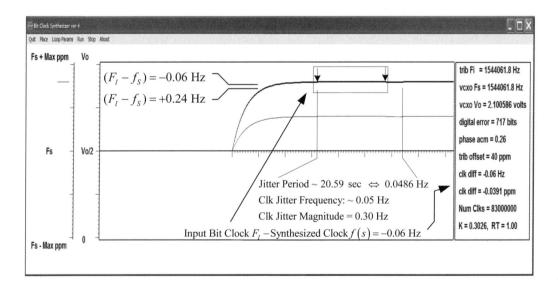

**Figure 11.17**   $B = 10$; frequency deviation = +40 ppm (1)

frequency. Figure 11.17 shows that the jitter period is measured to be approximately 20.59 seconds. The jitter frequency is 0.05 Hz and the jitter magnitude is equal to $(0.06 + 0.24)\,\text{Hz} = 0.30$ Hz. This also is the value for our loop time constant $k$. It shouldn't be a surprise that $k$ is also the value of the loop jitter. The loop error signal oscillating between two adjacent levels of $\varepsilon_D$ and $k$ is equal to the frequency difference between these adjacent values.

So this little constant $k$ has just taken on another huge role in the behavior of the DLL. Let's see if we can count all the fun things $k$ contributes to the DLL performance:

1.  $k$ is the frequency resolution of the DLL.
2.  $k$ is the circuit time constant that determines the attack rate of the loop.
3.  $k$ is the magnitude of the jitter in the loop synthesized frequency.

It turns out that this constant $k$ is an extremely important player in the performance of our DLL. The jitter we see for the $B = 10$ bit case doesn't sound like a big deal here, but we will show some examples in the next section where this jitter for smaller values of $B$ becomes very pronounced.

One other thing we need to mention here is the rise time of the exponential response. The rise time is defined to be the time required for the curve to rise from 10% to 90% of its final value. The rise time is dependent on the value of $k$. We will look at the computation of the rise time figure in greater

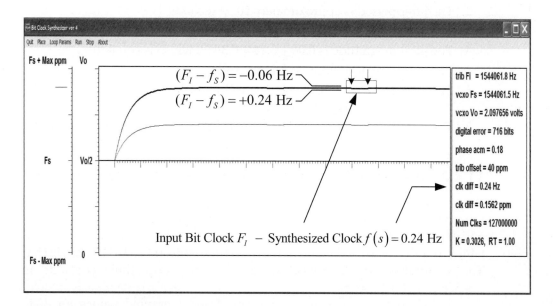

**Figure 11.18**   $B = 10$; frequency deviation = +40 ppm (2)

detail when we compare the simulation results with the DLL transient response results we calculate from either of the two expressions presented in Equation 11.16. For now, all we need to know is that the computed rise time for this 10-bit DLL is 7.28 seconds.

We mentioned several times that the clock jitter becomes more pronounced and the transient time of the circuit decreases as the value of the circuit time constant $k$ increases. This makes sense since the units of $k$ are Hz/bit. As the time constant $k$ gets larger, the frequency resolution decreases, which allows the loop to acquire a step in input frequency more quickly, but because its frequency resolution is decreased, once locked to in input frequency, the loop cannot track with as much precision. The opposite is true when the value of $k$ decreases. The next few simulation cases will demonstrate this very clearly as we take a look at the simulation examples for $B = 8$, 9, 11, and 12 bits. Once we acquire an understanding of these examples, we will look at some examples of how loop performance degrades when we set $B$ to lower values such as 6 and 7. These two examples are shown primarily to allow the reader to see what problems can arise if loop parameters are not selected with care.

The following are a few things to note about the 10-bit DLL driven by an input clock of $1.544\,\text{MHz} \pm 50\,\text{ppm}$ and its response to a $\pm 40\,\text{ppm}$ input frequency step:

1. The calculated value of the circuit time constant $k$ is 0.3016.
2. The jitter period is approximately 20.59 seconds.
3. The jitter frequency is approximately 0.0486 Hz.
4. The jitter magnitude is $0.06 + 0.24 = 0.3$ Hz, which is equal to $k$.
5. The exponential rise time is 7.28 seconds.
6. The digital error switches between 716 and 717 bits when tracking the high deviation +50 ppm input.
7. The digital error switches between 307 and 308 bits when tracking the low deviation −50 ppm input.

### 11.4.3   DLL Simulation Case II ⇒ *B* = 8 bits, *k* = 1.2062 Hz/bit

We see in Figure 11.19 the DLL transient response to an input bit clock whose frequency abruptly changes first by +40 ppm above center frequency and then by −40 ppm below center frequency. In this example, $B$ is set to 8 bits, and the measured value of the loop time constant is $k = 1.2166\,\text{Hz/bit}$. The computed or theoretical value for the loop time constant is $k = 308.8 / 256 = 1.20625$.

The difference between the theoretical and measured values for $k$ is a result of the limited number of simulation samples available during the

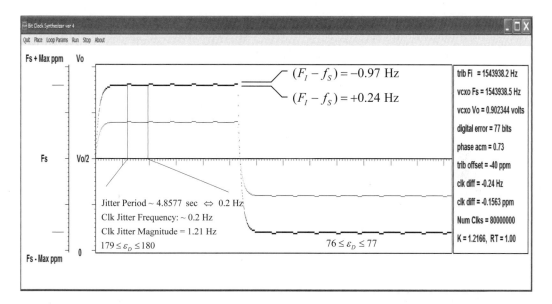

**Figure 11.19** $B = 8$; frequency deviation = ±40 ppm

exponential rise and fall times of the simulation waveform, and the corresponding error in determining the exact 10% and 90% rise/fall time points. Using the measured value of $k$, we can say that the loop can synthesize 256 discrete frequencies that are spaced 1.2166 Hz apart. If we look at the frequency curve, it's obvious that the synthesized frequency is bouncing above and below the desired solution of $F_I$ by 0.97 and 0.24 Hz. The sum of these two frequency offsets is $0.97 + 0.24 = 1.21$ Hz, which is what we would expect since the measured value of $k$ is 1.2166. This bouncing between two adjacent loop frequencies is the jitter in the synthesized clock. The elapsed time between minor ticks on the time axis is equivalent to $10^6$ input bit clock periods. From the plot, a quick approximation of the time for one jitter cycle is about 7.5 million clocks. The input clock at the end of the simulation is $1,543,938.2$ Hz, so we can approximate the jitter period to be

$$\approx \left(7.5 \times 10^6 \text{clocks}\right)\left(\frac{1 \text{ second}}{1543938.2 \text{ clocks}}\right) = 4.8577 \text{ second}$$

from which the jitter frequency is approximately $\left(1/4.8577\right)$ second $= 0.2059$ Hz.

On closer inspection of Figure 11.19, we notice that during the time when the loop is tracking the input bit clock at center frequency +40 ppm, the value of the digital error signal is within the range $179 \leq \varepsilon_D \leq 180$ bits. During the time when the loop is tracking the input bit clock at center frequency −40 ppm, the value of the digital error signal is within the range $76 \leq \varepsilon_D \leq 77$ bits. The loop is tracking the target frequency and is trying to generate a perfect solution. Because the loop can only synthesize 256 discrete frequencies and the target frequency is not one of those, it is forced to bracket the target frequency and play the error accumulation, error decay game. This is evident by looking at the fact that the digital error signal is alternating between 179 and 180 at the high deviation and alternating between 76 and 77 at the low deviation. This is the best the loop can do. We can make the loop perform better if we increase the value of $B$, which lowers the value of $k$ and in turn increases the frequency resolution. Keep in mind though that an increase in $B$ by 1 bit doubles the size of the loop elastic store memory.

Enough already! We have stated over and over the effect that $k$ has on the performance of the DLL. Hopefully by now this has become firmly entrenched in your memory. So for the remainder of the simulation examples, we will mention $k$ only to report its value.

The following is a list of things to note about this 8-bit DLL and its response to a ±40 ppm input frequency step:

1. The calculated value of the circuit time constant $k$ is 1.20635.
2. The jitter period is approximately 4.8577 seconds.
3. The jitter frequency is approximately 0.2059 Hz.
4. The jitter magnitude is $0.97 + 0.24 = 1.21$ Hz, which equals the rounded value of $k$.
5. The exponential rise time is 1.82 seconds.
6. The digital error switches between 179 and 180 when tracking the high deviation input frequency.
7. The digital error switches between 76 and 77 when tracking the low deviation input frequency.

### 11.4.4  DLL Simulation Case III => B = 9 bits, k = 0.6031 Hz/bit

Figure 11.20 illustrates the performance of the DLL for the case where $B = 9$ bits, the elastic store memory is 512 bits deep, and the value of $k = 0.6052$ Hz/bit. In this example, the loop can synthesize 512 discrete frequencies that are spaced 0.6052 Hz apart.

A quick comparison with the $B = 8$ bits case shows some interesting performance differences:

1. The calculated value of the circuit time constant $k$ is now 0.6031.
2. The jitter period has increased to approximately 7 seconds.
3. The jitter frequency has decreased to approximately 0.14 Hz.

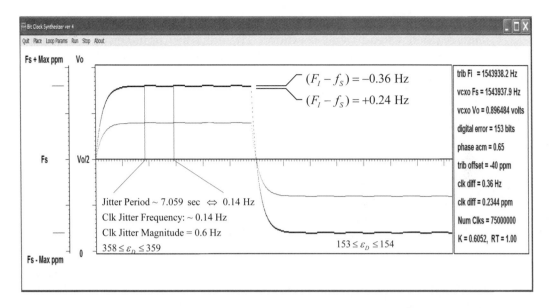

**Figure 11.20** $B = 9$; frequency deviation = center freq ±40 ppm

4. The magnitude of the jitter has decreased to 0.6 Hz, which equals the value of $k$.

5. The exponential rise time is 3.64 seconds twice as long as for the 8-bit DLL.

6. The digital error switches between 358 and 359 when tracking the high deviation input frequency.

7. The digital error switches between 153 and 154 when tracking the low deviation input frequency.

All these items depend on the value of $k$ and all were discussed in previous sections. There is not much else to mention here, except that the performance of the $B = 9$ bit DLL is better than the $B = 8$ bit DLL.

### 11.4.5 DLL Simulation Case IV ⇒ *B* = 11 bits, *k* = 0.1508 Hz/bit

Figure 11.21 illustrates the performance of the DLL for the case where $B = 11$ bits and the elastic store memory is 2048 bits deep. The loop can synthesize 2048 discrete frequencies that are spaced 0.1509 Hz apart. Here we see a significant increase in performance over the $B = 9$ bit DLL. The frequency resolution is small enough that it's difficult to detect the frequency jitter from looking at the plot.

The reader should note that the plot was stopped just prior to when the DLL reached a steady state condition. Therefore the digital error listed in the statistics column shows that $\varepsilon_D = 617$ as opposed to the actual steady state value of $\varepsilon_D = 614\,/\,615$.

Parameters to note in this example are as follows:

1. The calculated value of the circuit time constant $k$ is now 0.1508.
2. The magnitude of the jitter has decreased to 0.15 Hz, the same value as $k$.
3. The exponential rise time is 14.57 seconds, twice as long as the 10-bit DLL.
4. The digital error switches between 1433 and 1434 when tracking the high deviation input frequency.
5. The digital error switches between 614 and 615 when tracking the low deviation input frequency.

The major trade-off decision for this case is whether or not the long exponential transient response is acceptable over that of a $B = 10$ bit DLL. The $B = 11$ bit DLL gives us twice the frequency resolution, but we are only talking about a 0.1508 Hz per bit increase, and the price we pay is doubling the elastic store memory size and doubling the response rise time. Increasing the elastic store memory size may or may not play a part in overall circuit speed. When synthesizing much higher frequencies, the speed of the memory will become an issue.

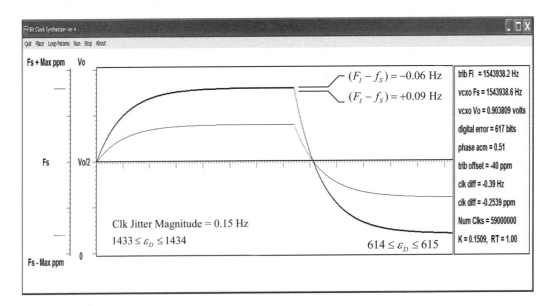

**Figure 11.21**   $B = 11$; frequency deviation = center freq ±40 ppm

As mentioned earlier, the rise time of the loop transient response may not be a big factor in a trade-off study. The reason is the bit clock the loop is tracking is always going to be well behaved, in that it will drift in frequency but will do so very slowly. Unless the write clock oscillator is defective, the ±40 ppm steps in frequency we see in the simulations will probably never occur.

### 11.4.6  DLL Simulation Case V => *B* = 12 bits, *k* = 0.0754 Hz/bit

Figure 11.22 illustrates the $B = 12$ case where the DLL utilizes a 4096-bit elastic store memory. The loop can synthesize 4096 discrete frequencies that are spaced 0.0755 Hz apart. In order to preserve the exponential portion of the transient response, the simulation for this DLL was stopped just prior to reaching the steady state condition. Parameters to note are as follows:

1. The calculated value of the circuit time constant $k$ is now 0.0755.
2. The jitter magnitude has decreased to 0.08 Hz, approximately equal to $k$.
3. The exponential rise time is 29.14 seconds, twice that of the 11-bit DLL.
4. The digital error switches between 2867 and 2868 when tracking the high deviation input frequency.
5. The digital error switches between 1228 and 1229 when tracking the low deviation input frequency.

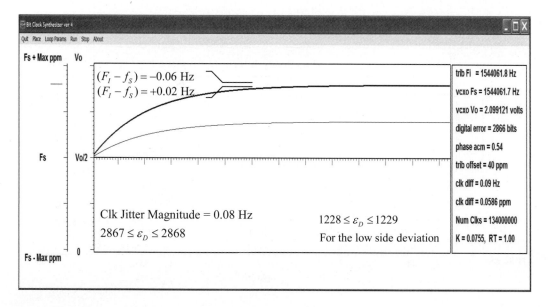

**Figure 11.22**  $B = 12$; frequency deviation = +40 ppm

The performance parameters for the $B = 12$ case speak for themselves. Of all the DLLs we have looked at, this provides the best frequency resolution, but it also has the longest response time. It is the most sluggish of them all. As $B$ continues to increase, the degree of circuit response sluggishness will continue to increase as well.

It should be crystal clear by now that increasing the value of $B$ by 1 halves the value of the time constant $k$, doubles the loop frequency resolution, doubles the elastic store memory size, and doubles the rise time. This is the DLL gospel.

### 11.4.7 DLL Simulation Case VI => B = 6,7,8,9 bits

Figure 11.23 shows a composite of responses for 6-, 7-, 8-, and 9-bit DLLs. The symbol

refers to the low and high amplitudes of the steady state frequency response curve. Listed below each curve are parameters of interest for that particular DLL. The main point of this figure is to visually summarize both the transient and steady state response of these four DLLs. Side-by-side observation of the response plots for all four DLLs allows us to see the relevant changes in DLL performance as the value of $B$ is increased from 6 to 9.

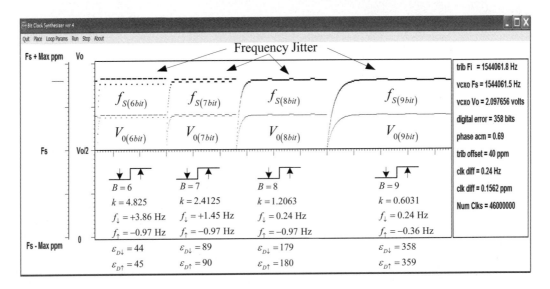

**Figure 11.23**   $B = 6, 7, 8, 9$ bits; frequency deviation $= +40$ ppm

Important points to remember are as follows:

1. The synthesized frequency jitter amplitude is always equal to $k$.
2. The jitter amplitude decreases with increasing $B$ and decreasing $k$.
3. The jitter frequency decreases with increasing $B$ and decreasing $k$.
4. The transient response times increases with increasing $B$ and decreasing $k$.
5. The jitter amplitude is confined to the interval between two adjacent values of $\varepsilon_D$.
6. There is a one for one correlation between the changes in $V_O$ and changes in $f_S$.

In general, the DLL performance improves as we increase the value of $B$. The limiting factor in the selection of the value for $B$ will be the rise time or attack time of the loop and the speed constraints placed on the loop by the size of its elastic store memory.

### 11.4.8  Simulation versus the Derived Transient Response Equation

The two expressions in Equation 11.16 give us a mathematical method of determining the transient response of a DLL based upon its computed time constant $k$. These expressions are repeated here for clarity:

$$f_S(nT) = F_I + (F_S(0) - F_I)e^{-knT}$$

$$f_S(nT) = F_S(0) + (F_I - F_S(0))(1 - e^{-knT})$$

Figure 11.24 shows a composite of transient responses for 8-, 9-, 10-, 11-, and 12-bit DLLs. We can easily see that the duration of the transient response increases with an increase in the value of $B$. Our goal here is twofold: First we would like to compare the results of Equation 11.16 with the transient response results obtained from the DLL bit-level simulation. Second, we want to use Equation 11.16 to calculate the DLL rise time. In both cases, we will use the second of the two expressions in Equation 11.16.

#### 11.4.8.1  Derived Transient Response

The parameters on the right side of Figure 11.24 are the measurements that correspond to the time when the last data point was calculated and plotted. Therefore the measurements we see all correspond to the transient response of the 12-bit DLL. We can see from the measurement box that the loop is trying to acquire an input bit clock frequency that has stepped up by +40 ppm. The synthesized bit clock frequency is trying to reach a target of $1,544,061.8$ Hz.

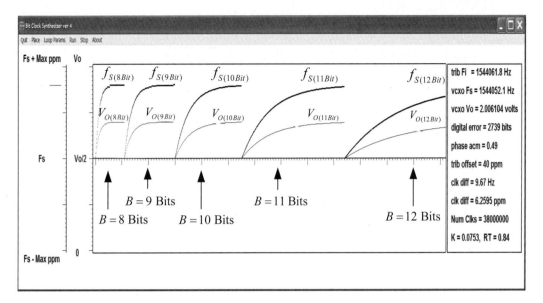

**Figure 11.24**  DLL transient rise times

After 38 million clock periods, the bit-level simulator says that the 12-bit DLL output frequency has attained the value of 1544052.1 Hz.

Let's see if we can get the same results from the use of our derived expressions for the DLL transient response. Remember that the simulator performs a bit-level simulation of the actual DLL hardware. It has no idea of the existence of the expressions in Equation 11.16.

So let's use the derived expression, fill in the values, and see what we come up with. We will use the calculated value for the DLL time constant $k$, and we will get the number of clock periods $n$ since time $t = 0$ from the simulator. We know the initial value of $f_S$ to be $F_S(0)$ or the loop center frequency, and we can calculate the target frequency as being the center frequency +40 ppm or 1,544,061.8 Hz. We can list these values separately here as

$$k = 0.0753$$
$$n = 38 \times 10^6$$
$$F_S(0) = 1.544 \times 10^6 \, \text{Hz}$$
$$F_I = 1.544061.8 \times 10^6 \, \text{Hz}$$

The value for $T$ is just the reciprocal of the input bit clock frequency and is

$$T = \frac{1}{1544061.8 \, \text{Hz}} = 6.4764 \times 10^{-7} \, \text{seconds}$$

We wish to know what the value of $f_s(nT)$ is; therefore we compute

$$knT = (0.0753)(38 \times 10^6)(6.4764 \times 10^{-7}) = 1.8532$$

Our transient response equation is

$$f_s(nT) = F_s(0) + (F_I - F_s(0))(1 - e^{-knT})$$

If we replace the terms in the equation with our known values, we get

$$f_s(nT) = 1.544 \times 10^6 + (1,544,061.8 - 1,544,000)(1 - e^{-1.8532})$$

$$f_s(nT) = 1.544 \times 10^6 + (61.8)(1 - 0.1567)$$
$$f_s(nT) = 1.544 \times 10^6 + (61.8)(0.8433)$$
$$f_s(nT) = 1.544 \times 10^6 + 52.1134$$
$$f_s(nT) = 1,544052.1134 \text{ Hz}$$

The simulation computed $f_s(nT)$ to be $1,544,052.1$ Hz, and Equation 11.16 predicted $f_s(nT)$ to be equal to $1,544,052.1134$ Hz or $1,544,052.1$ Hz, if we round up to the nearest tenth of a hertz as the simulator does. In this case, the simulated result and the calculated result are identical. This tends to validate both the predicted and simulated methods of determining DLL circuit performance.

### 11.4.8.2   Calculation of DLL Transient Response Rise Time

We will define the rise time of an exponential waveform as the time required for the waveform to rise from 10% to 90% of its final value. These two points occur at times $t_1$ and $t_2$, respectively. The portion of the curve that we are interested in is defined by the bold lines on the plot axis of Figure 11.25. This definition provides us with a uniform and consistent method to measure and compare the rise time of one exponential curve to any another.

We will begin with the transient response defined by the second expression of Equation 11.16:

$$f_s(t) = F_s(0) + (F_I - F_s(0))(1 - e^{-kt})$$

We can think of the transient response to be a sum of a DC component and an AC component. The $F_s(0)$ term is a constant and can be considered to take on the role of the steady state DC component that exists prior to the input clock

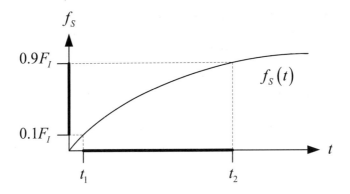

**Figure 11.25**   Exponential rise time

step in frequency. The $\left(F_I - F_S(0)\right)\left(1 - e^{-kt}\right)$ term is time varying transient and can be considered to take on the role of the AC component that occurs after the input clock step in frequency. We are only interested in measuring the rise time of the transient or AC component, so we subtract $F_S(0)$ from both sides of the equation and set the result equal to $z_S$:

$$z_S = f_S(t) - F_S(0) = \left(F_I - F_S(0)\right)\left(1 - e^{-kt}\right)$$

Now we wish to compute the value of $t_1$ at the 10% point in the curve. That is when

$$z_S = 0.1\left(F_I - F_S(0)\right)$$

and we wish to compute the value of $t_2$ at the 90% point in the curve, when

$$z_S = 0.9\left(F_I - F_S(0)\right)$$

We can compute the rise time by subtracting $t_1$ from $t_2$. Let's make the substitution of $\Delta F = F_I - F_S(0)$ to help simplify our notation. So now we have

$$z_S = \Delta F\left(1 - e^{-kt}\right)$$

Now let's solve for $t$:

$$e^{-kt} = 1 - \frac{z_s}{\Delta F}$$

$$-kt = \ln\left(1 - \frac{z_s}{\Delta F}\right)$$

$$t = -\frac{1}{k}\ln\left(1 - \frac{z_s}{\Delta F}\right)$$

At time $t_1$, $z_s = 0.1(\Delta F)$, and at time $t_2$, $z_s = 0.9(\Delta F)$, so

$$t_2 - t_1 = \left\{-\frac{1}{k}\ln\left(1 - \frac{0.9(\Delta F)}{\Delta F}\right)\right\} - \left\{-\frac{1}{k}\ln\left(1 - \frac{0.1(\Delta F)}{\Delta F}\right)\right\}$$

$$t_2 - t_1 = -\frac{1}{k}\left\{\ln(1 - 0.9) - \ln(1 - 0.1)\right\}$$

$$t_2 - t_1 = -\frac{1}{k}\left\{\ln(0.1) - \ln(0.9)\right\}$$

$$t_2 - t_1 = -\frac{1}{k}\ln\left(\frac{0.1}{0.9}\right)$$

$$t_2 - t_1 = \frac{1}{k}(2.1972)$$

We can rewrite this result with additional bits of precision as

$$\text{DLL rise time} = \frac{1}{k}(2.19722458)$$

**Equation 11.21: DLL rise time**

So once again our circuit time constant $k$ pops up in an important relation. The value of the circuit time constant $k$ is the only circuit parameter utilized in determining the rise time of our DLL. The reader should immediately
see that the DLL rise time is inversely proportional to the value of $k$ and that
when $k$ is halved, the rise time doubles, and when $k$ is doubled, the rise time
is halved.

Table 11.6 lists the rise time for 5-bit through 12-bit DLLs. It is interesting to note from this table that when $B$ increases by 1, the time constant $k$ is
halved, and the rise time is doubled. It is indeed very fascinating to see these
circuit parameters in action.

**Table 11.6**   DLL Transient Response Rise Time versus *B*

| Error bit width ($B$) | Time constant ($k$) | Computed rise time $t_2 - t_1$ (sec) |
|:---:|:---:|:---:|
| 5 | 9.6500000000 | 0.2276916660 |
| 6 | 4.8250000000 | 0.4553833321 |
| 7 | 2.4125000000 | 0.9107666642 |
| 8 | 1.2062500000 | 1.8215333284 |
| 9 | 0.6031250000 | 3.6430666567 |
| 10 | 0.3015625000 | 7.2861333134 |
| 11 | 0.1507812500 | 14.5722666269 |
| 12 | 0.0753906250 | 29.1445332538 |

## 11.4.9   Simulation versus the Derived Steady State Equations

We have just one more task to complete before we are finished and that is to compare the results obtained from the simulation with the results obtained from the derived steady state equations. We compare the simulation result of Figure 11.26 with the steady state VCXO frequency versus voltage response in Equation 11.4, and the VCXO frequency versus error signal response in Equation 11.5.

The two equations are repeated below for clarity:

$$f_O = \left[ \frac{F_{VCXOMax} - F_{VCXOMin}}{V_{ref}} \right] V_O + F_{VCXOMin}$$

**Equation 11.4: Ideal VCXO voltage versus frequency response**

$$f_O = \left[ \frac{F_{VCXOMax} - F_{VCXOMin}}{2^B} \right] \varepsilon_D + F_{VCXOMin}$$

**Equation 11.5: VCXO frequency response as a function of the digital error signal**

### 11.4.9.1   Comparison between the Simulation and Equation 11.5

Figure 11.26 shows the transient and steady state response for a 10-bit DLL that is locked to an input bit clock frequency −38 ppm below the center frequency. The simulation measurement box shows the value of the VCXO error voltage to be $V_O = 0.931641$ volts, the synthesized output frequency to be $f_s = 1,543,941.5$ Hz, and the digital error to be 318 bits.

Let's begin our comparison by calculating the digital error signal $\varepsilon_D$. Remember that

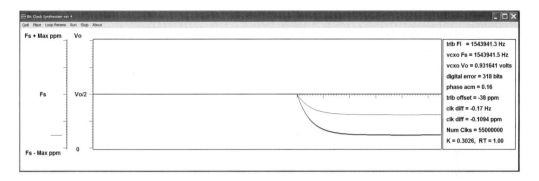

**Figure 11.26**   Ten-bit DLL input frequency deviation = –38 ppm

$$\varepsilon_D = \left( f_O - F_{VCXOMin} \right) \left[ \frac{2^B}{F_{VCXOMax} - F_{VCXOMin}} \right]$$

We calculate the ideal continuous valued VCXO output frequency to be

$$f_O - 38 \text{ ppm} = \left( 1.544 \text{ MHz} - 58.6720 \text{ Hz} \right) = 1,543,941.3280 \text{ Hz}$$

Substituting this value into the equation for $\varepsilon_D$ and solving, we obtain

$$\varepsilon_D = \left( 1,543,941.3280 - 1,543845.6 \right) \left[ \frac{1024}{308.8} \right]$$

$$\varepsilon_D = (95.728)(3.3161)$$

$$\varepsilon_D = 317.44$$

Since the value for $\varepsilon_D$ is not an integer, the loop will bracket this computed error signal value with integer values of 317 and 318 bits. Since the current value $\varepsilon_D$ in the simulation measurement box is 318, we will use that value. Let's use this value of $\varepsilon_D$ to compute the value of the loop synthesized frequency using Equation 11.5:

$$f_O = \left[ \frac{F_{VCXOMax} - F_{VCXOMin}}{2^B} \right] \varepsilon_D + F_{VCXOMin}$$

$$f_O = \left[ \frac{308.8}{1024} \right] (318) + 1,543,845.6$$

$$f_O = (0.3016)(318) + 1,543,845.6$$

$$f_O = 1,543,941.4969$$

$$f_{O \text{ Rounded}} = 1,543,941.5$$

The rounded result obtained from the equation agrees with the simulated results. So the two methods are identical to one another. The reader should keep in mind that we are not trying to verify the results of the simulator here. We are trying to verify the correctness of the derived equations through the comparisons made between the two.

### 11.4.9.2  Comparison between the Simulation and Equation 11.4

Now let's compare the simulation results with Equation 11.4. We can calculate the value of $V_O$ from

$$V_O = \left(\frac{V_{ref}}{2^B}\right)\varepsilon_D = \left(\frac{3 \text{ volts}}{1024 \text{ bits}}\right)318 \text{ bits} = 0.931640625 \text{ volts}$$

We round this value for $V_O$ to six significant figures, and we end up with the same value that the simulation produced:

$$V_{O\ Rounded} = 0.931641 \text{ volts}$$

Then, by substitution,

$$f_O = \left[\frac{F_{VCXOMax} - F_{VCXOMin}}{V_{ref}}\right]V_O + F_{VCXOMin}$$

$$f_O = \left[\frac{308.8 \text{ Hz}}{3 \text{ volts}}\right]0.931461 \text{ volts} + (1,544,000 - 154.4) \text{ Hz}$$

$$f_O = (102.9333)(0.931461) + (1,544,000 - 154.4) \text{ Hz}$$

$$f_O = 95.8784 + 1,544,000 - 154.4 = 1,543,941.4784 \text{ Hz}$$

$$f_{O\ Rounded} = 1,543,941.5 \text{ Hz}$$

The simulation measurements showed that $f_O = 1,543,941.5 \text{ Hz}$ when rounded to the nearest tenth of a hertz. The results obtained from the two methods are identical.

Remember that the software simulation actually operates at the bit level to simulate a DLL hardware architecture identical to the one we discussed in this chapter. The simulation has absolutely no idea that the equations we derived exist. The fact that the results from the bit-level simulation match those obtained from the theoretical calculations validates all the equations that we developed in this chapter.

## 11.5  ENGINEERING NOTE

The reader should take note that all simulations used in this chapter illustrated the DLL response to extreme cases of input signal bit rate perturbations of ±40 ppm or more. The chances of an engineer ever processing a signal like this are minimal. Most received signals are well behaved, mostly because the signal transmission hardware is built to very strict specifications. If an engineer were ever to see a communications signal with a sudden step in bit rate of ±40 ppm, he is probably right to assume that the clock oscillator in the transmission hardware has failed.

The design engineer can make the correct assumption that the bit rate of the signal he or she is processing will be constrained to operate within a strict range of values and that any drift within this range due to the effects of temperature and aging will be a very slow process. With this in mind, it would be quite correct to use much larger error signals beyond the 12-bit signals discussed in this chapter. This is because the DLL rise time in tracking a slowly varying input clock rate would not be of great concern.

## 11.6  SUMMARY OF USEFUL EQUATIONS

$B$ = the bit width of the digital error signal

$B$ = the size of the DLL elastic store memory $\left(\text{i.e., } 2^B \times 1 \text{ bits}\right)$

$F_{VCXOMax}$ = the maximum VCXO frequency deviation above center frequency

$F_{VCXOMin}$ = the maximum VCXO frequency deviation below center frequency

$\varepsilon_D$ = the value of the digital error signal

$V_O$ = the loop error voltage produced by the MDAC and applied to the VCXO

$V_{ref}$ = the reference voltage applied to the MDAC

$k$ = the DLL circuit time constant

$F_I$ = the value of the DLL input bit clock frequency, sometimes called the target frequency

$F_S(0)$ = the value of the steady state VCXO synthesized clock frequency at time $t = 0$

$f_S = f_S(t)$ = used interchangeably, the value of the VCXO synthesized frequency

$T$ = the period of the input bit clock = $1/F_I$

$n$ = the number of input bit clock periods since time zero $\left(t_0\right)$

Rise time = the time required for the exponential synthesized frequency response
to rise from 10% to 90% of its final value

$$\varepsilon_D = \left[ wPtr * -rPtr * \right] \mathrm{mod} \left( 2^B \right)$$

$$\varepsilon_D = \left( f_O - F_{VCXOMin} \right) \left[ \frac{2^B}{F_{VCXOMax} - F_{VCXOMin}} \right]$$

**Equation 11.1**

$$V_O = \left( \frac{V_{ref}}{2^B} \right) \varepsilon_D$$

**Equation 11.3**

$$f_O = \left[ \frac{F_{VCXOMax} - F_{VCXOMin}}{V_{ref}} \right] V_O + F_{VCXOMin}$$

**Equation 11.4**

$$f_O = \left[ \frac{F_{VCXOMax} - F_{VCXOMin}}{2^B} \right] \varepsilon_D + F_{VCXOMin}$$

**Equation 11.5**

$$f_S \left( nT \right) = F_I + \left( F_S \left( 0 \right) - F_I \right) e^{-knT}$$

$$f_S \left( nT \right) = F_S \left( 0 \right) + \left( F_I - F_S \left( 0 \right) \right) \left( 1 - e^{-knT} \right)$$

**Equation 11.16**

$$k = -\frac{1}{nT} \ln \left[ 1 - \frac{f_S \left( nT \right) - F_S \left( 0 \right)}{F_I - F_S \left( 0 \right)} \right]$$

**Equation 11.17**

$$k = \left[ \frac{F_{VCXOMax} - F_{VCXOMin}}{2^B} \right]$$

**Equation 11.20**

$$\text{DLL rise time} = \frac{1}{k}\left(2.1972\right)$$

**Equation 11.21**

## 11.7 REFERENCES

[1] David F. Stout and Milton Kaufman. *Handbook of Operational Amplifier Circuit Design*. New York: McGraw-Hill, 1976.

[2] Analog Devices Inc. *Analog–Digital Conversion Handbook*. 3rd ed. Englewood Cliffs, NJ: Prentice Hall, 1986.

[3] C5260 VCXO Data Sheet. Vectron International, New Hampshire.

[4] AD7837/AD7847 Dual 12-Bit MDAC Data Sheet. Analog Devices, Norwood, MA.

# CHAPTER TWELVE

# Channelized Filter Bank

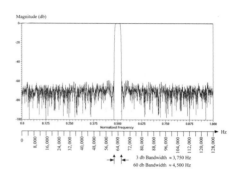

Many years ago, I devoted a great deal of time researching the concepts and mathematics of digital channelizers. The technical papers available at the time were highly theoretical and didn't really provide much in the way of useful information on the actual implementation of a channelizer. It wasn't until 1991 when I came upon a brilliantly written paper, "An Introduction to the FDM-TDM Digital Transmultiplexer" by Dr. John Treichler at Applied Signal Technology [4], that I finally understood how a channelizer (transmultiplexer) actually worked. To me it was as if the clouds suddenly parted and I could see all the channelizer wheels turning and all the gears grinding. I immediately sat down and wrote a software channelizer simulator and spent several days analyzing the simulation results. It worked to perfection, and I learned a great deal from it. Every channelizer design I have ever done since then, and the derivation of the stand-alone channelizer equation discussed in the second section of this chapter, has roots deeply embedded in Dr. Treichler's original paper.

## 12.1 INTRODUCTORY DESCRIPTION

This chapter deals with the filter bank method of converting a wideband input signal into multiple adjacent narrowband base band signals of equal bandwidth. This type of filter bank is called a *channelized filter bank* (CFB). This versatile circuit is found in many applications, such as frequency division multiplex (FDM) to time division multiplex (TDM) conversion, mixing consoles, wideband scanners, and the processing of wideband intercepts in radio astronomy, to name just a few. The channelizer can easily replace hundreds of receivers with not much more than a single integrated circuit. A channelizer utilized in an FDM to TDM conversion or in a TDM to FDM conversion application is oftentimes referred to in the literature as a *transmultiplexer*. In this discussion, we will use the term *channelizer* to reflect the fact that the filter bank we implement here is not limited in scope to only FDM to TDM conversion applications. The real-world FDM to TDM application, however, lends itself well to the development of the channelizer architecture. For this reason, we will use this application in this chapter as a means to develop the channelizer algorithm and to design a meaningful hardware architecture. Therefore we will need to spend some time discussing exactly what an FDM signal is.

An FDM signal consists of multiple independent channels of equal bandwidth that occupy adjacent spectral bands. We will utilize the symbol shown in Figure 12.1, part B, to represent a contiguous band of frequencies associated with a single FDM channel. The letter within the symbol identifies the channel. As is the custom, the symbol of the channel spectrum is drawn as a triangle and is intended to both illustrate the bandwidth of a single channel and differentiate between its lower and upper band edges.

Figure 12.1, part A, is an illustration of a functional signal translator usually referred to as a *mixer* by analog engineers. If a narrow band signal of bandwidth $f_B$ is input to a translator and multiplied by a tuning frequency $f_0$, then the spectrum of that signal will be up converted so that its spectrum is centered at some frequency $f_0$. In doing so, we end up with a signal whose spectrum is composed of a lower sideband and an upper sideband that are mirror images of one another. This is illustrated in part C of the figure. The signal is then passed through a filter to isolate the lower sideband, as illustrated in part D of the figure.

If we perform this translation on several independent signals, all of bandwidth $f_B$, using evenly spaced tuning frequencies $f_0, f_1, f_2 \cdots$ and then sum the outputs, we could build a multiplex structure similar to that shown in Figure 12.2. This structure represents a composite FDM signal composed of a group of adjacent FDM channels. Each channel occupies its own band of frequencies within the multiplex signal and is assumed to be distinct from all others. This symbol graphically illustrates a FDM signal.

An FDM signal consists of multiple independent channels of equal bandwidth that occupy adjacent spectral bands. Some engineering disciplines refer to this type of FDM signal as a *channel stack*. An FDM signal allows for the transmission of many signals on a single medium such as a copper conductor. Until the advent and maturation of modern digital technology, FDM was an efficient method of signal transmission because, depending on individual channel bandwidth, it enabled up to several hundred independent data channels to utilize the same transmission medium. This multiplex scheme saved a whole lot of copper wire during its day, and even though it is old technology, it is still in use today.

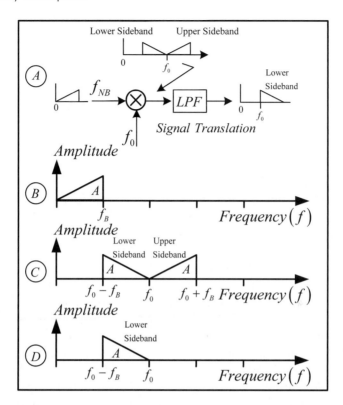

**Figure 12.1**    Representation of a single FDM channel

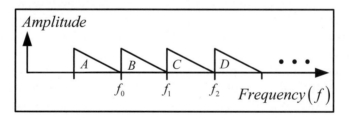

**Figure 12.2**    Representation of several FDM channels

Today's high-speed digital electronics and fiber-optic transmission lines have given rise to a newer multiplex scheme called *time division multiplex*, or TDM. TDM has the capability to carry thousands of independent channels on the same medium, such as a fiber-optic cable. This number far exceeds the bandwidth of the older FDM systems.

A typical TDM system interleaves samples from several independent channels. A group of $N$ multiplexed samples, composed of one sample per each channel, forms a frame of data. We will use the symbol shown in Figure 12.3 to

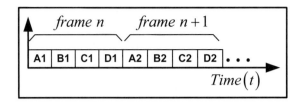

**Figure 12.3**  Representation of a TDM signal

represent a group of interleaved digital samples. Each sample is $B$ bits wide. Samples are labeled according to the input channel A, B, C, D… and according to the sequence number of each sample 1, 2, 3, 4…. This symbol graphically illustrates a four-channel TDM signal.

TDM systems have the advantage of being less expensive. They require less power, have far greater immunity to transmission noise and interference, and have a greater channel carrying capacity on a single medium. In addition, the multiplex/demultiplex equipment that performs channel add and drop services at each end of a fiber-optic transmission line is a great deal simpler and more efficient.

A channelizer can be thought of as a bank of adjacent band pass filters. In an FDM demultiplex application, we could use each of these filters to isolate a single channel for further processing. Another example would be to use a channelizer to segment the spectrum of a wideband signal into narrower bandwidths for processing and analysis. The applications are many. This chapter will develop an efficient method of implementing a digital channelizer filter bank (CFB) that can be used to convert an FDM signal into a TDM signal. This is certainly not the only application of a digital channelizer, but this application will give us a solid focus from which we can illustrate and demonstrate our development concepts. If this particular application is not in your future, this chapter is still a valuable resource because it discusses the properties of a poly phase filter (PPF) and a discrete Fourier transform (DFT) combination that allows us to design and efficiently implement a bank of identical band pass filters. The FDM to TDM conversion is used here, as a real-world application, only to help us better understand the design and implementation of these channelizer band pass filters.

This chapter is divided into four sections:

1. Section 12.2 is a general discussion and a top-level functional overview of the form, function, and high-level mathematics of a channelizer. This section will describe both the PPF and the DFT in the context of mini subsystems. We will introduce the basic mathematical concepts and use the resultant mathematical expressions to generate descriptive system

block diagrams. The material contained in this section is critical to understanding the overall channelizer operation. It will provide us with a solid platform from which to treat the critical design details presented in section 12.3.

2.  Section 12.3 is a detailed discussion of the efficient design of a channelizer. In this section, we will discuss the technical issues associated with the design. We will derive a standalone equation that completely describes a channelizer. This equation will give us an efficient method of directly implementing a channelizer, either in hardware or in software. As we proceed with the step-by-step development of the stand-alone channelizer equation, we will develop in a step-by-step fashion the block diagram of the channelizer architecture.

3.  Section 12.4 presents the results obtained from a channelizer software simulator. The simulator will provide us with graphical illustrations from which we can view and better understand the actual operation and performance of a typical channelizer.

4.  Section 12.5 takes a look at a real-world hardware implementation of a circuit card that implements several channelizers in parallel to process very wide bandwidth signals. This should give the reader some insight into the potential and power of channelizer-based signal processing.

## 12.2  CHANNELIZER FUNCTIONAL OVERVIEW

### 12.2.1  What Is a Channelizer?

The system block diagram in Figure 12.4 consists of two halves. The top half functionally depicts a four-channel FDM system. The bottom half functionally represents a four-channel TDM system. In this diagram, both systems input the exact same signals, labeled channel A through channel D. Also shown is a channelizer pathway between the two systems that converts the FDM signal format to the TDM signal format. The pathway is the signal processing topic we wish to discuss here, but before we begin, it makes sense to start by briefly describing both the FDM and TDM systems.

#### 12.2.1.1  FDM System Description

The FDM system illustrated in Figure 12.4 is shown as having four independent band limited analog inputs, labeled channel A through channel D. For the purposes of our design we have chosen the bandwidth of all the input signals to be 4 KHz. This choice is purely arbitrary. The bandwidths could have been any value, like 64 KHz or 23 MHz and so on. The point here is the bandwidth is determined solely by the application.

In the top half of the block diagram each input signal is represented as an analog function of time and by its analog spectrum. At the input all signals are at base band with bandwidths that range from 0 Hz to 4 KHz. The block diagram shows all signals except channel A being translated up in frequency by multiples of 4 KHz, so that the channel bands occupy adjacent 4 KHz bandwidths. The translator function is represented as a simple multiplier, but it performs the same operation as illustrated in Figure 12.1. In this

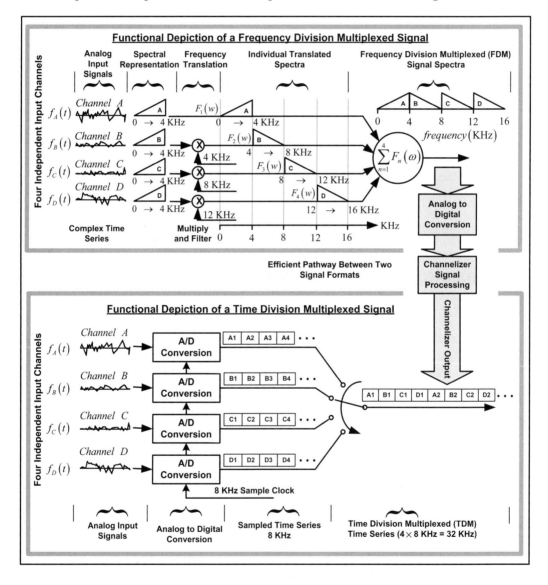

**Figure 12.4**  FDM and TDM signals and the pathway between the two

example we choose to keep the lower sideband of all the translated signals. We also note that the spectrum associated with channel A was not translated in frequency, and therefore its spectrum was not inverted. This is intended to keep this discussion in more general terms and to call attention to the fact that in an FDM system some channels are transmitted inverted and others are not. All four signals are then summed and output. The resulting output is an FDM signal consisting of four independent 4 KHz channels, occupying a total bandwidth of 16 KHz.

Under normal circumstances this FDM signal would be transmitted to a remote receiver where individual channels would be separated from the multiplex by first translating the desired channels to base band, low pass filtering, and then delivering them to their respective destinations. The demultiplexing process would be essentially the reverse of the multiplexing process.

### 12.2.1.2 TDM System Description

The TDM system illustrated in the bottom half of Figure 12.4 is shown as having the same input, consisting of four independent channels labeled channel A through channel D. The first step in the process is to apply each input channel to an analog to digital converter. In this particular example, all the A/D converters are sampling the input channels at an 8 KHz sample rate. The output of each A/D converter is a stream of digital samples that are $B$ bits wide. The samples are identified by their channel ID (A, B, C, or D) and by the successive sample index (1, 2, 3, 4…). The digitized samples are then multiplexed sample by sample to form an interleaved bit stream. The output sample rate of the interleaved stream is computed to be $4 \times 8K$ samples/second. The aggregate sample rate for this simple example is $32K$ samples/second. The output bit rate of the interleaved stream is computed by multiplying the sample rate by the number of bits in the sample, $4 \times 8K$ sample/second $\times B$ bits/sample $= B \times 32K$ bits/second.

The purpose of this chapter is to develop both the algorithm and the channelizer equation that describe in detail the process of converting the FDM signal represented in the top half of Figure 12.4 directly to the TDM signal represented in the bottom of Figure 12.4. This conversion process is represented in the figure by the box labeled "Channelizer Signal Processing" and the arrow labeled "Channelizer Output." In our development, we will not restrict ourselves to just four channels. Our objective is to produce a design equation that can be utilized for all applications, each with their own unique sample rates, channel capacities, channel bandwidths, and so on.

## 12.2.2 Basic Channelizer Block Diagram

A high-level block diagram of a channelizer is shown in Figure 12.5. At the top of the figure, we can see that the input to the channelizer is a wideband

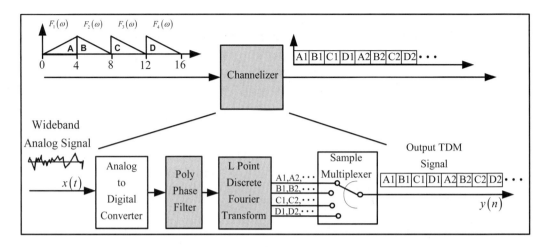

**Figure 12.5**  Channelizer functional block diagram

signal that is modeled as an FDM signal. The channelizer output is modeled as a set of base band time series that can be time division multiplexed
to produce a TDM signal, suitable for efficient transmission to some remote location.

It needs to be repeated here for clarity that the input signal need not
be an FDM signal. It could just as easily be a wideband input signal that we
wish to process in smaller bandwidth chunks. It also should be noted that the
output of the channelizer is not constrained to be a multiplexed TDM signal.
It could just as easily be a set of independent narrow band signals extracted
from the input. Don't forget that the FDM to TDM application we are discussing here is used solely to give us a concrete application to focus on. There are
countless other applications suitable for channelizer processing.

The channelizer block is broken down to its major components in the
bottom of the Figure 12.5. The block diagram shows that a wideband analog
signal is applied to an A/D converter, where it is digitized to samples that are
$B$ bits wide. The sampled data stream is then passed to a poly phase filter
and from there to an $L$ point DFT. The serial combination of a PPF and an
$L$ point discrete Fourier transform will produce $L$ equal bandwidth channelized band pass signals, all of which are down converted to base band. The
example input wideband signal is the same as the example FDM signal in
Figure 12.4 only to show that a properly designed channelizer can processes
signals with inverted spectra as well as signals with erect spectra.

The $L$ point DFT output consists of a vector $L$ points in length. Each
point is a single complex sample for the $L$th channel of the composite signal.
Successive DFTs will produce an output time series for each of the $L$ channels. We can think of the DFT as being a set of $L$ complex tuners that translate

the spectrum of the input signal down to base band in increments of the DFT bin width. It then isolates the spectral components of each adjacent bin with a low pass filter of fairly poor quality. The job of the PPF is to enhance the characteristics of the filter inherent to the DFT.

The final block shows a sample multiplexer that simply interleaves one sample per channel to form a continuous TDM data stream. The inclusion of this multiplexer is application dependent. Many applications do not use it simply because the intent is to segment the wideband input into $L$ independent time series. The PPF and the DFT blocks are highlighted to indicate that they are the heart of the channelizer system. The remainder of this chapter will deal specifically with these two functions. Initially we will break each of the two blocks down into their functional components and develop some high-level descriptive mathematics to characterize them. Later we will dig deeper to produce a detailed channelizer equation that can be implemented in both hardware and software.

### 12.2.3 System View of the DFT

Equation 12.1 mathematically describes an $L$ point DFT:

$$Y(k) = \sum_{n=o}^{L-1} x(n)e^{-j\left(2\pi\frac{k}{L}\right)n} \quad k = 0,1,2,\cdots L-1$$

**Equation 12.1**

The DFT processes an $L$ point input sequence $x(n)$ and produces an $L$ point output sequence $Y(k)$. Each point in the output is referred to as the output of a DFT bin. There are $L$ bins in a DFT, each of which are $(2\pi/L)f_s$ Hz wide, or in normalized notation, each bin has a width of $2\pi/L$.

The input sequence $x(n)$ is correlated against $L$ complex sinusoids, sometimes referred to as *analysis frequencies*. The $k$th correlation score is assigned to the $k$th DFT bin and given the notation $Y(k)$. The correlation score for the $k$th DFT bin is a measure of how close the frequency content in the input signal matches the $k$th analysis frequency. We can also think of the exponential term in Equation 12.1 as being a collection of $k$ complex tuners that translates the spectrum of $x(n)$ to base band in increments of the DFT bin widths. The base band spectrum is then low pass filtered and output as an estimate of the frequency content of the $k$th DFT bin. The collection of all $k$ bins is tabulated as $Y(k)$. The low pass filter function is the result of the summation operation.

In our discussions of a DFT, we will often refer to the internals of a DFT as a collection of complex tuners and a single low pass filter. It is also acceptable to view the internals of a DFT to be a collection of adjacent band pass filters that span the width of the input signal spectrum. Depending on the context of the discussion, both views are useful and will be used interchangeably. There are two ways to view the operation of a DFT:

1. The conventional view is that an $L$ point DFT is nothing more than a bank of $L$ identical band pass filters spectrally positioned adjacent to one another so that, taken together, they span the frequency range from 0 Hz to $f_s$ Hz, or from $-f_s/2$ Hz to $+f_s/2$ Hz, depending on the interpretation of the output.

   If the input to the DFT is a single $L$ point real or complex data vector, the output of the DFT will be an $L$ point complex vector. The $k$th complex point in the output vector represents an estimate of the instantaneous magnitude and phase of that portion of the input signal that lies within the bandwidth of the $k$th DFT band pass filter.

   In this view, the $L$ output points of the DFT provide us with a snapshot of the estimated input signal spectra across the entire signal bandwidth. It is customary to refer to each band pass output of a DFT as a *bin*. An $L$ point DFT will have $L$ output bins. The ideal spectral resolution of the DFT output would be the bandwidth of a single bin. Without some additional processing enhancement in the form of a PPF, this ideal resolution does not occur. In reality, for an unmodified DFT, the bandwidth of each bin's band pass filter overlaps into adjacent bins.

2. A second view of the DFT is similar to the first view, except we treat the DFT as a bank of complex tuners and a single low pass filter. Each output bin of the DFT is the base band version of the band pass signal in view 1. In this view, instead of applying a single input data vector, we apply a continuous sequence of input vectors. We can sequentially index each $L$ point input vector so that they take on the numbered sequence $0,1,2,3,\cdots,m,m+1,m+2,\cdots,\infty$. Thus the output of each of the $L$ DFT bins can be written as

$$Y(k,m) \quad \text{for} \quad \left\{ \begin{array}{ll} 0 \le k \le L-1 & L = \text{the number of complex tuners} \\ m = 0,1,2,\cdots\infty & \text{sample index of filter output sequence} \end{array} \right\}$$

   where the index $k$ refers to the output of the $k$th DFT bin and the index $m$ refers to the $m$th sample of the bin output sequence. For simplicity, we usually drop the sample index notation $m$ and refer to the $k$th bin output simply as $Y(k)$, with the understanding that the output consists of a sequence of length $m$ samples. For this view, the output of each DFT bin will be a steady sequence of complex samples that represent the time varying magnitude and phase of the input signal that falls within the bandwidth of that bin's filter.

Of the two views, view number 2 is the one that we will utilize for our channelizer. That is, the DFT is simply a bank of adjacent tuners and a single base band filter that isolates unique bands of frequencies within the input wideband signal. We interpret the steady output of an $L$ point DFT as an $L$ independent base band complex time series.

Figure 12.6 shows an example of processed samples output from the $k$th DFT bin. The input to the DFT is presented with a steady stream of serial data labeled $x(n)$. The data sequence is formed into $L$ sample vectors and applied to the input of an $L$ point DFT. Every time an input vector is processed, each bin of the DFT will output a single complex sample. The $m$th input vector produces the $m$th output sample. In this example, the input vectors are continuously processed, one after another, producing a sequence of output samples from each DFT bin labeled $Y(k,m)$. The output of each of the $L$ bins of the DFT can be thought of as a complex time series that represents the signal content that falls within that bin's bandwidth. In order to make this figure more descriptive, the output sample sequence is shown as being the magnitude of each complex sample in the output time series. The other DFT bins can be viewed as producing similar time series.

Each DFT channel or bin processes the same input sequence. The only difference is that the DFT frequency index $k$ takes on a unique value of 0 through $L-1$, corresponding to the $L$ bins of the DFT. So the tuner in successive bins shifts the input signal spectrum down in frequency by increments of

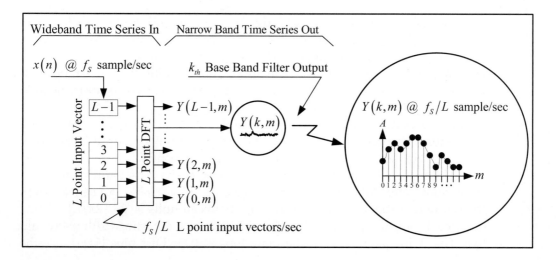

**Figure 12.6**  Output time series of the DFT *k*th channel

$2\pi/L$ Hz, which is the ideal bandwidth of a single DFT bin, prior to presenting them to the summer low pass filter.

We can visualize this process by looking at Figure 12.7. This figure illustrates the summation low pass filter superimposed on the shifted input signal spectrum for the DFT channels 0, 1, and 2. We can see that each time the value of $k$ is incremented the input signal is shifted down in frequency by an additional increment of $2\pi/L$. The DFT implements its bank of base band filters by incrementally down converting the input signal spectrum and then isolating a portion of that spectrum via the low pass filter properties of the summation operator. An $L$ point DFT will do this $L$ times to produce $L$ separate base band outputs represented by the notation

$$\{Y(0), Y(1), Y(2), \cdots Y(L-1)\}$$

In Chapter 3, "Review of the Fourier Transform," the frequency response of this filter inherent in the DFT was derived. The magnitude of the filter response is given by

$$\left|W\left(e^{j\omega}\right)\right| = \frac{\sin\left(\dfrac{\omega L}{2}\right)}{\sin\left(\dfrac{\omega}{2}\right)}$$

**Equation 12.2**

The plot of the magnitude in decibels for the DFT low pass filter is illustrated in Figure 12.8. There are two frequency axes shown: the top axis is calibrated in normalized radians, and the bottom axis is calibrated in normalized frequency. We can see from this figure that the filter response repeats itself every $2\pi$ radians or every hertz when the frequency axis is normalized to the sample rate $f_s$. If the reader prefers not to use normalized frequency, then the filter response repeats itself at the sample rate or every $f_s$ Hz. Depending on the reader's preference, the frequency response nulls occur at intervals of $1/L$, $f_s/L$ Hz, $2\pi/L$ radians, or $2\pi f_s/L$ radians/second.

The main lobe in the DFT filter response is the pass band of the DFT summation filter. All other lobes are referred to as *side lobes*. We can see from the figure that the main lobe bandwidth between nulls is $2 \times (2\pi/L) = 4\pi/L$, or twice the width of a single DFT bin. This is twice the bandwidth we would like. The 3 db bandwidth of the main lobe is 0.89 DFT bins [11]. Since the

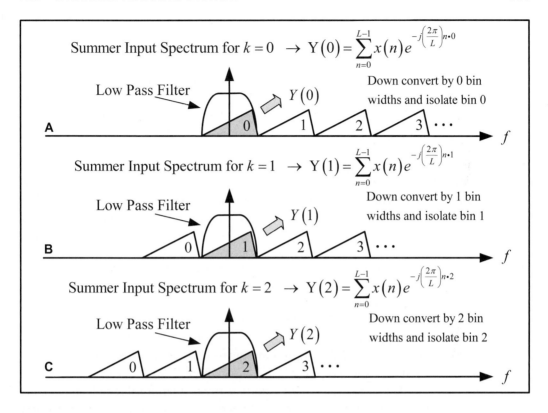

**Figure 12.7** Incremental tuning of the input signal spectrum

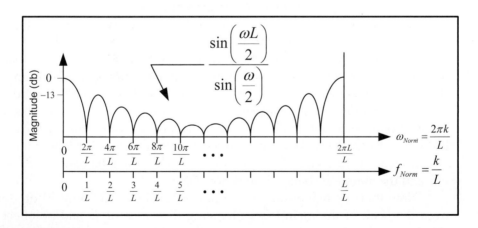

**Figure 12.8** Plot of the frequency response of the DFT filter

width in frequency of a DFT bin is equal to $f_s/L$ Hz, the 3 db bandwidth is given by $0.89 f_s/L$ Hz. The side lobe falloff from the main lobe occurs at a rate of approximately 6 db/octave.

We can also see from Figure 12.8 that the highest side lobe is only 13 db down from the main lobe. This indicates that the out of band rejection for this filter is not very impressive. We probably should be thankful for what we get, since we didn't expend any real effort to achieve this filter. This $\sin x/x$ like filter is the result of the summation operation and is inherent to the DFT.

We get this filter for free, and for many applications, this filter is sufficient for our needs. Nevertheless, if we want superior performance, as we will when we design the channelizer, we will need to improve upon this filter. We will discuss the method for doing this a bit later. It is very interesting to view Figure 12.8 in our minds and consider what happens as the length $L$ of the DFT increases. As $L$ gets larger, the width of all the response lobes gets narrower. This means the main lobe pass band gets narrower and the filter becomes more selective. An increase in DFT length, however, does nothing to increase the attenuation of the first side lobe. That's going to be stuck at 13 db, no matter how large $L$ becomes.

Figure 12.8 is an artist's conception. An actual plot of the filter frequency response of Equation 12.2 is shown in Figure 12.9. This figure shows the 1024-point frequency response plot for the case where $L = 16$ and the sample frequency $f_s$ is 8000 Hz. This plot contains a great deal of information. To start, it has three separate frequency axes. Axis 1 is calibrated in terms of normalized radians and has the range of 0 to $2\pi$. Axis 2 is calibrated in terms of normalized frequency and ranges from 0 to 1. Axis 3 is calibrated in hertz and has a range of $0 \text{ to } f_s$ Hz, where $f_s$ is the sample rate.

As an example, for the case where the sample rate $f_s = 8000$ Hz and $L = 16$, the side lobe bandwidth of the filter shown in Figure 12.9 is given by $f_s/L = 8000 \text{ Hz}/16 = 500$ Hz. The main lobe bandwidth is twice that, or 1000 Hz. We can clearly see from Figure 12.9 that the first side lobe is down 13 db from the main lobe and that successive side lobes have a pretty slow roll-off.

What happens if we increase the length of the DFT? Figure 12.10 shows the 1024-point summation filter frequency response plot for the case where $L$ is doubled to 32. We should expect to see the width of the side lobes and main lobe decrease by a factor of two, and we do. The side lobe bandwidth is 250 Hz, and the main lobe bandwidth is now 500 Hz.

Note that the magnitude of the first side lobe is still only 13 db down from the peak of the main lobe. Changing the value of $L$ had no effect on side lobe attenuation. It is important for the reader to remember that the plot illustrated in Figure 12.9 and Figure 12.10 is the frequency response of the low pass filter inherent to the unmodified DFT.

### 12.2.3.1 DFT Complex Exponential Term

The next term in the DFT Equation 12.1 that we need to look at is the complex exponential

$$e^{-j\left(2\pi\frac{k}{L}\right)n} \quad \left\{ \begin{array}{l} n = 0,1,2,3,\cdots,L-1 \\ k = 0,1,2,3,\cdots,L-1 \end{array} \right\}$$

**Equation 12.3: DFT phasor term**

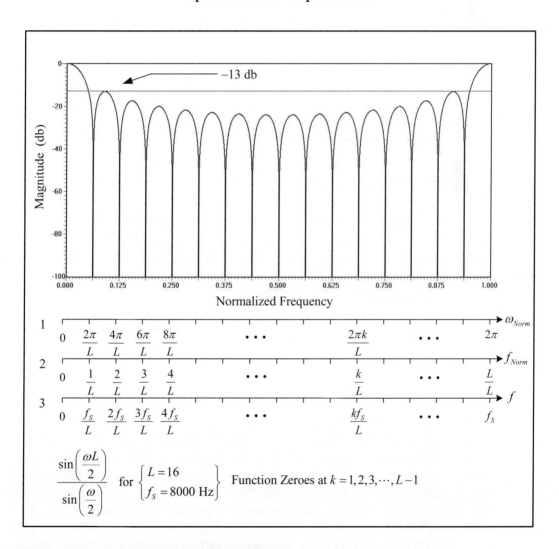

**Figure 12.9** Detailed plot of the summation filter for $L = 16$

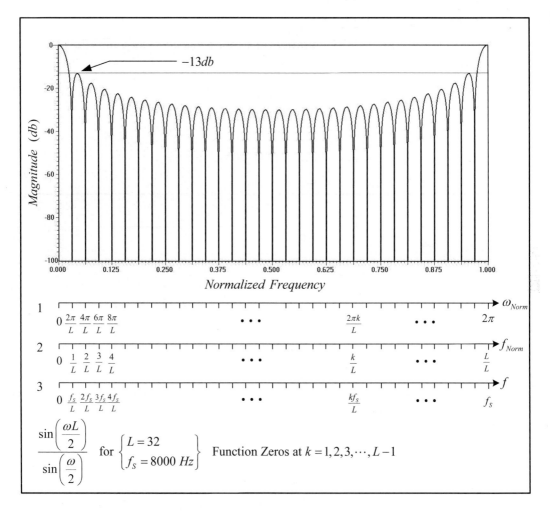

**Figure 12.10**  Detailed plot of the summation filter for $L = 32$

This term is an interesting and an extremely compact and versatile mathematical expression. We can think of this term as a complex phasor that rotates around the unit circle at specific frequencies. As illustrated in Figure 12.11, we see that Equation 12.3 divides the unit circle into $L$ equal arc lengths. Each arc length is shown between adjacent black dots in the figure and is determined by the size of the $2\pi/L$ radian chunk. The rate at which the phasor rotates is determined by the value of the DFT bin index $k$ and the data sample rate $f_s$.

The phasor will rotate at a normalized radian frequency of $2\pi k/L$. What causes the phasor to rotate? It's the index $n$ that increments at the sample

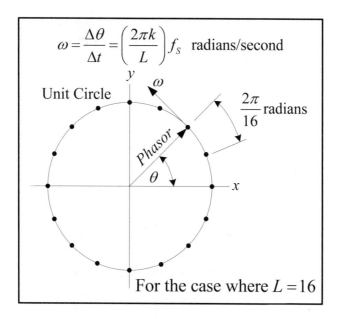

**Figure 12.11**   Complex phasor rotation on the unit circle

rate $f_s$. So we can say that the complex sinusoid $e^{-j\frac{2\pi k}{L}n}$ rotates around the unit circle at a rate of $(2\pi k/L)f_s$ radians/seconds.

The period of the sample clock is given by $T = 1/f_s$. We note from the figure that at any given sample time $nT$, the rotating phasor forms an angle $\theta$ with respect to the positive $x$-axis. The angular rate of phasor rotation is equal to the change in the angle $\theta$ with respect to time or

$$\omega = \frac{\Delta\theta}{\Delta t}$$

If $k$ is set to 0 then the phasor assumes the constant value of unity and does not rotate. This is equivalent to a DC waveform. If $k$ is set to 1, then the phasor will rotate around the unit circle taking on complex values at every possible radian chunk and the normalized rotation rate will be $2\pi/L$. If $k$ is set to 2, the phasor will rotate twice as fast and will take on complex values at every second radian chunk and the normalized rotation rate will be $4\pi/L$. The phasor will rotate at an incrementally faster rate as the value of $k$ is stepped through its range of $0 \leq k \leq L-1$. The beauty of Equation 12.3 is that its spectrum is simply an impulse at its normalized frequency. For example, the spectrum of the complex sinusoid $e^{j2\pi\left(\frac{m}{L}\right)n}$ is the single impulse at the

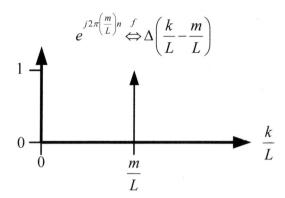

**Figure 12.12**  Spectrum of a positive complex sinusoid

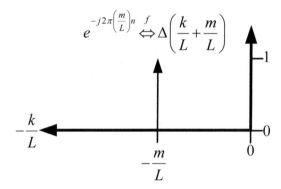

**Figure 12.13**  Spectrum of a negative complex sinusoid

normalized frequency $m/L$, as illustrated by the Fourier transform pair in Figure 12.12. Similarly the spectrum of the complex sinusoid $e^{-j2\pi\left(\frac{m}{L}\right)n}$ is illustrated by the Fourier transform pair in Figure 12.13.

Each complex exponential in the DFT is referred to as an *analysis frequency*. This is because the signal sequence input to a DFT is analyzed or correlated against all $L$ analysis frequencies. The output of each DFT bin is the correlation score of the input sequence and the analysis frequency corresponding to that bin.

So what does the spectrum of the sum of complex exponential terms in the DFT look like? Well, we can take the Fourier transform of all the negative exponential terms in Equation 12.3 to get

$$f\left\{\sum_{n=0}^{L-1} e^{-j2\pi\left(\frac{k}{L}\right)n}\right\} \text{ for } k = 0,1,\cdots,L-1 \;=\; \sum_{k_0=0}^{L-1} \Delta\left(\frac{k}{L}+\frac{k_0}{L}\right)$$

**Equation 12.4**

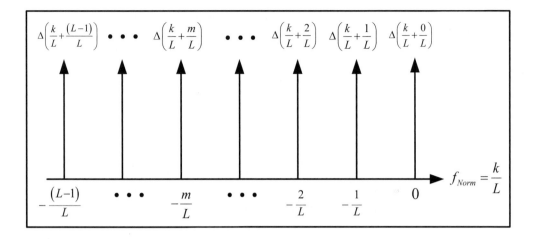

**Figure 12.14**  Spectrum of the sum of negative exponentials

Figure 12.14 illustrates the spectrum of the $L$ complex sinusoids described by Equation 12.4.

Each complex sinusoid is represented by an impulse in the frequency domain. The frequency axis is calibrated in normalized frequency. From the figure, we can clearly see that the spectrum is composed of a series of frequency impulses, all of which are to the left of the origin. The frequency of each successive phasor is centered at $-k_0/L$, where $k_0 = 0, 1, 2, \cdots, L-1$.

Now let's assume for a moment that we have a sum of positive exponentials given by the expression

$$\sum_{n=0}^{L-1} e^{+j2\pi\left(\frac{k}{L}\right)n}$$

The spectrum for this exponential sum is computed to be the sum of impulses given by

$$f\left\{\sum_{n=0}^{L-1} e^{+j2\pi\left(\frac{k}{L}\right)n}\right\} \text{ for } k = 0,1,\cdots,L-1 \;\; = \sum_{k_0=0}^{L-1} \Delta\left(\frac{k}{L} - \frac{k_0}{L}\right)$$

The spectral plot of this sum of positive exponentials is illustrated in Figure 12.15. Notice in this case that all the impulse spectra are located to the right of the plot origin.

Keep the line spectra in Figure 12.14 and Figure 12.15 in mind for a moment because they are important to the following discussion. It is important to note that for the band pass view of the DFT, the location of the impulses in Figure 12.15 mark the center of the $L$ band pass filters or $L$ bins of an $L$ point DFT.

Now let's take a look at an artist's conception of the main lobe of the DFT low pass filter plotted against normalized frequency, as illustrated in Figure 12.16. We can get a visual idea of just how good the DFT band pass filter is if we look at it with respect to the centers of the DFT bins shown in Figure 12.15.

The convolution of the spectrum of the DFT filter with the spectrum of the exponential $e^{j2\pi\left(\frac{k_O}{L}\right)n}$ results in a frequency translation of the low pass filter to a band pass filter centered at the frequency $k_O/L$. This is illustrated in Figure 12.17 for the case where the angular frequency of the phasor is set to $2\pi k_0/L$. In the figure, we can see that the spectrum of the main filter lobe is plotted relative to the normalized radian frequency $\omega/f_s$ and relative to the normalized linear frequency $f/f_s$. We see that the filter main lobe has been translated to

$$\omega/f_S = \frac{2\pi k_0}{L}$$

or

$$f/f_S = \frac{k_0}{L}$$

Now let's imagine that each of the line spectra shown in Figure 12.15 represents the center frequency of an individual channel in a wideband FDM signal. That is, each of the line spectra represents the center of a DFT bin. We

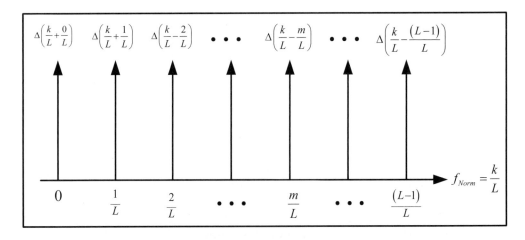

**Figure 12.15**   Spectrum of the sum of positive exponentials

can generate the graphical band pass filter version of the DFT by convolving the low pass frequency response of the DFT with each of the impulses located at the bin centers. If we superimpose the main lobe of the DFT filter on each bin center, we end up with the spectral plot illustrated in Figure 12.18. This plot is a graphical representation of the main lobes in the $L$ adjacent band pass filter view of the DFT. This interpretation of a DFT from a band pass point of view provides a great deal of information.

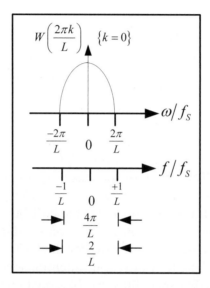

**Figure 12.16**  DFT summation filter main lobe

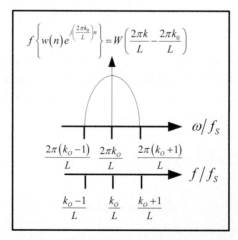

**Figure 12.17**  Translated DFT summation filter main lobe

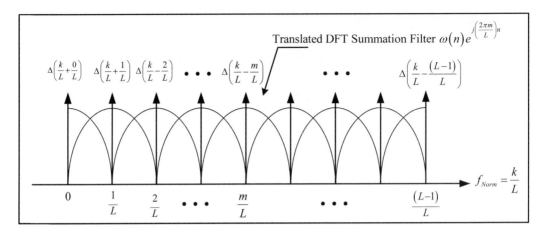

**Figure 12.18**  DFT summation band pass filter bank

Figure 12.18 shows that the translated filters are all identical, and each has a main lobe null-to-null width of $2/L$. Each filter pass band has a 50% overlap with the band pass of the filters adjacent to it. This plot is representative of the performance of the bank of band pass filters inherent in the unmodified DFT. If we remember that the DFT bins are defined to be at multiples of $1/L$, we can see that each filter is centered on a DFT bin.

Let's now revisit the expression for the DFT, as given in Equation 12.1, which is repeated here for clarity. Our intention is to graphically interpret the system operations that are defined by this equation:

$$Y(k) = \sum_{n=0}^{L-1} x(n) e^{-j2\pi\left(\frac{k}{L}\right)n} \quad k = 0, 1, 2, 3, \cdots, L-1$$

We begin by using brackets to separate terms in the DFT equation. This separation is shown in Equation 12.5. We can visualize the DFT as a sequence of system events, with the events within the brackets occurring first, followed by the events external to the brackets:

$$Y(k) = \sum_{n=0}^{L-1} \left[ x(n) e^{-j2\pi\left(\frac{k}{L}\right)n} \right] \quad k = 0, 1, 2, 3, \cdots, L-1$$

**Equation 12.5**

Let's now look at the DFT from a tune and low pass filter representation. In doing so, we can visualize the system operation that occurs within the brackets as an input time series denoted by $x(n)$ being tuned or down converted in frequency by the complex exponential

$$e^{-j2\pi\left(\frac{k}{L}\right)n}$$

Remember from Figure 12.14 that the complex exponential takes on $L$ different sinusoidal frequencies ranging from 0 Hz to $(L-1)/L$ Hz in multiples of the normalized frequency $1/L$. Each of these $L$ frequencies corresponds to a DFT bin. Each time the index $k$ is incremented, the spectrum of the input sequence $x(n)$ is down converted by another increment of $1/L$. Each increment will down convert the next adjacent frequency bin to base band so that the bin will be centered at 0 Hz.

The down converted signal is then summed sample by sample, which is the equivalent of processing the translated signal by a block-averaging filter that has the frequency response given by $\sin(\omega L/2)/\sin(\omega/2)$. As discussed previously, the main lobe of the filter has a low pass characteristic with nulls that span exactly two DFT bin widths. The filter main lobe from null to null spans a $2/L$ wide signal bandwidth centered at a DFT bin. This process is repeated $L$ times, producing DFT outputs consisting of $L$ complex numbers, each of which represents the magnitude and phase of the correlation score of the input signal within each of the $L$ adjacent bands.

If we process a single DFT, we get an output vector of $L$ numbers that can be interpreted as the estimated magnitude and phase of the spectrum of the input signal $x(n)$ taken at adjacent frequency bands. If we process $m$ DFTs in succession, the output of the DFT will take on the appearance of $L$ separate time series, each of which is $m$ complex samples in length. Each of the $L$ series can be visualized as the $m$ sample time history of the magnitude and phase of the band limited signal passed by the DFT filter.

Let's visualize the processing that goes on inside a DFT from a systems point of view. Snapshots of the internal processing of a DFT are illustrated in Figure 12.19. In part A of the figure, we see the spectrum $X(k/L)$ of some input signal $x(n)$. We also see the superimposed frequency impulses

$$\Delta\left(\frac{k}{L}-\frac{k_0}{L}\right) \quad \{\text{for } k_0 = 0,1,2,\cdots,L-1\}$$

that represent the center frequencies of sequential DFT bins. In addition, we also see the low pass frequency response of the main lobe of the DFT filter. The first operation the DFT performs is multiplying the input time series $x(n)$ by the complex exponential $e^{-j2\pi\left(\frac{0}{L}\right)n}$, where $k = 0$. This is the equivalent of translating by 0 Hz. In this case, the spectrum of the input signal is not translated and the portion of the spectrum that falls under the low pass response of the DFT is filtered from the remainder of the input signal spectrum. Depending on the notation the reader uses, the DFT output for this operation will be a single complex point associated with the bin

$$Y(k) = Y(0)$$

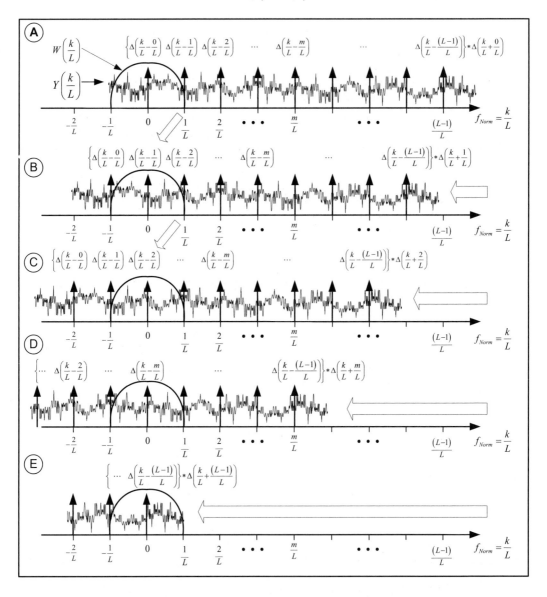

**Figure 12.19** System interpretation of the signal processing performed by a DFT

or

$$Y\left(\frac{k}{L}\right) = Y\left(\frac{0}{L}\right)$$

that represents an estimate of the magnitude and phase of the signal that falls within the band pass of the filter and is associated with the DFT bin 0.

The second operation the DFT performs is illustrated in part B of the figure. Here we see that the same time series $x(n)$ is multiplied by the complex exponential $e^{-j2\pi\left(\frac{1}{L}\right)n}$, where $k = 1$. This is the equivalent of translating the input signal spectrum by one DFT bin width. We note that the entire input signal spectrum is tuned or down shifted by exactly one DFT bin and passed through the low pass filter response of the DFT. The DFT output for this operation will be a single complex point associated with the bin

$$Y(k) = Y(1)$$

or

$$Y\left(\frac{k}{L}\right) = Y\left(\frac{1}{L}\right)$$

that represents an estimate of the magnitude and phase of the signal that falls within the filter bandwidth and is associated with DFT bin 1.

The third operation the DFT performs is illustrated in part C of the figure. Here we see that the same time series $x(n)$ is multiplied by the complex exponential $e^{-j2\pi\left(\frac{2}{L}\right)n}$, where $k = 2$. This is the equivalent of translating the input signal spectrum by two DFT bin widths. We note that the entire input signal spectrum is tuned or down shifted by exactly two DFT bins and passed through the low pass filter response of the DFT. The DFT output for this operation will be a single complex point associated with the bin

$$Y(k) = Y(2)$$

or

$$Y\left(\frac{k}{L}\right) = Y\left(\frac{2}{L}\right)$$

that represents the estimate of the magnitude and phase of the signal that falls within the filter bandwidth and is associated with DFT bin 2.

This process is repeated $L$ times: once for each of the $L$ DFT bins. In part D of the figure, we see the processing of the $m$th bin of the DFT, and in part E we see the processing for the $(L-1)$th DFT bin.

From Figure 12.19, we can see that the DFT is nothing more than a collection of $L$ tuners and a fairly poor quality low pass filter. There is no getting around it. That is, the DFT is merely a collection of tuners and a poor low pass filter. Clearly these filters are not ideal for our channelizer application, since each filter contains spectral leakage from adjacent DFT bins. When the high side lobes of the window filter are considered, the leakage problem only

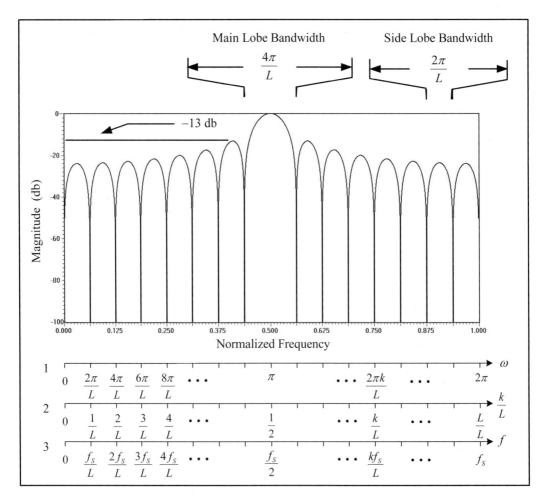

**Figure 12.20** DFT summation filter for $L = 16$

becomes worse, since the filter side lobes extend into all of the DFT bins and therefore the content of all the bins make a contribution to the value computed for any particular bin. Clearly this is not good for any application that requires precise band pass filtering.

To get a clear picture of the frequency response of the filter inherent to the DFT, we choose to view the DFT as a bank of band pass filters. A translated DFT filter for the case, where $L = 16$ and $k = 8$, is shown in Figure 12.20. The DFT low pass filter has been multiplied by the complex exponential

$$e^{j2\pi\left(\frac{k}{L}\right)n} = e^{j2\pi\left(\frac{L/2}{L}\right)n} = e^{j\pi n}$$

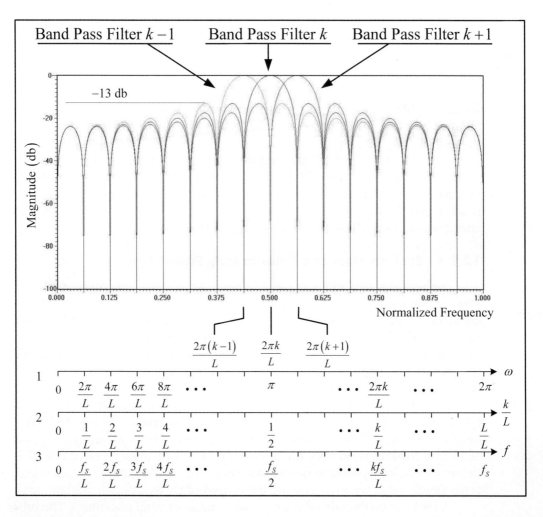

**Figure 12.21** Spectral leakage from adjacent band pass filters

Depending on our preference, the exponential has centered the main lobe of the band pass filter at either $\pi$ radians, 0.5, or $f_s/2$ Hz.

Figure 12.21 illustrates the same filter but this time with two adjacent filters superimposed on one another. In the figure we can see the filter response for the $k$th, $(k+1)$th, and $(k-1)$th DFT band pass filters. Note that the filter pass band or main lobe is two DFT bins wide at its nulls and overlaps the adjacent filter band pass by 50%. Also note that the relatively high side lobes of each filter extend into all other DFT bins. Therefore the signal content at the output of each individual DFT bin contains not only signal spectral components from that bin but also contaminating spectral components from adjacent bins and, to a lesser degree, from all other bins as well. Depending on the application, this spectral leakage can be a serious problem.

The low pass filter inherent to the DFT is sufficient for many applications. However, the $\sin x/x$ "like" filters that we get for free from the DFT are not the optimum set of band pass filters, and they are not suitable for narrow band channelizer applications. It makes sense for the channelizer application if the band pass of the DFT filter is defined to be equal to the bin width of the DFT. Therefore, in order to design our channelizer, we will need to change the bandwidth of the filter inherent to the DFT so that it is equal to a single DFT bin width.

We will vastly improve the quality of the DFT filter when we add a separate filter to the DFT. We will replace the sloppy DFT filter by a narrow band low pass filter with an impulse response $h(n)$ and in the process realize a much more efficient channelizer. Before we begin this discussion, we need to spend some time getting familiar with a PPF from a systems viewpoint.

### 12.2.4  System View of a Filter in Poly Phase Form

The mathematics of a PPF can get a little cumbersome. Sometimes math can obscure the overall picture of a system or function. Rather than cloud our discussion with a lot of Greek and squiggly characters, we are going to continue with the functional overview intent of this section, which means we will minimize the math and utilize descriptive graphics wherever possible. We will treat the mathematics later, after we get a firm understanding of the functional concepts.

Our objective now is to enhance the filter response characteristics of the DFT to reduce spectral leakage and to isolate the spectral content of each DFT bin from all other bins. In doing so, we will create, from a band pass point of view, a bank of band pass filters that uniquely isolate each bin of the DFT. Before we begin, we should define some common mathematical notation that is frequently used when discussing the output of DFTs.

A DFT can be thought of as a bank of $L$ adjacent band pass filters. The processed data out of each filter can be thought of as being placed in an output bin.

The $L$th filter output is referred to the $L$th bin of the DFT. The DFT output can be indexed in terms of these bins by using the index $k$, for $k = 0, 1, 2, \cdots, L-1$. If the DFT input sequence notation is $x(n)$, for $n = 0, 1, 2, \cdots, L-1$, then the DFT output sequence notation can be written as $Y(k)$, for $k = 0, 1, 2, \cdots, L-1$. In this representation, the index $k$ refers to the DFT bin number. We can also index the DFT output by its incremental normalized frequency of $k/L$. That is, for the same input sequence $x(n)$ the DFT output sequence is indexed as $Y(k/L)$ for $k = 0, 1, 2, \cdots, L-1$. Both of the two index methods are correct, and we shall use both interchangeably in this chapter.

We originally defined the DFT way back in Equation 12.1, which is repeated here for clarity:

$$Y(k) = \sum_{n=0}^{L-1} x(n)e^{-j2\pi\left(\frac{k}{L}\right)n} \qquad \text{for } k = 0, 1, 2, 3, \cdots, L-1$$

This definition can be modified by the addition of a filter impulse response $h(n)$ to produce

$$Y(k) = \sum_{n=0}^{L-1} h(n)x(n)e^{-j2\pi\left(\frac{k}{L}\right)n} \qquad \begin{array}{l} \text{for } k = 0, 1, 2, 3, \cdots, L-1 \\ \text{for } L = \text{ The number of bins in the DFT} \end{array}$$

### Equation 12.6

For our immediate purpose, we can visualize the processing described by Equation 12.6 as the spectrum of the input signal $X(k/L)$ being translated down in frequency by an amount $k_0/L$ and then multiplied by the frequency response of the DFT filter $H(k/L)$. This would occur for each of the $k$ bins of the DFT.

The frequency response of the new filter is a vast improvement over the $\sin x/x$ like response of the filter inherent to the naked DFT. The filter is a standard equal ripple finite impulse response (FIR) filter that has been designed so that it is centered at 0 Hz and has a bandwidth that approaches the spectral width of a single DFT bin or $2\pi/L$. Remember that the filter inherent to the DFT had a main lobe bandwidth of $4\pi/L$, the equivalent of two DFT bins. The width of the transition band of $H(k/L)$ is designed as narrow as possible and the stop band attenuation is on the order of 60 db or more to eliminate spectral leakage from other DFT bins—a remarkable improvement over the 13-db side lobe level of the inherent DFT filter.

We will show later that the length of the filter impulse response can be very long, up to $N$ coefficients, and the width of the DFT can be much shorter, up to $L$ bins, where $N$ is much greater than $L$.

From this point forward in our functional development, we will refer to the filter $h(n)$ as a poly phase filter (PPF) or as an enhanced DFT filter or simply as a filter. All three definitions are interchangeable. A PPF is just a fancy name for a common, ordinary, everyday FIR filter. The only difference between a poly phase FIR and an everyday run of the mill FIR is that the coefficients of a PPF are ordered in a slightly different fashion. This ordering will be discussed a bit later. It will be shown that we can implement the PPF with an impulse response $h(n)$ that can be much larger than the width of the DFT.

The concept that is important to understand at this point is that the data applied to the filter enhanced DFT are a real time continuous sequence. Figure 12.22 shows the continuous sequence of serial input data samples labeled $a(n)$ being applied to the input of the poly phase filter.

Sequences of length $L$ are carved out of $a(n)$ and formed into a steady stream of $L$ point vectors labeled $x(n)$, where $n = 0,1,2,3,\cdots L-1$. The poly phase filter processes these vectors to produce a continuous stream of $L$ sample output vectors $z(n)$, where $z(n)$ is just the product of the filter coefficient vector $h(n)$ and the input data vector $x(n)$—that is, $z(n) = x(n)h(n)$. The DFT then processes the $L$ point product vector $z(n)$ and produces an $L$ point output vector

$$Y(k) = \sum_{l=0}^{L-1} z(n)e^{-j2\pi\frac{k}{L}n} = \left\{Y(0),Y(1),\cdots,Y(k),\cdots,Y(L-1)\right\}$$

Each of the $L$ elements in the output vector represents a single complex point for each of the $L$ DFT bins. As a continuous stream of successive input vectors continue to be processed, we can view the output of each individual

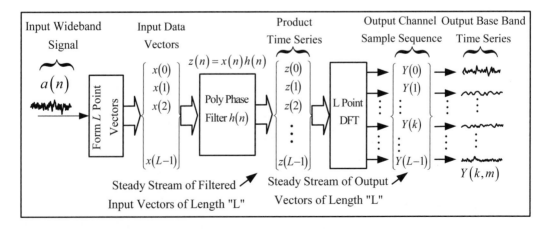

**Figure 12.22**  Poly phase enhanced DFT

DFT bin as being a continuous stream of complex samples that form a filtered narrow band time series that can be represented by

$$Y(k,m) = \{Y(0,m), Y(1,m), \cdots, Y(k,m), \cdots, Y(L-1,m)\}$$

where $m$ is the sample index for each channel that can take on the range of $m = 0, 1, 2, 3, \cdots, \infty$. For clarity, the output complex time series for each DFT bin in Figure 12.22 is drawn as though it is the computed magnitude of the series.

Even though we are using an $L$ point DFT, we will not constrain the length of the input time series $x(n)$ or the length of the filter impulse response $h(n)$ to be $L$ samples in length. In general, the filter characteristics of $h(n)$ improve as the length of the filter's impulse response increases. For example, we can attain narrower filter transition bands and greater out of band attenuation by increasing the number of coefficients that comprise the filter's impulse response. Narrow band filters with minimal spectral leakage are critical to the implementation of a channelizer. Digital filters with extremely tight specifications such as these usually have a long impulse response. For reasons we will discuss later, the length of the filter impulse response $h(n)$ and the length of the data series $x(n)$ will be chosen so that they are equal and so that their length is an integer multiple of the DFT size $L$.

The filter is normally designed using one of the many available filter design software packages that utilize the Parks-McClellan algorithm for the design of equal ripple FIR filters. These design techniques are discussed in Chapter 5, "Finite Impulse Response Digital Filtering," and in Appendix A, "Mixed Language C/C++ FORTRAN Programming." The design of these types of filters is discussed in Chapter 6, "Multirate Finite Impulse Response Filter Design."

We have spent a great deal of time talking about the use of a PPF. Now it's time to take a look at the filter's frequency response. Figure 12.23 illustrates an example filter that we might use in a channelizer application. A

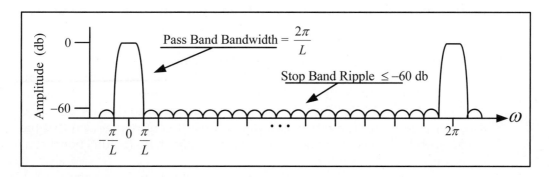

**Figure 12.23**  Frequency response of an enhanced DFT filter

premium filter such as the one shown in this figure is designed so that the bandwidth of the filter closely approximates the spectral width of a single DFT bin. If, for example, the sample rate of the input data was $f_s$ and the DFT was $L$ points in length, then the spectral width of each DFT bin is simply $f_s/L$ Hz or $2\pi f_s/L$ radians/second, depending on which notation you are using. If you prefer to work with normalized frequency notation where the sample rate is normalized to 1, then the spectral width of each DFT bin is simply $1/L$ or $2\pi/L$ radians.

The main trade-offs to consider in the design of the enhanced DFT filter are the sharpness of the filter transition bands versus the number of filter coefficients. The sharper the roll-off, the longer the filter impulse response. The amount of in band ripple and the amount of stop band attenuation will also affect the length of the filter impulse response. We will discuss these attributes later, when we actually design a channelizer.

A few things to take note of in Figure 12.23 are that the filter was designed with a null-to-null bandwidth very close to the DFT bin width of $2\pi/L$. The minimum stop band attenuation is 60 db. The equal ripple amplitude is the result of using the Parks-McClellan design algorithm. It is clear that this filter is significantly better than the $\sin(x)/x$ type of filter inherent in the stand-alone DFT. Keep in mind that the filter in Figure 12.23 is not set in concrete. We are free to design any filter we choose to best fit our channelizer application.

Now let's take a look at the conceptual band pass filter response of the DFT that has been enhanced by the addition of the PPF. The DFT band pass concept is illustrated in Figure 12.24. The frequency response is shown relative to three different frequency axes. The topmost axis is calibrated in normalized radians. The middle axis is calibrated in hertz where $f_s$ is the sample frequency. The lower axis is calibrated in normalized frequency.

The enhanced band pass filter bank illustrated in Figure 12.24 is a clear improvement over the naked DFT $\sin x/x$ like band pass filter bank of Figure 12.18. Notice that the filter bandwidth is very close to the DFT bin width, there is minimal overlap of bin filters, and the side lobes are suppressed below 60 db or more. All of these enhancements significantly reduce spectral leakage. Ideally the signal spectrum that falls within each DFT bin is effectively isolated from the spectra in all other DFT bins.

Now it's time to get into a bit more detail of the mechanics of the enhanced DFT filter. To do this, we will need to spend some time describing the content of Figure 12.25. The concepts that we are about to discuss, although simple, can seem to be a bit complicated at first. It is recommended that the reader spend whatever time necessary to fully understand the functional concepts illustrated in this figure.

At the top of the figure is the composite filter impulse response $h(n)$. The sequence $h(n)$ is more than likely obtained from a filter design software

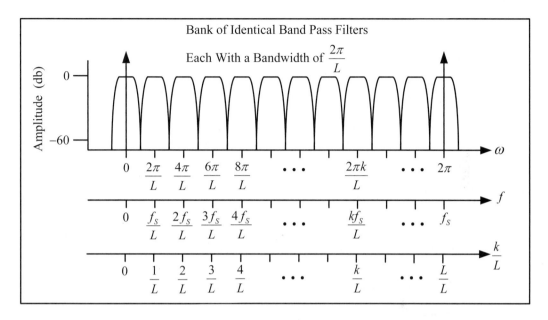

**Figure 12.24**   Enhanced DFT band pass filter response

package that inputs a user-defined set of specifications and spits out the filter impulse response. The impulse response tends to take on a shape similar to a $\sin(x)/x$ function, as you would expect since we would like the filter frequency response to be as close to a rectangular window as possible. The filter impulse response is $N$ coefficients long. Just below the impulse response is a graphical depiction of the $N$ coefficients that make up $h(n)$. The coefficients are $B$ bits wide. The number of coefficients $N$ is chosen to be some integer multiple of the width of the $L$ point DFT—that is, $N = qL$, where $q$ is an integer.

Since $N$ is a multiple of $L$, we segment the $N$ coefficient impulse response into $L$ shorter sequences, each of which will be $q = N/L$ coefficients in length. Each of these smaller sequences will be called a *phase* of the PPF structure and labeled

$$h_0(n), h_1(n) h_2(n) \quad \cdots \quad h_{L-1}(n)$$

The method by which we select the coefficients for each phase of the filter is straightforward. For filter $h_0(n)$, we take every $L$th coefficient of $h(n)$, beginning with coefficient $h(0)$. For filter $h_1(n)$, we take every $L$th coefficient of $h(n)$, beginning with coefficient $h(1)$. This continues until we get to filter $h_{L-1}(n)$, where once again we take every $L$th coefficient of $h(n)$, this time beginning with coefficient $h(L-1)$. When we are done, we end up with $L$

smaller filter impulse response sequences called *phases* that make up a PPF. Each phase of the filter has $q = N/L$ coefficients. Each phase of the PPF will produce an output labeled $z(p)$, where $p = 0, 1, 2, 3, \cdots, L-1$.

Figure 12.25 shows us that conceptually the $L$ filters are stacked up in a vertical array. The $L$ sample vector $x(n)$ is applied to the input of the filter

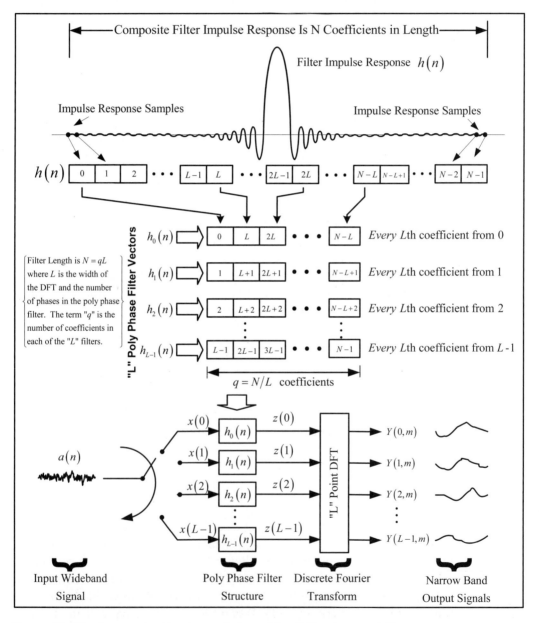

**Figure 12.25** Conceptual implementation of the DFT enhanced filter

stack. Conceptually the input data sequence is sequentially applied to (or wiped on) to each filter in the stack.

Every time the wiper arm completes one cycle, it has wiped on $L$ samples of the input signal $a(n)$. To distinguish these samples from the input signal, the $L$ samples are collectively labeled $x(n)$. For every cycle of the wiper arm, the filter stack produces an $L$ sample output vector $z(p)$. The vector $z(p)$ contains the elements $\{z(0), z(1), z(2), \cdots, z(L-1)\}$. As input data vectors are continually applied to the input of the filter stack, sequential $z(p)$ vectors are continuously computed by the filter stack and applied to the DFT. Since each phase of the filter is essentially a smaller FIR filter, when the $L$ newest input samples are wiped on, the oldest $L$ samples in the filter stack pipeline fall off and disappear.

If the input time series $a(n)$ is sampled at a rate of $f_s$ Hz, then it can be seen that the succession of vectors $x(n)$ is applied to the input of the DFT at a rate equal to $f_s/L$ Hz. This means that the succession of DFT output vectors $Y(k,m)$ also occurs at a rate equal to $f_s/L$ Hz.

This brings up an interesting point: Operationally, the bank of DFT filters is implemented by a single low pass filter. The input signal is down shifted in frequency in increments of $f_s/L$ Hz and then filtered by the low pass filter. The bandwidth of the filtered output signal is reduced to $f_s/L$ Hz. The reduced bandwidth signal no longer needs to be sampled at the high rate of $f_s$. For a complex signal, the sample rate needs to be at least equal to the highest frequency in the band. The output of each DFT bin is a complex time series of bandwidth $f_s/L$, so the minimal sample rate for that time series is also $f_s/L$. This means that this architecture takes care of not only the frequency down conversion and filtering but also the sample rate decimation. For this particular example, the decimation in output channel sample rate was a factor of $L$, and the reduction in the overall bandwidth of each channel was also a factor of $L$.

### 12.2.5  Channelizer Data Processing Concept

Now we will discuss the channelizer from a data processing point of view. The concept of data flow within the channelizer will be illustrated graphically. During the course of this dialogue, we will introduce some new concepts critical to the detail design discussions that follow.

The implementation of the enhanced DFT filter that we are about to discuss may not look like the implementation shown in Figure 12.25, but it is identical to it. The filter implementation in Figure 12.25 was conceptual and was intended as a tool to introduce the segmentation of filter coefficients that form the different phases of the PPF and the processing internal to the channelizer.

This method would work if it were implemented in either hardware or software; however, the method we will demonstrate next has the advantage of being simpler and more efficient to implement because it is less hardware and software intensive. The reader is encouraged to compare the two discussions on the enhanced DFT filter implementation and verify that both are identical. In doing so, you will benefit from a more thorough understanding of the PPF architecture.

With that said, let's take a look at Figure 12.26. This figure is intended to illustrate the simple data processing techniques that can be used to efficiently implement a channelizer. At the top of the figure, we show a segment of an infinite length data series, which we will call $a(n)$, that is sampled at a rate of $f_{S(IN)}$. Since our $L$ point DFT has a finite length, our algorithm calls for the processing of this data series in sequential blocks of $N$ samples, where $N$ is chosen to be an integer multiple of $L$.

To show how the data is collected, the figure illustrates the selection of four sequential $N$ sample sequences cut from the $a(n)$ input series. These sequential sequences are offset from one another by $M$ samples. The reason for this $M$ sample offset will be discussed shortly. Once a sequence of $N$ samples is selected, we will call it the channelizer input time series and label it $x(n)$. Every sequential sequence that is selected is labeled $x(n)$. Each sequence is multiplied sample for sample with the $N$ coefficients of the filter impulse response $h(n)$. The result, as shown in Figure 12.26, is an $N$ sample product series $x(n)h(n)$. The next process the channelizer performs is to segment the product series $\left[h(n)x(n)\right]$ into blocks that are $L$ samples in length. For the simple example in Figure 12.26, the product series is segmented into four blocks of $L$ samples each. These blocks are labeled $B_0(p), B_1(p), B_2(p)$, and $B_3(p)$. The four blocks are then summed element by element to produce a single block $z(p)$ that is also $L$ samples in length. Remember that the number of coefficients in each phase of the filter is given by $q = N/L$. We can see from the figure that each element of $z(p)$ contains exactly $q$, $h(n)x(n)$ product terms.

For the example in Figure 12.26 where we have four sample blocks, this summation is expressed mathematically by

$$z(p) = \sum_{i=0}^{3} B_i(p) \quad \text{for } p = 0, 1, 2, \cdots, L-1$$

The block of summed data $z(p)$ is an $L$ element summed product series, where each element is a partial sum of the $x(n)h(n)$ product vector. In other words, each element of $z(p)$ is the sum of $N/L$ product series elements. It is easily seen from Figure 12.26 that the first element of $z(p)$ is given by

$$z(0) = x(0)h(0) + x(L)h(L) + x(2L)h(2L) + \cdots + x(N-L)h(N-L)$$

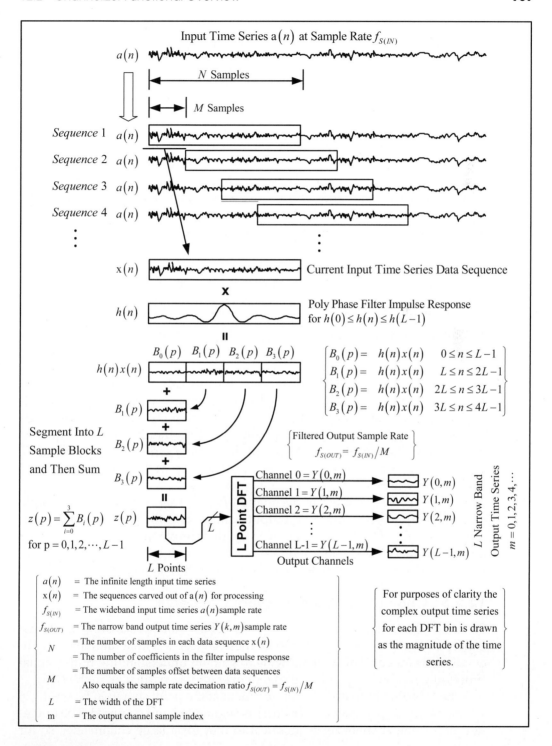

**Figure 12.26**   Channelizer data processing paths

This is identical to the sum of products for $z(0)$, illustrated in the conceptual block diagram of Figure 12.25. All the remaining $z(p)$ sum of products are computed in the same manner.

The $L$ element series $z(p)$ is now input to the $L$ point DFT. The DFT processes the $z(p)$ series and outputs an $L$ point series $Y(k,m)$. The notation $Y(k,m)$ represents the $m$th point in the output sequence of the $k$th bin of the DFT.

Let's go back now and explain the reason for the $M$ sample offset between successive input time series sequences $x(n)$. We define the sample rate of $a(n)$ to be the input sample rate $f_{S(IN)}$ and the sample rate of each of the DFT channelized time series as $f_{S(OUT)}$. If we set the value of $M$ to unity, successive $x(n)$ input time series sequences will be offset by one sample. This means that the channelizer will process one input sequence for every new input sample of the series $a(n)$. If the sequence length of $x(n)$ is $N$ samples, then successive sequences will be overlapped by $N-1$ samples. For this case, the sequence rate of the data processed by the DFT is equal to $f_{S(IN)}$ sequences/second, and the sample rate of each of the output $L$ channel time series will be $f_{S(OUT)} = f_{S(IN)}$ samples/second. Since the complex output of each DFT bin is band limited to $f_{S(IN)}/L$ Hz, an output sample rate of $f_{S(IN)}$ is clearly a case of oversampling. As a matter of fact, it is oversampled by a factor of $L$.

Now let's set the value of $M$ to 2. For this case successive $x(n)$ input time series, sequences will be offset by two samples. Therefore the channelizer will process one input sequence for every two input samples of the series $a(n)$. The sequence rate of the data processed by the DFT is now equal to $f_{S(IN)}/2$, and the sample rate of each of the $L$ time series output by the DFT will be $f_{S(OUT)} = f_{S(IN)}/2$ samples/second. If we set $M$ to 4, then the output sample rate will be equal to $f_{S(OUT)} = f_{S(IN)}/4$, and so on.

So $M$ is the sample rate decimation factor. That is, if the input sample rate for $a(n)$ is $f_{S(IN)}$, then the output sample rate for each of the $Y(k,m)$ channels is $f_{S(IN)}/M$. We can use $M$ to set the sample rate for all channelizer output channels. This is an important relation, and it will be a key design parameter when we discuss the detailed design of the channelizer.

## 12.2.6  Channelizer Mathematical Concepts

So far we have discussed the basic channelizer data processing functional concepts. The next step is to reinforce these concepts with a high-level mathematical treatment. This will allow us to gain an understanding of the signal processing that's going on in the channelizer and, in particular, in the DFT.

Let's begin our mathematical treatment at the point shown in Figure 12.26 where the product sequence $x(n)h(n)$ is segmented and summed. The operation of a small channelizer is identical to that for a large channelizer, so for this discussion, in order to minimize the volume of repetitive math, we will utilize, as an example, a channelizer that operates on only four channels.

For this discussion, we are going to assume the following:

- The length $L$ of the DFT is 4.
- The length $N$ of the input data series block $x(n)$ is 16.
- The length of the composite filter impulse response $h(n)$ is also 16.
- The number of coefficients in each phase of the filter is $q = N/L = 16/4 = 4$.
- We are only going to process a single data block, so the value of $M$ is not important in this example.

The input time series sequence $x(n)$ and the filter impulse response $h(n)$ are shown aligned with one another, as illustrated in Figure 12.27. The two are multiplied element for element to produce the product sequence $x(n)h(n)$.

In this example, after the multiplication is performed, the product series is segmented into four blocks of $q = N/L = 16/4 = 4$ elements each. The blocks are aligned and summed according to the relation

$$z(p) = \sum_{n=0}^{q-1} x(nL+p)h(nL+p) \quad \left\{\text{for } p = 0,1,2,\cdots(L-1)\right\}$$

**Equation 12.7**

For this simple example, $0 \le p < 4$ and $q = 4$. Plugging the values for the data vector $x(n)$ and the coefficient vector $h(n)$ into Equation 12.7, we compute the values of $z(p)$ to be

$$
\begin{aligned}
z(0) &= x(0)h(0) \quad +x(4)h(4) \quad +x(8)h(8) \quad +x(12)h(12)\\
z(1) &= x(1)h(1) \quad +x(5)h(5) \quad +x(9)h(9) \quad +x(13)h(13)\\
z(2) &= x(2)h(2) \quad +x(6)h(6) \quad +x(10)h(10) \quad +x(14)h(14)\\
z(3) &= x(3)h(3) \quad +x(7)h(7) \quad +x(11)h(11) \quad +x(15)h(15)
\end{aligned}
$$

**Equation 12.8: $z(p)$ Series Expansion**

| Align and Multiply the Data $x(n)$ Vector by the Coefficient Vector $h(n)$ | | | | | | | | | | | | | | | |
|---|---|---|---|---|---|---|---|---|---|---|---|---|---|---|---|
| $x(0)$ | $x(1)$ | $x(2)$ | $x(3)$ | $x(4)$ | $x(5)$ | $x(6)$ | $x(7)$ | $x(8)$ | $x(9)$ | $x(10)$ | $x(11)$ | $x(12)$ | $x(13)$ | $x(14)$ | $x(15)$ |
| $h(0)$ | $h(1)$ | $h(2)$ | $h(3)$ | $h(4)$ | $h(5)$ | $h(6)$ | $h(7)$ | $h(8)$ | $h(9)$ | $h(10)$ | $h(11)$ | $h(12)$ | $h(13)$ | $h(14)$ | $h(15)$ |

**Figure 12.27** Segmented product vector

This process, by which $z(p)$ is computed, is illustrated graphically in Figure 12.28. The $z(p)$ series is then input directly to the $L=4$ point DFT to produce an output vector $\{Y(0),Y(1),Y(2),Y(3)\}$. Our objective from this point on is to mathematically track the individual product samples $h(n)x(n)$ through the DFT to get a good understanding of the processing involved and how the $L$ output data series are computed.

Let's stop here for a moment and compare Equation 12.8 with the graphical representation of the PPF in Figure 12.25 and Figure 12.26. It is instructive to note that the filter coefficients in the series $z(0)$ are identical to the coefficients of $h_0$, the phase 0 filter of the PPF structure illustrated in both Figure 12.25 and Figure 12.26. In addition, the input signal samples are identical to those that are input to $h_0$ by the wiper arm in Figure 12.25. In other words, the value of $z(0)$ in the previous equation is identical to the value of $z(0)$ illustrated in both figures. It is left to the reader to verify that there is a one-to-one correspondence between the vectors $z(1)$, $z(2)$, and $z(3)$ detailed in Equation 12.8 and the filter coefficients and data samples graphically illustrated in Figure 12.25 and Figure 12.26. The reader can determine by

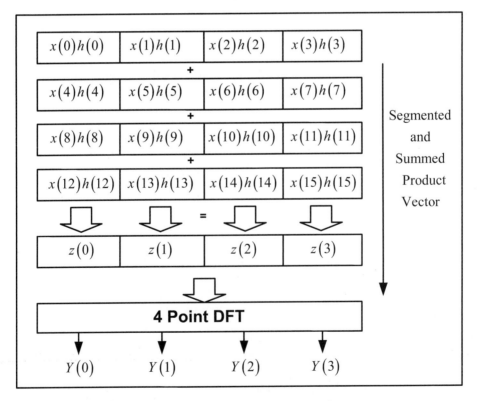

**Figure 12.28**  Segmentation and summing of the product vector

inspection that the segmentation and sum operation produces a DFT input vector $z(p)$ identical to the vector produced by the PPF shown in both figures.

The vector $z(p)$ is the input to the $L$ point DFT. The operation the DFT performs on $z(p)$ is described by

$$Y(k) = \sum_{p=0}^{L-1} z(p) e^{-j2\pi\left(\frac{k}{L}\right)p} \quad \left\{ \begin{array}{l} L = 4 \\ k = 0,1,2,3 \end{array} \right\}$$

**Equation 12.9: Four channel DFT equation**

The computed values for the DFT output channels $Y(k)$ are

$$Y(0) = z(0) + z(1)e^{-j(0)} + z(2)e^{-j(0)} + z(3)e^{-j(0)}$$

$$Y(1) = z(0) + z(1)e^{-j\left(\frac{\pi}{2}\right)} + z(2)e^{-j\left(\frac{2\pi}{2}\right)} + z(3)e^{-j\left(\frac{3\pi}{2}\right)}$$

$$Y(2) = z(0) + z(1)e^{-j\left(\frac{2\pi}{2}\right)} + z(2)e^{-j\left(\frac{4\pi}{2}\right)} + z(3)e^{-j\left(\frac{6\pi}{2}\right)}$$

$$Y(3) = z(0) + z(1)e^{-j\left(\frac{3\pi}{2}\right)} + z(2)e^{-j\left(\frac{6\pi}{2}\right)} + z(3)e^{-j\left(\frac{9\pi}{2}\right)}$$

**Equation 12.10: Channelizer DFT output**

### 12.2.6.1 Computing the DFT Output Y(0)

If we substitute the values for $z(p)$ documented in Equation 12.8 back into Equation 12.10, we can observe the signal processing that was achieved by the DFT. Let's start by looking at the DFT output channel $Y(0)$. When we substitute the values for $z(p)$ into the equation for $Y(0)$, we see that $Y(0)$ unfolds into the original product series. This process is illustrated very clearly in Equation 12.11. The value of $k$ in Equation 12.9 is 0, so the exponential term is equal to unity and the summation is simply an element-by-element multiplication of the input data by the filter coefficients.

$$Y(0) = z(0)e^{-j(0)} + z(1)e^{-j(0)} + z(2)e^{-j(0)} + z(3)e^{-j(0)}$$

$$\Downarrow \quad \Downarrow \quad \Downarrow \quad \Downarrow$$

$$\begin{array}{llll}
Y(0) = & h(0)x(0) & +h(1)x(1) & +h(2)x(2) & +h(3)x(3) \\
& +h(4)x(4) & +h(5)x(5) & +h(6)x(6) & +h(7)x(7) \\
& +h(8)x(8) & +h(9)x(9) & +h(10)x(10) & +h(11)x(11) \\
& +h(12)x(12) & +h(13)x(13) & +h(14)x(14) & +h(15)x(15)
\end{array}$$

**Equation 12.11: DFT Output Y(0)**

All the segmented product series are summed to produce a single output sample from channel $Y(0)$. Taken by itself, this doesn't say too much, but think of this: Sequential input sequences $x(n)$ will be shifted in time relative to the previous sequence. The next input data sequence $x(n)$ will be shifted in time relative to the previous sequence and multiplied by the same set of filter coefficients to produce a second output point. The third input data sequence $x(n)$ will be shifted in time and multiplied by the same set of filter coefficients to produce a third output point and so on. This process repeats itself for each successive input data sequence $x(n)$ to generate a series of output points that form the base band filtered output time series $Y(0,m)$. The sample index $m$ ranges theoretically from 0 to $\infty$, or less if the channelizer should be unplugged at some point.

You can visualize the time series data contained in the sequential input sequences $x(n)$ as sliding across the coefficients of the fixed filter $h(n)$. This is the same as convolving the data sequence with the filter sequence. If both $x(n)$ and $h(n)$ are $N$ point sequences, and we assign $y(n)$ to be the convolutional sum, then we can write

$$y(n) = \sum_{r=0}^{N-1} h(r)x(n-r)$$

If we are dealing with an $N$ point time series, we can use the Fourier transform pairs derived in Chapter 3, "Review of the Fourier Transform," to compute the Fourier transform of $y(n)$ in terms of DFT bins as

$$Y(k) = \sum_{n=0}^{N-1}\left[\sum_{r=0}^{N-1} h(r)x(n-r)\right]e^{-j2\pi\left(\frac{k}{N}\right)n} = H(k)X(k)$$

In terms of normalized frequency, we could write

$$Y\left(\frac{k}{N}\right) = H\left(\frac{k}{N}\right)X\left(\frac{k}{N}\right)$$

This result just means that we are convolving the input signal with the impulse response of the enhanced DFT filter, which equates to multiplying the frequency response of the filter $H(k)$ by the spectrum of the input signal $X(k)$. The filter frequency response will carve out the base band frequencies from the spectrum of the input signal and use them to form an estimate of the frequencies residing in the bandwidth of the DFT bin 0.

Expressed another way, since the filter is at base band, the data that is being filtered is the data that has been down converted to base band. For the

DFT output at port $Y(0,m)$, the value of $k=0$ in Equation 12.10, so the exponential term is unity and there is no down conversion, and the frequencies passed by the filter are the original input signal base band frequencies. If the bandwidth of $h(n)$ is equal to a single DFT bin width and the filter is at base band, then the filtered time series $Y(0,m)$  $m=0,1,2,\cdots$ can be thought of as being composed of the spectra associated with the 0th bin of the DFT.

### 12.2.6.2  Computing the DFT Output Y(1)

Now let's see what happens at the DFT output channel $Y(1)$. When we substitute the values of $z(p)$ into the equation for $Y(1)$, we see the equation unfold as illustrated in Equation 12.12. The equation for $Y(1)$ contains the exact same product series as the equation for $Y(0)$. This time, however, the DFT bin index $k=1$, and each product term is multiplied by the DFT exponential term, which is a rotating phasor. For our simple example, where the length of the DFT is 4, the phasor is moving around the unit circle at the sample rate in chunks of $\pi/2$ radians. This rotating phasor is nothing more than a complex digital tuner that is shifting the spectrum of the input data down in frequency so that the spectrum associated with the DFT channel 1, or bin 1, is tuned to base band, where it is multiplied with the base band filter impulse response $h(n)$ to produce an output sample for channel 1, labeled $Y(1,m)$. The DFT will continue to sequentially input data sequences $x(n)$, translate the sequence spectrum down by one DFT bin width, and multiply it with the frequency response of the PPF to compute sequential output samples for channel 1, resulting in the generation of a base band time series that can be represented by the sequence $Y(1,m)$  $m=0,1,2,\cdots$.

$$Y(1) = \quad z(0)e^{-j(0)} \qquad\qquad +z(1)e^{-j\left(\frac{\pi}{2}\right)} \qquad\qquad +z(2)e^{-j\left(\frac{2\pi}{2}\right)} \qquad\qquad +z(3)e^{-j\left(\frac{3\pi}{2}\right)}$$

$$\Downarrow \qquad\qquad\qquad \Downarrow \qquad\qquad\qquad \Downarrow \qquad\qquad\qquad \Downarrow$$

$$Y(1) = \quad h(0)x(0)e^{-j(0)} \quad +h(1)x(1)e^{-j\left(\frac{\pi}{2}\right)} \quad +h(2)x(2)e^{-j\left(\frac{2\pi}{2}\right)} \quad +h(3)x(3)e^{-j\left(\frac{3\pi}{2}\right)}$$

$$+h(4)x(4)e^{-j(0)} \quad +h(5)x(5)e^{-j\left(\frac{\pi}{2}\right)} \quad +h(6)x(6)e^{-j\left(\frac{2\pi}{2}\right)} \quad +h(7)x(7)e^{-j\left(\frac{3\pi}{2}\right)}$$

$$+h(8)x(8)e^{-j(0)} \quad +h(9)x(9)e^{-j\left(\frac{\pi}{2}\right)} \quad +h(10)x(10)e^{-j\left(\frac{2\pi}{2}\right)} \quad +h(11)x(11)e^{-j\left(\frac{3\pi}{2}\right)}$$

$$+h(12)x(12)e^{-j(0)} \quad +h(13)x(13)e^{-j\left(\frac{\pi}{2}\right)} \quad +h(14)x(14)e^{-j\left(\frac{2\pi}{2}\right)} \quad +h(15)x(15)e^{-j\left(\frac{3\pi}{2}\right)}$$

**Equation 12.12: DFT output $Y(1)$**

We can see that as sequential input sequences are computed, Equation 12.12 is just a convolutional sum of the filter impulse response $h(n)$ and a translated sequence $x(n)e^{-j2\pi\left(\frac{k_0}{L}\right)r}$, where $k_0 = 1$ and $0 \le r < N-1$. If both $x(n)$ and $h(n)$ are $N$ point sequences, and we assign $y(n)$ be the convolutional sum, then we can write

$$y(n) = \sum_{r=0}^{N-1} h(r)x(n-r)e^{-j2\pi\left(\frac{k_0}{L}\right)r}$$

**Equation 12.13**

We have a bit of an apples and oranges operation here. We are processing two $N$ sample sequences and translating one of the sequences by an exponential that is relative to the $L$ point DFT. In completely unscientific terms, we are mixing the "N domain" with the "L domain." So let's change the "L domain" to make it compatible. We know that $L = N/q$, so we can refer the complex exponential and the translation frequency to the "N domain" by substituting this value of $L$ into Equation 12.13 to produce

$$y(n) = \sum_{r=0}^{N-1} h(r)x(n-r)e^{-j2\pi\left(\frac{qk_0}{N}\right)r}$$

**Equation 12.14**

With the help of the DFT pairs developed in Chapter 3, "Review of the Fourier Transform," we can compute the DFT of $y(n)$ in terms of normalized frequency as

$$Y\left(\frac{k}{N}\right) = \sum_{n=0}^{N-1}\left[\sum_{r=0}^{N-1} h(r)x(n-r)e^{-j2\pi\left(\frac{qk_0}{N}\right)r}\right]e^{-j2\pi\left(\frac{k}{N}\right)n} = H\left(\frac{k}{N}\right)X\left(\frac{k}{N}+\frac{qk_0}{N}\right)$$

**Equation 12.15**

This result simply means that the convolution of the translated input sequence with the filter impulse response is just the multiplication of the low pass frequency response of the PPF $H(k/N)$ by the spectrum of the input signal $X(k/N)$ that has been translated down in frequency by $qk_0/N$. In this example, $q = N/L = 16/4 = 4$. Relative to the $L = 4$ point DFT, the normalized translation frequency is

$$k_0/L = qk_0/N = (4)(1)/16 = 1/4$$

or one $L$ point DFT bin.

The filter frequency response will carve out the base band frequencies from the spectrum of the input signal that has been translated down by a single bin width. The DFT will use this as the estimate of the signal content of bin 1. It's pretty cool to see how this stuff falls out so nicely.

### 12.2.6.3  Computing the DFT Output Y(2)

Let's take a look at the DFT output channel $Y(2)$. Once again, when we replace the series $z(p)$ with its element values, we get the unfolded equation for the DFT channel $Y(2)$ shown in Equation 12.16. We are starting to see a pattern here.

$$Y(2) = \qquad z(0)e^{-j(0)} \qquad\qquad +z(1)e^{-j(\pi)} \qquad\qquad +z(2)e^{-j(2\pi)} \qquad\qquad +z(3)e^{-j(3\pi)}$$

$$\Downarrow \qquad\qquad \Downarrow \qquad\qquad \Downarrow \qquad\qquad \Downarrow$$

$$
\begin{aligned}
Y(2) = \quad & h(0)x(0)e^{-j(0)} && +h(1)x(1)e^{-j(\pi)} && +h(2)x(2)e^{-j(2\pi)} && +h(3)x(3)e^{-j(3\pi)} \\
& +h(4)x(4)e^{-j(0)} && +h(5)x(5)e^{-j(\pi)} && +h(6)x(6)e^{-j(2\pi)} && +h(7)x(7)e^{-j(3\pi)} \\
& +h(8)x(8)e^{-j(0)} && +h(9)x(9)e^{-j(\pi)} && +h(10)x(10)e^{-j(2\pi)} && +h(11)x(11)e^{-j(3\pi)} \\
& +h(12)x(12)e^{-j(0)} && +h(13)x(13)e^{-j(\pi)} && +h(14)x(14)e^{-j(2\pi)} && +h(15)x(15)e^{-j(3\pi)}
\end{aligned}
$$

**Equation 12.16: DFT Output $Y(2)$**

We get the same product series as before, but this time the DFT bin index $k = 2$ and each element is multiplied by a rotating phasor that is moving around the unit circle at the sample rate in chunks of $\pi$ radians. The rotating phasor down converts the spectrum of the input time series, shifting the frequency band associated with the DFT channel 2 to base band, where it is filtered by the base band filter $h(n)$. The time series sequence that results from processing sequential input data series blocks can be represented by the sequence $Y(2, m)$ $m = 0, 1, 2, \cdots$.

The math for this translate-and-filter operation is identical to what we discussed for $Y(1)$, only this time the value of $k_0 = 2$. Using the same Equation 12.15 and substituting this value produces

$$Y\left(\frac{k}{N}\right) = \sum_{n=0}^{N-1}\left[\sum_{r=0}^{N-1} h(r)x(n-r)e^{-j2\pi\left(\frac{qk_0}{N}\right)r}\right]e^{-j2\pi\left(\frac{k}{N}\right)n} = H\left(\frac{k}{N}\right)X\left(\frac{k}{N} + \frac{qk_0}{N}\right)$$

Relative to the $L = 4$ point DFT, the normalized translation frequency is

$$k_0/L = qk_0/N = (4)(2)/16 = 2/4$$

or two $L$ point DFT bins.

### 12.2.6.4  Computing the DFT Output Y(3)

Finally let's take a look at the DFT output channel $Y(3)$. As shown in Equation 12.17, the DFT bin index $k = 3$, and we get the same product series, multiplied by a rotating phasor that moves around the unit circle at the sample rate in radian chunks of $3\pi/2$. This phasor down shifts the spectrum of the input data so that the spectrum associated with the DFT bin 3 is translated to base band, where it is filtered by the filter $h(n)$, producing an output point at $Y(3)$. The time series that results from processing sequential input data sequences can be represented by the sequence $Y(3,m)$ $m = 0,1,2,\cdots$.

$$Y(3) = \quad z(0)e^{-j(0)} \qquad +z(1)e^{-j\left(\frac{3\pi}{2}\right)} \qquad +z(2)e^{-j\left(\frac{6\pi}{2}\right)} \qquad +z(3)e^{-j\left(\frac{9\pi}{2}\right)}$$

$$
\begin{aligned}
Y(3) = \quad &h(0)x(0)e^{-j(0)} &&+h(1)x(1)e^{-j\left(\frac{3\pi}{2}\right)} &&+h(2)x(2)e^{-j\left(\frac{6\pi}{2}\right)} &&+h(3)x(3)e^{-j\left(\frac{9\pi}{2}\right)} \\
&+h(4)x(4)e^{-j(0)} &&+h(5)x(5)e^{-j\left(\frac{3\pi}{2}\right)} &&+h(6)x(6)e^{-j\left(\frac{6\pi}{2}\right)} &&+h(7)x(7)e^{-j\left(\frac{9\pi}{2}\right)} \\
&+h(8)x(8)e^{-j(0)} &&+h(9)x(9)e^{-j\left(\frac{3\pi}{2}\right)} &&+h(10)x(10)e^{-j\left(\frac{6\pi}{2}\right)} &&+h(11)x(11)e^{-j\left(\frac{9\pi}{2}\right)} \\
&+h(12)x(12)e^{-j(0)} &&+h(13)x(13)e^{-j\left(\frac{3\pi}{2}\right)} &&+h(14)x(14)e^{-j\left(\frac{6\pi}{2}\right)} &&+h(15)x(15)e^{-j\left(\frac{9\pi}{2}\right)}
\end{aligned}
$$

**Equation 12.17: DFT Output Y(3)**

The math for this translate-and-filter operation is identical to that discussed for $Y(1)$ and $Y(2)$, only this time the value of $k_0 = 3$. Substituting this value into Equation 12.15 produces

$$Y\left(\frac{k}{N}\right) = \sum_{n=0}^{N-1}\left[\sum_{r=0}^{N-1} h(r)x(n-r)e^{-j2\pi\left(\frac{qk_0}{N}\right)r}\right]e^{-j2\pi\left(\frac{k}{N}\right)n} = H\left(\frac{k}{N}\right)X\left(\frac{k}{N} + \frac{qk_0}{N}\right)$$

Relative to the $L = 4$ point DFT, the normalized translation frequency is

$$k_0/L = qk_0/N = (4)(3)/16 = 3/4$$

or three $L$ point DFT bins.

This stuff is fascinating. If we obey the laws of physics and the laws of mathematics, we can compute a whole lot of stuff that lets us peek beneath the surface of a complex problem and view all the inner wheels turning and all the hidden gears grinding to give us a detailed picture of how things work. I have often wondered what a mathematician visualizes in his mind when he writes an equation on a white board that is replete with squiggly lines, Greek characters, and arcane symbols. I believe that these people see in their minds a very clear picture of a functional system, where all the components of that system are neatly described by the individual terms in their equations. Following this belief, we have attempted to convert the complex concept of a channelizer into a functional system that can be easily visualized, where the individual components of the system are described by the terms in the channelizer descriptive equations. We have painted a systems picture of how the DFT based channelizer operates using graphics and a minimal amount of mathematics. This is the picture of the channelizer system that I visualize in my mind. By now, after all this work, you should have a fairly complete systems view of the inner workings of a channelizer. Hopefully this view is not mistaken as a hallucination!

## 12.3 CHANNELIZER DETAILED DESIGN CONCEPTS

At this point, the reader should have a good idea of the functional concepts and architecture of a basic channelizer. Now it is time to move forward and discuss the channelizer design parameters and detailed design concepts. Our intention in this section to derive a stand-alone equation that can be implemented directly in either hardware or software to produce a working channelizer that is suited for most reader applications. Along the way, we will introduce a few critical design considerations that are necessary to successfully implement a channelizer design. It is my intention to make this derivation as illustrative as possible. So we will derive the channelizer stand-alone equation piece by piece, in parallel with the development of the channelizer processing block diagram.

### 12.3.1 Definition of Channelizer Operational Parameters

Before we begin, let's define terms that will be used to describe and quantify the signal processing parameters inherent to a channelizer. Table 12.1 defines all the parameters we will utilize to mathematically derive the stand-alone channelizer equation.

### 12.3.2 Channelizer Input Signal Processing

Let's start at the beginning. For us, this will be the point where we select the input data to the channelizer. This process is illustrated in Figure 12.29. At the top of the figure we see a theoretically infinite data time series $a(n)$ sampled at a frequency of $f_{S(IN)}$. This is the data stream that we wish to process. The channelizer processes this data in blocks of $N$ samples at a time. Whenever we carve out an $N$ sample block from the series $a(n)$, we label it $x(n)$ and send it off to the channelizer for processing.

**Table 12.1**   Channelizer Design Parameters

| Parameter | Definition |
|---|---|
| $f_{S(IN)}$ | Channelizer wideband input signal sample rate |
| $f_{S(OUT)}$ | Channelizer narrow band output signal sample rate |
| $B$ | The bandwidth of each channelizer DFT bin |
| $N$ | The length of the enhanced DFT filter impulse response $h(n)$ <br> The length of the input data sequence $x(n)$ |
| $L$ | The size of the channelizer DFT |
| $M$ | The input signal sample rate decimation factor <br> The sample offset between successive $x(n)$ input signal sequence vectors |
| $Q$ | The number of coefficients in each phase of the enhanced DFT filter implemented in its poly phase form $Q = N/L$ <br> The number of rows in the $QxL$ folded product vector $h(n)x(n)$ |
| $n$ | The input data sample index $n = 0,1,2,\cdots\infty$ |
| $m$ | The decimated output signal sample index $m = 0,1,2,\cdots\infty$ |
| $k$ | The channelizer DFT channel or bin index $k = 0,1,2,\cdots L-1$ |
| $p$ | The index of the phases of the PPF $p = 0,1,2,\cdots L-1$ |
| $q$ | The index of the folded data and filter impulse response product term vectors used to compute $z(p)$ <br> $q = 0,1,2,\cdots Q-1$ |
| $a(n)$ | The wideband data time series to be processed |
| $x(n)$ | The sequence vector extracted from $a(n)$ for channelizer processing |
| $h(n)$ | The enhanced DFT filter impulse response |
| $z(p)$ | The folded and summed version of the product vector $h(n)x(n)$ |

Each of these data carve-outs are called *blocks*. In the figure, they are labeled Block 0, Block 1, Block 2, and so on. For clarity, we have given the data blocks a subscript to index the sequence in which they were created. From the figure, we can see that the sequential blocks of data are labeled $x_0(n), x_1(n), x_2(n), \cdots, x_m(n), \cdots$. These subscripts are useful when visualizing the order in which these sequential data blocks are processed by the channelizer.

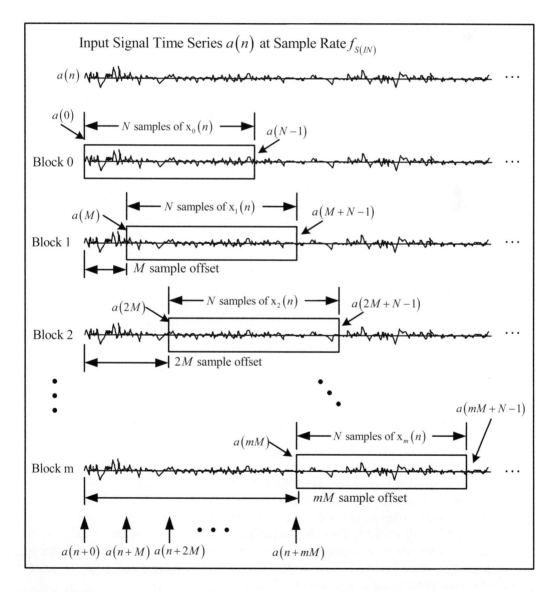

**Figure 12.29**  Channelizer input data

Figure 12.29 illustrates two very important concepts that we will deal with later during our mathematical derivations:

1. Each sequential block of data $x_i(n)$ is extracted from the infinitely long data sequence $a(n)$ and is offset from the previous block by $M$ samples.
2. Sequential blocks are shifted in time by $mM$ samples where the index $m = 0,1,2,3,\cdots$; therefore $m$ can be considered to be the block index.

These two items are extremely important and are critical to the efficient channelizer design. We will introduce the terms $m$ and $M$ mathematically later, but first we need to cement in our minds the visual concept of the channelizer input. Right now it is important for the reader to retain a picture of Figure 12.30 in his or her mind that illustrates the format and notation of the $m$th data block input to the channelizer.

Since the channelizer is processing individual blocks of data at a time, it needs to accumulate and store a current block of data while it is processing the previous block. To accomplish this, we can use a pair of memories that double buffer the input data samples. Figure 12.31 illustrates a double buffer built out of two memories, each of which is $N$ samples deep. Notice from the figure that a block of new data samples from the input series $a(n)$ are written into one memory, while the previously written block of data $x(n)$ is read from the second memory and sent to the channelizer. After one memory is filled with input samples and the second memory has emptied its output samples, the two memories reverse roles, where the second memory now stores input data and the first memory dumps its contents to the channelizer processing functions. This "ping-pong" action continues for as long as the channelizer is in operation. The main consideration to be made here is that the data labeled $x(n)$ needs to be read out of memory and processed in less time than it takes to clock through the $M$ samples of offset and then store the next $N$ samples of input data. This means that the channelizer has to complete all of its operations in a time frame that is shorter than

$$(N+M \text{ samples})/\left(f_{S(IN)} \text{ samples/seconds}\right) = \left[(N+M)/f_{S(IN)}\right] \text{ seconds}$$

If the channelizer processing takes longer than $(N+M)/f_{S(IN)}$ seconds, then the double buffer will drop data. In this case, either the system processing clock needs to be increased or some other processing architecture will have to be implemented such as a pipe line processing to speedup computations. We will slightly alter the architecture of this double buffer scheme to take advantage of a signal processing symmetry, which will be introduced later. The modified architecture that we will discuss later will allow us to reduce the processing clock rate by a factor of two.

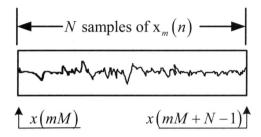

**Figure 12.30** *m*th input data block

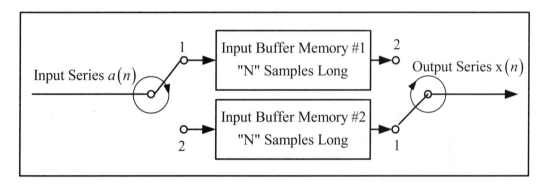

**Figure 12.31** Channelizer input data double buffer

For now, let's continue with this discussion using the functional input double buffer but with the addition of a third memory buffer used to hold and recirculate the enhanced DFT filter coefficients $h(n)$. The block diagram for this new structure is illustrated in Figure 12.32. Here we see the same functional double buffer as before, but this time the output series $x(n)$ is multiplied by the filter coefficients $h(n)$ to produce the product series $h(n)x(n)$. The reader should be able to picture in his or her mind the continuous operation of the memory buffers and the data flowing through them.

### 12.3.3 Segment Folding Memory

Once again, suppose we have an enhanced DFT filter impulse response $h(n)$ that is $N = 16$ coefficients in length and an input data block $x(n)$ that is $N = 16$ samples long. In this simple example, if our DFT is of length $L = 4$, we will break the filter into four phases of a PPF, and each phase will have $Q = N/L = 16/4 = 4$ coefficients. We can visualize folded data as being an array of $x(n)h(n)$ product terms organized as a $Q$ row by $L$ column array. We can also visualize the $z(p)$ terms as being the sum of the individual columns, where the index $p$ falls within the range $0 \le p \le (L-1)$ and represents the phase of the PPF. This data array is illustrated in Figure 12.33.

In Figure 12.33, the data is organized as an array of four columns: one column per filter phase $p$, where $p = 0, 1, 2, \cdots, (L-1)$, and $Q = 4$ rows. Each row is indexed by the variable $q$, where $q = 0, 1, 2, \cdots, (Q-1)$. We can compute the polyphase sum terms $z(p)$ using Equation 12.18:

$$z(p) = \sum_{q=0}^{Q-1} x(qL + p) h(qL + p) \quad \text{for} \quad \left\{ \begin{array}{l} p = 0, 1, 2, \cdots (L-1) \\ Q = N/L \end{array} \right\}$$

**Equation 12.18**

This equation represents a single sum of products vector for a single input data block. It does not account for subsequent input data blocks. We will modify Equation 12.18 later to include the notation that describes the computation of $z(p)$ when processing sequential input data blocks.

It should be noted that for longer data sequences, the same folded sum of products structure would apply. If the number of samples processed per block $N$ were increased to 128 and if the DFT size $L$ were increased to 16, then $Q = N/L = 128/16 = 8$. The folded data array would have 16 columns and 8 rows. The enhanced DFT filter would be implemented in a poly phase form with 16 phases and 8 coefficients per phase. The range of the index $p$ would be $0 \le p \le 15$, the range of the index $q$ would be $0 \le q \le 7$, and the DFT would be of width $L = 16$.

The question we should ask now is, how do we actually implement the folding and summation operations on the product series? We can implement the fold and summation operation in software simply by using Equation 12.18. The simplest method to implement the fold and sum operation in hardware is to use a memory that is $L$ samples in length that is configured to recirculate its contents like that illustrated in Figure 12.34.

Conceptually, the memory can be thought of as an $L$ stage shift register that is clocking data through its registers from one end to the other with its output fed back to its input through a summer. Initially the memory is filled with zeros. The input product series $h(n)x(n)$ is applied to the summer, whose output is fed to the input port of the recirculating memory. The output of the memory is then fed back to the second input of the summer. The memory content circulates in lock step with the input data stream and eventually produces a sum of products vector $z(p)$ of length $L$.

Figure 12.35 illustrates the content of the folding memory for our $N = 16$ element example during the fold and summation operation:

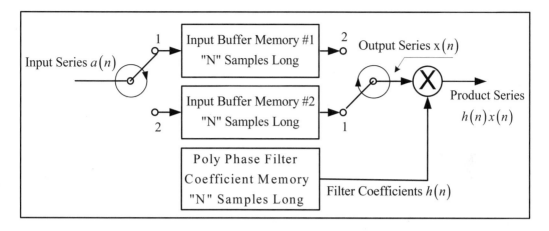

**Figure 12.32**   Input data double buffer with coefficient memory

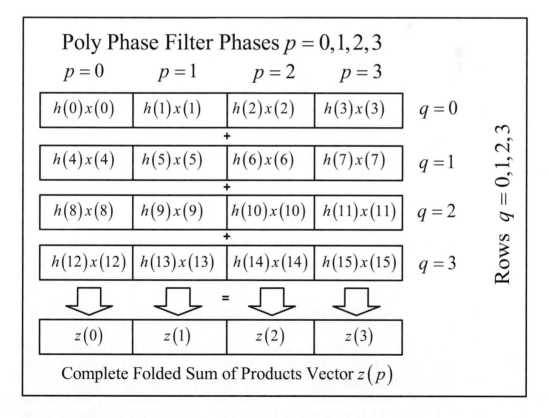

**Figure 12.33**   Fold and sum operation

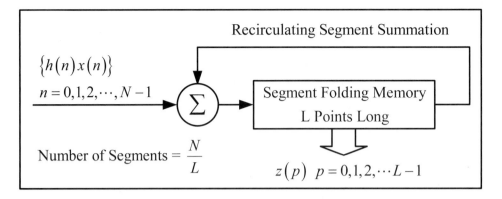

**Figure 12.34**  Folding memory block diagram

- In part A of the figure, the content of memory has been initialized to all zeros.
- Four clock ticks later, we see in part B that the memory has stored four consecutive product series samples producing an interim sum of product vector $z_0(p)$.
- In part C, four clock ticks later, we see that the content of memory has been recirculated and summed with the next four input sum of product series samples producing an interim sum of products vector $z_1(p)$.
- In part D, four clock ticks later, the content of memory has been recirculated and summed again, producing another interim vector $z_2(p)$.
- Finally, in part E, the last recirculation occurs and the final sum of product vector $z_3(p)$ is generated. This is the final vector we call $z(p)$ that is passed on to the DFT for further processing.

So we can see that the simple memory with feedback to a summer serially folds, accumulates, and stores the serial input sum of products data. At the beginning of each sequential input data block, the memory can be initialized either by loading all storage locations with zeros or by simply breaking the feedback path to the summer during the cycle shown in Figure 12.35, part B. In this case, the previous final sum of products data will be overwritten by fresh input data. For comparison purposes, the rows and columns of the sum of products illustrated in Figure 12.35, part E, are annotated with the same row and column notation $(p,q)$ that was used in Figure 12.33.

We now include the folding memory in the channelizer input block diagram of Figure 12.32 to produce the block diagram shown in Figure 12.36.

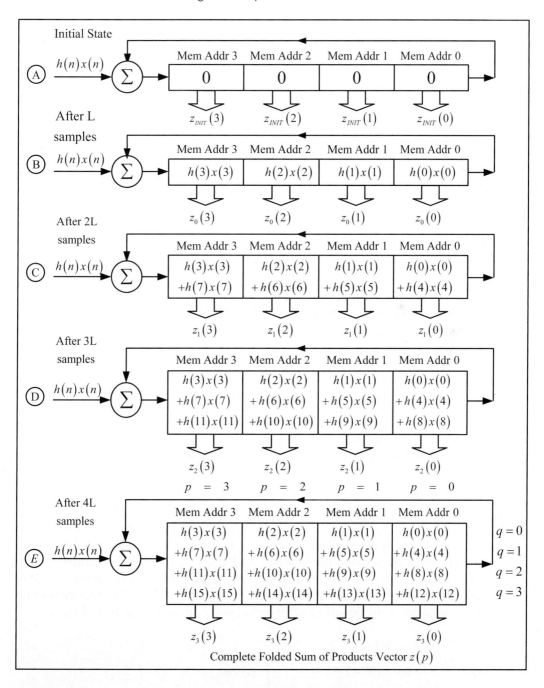

**Figure 12.35** Folding memory computational example

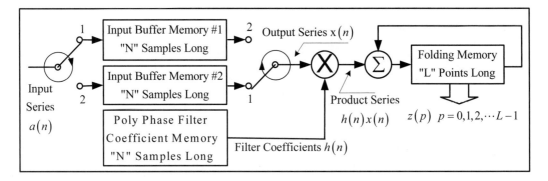

**Figure 12.36**  Input double buffer with folding memory

Remember that this block diagram of the channelizer input processing functions is still a preliminary block diagram. We will make some modifications to this diagram as we continue to introduce new concepts in our pursuit to derive the stand-alone channelizer equation.

Let's assume for the moment that the sum of products vector is not folded. Think of it as a single vector of length $N$ given by

$$z(n) = \sum_{n=0}^{N-1} h(n)x(n)$$

**Equation 12.19**

The Fourier transform of $z(n)$ can then be represented by

$$Y(k) = \sum_{n=0}^{N-1} z(n)e^{-j2\pi\left(\frac{k}{L}\right)n}$$

We know from previous discussion that $N = QL$ and that $z(n)$ is folded into $Q$ rows of $L$ columns. Remember that $Q$ also equals the number of coefficients in each phase of the enhanced DFT filter in its poly phase form. We will define $q$ and $p$ so that $0 \le q \le Q-1$ and $0 \le p \le L-1$. We can then redefine the index $n$ in Equation 12.19 by substitution of $q$ and $p$, as follows:

$$
\begin{aligned}
&0 \le n \le N-1 \\
&0 \le n \le QL-1 \\
&0 \le n \le (Q-1)L+(L-1) \\
&0 \le n \le qL+p
\end{aligned}
\left.
\begin{aligned}
\\
\\
\end{aligned}
\right\}
\quad \text{where} \quad
\left.
\begin{aligned}
0 \le q \le Q-1 \\
0 \le p \le L-1
\end{aligned}
\right\}
$$

We can make the last statement because of the range of the variable $q$ and the variable $p$. We can now rewrite Equation 12.19 as

$$z(p) = \sum_{q=0}^{Q-1} h(qL+p) x(qL+p) \quad \text{for} \quad \left\{ \begin{array}{l} p = 0,1,2,\cdots L-1 \\ Q = N/L \end{array} \right\}$$

**Equation 12.20**

where the index $n$ in $h(n)$ and $q(n)$ has been replaced by $qL+p$. If the summation index $n$ is at its minimum value (i.e., when $n=0$), then $0 = qL + p_{MIN}$, where $p_{MIN} = 0$. Thus $q=0$. When the summation index $n$ is at is at its maximum value (i.e., when $n = N-1$), then we have the condition where $N-1 = qL + p_{MAX}$. Since $p_{MAX} = L-1$, we end up with

$$
\begin{aligned}
qL + p_{MAX} &= N-1 \\
qL + L - 1 &= N-1 \\
q &= \{(N-1)-(L-1)\}/L \\
q &= (N-L)/L \\
q &= N/L - 1 \\
q &= Q-1
\end{aligned}
$$

Therefore the summation index $q$ sums over the range of $0$ to $Q-1$. This is identical to Equation 12.18, which we derived earlier by inspection from Figure 12.33.

The Fourier transform of $z(p)$ can now be written as

$$Y(k) = \sum_{p=0}^{L-1} \left[ \sum_{q=0}^{Q-1} h(qL+p) x(qL+p) \right] e^{-j2\pi\left(\frac{k}{L}\right)(qL+p)} \quad \text{for } k = 0,1,2,\cdots L-1$$

**Equation 12.21**

For clarity, we will temporarily collapse Equation 12.21 and rewrite it as

$$Y(k) = \sum_{p=0}^{L-1} z(p) e^{-j2\pi\left(\frac{k}{L}\right)(qL+p)} \quad \text{for } k = 0,1,2,\cdots L-1$$

where

$$z(p) = \sum_{q=0}^{Q-1} h(qL + p)x(qL + p) \quad \text{for} \quad \left\{ \begin{array}{l} p = 0, 1, 2, \cdots L - 1 \\ Q = N/L \end{array} \right\}$$

**Equation 12.22**

The expression for $Y(k)$ and $z(p)$ are a pair of relationships that are identified as Equation 12.22 and will serve as the starting point in our quest to derive a standalone channelizer equation.

## 12.3.4  Translation of Input Signal Spectra Bin Centers to 0 Hz

Let's begin by looking at an example frequency response $H(\omega)$ of the enhanced DFT filter shown in Figure 12.37. Here we see a filter centered at 0 Hz with a null-to-null bandwidth equal to $2\pi/L$, which is equal to the width of a single DFT bin. It has very narrow transition bands and a stop band of –60 db. Although this is very close to an ideal frequency response, it is by no means difficult to achieve. Filters like these can be easily designed using the techniques discussed in Chapter 5, "Finite Impulse Response Digital Filtering," and Chapter 6, "Multirate Finite Impulse Response Filter Design." A complete listing of the Parks-McClellan code to design these types of filters is included in Appendix A, "Mixed Language C/C++ FORTRAN Programming."

This is the filter frequency response that is representative of the filter impulse sequence $h(n)$. The spectrum of the input signal is illustrated in part A of Figure 12.38. Here we show the signal spectra with the superimposed frequency response of the enhanced DFT filter. We need to frequency shift the input spectra by a half bin width so that the spectrum of the signal in bin 0 and successive translated bins will be centered at 0 Hz. The result of the frequency down conversion is illustrated in part B of the figure. This half bin down conversion in frequency is sometimes referred to as a *spectral offset*.

The reader should keep in mind that this offset is particular to the channelizer development in this book and is inserted here only to illustrate how it is achieved. Other applications in the industry may require different values of offsets. If so, then the method given here can be used as a guide and the overall channelizer processing equation can be modified accordingly.

With that said, the next area we need to address in the evolution of our stand-alone channelizer equation is the implementation of this spectral offset. We can add a permanent frequency offset of $\pi/L$ to our equation if we replace the DFT bin index $k$ with the new index $k+1/2$. Once again, the DFT of the folded sum of products sequence $z(p)$ is given by

$$Y(k) = \sum_{p=0}^{L-1} z(p)e^{-j2\pi\left(\frac{k}{L}\right)(qL+p)} \quad \text{for } k = 0, 1, 2, \cdots L - 1$$

**Equation 12.23**

If we replace $k$ with $k+1/2$ in Equation 12.23, we arrive at Equation 12.24, which is the next stop in our derivation of the stand-alone channelizer equation:

$$Y(k) = \sum_{p=0}^{L-1} z(p) e^{-j\left(\frac{2\pi}{L}\right)(qL+p)\left(k+\frac{1}{2}\right)} \quad \text{for } k = 0,1,2,\cdots L-1$$

**Equation 12.24**

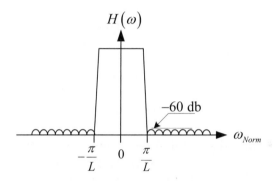

**Figure 12.37**  Enhanced DFT filter response

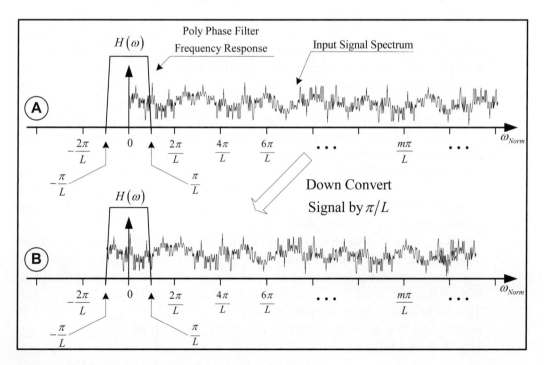

**Figure 12.38**  Input signal spectrum half bin offset

### 12.3.5 Sample Rate Decimation

In section 12.3.2 we emphasized the importance of the channelizer parameters $m$ and $M$. It was pointed out that sequential input signal time series blocks $x(n)$ were offset from one another by $M$ samples and that $m$ was the sequential block index. Every input block of data samples processed by the channelizer results in a single complex sample output in each of the $L$ bins of the channelizer. From the perspective of an output channel, the sample rate of the signal has been reduced or decimated by a factor of $M$. That is,

$$f_{S(OUT)} = \frac{f_{S(IN)}}{M}$$

The variable $m$ is the index of the sequential blocks of data $x(n)$ input to the channelizer. Since a single input block produces a single output point for each of the channelized outputs, it is easy to see that $m$ is also the index of the channelized data output samples.

Now consider the fact that the channelized output of an enhanced DFT is a complex signal with an ideal spectral bandwidth of $f_{S(IN)}/L$ Hz. In most applications, it is desirable to convert the complex signal output from each channel to a real signal. We will implement the complex to real conversion in section 12.3.7, but before we do that we need to discuss how we decimate the output channel sample rate.

In a typical application where a complex signal sampled at the Nyquist rate is converted to a real signal, the sample rate of the signal is increased or interpolated by a factor of two. The signal is then tuned so that its spectrum is centered at one-fourth the new sample rate. It is then filtered by a low pass image rejection filter, and the complex to real conversion is performed. This process is discussed in detail in Chapter 7, "Complex to Real Conversion."

In the derivation of our stand-alone equation, it makes sense to decimate the sample rate of the input signal by a factor of $M$ so that the channelized sample rate is twice the Nyquist rate for a complex signal but equal to the Nyquist rate for a real signal. This eliminates the need for an interpolation or image rejection filter in the complex to real conversion and simplifies our end-to-end design considerably.

We know that the Nyquist rate for the complex channelized time series is simply the width of the channel, or $f_{S(IN)}/L$ Hz. Since we are going to convert the channelized output signals from complex to real, we would like to select a value for $M$ so that the individual channel sample rate is $2f_{S(IN)}/L$ Hz. The decimated output sample rate is given by $f_{S(OUT)} = f_{S(IN)}/M$, so we can accomplish this by selecting the value of $M$ to be equal to $L/2$ or half the length of the DFT.

The selection of $M$ is such an important consideration that it is generally fixed at some optimum value. For a channelizer that outputs real data, the value of $L/2$ is considered to be optimum, so our stand-alone channelizer equation will fix the value of $M$ to be equal to $L/2$.

## 12.3.6   Time Varying Phase Terms

There are countless times in engineering where an engineer solves a systems problem by calculating the optimum value for some variable only to see that the results of that solution adversely affect some other system function. It is analogous to squeezing a balloon. By squeezing, the engineer solves a balloon diameter problem at one end, only to see a similar problem appear at the other end. Unfortunately this is the case with the selection of the sample rate decimation factor $M$. It turns out that selecting the value of $M$ to optimize the channelizer output sample rate creates a signal processing problem downstream in the form of undesirable time varying phase terms. This leaves us in an uncomfortable situation, but being good engineers, our first impulse is to identify, study, and solve this problem. This is exactly what we are going to do next.

Let's go back and consider the raw signal $a(n)$ that we wish to process. We can see from Figure 12.29 that the sequential data blocks we process are shifted in time by $mM$ samples, where $m$ is the block index and takes on the values $m = 0, 1, 2, 3, \cdots$. The sampled data within the blocks that we process can be represented by the notation

$$x(n + mM) \ \text{ for } 0 \le n < N, \text{ and } m = 0, 1, 2, \cdots$$

Every input data block is composed of the time series shifted by $mM$ samples. Let's take a look at how this shift in time affects the spectra of the data we are processing. To do this, we define the time series $x(n)$ and the spectral series $Y(k)$ to be Fourier transform pairs such that

$$x(n) \ \overset{f}{\leftrightarrow} \ Y(k)$$

The time series $x(n)$ is computed from its spectrum $Y(k)$ with the help of an inverse discrete Fourier transform (IDFT), given by

$$x(n) = \frac{1}{L} \sum_{k=0}^{L-1} Y(k) e^{j2\pi\left(\frac{k}{L}\right)n}$$

**Equation 12.25**

The factor $1/L$ is included to negate the factor of $L$ gain introduced by the forward DFT used to compute $Y(k)$ originally. Equation 12.25 gives us the inverse transform for a single $Y(k)$ vector that is $L$ elements long. Now envision Equation 12.25 not just processing a single input block but being used to process a steady stream of sequential blocks. Sequential blocks of data that are incrementally shifted in time by chunks of $M$ samples can be represented by the shifted time series notation $x(n+mM)$. Therefore, by substitution, we get

$$x(n+mM) = \frac{1}{L}\sum_{k=0}^{L-1} Y(k) e^{j2\pi\left(\frac{k}{L}\right)(n+mM)}$$

We can expand the exponential term to get

$$x(n+mM) = \frac{1}{L}\sum_{k=0}^{L-1} Y(k) e^{j2\pi\left(\frac{k}{L}\right)n} e^{+j2\pi\left(\frac{k}{L}\right)mM}$$

After rearranging terms we get

$$x(n+mM) = \frac{1}{L}\sum_{k=0}^{L-1} \left[ Y(k) e^{j2\pi\left(\frac{k}{L}\right)mM} \right] e^{j2\pi\left(\frac{k}{L}\right)n}$$

We can see that the right side of the equation is just the inverse Fourier transform of the term enclosed in brackets. The bracketed term is the original spectrum of $x(n)$ multiplied by the complex phasor

$$e^{j2\pi\left(\frac{k}{L}\right)mM}$$

Therefore the transform pair for the shifted time series is given by

$$x(n+mM) \overset{f}{\leftrightarrow} Y(k) e^{j2\pi\left(\frac{k}{L}\right)mM} \quad \text{for } m = 0,1,2,\cdots$$

**Equation 12.26**

In many applications, we wouldn't be too concerned about a static phase shift incurred through the use of a single DFT. But Equation 12.26 represents a continuous sequence of DFTs. The incrementing value of the index $m$ transforms the static phase shift into a *time varying phase shift*, which is another term for a phasor that is moving around the unit circle at the deci-mated sample rate, in chunks of $2\pi\frac{k}{L}M$ radians. The term $e^{j2\pi\left(\frac{k}{L}\right)mM}$ looks like a complex tuner and smells like a complex tuner, so there is a darn good chance that it really is a complex tuner. The $M$ sample time shift that served us so well in decimating the sample rate has introduced an unwanted tuner into each output channel. This tuner up converts each channel to a normalized center frequency given by

$$\omega_0 = 2\pi \left(\frac{k}{L}\right) M \text{ for } k = 0,1,2,\cdots,L-1$$

The tuned frequency increases for each channel as the DFT bin index $k$ increases from $0$ to $L-1$.

In section 12.3.5 we were able to decimate the output channel sample rate by introducing the time shift variable $M$, which was a good thing. It seemed simple enough at the time, but now we discover it has affected us in a not so good way by introducing an unwanted complex tuner in each of the $L$ channels. Let's take a closer look at this problem and see if we can use our engineering skills to mitigate its effects. We begin where we left off with Equation 12.24, our channelizer processing equation, which is repeated here for clarity:

$$Y(k) = \sum_{p=0}^{L-1} z(p) e^{-j\left(\frac{2\pi}{L}\right)(qL+p)\left(k+\frac{1}{2}\right)} \quad \text{for } k = 0,1,2,\cdots L-1$$

We know that $z(p)$ is the folded and summed product vector. If we recognize that the input data is shifted by $M$ samples every input data block, and we know that $Q = N/L$, then we can rewrite Equation 12.20 for $z(p)$ to reflect this incremental shift. Equation 12.20 is repeated here for clarity:

$$z(p) = \sum_{q=0}^{Q-1} h(qL+p)x(qL+p) \quad \text{for } \left\{ \begin{array}{l} p = 0,1,2,\cdots L-1 \\ Q = N/L \end{array} \right\}$$

If we incorporate these changes, we end up with

$$z(p) = \sum_{q=0}^{Q-1} h(qL+p)x(qL+p+mM) \quad \text{for } \left\{ \begin{array}{l} p = 0,1,2,\cdots L-1 \\ m = 0,1,2,\cdots \\ Q = N/L \end{array} \right\}$$

This shift in time alters the Fourier transform of $z(p)$ so that

$$Y(k,m) = \sum_{p=0}^{L-1} z(p) e^{-j\left(\frac{2\pi}{L}\right)(qL+p+mM)\left(k+\frac{1}{2}\right)} \quad \text{for } k = 0,1,2,\cdots L-1$$

We have added the index $m$ to $Y(k)$ to indicate that $Y(k,m)$ is the $m$th output sample of the $k$th channel. We can simplify this equation a little bit by separating the exponential terms

$$Y(k,m) = \sum_{p=0}^{L-1} z(p) \; e^{-j\left(\frac{2\pi}{L}\right)\left(k+\frac{1}{2}\right)qL} \; e^{-j\left(\frac{2\pi}{L}\right)\left(k+\frac{1}{2}\right)p} \; e^{-j\left(\frac{2\pi}{L}\right)\left(k+\frac{1}{2}\right)mM}$$

**Equation 12.27**

The first exponential term in Equation 12.27 $e^{-j\left(\frac{2\pi}{L}\right)\left(k+\frac{1}{2}\right)qL}$ can be simplified by noting

$$e^{-j\left(\frac{2\pi}{L}\right)\left(k+\frac{1}{2}\right)qL} = e^{-j\left(\frac{2\pi}{L}\right)kqL} \; e^{-j\left(\frac{2\pi}{L}\right)\left(\frac{1}{2}\right)qL} = e^{-j(2\pi)kq}e^{-j(\pi)q} = (1)(-1)^{q} = (-1)^{q}$$

This allows us to simplify Equation 12.27 to give us

$$Y(k,m) = \sum_{p=0}^{L-1} z(p) \; (-1)^{q} \, e^{-j\left(\frac{2\pi}{L}\right)\left(k+\frac{1}{2}\right)p} \; e^{-j\left(\frac{2\pi}{L}\right)\left(k+\frac{1}{2}\right)mM}$$

**Equation 12.28**

The last exponential term in Equation 12.28 $e^{-j\left(\frac{2\pi}{L}\right)\left(k+\frac{1}{2}\right)mM}$ can be separated and rewritten as

$$e^{-j\left(\frac{2\pi}{L}\right)\left(k+\frac{1}{2}\right)mM} = e^{-j\left(\frac{2\pi}{L}\right)kmM} \; e^{-j\left(\frac{\pi}{L}\right)mM}$$

Since the rewritten terms are not a function of the index $p$, they can be brought out in front of the summation, which produces

$$Y(k,m) = e^{-j\left(\frac{2\pi}{L}\right)kmM} \; e^{-j\left(\frac{\pi}{L}\right)mM} \sum_{p=0}^{L-1} z(p) \; (-1)^{q} \, e^{-j\left(\frac{2\pi}{L}\right)\left(k+\frac{1}{2}\right)p}$$

**Equation 12.29**

Since the $(-1)^{q}$ term is a function of the coefficient index $q$, it makes sense to encapsulate it in the folded vector $z(p)$, which is also a function of $q$. This action modifies the equation for $z(p)$ to give

$$z(p) = \sum_{q=0}^{Q-1} h(qL+p)x(qL+p+mM)(-1)^{q} \quad \text{for } p = 0,1,2,\cdots L-1$$

**Equation 12.30**

So now we can rewrite Equation 12.29 as

$$Y(k,m) = e^{-j\left(\frac{2\pi}{L}\right)kmM} e^{-j\left(\frac{\pi}{L}\right)mM} \sum_{p=0}^{L-1} z(p)\, e^{-j\left(\frac{2\pi}{L}\right)\left(k+\frac{1}{2}\right)p}$$

**Equation 12.31**

Back in section 12.3.5, we decided that the optimum value for $M$ was $L/2$ or half the width of the DFT. If we plug this value into Equation 12.31, we end up with

$$Y(k,m) = e^{-j\left(\frac{2\pi}{L}\right)km\frac{L}{2}} e^{-j\left(\frac{\pi}{L}\right)m\frac{L}{2}} \sum_{p=0}^{L-1} z(p)\, e^{-j\left(\frac{2\pi}{L}\right)\left(k+\frac{1}{2}\right)p}$$

**Equation 12.32**

The $L/2$ and $2/L$ terms in the first exponential and the $L$ and $1/L$ terms in the second exponential cancel one another, simplifying the expression to

$$Y(k,m) = e^{-j(\pi)km} e^{-j\left(\frac{\pi}{2}\right)m} \sum_{p=0}^{L-1} z(p)\, e^{-j\left(\frac{2\pi}{L}\right)\left(k+\frac{1}{2}\right)p} \quad \text{for} \left\{ \begin{array}{l} k = 0,1,2,\cdots,L-1 \\ m = 0,1,2,\cdots \end{array} \right\}$$

**Equation 12.33**

We can simplify this expression even further if we recognize that the term $e^{-j\pi mk}$ reduces to $(-1)^{mk}$, which allows us to rewrite the expression as

$$Y(k,m) = (-1)^{mk} e^{-j\left(\frac{\pi}{2}\right)m} \sum_{p=0}^{L-1} z(p)\, e^{-j\left(\frac{2\pi}{L}\right)\left(k+\frac{1}{2}\right)p} \quad \text{for} \left\{ \begin{array}{l} k = 0,1,2,\cdots,L-1 \\ m = 0,1,2,\cdots \end{array} \right\}$$

**Equation 12.34**

Let's stop for a moment and take a look at the two terms we pulled out of the summation $(-1)^{mk}$ and $e^{-j\left(\frac{\pi}{2}\right)m}$. The term $e^{-j\left(\frac{\pi}{2}\right)m}$ is the result of replacing the frequency bin index $k$ with $k+1/2$. This is a phasor that is rotating around the unit circle in chunks of $\pi/2$ radians at the output sample rate $f_{S(OUT)}$. This equates to a complex tuning frequency of $f_{S(OUT)}/4$.

The $(-1)^{mk}$ term tells us that for even DFT channels $k = 0,2,4,6,\cdots$, the channel spectra are multiplied by the unity sequence of $(1,1,1,1,\cdots)$, which does not modify their spectra. However, for odd DFT channels, $k = 1,3,5,7,\cdots$, the channel time series is multiplied by a unity sequence with alternating signs $(1,-1,\ 1,-1,\cdots)$. This is the equivalent of multiplying the channel time series by the complex exponential $e^{j\pi m}$, which is also the equivalent of the frequency $f_{s(OUT)}/2$. Multiplying a sequence by $f_s/2$ effectively inverts the spectrum of that sequence. This means that upon output, all the odd channels will have their input spectrum inverted.

We illustrate this frequency inversion phenomenon in Figure 12.39. In part A of the figure, we see the spectrum of a signal $x(n)$ that is represented by a series of triangles. Since this is the spectra of a digital signal, the spectrum repeats itself at the sample rate or every $f_s$ Hz. The bandwidth of each triangle in the diagram is $f_s/2$ Hz wide. In the digital world, the spectrum at frequencies above the Nyquist rate or $f_s/2$ Hz are identical to the spectrum at the frequencies below the Nyquist rate, with the exception that they are mirror images of one another. This is why we used triangles to represent the two halves of the digital spectrum.

In the figure, the spectrum has been labeled $A\uparrow$ for the "erect" or positive frequencies and $B\downarrow$ for the "inverted" or negative frequencies. The $B\downarrow$ portion of the spectrum is the mirror image of the $A\uparrow$ portion of the spectrum. It is as if the $A\uparrow$ portion was rotated or folded about the $f_s/2$ axis.

In part B of the figure, we see the same spectrum, but this time the signal $x(n)$ has been multiplied by the complex exponential $e^{j\pi n}$ or $f_s/2$ Hz.

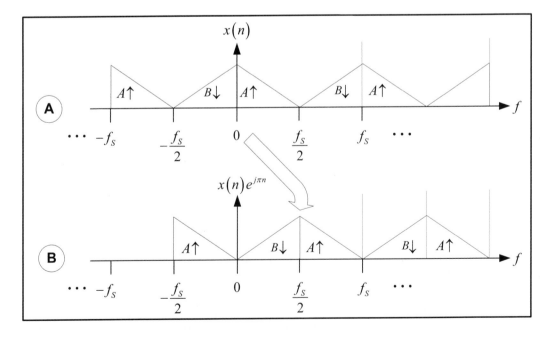

**Figure 12.39**  Odd channel spectrum inversion

This multiplication has translated the entire spectrum up in frequency by $f_s/2$ Hz. When this happens, negative frequency bands $B\downarrow$ and positive frequency bands $A\uparrow$ appear to be swapped. Furthermore the content of the frequency bands have been rotated so that the high frequencies now occupy the spectrum previously held by the low frequencies and vice versa. If this were to happen to a band of frequencies that was used to transport a simple telephone conversation, imagine what the caller's voice would sound like.

When this frequency switcheroo occurs, we say that the spectrum has been inverted. In order to visualize this more clearly, let's take another look at the channelizer output for the case illustrated way back in Equation 12.10. This equation is repeated here for clarity. In this case, both $h(n)$ and $x(n)$ were of length $N = 16$ and the DFT length was $L = 4$:

$$Y(0) = z(0) \quad +z(1)e^{-j(0)} \quad +z(2)e^{-j(0)} \quad +z(3)e^{-j(0)}$$

$$Y(1) = z(0) \quad +z(1)e^{-j\left(\frac{\pi}{2}\right)} \quad +z(2)e^{-j\left(\frac{2\pi}{2}\right)} \quad +z(3)e^{-j\left(\frac{3\pi}{2}\right)}$$

$$Y(2) = z(0) \quad +z(1)e^{-j(\pi)} \quad +z(2)e^{-j(2\pi)} \quad +z(3)e^{-j(3\pi)}$$

$$Y(3) = z(0) \quad +z(1)e^{-j\left(\frac{3\pi}{2}\right)} \quad +z(2)e^{-j\left(\frac{6\pi}{2}\right)} \quad +z(3)e^{-j\left(\frac{9\pi}{2}\right)}$$

We can insert the $(-1)^{mk}$ term introduced in Equation 12.34 into this equation and arrive at

$$Y(0,m) = z(0)(-1)^{0m} \quad +z(1)(-1)^{0m}e^{-j(0)} \quad +z(2)(-1)^{0m}e^{-j(0)} \quad +z(3)(-1)^{0m}e^{-j(0)}$$

$$Y(1,m) = z(0)(-1)^{1m} \quad +z(1)(-1)^{1m}e^{-j\left(\frac{\pi}{2}\right)} \quad +z(2)(-1)^{1m}e^{-j\left(\frac{2\pi}{2}\right)} \quad +z(3)(-1)^{1m}e^{-j\left(\frac{3\pi}{2}\right)}$$

$$Y(2,m) = z(0)(-1)^{2m} \quad +z(1)(-1)^{2m}e^{-j(\pi)} \quad +z(2)(-1)^{2m}e^{-j(2\pi)} \quad +z(3)(-1)^{2m}e^{-j(3\pi)}$$

$$Y(3,m) = z(0)(-1)^{3m} \quad +z(1)(-1)^{3m}e^{-j\left(\frac{3\pi}{2}\right)} \quad +z(2)(-1)^{3m}e^{-j\left(\frac{6\pi}{2}\right)} \quad +z(3)(-1)^{3m}e^{-j\left(\frac{9\pi}{2}\right)}$$

**Equation 12.35: DFT channel outputs for shifted input data**

The $Y(k,m)$ notation in this equation is meant to identify the $m$th output sample of the $k$th channel. Remember that the $m$th input block of data produces the $m$th output channelized sample.

We can simplify Equation 12.35 by factoring out the common terms to produce Equation 12.36. This helps us to better visualize the effect of the time varying phase terms:

$$Y(0,m) = (-1)^{0m} \{z(0) \ +z(1)e^{-j(0)} \ +z(2)e^{-j(0)} \ +z(3)e^{-j(0)} \ \}$$

$$Y(1,m) = (-1)^{1m} \{z(0) \ +z(1)e^{-j\left(\frac{\pi}{2}\right)} \ +z(2)e^{-j\left(\frac{2\pi}{2}\right)} \ +z(3)e^{-j\left(\frac{3\pi}{2}\right)}\}$$

$$Y(2,m) = (-1)^{2m} \{z(0) \ +z(1)e^{-j(\pi)} \ +z(2)e^{-j(2\pi)} \ +z(3)e^{-j(3\pi)} \ \}$$

$$Y(3,m) = (-1)^{3m} \{z(0) \ +z(1)e^{-j\left(\frac{3\pi}{2}\right)} \ +z(2)e^{-j\left(\frac{6\pi}{2}\right)} \ +z(3)e^{-j\left(\frac{9\pi}{2}\right)}\}$$

**Equation 12.36: DFT channel outputs in factored form**

The equation tells us that the time varying phase terms introduced by the decimation in the sample rate by a factor of $M = L/2$ have caused the even channels 0 and 2 to be multiplied by a factor of $(-1)^{0m} = 1$ and $(-1)^{2m} = 1$, respectively, which does not alter the orientation of their spectra. The odd channels 1 and 3, on the other hand, are multiplied by the factors $(-1)^{1m}$ and $(-1)^{3m}$, respectively, which is a unity time series with alternating signs $+1, -1, +1, -1, \cdots$. This time series is the equivalent of multiplying the outputs of those channels by half the sampling frequency or $f_{S(OUT)}/2$, which inverts the orientation of their spectra. If we look at successive samples out of each channel, as shown in Equation 12.37, we can clearly see the alternating sign sequence that occurs in the odd output channels:

$$Y(0,m) = Y(0,0), \ +Y(0,1), \ +Y(0,2), \ +Y(0,3), \ +Y(0,4), \ \cdots \quad \text{Channel 0}$$
$$Y(1,m) = Y(1,0), \ -Y(1,1), \ +Y(1,2), \ -Y(1,3), \ +Y(1,4), \ \cdots \quad \text{Channel 1}$$
$$Y(2,m) = Y(2,0), \ +Y(2,1), \ +Y(2,2), \ +Y(2,3), \ +Y(2,4), \ \cdots \quad \text{Channel 2}$$
$$Y(3,m) = Y(3,0), \ -Y(3,1), \ +Y(3,2), \ -Y(3,3), \ +Y(3,4), \ \cdots \quad \text{Channel 3}$$

**Equation 12.37: Signs of the output channel samples**

The reader is invited to expand Equation 12.35 by replacing $z(p)$ with its element values to further verify the odd channel spectral inversion at the product sample level.

This phenomenon of odd channel spectral inversion is not a good thing. The casual observer probably would have never expected to see this. However, if we take an optimistic look at the problem, it may not be such a bad thing either. This is because in real-life applications, there are more than likely some individual channels whose spectra we may actually want to see inverted. This is because in many applications, some channels in an FDM stack, for example, are transmitted with inverted spectra. If our channelizer reinverts these inverted channels, then all is well.

For this reason, we would like to include in our design the capability to control the spectral orientation of each channel. The way we are going to address this problem is to add the capability to program the channelizer to selectively invert or not invert the spectrum of each individual channel. It turns out that this is a fairly simple process. For the time being, let's assume that we have a memory that is one bit wide and $L$ bits in length. Each bit in the memory corresponds to a particular DFT bin or output channel. That is, bit $k$ in the memory corresponds to the $k$th channel. Each bit in memory can take on either a 0 or a 1, which is defined as follows:

- When the $k$th bit in memory is a 0, the $k$th channel spectrum will not be altered from whatever spectral orientation it may be in.
- When the $k$th bit in memory is a 1, the $k$th channel spectrum will be inverted from whatever spectral orientation it may be in.

For convenience in notation, we will name this memory $I$ to indicate that it is the invert memory control, and note that $I(k)$ represents the $k$th storage location in memory corresponding to the $k$th output channel. When a particular channel's $I$ bit is programmed to a 1, the channel output time series is multiplied by $f_{S(OUT)}/2$ to invert its spectrum. We can express $f_{S(OUT)}/2$ as $e^{j(\pi m)}$. We can control this term either turning it on or off by adding the memory $I$ to produce $e^{j(\pi m)I(k)}$. Now we can include this inverting function in our evolving channelizer equation to produce

$$Y(k,m) = (-1)^{mk}\, e^{-j\left(\frac{\pi}{2}\right)m}\, e^{j(\pi m)I(k)} \sum_{p=0}^{L-1} z(p)\, e^{-j\left(\frac{2\pi}{L}\right)\left(k+\frac{1}{2}\right)p} \quad \text{for} \quad \left\{ \begin{array}{l} k = 0,1,2,\cdots,L-1 \\ m = 0,1,2,\cdots \end{array} \right\}$$

**Equation 12.38**

In normal operation, the user simply preloads the $L$ locations in memory with either a logic 1 or 0 and then begins processing. Although we will not implement the spectral inversion control in hardware with an actual multiplier, we can visualize the operation of this inversion control in the block diagram of Figure 12.40.

Now is a good time to mention what the reader is probably already thinking: "Boy this evolving equation is getting pretty hairy!" Yes, it is getting a bit complicated, but if we understand the origin and function of each term in the equation, all should be clear at this point. As an incentive for the reader to press on, there is good news ahead. Later we will take advantage of some signal processing symmetry and tricks to substantially collapse Equation 12.38 into a more concise and less hairy form.

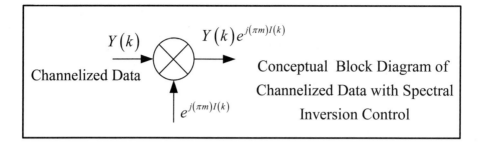

**Figure 12.40** Functional block diagram of the spectral inversion control

Looking back on this section, can you imagine the complexity of this problem if we had chosen $M$ to be some other value? I encourage you to set $M$ to values like $1$, $L/4$, $L/8$, or even something more interesting like $3L/4$ and see how it manifests itself in terms of unwanted channel tuning. If, for example, we had selected $M = L$, then the time varying phase term would vanish since $e^{j\left(\frac{2\pi}{L}\right)mMk} \Rightarrow e^{j\left(\frac{2\pi}{L}\right)mLk} \Rightarrow e^{j2\pi mk} = 1$ for all $m$ and $k$. If we set $M = L$, then this unwanted tuner problem would not have occurred but the output channel sample rate would have been decimated by a factor of $L$, which would have produced an output sample rate equal to half the Nyquist rate for real signals. As a consequence, we would have been forced to add interpolation filters to perform the complex to real conversion on each of the $L$ channel outputs, and the design complexity would have increased considerably. So there are trade-offs to be made when selecting the value of $M$.

### 12.3.7  Channel Complex to Real Conversion

At this point our stand-alone channelizer equation has taken on the look of something evil. At first glance it looks pretty formidable. Since we incrementally developed the equation term by term, the reader should be able to see how simple it really is. Every term in the equation tells a story. Every term is a component in the channelizer system. Hopefully the reader can look at this equation and form a picture of the overall system in his or her head.

We've spent a great deal of time building this equation term by term. Now it is time to start collapsing the equation to its final version by combining terms and taking advantage of symmetries that we will cover next in our discussion of channel complex to real conversion. If you have not already done so, please read Chapter 7, "Complex to Real Conversion," to better understand the logic in the following discussion.

The process of complex to real conversion, as detailed in Chapter 7, makes the assumption that the signal to be converted is sampled at the complex

Nyquist rate and requires us to first interpolate the sample rate of the signal by a factor of two and then quarter band translate the signal spectrum.

Because of the choice we made for $M$ we do not have to use interpolation filters for each DFT channel, but we do have to quarter band translate the signal so that it is centered at $f_{S(OUT)}/4$. Therefore it is necessary to multiply Equation 12.38 by $f_{S(OUT)}/4$ to center the output channel band between 0 Hz and $f_{S(OUT)}/2$ Hz and then take the real part.

Fortunately we have done our homework and decimated the sample rate of the input signal so that the sample rate of the channelizer output signals are all at the Nyquist rate for real signals. Therefore we will not need to invest the extra resources to implement $L$ interpolate by two filters. All we really need to do is translate the spectrum of the output channels by $f_{SOUT}/4$.

This is the same as multiplying by $e^{j\left(\frac{\pi}{2}\right)m}$. Then we will take the real part, resulting in

$$Y(k,m) = \mathrm{re}\left\{ e^{j\left(\frac{\pi}{2}\right)m} (-1)^{mk} e^{-j\left(\frac{\pi}{2}\right)m} e^{j(\pi m)I(k)} \sum_{p=0}^{L-1} z(p)\, e^{-j\left(\frac{2\pi}{L}\right)\left(k+\frac{1}{2}\right)p} \right\} \quad \text{for} \quad \left\{ \begin{array}{l} k = 0,1,2,\cdots,L-1 \\ m = 0,1,2,\cdots \end{array} \right\}$$

**Equation 12.39**

The $e^{j\left(\frac{\pi}{2}\right)m}$ and $e^{-j\left(\frac{\pi}{2}\right)m}$ terms cancel, and Equation 12.39 is immediately simplified to

$$Y(k,m) = \mathrm{re}\left\{ (-1)^{mk} e^{j(\pi m)I(k)} \sum_{p=0}^{L-1} z(p)\, e^{-j\left(\frac{2\pi}{L}\right)\left(k+\frac{1}{2}\right)p} \right\} \quad \text{for} \quad \left\{ \begin{array}{l} k = 0,1,2,\cdots,L-1 \\ m = 0,1,2,\cdots \end{array} \right\}$$

**Equation 12.40**

The term $e^{j(\pi)}$ simplifies to $(-1)$. Therefore the term $e^{j(\pi m)I(k)}$ can be reduced to $(-1)^{m\,I(k)}$. The two terms in Equation 12.40 involving $(-1)$ can be combined to form $(-1)^{m(k+I(k))}$. This further reduces the equation to

$$Y(k,m) = \mathrm{re}\left\{ (-1)^{m(k+I(k))} \sum_{p=0}^{L-1} z(p)\, e^{-j\left(\frac{2\pi}{L}\right)\left(k+\frac{1}{2}\right)p} \right\} \quad \text{for} \quad \left\{ \begin{array}{l} k = 0,1,2,\cdots,L-1 \\ m = 0,1,2,\cdots \end{array} \right\}$$

**Equation 12.41**

The real part of Equation 12.41 is given by

$$Y(k,m)_{REAL} = (-1)^{m(k+I(k))} \sum_{p=0}^{L-1} z(p) \cos\left[\left(\frac{2\pi}{L}\right)\left(k+\frac{1}{2}\right)p\right] \text{ for } \left\{ \begin{array}{l} k = 0,1,2,\cdots,L-1 \\ m = 0,1,2,\cdots \end{array} \right\}$$

**Equation 12.42**

The cosine function is magnitude periodic in $\pi$ radians and is negative symmetric about $L/2$. Therefore we can express this symmetry by writing

$$\begin{aligned} \cos\left[\left(\frac{2\pi}{L}\right)\left(k+\frac{1}{2}\right)p\right] &= -\cos\left[\left(\frac{2\pi}{L}\right)\left(k+\frac{1}{2}\right)(L-p)\right] && \text{for } p = 1 \text{ through } \frac{L}{2} \\ &= 1 && \text{for } p = 0 \\ &= 0 && \text{for } p = L/2 \end{aligned}$$

**Equation 12.43**

By recognizing this cosine symmetry, the upper limit of summation can be reduced from $L-1$ to $L/2-1$. The folded sum of products terms $z(p)$ and $z(L-p)$ can be combined into the term $z(p)-z(L-p)$, reducing the number of computational multiplies by a factor of two. In addition, since the cosine term is 1 for $p=0$, there is no need to perform a cosine multiplication on $z(0)$. Also since the cosine term for $p = L/2$ is equal to zero, there is no need to perform the multiplication on the $z(L/2)$ term. Therefore the summation portion of Equation 12.42 can be modified to

$$z(0) + \sum_{p=1}^{\frac{L}{2}-1} \left[z(p) - z(L-p)\right] \cos\left[\left(\frac{2\pi}{L}\right)\left(k+\frac{1}{2}\right)p\right]$$

**Equation 12.44**

For example, if we have a simple channelizer where $L=8$, the summation in Equation 12.44 can be written as

$$\begin{aligned} z(0) \ &+ \left[z(1) - z(7)\right]\cos\left[\left(\frac{\pi}{4}\right)\left(k+\frac{1}{2}\right)1\right] \\ &+ \left[z(2) - z(6)\right]\cos\left[\left(\frac{\pi}{4}\right)\left(k+\frac{1}{2}\right)2\right] \\ &+ \left[z(3) - z(5)\right]\cos\left[\left(\frac{\pi}{4}\right)\left(k+\frac{1}{2}\right)3\right] \end{aligned}$$

Notice that the $z(0)$ term is sitting all by itself and that the $z(L/2) = z(4)$ term is missing. There is no need to waste computational horsepower by multiplying $z(0)$ by one. There is also no need to include the $z(4)$ in the summation since it is multiplied by zero. If we incorporate this cosine symmetry into Equation 12.42, we arrive at the final version of the stand-alone channelizer equation given by

$$Y(k,m) = (-1)^{m\,(k+I(k))} \left\{ z(0) + \sum_{p=1}^{\frac{L}{2}-1} \left[ z(p) - z(L-p) \right] \cos\left[ \left( \frac{2\pi}{L} \right) \left( k + \frac{1}{2} \right) p \right] \right\} \text{ for } \left\{ \begin{matrix} k = 0,1,2,\cdots,L-1 \\ m = 0,1,2,\cdots \end{matrix} \right\}$$

where the $z(p)$ term is defined as

$$z(p) = \sum_{q=0}^{Q-1} h(qL+p) x \left( qL + p + m\frac{L}{2} \right) (-1)^q \quad \text{for } \left\{ \begin{matrix} p = 0,1,2,\cdots L-1 \\ Q = N/L \end{matrix} \right\}$$

### Equation 12.45: Channelizer stand-alone equation

Equation 12.45 represents the final stop in our incremental development of the stand-alone channelizer equation. We will utilize this equation in section 12.4, where we demonstrate the channelizer software simulation, and in section 12.5, where we demonstrate a channelizer hardware design example.

We need to stop for a moment and adjust our evolving channelizer block diagram to include the addition of the $(-1)^q$ term in the computation of $z(p)$. Equation 12.45 says that we compute $z(p)$ as follows:

$$z(p) = \sum_{q=0}^{Q-1} h(qL+p) x \left( qL + p + m\frac{L}{2} \right) (-1)^q \quad \text{for } p = 0,1,2,\cdots L-1$$

### Equation 12.46

Let's stick with the example channelizer that we have been playing with throughout the chapter where $N = 16$, $L = 4$, and $Q = N/L = 4$. Each successive data block $x(n)$ is offset from the previous data block by the $M = L/2$. For the very first input data block we process, the block index is $m = 0$. If we plug these values into Equation 12.46, we get

$$z(p) = \sum_{q=0}^{3} h(q4+p) x(q4+p+0)(-1)^q \quad \text{for } p = 0,1,2,3$$

### Equation 12.47

If we expand Equation 12.47, we arrive at

$$
\begin{array}{cccc}
q = 0 & q = 1 & q = 2 & q = 3 \\
z(0) = \ h(0)x(0)(-1)^0 & +h(4)x(4)(-1)^1 & +h(8)x(8)(-1)^2 & +h(12)x(12)(-1)^3 \\
z(1) = \ h(1)x(1)(-1)^0 & +h(5)x(5)(-1)^1 & +h(9)x(9)(-1)^2 & +h(13)x(13)(-1)^3 \\
z(2) = \ h(2)x(2)(-1)^0 & +h(6)x(6)(-1)^1 & +h(10)x(10)(-1)^2 & +h(14)x(14)(-1)^3 \\
z(3) = \ h(3)x(3)(-1)^0 & +h(7)x(7)(-1)^1 & +h(11)x(11)(-1)^2 & +h(15)x(15)(-1)^3
\end{array}
$$

**Equation 12.48**

The reader should note that the columns of Equation 12.48 are labeled $q = 0, 1, 2,$ and $3$ and correspond to the individual terms for each $z(p)$. We can see that the $(-1)^q$ term multiplies every odd term in each $z(p)$ by a negative one, and it multiplies every even term by unity.

For the second input data block, the block index $m = 1$ and $mL/2 = 2$. For this case, Equation 12.47 and Equation 12.48 can be written as

$$
z(p) = \sum_{q=0}^{3} h(q4 + p)x(q4 + p + 2)(-1)^q \quad \text{for } p = 0, 1, 2, 3
$$

$$
\begin{array}{cccc}
q = 0 & q = 1 & q = 2 & q = 3 \\
z(0) = \ h(0)x(2)(-1)^0 & +h(4)x(6)(-1)^1 & +h(8)x(10)(-1)^2 & +h(12)x(14)(-1)^3 \\
z(1) = \ h(1)x(3)(-1)^0 & +h(5)x(7)(-1)^1 & +h(9)x(11)(-1)^2 & +h(13)x(15)(-1)^3 \\
z(2) = \ h(2)x(4)(-1)^0 & +h(6)x(8)(-1)^1 & +h(10)x(12)(-1)^2 & +h(14)x(16)(-1)^3 \\
z(3) = \ h(3)x(5)(-1)^0 & +h(7)x(9)(-1)^1 & +h(11)x(13)(-1)^2 & +h(15)x(17)(-1)^3
\end{array}
$$

**Equation 12.49**

For this simple example, the third input block to the channelizer the block index $m = 2$ and $mL/2 = 4$. We can easily see that each successive data block is offset from the previous block by $L/2$ samples. This process will continue until the channelizer is turned off.

At this point the reader should be wondering, "How might we implement this crazy equation in software or in hardware?" In a software implementation of the channelizer, we would store the $N$ point product series $h(n)x(n)$ in heap memory pointed to by *pData. The process of selecting the data for storage into heap memory would have taken care of the $mL/2$ input data block offset and, as far as the software knows, the data in heap memory is indexed as $h(ql + p)x(ql + p)$. The folding memory would be implemented on the heap and is addressed via the pointer *pFold*. Both memories would be allocated as type "double." Equation 12.30 is easily implemented in the following code snippet:

```
for(p=0; p<L; p++){
    for(q=0; q<Q; q++){
        *(pFold+p) += *(pData +q*L +p))*pow((double)-1,(double)q);
    }
}
```

In hardware we need to alternate the signs of sequential terms in the summation of each $z(p)$. We need to modify the folding memory block diagram of Figure 12.34 to produce Figure 12.41.

In this figure, we show a multiplier in series with the summer and folding memory. One port of the multiplier inputs the $h(n)x(n)$ product series and the second input is the $(-1)^q$ term. For this simple folding memory function, we certainly would not want to increase the complexity by adding a multiplier just to change the sign of selected data samples. It's a good assumption that the design engineer will be using two's complement arithmetic. So we cannot just simply invert the sign of a number to negate it. So let's do something that's almost as simple. Let's go back to our ongoing example where the value of $L = 4$. As the data is recirculated in memory, let's look at the sign sequence we need to generate. The sequence for $q$ and $(-1)^q$ is illustrated in Table 12.2. The sign sequence is constant relative to each group of $L$ filter coefficients $h(n)$. The first row of the table lists each sequential set of $Q = N/L$ coefficients. The second row illustrates the value of $q$ for each set of $L$ coefficients, and the third row illustrates corresponding signs. Here we see

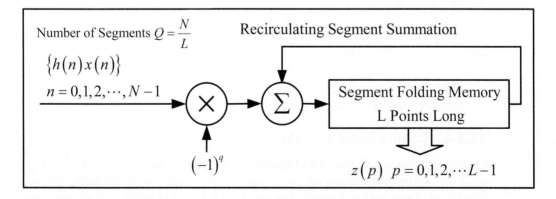

**Figure 12.41** Folding memory with inversion term

**Table 12.2**  Folding Memory Sign Sequence for $L = 4$

| | $\leftarrow h(0) \quad h(3) \rightarrow$ | $\leftarrow h(4) \quad h(7) \rightarrow$ | $\leftarrow h(8) \quad h(11) \rightarrow$ | $\leftarrow h(12) \quad h(15) \rightarrow$ |
|---|---|---|---|---|
| $q$ | 0  0  0  0 | 1  1  1  1 | 2  2  2  2 | 3  3  3  3 |
| $(-1)^q$ | +  +  +  + | −  −  −  − | +  +  +  + | −  −  −  − |

that the value of $q$ is held constant for $L$ consecutive input samples before it is incremented. The sign of the $(-1)^q$ term is also shown for each of the $L$ samples. Note that the sign is positive for the first $L$ input samples, negative for the second $L$ input samples, and so on. This behavior will hold true for all values of $L$.

Let's begin by replacing the multiplier with a cheap low power arithmetic logic unit (ALU), as illustrated in Figure 12.42. The input product sequence $h(n)x(n)$ is applied to the $B$ input port of the ALU, and a logic zero or ground is applied to the $A$ input port.

We have also expanded the width our folding memory by one bit to form a string of $L$ bits in the recirculating memory. These bits will be used to control the operation of the ALU. Initially these control bits are all set to logic 0. They recirculate in lock step with the folded data. The bits are inverted as they are recirculated, so they alternate between strings of logic 0 and strings of logic 1. If the bit is logic 0, it instructs the ALU to perform an addition of $(A + B) = (0 + B)$, effectively passing the input product sequence through to memory unchanged. When the bit is logic 1, it instructs the ALU to perform the subtraction $(A - B) = (0 - B)$, which negates the input product sequence sample into memory. The recirculating control bits alternate in sign every $L$ samples, duplicating the sign structure shown in Table 12.2. The fold and sum terms $z(p)$ obtained from the hardware architecture in Figure 12.42 are identical to the $z(p)$ illustrated in Equation 12.48.

This is a simple "no-hands" control mechanism that does not require a great deal of engineering thought.

### 12.3.8  Cosine Look Up Table

Our development of the stand-alone channelizer equation is not complete until we address the issue of the design of the cosine look up table. On the surface, it looks pretty simple. We just build a cosine table and plug it into our software or hardware channelizer design. Wrong! If we were to plunge

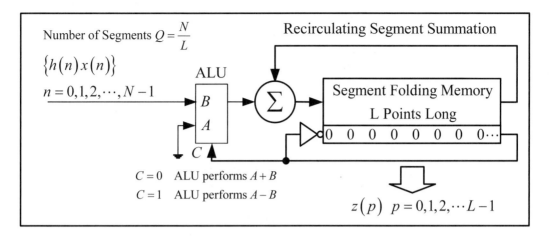

**Figure 12.42**   Modified folding memory architecture

headfirst into the table generation, we could very well end up with a rat's nest of a design. We could use brute force and fall into the trap of building a cosine look up table that contained a full period of all $L$ possible frequencies. If we succumbed to this design impulse, then a hardware implementation could end up with a huge, slow, and power-hungry look up table memory. In a software implementation, we could end up with a multidimensional array, or worse yet, we might end up performing the cosine calculation each time it's needed. Repetitive calculation of the cosine term would really slow the performance of a software channelizer. It is much better to do the calculations up front, one time, generate a look up table, and store it in heap memory. Access to the table would be done with pointer arithmetic, which is much faster than using an array with index addressing. In addition, using an indexed array might cause the compiler to implement the table in stack memory. If the look up table were large relative to the size of the stack, then the chances are highly likely that we would encounter a run time stack overflow error. These are not good things to have happen.

Enough said about what bad things might happen—let's discuss how we can efficiently implement this cosine look up table in either hardware or in software. First let's look at the cosine function we would like to compute:

$$\cos\left[\left(\frac{2\pi}{L}\right)\left(k+\frac{1}{2}\right)n\right] \quad \text{for} \quad \left\{ \begin{array}{l} k=0,1,2,\cdots,L-1 \\ n=0,1,2,\cdots,2L-1 \end{array} \right\}$$

**Equation 12.50**

The argument of the cosine is a function of both the indices $k$ and $n$. Table 12.3 illustrates the form the cosine argument reduces to for sequential values of $k$.

The lowest frequency generated by the cosine function occurs when $k = 0$. It is clear from the table that a full period of the lowest frequency cosine waveform contains the maximum number of unique values. A full period of all other frequencies when $k = 1, 2, 3, \cdots$ contains a subset of these values. For example, the argument sequence for the $k = 0$ case is given by

$$0, \frac{\pi}{L}, \frac{2\pi}{L}, \frac{3\pi}{L}, \frac{4\pi}{L}, \frac{5\pi}{L}, \frac{6\pi}{L}, \frac{7\pi}{L}, \frac{8\pi}{L}, \frac{9\pi}{L}, \frac{10\pi}{L} \cdots \frac{n\pi}{L} \cdots$$

**Equation 12.51**

The argument sequence for the $k = 1$ sequence is given by

$$0, \frac{3\pi}{L}, \frac{6\pi}{L}, \frac{9\pi}{L}, \frac{12\pi}{L} \cdots \frac{n3\pi}{L} \cdots$$

We can clearly see that the $k = 1$ sequence consists of every third element of the $k = 0$ sequence in Equation 12.51. Similarly the $k = 2$ sequence consists of every fifth element. This sequence decimation is similar for all values of $k = 0, 1, 2, \cdots, L-1$. This is a good thing because it means that we only need to build a look up table that can store one complete period of the lowest or $k = 0$

**Table 12.3** Breakdown of the Cosine Argument

| Index | | Base argument | Valid for |
|---|---|---|---|
| $k = 0$ | reduces to | $\dfrac{\pi}{L} n$ | for $n = 0, 1, 2, \cdots 2L - 1$ |
| $k = 1$ | reduces to | $\dfrac{3\pi}{L} n$ | for $n = 0, 1, 2, \cdots 2L - 1$ |
| $k = 2$ | reduces to | $\dfrac{5\pi}{L} n$ | for $n = 0, 1, 2, \cdots 2L - 1$ |
| $\vdots$ | | | for $n = 0, 1, 2, \cdots 2L - 1$ |
| $k = L - 1$ | reduces to | $\dfrac{(2L-1)\pi}{L} n$ | for $n = 0, 1, 2, \cdots 2L - 1$ |

cosine frequency. In order to generate the table, we will compute and store every possible value of a single period of the cosine waveform at its lowest frequency. All other frequencies of the cosine waveform will contain a subset of these values. So the number of samples computed for the lowest frequency cosine waveform will determine the length of the table.

Because of the $k+1/2$ term in Equation 12.50, a table of length $L$ will only be able to store a half period of sinusoid when $k=0$. Therefore it will be necessary to construct a table that is $2L$ samples in length, and the range of the index $n$ will be $n = 0, 1, 2, \cdots 2L - 1$.

Let's observe an example cosine look up table for $k=0$ for the case where $L = 16$, as shown in Table 12.4. The table shows the index $n$ ranging from 0 to 31. The table is $2L$ or 32 samples long.

The table entries are computed using Equation 12.50, which is repeated here for clarity:

$$\cos\left[\left(\frac{2\pi}{L}\right)\left(k+\frac{1}{2}\right)n\right] \quad \text{for} \quad \left\{ \begin{array}{l} k = 0, 1, 2, \cdots, L-1 \\ n = 0, 1, 2, \cdots, 2L-1 \end{array} \right\}$$

When we set $k=0$ and $L=16$, the cosine argument collapses and we end up with

$$\cos\left[\left(\frac{\pi}{16}\right)n\right] \quad \text{for} \quad n = 0, 1, 2, \cdots, 2L-1$$

**Equation 12.52**

**Table 12.4**  Cosine Look Up Table for $k = 0$

| $n$ | cos() | $n$ | cos() | $n$ | cos() | $n$ | cos() |
|---|---|---|---|---|---|---|---|
| 0 | 1.000 | 8 | 0.000 | 16 | -1.000 | 24 | 0.000 |
| 1 | 0.981 | 9 | -0.195 | 17 | -0.981 | 25 | 0.195 |
| 2 | 0.924 | 10 | -0.383 | 18 | -0.924 | 26 | 0.383 |
| 3 | 0.831 | 11 | -0.556 | 19 | -0.831 | 27 | 0.556 |
| 4 | 0.707 | 12 | -0.707 | 20 | -0.707 | 28 | 0.707 |
| 5 | 0.556 | 13 | -0.831 | 21 | -0.556 | 29 | 0.831 |
| 6 | 0.383 | 14 | -0.924 | 22 | -0.383 | 30 | 0.924 |
| 7 | 0.195 | 15 | -0.981 | 23 | -0.195 | 31 | 0.981 |

Now that we have constructed the table, we need to know how to use it. We can easily extract the samples necessary for the $k = 0$ sinusoid. All we have to do is linearly address the table at the indices $n = 0, 1, 2, \cdots 31$. How, you might ask, do we get the samples for the other 15 frequencies? The answer is in the method with which we address the table. If we compute the sample values for the remaining 15 cosine frequencies and place them in columns adjacent to the $k = 0$ frequency column in Table 12.4, we can see by inspection a fairly simple address pattern. The addressed samples for any of the $L$ sinusoidal frequencies are obtained from the addresses pointed to by

$$\left[ (2k+1)n \ \ Mod(2L) \right] \text{ for } k = 0, 1, 2, \cdots, L-1$$

**Equation 12.53: Cosine look up table address map**

For example, if the table is implemented in software in an array called DFTCoeff [ ], we would address the table as

```
Coeff = DFTCoeff [((2*k+1)*n)%(2*L)];
```

Now let's take a look at a complete set of cosine waveforms for the case where $L = 16$. The waveforms are tabulated in the Excel worksheet shown in Table 12.5. Columns 0 through 15 of the worksheet correspond to the waveform samples for each of the 16 frequencies, $k = 0$ through 15, versus the index $n$, which is listed in the far left column.

We can equate the entries in each column to the entries in the 32-sample look up Table 12.4 for $k = 0$ via the address map in Equation 12.53. The reader is strongly urged to use Equation 12.53 to map several of the cosine look up table entries from Table 12.4 to Table 12.5 to fully understand the mapping equation.

An Excel plot of the first two table frequencies for $k = 0$ and $k = 1$ is illustrated in Figure 12.43. Note that for $k = 0$ the waveform is a full period in length, and for $k = 1$ the waveform is three periods in length. Why do you suppose this is? Hint: go back and examine Equation 12.50 and remember that when $k = 1$, every third sample of the table is addressed; therefore its frequency should be three times as fast. We should note that, as predicted by Equation 12.43, the table entries are all equal to 1 for $n = 0$, they are equal to 0 for $n = L/2$ and for $n = 3L/2$, and they are all equal to $-1$ for $n = L$.

The reader should understand that although the Table 12.5 contains entries for all values of $k$, if Equation 12.53 is used to compute the cosine table address, the only entries that are necessary for proper operation of the channelizer are the ones in the column for $k = 0$. Therefore, for proper operation, the complete look up table in this example needs to only be 32 samples in length.

**Table 12.5**   Cosine Waveform Samples for $L = 16$ and $k = 0$ through $k = 15$

| k \ n | 0 | 1 | 2 | 3 | 4 | 5 | 6 | 7 | 8 | 9 | 10 | 11 | 12 | 13 | 14 | 15 |
|---|---|---|---|---|---|---|---|---|---|---|---|---|---|---|---|---|
| 0 | 1.000 | 1.000 | 1.000 | 1.000 | 1.000 | 1.000 | 1.000 | 1.000 | 1.000 | 1.000 | 1.000 | 1.000 | 1.000 | 1.000 | 1.000 | 1.000 |
| 1 | 0.981 | 0.831 | 0.556 | 0.195 | -0.195 | -0.556 | -0.831 | -0.981 | -0.981 | -0.831 | -0.556 | -0.195 | 0.195 | 0.556 | 0.831 | 0.981 |
| 2 | 0.924 | 0.383 | -0.383 | -0.924 | -0.924 | -0.383 | 0.383 | 0.924 | 0.924 | 0.383 | -0.383 | -0.924 | -0.924 | -0.383 | 0.383 | 0.924 |
| 3 | 0.831 | -0.195 | -0.981 | -0.556 | 0.556 | 0.981 | 0.195 | -0.831 | -0.831 | 0.195 | 0.981 | 0.556 | -0.556 | -0.981 | -0.195 | 0.831 |
| 4 | 0.707 | -0.707 | -0.707 | 0.707 | 0.707 | -0.707 | -0.707 | 0.707 | 0.707 | -0.707 | -0.707 | 0.707 | 0.707 | -0.707 | -0.707 | 0.707 |
| 5 | 0.556 | -0.981 | 0.195 | 0.831 | -0.831 | -0.195 | 0.981 | -0.556 | -0.556 | 0.981 | -0.195 | -0.831 | 0.831 | 0.195 | -0.981 | 0.556 |
| 6 | 0.383 | -0.924 | 0.924 | -0.383 | -0.383 | 0.924 | -0.924 | 0.383 | 0.383 | -0.924 | 0.924 | -0.383 | -0.383 | 0.924 | -0.924 | 0.383 |
| 7 | 0.195 | -0.556 | 0.831 | -0.981 | 0.981 | -0.831 | 0.556 | -0.195 | -0.195 | 0.556 | -0.831 | 0.981 | -0.981 | 0.831 | -0.556 | 0.195 |
| 8 | 0.000 | 0.000 | 0.000 | 0.000 | 0.000 | 0.000 | 0.000 | 0.000 | 0.000 | 0.000 | 0.000 | 0.000 | 0.000 | 0.000 | 0.000 | 0.000 |
| 9 | -0.195 | 0.556 | -0.831 | 0.981 | -0.981 | 0.831 | -0.556 | 0.195 | 0.195 | -0.556 | 0.831 | -0.981 | 0.981 | -0.831 | 0.556 | -0.195 |
| 10 | -0.383 | 0.924 | -0.924 | 0.383 | 0.383 | -0.924 | 0.924 | -0.383 | -0.383 | 0.924 | -0.924 | 0.383 | 0.383 | -0.924 | 0.924 | -0.383 |
| 11 | -0.556 | 0.981 | -0.195 | -0.831 | 0.831 | 0.195 | -0.981 | 0.556 | 0.556 | -0.981 | 0.195 | 0.831 | -0.831 | -0.195 | 0.981 | -0.556 |
| 12 | -0.707 | 0.707 | 0.707 | -0.707 | -0.707 | 0.707 | 0.707 | -0.707 | -0.707 | 0.707 | 0.707 | -0.707 | -0.707 | 0.707 | 0.707 | -0.707 |
| 13 | -0.831 | 0.195 | 0.981 | 0.556 | -0.556 | -0.981 | -0.195 | 0.831 | 0.831 | -0.195 | -0.981 | -0.556 | 0.556 | 0.981 | 0.195 | -0.831 |
| 14 | -0.924 | -0.383 | 0.383 | 0.924 | 0.924 | 0.383 | -0.383 | -0.924 | -0.924 | -0.383 | 0.383 | 0.924 | 0.924 | 0.383 | -0.383 | -0.924 |
| 15 | -0.981 | -0.831 | -0.556 | -0.195 | 0.195 | 0.556 | 0.831 | 0.981 | 0.981 | 0.831 | 0.556 | 0.195 | -0.195 | -0.556 | -0.831 | -0.981 |
| 16 | -1.000 | -1.000 | -1.000 | -1.000 | -1.000 | -1.000 | -1.000 | -1.000 | -1.000 | -1.000 | -1.000 | -1.000 | -1.000 | -1.000 | -1.000 | -1.000 |
| 17 | -0.981 | -0.831 | -0.556 | -0.195 | 0.195 | 0.556 | 0.831 | 0.981 | 0.981 | 0.831 | 0.556 | 0.195 | -0.195 | -0.556 | -0.831 | -0.981 |
| 18 | -0.924 | -0.383 | 0.383 | 0.924 | 0.924 | 0.383 | -0.383 | -0.924 | -0.924 | -0.383 | 0.383 | 0.924 | 0.924 | 0.383 | -0.383 | -0.924 |
| 19 | -0.831 | 0.195 | 0.981 | 0.556 | -0.556 | -0.981 | -0.195 | 0.831 | 0.831 | -0.195 | -0.981 | -0.556 | 0.556 | 0.981 | 0.195 | -0.831 |
| 20 | -0.707 | 0.707 | 0.707 | -0.707 | -0.707 | 0.707 | 0.707 | -0.707 | -0.707 | 0.707 | 0.707 | -0.707 | -0.707 | 0.707 | 0.707 | -0.707 |
| 21 | -0.556 | 0.981 | -0.195 | -0.831 | 0.831 | 0.195 | -0.981 | 0.556 | 0.556 | -0.981 | 0.195 | 0.831 | -0.831 | -0.195 | 0.981 | -0.556 |

(continued on next page)

**Table 12.5** Cosine Waveform Samples for $L = 16$ and $k = 0$ through $k = 15$ (continued)

| $k$ | 0 | 1 | 2 | 3 | 4 | 5 | 6 | 7 | 8 | 9 | 10 | 11 | 12 | 13 | 14 | 15 |
|---|---|---|---|---|---|---|---|---|---|---|---|---|---|---|---|---|
| $n$ | | | | | | | | | | | | | | | | |
| 22 | −0.383 | 0.924 | −0.924 | 0.383 | 0.383 | −0.924 | 0.924 | −0.383 | −0.383 | 0.924 | −0.924 | 0.383 | 0.383 | −0.924 | 0.924 | −0.383 |
| 23 | −0.195 | 0.556 | −0.831 | 0.981 | −0.981 | 0.831 | −0.556 | 0.195 | 0.195 | −0.556 | 0.831 | −0.981 | 0.981 | −0.831 | 0.556 | −0.195 |
| 24 | 0.000 | 0.000 | 0.000 | 0.000 | 0.000 | 0.000 | 0.000 | 0.000 | 0.000 | 0.000 | 0.000 | 0.000 | 0.000 | 0.000 | 0.000 | 0.000 |
| 25 | 0.195 | −0.556 | 0.831 | −0.981 | 0.981 | −0.831 | 0.556 | −0.195 | −0.195 | 0.556 | −0.831 | 0.981 | −0.981 | 0.831 | −0.556 | 0.195 |
| 26 | 0.383 | −0.924 | 0.924 | −0.383 | −0.383 | 0.924 | −0.924 | 0.383 | 0.383 | −0.924 | 0.924 | −0.383 | −0.383 | 0.924 | −0.924 | 0.383 |
| 27 | 0.556 | −0.981 | 0.195 | 0.831 | −0.831 | −0.195 | 0.981 | −0.556 | −0.556 | 0.981 | −0.195 | −0.831 | 0.831 | 0.195 | −0.981 | 0.556 |
| 28 | 0.707 | −0.707 | −0.707 | 0.707 | 0.707 | −0.707 | −0.707 | 0.707 | 0.707 | −0.707 | −0.707 | 0.707 | 0.707 | −0.707 | −0.707 | 0.707 |
| 29 | 0.831 | −0.195 | −0.981 | −0.556 | 0.556 | 0.981 | 0.195 | −0.831 | −0.831 | 0.195 | 0.981 | 0.556 | −0.556 | −0.981 | −0.195 | 0.831 |
| 30 | 0.924 | 0.383 | −0.383 | −0.924 | −0.924 | −0.383 | 0.383 | 0.924 | 0.924 | 0.383 | −0.383 | −0.924 | −0.924 | −0.383 | 0.383 | 0.924 |
| 31 | 0.981 | 0.831 | 0.556 | 0.195 | −0.195 | −0.556 | −0.831 | −0.981 | −0.981 | −0.831 | −0.556 | −0.195 | 0.195 | 0.556 | 0.831 | 0.981 |

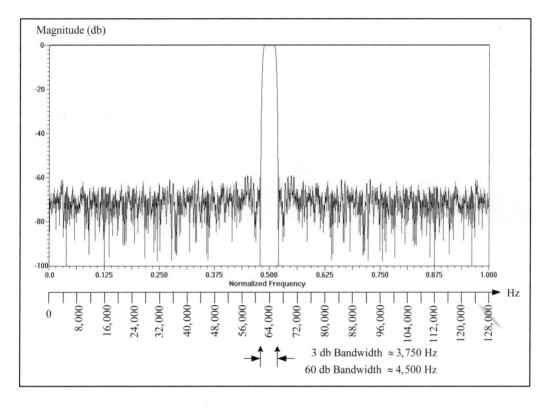

**Figure 12.45**   Enhanced DFT filter response centered at half the sample rate

width of a single channelizer DFT bin, which in the software simulation we discuss next is 4 KHz. This is clearly illustrated in Figure 12.45, where the filter response has been translated to half the sample rate.

Figure 12.46 illustrates the channelizer band pass filter concept for the first 16 of a possible 32 band pass filters. This figure shows the filter response tuned to the first 16 DFT bin centers and overlaid on one another. This bank of band pass filters view is an accurate representation of a channelizer, and it provides a bit more intuitive picture of the channelizer processing. The reader should readily agree that the enhanced DFT filter illustrated in Figure 12.46 is a significant improvement over the regular DFT.

Looking at Figure 12.46, one can think of many potential applications. For example, a channelizer similar to this could be a major component in the design of a mixing console for use in the audio industry. In the world of telephony, a channelizer similar to this can be used to convert the older but still existing FDM technology to the modern TDM technology for the efficient transmission of telephone signals. The output of each one of the band pass filters in the figure could represent a single demultiplexed FDM

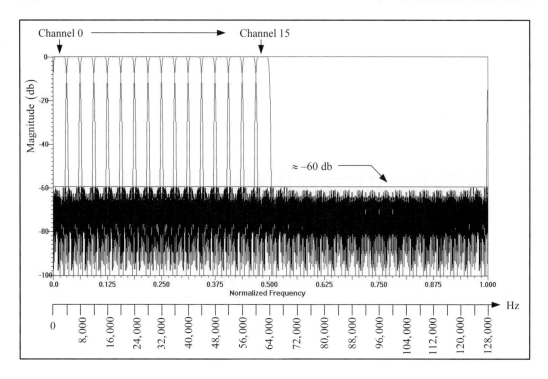

**Figure 12.46**  The first 16 channels of a 32-channel channelizer

telephone channel. A channelizer such as this would allow the older FDM systems to upgrade from old copper transmission lines to today's newer fiber-optic transmission lines. Radio astronomers could use a channelizer to separate and catalog the component frequencies of wideband signal intercepts. A similar application would be the separation and detection of various radio channels received by a wideband scanner. The list of potential applications is long. The advantage in using a channelizer in these applications and many more like them is that all channels would be available simultaneously.

### 12.3.10  Channelizer Block Diagram

We have finally reached the point where we can finish our channelizer system block diagram. The first thing we will do is to modify for the second time the folding memory architecture of Figure 12.42. The modification we will make is illustrated in Figure 12.47.

Basically all we have done is to convert the single buffer folding memory into a split buffer folding memory. Each buffer will hold half of the $z(p)$ terms. In the figure, we show the top memory holding the

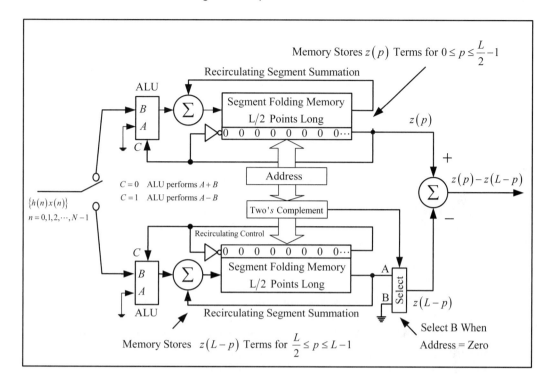

**Figure 12.47**   Split buffer folding memory

$z(p)$ terms for $0 \leq p \leq L/2 - 1$. The bottom memory holds the $z(p)$ terms for $L/2 \leq p \leq L - 1$. In order to speed up the processing, we have placed all the terms that will be processed as $z(p)$ in the top memory and all the terms that will be processed as $z(L-p)$ in the bottom memory. Now, as the figure shows, we can perform the computation of $z(p) - z(L-p)$ in a single memory read cycle, as opposed to two cycles for the single buffer architecture. The split buffer is essentially two half buffers operating in parallel that use the same memory address generator but with two minor changes.

The first change is the two's complement conversion of the bottom memory address. The memory address of the buffer that retrieves the $z(L-p)$ terms is the two's complement of the address used to retrieve the $z(p)$ terms. Let's look at the summation of the $z(p)$ terms for the simple case where $L = 8$. This summation is shown in Equation 12.54. We can see that the memory address ranges from 0 to 7. We can also see that the address for the $z(L-p)$ is indeed the two's complement of the address for the $z(p)$ terms. This is a pretty neat deal because it relieves us of the burden of having to do a lot of complicated bookkeeping:

$$z(0) \quad + \left[ z(1) - z(7) \right] \cos \left[ \left( \frac{\pi}{4} \right) \left( k + \frac{1}{2} \right) \right]$$

$$+ \left[ z(2) - z(6) \right] \cos \left[ \left( \frac{\pi}{4} \right) \left( k + \frac{1}{2} \right) 2 \right]$$

$$+ \left[ z(3) - z(5) \right] \cos \left[ \left( \frac{\pi}{4} \right) \left( k + \frac{1}{2} \right) 3 \right]$$

**Equation 12.54**

The second change is the inclusion of a two input multiplexer that will select either the output of the lower memory or logic 0. This comes in handy when summing the first $z(0)$ term. On output, the multiplexer simply selects the B port, replacing the $z(L/2)$ term with a logic 0. In this example, the $z(L/2)$ term is $z(4)$. This selection occurs when the memory address is zero. All other addresses cause the selection of the A port. The resulting sum for address 0 is then $z(0) - 0$. This is the more general solution. An alternative solution is to simply load a zero in address location 0 of the bottom memory. Figure 12.48 illustrates a split buffer for the simple case where $L = 8$. Buffer 1 holds the folded and summed values for $z(0)$ through $z(3)$. Buffer 2 holds the folded and summed values for $z(4)$ through $z(7)$.

The address for buffer 2 is the two's complement of the buffer 1 address. On output, the "select A or B" multiplexer selects the B input port when the address to the buffer is zero. For all other addresses, the multiplexer selects

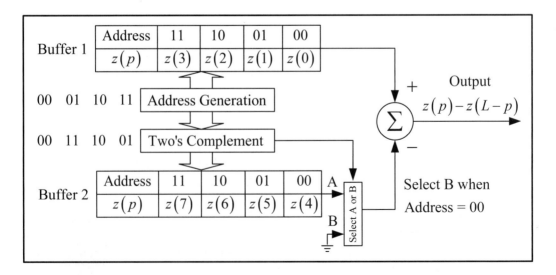

**Figure 12.48**  Split buffer example for $L = 8$

the A port. The address generation and the resultant output are tabulated in Table 12.6.

The odds are astronomical that a channelizer built in hardware will be designed using the VHSIC hardware description language (VHDL), or something that has evolved beyond that, and implemented on an application-specific integrated circuit (ASIC) or some newer technology. The split buffer in Figure 12.47 is loaded with symmetry and can be easily implemented and stored as a design specific library component. Now the designer can use this component many times by simply pointing and clicking, so to speak. This turns out to be a good thing because we are going to use this split buffer folding memory in a double buffer configuration so we can increase the processing speed of the hardware.

The final channelizer block diagram is illustrated in Figure 12.49. Here we see the input double buffer memory that alternately stores and dumps $N$ samples of data. The double buffer allows the channelizer to process the previous buffer of data during the time that the second buffer is filling. The double buffer concept is intended only to speed up the processing of data for real time applications and is therefore not a specific term in the channelizer equation.

The samples dumped from memory are multiplied by the enhanced DFT filter coefficients $h(n)$ and then input to a double buffered folding memory. For simplicity, the folding memory is illustrated as a simple block in the diagram, but the contents of that block are identical to that shown in Figure 12.47. The data stored in one of the folding memories is processed while the second folding memory fills with new data. Once again, this is done to increase the processing speed for real time applications.

The output of one of the two folding memories is selected as an input to the summer where the $z(p) - z(L - p)$ term is computed. Once computed, the

**Table 12.6**   Split Buffer Addresses and Output Example for $L = 8$

| Addr $z(p)$ buffer 1 | | Addr $z(L-p)$ buffer 2 | | Two's complement (addr) buffer 2 | Output $z(p) - z(L-p)$ |
|---|---|---|---|---|---|
| 00 | $z(0)$ | 00 | $z(4)$ | 00 | $z(0) - 0$ |
| 01 | $z(1)$ | 01 | $z(5)$ | 11 | $z(1) - z(7)$ |
| 10 | $z(2)$ | 10 | $z(6)$ | 10 | $z(2) - z(6)$ |
| 11 | $z(3)$ | 11 | $z(7)$ | 01 | $z(3) - z(5)$ |

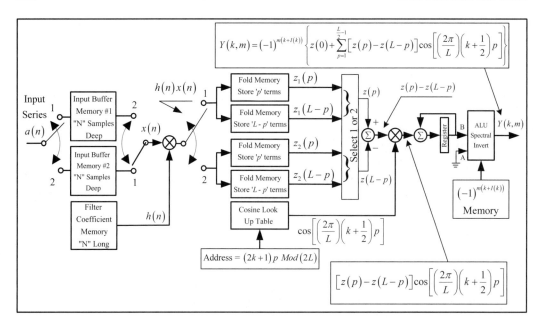

**Figure 12.49** Channelizer processing block diagram

$z(p)$ terms are multiplied by the contents of the cosine look up table and then fed to an accumulator, which implements the summation operation:

$$z(0) + \sum_{p=1}^{\frac{L}{2}-1} [z(p) - z(L-p)] \cos\left[\left(\frac{2\pi}{L}\right)\left(k + \frac{1}{2}\right)p\right]$$

Finally the summed output is fed to an ALU where its spectrum is inverted or not, depending on the contents of the $(-1)^{m(k+I(k))}$ memory.

The whole process is fairly simple if you think about it. And the good news is the architecture is identical for any compatible values of $N$ and $L$. The example hardware implementation that we will discuss in section 12.5 will dramatically increase the scale of processing without modification of the basic architecture.

## 12.4 CHANNELIZER SOFTWARE SIMULATION RESULTS

In section 12.1 we spent some time discussing a potential application of a channelizer. This was done only to provide some focus for the sections that followed. This is only one of hundreds of applications that are relevant to channelizer processing.

In section 12.2 we spent a great deal of time developing the functional overview of a channelizer in order to build a solid conceptual foundation. The intent of this early section was to introduce the subject and provide enough high-level discussion to turn on a light bulb in the reader's mind.

In section 12.3 we introduced and discussed all the technical issues associated with the detailed design of a channelizer. The intent of this section was to develop the mathematics of the channelizer to the point where we had derived a stand-alone equation that could be implemented and utilized in either software or hardware applications.

In section 12.4 we will use a software simulation written by the author to demonstrate the actual signal processing of a representative channelizer. The intent of this section is to provide a graphical example to further ingrain into the reader's mind the relevance of the previous discussions and hopefully make that light bulb burn a little brighter.

Let's begin by defining the application. We want to use a channelizer to process a test signal that is both representative of a real-world signal and is simple enough to provide an easily comprehensible demonstration. The spectrum of the input test signal that we will process is illustrated in Figure 12.50.

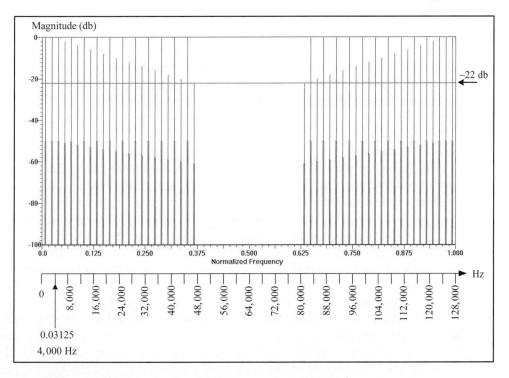

**Figure 12.50**  Channelizer input test signal

The wideband test signal consists of an FDM stack of 12 independent narrow band signals. Our objective is to use the channelizer to simultaneously demultiplex these narrow band signals into 12 independent base band channels. The bandwidth of each channel is 4 KHz. The signal in each channel consists of two pure sinusoidal tones positioned at 1 KHz and 3 KHz relative to that channel's lower band edge. The two sinusoids define the band edges of the signal within a channel. The 2-KHz difference between the two sinusoids defines the bandwidth of the signal within a channel. The sinusoid at 1 KHz is programmed to have a magnitude of 0 db and is used as a channel reference tone. The tone at 3 KHz is intended to represent the channel identification tone. Successive channel 3 KHz identification tones are programmed to decrease by approximately 2 db from the level of the previous channel ID tone. The ID tones range from 0 db for channel 0 down to –22 db for channel 11. The ID tone allows us to readily identify the demultiplexed narrow band signal within the output channels. The amplitude cursor in Figure 12.50 clearly shows the level of the ID sinusoid for channel 11 is –22 db.

The test signal spectrum in the figure is the Fourier transform of the time series $a(n)$ that we have addressed in our previous discussions. The sample rate of the test signal time series is 128 KHz. The bandwidth of each of the 12 individual FDM channels is 4 KHz. We want to process this test signal to produce 12 independent real output base band channels, each with a bandwidth of 4 KHz at a sample rate of 8 KHz.

With this information in hand, let's list all the critical design parameters that we will need to design and simulate our channelizer in Table 12.7. We can readily write down the obvious like the input and output sample rates, the channel bandwidth, and the sample rate decimation factor. The size of the channelizer DFT is simply the input sample rate divided by the output channel bandwidth. In this simulation, this is determined by L = 128 KHz / 4 KHz = 32 bins. The value of $N$ is determined by the length of the enhanced DFT filter or PPF. There are two major criteria to meet in the choice of $N$: The first is that the number of coefficients in the filter must be a multiple of the DFT width $L$—in our case a multiple of 32. The second is that the length of the filter impulse response must be long enough to produce a filter that meets the application dictated frequency response characteristics, such as the width of the pass band, stop band attenuation, transition bandwidth, and pass band ripple.

We will use the filter illustrated in Figure 12.44 in the software simulation. The specifications for that filter are repeated here for clarity. The filter impulse response is 384 coefficients in length. The coefficients are quantized to 16-bit integer. It was designed to filter a signal sampled at 128 KHz with less than 0.01 db pass band ripple and 60 db of stop band attenuation. Our channelizer software simulator will process signals that occupy adjacent 4 KHz channels. The target signal within each band ranges from 1000 Hz to 3000 Hz, giving us a 1000 Hz in-band guard band on either side. This guard band

**Table 12.7**   Channelizer Software Simulation Design Parameters

| Parameter | Description | Value |
|---|---|---|
| $f_{S(IN)}$ | Input signal sample rate | 128 KHz |
| $f_{S(OUT)}$ | Output channelized signal sample rate | 8 KHz |
| $B$ | Channel bandwidth | 4 KHz |
| $L = f_{S(IN)}/B$ | Number of bins in the DFT | 32 |
| $M = f_{S(IN)}/f_{S(OUT)} = L/2$ | Decimation factor | 16 |
| $N$ | Length of PPF $h(n)$ <br> Length of input data vector $x(n)$ | 384 |
| $Q = N/L = 384/32$ | Number of coefficients in each phase of the PPF | 12 |

allows the transition bands of the filter to overlap one another with no significant leakage from adjacent bands. The filter 3 db bandwidth is approximately 3750 Hz, and the filter 60 db bandwidth is approximately 4500 Hz. At 60 db, the filter responses overlap by 500 Hz or 1/2 the allowed guard band.

The software simulation implements the channelizer stand-alone equation documented in Equation 12.45. The simulation opted to invert all odd channels to account for the odd channel spectral inversion phenomenon discussed previously. Therefore the outputs of all channels will be spectrally erect.

The channelizer output spectrum for channel 0 is illustrated in Figure 12.51. We can identify it as being that of channel 0 because the level of the ID tone at 3 KHz is 0 db. As we can see from the figure the output sample rate is $f_{S(IN)}/M = 128$ KHz/16=8 KHz. The channel reference tone is sitting at 1 KHz and the channel ID tone is sitting at 3 KHz. Since the channel signal is real, we expect to see the positive channel frequencies plotted from 0 Hz to $f_S/2$ Hz and the negative frequencies plotted from $f_S/2$ Hz to $f_S$ Hz. The reader should note that because of the bandwidth of each channel's test signal and the design of the DFT filter bandwidth, there is no apparent leakage. The combination of the 1000 Hz signal guard band on either side of the channelized test signal and the narrow transition bands of the filter prevents spectral leakage from adjacent channels. As the plot shows, we have at least 100 db of pure tone with no evidence of spurs or adjacent channel interference.

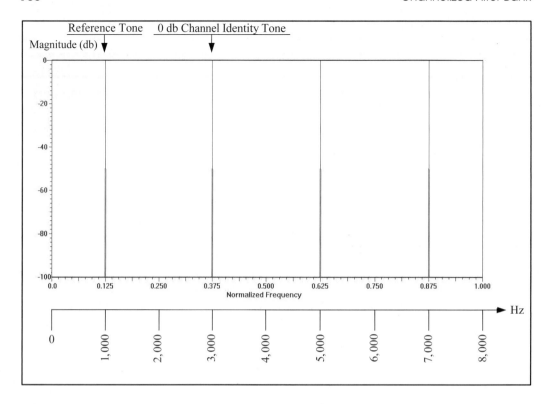

**Figure 12.51**   Channelizer output: Channel 0

The spectrum of channel 1 is illustrated in Figure 12.52. Note that the channel ID tone is 2 db down from the channel reference tone, which clearly identifies this as channel 1.

The spectral plot of channel 2 is illustrated in Figure 12.53. As expected, the reference tone and ID tone are at 1 KHz and 3 KHz, respectively, and the ID tone is 4 db below the reference tone. Figure 12.54 and Figure 12.55 show the spectrum for channel 3 and channel 11, respectively. Note that the ID tone is 6 db down for channel 3 and 22 db down for channel 11.

It is clear from the figures illustrating the FDM input signal and the channelized output signals that the channelizer performance is very impressive. The software simulation illustrates a very realistic application. The two-tone channel signals we used for demonstration purposes could very well have been 4-KHz telephone channels. For example, the bandwidth of a telephone channel ranges from 300 Hz to 3400 Hz, and the first level of a FDM signal is defined by the International Telephone and Telegraph Consultative Committee (CCITT) as a group that consists of 12 telephone channels. The same channelizer we simulated here could easily have demultiplexed an FDM

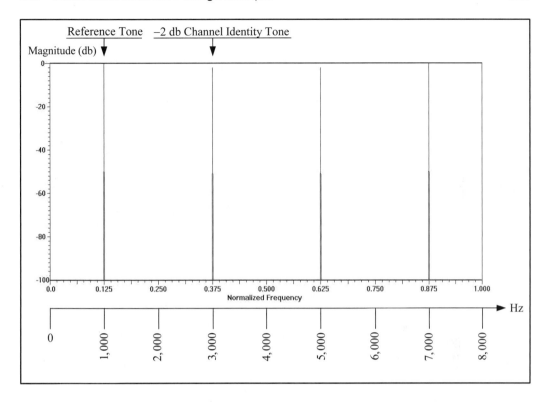

**Figure 12.52**   Channelizer output: Channel 1

group. The same design can be easily scaled up to process even wider band FDM signals such as a CCITT-defined 60-telephone channel super group. This is only one of the many applications that lend themselves to channelizer processing.

## 12.5   CHANNELIZER HARDWARE DESIGN EXAMPLE

In the previous sections, we used a simple four-channel example to demonstrate the many concepts and design issues associated with a channelizer. Our software simulation raised the bar a bit to 32 channels, which we used to process a 12-channel FDM signal. Now it is time to increase the scale of the channelizer dramatically in order to demonstrate a practical example of a realistic system. It is important for the reader to keep in mind that the expanded hardware example presented here utilizes the exact same stand-alone equation that we developed in section 12.3 and presented as Equation 12.45.

Suppose we wanted to build an interface between an existing telephone company FDM system and a newly installed telephone company TDM

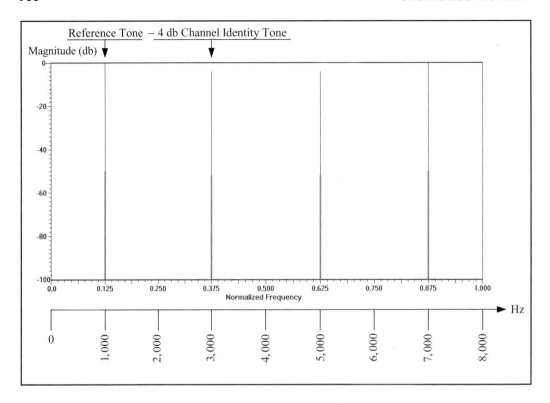

**Figure 12.53**  Channelizer output: Channel 2

system. Our motivation is to get the telephone channels off of the old copper transmission lines and on to the more modern fiber-optic transmission lines without trashing all the existing FDM equipment.

Our task is to convert two telephone company super master groups each transporting 900 independent 4-KHz FDM analog telephone channels to a single TDM digital multiplex called a *STM-1*, which is capable of transporting up to 1920 telephone channels on a fiber-optic cable.

The term *STM* has not been discussed previously. It is the abbreviation used for synchronous transport module. The STM is a specification for a family of multiplex formats and protocols denoted by the term *STM-N*. They are associated with the synchronous digital hierarchy (SDH) and synchronous optical network (SONET) specifications for high-speed digital pulse code modulation (PCM) transmission. Detailed specification for these protocols is given in the International Telecommunications Union (ITU-T) G.707 specification [6] in the references at the end of this chapter. For the purposes of this example, it is not necessary for the reader to know or understand these

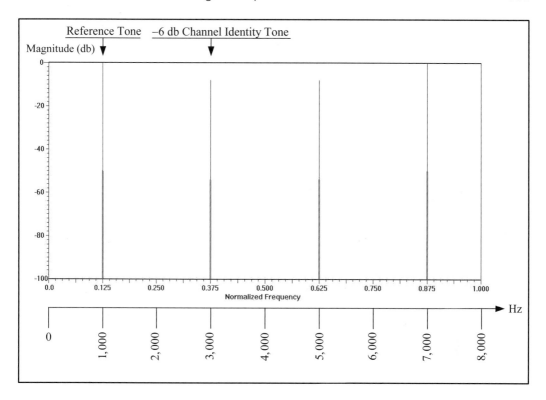

**Figure 12.54** Channelizer output: Channel 3

specifications. They are included only to make this a realistic application that can be implemented and used in industry.

Suppose further that we want to format our channelized output so that it can be transmitted to some remote location via high-speed fiber-optic cable. For a more general approach, we will also assume that the two FDM signals are constructed with a mix of inverted and noninverted channel spectra. The hardware channelizer design example that we present here efficiently performs this FDM to TDM conversion. The reader should note that the terms *super master group* and *STM-1* are not the focus of this design example. We don't want to get lost in terminology that we haven't discussed previously. So, for our purposes, these terms merely represent an industry standard formatted input to and formatted output from our channelizer. The focus in this example application is specific to the fact that we have a total of 1800 FDM telephone channels, each of which occupies 4 KHz of bandwidth. We need to demultiplex these 1800 channels using a channelizer identical to the one that we spent so much time designing in this chapter.

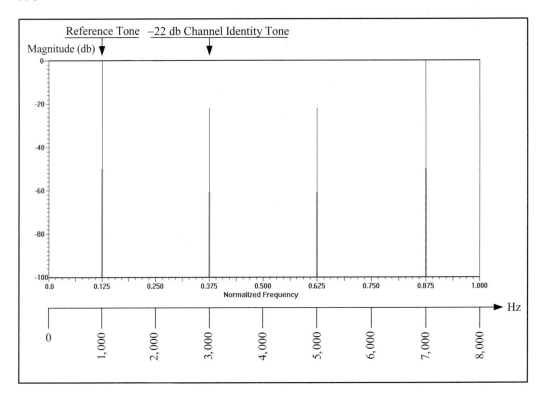

**Figure 12.55**  Channelizer output: Channel 11

The block diagram of the hardware design example is shown in Figure 12.56. Let's now step through the diagram one block at a time and discuss the processing that takes place. On the left side of the diagram, we see that both of the input wideband analog super master group FDM signals are applied to an analog to digital converter. Each of these two input signals contains 900 telephone channels that occupy a bandwidth of 3.872 MHz. It is assumed that these signals have been down converted to base band prior to their input to the A/Ds.

The A/D block contains all the necessary components, like an antialiasing filter, sample and hold logic, gain control, and other analog gizmos, to make this block practical. Since this is an analog-processing block, it would be implemented on its own circuit card and shielded as much as possible from the noisy digital ground lines and digital processing components.

The output of each A/D converter is applied to an array of four quad digital tuner chips. Tuner chips like these are manufactured by several

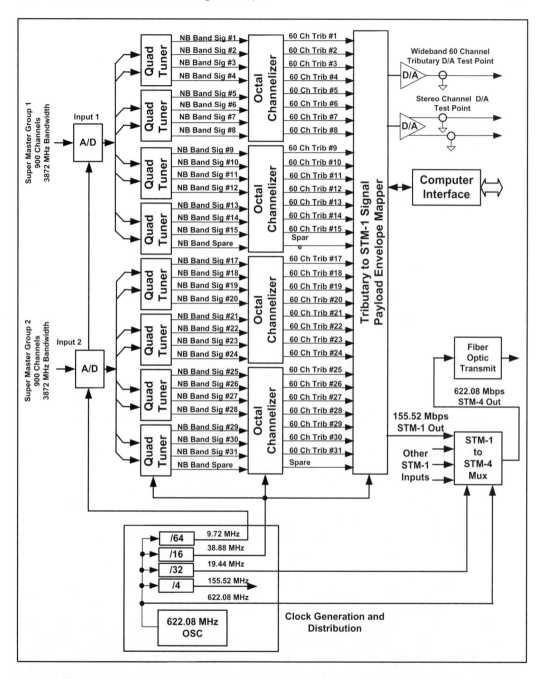

**Figure 12.56**   Channelizer hardware design example

companies and are readily available as off-the-shelf components. In addition these, tuner functions are available as field programmable gate array (FPGA) cores that can be inserted into an FPGA design with the click of the mouse. We show these digital tuners as separate components purely for the functional clarity of this design example.

All the digital tuners are configured to operate in synchronism with one another so that all their outputs will be clock synchronous. The tuners are programmed to output real signals. The input super master group signal FDM multiplex contains 900 telephone channels. Each super master group is composed of fifteen 60-channel super group multiplex signals. Each tuner down converts and isolates a particular 60-channel super group from the wideband 900-channel input and then decimates the sample rate to produce a narrow band 60-channel FDM output. As shown in Figure 12.56, there are 16 tuner outputs for each set of 4 tuners. Of those 16 outputs, 15 are narrow band super group signals. One output is set aside as a spare.

There are four *octal channelizer* FPGAs that have been programmed so that each chip implements eight clock synchronous channelizers. Each of the eight channelizers in an octal channelizer FPGA will process and demultiplex 60 channels of one super group. The beauty of FPGA design is that once the design of a single 60-channel channelizer is complete, it can be stored as a development system library component and then replicated eight times per chip. If assigning the I/O pins is not considered, the replication isn't much more than a point-and-click operation. The high density of most available FPGAs provides for sufficient real estate to implement eight independent channelizer circuits with enough real estate leftover for additional circuitry if necessary. The decision made in this example was to not fully pack the chip with circuitry in order to leave some design headroom for future modifications.

To quickly review, the analog bandwidth of each wideband input signal is 3.872 MHz. The sample clock to each A/D is 9.720 MHz. Each of the sampled wideband signals is applied to the inputs of a tuner group consisting of four quad digital tuners. Fifteen of the sixteen tuners in each group will be utilized for processing, while the sixteenth tuner will be held in standby as a spare.

Each tuner down converts and filters a particular 324 KHz band, which is the equivalent of 81 4 KHz channels. This is a convenient set of values because it results in an easily generated set of sample clocks in the clock generation tree, which is described later. Each tuner then decimates the original 9.720 MHz sample rate by a factor of 15 to 648 KHz, which is 8 KHz per channel, to produce a narrow band output containing the equivalent of 81 4 KHz bandwidth channels. As we can see from the block diagram, each of the two quad tuner banks output 15 narrow band signals, each of which contain the equivalent of an 81-channel stack.

The bandwidth of each narrow band output is the equivalent of 81 adjacent 4-KHz channels. The band edges of adjacent narrow band outputs can overlap one another. This is fine because the channelizer will be programmed to only process a set of 60 contiguous channels in the middle of the 81-channel band. This provides an approximate 10-channel guard band on either edge of the 60 channels and eliminates the edge effects of the tuner filter.

The 81-channel narrow band signals are applied to the input of the octal channelizer FPGAs. Each channelizer processes only the middle 60 channels, beginning with some programmed channel such as channel 9, 10, or 11. This will provide a sufficiently wide guard band on the low end and the upper end of the narrow band signal. The 32 outputs of the 4 octal channelizer FPGAs are composed of 30 60-channel TDM outputs and 2 spares, for a total of 1800 active channels and 120 spare channels. The spare channels can be utilized for redundancy, special purpose signal transport, or any number of other things. If the spare channels are taken into account, the four octal channelizers output a total of 1920 channels

Since the tuners are synchronized with one another and since every processing chip uses the same system clock or derivative of the system clock, the 30 channelizer active outputs and two spares will be clock synchronous with one another as well.

The tributary payload envelope mapper FPGA encapsulates the circuitry to multiplex 1920 channels into a single-byte-wide data stream. When protocol overhead bits are added, the clock rate of the composite data streams is 19.44 MHz. The channels are packed into a STM-1 multiplex format. The 19.44 MHz byte rate of the multiplexed output signal is equivalent to a 155.52 Mbit/second serial bit rate.

The STM-1 formatted output signal is then applied to the input of a STM-1 to STM-4 multiplex chip, which is also readily available on the market. The STM-1 to STM-4 multiplexes up to four STM-1 signals to produce a 622.08 Mbps serial output suitable for transmission over a fiber-optic cable. If all four inputs to the STM-4 multiplex are utilized, then we can easily transport up to 7680 channels on the same fiber-optic transmission line. This is huge when compared to the capacity of a copper wire, but it is still small in comparison with higher-level fiber-optic multiplexes.

Also shown attached to the mapper FPGA are two digital to analog converters used as an aid in the stand-alone debug and testing of this channelizer. The first D/A is used to select and convert any of the 30 narrow band 60-channel super group tributaries for display on an oscilloscope or spectrum analyzer or for transmission to other analog equipment that may be part of a companion system. The second D/A is used to convert any 2 of the 1920 channels to drive a spectrum analyzer or a set of stereo headphones for audio testing purposes. In addition, there is a computer interface shown that allows the channelizer to connect to the mother system's computer control or a local test station computer. This interface is implemented in the mapper FPGA as well.

The clock tree is included in the block diagram for the sake of being thorough. The channelizer mother clock is generated from a 622.08 MHz, 50% duty cycle oscillator. This is standard frequency and is an off-the-shelf product for most oscillator manufacturers. The divide-down tree inputs the mother clock and generates all the necessary channelizer processing clocks. It goes without saying that great care must be taken in the design and layout of the high-speed clock tree and clock distribution network.

Overall, this is a fairly safe design that can be easily implemented on a single circuit card or two cards if the analog circuitry needs to be isolated from the digital logic. Other than the concern for the high-speed clocking there is not a great deal of design risk associated with this channelizer.

## 12.6  SUMMARY OF USEFUL EQUATIONS

$$Y(k) = \sum_{n=0}^{L-1} x(n) e^{-j2\pi\left(\frac{k}{L}\right)n} \quad k = 0,1,2,3,\cdots,L-1$$

**Equation 12.1**

$$\left|W\left(e^{j\omega}\right)\right| = \frac{1}{L}\left[\frac{\sin\left(\frac{\omega L}{2}\right)}{\sin\left(\frac{\omega}{2}\right)}\right] \quad \left\{\begin{array}{l} = 1 \text{ for } \omega = 0 \\ <1 \text{ for } \omega > 0 \end{array}\right\}$$

**Equation 12.2: DFT inherent frequency response**

$$Y(k,m) = (-1)^{m(k+I(k))}\left\{z(0)+\sum_{p=1}^{\frac{L}{2}-1}\left[z(p)-z(L-p)\right]\cos\left[\left(\frac{2\pi}{L}\right)\left(k+\frac{1}{2}\right)p\right]\right\} \text{ for } \left\{\begin{array}{l} k=0,1,2,\cdots,L-1 \\ m=0,1,2,\cdots \end{array}\right\}$$

**Equation 12.45A: Channelizer stand-alone equation**

$$z(p) = \sum_{q=0}^{Q-1} h(qL+p)x(qL+p+mM)(-1)^q \quad \text{for } \left\{\begin{array}{l} p=0,1,2,\cdots L-1 \\ R=N/L \end{array}\right\}$$

**Equation 12.45B: Expanded equation for $z(p)$**

$$\left[(2k+1)n\ Mod(2L)\right] \text{ for } k = 0,1,2,\cdots,L-1$$

**Equation 12.53: Cosine look up table address map**

## 12.7 REFERENCES

[1] John Treichler. *An Introduction to the FDM-TDM Digital Transmultiplexer*. Sunnyvale, CA: Applied Signal Technology, 1991.

[2] John Treichler. *An Introduction to the FDM-TDM Digital Transmultiplexer*. Sunnyvale, CA: Applied Signal Technology, 2003.

[3] James H. McClellan, Thomas Parks, and Lawrence R. Rabiner. "A Computer Program for Designing Optimum FIR Linear Phase Digital Filters." *IEEE Transactions on Audio and Electroacoustics*, vol. AU-21, no. 6 (December 1973).

[4] Ronald E. Crochiere and Lawrence R. Rabiner. *Multirate Digital Signal Processing*. Englewood Cliffs, NJ: Prentice Hall, 1983.

[5] Oran Brigham. *The Fast Fourier Transform*. Englewood Cliffs, NJ: Prentice Hall, 1974.

[6] David R. Smith. *Digital Transmission Systems*. 2nd ed. New York: Chapman and Hall, 1993.

[7] International Telecommunication Union, Telecommunication Standardization Sector (ITU-T). Recommendation G.707. March 1996. Network Node Interface for the Synchronous Digital Hierarchy (SDH).

[8] P. P. Vaidyanathan. *Handbook of Digital Signal Processing Engineering Applications*. Edited by Douglas Elliott. San Diego, CA: Academic Press, 1987.

[9] Alan V. Oppenheim and Ronald W. Schafer. *Discrete Time Signal Processing*. Englewood Cliffs, NJ: Prentice Hall.

[10] Roger L. Freeman. *Telecommunication Transmission Handbook*. 3rd ed. New York: John Wiley and Sons, 1991.

[11] Fredric J Harris. *Handbook of Digital Signal Processing Engineering Applications*. Edited by Douglas Elliott. San Diego, CA: Academic Press.

# CHAPTER THIRTEEN

# Digital Automatic Gain Control

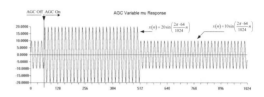

In many electronic systems, one of the most important functions within that system is its automatic gain control (AGC). The design of an AGC algorithm is 100% application dependent. Complete knowledge of the statistics of the input signal whose gain is to be monitored and controlled are the key to the successful design of an AGC circuit. There are several tried-and-true gain control algorithms, and typically a design engineer will choose one of these methods and modify its operational characteristics to fit his or her system specifications.

Generally the AGC algorithm is dependent to a large extent upon the properties of the input signal. For example, suppose the input signal to an AGC circuit is supposed to ideally have constant amplitude. Suppose further that imperfections in the transmission channel modulate the signal in such a manner as to cause the signal amplitude to slowly vary over time. Correcting this amplitude variance is a good application for one type of AGC circuit. In this example the AGC would track the input and compensate for the signal gain fluctuations to produce a constant amplitude output signal. Ideally the AGC would restore the signal to its pretransmission quality.

Another application might be the boosting of signal gain during periods where the signal begins to fade and decreasing the signal gain during periods where the signal begins to grow. Although a bit out of date these days, but still a very good descriptive application, would be the play back of data from a digital tape recorder. A digital tape recorder lays down data streams on several independent data tracks called *channels*. Ideally, on playback, each channel of a track tape recorder would have a digital AGC (dAGC) following the read head amplifier to compensate for gain changes due to tape wear

**977**

caused by the burnishing of the tape as it passes over the record and playback heads. Typically the amplitude of a signal recorded on an individual channel is reduced somewhat each time the tape is played back. Over time, this gain degradation can accumulate to the point where the reproduce logic begins to generate bit errors. As the bit error rate increases, the signal quality diminishes. A dAGC circuit imbedded within the reproduce logic can mitigate many of the problems typically associated with degraded amplitude signals.

Another example might be the gain control on a typical car radio that compensates for signal fade and signal gain as the car is driven in areas such as mountainous regions, where the signal is partially attenuated on occasion.

When an AGC boosts the gain of a signal, the circuit will have certain "attack" parameters that will determine the amount of incremental gain change that occurs over a specific amount of time or $(+\Delta g/\Delta t)$. This is usually referred to as the *attack rate* of the circuit. When an AGC reduces the gain of a signal, the circuit will have certain "decay" parameters that will determine the amount of gain change per unit time or $(-\Delta g/\Delta t)$. This is referred to as the *decay rate* of the circuit. Many times the attack rate and decay rate of an AGC circuit are purposely designed so they are not the same. In many implementations of a simple dAGC, the response time of the circuit is usually determined by setting constant values for both $\Delta g$ and $\Delta t$.

Since the AGC modifies the gain of an input signal, the designer needs to think about the response envelope of the circuit. That is, over what range of input signal amplitude is the AGC active? No circuit is perfect. If the AGC cannot precisely modify the gain of an input signal to produce a specific output signal amplitude, then it will usually end up oscillating between the next higher gain level and next lower gain level in a never-ending quest to control the output signal amplitude. This constant gain oscillation, if significantly large, will result in an unwanted modulation or amplitude jitter of the output signal. In this case, the AGC probably has done more harm than good. This phenomenon is particularly true for dAGC circuits that have finite precision due to the fact that they deal with quantized input signals and produce quantized gain values. For this reason, many AGC circuits usually employ a dead band where they remain idle until the input signal either increases above some preset upper threshold or decreases below some other preset lower threshold. In between these high and low thresholds, the AGC maintains a fixed gain setting. This helps prevent the gain oscillation problems that can wreak havoc on any follow on processing logic.

The AGC circuit that strives to maintain a constant output signal amplitude is sometimes referred to as a type I AGC. A typical response curve for a type I AGC is illustrated in Figure 13.1. Here the AGC attempts to keep the signal amplitude at some constant level specified by an amplitude threshold $T$. If the input signal should experience a positive step in amplitude, the output signal amplitude will initially experience a similar step. This step

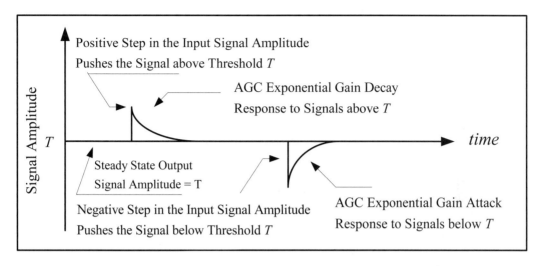

**Figure 13.1** Type I AGC amplitude response

effectively increases the signal gain to some level above the desired threshold $T$. This effective gain step is illustrated in the figure. The AGC response is to reduce the circuit gain over some period of time until the output signal amplitude is again at the desired threshold level. In this example, the gain exponentially decays until the signal amplitude is again equal to the preset threshold $T$. Should the input signal experience a negative step in amplitude, the AGC will respond by increasing the gain until the output signal amplitude is increased back to the prescribed threshold level. The AGC attack response in this example is shown to be exponential. A type I AGC circuit is useful primarily in applications involving DC signals whose values, for some reason or another, vary slowly over time, or in applications involving constant amplitude sinusoidal carrier signals.

The type II AGC circuit utilizes two thresholds: an upper threshold $T_H$ and a lower threshold $T_L$. The gain response for this type is illustrated in Figure 13.2. This type of circuit performs exactly the same as a type I AGC when the amplitude of the signal increases above the upper threshold $T_H$ or when the amplitude of the signal decreases below the lower threshold $T_L$. The gain region between these two thresholds is referred to as the *AGC dead band*. When the signal amplitude falls within the dead band, the AGC gain remains constant. In other words, the signal is free to wander in amplitude anywhere within the dead band without AGC correction. A type II AGC is more suitable to applications that process wideband signals. One example of this is a system that employs amplitude modulation. A type I AGC would attempt to strip off any amplitude modulation, while the type II AGC would simply scale the modulation so that it remains within acceptable bounds.

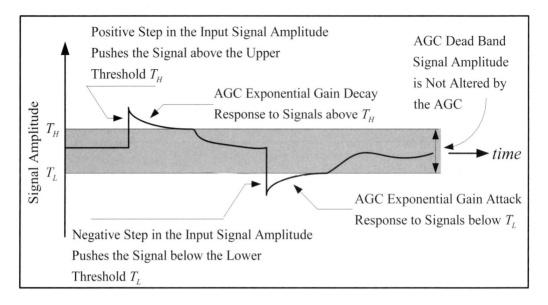

**Figure 13.2**   Type II AGC amplitude response

In this chapter we will first look at a type I AGC circuit that modifies the gain of the input signal based on measurements obtained on the root mean square (RMS) power of the output signal. This AGC is a feedback control circuit and the mechanics of its operation can be fascinating to those interested in adaptive circuits. We will then use the principles developed for the type I AGC to briefly discuss the type II AGC. This is possible because the type II AGC is a simple extension of the type I, and with the exception of some additional hardware, the two types are basically identical to one another. The rest of this chapter will be spent in the development of the AGC mathematical relationships. We will then design an AGC circuit and analyze its performance. We will reinforce the mathematics and design concepts with the presentation of results obtained from software simulations.

Typically the design of an AGC circuit borrows heavily from the teachings of feedback control theory. The principles of this theory can fill an entire book. Rather than get bogged down in the mathematics of control theory, this chapter will develop the AGC design using simple reasoning and straightforward mathematical relationships. If the reader is interested in pursuing this topic further, he or she can use the principles of feedback control theory to expand on the models developed here. The benefits gained would be a more precise study of overall system stability through the use of pole/zero analysis.

## 13.1   DESIGN OF A TYPE I RMS AGC CIRCUIT

### 13.1.1   AGC RMS Computation

In a digital system, we compute an estimate of the RMS value of a signal $x(n)$ by first summing a sequence of $N$ successive squared samples then dividing the sum by $N$ to produce what is termed the *mean square* of the signal. This process is demonstrated as follows:

$$x(n)_{Mean\ Square} = \frac{1}{N}\sum_{n=0}^{N-1}x^2(n) = \frac{x^2(0)+x^2(1)+x^2(2)+\cdots+x^2(N-1)}{N}$$

### Equation 13.1

Following that, all we need do is to take the square root of the mean square value to get the RMS value of the sequence:

$$x(n)_{RMS} = \sqrt{\frac{1}{N}\sum_{n=0}^{N-1}x^2(n)} = \sqrt{\frac{x^2(0)+x^2(1)+x^2(2)+\cdots+x^2(N-1)}{N}}$$

### Equation 13.2

The block diagram of three different circuits that can be used to implement this RMS computation is illustrated in Figure 13.3. Part A of the figure utilizes a simple tapped delay line and summer, part B utilizes a recursive accumulator, and part C uses a combination of both.

The delay line approach has the advantage in that a new RMS estimate can be obtained on each tick of the data sample clock. This provides for a continuous stream of RMS estimates at the system sample rate. For long sequences, however, the delay line method can require a fair amount of hardware. The recursive accumulator has the advantage of using less hardware because it collapses the $N$ register delay line and the $N$ summers into a single register and summer. However, since the sum of squares requires the accumulation of $N$ sequential samples, it can only produce an RMS estimate every $N$ sample clock periods. In addition, some timing is required to clear the accumulator and clock the output register every $N$ samples. The combination of the two eliminates the large summation tree by using the recursive accumulator and a delay line in parallel. In operation, the last sample out of the delay line is subtracted from the recursive sum to truncate the sum to $N$ samples. This configuration will produce RMS estimates at each tick of the sample clock, and it does not require a large summation tree.

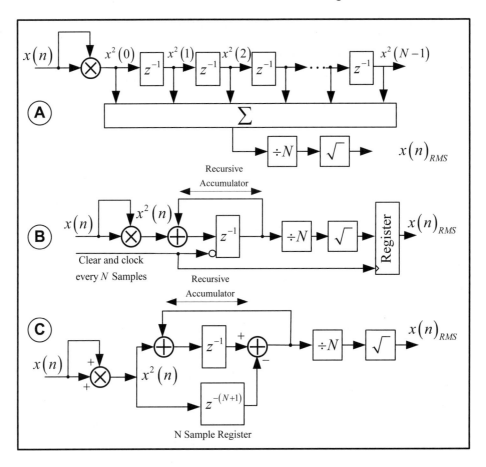

**Figure 13.3**  RMS implementation block diagram

In all three cases, the number of squared samples to be summed is usually chosen to be a power of two. This allows the divide by $N$ function to be implemented by a simple shift operation.

In all cases, the computed RMS value effectively lags the input signal sequence by $N$ sample periods. Therefore computed RMS estimates will lag any perturbations in the input signal. For well-behaved signals, the tapped delay line can approximate accurate RMS values each clock cycle. However, for signals that experience random or large gain fluctuations, the contributions of the first few samples input to the delay line are swamped by the contribution of the many samples remaining in the pipeline. In this case, accurate RMS estimates will lag the initial gain perturbation. This lag in the RMS estimate of the signal can result in the AGC overcorrecting the gain estimates. This overcorrection results in a ringing transient response to input signal amplitude steps.

This issue and its solution will be addressed later. For the remainder of this discussion, we will assume the use of the delay line architecture mainly because we will want to compute a new RMS estimate every sample clock. Furthermore, we will represent the mathematical operation performed by the RMS delay line by the following notation:

$$x(n)_{RMS} = \sqrt{\frac{1}{N}\sum_{k=0}^{N-1} x^2(k)} \quad \text{where} \quad \left\{ \begin{array}{ll} n = & \text{the index of the current RMS measurement} \\ k = & \text{the index of the } k\text{th element in the } N \text{ element RMS delay} \end{array} \right\}$$

**Equation 13.3**

### 13.1.2   RMS AGC Circuit Signal Flow

Let's begin our discussion on RMS AGC circuits by looking at the basic signal flow through the circuit. In its most basic terms, the input sequence represented by $x(n)$ is multiplied by a gain sequence $g(n)$ to produce an output sequence $y(n) = g(n)x(n)$. The gain term $g(n)$ is computed at each sample time. It is used to modify the amplitude of the samples in the sequence $x(n)$ to form a new sequence $y(n)$. This modification is based on some user-supplied specifications, which we will discuss later. This flow is illustrated in Figure 13.4. All we need to do now is come up with an algorithm that determines how the value of the gain term $g(n)$ is computed.

### 13.1.3   Design of the Type I RMS AGC Circuit

This section will discuss the fundamental design of a core AGC circuit. We will then demonstrate and analyze the performance of this core circuit. This AGC is a feedback circuit and, therefore, is subject to stability considerations. The intent in this section is to use the design to acquaint the reader with issues that should be considered in the development of a working AGC.

The core circuit we design can serve as a baseline AGC that, with specific modifications and/or enhancements, can be made compatible with both the input signal statistics and the performance requirements of a broad range of applications. In other words, this core AGC is not necessarily the whole enchilada. Depending on the application, however, it could serve as a sizeable portion of that enchilada. With that said, let's begin.

Our design goal is to force the AGC to maintain a user-specified output signal RMS level. Therefore we will need to utilize the measured RMS value of the output signal sequence $y(n)$ as an input to some feedback/control network that will allow us to iteratively generate a stable gain sequence $g(n)$. Typically a feedback circuit is designed to minimize the error measured between the actual circuit output signal value and some user-specified desired output value. In order to design this type of feedback circuit, we need to first

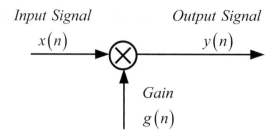

**Figure 13.4**  AGC signal flow

identify a meaningful error signal. Once our error signal is determined, the objective of the adaptive feedback design will be to minimize this error to produce an optimum gain setting.

In our design, we want to be able to externally specify the desired RMS level of the output signal. Therefore it seems natural that the error signal should provide the AGC circuit with information concerning both the magnitude and the polarity of the difference between the actual output signal RMS measurement and the user-supplied or desired signal RMS value. The circuit will be designed to iteratively modify the gain until the error is minimized or, in the ideal case, until the error is forced to zero. In doing so, the AGC feedback loop will attempt to drive the RMS value of the output signal to the user-specified RMS value.

In this chapter we will identify this error signal as $\varepsilon$. Since we are dealing with a digital signal, we will also use the equivalent notation of $\varepsilon(n)$ to indicate that each successive estimate of the error is a function of the sample index $n$. Similarly, the measured RMS value of the output signal will be expressed as either $y_{RMS}$ or $y(n)_{RMS}$. Both notations are correct and, for purposes of clarity, both will be used interchangeably, depending on the context of the discussion. The first thing we need to accomplish in the design of this AGC circuit is to identify an error signal that is easily computed and then determine just how we can use this signal to achieve our circuit goals.

Let's begin by defining our error signal to be the difference between the desired output signal RMS value represented by $d_{RMS}$ and the actual or measured RMS of the output signal represented by $y_{RMS}$. In our circuit, the operator will simply specify a programmable value for $d_{RMS}$, and the AGC will then iteratively compute the gain so that over some time interval the RMS value of the output signal approaches and ideally equals the desired value. In this design, we determine that the error signal is given by

$$\varepsilon = d_{RMS} - y_{RMS}$$

It is understood that the symbol $d$ and $d_{RMS}$ refer to the same quantity and therefore, for simplicity, we will drop the $RMS$ subscript and refer to our error signal in either of the two forms:

$$\varepsilon \quad\;\; = d - y_{RMS}$$
$$\varepsilon(n) \;\; = d - y(n)_{RMS}$$

The first representation is merely a shorthand notation for the error signal, while the second representation includes the sample index $n$, which clearly indicates that the desired RMS value $d$ is a constant and that both the RMS feedback signal and the error signal are computed every sample instant.

Let's get started. The first thing our AGC circuit design needs to do is compute the RMS value of the output signal and then feed it back as one input to the error signal. This computation is illustrated in the block diagram in Figure 13.5.

The second thing our AGC design needs to include is the computation of the error signal $\varepsilon = d - y_{RMS}$. Figure 13.6 illustrates the addition of this function in the circuit block diagram.

If we take a look at the error signal, we can see that depending on the value of the desired RMS and the measured RMS output, the error signal can be quite large. We do not wish our AGC circuit to have instantaneously large changes in gain. We would much rather see the iterative steps of gain change be small so that the overall circuit gain changes slowly over time. This avoids the potential of huge jumps or even oscillation in signal gain, which will cause havoc in downstream processors. For this reason, we add a scale factor to the error signal that we will designate as $\mu$. It is the function of this scale factor to attenuate large gain changes and thus buffer instantaneous signal amplitude variations. Because of the function it performs, $\mu$ is often referred to as the *circuit scale factor* or the *circuit feedback coefficient*. We will use the term *feedback coefficient* in this chapter. The incorporation of the feedback coefficient $\mu$ modifies the feedback error signal such that $\mu\varepsilon = \mu[d - y_{RMS}]$. The block diagram that illustrates the incorporation of the feedback coefficient is shown in Figure 13.7.

Now we are in a position to define the incremental change in circuit gain to be nothing more than the scaled feedback error signal or

$$\Delta g(n) = \mu\varepsilon(n) = \mu\left[d - y(n)_{RMS}\right]$$

**Equation 13.4**

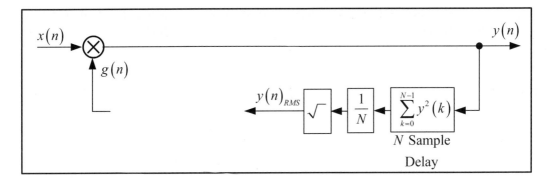

**Figure 13.5**  AGC feedback RMS signal computation block diagram

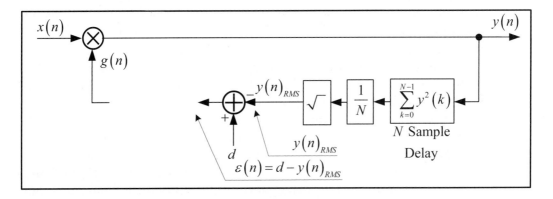

**Figure 13.6**  AGC feedback error signal computation block diagram

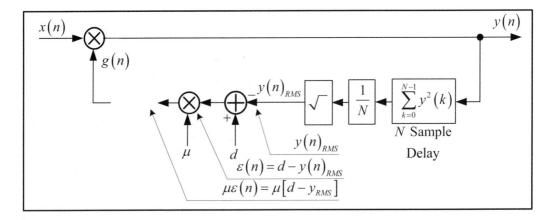

**Figure 13.7**  Feedback scaled error signal computation block diagram

We generate the next value for circuit loop gain $g(n)$ by adding the computed incremental gain to the last value of the loop gain $g(n-1)$. Doing so gives us the relation

$$
\begin{aligned}
g(n) &= g(n-1) + \Delta g(n) \\
&= g(n-1) + \mu \varepsilon(n) \\
&= g(n-1) + \mu \left[ d - y(n)_{RMS} \right]
\end{aligned}
$$

**Equation 13.5**

Now we can say the relationship between the output and the input signals of the AGC circuit is given by

$$
\begin{aligned}
y(n) &= \left\{ g(n-1) + \mu \left[ d - y(n)_{RMS} \right] \right\} x(n) \\
&= \left\{ g(n-1) + \Delta g(n) \right\} x(n) \\
&= g(n) x(n)
\end{aligned}
$$

**Equation 13.6**

The reader can see from Equation 13.6 that the current value of the output signal $y(n)$ depends on the previous $N$ samples of $y(n)$ that are necessary to compute $y(n)_{RMS}$. This is the reason why the change in the signal gain is effectively delayed by $N$ samples. The block diagram of the complete Type I RMS AGC circuit is illustrated in Figure 13.8.

The operation of the AGC circuit is functionally very simple. We first compute an error signal $\varepsilon(n) = \left[ d - y(n)_{RMS} \right]$, which is then scaled to a small number by the feedback coefficient $\mu$. If the computed RMS value of the output signal happens to be larger than the desired RMS value, the error signal and thus the incremental gain $\Delta g$ will be negative. This negative incremental gain reduces the circuit gain by some small amount. The reduced gain lowers the measured RMS of the output signal, which in turn reduces the magnitude of the error signal. Ideally this scenario will continue at each iteration, and the error signal will get smaller, which in turn makes the negative gain change smaller. In the limit for an infinitely precise system, the error signal will be driven to zero, the gain will remain constant, and the circuit will operate in the steady state mode given by

$$
\lim_{y_{RMS} \to d} y(n) = \lim_{y_{RMS} \to d} \left\{ g(n) + \mu \left[ d - y_{RMS} \right] \right\} x(n) = \left\{ g(n) + \mu[0] \right\} x(n) = g(n) x(n)
$$

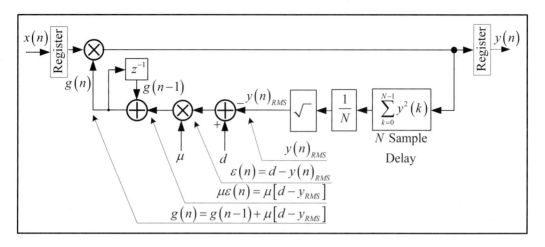

**Figure 13.8**   Complete RMS based AGC circuit block diagram

If the computed RMS value of the output signal happens to be smaller than the desired RMS value, the error signal and thus the incremental gain $\Delta g$ will be positive. This positive incremental gain increases the circuit gain by some small amount. The increased circuit gain increases the measured RMS of the output signal and, as before, reduces the loop error. Over time, the gain will be incrementally increased, which amplifies the output signal and continues to increase the output signal RMS and reduce the error signal. Ideally the error signal will eventually reach zero, and the circuit transient response will settle and the circuit will reach a stable steady state.

### 13.1.4   Determination of the Feedback Coefficient $\mu$

The heart of the AGC design is the determination of the scale factor or feedback coefficient $\mu$. This is not a trivial design issue, nor is it an exact science. In this section, we will introduce several key design issues that the engineer needs to deal with when choosing the circuit feedback coefficient. There are many methods that can be utilized to determine the value of $\mu$, limited only by the ingenuity of the design engineer. We will limit our discussion to the two possible methods listed here:

- The feedback coefficient $\mu$ is a fixed constant.
- The feedback coefficient $\mu$ is time varying.

#### 13.1.4.1   The Feedback Coefficient $\mu$ Is a Fixed Constant

In order to make the feedback gain equation given by $g(n)$ work, we need to define the value for the feedback coefficient $\mu$. This is the one area in the design of a dAGC that can benefit from the study of feedback control theory. Specifically, this benefit comes from the derivation of the circuit poles and zeros and in the determination of system stability. As mentioned earlier, feedback and control theory is a topic that requires an entire book; therefore we will not delve into that subject in this short chapter. Instead we will develop the mathematics necessary to understand and define a value for $\mu$ that may not be the optimum value (as would be determined from control theory) but is still sufficient for many designs.

Generally the statistics of the input signal and the architecture of the AGC circuit will play a significant role the determination for the value of $\mu$. When we compute the value of $\mu$, there are several considerations we must take into account. At every iteration of the gain computation, the previous value of loop gain is changed by adding or subtracting the incremental gain $\Delta g(n) = \mu[d - Y_{RMS}(n)]$. Depending on the relative magnitude of both the $d$ and the $Y_{RMS}(n)$ terms, there are three possible results that can be obtained when computing the incremental gain $\Delta g$:

1. $d > Y_{RMS}(n)$   and the gain $g(n)$ incrementally increases since        $\mu[d - Y_{RMS}(n)] = +\Delta g$

2. $d = Y_{RMS}(n)$   and the gain $g(n)$ remains constant since        $\mu[d - Y_{RMS}(n)] = 0$

3. $d < Y_{RMS}(n)$   and the gain $g(n)$ incrementally decreases since        $\mu[d - Y_{RMS}(n)] = -\Delta g$

It's also common knowledge that the value of the overall circuit gain $g(n)$ should never be negative. After all, multiplying the input signal by a negative gain will invert the signal amplitude. Ideally, since the overall gain is nothing more than a scale factor, it should range from some very small fractional value greater than 0 to some maximum value $G$. Therefore we can initially state that $g(n)$ must always remain greater than zero or

$$g(n) > 0$$

or equivalently

$$g(n) = g(n-1) + \mu[d - y_{RMS}(n)] > 0$$

It is OK if the quantity $[d - Y_{RMS}(n)]$ is a negative number because that is how we get the negative gain increment $-\Delta g$. However, the magnitude of $-\Delta g$ must not exceed $g(n-1)$. If it does, then the overall gain will become negative and the AGC circuit will become unstable. The only way the gain can become negative is if the error is both negative and larger in magnitude than the previous gain value. Therefore we want to make sure that

$$\mu \left[ d - y_{RMS} \right] < \left| g(n-1) \right|$$

or

$$\mu < \left| \frac{g(n-1)}{\left[ d - y_{RMS} \right]} \right|$$

**Equation 13.7**

This gives us an upper bound on value of the feedback coefficient $\mu$. This upper bound in Equation 13.7 is computed using only the initial gain value $g(n-1)$ and therefore is only valid for the first gain iteration. The upper bound of the feedback coefficient may be substantially different for the second, third, fourth,… iterations. That is, if we were to replace $g(n-1)$ with $g(n), g(n+1), g(n+2)\cdots$, we would get a different upper bound for $\mu$.

In some applications, we would like to set $\mu$ to some constant value and leave it alone for the duration of circuit operation. Because of the changes that occur during the transient response of the circuit, using each newly computed value of $y(n)_{RMS}$ would cause the value of $\mu$ to vary, according to the relationship in Equation 13.7. The reader may think, "If we want a constant $\mu$, which particular value of the potentially large number of $y(n)_{RMS}$ values do we use to compute a constant value for the feedback coefficient?" Depending on the system parameters, the answer can simple or it can be difficult.

Let's deal with the simple case first. We design the AGC armed with a good knowledge of the input signal statistics. Therefore we can predict with some certainty the behavior of the input signal. The worst-case scenario would be for the input signal to have large step changes in amplitude. For example, the input signal amplitude could suddenly increase or decrease by 3 db. This would result in a drastic change in the feedback error signal and could very easily cause the AGC loop gain to oscillate if the value of the feedback coefficient $\mu$ were too large. Without the benefit of feedback and control theory, the methods used to calculate the feedback coefficient $\mu$ are many, and the result is more than likely not the optimum value. We can, however, determine values that are acceptable for use in most AGC applications. There are many considerations to be made when determining the value of $\mu$. A few of these considerations and the resultant computational methods are discussed next.

**Method 1**. One method of dealing with this issue is to set the initial value of $g(n-1)$ to unity and then normalize the error signal to the largest possible expected error, which gives us

$$\mu < \left| \frac{1}{\left[ d - y_{RMS} \right]_{MAX}} \right|$$

**Equation 13.8**

In addition, it is common to scale the value of $\mu$ by a factor of $1/(2N)$, where $N$ is the length of the RMS delay line. Therefore we end up with the expression

$$\mu < \frac{1}{2N} \left| \frac{1}{\left[ d - y_{RMS} \right]_{MAX}} \right|$$

**Equation 13.9**

If we substitute the value of $\mu$ in Equation 13.9 into the loop gain equation, we get

$$
\begin{aligned}
g(n) &= g(n-1) + \mu \left[ d - y_{RMS} \right] \\
&= g(n-1) + \frac{1}{2N} \frac{\left[ d - y_{RMS} \right]}{\left| \left[ d - y_{RMS} \right]_{MAX} \right|} \\
&= g(n-1) + \Delta g
\end{aligned}
$$

**Equation 13.10**

In this chapter we will always refer to the gain increment $\Delta g$ as the small amount of gain that is added to the previous value of the gain $g(n-1)$ to produce the current value of the circuit gain $g(n)$. This will be true no matter how $\Delta g$ is computed. In the case of Equation 13.10 $\Delta g$ is given by

$$\Delta g = \frac{1}{2N} \frac{\left[ d - y_{RMS} \right]}{\left| \left[ d - y_{RMS} \right]_{MAX} \right|}$$

Equation 13.9 gives us a very good first estimate for a usable constant $\mu$. This is true because the error term $\varepsilon$ is normalized to unity by dividing it by the maximum possible error $\varepsilon_{MAX}$ or

$$\frac{\varepsilon}{\varepsilon_{MAX}} = \frac{\left[d - y_{RMS}\right]}{\left|\left[d - y_{RMS}\right]_{MAX}\right|}$$

therefore the incremental gain $\Delta g$ is never going to be larger than $1/(2N)$.

Just because Equation 13.9 appears to be a good first estimate, that doesn't mean it is the optimum one. It does not guarantee that the feedback circuit will be stable. As is always the case, the design engineer should simulate the AGC design using his or her choice of $\mu$ to verify its performance. The simulation can be a fairly fancy and rigorous piece of custom software with graphics, or it can be as simple as an Excel spreadsheet with data plots. Either way, one should remember that this is a feedback circuit, and therefore it can encounter stability problems if its operational parameters are not correct. Therefore the circuit should be fully verified through simulation.

Let's take a look at a quick example: Suppose we are processing a 20 volt peak-to-peak sinusoidal signal, and we know this signal can abruptly change in amplitude by as much as 3 db. This is a very crazy signal indeed, but sometimes an exaggerated example is the clearest to understand. We would therefore expect that, over the lifetime of the AGC, we will see sudden steps in amplitude from 20 volts down to 10 volts and from 20 volts up to 40 volts. If our desired output signal RMS value is a constant $d = 14$, then we will expect a maximum possible error signal of

$$\varepsilon_{MAX} = \left[d - y_{RMS}\right]_{MAX} = \left[14 - 40/\sqrt{2}\right] = \left[-14.28\right]$$

If our AGC RMS delay is $N = 16$ samples long, then we would compute our constant $\mu$ to be

$$\mu < \frac{1}{2N}\left|\frac{1}{\left[d - y_{RMS}\right]_{MAX}}\right| = \frac{1}{2(16)}\left|\frac{1}{\left[14 - 40/\sqrt{2}\right]}\right| = 0.0022$$

A rule of thumb is "a small $\mu$ is a good $\mu$." The value of 0.0022 is small, so the probability is high that it is also a good value. A simple Excel simulation will verify if this is a good choice for our AGC, and it will give us the tools to tweak our way into deriving a more optimum value.

We mentioned earlier that the determination of $\mu$ can be simple or it can be difficult. We just discussed a simple case. Now let's warn the reader of the issue that may make this computation a little more difficult. Suppose that the input signal is well behaved and that the maximum expected RMS value for the input signal is close to the desired RMS output value. That is, suppose that the quantity $\left[d - y_{RMS}\right]_{MAX}$ is close to zero. Even though the error is normalized to the max possible error value, as shown in Equation 13.10, the value computed for $\mu$ can hover near a value close to $1/2N$ and therefore the loop gain can become unstable. If our AGC circuit design should face this problem, then what method can we use to solve it? This leads us into method 2.

**Method 2.** Once again, if we know the statistics of the input signal, and if we know the bit width and mathematical format of the processing hardware, we can replace the computed value $y_{RMS}$ with a constant equal to the maximum expected RMS that can ever occur. For example, if the bit width of the digital signal being processed by the AGC is 8 bits and the samples are encoded in two's complement format, then the maximum signal magnitude is $2^7 = 128$. The maximum possible RMS value would then be $128/\sqrt{2} = 90.5$. If we know that our desired RMS value $d$ is always going to be substantially smaller than the maximum value of $y_{RMS}$, then the quantity $\left[d - y_{RMS(MAX)}\right]$ will never result in a number close to zero.

**Method 3.** We can expand on method 2 by realizing that there is nothing preventing us from replacing the computed $y_{RMS}$ term with any value. We could replace it with any reasonable constant that we label $y_{MAX}$. For example, in method 2, we could replace the variable $y_{RMS}$ with the constant $y_{MAX} = 128$. Now because the largest value we can set our desired RMS value to is $d = 90.5$, we are assured that the term $\left[d - y_{MAX}\right]$ will never be close to zero. In many applications, we would like the amplitude of the AGC output signal $y(n)$ to initially equal the amplitude of the input signal. Therefore we usually set the initial gain term $g(n-1)$ to unity. For this example, if we assign a value of unity to the $g(n-1)$ term and we set $y_{RMS}$ to $y_{MAX} = 128$, then we can compute $\mu$ for an AGC with a 16-sample RMS delay:

$$\mu < \frac{1}{2N}\left|\frac{g(n-1)}{[d-y_{RMS}]}\right| \Rightarrow \frac{1}{2N}\left|\frac{g(n-1)}{[d-y_{MAX}]}\right| = \frac{1}{2(16)}\left|\frac{1}{[d-128]}\right| = \frac{1}{32}\left|\frac{1}{[d-128]}\right|$$

**Equation 13.11**

In our example, the desired value $d$ can never be set larger than 90.5 unless we intentionally want the circuit to go unstable. Therefore the upper bound for $\mu$ in this example would be

$$\mu < \frac{1}{32}\left|\frac{1}{[d-128]}\right| = \frac{1}{32}\left|\frac{1}{[90.5-128]}\right| = 0.0008$$

This is a simple example and the 8-bit data path we used is small by today's standards. It does, however, illustrate another method of determining the value of feedback coefficient. The object is to prevent the $[d - y_{RMS}]_{MAX}$ term from becoming small or compensate for it if it does.

Using this method for determining a fixed value for the feedback coefficient, we would simply set the value we computed for $y_{MAX}$ to some value that is much larger than the largest possible value of the desired RMS value $d$. If, for example, the desired RMS term $d$ is restricted to some range, say $r_1 \leq d \leq r_2$, then the $y_{MAX}$ term can be arbitrarily set to some value much larger than $r_2$.

For example, suppose we were tasked to stabilize the gain of a slowly varying input DC signal whose RMS value slowly wanders about $y_{RMS} = 7.0$ so that the output signal RMS level is rock solid at $d = 7.1$. Stabilizing this amplitude variance might be a good application for a simple AGC circuit. Since the signal amplitude slowly varies with time, we can approximate it as being a DC signal and use an AGC with an RMS delay whose length is 1.

In this very realistic application, the value of the feedback coefficient is calculated as

$$\mu < \frac{1}{2N} \left| \frac{g(n-1)}{[d - y_{RMS}]_{MAX}} \right| = \frac{1}{2(1)} \left| \frac{1}{[7.1 - 7.0]} \right| = 5.0$$

This large value for $\mu$ is going to multiply the feedback error by a factor of 5. This is a huge value and will almost certainly guarantee instability. For comparison purposes, if we replace the $y_{RMS}$ term with the $y_{MAX}$ term for the 8-bit processing path in the previous equation, we can alternatively compute the feedback coefficient as

$$\mu < \frac{1}{2N} \left| \frac{g(n-1)}{[d - y_{MAX}]} \right| = \frac{1}{2} \left| \frac{1}{[7.1 - 128]} \right| = 0.0041$$

This value for $\mu$ is greater than zero and less than unity and represents a fairly common value. Remember that $\mu$ should be small. We just need to get a good first estimate on what "small" means. Later in our simulations, we can tweak the value of $\mu$ until we get a more optimum solution.

**Method 4.** Another method of calculating a good first estimate for $\mu$ is the option of setting the $g(n-1)$ term to unity and then removing the desired RMS value from the denominator entirely to get the expression

$$\mu < \frac{1}{2N} \left| \frac{g(n-1)}{[y_{MAX}]} \right| = \frac{1}{2N} \left| \frac{1}{[y_{MAX}]} \right| = \frac{1}{2N[y_{MAX}]}$$

**Equation 13.12**

The absolute value function is removed because $y_{MAX}$ is always a positive number. Now if we substitute Equation 13.12 into the loop gain equation, we end up with

$$
\begin{aligned}
g(n) &= g(n-1) + \mu\left[d - y_{RMS}\right] \\
&= g(n-1) + \frac{\left[d - y_{RMS}\right]}{2N\left[y_{MAX}\right]}
\end{aligned}
$$

**Equation 13.13**

We can see from Equation 13.13 that the ratio of the error signal to the maximum value of the output signal $\left[d - y_{RMS}\right]/y_{MAX}$ will always be less than unity, and therefore the magnitude of $\Delta g$ will always be less than unity. Why? Because with the exception of a DC signal, $y_{RMS} < y_{MAX}$. It is important to note that the procedure for determining the optimum $\mu$ is not a hard and fast rule. The design engineer is free to choose any of the numerous methods to determine a working value for $\mu$.

**Method 5**. Another and most commonly used method is to replace the $y_{MAX}$ term in Equation 13.12 with a scale factor $S$ that isn't dependent on anything except the designer's imagination. In this case, $\mu = 1/S$ and the gain equation is given by

$$
g(n) = g(n-1) + \left[\frac{d - y_{RMS}}{S}\right]
$$

**Equation 13.14**

This method is illustrated in the simulations that follow.

The reader might ask the question, "Why in the heck would we ever want to use an AGC on some DC signal whose amplitude is slowly time varying?" On the surface, the answer to this question sure looks like "We wouldn't!" But let's take a look below the surface, so to speak, and see if we can find another use for this simple AGC circuit. Perhaps our application is to track the amplitude variance of this DC "like" signal and provide a real time precision measurement of this variance to some instrumentation logic located elsewhere within the system. For example, suppose the slowly varying DC signal is representative of the slight fluctuations of a remote temperature sensor located on a critical piece of machinery. In this case, the variance of the DC signal contains useful information. As we will see later when we study the plots of the AGC internal waveforms, the loop gain precisely tracks the amplitude fluctuations of the input signal and could very easily be used as the precision instrumentation signal we need.

In this application, the AGC could be used as a very simple but very accurate amplitude demodulator. The length of the RMS delay line $N$ could be lengthened from 1 to some larger number such as 16 to smooth out and eliminate any additive noise that may be present on the input signal. The reader might ask, "Why would increasing the RMS delay length eliminate additive noise?" The short answer is that the RMS delay line is nothing more than a digital filter with unity coefficients and, as such, acts like a $\sin x/x$ filter. The longer the delay line, the narrower the bandwidth of the filter main lobe, which helps eliminate high-frequency noise from the loop gain path.

While we are on this subject, it is fun to think of the possibilities we have at our disposal if the RMS delay line were not constrained to use unity coefficients. What if the delay line was an actual digital low pass filter of some bandwidth B, with coefficients defined by the filter impulse response $h(n)$? There is absolutely nothing wrong with this approach. In fact, it makes this simple AGC a great deal more robust and a great deal more applicable for instrumentation applications.

We mentioned earlier that we scaled the value of $\mu$ by a factor of $1/(2N)$ . Now it's time to discuss the reason why. Let's visualize an RMS delay line of length $N$ in the feedback path of the AGC circuit. We should be able to immediately visualize two things. First, for wideband input signals, larger values of $N$ should produce more accurate estimates of the output signal RMS value. Second, larger values of $N$ will also increase the delay between a change in the input signal amplitude and the corresponding change in circuit gain. That is, the gain changes will lag the input signal by a length of time on the order of $N$ signal samples. To compensate for this lag in gain and the potential instability it may cause, we need to reduce the size of the feedback coefficient, thereby reducing the size of the incremental gain steps $\Delta g$. This will reduce the overshoot and undershoot in the transient response to input signal gain changes.

Let's illustrate this point with a simple example. Suppose the input signal to our AGC experiences a large positive step in amplitude. The change in circuit gain in response to this input signal step will not be instantaneous. Therefore the output signal will also experience a large step in amplitude. It will take a few sample periods for the new output samples to move through the RMS delay pipe and make their contribution to the overall RMS measurement.

The first output sample after the input step will contribute roughly $(1/N)$th to the overall sum of squares in the RMS computation. As the second, third, fourth, and fifth samples after the input step are clocked into the RMS pipe, the measured RMS value will slowly increase. This increase in RMS will cause the circuit error to increase and the loop gain to slowly decrease. This

lower gain then lowers the output signal RMS value. Keep in mind that the effective error signal $\varepsilon(n)$ lags the input by approximately $N$ sample periods.

If the value of the feedback coefficient $\mu$ is large, the resulting negative incremental gain $\mu\varepsilon = -\Delta g$ will also be large and will also be lagging the
input signal by $N$ sample periods. As RMS samples are passed through the
pipeline, the negative incremental gain will continue to grow. The high negative incremental gain will cause the circuit gain to undershoot its optimum
value. This undershoot will attenuate the input signal amplitude below the
desired level.

Eventually this undershoot will be detected in the error circuit and the
feedback logic will attempt to correct things by increasing the incremental gain. Because of the delay, the same scenario occurs again, only this time
the gain overshoots the optimum value. This oscillation between successive
undershoots and overshoots can slowly die out, causing the AGC circuit to
eventually stabilize, or it can increase in amplitude, causing runaway oscillation and numerical overflow. Either way, this transitory oscillation is not a
desired circuit behavior. In a nutshell, the longer the RMS delay, the slower or
more lethargic the AGC circuit becomes. It becomes less responsive to abrupt
input signal changes, and it can over correct in its attempt to iterate to the
optimum gain. The solution is to make the incremental gain steps smaller to
slow down and prevent the cyclic undershoot and overshoot problem. We accomplish this by reducing the value of the feedback coefficient $\mu$ by some
scale factor to compensate for the $N$ sample feedback delay. The value of
the scale factor has been determined empirically to be equal to the reciprocal
of twice the length of the delay line, or $2N$. Scaling the feedback coefficient
slows the adaption process and therefore attenuates or even prevents overcorrection due to input steps in amplitude. This overcorrection problem will
be illustrated later when we perform some example circuit simulations. The
reader is reminded that this oscillatory behavior that can lead to an unstable
circuit can easily be predicted and corrected during circuit design through the
use of pole placement in the system pole-zero diagram.

The engineer should take note that the length of the RMS delay line is an
important design trade-off issue: the longer delay produces a more accurate
RMS estimate, but it slows the transient gain response of the AGC. For single
frequency input signals, the length $N$ only needs to be as long as a single period. Anything longer will be inefficient. In fact, in this case, anything longer
than an integer number of periods will produce a degraded estimate. For narrow band signals, the length $N$ might be as long as the period of the lowest
frequency in the pass band.

### 13.1.4.2   The Feedback Coefficient $\mu$ Is Time Varying

One of the big issues we had with determining a fixed constant $\mu$ was trying to determine a single value that would work for a wide variance of input

signal amplitudes and one that would work for all time. Although the use of a fixed constant $\mu$ is valid, there are always efficiency problems associated with a "one-size-fits-all" approach. In this paragraph, we will present a method that provides us with a much more optimum feedback coefficient. We achieve this enhancement at the expense of some additional circuit computations. Our approach is to derive a time varying feedback coefficient, which we will denote as $\mu(n)$ to indicate that it is a function of the sample index. In this approach, we will compute a new feedback coefficient every sample period using the same relation as we derived in Equation 13.7, repeated here for clarity:

$$\mu(n) < \left| \frac{g(n-1)}{[d - y_{RMS}]} \right|$$

This time we will use at each sample instant the last value of the circuit loop gain $g(n-1)$ to compute the next value of $\mu(n)$ and hence the next incremental gain $\Delta g(n)$. We still have the problem of the denominator approaching zero, which will happen as the AGC iteratively adapts. Using the same arguments as we did for the constant feedback coefficient case, we arrive at the time varying relations

$$\mu(n) \le \frac{1}{2N} \left| \frac{g(n-1)}{[d - y_{RMS}]_{(MAX)}} \right|$$

or

$$\mu(n) \le \frac{1}{2N} \left| \frac{g(n-1)}{y_{MAX}} \right|$$

or

$$\mu(n) \le \left| \frac{g(n-1)}{S} \right|$$

**Equation 13.15**

In the time varying computation of the feedback coefficient, the value of $g(n-1)$ is the actual value of the previous loop gain, as opposed to unity. The use of this time varying $\mu$ will be illustrated in the simulations that follow.

### 13.1.5   AGC Simulation: A DC Signal Using a Constant $\mu$

For this example, since our input signal is a DC voltage with a small amplitude variance, we will make the assumption that we can estimate the RMS value of the output signal $y_{RMS}$ using only the previous output sample $y(n-1)$. That is, we set $N = 1$. On the surface this sounds a bit odd. The reader might ask, "How can we compute the RMS value of a digital sequence using only a single sample?" The answer has much to do with the statistics of the input signal. For example, if our AGC circuit application is to control the level of a DC signal or perhaps a very slow time varying signal, then a single sample estimate of the RMS value can be sufficient.

This is because the RMS value of a waveform is a measured average of that waveform. The RMS value of a DC waveform is simply the value of the DC waveform. That is, if the signal $x(n)$ is a DC signal, then $x(n)_{RMS} = x(n)$. For example, the RMS computation in Equation 13.2 says

$$x(n)_{RMS} = \sqrt{\frac{1}{N}\sum_{n=0}^{N-1} x^2(n)} = \sqrt{\frac{x^2(0) + x^2(1) + x^2(2) + \cdots + x^2(N-1)}{N}}$$

If the signal $x(n)$ is the constant DC value $v$, then

$$x(n)_{RMS} = \sqrt{\frac{1}{N}\sum_{n=0}^{N-1} v^2(n)} = \sqrt{\frac{v^2(0) + v^2(1) + v^2(2) + \cdots + v^2(N-1)}{N}} = \sqrt{\frac{Nv^2}{N}} = v$$

For other signals, such as narrow band signals or narrow band signals plus noise, we will get a more accurate estimate of the signal RMS by using a much longer delay line.

Seeing how the AGC responds to a DC signal will help us better understand its response to a sinusoidal or narrow band signal. Therefore, for instructional purposes, we will begin our simulation examples by showing the AGC response to the input stepped DC signal shown in Figure 13.9.

Before we get started, we should define the annotation at the bottom of each signal simulation plot. The simulation processes 1024 input signal samples, and therefore the plots are 1024-points long. The sample space is divided into three sections called *amplitude profiles*, abbreviated as *Ap* in the notation. The first profile extends from sample 0 to sample 63, the second profile ranges from sample 64 to sample 511, and the third profile ranges from sample 512 to sample 1023. The amplitude of the input signal can change at the boundary of any one of these profiles. The simulation disables the AGC for the first 64 samples, which is the duration of profile 1, and enables the AGC at the beginning of profile 2. The desired value of the output signal RMS is listed in the plot annotation as $d$. In this simulation of a DC input signal, this value is arbitrarily set to $d = 6.123$, and the amplitude profile is $20/20/10$. The plots of the input test signal display the values for the amplitude profile $Ap$ and desired RMS $d$, at the bottom of the plot, as illustrated in Figure 13.9.

The simulation output plots contain additional annotation, as illustrated in Figure 13.10. The value of the loop gain for each of the three profiles is annotated as $LGain = aa / bb / cc$. In this first simulation the loop gain is computed to be $LGain = 1.000 / 0.306 / 0.612$. Although the value of the loop gain is computed for each sample instant, the loop gain for each profile displayed in the plot annotation is that computed at the end of each profile segment. That is, at sample instants 63, 511 and 1023.

The measured RMS value of the AGC output signal is also calculated at each sample time, and the value measured at the end of each profile segment is

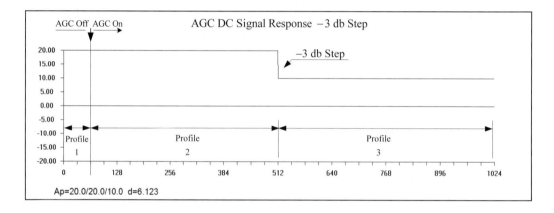

**Figure 13.9**   AGC simulation 1 DC input signal with a –3 db step

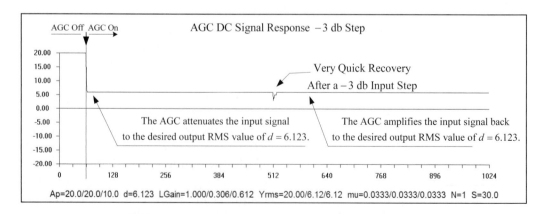

**Figure 13.10**   AGC simulation 1 processed DC output signal

listed in the annotation as $y_{RMS} = aa / bb / cc$. In the first simulation, the measured RMS values are $y_{RMS} = 20.00 / 6.12 / 6.12$. The value of the feedback coefficient $\mu$ for each profile segment is also contained in the annotation. In this constant $\mu$ simulation, the values are listed as $mu = 0.0333 / 0.0333 / 0.0333$. The length of the RMS delay line is listed as $N = 1$, and the feedback coefficient scale factor is listed as $S = 30.0$. The scale factor is the denominator in the expression used to compute the value of $\mu$, as determined by Equation 13.14. With that said, let's begin with the discussion of the AGC simulation 1. The test signal input to the AGC in this simulation is a DC signal with a −3 db step at sample $n = 512$. This signal is illustrated in Figure 13.9. The author has chosen to mark the boundaries of all three amplitude profile segments in this figure. Although not marked, all plots for all simulations that follow use the same profile distribution. The amplitude profile for this test input signal is $Ap = 20 / 20 / 10$. In this simulation, we computed the feedback coefficient as

$$\mu = \frac{1}{2N} \left| \frac{1}{\left[ d - y_{RMS} \right]_{MAX}} \right| = \frac{1}{2} \left| \frac{1}{\left[ 6.123 - 20 \right]_{MAX}} \right| = \frac{1}{2} \left| \frac{1}{-13.877} \right| = \left| \frac{1}{-27.754} \right| = 0.0360$$

The value of −27.754 was rounded to −30 to give us a feedback coefficient of

$$\mu = \left| \frac{1}{S} \right| = \left| \frac{1}{-30} \right| = 0.0333$$

The plot of the AGC output signal is illustrated in Figure 13.10. There are three things we immediately notice from this plot: The first thing we notice is that when the AGC is not enabled during the period of time defined by amplitude profile 1, the output signal follows the input signal. That is, the amplitude of the output DC signal is 20, just as we would expect. The second thing we notice is that immediately after the AGC is enabled, the RMS of the output signal decreases to the desired output RMS value of $d = 6.123$. The third thing we notice is how quickly the desired RMS output is achieved. From the plot, it looks instantaneous. Actually, the ACG required seven sample periods to iterate the gain to 0.306.

The output signal amplitude during segment two is given by $(0.306)(20) = 6.12$, which is within measurement error of the desired RMS value of $d = 6.123$. This is indeed a very responsive AGC. Its response time is due to the value of the feedback coefficient we chose, and it may be too quick for most applications. We will look at slower responses later. The plot also shows that there is a small variance in the output RMS level at sample 512 caused by the −3 db step in the input signal. The AGC quickly applies additional gain to boost the output signal back to the desired value of $d = 6.123$. The annotation at the bottom of the plot shows that the AGC doubled its gain from 0.306 in profile 2 to 0.612 in profile 3 in response to the halving of the input signal amplitude from 20 down to 10.

The track of the loop gain is illustrated in Figure 13.11. As before, we note the lightning fast gain response that occurs immediately after the AGC is enabled and at the point where the –3 db amplitude step occurs. Although not clearly shown in the gain plot, both of these response curves are exponential with a time constant that is controlled by the feedback coefficient $\mu$. We will address the mathematics of the exponential response later.

A more revealing plot is the track of the loop error illustrated in Figure 13.12. In this figure, we can see that the loop error is zero for the entire simulation, with the exception of the times when the AGC is initially enabled and the time when the negative step occurs. It is beneficial to look at the mathematics of the loop error to get a better understanding of how the AGC loop works.

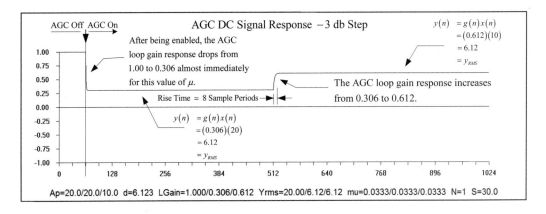

**Figure 13.11**   AGC simulation 1 track of the loop gain

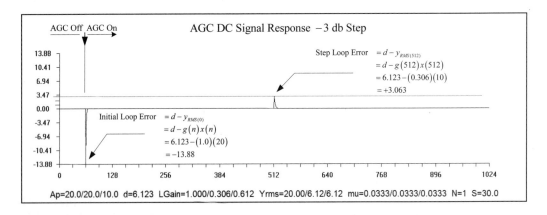

**Figure 13.12**   AGC simulation 1 track of the loop error

For the first 64 samples, the AGC is disabled. The input signal is 20 volts DC. The desired RMS of the output signal is $d = 6.123$. The loop gain is set to unity. Immediately after the AGC is enabled, the loop gain is still unity because it can't change instantaneously. Therefore the RMS value of the output signal $y_{RMS}$ is given by $y_{RMS} = g(n)x(n)$. The loop error immediately after the AGC is enabled is $[d - y_{RMS}] = [d - g(n)x(n)]$. Therefore the initial loop error is calculated by

$$\begin{aligned}
\text{Initial loop error} \quad &= d - y_{RMS} \\
&= d - g(n)x(n) \\
&= d - g(64)x(64) \\
&= 6.123 - (1)(20) \\
&= -13.877
\end{aligned}$$

If we look at the initial loop error at sample 64 in Figure 13.12, we see that the error takes an immediate negative spike all the way down to $\varepsilon = -13.88$, as predicted. Although it is not apparent in the figure, the recovery from the negative spike takes on an exponential shape.

At sample 512 the input signal steps down by 3 db. Now the AGC sees a much lower value for $y_{RMS}$. Right away it determines that the gain setting is too low, and it needs to increase it to compensate for the reduced amplitude of the input signal. How does it know this? It knows because the error signal $[d - y_{RMS}] = [d - g(n)x(n)]$ tells it so. The loop error signal in response to the −3 db input step at sample 512 is computed to be

$$\begin{aligned}
\text{Step loop error} \quad &= d - y_{RMS} \\
&= d - g(n)x(n) \\
&= d - g(512)x(512) \\
&= 6.123 - (0.306)(10) \\
&= +3.063
\end{aligned}$$

Remember that the loop gain does not change instantaneously, so the loop gain in the above error computation is equal to the last steady state zero error loop gain of $g(512) = 0.306$. These simulation results were obtained with a $\mu$ scale factor of $s = 30$. Now let's see what happens if, for the exact same input signal, we make the feedback coefficient smaller. In the following figures, we will compute the feedback coefficient using the method given by Equation 13.15, which is repeated here for clarity:

$$\mu(n) \le \left| \frac{g(n-1)}{S} \right| = \frac{1}{S}$$

In this case, we choose to set $S = 128$, which gives us a constant feedback coefficient $\mu = 1/128 = 0.0078$. The track of the loop gain with this smaller value of $\mu$ is illustrated in Figure 13.13. The smaller $\mu$ has increased the length of the loop gain transient response. The AGC response to an abrupt change in input signal amplitude is slower. The steady state values are identical to the previous case where $\mu = 0.0333$; the only difference is in the speed of the loop transient response.

The track of the loop error is illustrated in Figure 13.14. Once again, with the exception of the transient response, the track is identical to that of the larger $\mu$ case. The reader should note that the exponential behavior of the transient response becomes more apparent as the value of $\mu$ decreases and as the loop response slows. Finally, the AGC output $y(n)$ is illustrated in Figure 13.15. After the initial adaption, the output holds steady at 6.123 volts, which is equal to our desired RMS output. Remember that for DC signals, the amplitude and RMS values are the same.

So far we have taken a look at how the AGC processes and responds to a stepped DC input signal. This was the simplest method with which to demonstrate the track of the AGC loop gain and loop error signals. Let's complicate the problem a little bit with the use of a slowly time varying DC signal.

### 13.1.6  AGC Simulation: A Slowly Varying Signal with a Constant $\mu$

During this discussion we mentioned that the type I AGC could be used to process a DC signal to remove a slowly varying amplitude component. For lack of a better term, a signal such as this might be described as having "DC wander." We also mentioned previously that the signals internal to the AGC could be used externally as precise instrumentation signals to convey critical information about the input signal. The loop gain signal in this particular case could be utilized by external instrumentation to extract specific information contained within the input signal. We will discuss these two topics next.

Suppose we receive a DC signal that, for whatever reason, has a slight amplitude variance. Perhaps this variance is an attribute that we wish to eliminate. Perhaps this variance actually contains information such as a slowly varying temperature measured by some remote sensor, and we would like to extract this information and transmit it to some other processing system. There are a great many applications that come to mind. The time varying component might be the voltage fluctuations from an accelerometer. It might be a small voltage that represents the RPM of an engine under test, or it might

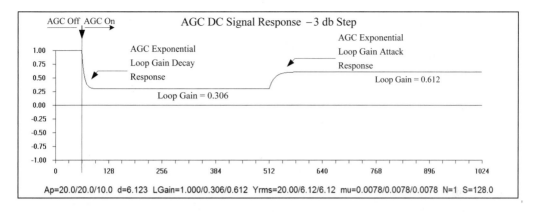

**Figure 13.13**   AGC simulation 1 track of the loop gain response with smaller μ

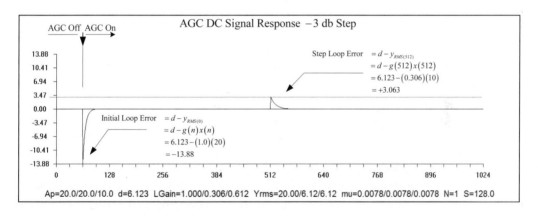

**Figure 13.14**   AGC simulation 1 track of the loop error response with smaller μ

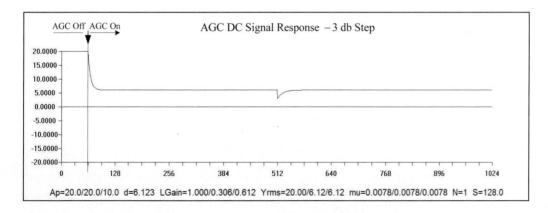

**Figure 13.15**   AGC simulation 1 output signal RMS with smaller μ

be a small voltage that represents the pressure changes in a propellant tank. The possibilities are endless.

Suppose we were given the task of processing a time varying DC voltage that was proportional to the fill level of a vat of highly evaporative chemicals. The DC voltage would slowly decrease as the chemicals evaporated and increase as the vat is replenished. Our assignment is to strip off the variance from the DC signal and pass it off to an instrumentation system that will analyze the variance and adjust the chemical mix accordingly. Or perhaps we are given the task of processing a DC signal whose variance was a measure of the voltage output by a spacecraft battery whose overall charge is affected by the attitude of the solar panels relative to the sun. These or thousands of other applications that transmit low bandwidth data could be digitized and transmitted once per second using a single shared time slot of a pulse code modulation (PCM) telemetry signal.

At the receiving end, the time varying DC could be processed by a type I AGC to separate the DC voltage from the time varying voltage. In any case, whether the amplitude variance contains intelligent information, or whether it is garbage and needs to be eliminated, our AGC circuit can handle both operations. Suppose our AGC inputs a time varying DC signal such as that illustrated in Figure 13.16.

The input signal is composed of a 10-volt DC term and a slowly varying amplitude term that is modeled in this example by a 1-volt sinusoid. In this example, the period of the sinusoid is 1024 samples, so we see exactly one cycle of variance across the entire simulation.

Our objective is twofold: First we wish to eliminate the amplitude variance and pass the DC term on for further processing, and second we would like to extract the time varying component and pass it off to other equipment,

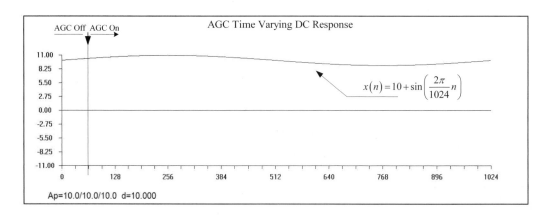

**Figure 13.16**    AGC simulation 2 time varying DC input signal

which will then process it and recover whatever information it contains. One solution might be to pass the composite signal through a band pass filter to extract the time varying component. But this method is impractical because the DC component and the time varying component are spaced extremely close in frequency. The filter would need to have a very narrow pass band and a very narrow transition band relative to the sample rate of the signal. The narrow transition band requirement alone would cause the impulse response of a finite impulse response (FIR) filter to grow to an unrealizable length. So the filter solution doesn't make the cut. The second solution might be to use the AGC circuit we just designed. It could easily perform the required operations and, at the same time, use minimal hardware if implemented in a field programmable gate array (FPGA) or minimal resources if implemented in software either on some digital signal processing (DSP) processor or some DSP core within a FPGA. So this is the method that we choose to use and demonstrate. Before we look at the AGC internals, let's jump right to the results to see if our type I AGC can remove the variance term. The processed signal out of the AGC is illustrated in Figure 13.17.

When the AGC is disabled, it defaults to a loop gain of unity so the output signal is a copy of the input signal. We can see the first 64 samples out of the AGC mirror the input sinusoidal variance. As soon as the AGC is enabled, it quickly generates a loop error signal and adjusts the loop gain so that the amplitude variance is removed and the output signal is the 10-volt DC component of the original input signal.

We derived the test signal in this simulation by setting the three amplitude profile segments to 10 $(Ap = 10,10,10)$, and then we added a 1-volt sinusoid at the normalized frequency

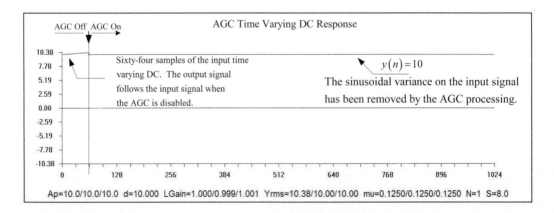

**Figure 13.17**   AGC simulation 2 time varying DC removal output signal

$$f_{NORM} = 1/1024 \cong 0.001$$

We set our desired output signal RMS to $d = 10$ and the RMS delay to $N = 1$. We want this loop to react quickly to small changes in the input signal, so we set our $\mu$ scale factor to the small number $S = 8$. We can do this because we know that the signal is well behaved and will not experience any large gain steps.

For this application, if the scale factor is set to a large number, the loop gain will lag the input and the track of the gain will be nothing more than a delayed or phase-shifted version of the input. Therefore we use Equation 13.14 to compute the feedback coefficient to be

$$\mu = (1/S)$$
$$= (1/8)$$
$$= 0.125$$

We can see from the plot that we achieved our goal in that the output signal RMS was measured to be 10 and 10 during the two profile segments where the AGC was enabled. Thus far, we have demonstrated that the type I AGC can track a time varying DC input and remove its amplitude variance.

Now how do we extract the variance term and pass it to some external instrumentation for further processing? The answer is simple. We have already done it. As illustrated in Figure 13.18, the track of the loop gain is proportional to the input signal variance. It is phase shifted by 180°, but it contains all the information that once rode on the back of the DC signal. In this example, the loop gain takes on the sinusoidal form with the same period as the input variance and with amplitude swings between 0.9091 and 1.1111.

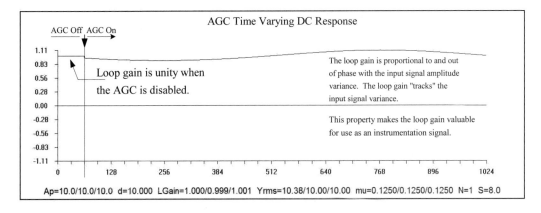

**Figure 13.18**   AGC simulation 2 track of the loop gain for the time varying input

The peak-to-peak amplitude swing is 0.2 volts, or 1/10 of the input signal variation. All we need to do is provide this signal to some other piece of equipment, which will then extract the information. The loop gain values listed in the plot annotation are the single valued snapshots of the gain taken at samples 63, 511, and 1023 and do not reflect the history of the time varying loop gain. More than likely, the time varying component will change only a fraction of a volt over the course of several hours or even days. No matter what, the track of that change is captured by the track of the AGC loop gain.

To give the reader a feeling for how precise this AGC can be, observe the track of the loop error illustrated in Figure 13.19. With the exception of the –0.3927-volt spike when the AGC is initially enabled, the magnitude of the error is exceedingly small.

On the plot, it is almost indistinguishable from zero. No matter how small the error is, as long as it is larger than the resolution of the floating point number representation in our DSP processor, the loop will detect it, and it will use it to adjust the loop gain. We can use the loop gain as a signal suitable for external processing purposes if we so desire. For completeness, the track of the output signal RMS value is illustrated in Figure 13.20. As we can see, the output RMS is a constant value equal to the desired RMS value of $d = 10$ that we entered in the simulation.

We have demonstrated that the loop can track a small amplitude variance, but we might ask, "Just how robust is this AGC? What will happen if we stress test the loop by exaggerating both the frequency and the magnitude of the signal variance?" Suppose we boost the amplitude of the variance to 20% of the 10-volt DC signal and further suppose that we increase the frequency fourfold. The input signal with these attributes is illustrated in Figure 13.21. In this figure, we see a 10-volt DC signal with four cycles of a 2-volt sinusoid riding on top. The input signal is given by

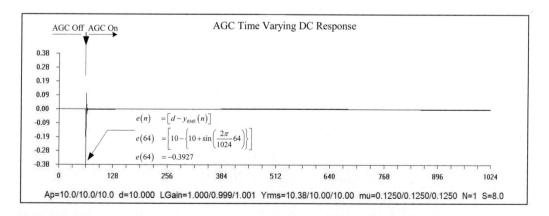

**Figure 13.19** AGC simulation 2 track of the loop error for a time varying DC input

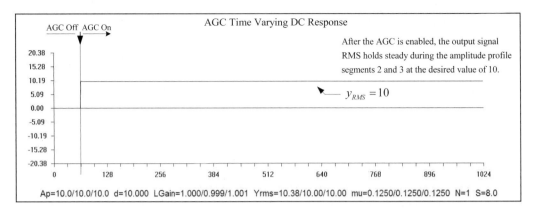

**Figure 13.20**   AGC simulation 2 track of the output signal RMS measurement

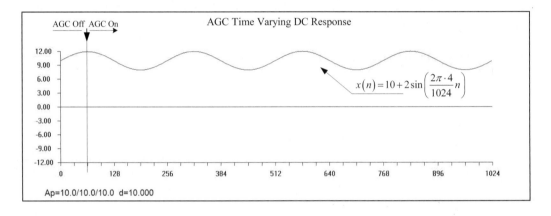

**Figure 13.21**   AGC simulation 3 time varying DC input signal

$$x(n) = 10 + 2\sin\left(\frac{2\pi \cdot 4}{1024}n\right)$$

This is not a small variance relative to the DC pedestal. The AGC loop is going to have to be fast in order to track this swiftly changing signal. We calculate the first estimate of the feedback coefficient to be

$$\mu = \frac{1}{2N}\left|\frac{1}{d - y_{RMS(MAX)}}\right| = \frac{1}{2}\left|\frac{1}{10 - 12}\right| = 0.250$$

This results in a very large $\mu$. Simulations show that for this value of $\mu$, the loop error, loop gain, and output signal suffered from an oscillatory

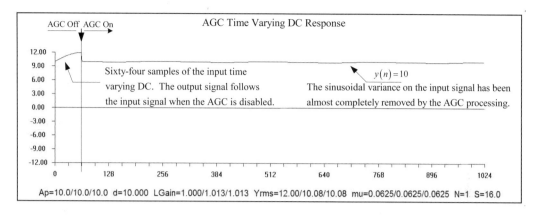

**Figure 13.22** AGC simulation 3 loop output signal

response immediately after the AGC was enabled. Although the oscillations decayed rapidly, the transition response was still unacceptable. Since the difference between $d$ and $y_{RMS(MAX)}$ is small, we opt to set the scale factor $S$ to 16 and use Equation 13.14 to compute the feedback coefficient. In doing so, we end up with

$$\mu = 1/S = 1/16 = 0.0625$$

Using this $\mu$, the AGC processed output signal is illustrated in Figure 13.22. We can see some slight undulation in the output, which suggests that we could increase the value of $\mu$ somewhat to reduce the loop response time and iron out the signal variance. Once again, since the AGC output signal tracks the input signal with a gain of unity when the AGC is disabled, we see at the beginning of the simulation the first 64 samples of the input sinusoidal waveform. When the AGC is enabled at sample 64, the output signal takes on the appearance of a DC level of 10 volts. We can see from the plot annotation that there is some variance left on the signal since the measured RMS is 10.08, as opposed to the desired 10.0.

Don't forget that we are treating this input signal with a huge ±2 volt deviation as if it were a DC signal with a very slow amplitude variance. This assumption allows us to closely approximate the signal RMS value as being the actual amplitude of the sample. In this example, we are stretching the limits of that assumption, which is a contributing factor in the residue undulations seen in the output waveform. Even so, the result is pretty impressive.

The track of the loop gain is illustrated in Figure 13.23. Once enabled, the loop gain tracks the input signal variance with a phase offset of 180°. This makes sense. If the loop gain is 180° out of phase with the input signal

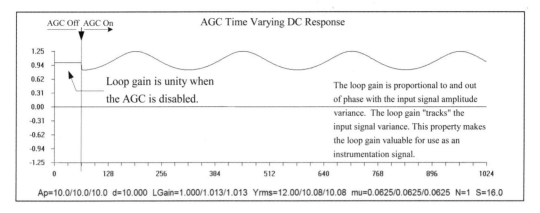

**Figure 13.23**   AGC simulation 3 track of the loop gain for a time varying input

variance, and if the magnitude of the gain is proportional to that of the input signal variance, then the product of the two should produce a constant amplitude output, effectively stripping the variance off the input signal. The loop gain shown in the figure annotation is the value at the end of each profile segment (i.e., at samples 63, 511, and 1023) and is not relevant here, since it does not represent the entire AC track of the gain.

The simulation printout shows the loop gain oscillating between 0.83333 at its lowest point and 1.25000 at its highest point. When lined up 180° out of phase, the products of the sinusoidal peaks, $12 \cdot 0.83333 = 10$ and $8 \cdot 1.25000 = 10$. All matching intermediate points should ideally have the same product.

The track of the loop error is illustrated in Figure 13.24, and the track of the AGC output RMS is illustrated in Figure 13.25. As the reader can see, we have taken some liberties in this last simulation when approximating the RMS value of a slowly varying DC to be equal to the DC value, when in fact the frequency of the variance was anything but slow and the amplitude of the of the variance was anything but small. This fact tends to illustrate the robustness of this AGC circuit.

### 13.1.7   AGC Simulation: Sinusoidal Input Signal Using a Constant $\mu$

Let's repeat the signal simulation 1, only this time instead of using a DC input signal, we will use a sinusoidal input signal. We choose to use as our input test signal a sinusoidal waveform given by

$$x(n) = A \sin\left( \frac{2\pi \cdot 64}{1024} n \right)$$

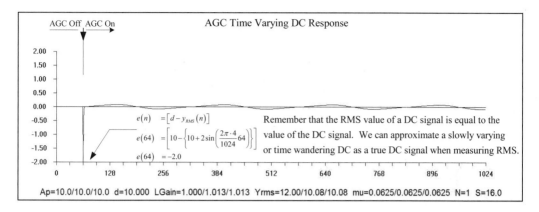

**Figure 13.24**   AGC simulation 3 track of the loop error for a time varying input

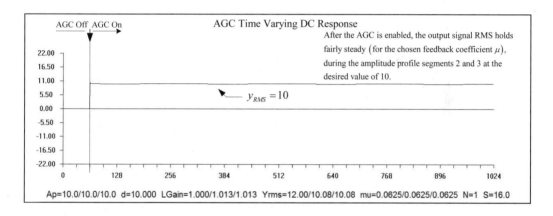

**Figure 13.25**   AGC simulation 3 track of the output signal RMS

The amplitude profile for this simulation is the same as used for the DC simulation 1 and is given by $A_p = 20, 20, 10$. The input signal is illustrated in Figure 13.26. It initially has an amplitude of 20 volts. At sample 512, the input signal experiences a –3 db step down to 10 volts. The normalized frequency of the input sinusoid is $f_n = 0.0625$. As before, we have chosen our desired output signal RMS to be $d = 6.123$. This RMS level is represented by the solid line superimposed on the figure. The first thing we as designers need to do is to compute the initial estimate of the feedback coefficient $\mu$. For illustration purposes, let's calculate $\mu$ using Equation 13.9.

$$\mu < \frac{1}{2N} \left| \frac{1}{\left[ d - y_{RMS} \right]_{MAX}} \right|$$

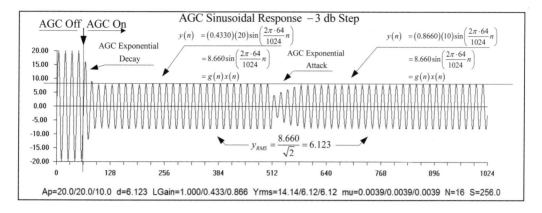

**Figure 13.27**   AGC simulation 4 processed sinusoidal output signal $N = 16$

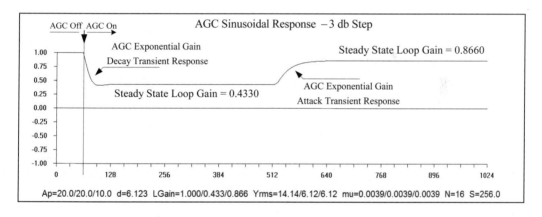

**Figure 13.28**   AGC simulation 4 track of the loop gain $N = 16$

decay rate depends on the value selected for $\mu$. This type of circuit response is definitely pleasing to the eye. The track of the loop gain in Figure 13.28 clearly shows the attack and decay transients and the value of steady state gain that occurs after the loop drives the error toward zero and the transient response dies. The plot annotation clearly shows the output RMS as being 14.14 volts in profile segment 1 before the AGC is enabled and 6.12 volts in profile segments 2 and 3 after the AGC is enabled. The annotation also shows the value of the feedback coefficient is constant at 0.0039 across all three profile segments. The RMS delay is 16 samples in length, and the scale factor for the chosen $\mu$ is given by $S = 1/0.0039 \approx 256$.

The track of the AGC loop error is illustrated in Figure 13.29. This loop error track is similar to the track for the case where the input signal was DC.

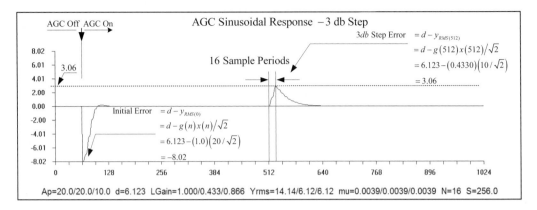

**Figure 13.29**   AGC simulation 4 track of the loop error $N = 16$

The one difference is that in this simulation is that the loop uses a 16-sample RMS delay line and therefore has a slower response. The reader can see the exponential shape of the error track. The initial error after the AGC is enabled is computed as follows. Since the initial gain is unity, we know that prior to the AGC being enabled the output signal will be a copy of the input signal. Therefore the initial error is computed as

$$\varepsilon = \left[d - y_{RMS}\right] = \left[d - x_{RMS}\right] = \left[6.123 - 20/\sqrt{2}\right] = -8.019$$

Looking at the error from a different angle, the initial loop error at sample 64 is computed by

$$
\begin{aligned}
\text{Initial loop error} \quad &= d - y_{RMS} && = d - g(n)x(n)/\sqrt{2} \\
&= d - g(64)x(64)/\sqrt{2} \quad &&= 6.123 - (1)\left(20/\sqrt{2}\right) \\
&= -8.019
\end{aligned}
$$

By the same reasoning and understanding that the loop gain cannot change instantaneously, we can compute the loop error response to the −3 db step at sample 512 as

$$
\begin{aligned}
\text{Step loop error} \quad &= d - y_{RMS} && = d - g(n)x(n)/\sqrt{2} \\
&= d - g(512)x(512)/\sqrt{2} \quad &&= 6.123 - (0.433)\left(10/\sqrt{2}\right) \\
&= +3.0613
\end{aligned}
$$

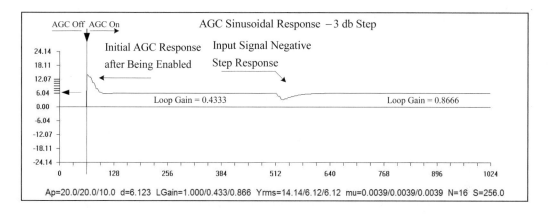

**Figure 13.30**   AGC simulation 4 track of the output signal RMS value

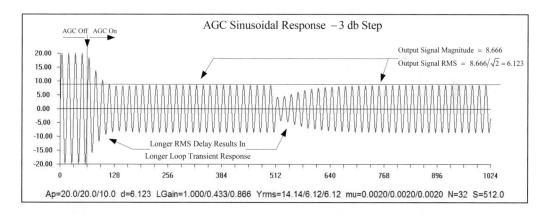

**Figure 13.31**   AGC simulation 4 processed sinusoidal output signal $N = 32$

The peak value of this step loop error is indicated by the overlaid dotted line. It is interesting to note that the complete rise time of the step error response is equal to 16 sample periods or the length of the RMS delay line. The track of the output signal RMS is illustrated in Figure 13.30.

Next we input the exact same test signal as illustrated in Figure 13.26 but process it with an AGC that has an RMS delay of $N = 32$. The only difference in the circuit setup was that we increased the scale factor $S$ from 256 to 512, which halved the feedback coefficient from $\mu = 0.0039$ to $\mu = 0.0020$. The resulting output signal is illustrated in Figure 13.31. Comparing this AGC output for $N = 32$ with that of the AGC output for $N = 16$, we can see that doubling the length of the RMS delay essentially halved the feedback coefficient and resulted in the increased length of the exponential transient response. The track of the loop gain and loop error is illustrated in Figure 13.32 and Figure 13.33, both of which reflect the increase in the loop RMS delay. It should be pointed out that the loop error response to the –3 db step at sample 512 has a total rise time equal to 32 sample periods. The measurement of the output RMS is illustrated in Figure 13.34.

The reader may be asking the question, "Why does the loop error drop immediately to –8.02 when the AGC in turned on but increase exponentially to its peak value when the input signal experiences a –3 db drop in amplitude?" The reason the AGC error drops almost instantaneously when enabled is that the loop error $\varepsilon = \left[ d - y_{RMS} \right] = \left[ d - x_{RMS} \right] = \left[ 6.123 - 20/\sqrt{2} \right] = -8.019$ is already computed at that point and is instantaneously applied when the AGC is enabled. When the input signal drops by –3 db at the end of amplitude profile 2, the error has not yet been computed. It will take N samples of input data for the error track to reach its maximum value before it exponentially decays toward zero.

## 13.1.8 Effects of Choosing Too Large a Value for $\mu$

We have just discussed a properly operating AGC, and we have viewed the track of its internal loop gain and loop error signals. Now let's play around a bit and see how the performance of the AGC loop degrades as the feedback coefficient gets larger and larger. This part of the simulation should provide some valuable insight as to the behavior of the AGC internals. For comparison purposes, this simulation uses the same sinusoidal input signal that is illustrated in Figure 13.26.

The signal tracks in Figure 13.35, Figure 13.36, and Figure 13.37 illustrate the track of the loop gain, loop error, and the AGC output signal for the case where the value selected for the feedback coefficient is too large. We mentioned earlier that too large of value for $\mu$ will pass a large loop error and therefore cause the loop gain to overcorrect. This overcorrection results in a succession of undershoot and overshoot oscillations that may or may not settle to the desired steady state values. In this case, we set the value of $\mu$ to be approximately two and a half times the previous value of $\mu = 0.0039$ to $\mu = 0.0104$. That is, we reduced the scale factor $S$ from 256 to 96. The first thing we notice in Figure 13.35 is a ringing in the loop gain transient response that in this case exponentially settles out to the desired steady state condition.

When the AGC is initially enabled, the error signal is large, and we see a very large ringing that occurs as the loop tries to iterate to a valid solution. We also see some overshoot of the loop gain when the AGC responds to the –3 db step of the input signal. The track of the loop error is illustrated in Figure 13.36. We note the same initial error peak of –8.02 followed by a ringing response that eventually settles out. How does this ringing affect the output signal? We

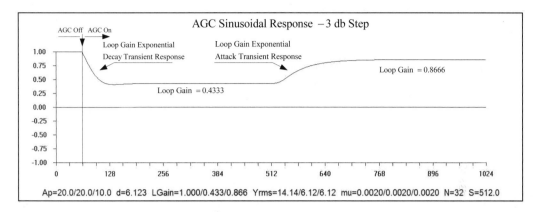

**Figure 13.32** AGC simulation 4 track of the loop gain $N$ = 32

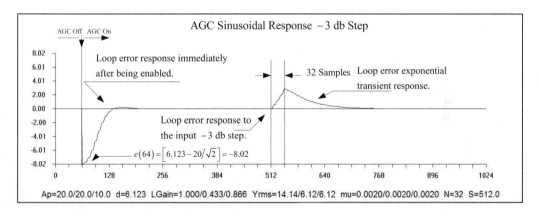

**Figure 13.33** AGC simulation 4 track of the loop error $N$ = 32

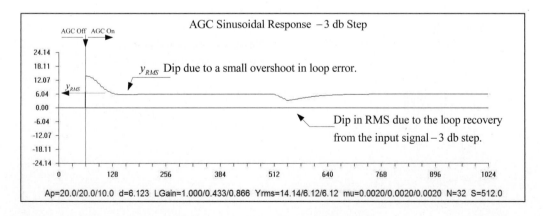

**Figure 13.34** AGC simulation 4 track of the output signal RMS $N$ = 32

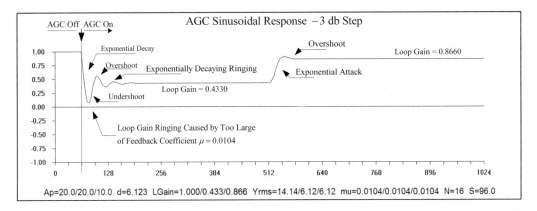

**Figure 13.35**   AGC simulation 4 track of an overvalued loop gain μ = 0.0104

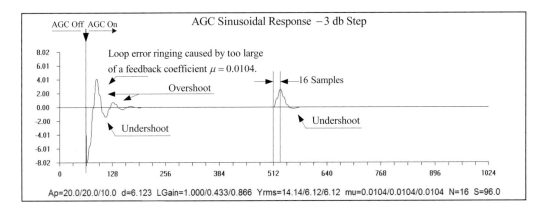

**Figure 13.36**   AGC simulation 4 track of the loop error with overvalued μ = 0.0104

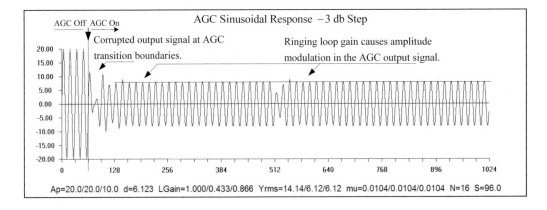

**Figure 13.37**   AGC simulation 4 processed output signal with overvalued μ = 0.0104

can see the effects to the output signal in Figure 13.37. The first thing that is obvious is the initial distortion in the signal when the AGC is enabled and initially responds to a fairly large error signal. The second thing we see, which is not so obvious to the eye, is the slight undulation or amplitude modulation of the output signal following both adaption transients. This may not look too bad when viewing the time domain representation of the signal, but the amplitude modulation creates some frequency components in the signal spectrum that may not be desired.

Let's increase the value of the feedback coefficient again and observe the additional performance degradation this action causes. Figure 13.38, Figure 13.39, and Figure 13.40 illustrate the track of the loop gain, loop error, and the output signal for the case where the value of the feedback coefficient increases to $\mu = 0.0179$, which correlates to a change in the scale factor $S$ from 96 to 56. Right away we can see from the track of the loop gain in Figure 13.38 that the performance of the AGC has degraded dramatically. The ringing of the loop gain transient response is much larger and lasts much longer. The interval between the enabling of the AGC and the negative step is $512 - 64 = 448$ sample periods. The desired loop gain during this profile segment should be 0.433. As can be seen from the plot annotation, the gain has only settled to 0.440 at sample 511. If the sample rate of the input signal is 40 MHz, then this says that the AGC transient response is longer than $(448)(25n$ seconds$)$ or $11.2\mu$ seconds. It is clear that the performance of this AGC is unacceptable for most if not all applications. We can clearly see the overshoot and ringing following the gain correction to the –3 db step.

Who wants to be on the receiving end of a signal processed by an AGC that has a step response that rings for more than $11\mu$ seconds? If an engineer presented this circuit at a major design review, he might be on the receiving end of a lengthy barrage of tomatoes and cabbages hurled from the technically astute audience.

The corresponding track of the loop error for this underperforming circuit is illustrated in Figure 13.39. The ringing in the loop error due to the loop over correction is clearly evident. The signal output by this AGC is illustrated in Figure 13.40. Here we can see that the large overcorrections in the loop gain have actually caused a phase reversal in the processed signal following the enabling of the AGC. In addition, the amplitude modulation of the processed signal caused by the oscillating transient response of the loop gain is clearly evident. This AGC is on the verge of being completely unstable. An unruly input signal or an increase in $\mu$ will more than likely cause the circuit to become unstable and reach an overflow condition.

To be fair, the major ringing occurs when the AGC is initially enabled and not in response to an input signal perturbation. This problem can be corrected by a modification of the AGC turn on procedure, such as slowing down the loop response for a period of time after the circuit is enabled. That

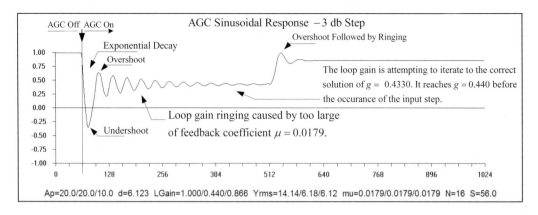

**Figure 13.38**   AGC simulation 4 track of the loop gain with overvalued μ = 0.0179

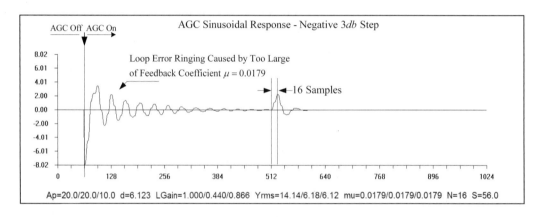

**Figure 13.39**   AGC simulation 4 track of the loop error with overvalued μ = 0.0179

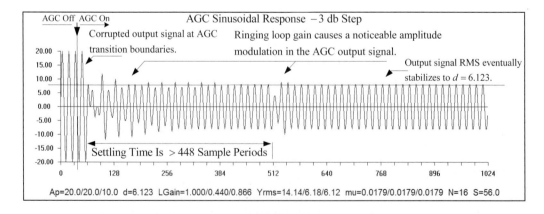

**Figure 13.40**   AGC simulation 4 processed output signal with overvalued μ = 0.0179

is, use a very small value $\mu$ for the first 500 milliseconds or so after circuit turn on, and then adjust the value of $\mu$ to its programmed operational value. This would accomplish two things:

- It would suppress the initial turn on transients of the loop.
- It would allow for the use of a larger valued $\mu$ during normal loop operation.

## 13.1.9   Loop Exponential Transient Response

We have stated and have graphically shown that the transient response of a stable RMS AGC circuit is exponential. Now it's time to derive the mathematical expression for the exponential transient response. Let's begin with the basic expression that describes the AGC in the continuous time domain:

$$y(t) = \left\{ g(t) + \mu \left[ d - y_{RMS}(t) \right] \right\} x(t)$$

We can interpret this relation to mean that the output signal is equal to the product of the current gain plus the change in gain times the input signal or

$$y(t) = \left\{ g(t) + \Delta g(t) \right\} x(t)$$

Therefore we can state that the change in the output signal is the result of the product of the input signal and the change in loop gain or

$$\begin{aligned} \Delta y(t) &= \Delta g(t) x(t) \\ &= \mu \left[ d - y_{RMS}(t) \right] x(t) \end{aligned}$$

**Equation 13.16**

We can also state that the change in the RMS value of the output signal is equal to the product of the change in loop gain and the RMS value of the input signal or

$$\Delta y_{RMS}(t) = \mu \left[ d - y_{RMS}(t) \right] x_{RMS}(t)$$

We make the assumption that the input $x(t)$ is a steady state signal that experiences an amplitude step at time $t = 0$ and then forever remains in a steady state condition. If this is the case, then from time $t = 0$ forward, $x_{RMS}(t)$ is a constant. This behavior is identical to the behavior of the input signals in all our previous discussions. Therefore the change in the RMS value of the output signal is proportional to the change in the loop gain or

$$\Delta y_{RMS}(t) = k\mu\big[d - y_{RMS}(t)\big]$$

**Equation 13.17**

where the $k$ term in Equation 13.17 is a proportionality constant. We would like to compute the change in the RMS of the output signal over some small interval of time. In other words, we wish to compute the measurement of $\Delta y_{RMS}(t)/\Delta t$. In order to achieve an accurate measurement, we let the incremental changes for $\Delta y_{RMS}(t)$ and $\Delta t$ shrink to infinitesimal size so that we can write

$$\lim_{\Delta t \to 0} \frac{\Delta y_{RMS}(t)}{\Delta t} = \frac{d}{dt} y_{RMS}(t)$$

For simplicity, we will drop the RMS subscript from the $y(t)$ term. We can now write

$$\frac{dy(t)}{dt} = k\mu\big[d - y(t)\big]$$

where $k$ is an unknown proportionality constant. So we know that the instantaneous change in the output $y$ with respect to $t$ is proportional to the current value of $y$. Now if we pull the negative sign out of the brackets on the right side of the equation, we get

$$\frac{dy(t)}{dt} = -k\mu\big[y(t) - d\big]$$

We divide both sides of the equation by the quantity $\big[y(t) - d\big]$ and multiply both sides by $dt$ to get

$$\frac{dy(t)}{\big[y(t) - d\big]} = -k\mu dt$$

We are interested in how this relation behaves over all time, so we integrate both sides to arrive at

$$\int \frac{dy(t)}{\big[y(t) - d\big]} = -k\mu \int dt$$

In order to perform the integration, we need to consider two cases: the case where the quantity $[y-d]$ is positive and the case where it is negative. We know from basic calculus that

$$\int \frac{du}{u} = \ln u + C \qquad \{\text{for } u > 0\}$$
$$= \ln(-u) + C \quad \{\text{for } u < 0\}$$

The second expression is valid because if the quantity $u$ is negative, then $-u$ is positive and

$$\int \frac{du}{u} = \int \frac{d(-u)}{-u} = \ln(-u) + C$$

The desired RMS term $d$ is a constant, so after performing the integration, we end up with

$$\ln\big[(y(t)-d)\big] = -k\mu t + c$$

We can dismiss the natural logarithm operator by using the relation $e^{\ln x} = x$ to get

$$e^{\ln(y(t)-d)} \qquad = e^{-k\mu t + c}$$
$$\big[y(t)-d\big] \quad = e^{-k\mu t}e^{c}$$
$$\big[y(t)-d\big] \quad = Ae^{-k\mu t} \quad \{\text{where } A = e^{c}\}$$

If the quantity $\big[y(t)-d\big]$ is positive, then

$$y(t) - d = Ae^{-k\mu t}$$

Rearranging terms, we end up with

$$y(t) = d + Ae^{-k\mu t}$$

At time $t = 0$, the value of $y(t) = y(0)$, so

$$y(0) = d + Ae^{-k\mu 0} = d + A$$

therefore $A = y(0) - d$. Making this substitution for $A$, we get

$$y(t) = d + (y(0) - d)e^{-k\mu t} \quad \left\{\text{for } \left[y(t) - d\right] > 0\right\}$$

### Equation 13.18

The exponential shape of this transient response is illustrated in Figure 13.41. We can see from the figure that given the initial condition of $y(0) > d$, the exponential decay of the initial loop error forces the loop gain to exponentially decrease over time until the initial signal offset $\left[y(0) - d\right]e^{-k\mu t} \to 0$ and $y(t) \to d$. When this occurs, the loop error also equals zero, or

$$e(t) = \left[d - y(t)\right] = 0$$

Using similar reasoning for the case, where the quantity $\left[y(t) - d\right]$ is negative, we get

$$\ln\left(-\left[y(t) - d\right]\right) = -k\mu t + c$$

$$-\left[y(t) - d\right] = e^{-k\mu t + c} = e^{-k\mu t}e^c = Ae^{-k\mu t} \quad \left\{\text{where } A = e^c\right\}$$

At $t = 0$, $y(t) = y(0)$. Therefore $A = \left[d - y(0)\right]$, and we end up with

$$y(t) = d - \left[d - y(0)\right]e^{-k\mu t} \quad \left\{\text{for } \left[y(t) - d\right] < 0\right\}$$

### Equation 13.19

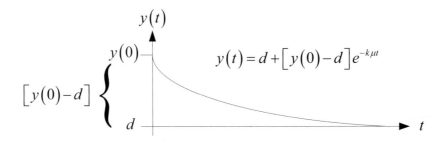

**Figure 13.41**   Loop exponential transient response

We can opt to bring the minus sign out of the brackets in Equation 13.19 to arrive at

$$y(t) = d + [y(0) - d]e^{-k\mu t} \qquad \left\{ \text{for} \quad \begin{array}{c} [y(t) - d] > 0 \\ [y(t) - d] < 0 \end{array} \right\}$$

**Equation 13.20**

We can see that Equation 13.20 and Equation 13.18 are identical, so we can use the same equation for the two cases where the quantity $[y(t) - d]$ is either positive or negative. The exponential shape of this transient response for the case where $[y(0) - d]$ is negative is illustrated in Figure 13.42. We can see from the figure that, given the initial condition $y(0) < d$, the exponential decay of the initial loop error forces the loop gain to exponentially increase over time until the initial signal offset $[y(0) - d]e^{-k\mu t} \to 0$ and $y(t) \to d$. When this occurs, the loop error also equals zero, or

$$e(t) = [d - y(t)] = 0$$

The only thing we need to do now is to determine the value of the proportionality constant $k$. We will begin the derivation using Equation 13.20 as the starting point. This equation is repeated here for clarity:

$$y(t) = d + [y(0) - d]e^{-k\mu t}$$

We begin by rearranging terms so that the exponential term is on the left side of the equation. When we do so, we arrive at

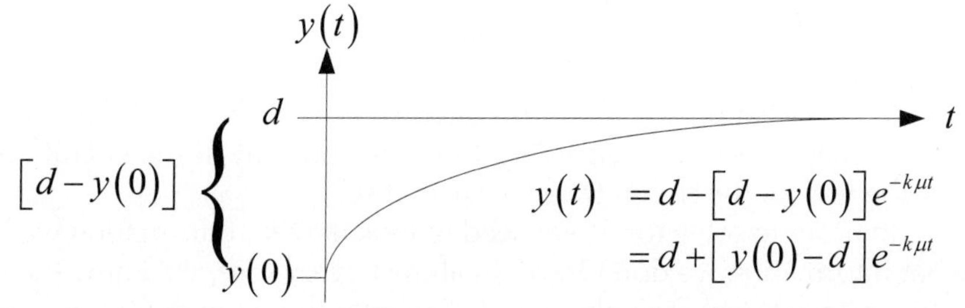

**Figure 13.42**   Loop exponential transient response

$$e^{-k\mu t} = \left[\frac{y(t)-d}{y(0)-d}\right]$$

We want to solve for $k$, so we take the natural log of both sides to get

$$-k\mu t = \ln\left[\frac{y(t)-d}{y(0)-d}\right]$$

If we divide both sides by the quantity $-\mu t$, we arrive at

$$k = \frac{-1}{\mu t}\ln\left[\frac{y(t)-d}{y(0)-d}\right]$$

We change our viewpoint from continuous time to discrete time by making a change in variables. We set $t = nT$, where $T$ is the discrete time system sample period and $n$ is the sample index. If we normalize the sample rate by setting $T = 1$, then we end up with the expression

$$k = \frac{-1}{\mu n}\ln\left[\frac{y(n)-d}{y(0)-d}\right]$$

**Equation 13.21**

We can also substitute $nT = t$ into Equation 13.20 and then normalize the sample period to $T = 1$ to get the discrete version of the continuous equation given by

$$y(n) = d + \left[y(0)-d\right]e^{-\mu kn} \quad \left\{\text{for} \quad \begin{array}{c} \left[y(n)-d\right]>0 \\ \left[y(n)-d\right]<0 \end{array}\right\}$$

**Equation 13.22**

From Equation 13.22, we can clearly see that the feedback coefficient $\mu$ plays a major role in determining the time constant of the circuit and therefore the time of the circuit transient response.

In order to solve for $k$, we need to look at what information we have and what information we don't have. In almost every case, we know the value for $y(0)$, and we know the value for $d$. The only thing we don't know is a value for the index $n$ and the corresponding value for $y(n)$.

One method of obtaining a value for $n$ and $y(n)$ is to create a simple Excel spreadsheet and generate a table such as the one illustrated in Table 13.1. The table is divided into two sections. The section on the left side of the table is titled *measured output*. The columns in this section tabulate the measured values obtained by incrementally computing the AGC output using Equation 13.6, repeated here for clarity:

$$y(n) = \left\{ g(n-1) + \mu \left[ d - y(n)_{RMS} \right] \right\} x(n)$$
$$= \left\{ g(n-1) + \Delta g(n) \right\} x(n)$$
$$= g(n) x(n)$$

The section on the right of the table is titled *calculated output*. The columns in this section tabulate the calculated values obtained through the use of the transient Equation 13.22. The intent of this table is to compare the results between the two sections and therefore between the two equations. In this simple example, the input to the AGC is a simple 18-volt DC signal. Our desired output is another DC signal at 20 volts. We choose a feedback coefficient of $\mu = 0.005$. All this information is shown under the heading *AGC circuit parameters* at the top of the table. The *measured output* section of the table contains five columns that tabulate sequential iterations of Equation 13.6. The first column is the sample index $n$, the second column is the RMS value of the output signal $y_{RMS}(n)$, the third column is the loop error $\varepsilon(n) = \left[ d - y_{RMS}(n) \right]$, the fourth column is the incremental gain or $\Delta g = \mu \varepsilon$, and the fifth column is the value of the overall loop gain. In this example, we arbitrarily select the sample index $n = 40$, which fixes the corresponding RMS value of the output to be $y_{RMS}(40) = 19.95401$. These two quantities are highlighted on the table. Plugging these two values into Equation 13.21, we can compute the value of the proportionality constant $k$:

$$k = \frac{-1}{\mu n} \ln \left[ \frac{y(n) - d}{y(0) - d} \right]$$
$$= \frac{-1}{(0.005)(40)} \ln \left[ \frac{y(40) - d}{y(0) - d} \right]$$
$$= \frac{-1}{(0.005)(40)} \ln \left[ \frac{19.95401 - 20}{18 - 20} \right]$$
$$= 18.862$$

**Table 13.1**   AGC Exponential Transient Response

| AGC Circuit Parameters | | | | | | |
|---|---|---|---|---|---|---|
| μ | d RMS | X peak | x RMS | | k | |
| 0.005 | 20 | 18 | 18 | | 18.8624 | |
| Measured Output | | | | | Calculated Output | |
| $y(n)=\left\{g(n-1)+\mu\left[d-y(n)_{RMS}\right]\right\}x(n)$ | | | | | $y(n)=d+\left[y(0)-d\right]e^{-\mu kn}$ | |

| n | y RMS | ε=[d-y RMS] | με | g(n) | | μkn | exp(-μkn) | y RMS | Difference |
|---|---|---|---|---|---|---|---|---|---|
| 0 | 18.0000 | | | 1.0000 | | 0.0000 | 1.0000 | 18.0000 | 0.0000 |
| 1 | 18.1800 | 2.0000 | 0.0100 | 1.0100 | | 0.0943 | 0.9100 | 18.1800 | 0.0000 |
| 2 | 18.3438 | 1.8200 | 0.0091 | 1.0191 | | 0.1886 | 0.8281 | 18.3438 | 0.0000 |
| 3 | 18.4929 | 1.6562 | 0.0083 | 1.0274 | | 0.2829 | 0.7536 | 18.4929 | 0.0000 |
| 4 | 18.6285 | 1.5071 | 0.0075 | 1.0349 | | 0.3772 | 0.6857 | 18.6285 | 0.0000 |
| 5 | 18.7519 | 1.3715 | 0.0069 | 1.0418 | | 0.4716 | 0.6240 | 18.7519 | 0.0000 |
| 6 | 18.8643 | 1.2481 | 0.0062 | 1.0480 | | 0.5659 | 0.5679 | 18.8643 | 0.0000 |
| 7 | 18.9665 | 1.1357 | 0.0057 | 1.0537 | | 0.6602 | 0.5168 | 18.9665 | 0.0000 |
| 8 | 19.0595 | 1.0335 | 0.0052 | 1.0589 | | 0.7545 | 0.4702 | 19.0595 | 0.0000 |
| 9 | 19.1441 | 0.9405 | 0.0047 | 1.0636 | | 0.8488 | 0.4279 | 19.1442 | 0.0000 |
| 10 | 19.2212 | 0.8559 | 0.0043 | 1.0678 | | 0.9431 | 0.3894 | 19.2212 | 0.0000 |
| 11 | 19.2913 | 0.7788 | 0.0039 | 1.0717 | | 1.0374 | 0.3544 | 19.2913 | 0.0000 |
| 12 | 19.3550 | 0.7087 | 0.0035 | 1.0753 | | 1.1317 | 0.3225 | 19.3551 | 0.0000 |
| 13 | 19.4131 | 0.6450 | 0.0032 | 1.0785 | | 1.2261 | 0.2934 | 19.4131 | 0.0000 |
| 14 | 19.4659 | 0.5869 | 0.0029 | 1.0814 | | 1.3204 | 0.2670 | 19.4659 | 0.0000 |
| 15 | 19.5140 | 0.5341 | 0.0027 | 1.0841 | | 1.4147 | 0.2430 | 19.5140 | 0.0000 |
| 16 | 19.5577 | 0.4860 | 0.0024 | 1.0865 | | 1.5090 | 0.2211 | 19.5577 | 0.0000 |
| 17 | 19.5975 | 0.4423 | 0.0022 | 1.0888 | | 1.6033 | 0.2012 | 19.5975 | 0.0000 |
| 18 | 19.6338 | 0.4025 | 0.0020 | 1.0908 | | 1.6976 | 0.1831 | 19.6338 | 0.0000 |
| 19 | 19.6667 | 0.3662 | 0.0018 | 1.0926 | | 1.7919 | 0.1666 | 19.6667 | 0.0000 |
| 20 | 19.6967 | 0.3333 | 0.0017 | 1.0943 | | 1.8862 | 0.1516 | 19.6967 | 0.0000 |
| 21 | 19.7240 | 0.3033 | 0.0015 | 1.0958 | | 1.9806 | 0.1380 | 19.7240 | 0.0000 |
| 22 | 19.7488 | 0.2760 | 0.0014 | 1.0972 | | 2.0749 | 0.1256 | 19.7489 | 0.0000 |
| 23 | 19.7714 | 0.2512 | 0.0013 | 1.0984 | | 2.1692 | 0.1143 | 19.7715 | 0.0000 |
| 24 | 19.7920 | 0.2286 | 0.0011 | 1.0996 | | 2.2635 | 0.1040 | 19.7920 | 0.0000 |
| 25 | 19.8107 | 0.2080 | 0.0010 | 1.1006 | | 2.3578 | 0.0946 | 19.8107 | 0.0000 |
| 26 | 19.8278 | 0.1893 | 0.0009 | 1.1015 | | 2.4521 | 0.0861 | 19.8278 | 0.0000 |
| 27 | 19.8433 | 0.1722 | 0.0009 | 1.1024 | | 2.5464 | 0.0784 | 19.8433 | 0.0000 |
| 28 | 19.8574 | 0.1567 | 0.0008 | 1.1032 | | 2.6407 | 0.0713 | 19.8574 | 0.0000 |
| 29 | 19.8702 | 0.1426 | 0.0007 | 1.1039 | | 2.7350 | 0.0649 | 19.8702 | 0.0000 |
| 30 | 19.8819 | 0.1298 | 0.0006 | 1.1045 | | 2.8294 | 0.0591 | 19.8819 | 0.0000 |
| 31 | 19.8925 | 0.1181 | 0.0006 | 1.1051 | | 2.9237 | 0.0537 | 19.8925 | 0.0000 |
| 32 | 19.9022 | 0.1075 | 0.0005 | 1.1057 | | 3.0180 | 0.0489 | 19.9022 | 0.0000 |
| 33 | 19.9110 | 0.0978 | 0.0005 | 1.1062 | | 3.1123 | 0.0445 | 19.9110 | 0.0000 |
| 34 | 19.9190 | 0.0890 | 0.0004 | 1.1066 | | 3.2066 | 0.0405 | 19.9190 | 0.0000 |
| 35 | 19.9263 | 0.0810 | 0.0004 | 1.1070 | | 3.3009 | 0.0368 | 19.9263 | 0.0000 |
| 36 | 19.9329 | 0.0737 | 0.0004 | 1.1074 | | 3.3952 | 0.0335 | 19.9329 | 0.0000 |
| 37 | 19.9390 | 0.0671 | 0.0003 | 1.1077 | | 3.4895 | 0.0305 | 19.9390 | 0.0000 |
| 38 | 19.9445 | 0.0610 | 0.0003 | 1.1080 | | 3.5839 | 0.0278 | 19.9445 | 0.0000 |
| 39 | 19.9495 | 0.0555 | 0.0003 | 1.1083 | | 3.6782 | 0.0253 | 19.9495 | 0.0000 |
| 40 | 19.9540 | 0.0505 | 0.0003 | 1.1086 | | 3.7725 | 0.0230 | 19.9540 | 0.0000 |
| 41 | 19.9581 | 0.0460 | 0.0002 | 1.1088 | | 3.8668 | 0.0209 | 19.9581 | 0.0000 |
| 42 | 19.9619 | 0.0419 | 0.0002 | 1.1090 | | 3.9611 | 0.0190 | 19.9619 | 0.0000 |
| 43 | 19.9653 | 0.0381 | 0.0002 | 1.1092 | | 4.0554 | 0.0173 | 19.9653 | 0.0000 |

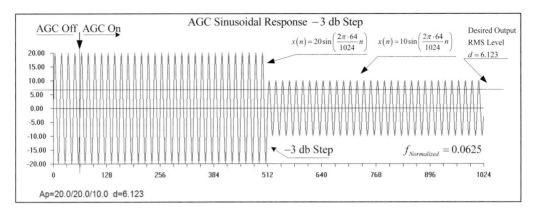

**Figure 13.26**  AGC simulation 4 sinusoidal input signal with a −3 db step

We know from the figure that $y_{RMS(MAX)} = 20/\sqrt{2} = 14.14$ and we are given the desired output signal RMS value as $d = 6.123$. We compute the period of the input sinusoid to be $1024/64 = 16$ samples, so we can set the value of our RMS delay line to be $N = 16$. This is simply the number of samples in a single cycle of the test sinusoid. Plugging these values into Equation 13.9 gives us

$$\mu \; < \frac{1}{2N}\left|\frac{1}{[d - y_{RMS}]_{MAX}}\right|$$

$$< \frac{1}{(2)(16)}\left|\frac{1}{6.123 - 14.14}\right|$$

$$< 0.0039$$

This value for $\mu$ is pretty close to $\mu = 1/S = 1/256$, so we will use in our simulation $S = 256$. Keep in mind that the engineer could have used any number of other criteria for determining the value for $\mu$. The method we used here is just one of many. The AGC processed output signal is illustrated in Figure 13.27. Here we can see that, for this particular value of $\mu$, the output signal is well behaved with a steady state amplitude of 8.66. The steady state signal amplitude is a constant value, as indicated by the straight line superimposed on the signal peaks. The RMS value for this steady state signal is $y_{RMS} = 8.66/\sqrt{2} = 6.123$, which of course is the desired output level that we specified. In Figure 13.28, we can see the results of the exponential decay of the loop gain immediately after the AGC is enabled, and we can see the exponential attack of the loop gain following the −3 db step in input amplitude. Once again, the transient response of this circuit and hence the attack and

Since the feedback coefficient $\mu$ is a constant, we can compute the constant product of $\mu k$:

$$\mu k = (0.005)(18.862) = 0.0943$$

Plugging this value of $\mu k$ in to Equation 13.22 gives us

$$
\begin{aligned}
y(n) &= d + [y(0) - d]e^{-\mu k n} \\
&= 20 + [18 - 20]e^{-0.0943n} \\
&= 20 - 2e^{-0.0943n}
\end{aligned}
$$

**Equation 13.23**

Equation 13.23 gives us the calculated exponential transient response of the AGC for the initial conditions of $y_{RMS}(0) = 18$, $\mu k = 0.0943$, and $d = 20$. The tabulation of the calculated transient response is illustrated in the *calculated output* section of Table 13.1. The first column of the right side of the table is the computed value of $\mu k n$, the second column is the computed value of $e^{-\mu k n}$, and the third column is the calculated value of the RMS output from the AGC. The fourth column is just the difference between the measured $y_{RMS}(n)$ and the calculated $y_{RMS}(n)$. That column is all zeros, indicating that the calculated method produces results identical those obtained by simulation of the circuit and measuring the actual output. Examples of the response curve for both methods are illustrated in Figure 13.43 and Figure 13.44.

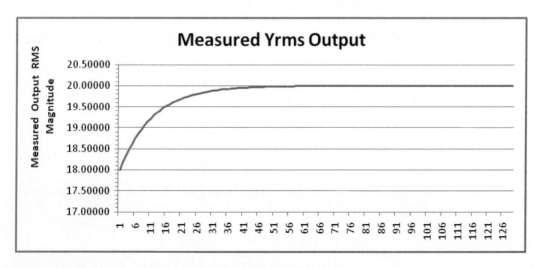

**Figure 13.43**   AGC RMS output measured by simulation

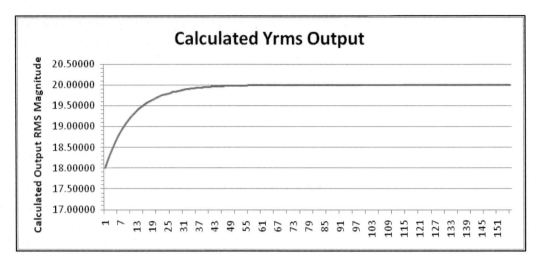

**Figure 13.44**   AGC RMS output predicted by equation

One other result is presented without derivation to further demonstrate the exponential response of this AGC circuit. The output signal transient response of an unstable AGC is illustrated in the Excel plot of Figure 13.45. In this example, $y_{RMS}(0) = 3$ and the desired output $d = 1.25$. The value of $\mu$ was chosen to make the circuit unstable during the transition period. This is evident from the output oscillations in the figure. Superimposed on the oscillating transient response is the computed transient response envelope. It is clear that both the measured signal and the computed signal envelope have an identical exponential transient response.

### 13.1.10   AGC Simulation Using a Time Varying Feedback Coefficient

All the simulations up to this point have used a constant feedback coefficient. There is another option and that is to use a time varying $\mu$ computed using Equation 13.15 and repeated here for clarity:

$$\mu(n) \le \frac{1}{2N} \left| \frac{g(n-1)}{[d - y_{RMS}]_{(MAX)}} \right|$$

The denominator terms can be modified to give the two other expressions previously discussed:

$$\mu(n) \le \frac{1}{2N} \left| \frac{g(n-1)}{y_{MAX}} \right| \quad \text{or} \quad \mu(n) \le \left| \frac{g(n-1)}{S} \right|$$

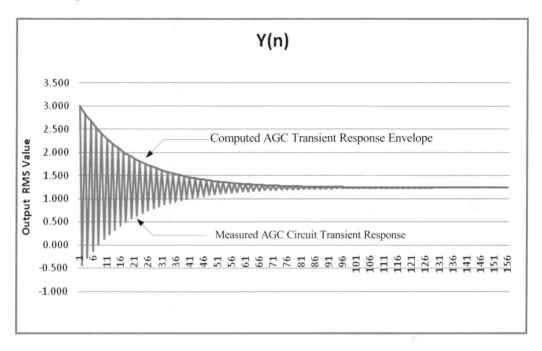

**Figure 13.45**   Unstable AGC measured output signal and computed transient envelope

The important thing to note here is that in the equations the terms $[d - y_{RMS}]_{(MAX)}$, $y_{MAX}$, and $S$ are all constants, and $\mu(n)$ is proportional to the value of the previous loop gain $g(n-1)$. In the following simulation plots, the annotation at the bottom refers to $2N(y_{MAX})$ as $S$. Let's begin by inputting a sinusoidal test signal to our AGC circuit that has both a positive and a negative 3 db step in amplitude, as illustrated in Figure 13.46. Here we see an input sinusoidal signal with an amplitude profile of $Ap = 10, 20, 10$ volts. As before, the AGC is turned off for the first 64 samples and is enabled from that point on. In this simulation, we have chosen a value for the scale factor $S = 128$, we set the initial gain to unity, and we set the desired RMS output level to be $d = 7.071$ volts, which corresponds to an output signal amplitude of 10 volts. The processed output signal is illustrated in Figure 13.47. The first thing we notice in the figure is the well-behaved output signal. The output signal amplitude, with the exception of the times when the input experiences a step in amplitude, is indeed held to a constant 10 volts. This is satisfying to see. The second thing we see is the value of the feedback coefficient $\mu$, captured at samples 63, 511, and 1023. These three values are listed in the plot annotation as $\mu = 0.0078, 0.0039, 0.0078$ and correspond to each of the three profile segments.

The next thing the reader may notice is the single cycle of output signal at sample 64 that jumps up almost to the 20-volt amplitude of the input signal and then attenuates rather quickly to the desired level of 10 volts. This occurs right at the beginning of the positive amplitude step. Why did this sinusoidal-like spike occur? The test frequency we input into the AGC has a period of 16 samples. The RMS delay in the loop feedback is also set to 16 samples. This means that the loop gain is essentially delayed by 16 samples relative to the input signal. In this case, the loop gain could not change fast enough to knock down the first 16-sample cycle of the positive step input signal. It is easily seen from the signal peaks that the signal is being attenuated as the stepped up samples of the output signal are shifting through the RMS delay line.

As the stepped up signal is shifted through the RMS delay line, the loop error steadily increases and the loop gain steadily decreases until the input

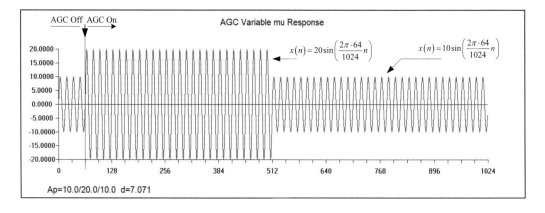

**Figure 13.46**   AGC simulation 5 sinusoidal input signal with a ±3 db step

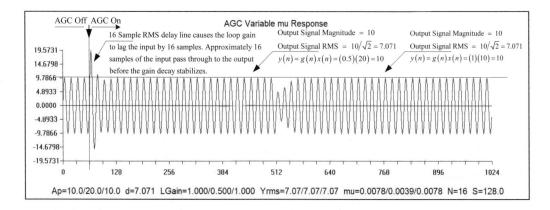

**Figure 13.47**   AGC simulation 5 processed output signal with variable μ

signal is attenuated by half, the loop error is zero, and the loop gain stabilizes at 0.5. The track of the loop gain is illustrated in Figure 13.48. Since the desired output signal amplitude is 10 volts, the loop gain begins at 1.0 in profile segment 1, exponentially decreases to 0.5 in profile segment 2, and exponentially increases back to 1.0 in the profile segment 3.

The track of the loop error signal is illustrated in Figure 13.49. It is easily seen in this figure that the error response has a delay equal to the length of the RMS delay, in this case 16 samples. The variable feedback coefficient has the effect of reducing the magnitude of the loop error swings. For a constant feedback coefficient, the initial loop error at sample 64 would have been

$$\varepsilon = \left[ d - y_{RMS} \right] = \left[ 7.071 - 14.14 \right] = -7.071 \text{ volts}$$

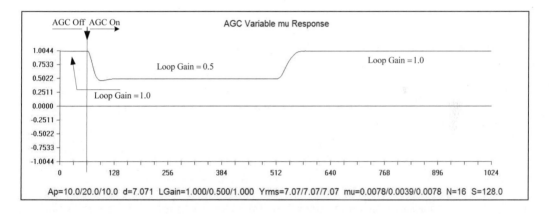

**Figure 13.48**   AGC simulation 5 track of the loop gain $N = 16$

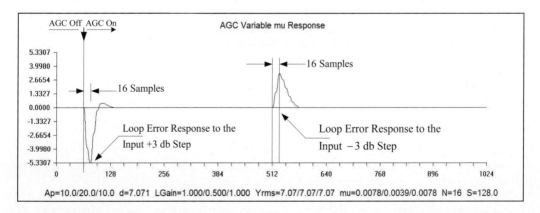

**Figure 13.49**   AGC simulation 5 track of the loop error $N = 16$

For the time varying $\mu$, we can see from the figure that the initial loop error is significantly less than $-7.071$ volts and that the maximum error eventually reaches approximately $-5.3$ volts after 16 samples of delay. It is also noticeable from other simulation plots that are not shown here that the positive overshoot following the initial negative error has been significantly reduced.

The track of the variable feedback coefficient $\mu$ is illustrated in Figure 13.50. The track is well behaved and tends to follow the track of the loop gain. This is not surprising since the feedback coefficient is proportional to the loop gain.

The track of the output signal RMS value is illustrated in Figure 13.51. It can be seen from this figure that, with the exception of the input signal perturbations, the RMS tracks true to the desired value of $d = 7.071$.

The advantage of using a variable feedback coefficient is that it speeds the loop response and can substantially attenuate the ringing that occurs due to an input signal step. The disadvantage of using a variable feedback

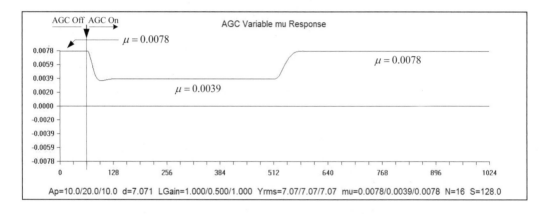

**Figure 13.50**   AGC simulation 5 track of the variable feedback coefficient $N = 16$

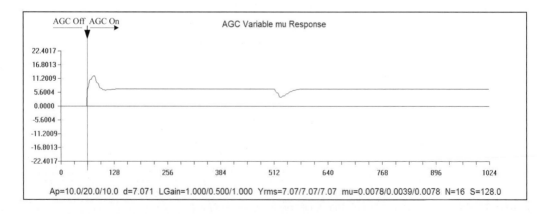

**Figure 13.51**   AGC simulation 5 track of the output signal RMS

coefficient is that since the circuit time constant is dependent on $\mu$, the exponential transient response also becomes time varying.

### 13.1.11  AGC Simulation Sinusoidal Input Signal with Noise

So far we have dealt only with pure signals—that is, signals without any additive noise. How does the AGC loop respond when processing a noisy input signal? In this paragraph, we will take a brief look at the behavior of the AGC internals when tasked to operate on an input signal corrupted with an additive noise component. The time domain plot of a pure sinusoidal input test signal is illustrated in Figure 13.52. Here we see a sinusoidal signal with a constant amplitude of 6 volts at a normalized frequency of $f_{NORM} = 64/1024 = 1/16 = 0.0625$. The spectrum of this pure tone computed by a 1024-point discrete Fourier transform (DFT) is illustrated in Figure 13.53.

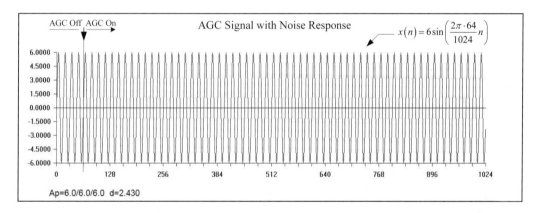

**Figure 13.52**   AGC simulation 6 pure sinusoidal input signal

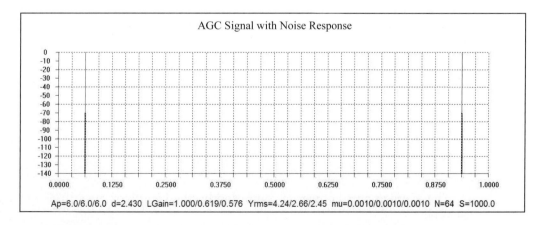

**Figure 13.53**   AGC simulation 6 pure sinusoidal input signal spectrum

The reader will note that we have chosen the frequency of the test signal so that it completes an integer number of cycles within the 1024-point simulation window. This eliminates the problems associated with spectral leakage and, for demonstration purposes, gives us an ideal test signal.

The simulation generates a noise component by using a standard programming function called *rand( )*. Successive calls to *rand( )* produce a random number that is somewhere within the range of 0 and *rand_max = 0x7FFF*. The random number is normalized to unity and multiplied by the desired maximum noise amplitude. Then the DC component is removed, as illustrated in Equation 13.24. For this simulation, we choose the maximum noise amplitude equal to 2, so we end up with

$$Nse(k) = (NoiseAmplitude)\left(\frac{(double)\,rand(\ )}{(double)\,RAND\_MAX}\right) - \left(\frac{NoiseAmplitude}{2}\right) \quad \{For\ 0 \le k < K\}$$

**Equation 13.24**

$$Nse(k) = 2.0\left(\frac{(double)\,rand(\ )}{(double)\,RAND\_MAX}\right) - 1.0 \quad \{For\ 0 \le k < K\}$$

**Equation 13.25**

The plot of the noisy input test signal is illustrated in Figure 13.54. We can see from the figure that the signal definitely has some hair on it. The amount of noise added is fairly excessive. We compute the *signal to noise ratio* or *SNR* of this signal by breaking the signal into its signal and noise components. The input signal is a pure sinusoid with an amplitude of 6 volts. We know from our study of Fourier series that a sinusoidal waveform can be represented by the sum of its Fourier coefficients $C_n$. In this case, there are two coefficients located at the normalized frequencies given by $\pm f_0 = \pm 0.0625$. Each component has the amplitude of $A/2$. The signal power can then be calculated by

$$Sig_{PWR} = \sum_{n=-\infty}^{\infty} |C_n|^2 = \left(\frac{A}{2}\right)^2 + \left(\frac{A}{2}\right)^2 = \frac{A^2}{2}$$

$$Sig_{PWR} = \frac{6^2}{2} = 18$$

The average noise power is computed by the simulator over 1024 samples to be

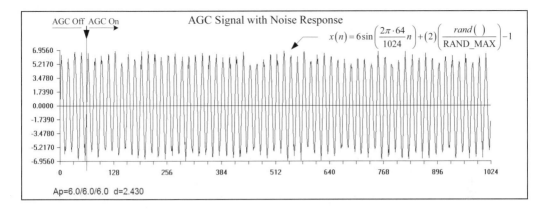

**Figure 13.54**   AGC simulation 6 sinusoidal input signal with noise

$$Nse_{PWR} = \frac{1}{K}\sum_{k=0}^{K-1} Nse^2 = \frac{1}{1024}\sum_{k=0}^{1023}\left[2\left(\frac{rand(\ )}{RAND\_MAX}\right)-1\right]^2 = 0.35$$

The SNR for this input signal is then computed to be

$$SNR = 10\mathrm{Log}_{10}\left(\frac{Sig_{PWR}}{Nse_{PWR}}\right) = 10\mathrm{Log}_{10}\left(\frac{18}{0.35}\right) \simeq 17.1 \text{ db}$$

**Equation 13.26**

The spectrum of the noisy input signal is computed with a 1024-point DFT and is illustrated in Figure 13.55. We can see the impulse-like spectra of the sinusoidal component of the input signal at the normalized frequencies 0.0625 and 0.9375. Since the DFT frequencies from beyond 0.5000 to 1.0000 are viewed as negative frequencies, the frequency at 0.9375 can be visualized as being the negative frequency $0.9375 - 1.0000 = -0.0625$. The DFT output is normalized to the largest bin magnitude, which sets the peak signal spectra at 0 db. The average noise level out of the DFT is plotted relative to the signal at –45.4 db. For this particular plot, we can state that the $SNR = 45.4$ db. This sure is a big change from the spectral plot of the pure sinusoid in Figure 13.53, where we essentially had an infinite SNR.

The reader is probably scratching his or her head right about now. We just stated that the SNR of the test signal was 17.1 db, and now we present a spectral plot obtained through the use of a 1024-point DFT that shows the SNR to be 45.4 db. What in the heck is going on here? The answer lies in the processing. Both the SNR of 17.1 db and 45.4 db are correct. The $SNR$ of the

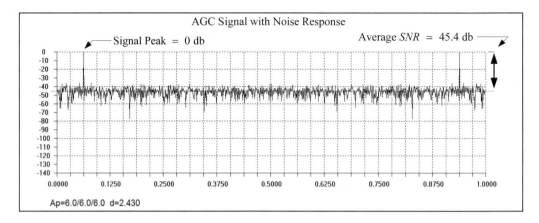

**Figure 13.55** AGC simulation 6 sinusoidal input signal with noise spectrum

signal input to the DFT is 17.1 db, and the *SNR* of the signal output from the DFT is 45.4 db. The difference between the *SNR* of the input and output signals is due to the huge factor of *N* processing gain we get from the *N* point DFT. This was discussed way back in Chapter 3, "Review of the Fourier Transform." For clarity, we will briefly summarize the mathematics used for computing the *SNR* of a signal processed by a DFT.

We consider the output of a DFT to be the frequency domain representation of a time domain signal. The output of a DFT is generally a string of *N* complex numbers. The peak value at the output of a DFT for a real sinusoid input signal that has an integer number of cycles within the *N* point DFT sample space is given by

$$S(f)_{PEAK} = \frac{AN}{2}$$

Notice the factor of *N* that was not present when we computed the value for $Sig_{PWR}$ in the time domain. This factor of *N* is the processing gain of the DFT. The noise in the signal processed by the DFT is not correlated with any of the *N* analysis frequencies used by the DFT. Therefore we do not see a factor of *N* performance enhancement for the noise samples. We can therefore state that for our $N = 1024$ point DFT the amplitude of the signal peak is given by

$$S(f)_{PEAK} = \frac{AN}{2} = \frac{(6)(1024)}{2} = 3072$$

We could have determined the exact same value if we would have searched every DFT output point to find the one with the maximum magnitude. In doing so, we would have used the following relation:

$$S(f)_{PEAK} = Max\left\{\sqrt{S^2(f)_{REAL} + S^2(f)_{IMAG}}\right\} \quad \{\text{For a } K \text{ point DFT}\}$$

In the same fashion, we can compute the average value of the spectral noise with the relation

$$N(f)_{AVG} = \frac{1}{K}\sum_{k=0}^{K-1}\sqrt{N^2(f)_{REAL} + N^2(f)_{IMAG}} \quad \{\text{For a } K \text{ point DFT}\}$$

Once we know the peak signal value and the average noise value, we can compute the SNR as

$$SNR_{AVG} = 20Log_{10}\left(\frac{S(f)_{PEAK}}{N(f)_{AVG}}\right) = 20Log_{10}\left(\frac{Max\left\{\sqrt{S^2(f)_{REAL} + S^2(f)_{IMAG}}\right\}}{\frac{1}{K}\sum_{k=0}^{K-1}\sqrt{N^2(f)_{REAL} + N^2(f)_{IMAG}}}\right)$$

The simulation produced the following results:

$$S(f)_{PEAK} = \frac{AN}{2} = \frac{(6)(1024)}{2} = 3072$$

$$N(f)_{AVG} = \frac{16941.59}{1024} = 16.5445$$

$$SNR_{AVG} = 20Log_{10}\left(\frac{3072}{16.5445}\right) \simeq 45.4 \text{ db}$$

The SNR of 45.4 db matches the simulation plot in Figure 13.55.

The reader should be cautioned here. The value of the SNR derived from the use of a DFT is indeed boosted by its processing gain. We could have used a 2048-point DFT, which is twice as long as the 1024-point DFT, and measured the SNR to be close to 48 db or approximately 3 db greater. So the SNR out of a DFT is not the actual *SNR* of the signal input to the DFT. It is the input *SNR* boosted by the DFT processing gain. This is one reason why the DFT is such an important signal-processing tool. It can be used to "pop" a periodic signal out of the noise for a great many signal-processing applications.

The processed signal out of the AGC is illustrated in Figure 13.56. The desired RMS value of the output signal was specified to be $d = 2.43$, which means the output signal amplitude should be equal to $(\sqrt{2})(2.43) = 3.43$. The horizontal line overlaid on the figure at approximately 3.43 volts gives us a visual idea of the deviation in output signal amplitude caused by the additive noise.

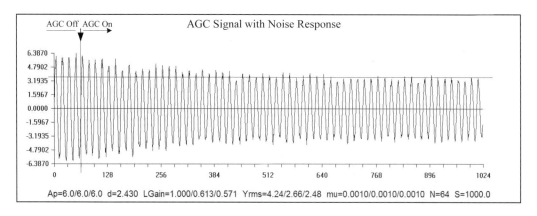

**Figure 13.56**   AGC simulation 6 processed output signal

One noteworthy observation is that the scaled loop gain has reduced the noise amplitude and therefore the output signal amplitude variation. A quick comparison of the input signal in Figure 13.54, and the processed output signal in Figure 13.56 shows a remarkable improvement in signal quality. The track of the loop gain for this example is illustrated in Figure 13.57.

The reader should note that loop response is very slow because of the long 64-sample RMS delay line and because of the very large value of $S = 1000$. The large $S$ forces the value of the feedback coefficient to be very small, in this case $\mu = 0.001$. The long RMS delay was used to help smooth the noise variance in the loop error computations.

The track of the loop error for the case where the input signal was not contaminated with additive noise is illustrated in Figure 13.58. The track of the loop error for the case where the input signal is corrupted with additive noise is illustrated in Figure 13.59. The noiseless error track is a smooth exponential transient from max error to essentially zero error, while the signal plus noise error track reflects the perturbations in the transient response due to noise in the RMS computations. The track resembles a noisy exponential that tries to adapt to the ideal solution of zero error. Because of the noise, however, the error signal will never maintain the ideal solution of zero. There will always be some variance about the zero mean.

Although not developed in this book, the AGC results for signal plus noise application such as the one we just discussed could be enhanced by replacing the RMS delay line that uses unity coefficients with a narrow band low pass filter coefficient vector $h(n)$. This would band limit the high-frequency additive noise and smooth the RMS estimate considerably. The relation we have used in this book to compute the RMS value of a signal is given by

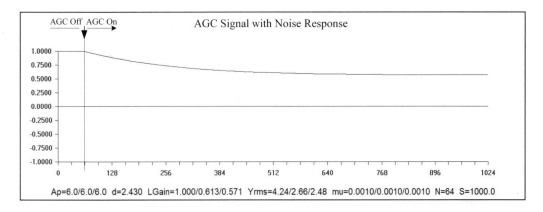

**Figure 13.57**   AGC simulation 6 track of the loop gain

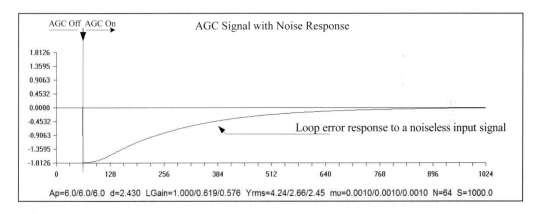

**Figure 13.58**   AGC simulation 6 track of the loop error without noise

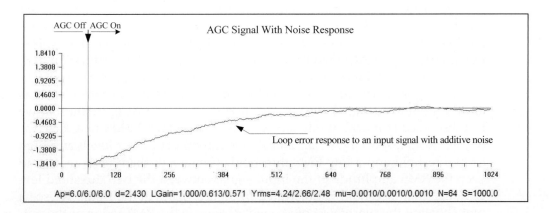

**Figure 13.59**   AGC simulation 6 track of the loop error with noise

$$y_{RMS} = \sqrt{\frac{1}{K}\sum_{k=0}^{K-1} y^2(k)}$$

**Equation 13.27**

It is clear that the $K$ samples $y(k)$ are multiplied by unity coefficients prior to being summed. We can enhance the performance of the RMS computation if we filter the samples $y(k)$. This would entail the multiplication of the input sequence by a low pass filter impulse response $h(n)$, as illustrated by

$$y_{RMS} = \sqrt{\frac{1}{K}\sum_{k=0}^{K-1}\left[h(h)y(k)\right]^2}$$

**Equation 13.28**

Use of a low pass filter would smooth the RMS estimates by eliminating high-frequency noise components. It should be noted that the length of the RMS delay line would be affected due to the length of the filter impulse response. However, for many applications, the overall performance of the AGC would be improved.

## 13.2  DESIGN OF A TYPE II RMS AGC CIRCUIT

In the previous discussion, we dealt with a type I AGC that iteratively adjusted the loop gain to force the RMS value of the output signal to equal the desired output RMS level $d$. The desired output $d$ was the optimum solution to the adaption problem that the AGC loop was tasked to solve. The type I AGC has its place in the world of AGC applications, but it is not the AGC of choice for many communications signals. An amplitude modulated signal is a good example where a type I AGC is not applicable. This is because the type I AGC would do its best to wipe off the modulation and, in the ideal world, leave only a constant amplitude carrier as the output signal. For signals that do have some legitimate amplitude variance, such as an AM modulated signal, we need to modify the type I AGC circuit that we just discussed to morph it into a type II AGC. The type II RMS AGC has two desired RMS levels labeled $d_H$ and $d_L$. We can think of these levels as a high RMS threshold and a low RMS threshold. The task of the type II AGC is to keep the RMS amplitude of the input signal between the two threshold levels. This dual threshold is illustrated in Figure 13.60. In the figure we see a plot showing the placement of $d_H$ and $d_L$ versus the RMS value of the output signal. The desired RMS levels $d_H$ and $d_L$ segment the plot into three

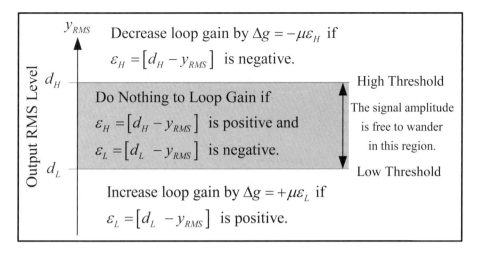

**Figure 13.60**  Type II RMS AGC thresholds

regions. The goal of the type II AGC is to keep the RMS value of the output signal within the middle or shaded region.

The type II AGC computes the error signals associated with $d_H$ and $d_L$:

$$\varepsilon_H = \left[ d_H - y_{RMS} \right]$$
$$\varepsilon_L = \left[ d_L - y_{RMS} \right]$$

The AGC monitors the sign of both error expressions and makes the following decision:

$$Sign(\varepsilon_H) = \quad - \quad \text{and} \quad y_{RMS} > d_H \quad \text{Decrease loop gain} \quad \Delta g = -\mu\varepsilon_H$$
$$Sign(\varepsilon_H\varepsilon_L) = \quad +- \quad \text{and} \quad d_L < y_{RMS} < d_H \quad \text{Do nothing} \quad \Delta g = 0$$
$$Sign(\varepsilon_L) = \quad + \quad \text{and} \quad y_{RMS} < d_L \quad \text{Increase loop gain} \quad \Delta g = +\mu\varepsilon_L$$

The block diagram of a type II AGC is illustrated in Figure 13.61. This figure represents only one of several possible implementations of a type II AGC. For example, the designer can choose to use two different values for $\mu$ (i.e., $\mu_H$ and $\mu_L$), so that independent control over the attack and decay time constants of the loop gain can be maintained. Many AGC circuits are designed with different attack and decay rates, and this would be one method of achieving that. Another option for the designer is to adjust the gain in fixed increments that are not dependent on the magnitude of the loop error—that is, to use incremental gain values that are fixed. For

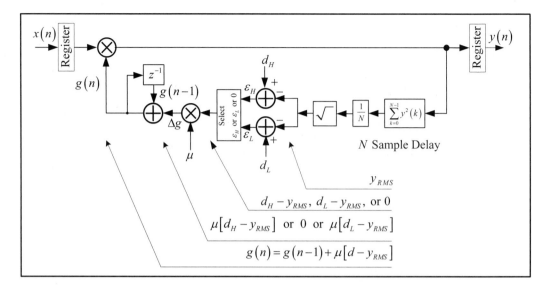

**Figure 13.61** Type II RMS AGC block diagram

example, one might choose the attack and decay incremental gain values to be $+\Delta g = 0.01$ and $-\Delta g = -0.015$.

As an example, suppose we wished to design an AGC circuit that restricts the RMS amplitude of an input signal to be between 13 and 23 volts. All we need do is program the two desired level thresholds to be $d_H = 23$, $d_L = 13$. As long as the input signal maintains an RMS voltage between those two limits, it is free to range anywhere in between without any change in circuit gain. Should the RMS voltage of the output signal exceed the $d_H$ threshold, the loop gain will be decreased, reducing the signal RMS amplitude until it falls within the shaded area of Figure 13.60. Should the RMS voltage of the output signal fall below the lower threshold set by $d_L$, the loop gain would be increased until the RMS amplitude of the signal is once again within the shaded region of the figure.

In summary, there are any number of design configurations for both the type I and type II AGC circuits, limited only by the imagination of the engineer doing the design. The design engineer should be cautious, however. Whenever designing a complex circuit, especially one that uses feedback, it is always a good idea to simulate the design before it is coded or fabricated. No matter how much you understand the circuit you're designing, there will always be some surprises that will cause a "rethink" of the design. It's always nice to catch these "design rethinks" before the design is set in concrete. A good rule of thumb that an engineer should remember is, "a one inch red line on a schematic can keep 20 support people employed for a week." Think of the effect that postfabrication changes can have on program cost, schedule, and ultimately your yearly evaluation.

The goal of this chapter was to avoid complex mathematics as much as possible and still convey enough information to provide an engineer with a baseline knowledge of dAGC design. Normally, feedback circuits such as a dAGC are studied using feedback control theory. Discussions on feedback control theory can fill an entire book. It is mathematically intensive and requires a strong familiarity with all the complex concepts involved. This chapter has provided examples of baseline AGC models. The reader can analyze these models in greater detail by using the tools provided by feedback control mathematics. For example, by using these tools, the reader would be able to derive the system poles and zeros and easily determine system stability and from that establish the optimum value of the feedback constant $\mu$. There are many excellent digital feedback control books on the market, some of which are included in this chapter's references.

## 13.3  REFERENCES

[1] Benjamin C. Kuo. *Digital Control Systems*. New York: Holt, Rinehart, and Winston, 1980.

[2] Charles Philips and H. Troy Nagle. *Digital Control System*. Englewood Cliffs, NJ: Prentice Hall, 1990.

[3] Gene Franklin and David Powell. *Digital Control*. Philippines: Addison-Wesley, 1980.

[4] Raymond Jacquot. *Modern Digital Control Systems*. New York: Marcel Dekker, 1981.

# APPENDIX A

# Mixed Language C/C++ FORTRAN Programming

```
extern "C"
{
  void __stdcall SUB1(int*, int*);
  int  __stdcall FUN1(int*, int*);
}
```

There is a good chance that engineers who have been in the business for a while have accumulated a few dusty old FORTRAN programs, functions, or subroutines that represent some pretty valuable legacy code. If these coded routines weren't considered valuable, the engineers more than likely would never have saved them. But typically these routines represent a treasure chest of tested, debugged, and proven code that is still relevant in today's engineering environment. The one big problem is that most of the software development today is in C or C++.

If this is the predicament that you find yourself in, there is some good news and some bad news. The good news is that you, the program manager, and rest of the design engineering staff have a wealth of proven FORTRAN code ready to be used. Incorporating this proven code into a project very well could result in a significant reduction in labor costs and a significant reduction in program schedule. The bad news, of course, is that C and C++ are pretty much the preferred languages in use today. Therefore writing deliverable code in FORTRAN is really not a viable option.

So what are you, the program manager, and the rest of the engineering staff to do in a situation such as this? One alternative is to build a mixed language program. The bulk of the program, including the main program, would be written in the language of choice, C/C++. Wherever necessary, the C main program could call legacy FORTRAN functions or subroutines. This approach incorporates the best attributes of both software worlds.

Mixed language programming is a reasonable, justifiable, and acceptable method of developing, testing, and delivering legitimate software products.

Because this method utilizes proven legacy code, it is the design choice of most forward-thinking program managers and program engineers.

This appendix presents a simple method for the writing, compiling, and linking of mixed language C/C++ / FORTRAN programs.

In Chapter 5, "Finite Impulse Response Digital Filtering," and Chapter 6, "Multirate Finite Impulse Response Filter Design," we spent a great deal of time discussing and using the legendary Parks-McClellan FORTRAN program used to develop optimum linear phase finite impulse response (FIR) digital filters.

Therefore it makes sense to spend some time in this appendix discussing how to incorporate this valuable legacy code into a C/C++ application. To support Chapters 5 and 6, this appendix contains a complete listing of the Parks-McClellan FORTRAN program that has been modified to convert it into a FORTRAN subroutine. At the end of this appendix, we will demonstrate a simple method for the compilation and linking of this valuable subroutine with a C/C++ main program.

There are two main areas we will cover:

- The first area has to do with the calling convention and argument passing between a C/C++ main and a FORTRAN function or subroutine. Argument types include fixed point variables, floating point variables, fixed and floating point arrays, and pointers to fixed and floating point variables and memory blocks. The procedure for passing character strings is a simple extension to this process and is left as an exercise for the reader.
- The second area has to do with the procedure for compiling the FORTRAN routines, linking these routines to the C/C++ main, and building the C/C++ main program. At this point, it is appropriate to state that the mixed language method described here is set of guidelines and procedures that have been researched and successfully implemented by the author. There exists a plethora of information loosely spread across the internet that addresses mixed language programming, but as far as can be determined, none of this information is condensed into a single standardized protocol. Microsoft provides some information on the subject, but it is sparse and distributed in bits and pieces throughout their C++ code development documentation.

With that said, the reader is cautioned that even though the procedure documented in this appendix enables the successful compilation and linkage of mixed C/C++ and FORTRAN code, the methods are not a standardized protocol.

The following guidelines and procedures have been tested and verified using the following tools:

- The C main programs are written using the Microsoft Visual C++ 6.0 Professional Compiler. The mixed language programs were compiled and linked by the same Visual C++ compiler.
- The FORTRAN functions and subroutines are written and/or compiled in Microsoft FORTRAN Power Station 4.0. The Parks-McClellan FORTRAN subroutine contained in this appendix was written way back in 1974 in FORTRAN IV.

## A.1  WRITING A C/C++ MAIN PROGRAM

For the most part, the C/C++ main program in a mixed language application isn't any different from any other standard, everyday, ordinary C/C++ main program. However, there are a few simple rules and procedures to follow in order to make the mixed language program compile and link successfully. The rules are listed here and their implementation is illustrated by the examples that follow:

1. The FORTRAN source code functions and subroutines should be compiled to produce object files that can be linked to the C/C++ main program.
2. The source code for these FORTRAN functions and subroutines should be centrally located in a separate file. In this appendix, that file is named **ftn_files.f90**.
3. The source file should be compiled in the FORTRAN complier to generate an object file, which in this appendix is named **ftn_files.obj**.
4. The FORTRAN object files can be accessed by the C main program only if these routines have been previously declared in the C main as having C linkage.
5. The FORTRAN object files should be in the same project workspace as the C/C++ source files.
6. The FORTRAN functions and subroutines must be prototyped outside the C/C++ main using the **extern "C"** syntax. Because of C++ name decoration of external symbols, the naming conventions for C++ are different than for C. The **extern "C"** syntax causes C++ to drop name decoration, making it possible for C++ to share data and routines with other languages.

## A.2  CALLING SUBROUTINES AND FUNCTIONS FROM A C/C++ MAIN

Suppose we have a file that we have named **ftn_files.f90**. Further suppose that this file contains the source code for all the FORTRAN functions and FORTRAN subroutines we intend to call from a C/C++ main program. And further suppose that this file is compiled to produce the object file **ftn_files. obj**. This file can be visualized as a kind of FORTRAN library. Now suppose this "library" file contained the object code for the FORTRAN subroutine,

```
SUBROUTINE SUB1(I,J)
```

where the arguments **I** and **J** are both integers. If this subroutine is to be linked to the C/C++ main program, we need to prototype this routine by utilizing an **extern** statement outside the C/C++ main program that has the form

```
extern "C" { void __stdcall SUB1(int*, int*);}
```

Note the use of "void" as the return type. This is necessary since the subroutine does not return a value. Also note that there are two underscore characters that precede the **stdcall** statement. Suppose the same "library" file contains the object code for the FORTRAN function,

```
INTEGER FUNCTION FUN1(I,J)
```

whose arguments **I** and **J** are both integers. We can prototype this function outside the C/C++ main program with an **extern** statement of the form

```
extern "C" { int __stdcall FUN1(int*, int*);}
```

Note the use of "**int**" as the return type since the function returns an integer value. If more than one subroutine or function is to be linked into the C main program, then the extern statements can be either written separately, as shown in Listing A.1, or combined into a single extern statement, as shown in Listing A.2.

## Listing A.1   Separate Extern Statements

```
extern "C" { void __stdcall SUB1(int*, int*);}
extern "C" { int  __stdcall FUN1(int*, int*);}
```

## Listing A.2   Combined Extern Statements

```
extern "C"
  {
    void __stdcall SUB1(int*, int*);
    int  __stdcall FUN1(int*, int*);
  }
```

The return type given in the function prototype must match the return type in declarations of the function outside the main C/C++ program.

The **extern "C"** syntax is only necessary when using a C++ compiler. To make the prototyping compatible with both C and C++ compilers, the **extern "C"** { } block can be enclosed in two short **#ifdef** blocks. For example, a prototyping block compatible with both C and C++ compilers can be written as shown in Listing A.3.

### Listing A.3   extern "C" syntax

```
#ifdef __cplusplus
extern "C" {
#endif

  void __stdcall SUB1(int*, int*);
  int  __stdcall FUN1(int*, int*);

#ifdef __cplusplus
}
#endif
```

For simplicity, all of the code listings from this point forward will be referred to in the general sense as being compatible with both C/C++ compilers and will continue to use the extern "C" statement

```
extern "C" { void __stdcall SUB1(int*, int*);}
```

with the understanding that it is equivalent to the **#ifdef** block shown in Listing A.3.

The C/C++ main program uses the **__stdcall** keyword to call the FORTRAN subroutines and functions with the correct calling convention. According to Microsoft, the C/C++ source code must use all uppercase names for the FORTRAN subroutines and functions.

Notice that in the function prototyping syntax, each argument passed in the call list has the form **<type * >**, such as **<int*>**. This specifies that the argument will be passed by reference. That is, the address of the argument is passed between C main and the FORTRAN subroutine or function. The reader should note that the **__stdcall** keyword looks to be compiler dependent, since it is incompatible with mixed language code tested using the Intel FORTRAN compiler.

The partial code of Listing A.4 shows how the FORTRAN subroutine SUB1 and function FUN1 are called from a C/C++ main program.

## Listing A.4   Calling a FORTRAN Subroutine and Function

```
// C/C++ main program    ***************************
extern "C" {void __stdcall SUB1(int*,int*);}
extern "C" {int  __stdcall FUN1(int*,int*);}

main( ) {
  ⋮
int i,j,c;
  ⋮
SUB1(&i,&j);
c=FUN1(&i,&j);
  ⋮
return;
}
```

This procedure is very straightforward. As the reader can see, there isn't a great deal of difference between a standard C/C++ main and a C/C++ main used in a mixed language environment.

## A.3  WRITING A FORTRAN SUBROUTINE

Coding a FORTRAN subroutine for mixed language programming that will be compiled and then linked to a C/C++ main program isn't any different than coding the same subroutine to be compiled and linked in a normal FORTRAN main program. For example, the simple subroutine SUB1 is coded as shown in Listing A.5.

## Listing A.5  Example FORTRAN Subroutine

```
C
C       FORTRAN SUBROUTINE SOURCE CODE
C
        SUBROUTINE SUB1(A,B)
        INTEGER A,B
        B=4*A
        RETURN
        END
```

In SUB1, an integer A is passed in the argument list. That integer is multiplied by four and then assigned to the integer B, which is returned to the calling program through the argument list. Note that both A and B are typed as INTEGER in the subroutine.

## A.4 WRITING A FORTRAN FUNCTION

Coding the FORTRAN Function for mixed language programming that will be compiled and then linked to a C/C++ main program isn't any different than coding the same function that is compiled and linked in a FORTRAN main program. For example, the simple FORTRAN function FUN1 is coded as shown in Listing A.6.

### Listing A.6   Example FORTRAN Function

```
      C
      C       FORTRAN FUNCTION SOURCE CODE
      C
              INTEGER FUNCTION FUN1(D,E)
              INTEGER D,E
              FUN1=10*D+E
              RETURN
              END
```

In FUN1, an integer D and an integer E are passed to the function in the argument list. The integer D is multiplied by 10 and summed with E. The quantity 10*D + E is assigned to FUN1 and returned as the computed value of the function. Once again, the arguments D and E are typed within the function as INTEGER. The value returned by the function is also typed INTEGER.

## A.5 PASSING INTEGER ARGUMENTS

Passing integer arguments from a C/C++ main to a FORTRAN function or subroutine is straightforward. For example, suppose our C main program calls the FORTRAN subroutine SUB1(A,B) shown in Listing A.5, where both A and B are integers.

The FORTRAN file containing the source code for function SUB1 would be added to the FORTRAN "library file," which we have arbitrarily defined to be **ftn_files.f90**. The source file would be compiled to produce the object file **ftn_files.obj** and then added to the C/C++ project workspace. The C/C++ main program could be coded as in Listing A.7.

### Listing A.7   Passing Integer Arguments to a FORTRAN Subroutine

```
// C/C++ main program ************************
extern "C" {void __stdcall SUB1(int*,int*);}
main( ) {
   ⋮
int a,b;
   ⋮
SUB1(&a,&b);
   ⋮
return;
}
```

Here are some important things to note in this listing:

- The FORTRAN subroutine is prototyped outside of the C/C++ main, using the **extern "C"** syntax, and both of the integer arguments are passed by reference.
- In the body of the C main program, the call to SUB1 prefixes each integer argument in the subroutine call list with the address operator **&**.

The exact same procedure is utilized when a C/C++ main calls a FORTRAN function. For example, suppose our C/C++ main program called the FORTRAN integer function FUN2 (A,B), coded as shown in Listing A.8.

### Listing A.8   Example FORTRAN Function with Integer Arguments

```
C
C        FORTRAN FUNCTION SOURCE CODE
C
         INTEGER FUNCTION FUN2(A,B)
         INTEGER A,B
         FUN2=A*B
         RETURN
         END
```

The FORTRAN file containing the source code for function FUN2 would be added to the FORTRAN library file **ftn_files.f90**. The source file would be compiled to produce the object file **ftn_files.obj** and then added to the C/C++ project workspace. The C/C++ main program listing could be written as shown in Listing A.9.

### Listing A.9   Passing Integer Arguments to a FORTRAN Function

```
// C/C++ main program ************************
extern "C" {int __stdcall FUN2(int*,int*);}
main( ) {
   ⋮
int a,b,c;
   ⋮
c=FUN2(&a,&b);
   ⋮
return;
}
```

The following are some important things to note about Listing A.9:

- The FORTRAN function is prototyped outside of the C/C++ main, using the **extern "C"** syntax, and both of the integer arguments are passed by reference.
- In the body of the main program, the call to FUN2 prefixes each integer argument with the address operator **&**.

As the reader can see, the procedure for calling FORTRAN subroutines and FORTRAN functions are exactly the same. The only difference is that FORTRAN functions return a value and therefore must be prototyped with a return type. FORTRAN subroutines do not return a value, except through the call argument list, and therefore always are prototyped with the return type **void**.

## A.6   PASSING FLOATING POINT ARGUMENTS

The procedure for the passing of floating point arguments to and from a called FORTRAN subroutine or function is identical to that described previously for the case where the argument list consists of integers. The only difference is in the change of variable types. For example, suppose our FORTRAN subroutine is coded as shown in Listing A.10.

### Listing A.10   Example FORTRAN Subroutine with Floating Point Arguments

```
C
C         FORTRAN SUBROUTINE SOURCE CODE
C
          SUBROUTINE SUB8(A, B, C, D)
          REAL*4 A,B
```

```
DOUBLE PRECISION C,D

B = 14.3 * A
D = 14.3 * C

RETURN
END
```

In Listing A.10 of the FORTRAN subroutine SUB8, the variables A and B are typed as **REAL*4**, declaring them to 4-byte words. Variables C and Dare typed as **DOUBLE PRECISION**, declaring them to 8-byte words. The equivalent types in C/C++ are **float** and **double**, respectively. The extern "C" syntax prototype is modified to reflect the type of the arguments and is given by

```
extern "C"{void __stdcall SUB8(float*,float*,double*,double*);}
```

This is the same procedure that is utilized for a C/C++ prototype of a FORTRAN function. The C/C++ main program code snippet shown Listing A.11 illustrates the relevant code necessary to call SUB8.

### Listing A.11  Passing Floating Point Arguments to a FORTRAN Subroutine

```
// C/C++ main program ******************************
extern "C" {void __stdcall SUB8(float*,float*,double*,double*);}

main( ) {
  ⋮
float a,b;
double c,d
a=3.3;
c=4.4;
  ⋮
SUB8(&a,&b,&c,&d);
  ⋮
return;
}
```

Once again note the **extern** statement is outside the main program, and all the variables in the SUB8 call list are preceded by the address operator **&**.

## A.7   PASSING ARRAY ARGUMENTS

Passing array arguments to a FORTRAN subroutine or function is just as simple as passing a variable. Suppose we have an array in the C/C++ main that we wish to pass to a FORTRAN subroutine called SUB9. Further suppose that this array contains **N** floating point values and we wish the subroutine to compute the mean value of all the elements in the array and return that value. There are two important issues to remember when passing an array as an argument:

- In C/C++, the name of the array is a pointer to the first element in the array.
- Arrays in FORTRAN begin at index 1, while arrays in C/C++ begin at index 0.

The FORTRAN subroutine SUB9 shown in Listing A.12 has three arguments. The array containing the **N** elements is named **ARRAY** and is typed as **DOUBLE PRECISION**. The number of elements in the array **N** is typed as **INTEGER**, and the calculated mean value **MEAN** is typed as **DOUBLE PRECISION** and returned to the calling program. The subroutine argument list is given by

```
SUBROUTINE SUB9(ARRAY,N,MEAN)
```

Notice in the FORTRAN subroutine that ARRAY is dimensioned as ARRAY(1). This maps the address of the first element in the FORTRAN array to the address of the first element in the C/C++ array. This mapping of addresses takes care of array starting index issue. Also note that we did a conversion from INTEGER to DOUBLE PRECISION when the variable D was equated to N. The listing for the FORTRAN subroutine is given in Listing A.12.

### Listing A.12   Example FORTRAN Subroutine with Array Arguments

```
C
C       FORTRAN SUBROUTINE SOURCE CODE
C
        SUBROUTINE SUB9(ARRAY,N,MEAN)
        DOUBLE PRECISION ARRAY(1), MEAN, SUM, D
        INTEGER N,J

        SUM = 0.0
        MEAN = 0.0
        D = N
```

```
         DO 100 J=1,N
     100 SUM = SUM + ARRAY(J)
         IF(N.GT.0) MEAN = SUM/D
         RETURN
         END
```

The FORTRAN subroutine is prototyped in the **extern "C"** syntax as

```
extern "C" { void __stdcall SUB9(double*, int*, double*);}
```

The C/C++ main program shown in Listing A.13 illustrates the relevant code snippets necessary to pass an array as a FORTRAN argument.

### Listing A.13   Passing Array Arguments to a FORTRAN Subroutine

```
// C/C++ main program *****************************
extern "C" {void __stdcall SUB9(double*,int*,double*);}

main( ) {
  ⋮
double dIN[ ] = {1.1, 2.2, 3.3, 4.4, 5.5}; // input array
double dMean;                    // calculated mean value
int    n=5;                      // number of array elements
  ⋮
SUB9(dIN, &n, &dMean);
  ⋮
return;
}
```

Notice that in the argument list in the SUB9 call statement the input array dIN[ ] is not prefixed with the address operator **&**. This is because the array name in C/C++ is a pointer to the address of the first element in the array.

## A.8   PASSING POINTER ARGUMENTS

Since the array name in C is a pointer to the first element in the array, passing an array as an argument should give us some insight into passing pointers to a FORTRAN subroutine or to a FORTRAN function. In this example, let's assume that we want to call a FORTRAN subroutine, input a block of double precision data labeled AIN, multiply each element in the block by two, and then return the result in another double precision data block labeled AOUT. The number of elements in the two arrays AIN and AOUT is passed as the integer N. The simple FORTRAN subroutine that accomplishes this feat is named SUB5 and is given in Listing A.14.

### Listing A.14   Example FORTRAN Subroutine with Array Arguments

```
C
C         FORTRAN SUBROUTINE SOURCE CODE
C
          SUBROUTINE SUB5(AIN,AOUT,N)
          DOUBLE PRECISION AIN(1),AOUT(1)
          INTEGER N, J

          DO 10 J   = 1,N
       10 AOUT(J) =   2*AIN(J)
          RETURN
          END
```

The source file for SUB5 is added to the FORTRAN source library file **ftn_files.f90**. The source file is compiled to produce an object file named **ftn_files.obj** and added to the C/C++ project workspace, as described previously. Once again we will use the **extern "C"** syntax outside the C/C++ main program to prototype the FORTRAN subroutine as

```
extern "C" { void __stdcall SUB5(double*,double*,int*);}
```

Note that once again all arguments are prototyped as being passed by reference. The relevant code in the C/C++ main program is illustrated in Listing A.15. The first line of code external to the C main is the **extern "C"** statement. The C/C++ main program declares two pointers of type double. These pointers will each point to a block of memory allocated from the heap.

The pointer *pAIN points to the heap memory that has been allocated to serve as our block of input memory, and the pointer *pAOUT points to the heap memory that has been allocated as our block of output memory. The next two major operations within the main body are to **calloc** enough memory from the heap to store **nElements** of type double for the memory pointed to by *pAIN and to **calloc** enough memory to store **nElements** of type double for the memory pointed to by *pAOUT. If **calloc** fails, the program should provide some sort of error report to the user. In this example, we inform the user via a "printf" statement. If calloc should fail for any reason, we free any memory previously allocated, display an error message, and then exit the program.

The next statement of interest is the initialization of the input memory. In this simple example, the memory is initialized to nothing more than an incremental count from 0 to nElements-1. Finally, the call is made to SUB5 to perform the simple memory-to-memory operation.

Note that the arguments **pAIN** and **pAOUT** are not preceded by the address operator **&**. This is because both of these arguments are pointers, and they are already pointing to the address of the first element in their respective data blocks.

Also note that the arrays within the FORTRAN subroutine are typed as DOUBLE PRECISION and that they are dimensioned AIN(1) and AOUT(1). This maps the address pointed to by **pAIN** to AIN(1) and the address pointed to by **pAOUT** to AOUT(1).

### Listing A.15   Passing Pointer Arguments to a FORTRAN Subroutine

```
extern "C" { void __stdcall SUB5(double*,double*,int*);}

main( )
    ⋮

double *pAIN,*pAOUT;  // pointers to input and output arrays
int i;                // general purpose index
int nElements = 32;   // number of elements in each buffer
    ⋮

//**** allocate heap memory for the input data buffer pointed
//**** to by *pAIN
if(NULL==(pAIN=(double*) calloc(nElements,sizeof(double)))){
printf("Could not allocate heap memory for pAIN")
return(0);
}

//**** allocate heap memory for the output data buffer pointed
//**** to by *pAOUT
if(NULL==(pAOUT=(double*) calloc(nElements,sizeof(double)))){
printf("Could not allocate heap memory for pAOUT")
free(pAIN);
return(0);
}
    ⋮

//**** initialize the input buffer

for(i=0;i<nElements;i++) *(pAIN + i) = (double) i;

SUB5(pAIN,pAOUT,&nElements);
```

```
        ⋮
//**** tidy up before leaving
 free(pAIN);
 free(pAOUT);
 return;
 }
```

## A.9   COMPILE/LINK MIXED LANGUAGE C/C++ FORTRAN PROGRAMS

The compilation and linkage of FORTRAN routines into a C main program depends, to some degree, on the manufacturer of the C and FORTRAN compilers and on the development environment the reader may employ. The author utilized the Microsoft Visual C++ 6.0 Professional compiler and the Microsoft FORTRAN Power Station 4.0 compiler. The procedure presented here is relevant to these two tools. The procedure may or may not have to be tweaked a bit for other development environments.

The Visual C++ compiler and the FORTRAN compiler are installed in their own separate directories. This separation of development tools presents some minor issues, probably considered to be negligible by some, in that it prevents the two tools from sharing resources. These problems are solved with a few extra keystrokes when it comes time to build the final application. With this in mind, let's proceed with an orderly set of steps that are necessary to build a mixed language C/C++ / FORTRAN program:

1. Using the Visual C development tools, create a C/C++ project and generate the C/C++ main program code.
2. Using the FORTRAN development tools, create a separate FORTRAN project and either generate the FORTRAN code or copy existing legacy FORTRAN code. Add this source code to the source library file, which in this appendix is named **ftn_files.f90**.
3. Add the following Microsoft Power Station library routines to the C/C++ project:
   **a.** libf.lib
   **b.** console.lib
   **c.** portlib.lib
4. In the Visual C++ 6.0 development environment, adding files is very simple:
   **a.** In the menu bar, click on "**Project**".
   **b.** Then in successive drop down menus click on "**Add To Project**", and then click on "**Files**".

   **c.** A "**Search for Files**" dialog box pops up that enables the user to search for the desired files to add to the C/C++ project.

   **d.** The Microsoft Power Station 4.0 library files are located by default in the path "**C:\MSDEV\LIB**". This path will be different if the reader installed power station in some other directory.

5. Try to keep all the FORTRAN source files collected in a single project file:

   **a.** For example, let's assume we named the collection file **<ftn_files. f90>**, where the file extension is determined by the development system.

   **b.** Use the FORTRAN development tools to compile <ftn_files.f90> to produce a FORTRAN object file most likely named <ftn_files.obj>.

6. Move the FORTRAN object file to the C/C++ workspace:

   **a.** Add the <ftn_files.obj> file to the C/C++ project using the same procedure outlined in step 4.

   **b.** Use the C/C++ development system to compile the C/C++ source code and link the FORTRAN object code.

7. Assuming the code for both languages is completely debugged, step 6 should produce a successful mixed language executable that will run without error.

## A.10 PARKS-MCCLELLAN FORTRAN SUBROUTINE CALLED FROM C MAIN

This section is written specifically to provide support for Chapter 5, "Finite Impulse Response Digital Filtering," and Chapter 6, "Multirate Finite Impulse Response Filter Design," which heavily depend on the use of the Parks-McClellan FORTRAN program to design optimum linear phase FIR filters. If the reader has access to other filter design application software and has no need to develop his or her own filter design code, then this section can be skipped.

   The Parks-McClellan FORTRAN program for computing optimum linear phase digital filters has been around in one form of legacy code or another since December 1973. It can be found in practically every signal processing engineer's toolbox. It's probably in the possession of a huge majority of engineers who were actively involved in designing digital filters back in the 1970s and 1980s. Most of the glossy, Windows-based filter design programs available today use either the Parks-McClellan program or some variant of it. Most of these programs also allow for the design of filters using less efficient and less productive methods such as the window method. These programs are pretty expensive, and most users will only utilize the Parks-McClellan portion of it.

An argument, therefore, can be made for the reader to write his or her own filter design program that utilizes the Parks-McClellan program.

A fair amount of engineers today feel that use of proven legacy FORTRAN code is not fashionable in today's engineering environment. This may be true for many cases, but it is not necessarily true for all. The FORTRAN version of the Parks-McClellan filter design program has been around for years. It is tried and true and is backed up by tons of documentation. This program is documented as an industry standard in just about every digital signal processing (DSP) book written since 1974. The only thing an engineer needs to do to utilize this valuable legacy code in a mixed language environment is to add one line of code to a C main program. This is nothing more than a simple extern "C" { } statement. The user may also opt to comment out the FORTRAN code in the output section of the program. Since this code is a contiguous set of lines, it takes less than a minute to do. The only investment made to use this valuable program is the time it takes to write one line of code and comment out or remove the program output section, as the author has done.

The alternative is to use a variant of this program written in C, found somewhere on the internet, that may or may not be reliable, has not been thoroughly reviewed, is not documented in any DSP text, and has not been extensively used. This can be quite a big risk. An engineer should look at the trade-offs prior to investing in unproven and uncorroborated code.

With that said, there is a C version of the Parks-McClellan filter design program available on the internet, authored by Jake Janovetz. It is free software found on the internet that can be redistributed and or modified under the terms of the GNU library General Public License, as published by the Free Software Foundation. This library is distributed without any warranty of merchantability or fitness for a particular purpose. The author has downloaded this code and tested it against the original FORTRAN program by Parks and McClellan. The C program produces exactly the same floating point coefficients as the FORTRAN program, but it does not provide any auxiliary information such as computed filter band deviations. The engineer who uses this C program would have to compute the deviation in each filter band separately in order to verify compliance with design specification.

The intent here is not to discuss the Parks-McClellan program but rather to demonstrate how it can be embedded in a mixed language program. We will demonstrate how to compile and link this valuable piece of legacy code to a modern day C program. A detailed discussion is presented on the use of the Parks-McClellan program in Chapter 5, "Finite Impulse Response Digital Filtering." There are many texts and papers available that delve deep into the physics of this historic program. Some of these sources are included in the references at the end of this appendix.

The reader should note that the included Parks-McClellan FORTRAN subroutine is vintage 1974 and therefore contains an output section that produces a line printer output. This section contains several FORTRAN WRITE and FORMAT statements. If the reader is inclined to include this routine in a mixed language program, it is suggested that all these WRITE and FORMAT statements be commented out, which is what this author has done. The data normally output by the subroutine can be passed back to the calling program through the argument list so that it may be output to a file, a display, or a printer either by standard C/C++ I/O or by the use of Windows application programming interfaces (APIs) for graphical user interface (GUI) applications. The 1974 version of the Parks- McClellan subroutine that outputs data via the line printer using FORTRAN WRITE statements originally had the call list

```
PARK(NFILT,JTYPE,NBANDS,LGRID,dEdge,Fx,Wtx,pH)
```

In this appendix, all the WRITE and FORMAT statements in the modified subroutine have been left in the routine but have been commented out. In their place, the design variables necessary to output a meaningful documentation disk file have been brought out via the call list so that they may be handled by standard C I/O. The modified subroutine call list formatted for use within a C main is shown here. The Parks-McClellan FORTRAN subroutine is illustrated in Listing A.16.

```
PARK(&iNFILT,&iJTYPE,&iNBANDS,&iLGRID,dEdge,dFx,dWtx,pH,
     &iNFCNS,&iNEG,&iNODD,&iNZ,&iKUP,&dDev,&iSTATUS);
```

Alternatively, instead of commenting out the subroutine I/O as we have done, the FORTRAN subroutine may be modified directly so that data within the routine can be output to a file directly.

For clarity, the C code used here to format the routine output for storage to disk file is presented in Listing A.17. If the reader only has the need for the filter coefficients produced by the Parks-McClellan routine and can do without the filter band information, then the call list arguments

```
&iNFCNS,&iNEG,&iNODD,&iNZ,&iKUP,&dDev,&iSTATUS
```

can be removed and their use in the C main provided at the end of this appendix can be deleted.

The method of function prototyping, rules of argument passing, and program compile and linkage procedures are identical to those described for the example programs discussed previously.

## A.10.1   The Parks-McClellan FORTRAN Subroutine

The Parks-McClellan FORTRAN algorithm for the design of optimum linear phase digital filters in the form of a subroutine is illustrated in Listing A.16. This listing is a modified version of the original Parks-McClellan program. The code was modified to convert the program into a FORTRAN subroutine and has been used in the industry since the mid-1970s. The modified code also has the original FORTRAN WRITE and FORMAT statements commented out. In its place, certain subroutine variables have been passed back to the calling program via the call list to aid in the generation of an output disk file that documents the filter design parameters and the filter coefficients.

The original Parks McClellan program was published in the paper titled "A Computer Program for Designing Optimum FIR Linear Phase Digital Filters" in 1973 by the IEEE [1]. It is reprinted here with permission from *IEEE Transactions on Audio and Electroacoustics*, vol. AU-21, no. 6, December 1973.

This subroutine is included for the support of Chapter 5, "Finite Impulse Response Digital Filtering," and Chapter 6, "Multirate Finite Impulse Response Filter Design."

**Listing A.16   Modified Parks-McClellan FORTRAN Subroutine**

```
      SUBROUTINE PARK(NFILT,JTYPE,NBANDS,LGRID,EDGE,FX,WTX,H,
    1                 ENFCNS,NEG,NODD,NZ,KUP,EDEV,STATUS)
C   PROGRAM FOR THE DESIGN OF LINEAR PHASE FINITE IMPULSE
C   RESPONSE (FIR) FILTERS USING THE REMEZ EXCHANGE ALGORITHM
C   JIM MCCLELLAN, RICE UNIVERSITY, APRIL 13, 1973
C   THREE TYPES OF FILTERS ARE INCLUDED--BANDPASS FILTERS
C   DIFFERENTIATORS, AND HILBERT TRANSFORM FILTERS
C
C   INPUT ARGUMENTS
C
C   NFILT  = FILTER LENGTH (I.E. NUMBER OF TAPS)
C   JTYPE  = FILTER TYPE  1 = MULTIPLE PASSBAND/STOPBAND
C                         2 = DIFFERENTIATOR
C                         3 = HILBERT TRANSFORMER
C   NBANDS = NUMBER OF BANDS IN THE TRANSFER FUNCTION
C   LGRID  = GRID DENSITY
C   EDGE   = BANDEDGES, LOWER AND UPPER EDGES FOR EACH
C            BAND WITH A MAXIMUM OF 10 BANDS.
C   FX     = DESIRED FUNCTION FOR A FIR FILTER, OR DESIRED
C            SLOPE FOR A DIFFERENTIATOR.
C   WTX    = WEIGHT FUNCTION IN EACH BAND. FOR A DIFFERENTIATOR
C            THE WEIGHT FUNCTION IS INVERSLY PROPORTIONAL TO F.
C
```

```
C   OUTPUT ARGUMENTS
C
C   H       =   THE ARRAY OF COMPUTED FILTER COEFFICIENTS
C   ENFCNS  =   THE NUMBER OF COSINE APPROXIMATION FUNCTIONS

C   THE FOLLOWING INPUT DATA SPECIFIES A LENGTH 32 BANDPASS
C   FILTER WITH STOPBANDS 0 TO 0.1 AND 0.425 TO 0.5, AND
C   PASSBAND FROM 0.2 TO 0.35 WITH WEIGHTING OF 10 IN THE
C   STOPBANDS AND 1 IN THE PASSBAND.  THE IMPULSE RESPONSE
C   WILL BE PUNCHED AND THE GRID DENSITY IS 32.
C
C   SAMPLE INPUT DATA SETUP
C
C   NFILT = 32
C   JTYPE = 1
C   NBANDS= 3
C   LGRID = 32
C   EDGE  = 0.0 , 0.1 , 0.2 , 0.35 , 0.425 , 0.5
C   FX    = 0,1,0
C   WTX   = 10,1,10
C
C   THE FOLLOWING INPUT DATA SPECIFIES A LENGTH 32 WIDEBAND
C   DIFFERENTIATOR WITH SLOPE 1 AND WEIGHTING OF 1/F.  THE
C   IMPULSE RESPONSE WILL NOT BE PUNCHED AND THE GRID
C   DENSITY IS ASSUMED TO BE 16.
C
C   NFILT = 32
C   JTYPE = 2
C   NBANDS= 1
C   LGRID = 16
C   EDGE  = 0.0 , 0.5
C   FX    = 1.0
C   WTX   = 1.0
C
        COMMON PI2,AD,DEV,X,Y,GRID,DES,WT,ALPHA,IEXT,NFCNS,NGRID
        DIMENSION IEXT(130),AD(130),ALPHA(130),X(130),Y(130)
        DIMENSION DES(2080),GRID(2080),WT(2080)
        DIMENSION EDGE(1),FX(1),WTX(1),H(1),DEVIAT(10)
        DOUBLE PRECISION FUP, DES,GRID,WT,ALPHA,AD,DEV,X,Y
        DOUBLE PRECISION PI2,PI
        DOUBLE PRECISION WATE,EFF
        DOUBLE PRECISION CHANGE,DELF,TEMP,H
```

```
         DOUBLE PRECISION EDEV
         INTEGER ENFCNS, STATUS

CC       DIMENSION H(256)
CC       DIMENSION DES(1045),GRID(1045),WT(1045)
CC       DIMENSION IEXT(66),AD(66),ALPHA(66),X(66),Y(66)
CC       INTEGER COEF(64),IFNM(6)
CC       DOUBLE PRECISION EDGE(20),FX(10),WTX(10)
         DOUBLE PRECISION H(256),DEVIAT(10)

         PI2=6.283185307179586D0
         PI=3.141592653589793D0
         LUO=6
         LUDIST =LUO
         STATUS = 0
C
C    THE PROGRAM IS SET UP FOR A MAXIMUM LENGTH OF 128, BUT
C    THIS UPPER LIMIT CAN BE CHANGED BY REDIMENSIONING THE
C    ARRAYS IEXT, AD, ALPHA, X, Y, H TO BE NFMAX/2 + 2.
C    THE ARRAYS DES, GRID, AND WT MUST DIMENSIONED
C    16(NFMAX/2 + 2).
C
         NFMAX=256
C
         IF(NFILT.GT.NFMAX.OR.NFILT.LT.3) CALL ERROR
         IF(JTYPE.LT.1.OR.JTYPE.GT.3)CALL ERROR
         IF(NBANDS.LE.0.OR.NBANDS.GT.10) CALL ERROR
C
C    GRID DENSITY IS ASSUMED TO BE 16 UNLESS SPECIFIED
C    OTHERWISE
C
         IF(LGRID.LE.0.0)LGRID = 16
         JB=2*NBANDS
         IF(JTYPE.EQ.0) CALL ERROR
         NEG=1
         IF(JTYPE.EQ.1) NEG=0
         NODD=NFILT/2
         NODD=NFILT-2*NODD
         NFCNS=NFILT/2
         IF(NODD.EQ.1.AND.NEG.EQ.0) NFCNS=NFCNS+1
C
C    SET UP THE DENSE GRID.  THE NUMBER OF POINTS IN
```

```
C   THE GRID IS (FILTER LENGTH + 1)*GRID DENSITY/2
C
      GRID(1)=DBLE(EDGE(1))
      DELF=DBLE(FLOAT(LGRID*NFCNS))
      DELF=0.5D0/DELF
      IF(NEG.EQ.0) GO TO 135
      IF(DBLE(EDGE(1)).LT.DELF) GRID(1)=DELF
135   CONTINUE
      J=1
      L=1
      LBAND=1
140   FUP=DBLE(EDGE(L+1))
145   TEMP=GRID(J)
C
C   CALCULATE THE DESIRED MAGNITUDE RESPONSE AND THE WEIGHT
C   FUNCTION ON THE GRID
C
      DES(J)=EFF(TEMP,FX,WTX,LBAND,JTYPE)
      WT(J)=WATE(TEMP,FX,WTX,LBAND,JTYPE)
      J=J+1
      GRID(J)=TEMP+DELF
      IF(GRID(J).GT.FUP) GO TO 150
      GO TO 145
  150 GRID(J-1)=FUP
      DES(J-1)=EFF(FUP,FX,WTX,LBAND,JTYPE)
      WT(J-1)=WATE(FUP,FX,WTX,LBAND,JTYPE)
      LBAND=LBAND+1
      L=L+2
      IF(LBAND.GT.NBANDS) GO TO 160
      GRID(J)=DBLE(EDGE(L))
      GO TO 140
  160 NGRID=J-1
      IF(NEG.NE.NODD) GO TO 165
      IF(GRID(NGRID).GT.(0.5D0-DELF)) NGRID=NGRID-1
  165 CONTINUE
C
C   SET UP A NEW APPROXIMATION PROBLEM WHICH IS EQUIVALENT
C   TO THE ORIGINAL PROBLEM
C
      IF(NEG) 170,170,180
  170 IF(NODD.EQ.1) GO TO 200
      DO 175 J=1,NGRID
      CHANGE=DCOS(PI*GRID(J))
```

```
          DES(J)=DES(J)/CHANGE
      175 WT(J)=WT(J)*CHANGE
          GO TO 200
      180 IF(NODD.EQ.1) GO TO 190
          DO 185 J=1,NGRID
          CHANGE=DSIN(PI*GRID(J))
          DES(J)=DES(J)/CHANGE
      185 WT(J)=WT(J)*CHANGE
          GO TO 200
      190 DO 195 J=1,NGRID
          CHANGE=DSIN(PI2*GRID(J))
          DES(J)=DES(J)/CHANGE
      195 WT(J)=WT(J)*CHANGE
C
C   INITIAL GUESS FOR THE EXTREMAL FREQUENCIES--EQUALLY
C   SPACED ALONG THE GRID
C
      200 TEMP=DBLE(FLOAT(NGRID-1))/DBLE(FLOAT(NFCNS))
          DO 210 J=1,NFCNS
      210 IEXT(J)=(J-1)*TEMP+1
          IEXT(NFCNS+1)=NGRID
          NM1=NFCNS-1
          NZ=NFCNS+1
C
C   CALL THE REMEZ EXCHANGE ALGORITHM TO DO THE
C   APPROXIMATION PROBLEM
C
          CALL REMEZ(EDGE,NBANDS)
C
C   CALCULATE THE IMPULSE RESPONSE.
C
          IF(NEG) 300,300,320
      300 IF(NODD.EQ.0) GO TO 310
          DO 305 J=1,NM1
      305 H(J)=0.5D0*ALPHA(NZ-J)
          H(NFCNS)=ALPHA(1)
          GO TO 350
      310 H(1)=0.25D0*ALPHA(NFCNS)
          DO 315 J=2,NM1
      315 H(J)=0.25D0*(ALPHA(NZ-J)+ALPHA(NFCNS+2-J))
          H(NFCNS)=0.5D0*ALPHA(1)+0.25D0*ALPHA(2)
          GO TO 350
      320 IF(NODD.EQ.0) GO TO 330
```

```
          H(1)=0.25D0*ALPHA(NFCNS)
          H(2)=0.25D0*ALPHA(NM1)
          DO 325 J=3,NM1
      325 H(J)=0.25D0*(ALPHA(NZ-J)-ALPHA(NFCNS+3-J))
          H(NFCNS)=0.5D0*ALPHA(1)-0.25D0*ALPHA(3)
          H(NZ)=0.0D0
          GO TO 350
      330 H(1)=0.25D0*ALPHA(NFCNS)
          DO 335 J=2,NM1
      335 H(J)=0.25D0*(ALPHA(NZ-J)-ALPHA(NFCNS+2-J))
          H(NFCNS)=0.5D0*ALPHA(1)-0.25D0*ALPHA(2)
C
C  PROGRAM OUTPUT SECTION.
C  THE FORTRAN LINE PRINTER OUTPUT SECTION IS COMMENTED OUT.
C  ALL OUTPUT DATA ARRAYS ARE PASSED TO THE CALLING PROGRAM
C  VIA THE SUBROUTINE CALL LIST.
C
C 350 WRITE(LUO,360)
C 360 FORMAT(1H1,70(1H*)//25X,'FINITE IMPULSE RESPONSE (FIR)'/
C     125X,'LINEAR PHASE DIGITAL FILTER DESIGN'/
C     225X,'REMEZ EXCHANGE ALGORITHM'/)
C     IF(JTYPE.EQ.1) WRITE(LUO,365)
C 365 FORMAT(25X,'BANDPASS FILTER'/)
C     IF(JTYPE.EQ.2) WRITE(LUO,370)
C 370 FORMAT(25X,'DIFFERENTIATOR'/)
C     IF(JTYPE.EQ.3) WRITE(LUO,375)
C 375 FORMAT(25X,'HILBERT TRANSFORMER'/)
C     WRITE(LUO,378) NFILT
C 378 FORMAT(15X,'FILTER LENGTH = ',I3/)
C     WRITE(LUO,380)
C 380 FORMAT(15X,'***** IMPULSE RESPONSE *****')
C     DO 381 J=1,NFCNS
C     K=NFILT+1-J
C     IF(NEG.EQ.0) WRITE(LUO,382) J,H(J),K
C     IF(NEG.EQ.1) WRITE(LUO,383) J,H(J),K
C 381 CONTINUE
C 382 FORMAT(20X,'H(',I3,') = ',E15.8,' = H(',I4,')')
C 383 FORMAT(20X,'H(',I3,') = ',E15.8,' = -H(',I4,')')
C      IF(NEG.EQ.1.AND.NODD.EQ.1) WRITE(LUO,384) NZ
C 384 FORMAT(20X,'H(',I3,') = 0.0')
C     DO 450 K=1,NBANDS,4
C     KUP=K+3
C     IF(KUP.GT.NBANDS) KUP=NBANDS
```

```
C      WRITE(LUO,385) (J,J=K,KUP)
C 385 FORMAT(/24X,4('BAND',I3,8X))
C      WRITE(LUO,390) (EDGE(2*J-1),J=K,KUP)
C 390 FORMAT(2X,'LOWER BAND EDGE',5F15.9)
C      WRITE(LUO,395) (EDGE(2*J),J=K,KUP)
C 395 FORMAT(2X,'UPPER BAND EDGE',5F15.9)
C      IF(JTYPE.NE.2) WRITE(LUO,400) (FX(J),J=K,KUP)
C 400 FORMAT(2X,'DESIRED VALUE',2X,5F15.9)
C      IF(JTYPE.EQ.2) WRITE(LUO,405) (FX(J),J=K,KUP)
C 405 FORMAT(2X,'DESIRED SLOPE',2X,5F15.9)
C      WRITE(LUO,410) (WTX(J),J=K,KUP)
C 410 FORMAT(2X,'WEIGHTING',6X,5F15.9)
C      DO 420 J=K,KUP
C 420 DEVIAT(J)=DEV/WTX(J)
C      WRITE(LUO,425) (DEVIAT(J),J=K,KUP)
C 425 FORMAT(2X,'DEVIATION',6X,5F15.9)
C      IF(JTYPE.NE.1) GO TO 450
C      DO 430 J=K,KUP
C 430 DEVIAT(J)=20.0*ALOG10(DEVIAT(J))
C      WRITE(LUO,435) (DEVIAT(J),J=K,KUP)
C 435 FORMAT(2X,'DEVIATION IN DB',5F15.9)
C 450 CONTINUE
C      WRITE(LUO,455) (GRID(IEXT(J)),J=1,NZ)
C 455 FORMAT(/2X,'EXTREMAL FREQUENCIES'/(2X,5F12.7))
C      WRITE(LUO,460)
C 460 FORMAT(/1X,70(1H*)/1H1)
  350 CONTINUE
      ENFCNS = NFCNS
      EDEV   = DEV

      NEWB=NFILT/2
      DO 470 ICOEF=1,NEWB
      LCOEF=NFILT-ICOEF+1
      IF(NEG.EQ.0.0)H(LCOEF)=H(ICOEF)
      IF(NEG.EQ.1.0)H(LCOEF)=-H(ICOEF)
  470 CONTINUE
      RETURN
      END

      DOUBLE PRECISION FUNCTION EFF(TEMP,FX,WTX,LBAND,JTYPE)
C
C  FUNCTION TO CALCULATE THE DESIRED MAGNITUDE RESPONSE
C  AS A FUNCTION OF FREQUENCY.
```

```
C
CC    DIMENSION FX(10),WTX(10)
      DIMENSION FX(1),WTX(1)
      DOUBLE PRECISION TEMP
      IF(JTYPE.EQ.2) GO TO 1
      EFF=DBLE(FX(LBAND))
      RETURN
    1 EFF=DBLE(FX(LBAND))*TEMP
      RETURN
      END
      DOUBLE PRECISION FUNCTION WATE(TEMP,FX,WTX,LBAND,JTYPE)
C
C  FUNCTION TO CALCULATE THE WEIGHT FUNCTION AS A FUNCTION
C  OF FREQUENCY.
C
CC    DIMENSION FX(10),WTX(10)
      DIMENSION FX(1),WTX(1)
      DOUBLE PRECISION TEMP
      IF(JTYPE.EQ.2) GO TO 1
      WATE=DBLE(WTX(LBAND))
      RETURN
    1 IF(DBLE(FX(LBAND)).LT.0.0001D0) GO TO 2
      WATE=DBLE(WTX(LBAND))/TEMP
      RETURN
    2 WATE=DBLE(WTX(LBAND))
      RETURN
      END


      SUBROUTINE ERROR
C     WRITE(LUDISP,1)
C   1 FORMAT(' ******** ERROR IN INPUT DATA ********')
      STOP
      END


      SUBROUTINE REMEZ(EDGE,NBANDS)
C
C  THIS SUBROUTINE IMPLEMENTS THE REMEZ EXCHANGE ALGORITHM
C  FOR THE WEIGHTED CHEBYCHEV APPROXIMATION OF A CONTINUOUS
C  FUNCTION WITH A SUM OF COSINES.  INPUTS TO THE SUBROUTINE
C  ARE A DENSE GRID WHICH REPLACES THE FREQUENCY AXIS, THE
C  DESIRED FUNCTION ON THIS GRID, THE WEIGHT FUNCTION ON THE
C  GRID, THE NUMBER OF COSINES, AND AN INITIAL GUESS OF THE
```

```fortran
C  EXTREMAL FREQUENCIES.  THE PROGRAM MINIMIZES THE CHEBYCHEV
C  ERROR BY DETERMINING THE BEST LOCATION OF THE EXTREMAL
C  FREQUENCIES (POINTS OF MAXIMUM ERROR) AND THEN CALCULATES
C  THE COEFFICIENTS OF THE BEST APPROXIMATION.
C
      COMMON PI2,AD,DEV,X,Y,GRID,DES,WT,ALPHA,IEXT,NFCNS,NGRID
      DIMENSION EDGE(1)
      DIMENSION IEXT(130),AD(130),ALPHA(130),X(130),Y(130)
      DIMENSION DES(2080),GRID(2080),WT(2080)
      DIMENSION A(130),P(130),Q(130)

      DOUBLE PRECISION DES,GRID,WT,ALPHA
      DOUBLE PRECISION CN,FSH
      DOUBLE PRECISION PI2,DNUM,DDEN,DTEMP,A,P,Q
      DOUBLE PRECISION AD,DEV,X,Y
      DOUBLE PRECISION D,ARCOS,GEE
      DOUBLE PRECISION AA,BB,COMP,DELF,DEVI,DEVL
      DOUBLE PRECISION ERR,FT,GTEMP,XE,XT,YNZ,Y1

CC    DIMENSION IEXT(66),AD(66),ALPHA(66),X(66),Y(66)
CC    DIMENSION DES(1045),GRID(1045),WT(1045)
C
C  THE PROGRAM ALLOWS A MAXIMUM NUMBER OF ITERATIONS OF 25
C
      ITRMAX=25
      DEVL=-1.0D0
      NZ=NFCNS+1
      NZZ=NFCNS+2
      NITER=0
 100  CONTINUE
      IEXT(NZZ)=NGRID+1
      NITER=NITER+1
      IF(NITER.GT.ITRMAX) GO TO 400
      DO 110 J=1,NZ
      DTEMP=GRID(IEXT(J))
      DTEMP=DCOS(DTEMP*PI2)
 110  X(J)=DTEMP
      JET=(NFCNS-1)/15+1
      DO 120 J=1,NZ
      JJ=J
 120  AD(J)=D(JJ,NZ,JET)
      DNUM=0.0D0
      DDEN=0.0D0
```

```
      K=1
      DO 130 J=1,NZ
      L=IEXT(J)
      DTEMP=AD(J)*DES(L)
      DNUM=DNUM+DTEMP
      DTEMP=DBLE(FLOAT(K))*AD(J)/WT(L)
      DDEN=DDEN+DTEMP
  130 K=-K
      DEV=DNUM/DDEN
      NU=1
      IF(DEV.GT.0.0D0) NU=-1
      DEV=DBLE(FLOAT(-NU))*DEV
      K=NU
      DO 140 J=1,NZ
      L=IEXT(J)
      DTEMP=DBLE(FLOAT(K))*DEV/WT(L)
      Y(J)=DES(L)+DTEMP
  140 K=-K
      IF(DEV.GE.DEVL) GO TO 150
C     CALL OUCH
      STATUS = 1
      GO TO 400
  150 DEVL=DEV
      JCHNGE=0
      K1=IEXT(1)
      KNZ=IEXT(NZ)
      KLOW=0
      NUT=-NU
      J=1
C
C  SEARCH FOR THE EXTREMAL FREQUENCIES OF THE BEST
C  APPROXIMATION
C
  200 IF(J.EQ.NZZ) YNZ=COMP
      IF(J.GE.NZZ) GO TO 300
      KUP=IEXT(J+1)
      L=IEXT(J)+1
      NUT=-NUT
      IF(J.EQ.2) Y1=COMP
      COMP=DEV
      IF(L.GE.KUP) GO TO 220
      ERR=GEE(L,NZ)
      ERR=(ERR-DES(L))*WT(L)
```

```
          DTEMP=DBLE(FLOAT(NUT))*ERR-COMP
          IF(DTEMP.LE.0.0D0) GO TO 220
          COMP=DBLE(FLOAT(NUT))*ERR
     210  L=L+1
          IF(L.GE.KUP) GO TO 215
          ERR=GEE(L,NZ)
          ERR=(ERR-DES(L))*WT(L)
          DTEMP=DBLE(FLOAT(NUT))*ERR-COMP
          IF(DTEMP.LE.0.0D0) GO TO 215
          COMP=DBLE(FLOAT(NUT))*ERR
          GO TO 210
     215  IEXT(J)=L-1
          J=J+1
          KLOW=L-1
          JCHNGE=JCHNGE+1
          GO TO 200
     220  L=L-1
     225  L=L-1
          IF(L.LE.KLOW) GO TO 250
          ERR=GEE(L,NZ)
          ERR=(ERR-DES(L))*WT(L)
          DTEMP=DBLE(FLOAT(NUT))*ERR-COMP
          IF(DTEMP.GT.0.0D0) GO TO 230
          IF(JCHNGE.LE.0) GO TO 225
          GO TO 260
     230  COMP=DBLE(FLOAT(NUT))*ERR
     235  L=L-1
          IF(L.LE.KLOW) GO TO 240
          ERR=GEE(L,NZ)
          ERR=(ERR-DES(L))*WT(L)
          DTEMP=DBLE(FLOAT(NUT))*ERR-COMP
          IF(DTEMP.LE.0.0D0) GO TO 240
          COMP=DBLE(FLOAT(NUT))*ERR
          GO TO 235
     240  KLOW=IEXT(J)
          IEXT(J)=L+1
          J=J+1
          JCHNGE=JCHNGE+1
          GO TO 200
     250  L=IEXT(J)+1
          IF(JCHNGE.GT.0) GO TO 215
     255  L=L+1
          IF(L.GE.KUP) GO TO 260
```

```
          ERR=GEE(L,NZ)
          ERR=(ERR-DES(L))*WT(L)
          DTEMP=DBLE(FLOAT(NUT))*ERR-COMP
          IF(DTEMP.LE.0.0D0) GO TO 255
          COMP=DBLE(FLOAT(NUT))*ERR
          GO TO 210
      260 KLOW=IEXT(J)
          J=J+1
          GO TO 200
      300 IF(J.GT.NZZ) GO TO 320
          IF(K1.GT.IEXT(1)) K1=IEXT(1)
          IF(KNZ.LT.IEXT(NZ)) KNZ=IEXT(NZ)
          NUT1=NUT
          NUT=-NU
          L=0
          KUP=K1
          COMP=YNZ*(1.00001D0)
          LUCK=1
      310 L=L+1
          IF(L.GE.KUP) GO TO 315
          ERR=GEE(L,NZ)
          ERR=(ERR-DES(L))*WT(L)
          DTEMP=DBLE(FLOAT(NUT))*ERR-COMP
          IF(DTEMP.LE.0.0D0) GO TO 310
          COMP=DBLE(FLOAT(NUT))*ERR
          J=NZZ
          GO TO 210
      315 LUCK=6
          GO TO 325
      320 IF(LUCK.GT.9) GO TO 350
          IF(COMP.GT.Y1) Y1=COMP
          K1=IEXT(NZZ)
      325 L=NGRID+1
          KLOW=KNZ
          NUT=-NUT1
          COMP=Y1*(1.00001D0)
      330 L=L-1
          IF(L.LE.KLOW) GO TO 340
          ERR=GEE(L,NZ)
          ERR=(ERR-DES(L))*WT(L)
          DTEMP=DBLE(FLOAT(NUT))*ERR-COMP
          IF(DTEMP.LE.0.0D0) GO TO 330
          J=NZZ
```

```
         COMP=DBLE(FLOAT(NUT))*ERR
         LUCK=LUCK+10
         GO TO 235
     340 IF(LUCK.EQ.6) GO TO 370
         DO 345 J=1,NFCNS
     345 IEXT(NZZ-J)=IEXT(NZ-J)
         IEXT(1)=K1
         GO TO 100
     350 KN=IEXT(NZZ)
         DO 360 J=1,NFCNS
     360 IEXT(J)=IEXT(J+1)
         IEXT(NZ)=KN
         GO TO 100
     370 IF(JCHNGE.GT.0) GO TO 100
C   CALCULATION OF THE COEFFICIENTS OF THE BEST APPROXIMATION
C   USING THE INVERSE DISCRETE FOURIER TRANSFORM
C
     400 CONTINUE
         NM1=NFCNS-1
         FSH=1.0D-06
         GTEMP=GRID(1)
         X(NZZ)=-2.0D0
         CN=DBLE(FLOAT(2*NFCNS-1))
         DELF=1.0D0/CN
         L=1
         KKK=0
         IF(EDGE(1).EQ.0.0.AND.EDGE(2*NBANDS).EQ.0.5) KKK=1
         IF(NFCNS.LE.3) KKK=1
         IF(KKK.EQ.1) GO TO 405
         DTEMP=DCOS(PI2*GRID(1))
         DNUM=DCOS(PI2*GRID(NGRID))
         AA=2.0D0/(DTEMP-DNUM)
         BB=-(DTEMP+DNUM)/(DTEMP-DNUM)
     405 CONTINUE
         DO 430 J=1,NFCNS
         FT=(J-1)*DELF
         XT=DCOS(PI2*FT)
         IF(KKK.EQ.1) GO TO 410
         XT=(XT-BB)/AA
         FT=ACOS(XT)/PI2
     410 XE=X(L)
         IF(XT.GT.XE) GO TO 420
         IF((XE-XT).LT.FSH) GO TO 415
```

```
      L=L+1
      GO TO 410
  415 A(J)=Y(L)
      GO TO 425
  420 IF((XT-XE).LT.FSH) GO TO 415
      GRID(1)=FT
      A(J)=GEE(1,NZ)
  425 CONTINUE
      IF(L.GT.1) L=L-1
  430 CONTINUE
      GRID(1)=GTEMP
      DDEN=PI2/CN
      DO 510 J=1,NFCNS
      DTEMP=0.0D0
      DNUM=DBLE(FLOAT(J-1))*DDEN
      IF(NM1.LT.1) GO TO 505
      DO 500 K=1,NM1
  500 DTEMP=DTEMP+A(K+1)*DCOS(DNUM*DBLE(FLOAT(K)))
  505 DTEMP=2.0D0*DTEMP+A(1)
  510 ALPHA(J)=DTEMP
      DO 550 J=2,NFCNS
  550 ALPHA(J)=2.0D0*ALPHA(J)/CN
      ALPHA(1)=ALPHA(1)/CN
      IF(KKK.EQ.1) GO TO 545
      P(1)=2.0D0*ALPHA(NFCNS)*BB+ALPHA(NM1)
      P(2)=2.0D0*AA*ALPHA(NFCNS)
      Q(1)=ALPHA(NFCNS-2)-ALPHA(NFCNS)
      DO 540 J=2,NM1
      IF(J.LT.NM1) GO TO 515
      AA=0.5D0*AA
      BB=0.5D0*BB
  515 CONTINUE
      P(J+1)=0.0D0
      DO 520 K=1,J
      A(K)=P(K)
  520 P(K)=2.0D0*BB*A(K)
      P(2)=P(2)+A(1)*2.0D0*AA
      JM1=J-1
      DO 525 K=1,JM1
  525 P(K)=P(K)+Q(K)+AA*A(K+1)
      JP1=J+1
      DO 530 K=3,JP1
  530 P(K)=P(K)+AA*A(K-1)
      IF(J.EQ.NM1) GO TO 540
```

```
      DO 535 K=1,J
  535 Q(K)=-A(K)
      Q(1)=Q(1)+ALPHA(NFCNS-1-J)
  540 CONTINUE
      DO 543 J=1,NFCNS
  543 ALPHA(J)=P(J)
  545 CONTINUE
      IF(NFCNS.GT.3) RETURN
      ALPHA(NFCNS+1)=0.0D0
      ALPHA(NFCNS+2)=0.0D0
      RETURN
      END


      DOUBLE PRECISION FUNCTION D(K,N,M)
C
C  FUNCTION TO CALCULATE THE LAGRANGE INTERPOLATION
C  COEFFICIENTS FOR USE IN THE FUNCTION GEE.
C
      COMMON PI2,AD,DEV,X,Y,GRID,DES,WT,ALPHA,IEXT,NFCNS,NGRID
      DIMENSION IEXT(130),AD(130),ALPHA(130),X(130),Y(130)
      DIMENSION DES(2080),GRID(2080),WT(2080)

      DOUBLE PRECISION DES,GRID,WT,ALPHA
      DOUBLE PRECISION AD,DEV,X,Y
      DOUBLE PRECISION Q
      DOUBLE PRECISION PI2

CC    DIMENSION IEXT(66),AD(66),ALPHA(66),X(66),Y(66)
CC    DIMENSION DES(1045),GRID(1045),WT(1045)

      D=1.0D0
      Q=X(K)
      DO 3 L=1,M
      DO 2 J=L,N,M
      IF(J-K)1,2,1
    1 D=2.0D0*D*(Q-X(J))
    2 CONTINUE
    3 CONTINUE
      D=1.0D0/D
      RETURN
      END


      DOUBLE PRECISION FUNCTION GEE(K,N)
C
```

```fortran
C   FUNCTION TO EVALUATE THE FREQUENCY RESPONSE USING THE
C   LAGRANGE INTERPOLATION FORMULA IN THE BARYCENTRIC FORM
C
        COMMON PI2,AD,DEV,X,Y,GRID,DES,WT,ALPHA,IEXT,NFCNS,NGRID
        DIMENSION IEXT(130),AD(130),ALPHA(130),X(130),Y(130)
        DIMENSION DES(2080),GRID(2080),WT(2080)

        DOUBLE PRECISION DES,GRID,WT,ALPHA
        DOUBLE PRECISION P,C,D,XF
        DOUBLE PRECISION PI2
        DOUBLE PRECISION AD,DEV,X,Y

CC      DIMENSION IEXT(66),AD(66),ALPHA(66),X(66),Y(66)
CC      DIMENSION DES(1045),GRID(1045),WT(1045)

        GEE=0.0D0
        IF(N.LE.0)RETURN
        P=0.0D0
        XF=GRID(K)
        XF=DCOS(PI2*XF)
        D=0.0D0
        DO 1 J=1,N
        C=XF-X(J)
        C=AD(J)/C
        D=D+C
      1 P=P+C*Y(J)
        IF(D.EQ.0.0D0) D=1.0D-24
        GEE=P/D
        RETURN
        END


        SUBROUTINE OUCH
C       WRITE(LUDISP,1)
C     1 FORMAT(' ********* FAILURE TO CONVERGE *********'/
C     1'PROBABLE CAUSE IS MACHINE ROUNDING ERROR'/
C     2'THE IMPULSE RESPONSE MAY BE CORRECT'/
C     3'CHECK WITH A FREQUENCY RESPONSE')
        RETURN
        END
```

## A.10.2   The C Main Program

The input arguments in the call list to the Parks-McClellan program are thoroughly documented in Chapter 5, "Finite Impulse Response Digital Filtering," and are not discussed here. This section is only concerned with the mixed language protocol necessary to compile and link a C main to the Parks-McClellan program. In this regard, we present a code snippet taken from a filter design program written by the author that runs in Windows and is written in C. The purpose of this program is to illustrate both the call to the PM subroutine and the code used to format the returned data for disk storage. The reader can use this C code directly or modify it specifically for his or her own application. The reader should keep in mind that this code snippet was cut from a much larger program and therefore may have some broken pipes in that the definition and initiation of some global variables is not traceable back to their original include files. The C code snippet illustrated in Listing A.17 illustrates the very basic format for this mixed language program.

There is no need at this point for the reader to be concerned about the definition of the filter specifications. The definition of all these specifications is thoroughly discussed in Chapter 5, "Finite Impulse Response Digital Filtering."

The reader should note that all the rules of mixed language programming that we discussed previously are implemented in this procedure. The first thing to note is the **extern** statement that prototypes the Parks-McClellan FORTRAN subroutine named PARK. This statement is placed outside the C main program.

The first thing the program does is **calloc** enough memory to hold the filter coefficients returned by the Parks-McClellan subroutine and pointed to *pH. It then opens a <dat> file to hold the binary formatted filter coefficients and a <txt> file to hold the filter documentation data. Following that, the program implements the user-entered parameters to call the FORTRAN subroutine PARK to compute the filter coefficients.

The remainder of the code is nothing more than formatting and storing the data produced by the subroutine to the two disk files. To help in understanding the code contained within this procedure, the text file that was generated by the C main is illustrated in Table A.1 for the example where the input arguments given as

| | |
|---|---|
| Number of filter coefficients | iNFILT = 34 |
| Filter type | iJTYPE = 1 |
| Number of filter bands | iNBANDS = 2 |

Size of grid                          iLGRID = 16
Filter band edges                     dEdge[4] = 0.00 0.20 0.30 0.50
Filter desired function               dFx[2] = 1.0 0.0
Filter weighting function             dWtx[2] = 1.00 1.20
Filter output coefficient             pH[32]

**Table A.1**   Example Filter Documentation Text File

```
            Finite Impulse Response FIR
            Linear Phase Digital Filter Design
            Remez Exchange Algorithm
            Band Pass Filter
            Filter Length = 34


            *****   Impulse Response   *****
    H(0)   =      0.00057940   =   H(33)   =        19
    H(1)   =     -0.00143848   =   H(32)   =       -46
    H(2)   =     -0.00199142   =   H(31)   =       -64
    H(3)   =      0.00300130   =   H(30)   =        98
    H(4)   =      0.00418997   =   H(29)   =       137
    H(5)   =     -0.00610421   =   H(28)   =      -199
    H(6)   =     -0.00802244   =   H(27)   =      -262
    H(7)   =      0.01107391   =   H(26)   =       363
    H(8)   =      0.01416825   =   H(25)   =       464
    H(9)   =     -0.01899166   =   H(24)   =      -621
    H(10)  =      0.02417187   =   H(23)   =      -791
    H(11)  =      0.03223908   =   H(22)   =      1056
    H(12)  =      0.04212628   =   H(21)   =      1380
    H(13)  =      0.05858603   =   H(20)   =     -1919
    H(14)  =      0.08527554   =   H(19)   =     -2793
    H(15)  =      0.14773008   =   H(18)   =      4841
    H(16)  =      0.44889525   =   H(17)   =     14709

                             Band 1          Band 2
    Lower Band Edge        0.0000000       0.3000000
    Upper Band Edge        0.2000000       0.5000000
    Desired Value          1.0000000       0.0000000
    Weighting              1.0000000       1.2000000
    Deviation              0.0011563       0.0009636
    Deviation(db)          0.0100375     -60.3223428
    Sum of Floating Point Coefficients     0.9988438
```

## Listing A.17   Call to the PM Subroutine and Disk File Output

```
/*****************************************************/
extern "C" {void__stdcall PARK(int*, int*, int*, int*, double*,
        double*, double*, double*, int*, int*, int*, int*,
                                    int*,double*,int*);}
/*****************************************************/
// C main program begin

int WINAPI WinMain(HINSTANCE hInst, // windows code
            HINSTANCE hPreviousInst,
            PSTR szCmdLine, int iCmdShow) {
:
//****calloc heap mem store filter coeffs pointed to by *pH
if(NULL==(pH= (double*)calloc(NFMAX,sizeof(double)))) {
MessageBox(hDlg,"Cannot calloc coeff mem",
                "Memory Error",MB_OK);
free(pH);
return 0;
}                               // end of if on pH mem calloc

//**** open .dat file to hold computed filter coefficients
sprintf(szMsg,"");
strcat(szMsg,szPath);
strcat(szMsg,szFirFileName);
strcat(szMsg,".dat");
if(NULL == (pmTxtFilterFile = fopen(szMsg,"w"))) {
sprintf(szMsg,"Cannot Open Coefficient Text Disk File\n"
            "DesFilterDlgProc Message");
//**** can't open disk file so inform user
MessageBox(hWnd,szMsg,"Error Message",MB_OK);
free(pH); // then tidy up and leave
return 0;
}                               //  end of if on .dat file open

//****  open .txt text file to hold filter documentation
sprintf(szMsg,"");
strcat(szMsg,szPath);
strcat(szMsg,szFirFileName);
strcat(szMsg,".txt");
if(NULL == (pmTxtFilterDocFile = fopen(szMsg,"w"))) {
sprintf(szMsg,"Cannot Open Filter Doc Text Disk File\n"
```

```
                "DesFilterDlgProc Message");
//**** can't open disk file so inform user
MessageBox(hWnd,szMsg,"Error Message",MB_OK);
free(pH); // and tidy up and leave
_fcloseall();
return 0;
}                                 // end of if on.txt file open


/*****************************************************/
/* Call the modified Parks-McClellan FORTRAN routine to  */
/* compute filter coefficients                           */
/*****************************************************/

PARK(&iNFILT,&iJTYPE,&iNBANDS,&iLGRID,dEdge,dFx,dWtx,pH,
      &iNFCNS,&iNEG,&iNODD,&iNZ,&iKUP,&dDev,&iSTATUS);

switch (iSTATUS) { // check PM status on output

case 0: //**** PM status is good so continue
break;

case 1: //**** PM status is bad send error message and abort
   sprintf(szMsg,"******* FAILURE TO CONVERGE *******\n"
                 " PROBABLE CAUSE IS MACHINE ROUNDING ERROR\n"
                 "    THE IMPULSE RESPONSE MAY BE CORRECT\n"
                 "       CHECK WITH A FREQUENCY RESPONSE");
   MessageBox(hWnd,"szMsg","Park McClellan Error",MB_OK);
   free(pH);                 // routine did not converge so
   fcloseall();              // send message to user and
   break;                    // tidy up and leave
   default:
break;
}                                 // end of switch on iSTATUS


/*****************************************************/
/* generate the binary representation of the integer     */
/* filter coefficients and output to .dat file           */
/*****************************************************/
sumf= 0;

rewind(pmTxtFilterFile); // rewind .dat file
fseek(pmTxtFilterFile,0L,SEEK_SET);// set text file ptr to
                    //start of file
```

```c
numcoeff = 0; // clear coefficient count

for(i=0;i<iNFILT; i++) { // convert dbl prec to 16 bit int
  c =(short int)(*(pH + i)*pow(2,15) + 0.5);//scale & rnd int
                              //coefficient
    sumf=sumf+ *(pH + i);  // accumulate sum of float coeff
    sprintf(szBuf," ");

    for(j=0;j<16;j++){      // compute binary representation
      d= (c ]*] (15-j));
      if((d & mask) == 0) {
      strcat(szBuf,"0");}
      if((d & mask) == 1) {strcat(szBuf,"1");}
      }                      // end of for loop on j

    strcat(szBuf,",\n");
    fwrite(szBuf,strlen(szBuf),1,pmTxtFilterFile); // write 16
                       // bit binary coefficient to file
    numcoeff +=1;        // increment number of coefficients
    }                      // end of for loop on i

sprintf(szBuf,"%f,\n",sumf );
fwrite(szBuf,strlen(szBuf),1,pmTxtFilterFile); // write
                   //normalized coeff sum to file

/*****************************************************/
/*  generate the filter print documentation .txt file  */
/*****************************************************/
rewind(pmTxtFilterDocFile); // rewind text file
fseek(pmTxtFilterDocFile,0L,SEEK_SET); // set text file ptr
                          // to start of file

//**** generate beginning of file header
sprintf(szBuf,"Finite Impulse Response FIR\n"
            "Linear Phase Digital Filter Design\n"
            "Remez Exchange Algorithm\n");
write(szBuf,strlen(szBuf),1,pmTxtFilterDocFile);

//**** write filter type
if(iJTYPE == 1) sprintf(szBuf,"Band Pass Filter\n");
if(iJTYPE == 2) sprintf(szBuf,"Differentiator\n");
if(iJTYPE == 3) sprintf(szBuf,"Hilbert Transformer\n");
fwrite(szBuf,strlen(szBuf),1,pmTxtFilterDocFile);
```

```
sprintf(szBuf,"Filter Length = %d",iNFILT);//write # of coeff
fwrite(szBuf,strlen(szBuf),1,pmTxtFilterDocFile);

//**** write end of file header
sprintf(szBuf,"\n***** Impulse Response *****\n");
fwrite(szBuf,strlen(szBuf),1,pmTxtFilterDocFile);

for(j=0;j<iNFCNS;j++){
   k=iNFILT -1 -j;

//****    compute 16 bit integer coefficients
   c =(short int)( *(pH + j)*pow(2,15) + 0.5);
   //**** output dbl precision and integer coefficients to
   //**** documentation text file
   if(iNEG == 0) {sprintf(szBuf,"\tH(%d)\t= %15.8f\t="
             "H(%d)\t=\t%d\n",j,*(pH+j),k,c);
   if(iNEG == 1) sprintf(szBuf,"\tH(%d)\t= %15.8f\t="
             "-H(%d)\t=\t %d\n",j,*(pH+j),k,c);
   fwrite(szBuf,strlen(szBuf),1,pmTxtFilterDocFile);
}                              //  end of for loop on j

if((iNEG ==1) && (iNODD == 1)) {
   sprintf(szBuf,"\tH(%d)\t=\t0.0\n",iNZ);
   fwrite(szBuf,strlen(szBuf),1,pmTxtFilterDocFile);
   }                           // end of if on iNEG

//**** Output filter design parameters to the documentation
//**** text file
for(k=1;k<iNBANDS+1;k+=4){// output filter band id's
   strcpy(szMsg,"\n\n\t\t\t");
   iKUP = k+3;
   if(iKUP > iNBANDS) iKUP = iNBANDS;
   for(j=k;j<iKUP+1;j++){
     sprintf(szBuf,"Band %d\t\t",j);
     strcat(szMsg,szBuf);
     }                         // end of for loop on j

   fwrite(szMsg,strlen(szMsg),1,pmTxtFilterDocFile);
   sprintf(szMsg,"\nLower Band Edge\t");//output lwr band edges
   for(j=k;j<iKUP+1;j++){sprintf(szBuf,"%15.7f\t",dEdge[2*j-2]);
     strcat(szMsg,szBuf);
     }                         // end of for loop on j

   fwrite(szMsg,strlen(szMsg),1,pmTxtFilterDocFile);
```

```
          sprintf(szMsg,"\nUpper Band Edge\t");// output upper band edges
          for(j=k;j<iKUP+1;j++){
             sprintf(szBuf,"%15.7f\t",dEdge[2*j-1]);
             strcat(szMsg,szBuf);
             }                          // end of for loop on j

       fwrite(szMsg,strlen(szMsg),1,pmTxtFilterDocFile);

       if(iJTYPE != 2) { // output filter desired values
            sprintf(szMsg,"\nDesired Value\t");
            for(j=k;j<iKUP+1;j++){
               sprintf(szBuf,"%15.7f\t",dFx[j-1]);
               strcat(szMsg,szBuf);
               }                          // end of for loop on j

            fwrite(szMsg,strlen(szMsg),1,pmTxtFilterDocFile);
       }                               // end of if on iJTYPE

         if(iJTYPE == 2) {          // output filter desired slope
            sprintf(szMsg,"\nDesired Slope\t");
            for(j=k;j<iKUP+1;j++){
               sprintf(szBuf,"%15.7f\t",dFx[j-1]);
               strcat(szMsg,szBuf);
               }                          // end of for loop on j
            fwrite(szMsg,strlen(szMsg),1,pmTxtFilterDocFile);
            }                          // end of if on iJTYPE

       sprintf(szMsg,"\nWeighting\t");// output filter weights
       for(j=k;j<iKUP+1;j++){
          sprintf(szBuf,"%15.7f\t",dWtx[j-1]);
          strcat(szMsg,szBuf);
          }                          // end of for loop on j
       fwrite(szMsg,strlen(szMsg),1,pmTxtFilterDocFile);

       sprintf(szMsg,"\nDeviation\t"); //output band deviation
       for(j=k;j<iKUP+1;j++){
          sprintf(szBuf,"%15.7f\t",dDev/dWtx[j-1]);
          strcat(szMsg,szBuf);
          }                          // end of loop on j
       fwrite(szMsg,strlen(szMsg),1,pmTxtFilterDocFile);

       if(iJTYPE == 1){           //  output band deviation in db
          sprintf(szMsg,"\nDeviation(db)\t");
```

```
      for(j=k;j<iKUP+1;j++){
        arg = dDev/dWtx[j-1];
        if(dFx[j-1] == 1) arg = arg+1;
        sprintf(szBuf,"%15.7f\t",20*log10(arg));
        strcat(szMsg,szBuf);
        }                       // end of loop on j
   fwrite(szMsg,strlen(szMsg),1,pmTxtFilterDocFile);

      sprintf(szMsg,"\nSum of Floating Point Coefficients ="
                                    " %15.7f" ,sumf);
      fwrite(szMsg,strlen(szMsg),1,pmTxtFilterDocFile);
      }                         // end of if on iJTYPE


 }                              // end of for loop on k

free(pH); // done so let us tidy up and leave
_fcloseall();

Return  DefWindowProc(hWnd,iMsg,wParam,lParam);// C main end
}
```

## A.10.3  Compilation and Linkage

The procedure is identical to that used in our previous examples:

1. Open a new FORTRAN project in Microsoft FORTRAN Power Station 4.0 called **pm**. Power Station will name the source file **pm.f90**.
2. Add the Parks-McClellan FORTRAN subroutine to the project.
3. Compile **pm.f90** and produce an object file named **pm.obj**.
4. Open a new C++ project in Microsoft Visual C++, and write the C/C++ main source code.
5. Add the **pm.obj** file to the C++ project.
6. Add the following Power Station libraries to the C++ project:
   **a.** console.lib
   **b.** portlib.lib
   **c.** libf.lib
7. Build the C++ program, compile and link the C/C++ source code and link the FORTRAN object code.
8. Run the program.

## A.11 REFERENCES

[1] James H. McClellan, Thomas W. Parks, and Lawrence R. Rabiner. "A Computer Program for Designing Optimum FIR Linear Phase Digital Filters." *IEEE Transactions on Audio and Electroacoutsics*, vol. AU-21, no. 6 (1973).

[2] Paul M. Embree and Bruce Kimble. *C Language Algorithms for Digital Signal Processing*. Englewood Cliffs, NJ: Prentice Hall, 1991.

[3] Emmanuel C. Ifeachor and Barrie W. Jervis. *Digital Signal Processing: A Practical Approach*. Workingham, UK: Addison-Wesley, 1993.

[4] Alan V. Oppenheim and Ronald W. Schafer. *Digital Signal Processing*. Englewood Cliffs, NJ: Prentice Hall, 1975.

[5] Steven A. Tretter. *Introduction to Discrete Time Signal Processing*. New York: John Wiley and Sons, 1976.

[6] P. P. Vaidyanathan. *Handbook of Digital Signal Processing Engineering Applications*. Edited by Douglas Elliott. San Diego, CA: Academic Press, 1987.

[7] Charles S. Williams. *Designing Digital Filters*. Englewood Cliffs, NJ: Prentice Hall, 1986.

[8] Lawrence R. Rabiner and Bernard Gold. *Theory and Applications of Digital Signal Processing*. Englewood Cliffs, NJ: Prentice Hall, 1975.

[9] G. J. Borse. *FORTRAN and Numerical Methods for Engineers*. Boston: PWS Engineering.

# Index

*Note:* The letters *f*, *e*, and *t* indicate that the entry refers to a page's figure, equation, or table, respectively.

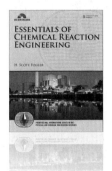  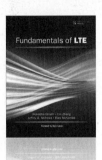  

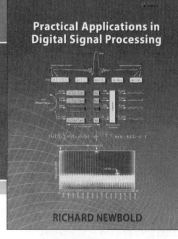

**Practical Applications in Digital Signal Processing**

RICHARD NEWBOLD

# FREE
# Online Edition

Your purchase of **Practical Applications in Digital Signal Processing** includes access to a free online edition for 45 days through the **Safari Books Online** subscription service. Nearly every Prentice Hall book is available online through **Safari Books Online**, along with thousands of books and videos from publishers such as Addison-Wesley Professional, Cisco Press, Exam Cram, IBM Press, O'Reilly Media, Que, Sams, and VMware Press.

**Safari Books Online** is a digital library providing searchable, on-demand access to thousands of technology, digital media, and professional development books and videos from leading publishers. With one monthly or yearly subscription price, you get unlimited access to learning tools and information on topics including mobile app and software development, tips and tricks on using your favorite gadgets, networking, project management, graphic design, and much more.

## Activate your FREE Online Edition at
## informit.com/safarifree

**STEP 1:**    Enter the coupon code: ZFFJDDB.

**STEP 2:**    New Safari users, complete the brief registration form.
             Safari subscribers, just log in.

If you have difficulty registering on Safari or accessing the online edition,
please e-mail customer-service@safaribooksonline.com